Nuclear Power Plants Worldwide

ISSN 1067-9979

Nuclear Power Plants Worldwide

First Edition

Peter D. Dresser, Editor

Gale Research Inc. • *DETROIT* • *WASHINGTON, D.C.* • *LONDON*

Editor: Peter D. Dresser
Contributing Editors: Karen Bellenir, Al Cook, Michael Richman, Mary Gillis, and Linda M. Ross
Translators: Eduardo Almaguer and Al Cook
Aided by: Charles Ceplecha, Yvette M. Costa, Danielle Des Rochers, and Sharon Miscavage

Gale Research Inc. Staff
Senior Editor: Amy Lucas
Project Coordinator: Annette Piccirelli

Aided by: Sara Burak, Ned Burels, Sandra Doran, Joyce Jakubiak, Kristin Kahrs, Jan Klisz, John Krol, Christine Maurer, Kathleen Lopez Nolan, Diane Sawinski, Christopher P. Scanlon, and Gwen E. Turecki

Supervisor of Editorial Programming Services: Theresa A. Rocklin
Programmer: Ida M. Wright

Data Entry Supervisor: Benita L. Spight
Data Entry Group Leader: Gwendolyn Tucker
Data Entry Associate: Civie Green

Production Director: Mary Beth Trimper
Production Assistant: Mary Kelley

Art Director: Cynthia Baldwin
Graphic Designers: Bernadette M. Gornie and Mary Krzewinski
Keyliner: C.J. Jonik
Map Illustrations by Nicholas Jakubiak
Cover Photograph by Robert J. Huffman

While every effort has been made to ensure the reliability of the information presented in this publication, Gale Research Inc. does not guarantee the accuracy of the data contained herein. Gale accepts no payment for listing; and inclusion in the publication of any organization, agency, institution, publication, service, or individual does not imply endorsement by the editors or publisher. Errors brought to the attention of the publisher and verified to the satisfaction of the publisher will be corrected in future editions.

∞™ This book is printed on acid-free paper that meets the minimum requirements of American National Standard for Information Sciences–Permanence Paper for Printed Library Materials, ANSI Z39.48-1984.

♻ This book is printed on recycled paper that meets Environmental Protection Agency Standards.

This publication is a creative work copyrighted by Gale Research Inc. and fully protected by all applicable copyright laws, as well as by misappropriation, trade secret, unfair competition, and other applicable laws. The authors and editors of this work have added value to the underlying factual material herein through one or more of the following: unique and original selection, coordination, expression, arrangement, and classification of the information.

Gale Research Inc. will vigorously defend all of its rights in this publication.

Copyright © 1993
Gale Research Inc.
835 Penobscot Bldg.
Detroit, MI 48226-4094

All rights reserved including the right of reproduction in whole or in any part in any form.

ISBN 0-8103-8880-4
ISSN 1067-9979

Published simultaneously in the United Kingdom
by Gale Research International Limited
(an affiliated company of Gale Research Inc.)

The trademark I(T)P™ is used under license.

Contents

Highlights vii
Introduction ix
User's Guide xi
Symbols and Abbreviations xvii

How a Nuclear Power Plant Works xix

Maps of Plant Locations xxix

    Map 1: Belgium, France, Netherlands, Spain, and United Kingdom xxx
    Map 2: Austria, Bulgaria, Czech Republic, Finland, Germany,
           Hungary, Italy, Poland, Romania, Slovenia, Sweden, and
           Switzerland xxxii
    Map 3: United States (western states) and Mexico xxxiv
    Map 4: United States (eastern states), Canada, and Cuba xxxvi
    Map 5: Japan and Republic of Korea xxxviii
    Map 6: China, India, Pakistan, Philippines, and Taiwan xl
    Map 7: Armenia, Kazakhstan, Lithuania, Russia, and Ukraine xlii
    Map 8: Argentina and Brazil xliii
    Map 9: South Africa xliii

Country Reports and Plant Profiles

Argentina 1
Armenia 5
Austria 9
Belgium 13
Brazil 19
Bulgaria 23
Canada 27
China   See People's Republic of China
Cuba 41
Czech Republic 45
Finland 53
France 59
Germany 79
Hungary 105
India 111
Indonesia 117
Italy 119
Japan 123
Kazakhstan 145
Korea   See Republic of Korea

Lithuania	149
Luxembourg	153
Mexico	157
Netherlands	161
Pakistan	165
People's Republic of China	169
Philippines	177
Poland	185
Republic of Korea	189
Republic of South Africa	203
Romania	207
Russia	211
Slovenia	227
South Africa See Republic of South Africa	
Spain	231
Sweden	239
Switzerland	247
Taiwan	251
Ukraine	257
Union of Soviet Socialist Republics	269
See also Armenia, Kazakhstan, Lithuania, Russia, Ukraine	
United Kingdom	277
United States	317
Yugoslavia See Slovenia	

Appendixes

Appendix I: NRC Report Cards	487
Appendix II: Glossary	525

Indexes

Technical Problems Index	533
Alphabetical Index	543

Highlights

Researchers, students, journalists, environmentalists, and others can turn to *Nuclear Power Plants Worldwide (NPPW)* for information on the commercial nuclear power industry. *NPPW* provides nontechnical profiles on all of the currently operating commercial nuclear plants in 39 countries, plus coverage of nonoperable plants, such as those that are under construction, shut down, or cancelled.

Essay and Maps Provide Industry Background

NPPW's opening essay, "How a Nuclear Power Plant Works," familiarizes the layperson with the different types of nuclear reactors and how they operate. The essay is followed by maps showing the locations of select power plants around the world.

Country Reports and Plant Profiles

Each country chapter begins with a report providing vital background information for the descriptive listings. Included in these country reports are data on the history, national issues, and statistics of the country's commercial nuclear industry. The profiles of each plant that follow provide information on a total of 741 nuclear power units at 310 sites, including plants that are:

- operable
- planned
- on order
- under construction
- cancelled
- indefinitely deferred
- shut down
- decommissioning

Profiles provide complete contact data, operational and status details, key dates, plant histories, and newsmaking events.

Appendixes Provide Plant Evaluations and Glossary

NPPW includes two appendixes: **NRC Report Cards**, which provide assessments prepared by the U.S. Nuclear Regulatory Commission for plants in the United States, and the **Glossary**, which defines more than 160 terms related to nuclear power plants.

Indexes Allow Convenient Access

Two indexes speed access to information in the country reports and plant profiles:

The **Technical Problems Index** categorizes units from the plant profiles that have had significant accidents, breakdowns, or technical failures by the main operating systems and components of a nuclear power plant.

The **Alphabetical Index** includes all listed nuclear power plants, utilities, organizations, agencies, companies, and contractors from the country reports and plant profiles in one alphabetic sequence.

Introduction

The numbers make it clear: nuclear power is an important energy source in the world today. By the end of 1991, 420 operating nuclear power units in 26 countries supplied approximately one-sixth of the world's electricity. Another 166 power plants were either planned, ordered, or under construction.

Nuclear Power Plants Worldwide (NPPW) is a comprehensive, nontechnical guide to the international commercial nuclear power industry—those utilities and plants that create significant amounts of electricity from nuclear reactors. *NPPW* provides information on all currently operating nuclear power plants spanning 39 countries. Covered are more than 740 plants that are:

- operable
- planned
- on order
- under construction
- cancelled
- indefinitely deferred
- shut down
- decommissioning

NPPW does not cover research reactors or reactors used in the production of nuclear weapons.

Introductory Materials Provide Background on the Industry

NPPW begins with two introductory sections that help to orient users to plant operations and locations.

- An opening essay, "How a Nuclear Power Plant Works," provides a nontechnical explanation of the mechanics of nuclear power plant operations, complete with diagrams to help general users. Included are descriptions of the different types of reactors in use today, such as pressurized water reactors and high temperature gas-cooled reactors, and a step-by-step description of a nuclear chain reaction.

- A series of Maps of Plant Locations features nine maps of the United States, South America, Africa, Europe, and Asia that show the countries and the approximate sites of all the world's operating nuclear power stations, plus other plants that are shut down, decommissioning, on order, planned, or indefinitely deferred.

Country Reports and Plant Profiles

NPPW is organized into 39 country chapters, each beginning with a narrative report that surveys the national issues related to that country's particular nuclear industry. Topics include how the country got started in nuclear energy, statistics about the country's power mix, new technologies, and national construction programs and major shutdowns.

Country reports are followed by plant profiles that offer an in-depth look at that country's individual plants. Profiles include contact information and details on the plants' status, technology, history, finances, regulation, and newsmaking events. A brief bibliography of the key sources used to compile each plant profile is also provided for users who wish to pursue further research.

Appendixes and Indexes Increase Accessibility

Two appendixes provide plant evaluations and definitions of terms used in the nuclear industry:

- The **NRC Report Cards** provide assessments of select U.S. nuclear power plants by the U.S. Nuclear Regulatory Commission, allowing users to quickly evaluate each plant's overall performance.
- The **Glossary** provides clear and nontechnical definitions of more than 160 terms related to nuclear power plants–from "alpha particle" to "whole-body exposure."

Two indexes speed users to information in the country reports and plant profiles:

- The **Technical Problems Index** provides an overview of significant accidents, breakdowns, or technical failures experienced by the plants listed. For example, users can spot at-a-glance all the plants that have had major releases of radioactive steam from their generator tubes.
- The **Alphabetical Index** provides quick access to all of the nuclear power plants, utilities, organizations, agencies, companies, and contractors listed in the country reports and plants profiles.

For more information on the arrangement, content, and indexes of *NPPW*, consult the User's Guide following this introduction.

Method of Compilation

NPPW was compiled from information obtained from a wide variety of sources representing a broad spectrum of views on the nuclear power industry. Included are data obtained from: nuclear power plants, association and industry publications, publications of antinuclear groups, U.S. government documents, and newspapers, magazines, and books.

Acknowledgments

The editor would like to thank the nuclear power plant managers and public information specialists around the world who provided information for this directory.

The editor also would like to thank the contributing editorial staff, without whom *NPPW* would not have been produced. Special thanks to Karen Bellenir, the editor's right-hand writer who really stood in the gap; Mike Richman, whose Washington, D.C., connections added significant editorial depth; and Al Cook, who tackled the French- and German-language documents and worked them into excellent English-language essays. Finally, warm thanks to Lisa, Benny, and Joey for their patience in the last month before deadline.

The Gale project coordinator would like to thank Jim Ottaviani, Engineering Librarian at the University of Michigan North Engineering Library, and Cindy Naegeli, Nuclear Information Supervisor at Detroit Edison, for lending their time and expertise to this project.

Comments Welcome

Suggestions for improving *NPPW*'s scope, content, or presentation are welcome. Please address all comments to:

Editor
Nuclear Power Plants Worldwide
Gale Research Inc.
835 Penobscot Bldg.
Detroit, MI 48226-4094
Phone: (313) 961-2242
Fax: (313) 961-6815
Toll-free: (800) 347-GALE

User's Guide

Nuclear Power Plants Worldwide (NPPW) is organized into the following sections:

- How a Nuclear Power Plant Works
- Maps of Plant Locations
- Country Reports and Plant Profiles
- Appendixes

and two indexes. Each section is described fully below.

How a Nuclear Power Plant Works

This essay explains, in nontechnical language, the mechanics of nuclear power plant operations and the various types of reactors currently in use today, including: boiling water reactors, gas-cooled reactors, heavy-water/light-water reactors, liquid metal fast breeder reactors, and more. Helpful diagrams are also included. Factual data for the essay was obtained from the U.S. Department of Energy.

Maps of Plant Locations

This collection of nine maps shows the locations of all currently operational commercial nuclear reactors worldwide, as well as the locations of commercial plants that are shut down, decommissioning, on order, planned, or indefinitely deferred; cancelled plants are not included. The corresponding plant profiles provide additional information about the operational status of each plant.

Country Reports and Plant Profiles

NPPW is organized into 39 country chapters as shown on the Contents pages. Each chapter begins with a country report covering the history, national issues, and statistics of the country's commercial nuclear industry. Each country report provides vital background information on the individual plant profiles that follow in the chapter.

Plant profiles offer a detailed look at the individual units, including contact information, key dates, and in-depth reports on plant histories and newsworthy events. For more detailed information about the plant profiles, refer to the Sample Plant Profile below.

Country reports and plant profiles were compiled using materials obtained directly from listed plants, nuclear association and industry publications, publications of antinuclear groups, and newspapers, books, and magazines.

Sample Plant Profile

Below is a fictitious entry in which each numbered section designates an item of information that might be included in a typical *NPPW* listing. The numbered sections are explained in the descriptive paragraphs following the sample.

---

① ⬚1002⬚ ● ②

③ **Braxton Nuclear Power Plant, Units 1-4**

④ *[See Also Braxton Nuclear Power Plant, Units 5-8]*

⑤ 103 Outer Lake Rd.
Braxton, NY 14519
⑦ Elizabeth Ross, Plant Mgr.

⑥ Phone: (215)333-3322
Fax: (215)333-3336
Telex: 27911 der

⑧ **Owning Utility**

Roberts Gas and Electric Co.
89 North Ave.
Riverton, NY 14701

Phone: (714)446-4700
Fax: (714)446-1510

⑨ Benjamin Peters, V.Pres.,
   Nuclear Production

⑩ **Contact**

Joseph Jameson, Pub.Aff.Mgr.
Roberts Gas and Electric Co.

Phone: (714)446-2727

⑪ **Basic Facts**

Units 1-4 *Status:* Operable. *Type of Reactor:* Pressurized Light Water Reactor. *Megawatts (electric):* 798. *Megawatts (thermal):* 1520. *Reactor System Supplier:* Westinghouse Electric Corp. (United States). *Generator Supplier:* Westinghouse Electric Corp. (United States). *Architect Engineer:* Gilbert Associates (United States). *Constructor:* Bechtel.

⑫ **Key Dates**

Units 1-4 *Ordered:* 1971. *Construction Began:* 1972. *Criticality:* November 1979. *First Power:* December 1979. *Commercial Operation:* March 1980. *Shut Down Expected:* April 2016.

⑬ **Operating Costs**

Units 1-4
1981: $22,481,000   1985: $31,609,000
1982: $29,570,000   1986: $37,389,000
1983: $26,955,000   1987: $37,763,000
1984: $32,679,000   1988: $44,427,000

⑭ **Plant Report**

Many owners. The Braxton Nuclear Power Plant is located approximately 19 miles southwest of Frankville, New York. Although the station was built and is operated by Roberts Gas & Electric, Roberts' nuclear program encountered financial difficulties during the early 1980s and the utility sold interests in the unit to regional electric corporations. As of December 31, 1990, Roberts retained only a 12.5 percent ownership interest. Other owners are: the Valley Municipal Power Agency (37.5 percent), the New York State Electric Membership Corporation (28.125 percent), the Newman Municipal Power Agency (12.5 percent), and the Saskaman River Electric Cooperative (9.375 percent).

Construction challenged. Roberts had estimated that demand for electric power would grow 10 percent per year when the company originally made plans for Braxton in 1971. The Northeastern States Environmental Study Group claimed that the additional capacity was not necessary and attempted to force cancellation of the plant. The New York Utilities Commission, however, ruled that the construction was "prudent and reasonable."

⑮ **Sources**

"New York Utility Sells Interests in A-Plant." *The Washington Post*, February 19, 1983.
"Roberts Gas & Electric Gets 100 percent of Braxton Cost in Rate Base" *Electrical Nation*, October 1986.
Roberts Gas & Electric Co. Annual Report 1990.
Roberts Gas & Electric Co. Press Release. Department of Public Affairs, August 1992.

⑯ —Sandra Lockhart

---

① **Entry Number.** Entries in *NPPW* are numbered sequentially, and the entry number (rather than the page number) is used in the indexes to refer to a specific entry.

② **Plant Status Symbol.** A symbol is used to show the current operational status of each plant. Plants may be:

- ● operable
- ☐ planned
- ■ on order
- ○ under construction
- ☆ cancelled
- △ indefinitely deferred
- ★ shut down
- ▲ decommissioning

A legend showing the symbols and their definitions appears at the bottom of each page.

These symbols also are used in the two indexes.

③ **Plant Name.** The complete name of the nuclear power plant is given, including specific units that are part of the plant complex. If details about different units at the same plant vary greatly, then individual units or groups of units are described as separate entries. All listed plant names appear in the Alphabetical Index.

④ **See Also Reference.** See Also references are used to direct users to related plant profiles.

⑤ **Plant Address.** The full mailing address of the nuclear power plant is given, including street address, city, state/province, country (for non-U.S. plants only), and zip or postal code.

⑥ **Plant Phone, Fax, and Telex Numbers.** These contact numbers for the nuclear power plant are provided when available.

⑦ **Plant Manager Name.** The name of the manager who oversees nuclear operations at the listed plant.

⑧ **Owning Utility.** The name, address, and phone, fax, and telex numbers of the investor-owned utility or government organization that owns and/or operates the plant. All listed owning utilities appear in the Alphabetical Index.

⑨ **Utility Manager Name.** The name of the manager who oversees nuclear operations at the listed utility.

⑩ **Contact.** The name of the person to contact for additional information about the listed nuclear power plant, often a public relations officer or other nuclear information specialist. Information about where the contact can be reached–at the utility or at the plant itself–is also given, along with additional address and phone data if the contact is not located at the plant or owning utility.

⑪ **Basic Facts.** Included in this paragraph are: plant status, type of reactor, electrical output, thermal output, and primary contractors (reactor supplier, generator supplier, architect or engineer, and constructor). See paragraphs below for detailed information about each of these categories. If there is more than one unit at a plant, and the details of the units differ greatly, then individual units or groups of units are described as separate Basic Facts paragraphs within the same entry.

- **Plant Status.** This item indicates the current operational status of the listed plant. Plants may be operable, under construction, on order, planned, indefinitely deferred, cancelled, decommissioning, or shut down.

- **Type of Reactor.** The type of reactor is indicated here. Reactor types include boiling water reactors, gas-cooled reactors, heavy-water/light-water reactors, high temperature gas reactors, light-water breeder reactors, light-water-cooled, graphite-moderated reactors, liquid metal fast breeder reactors, pressurized heavy water reactors, and pressurized light-water reactors. For more information about reactor types, refer to "How a Nuclear Power Plant Works" beginning on page xix.

- **Electrical Output.** Total amount of electricity produced by the plant, expressed in megawatts.

- **Thermal Output.** Total amount of heat produced by the plant, expressed in megawatts; equal to the amount of electricity generated by the plant plus the amount of waste heat generated by the plant.

- **Reactor Supplier.** The name of the company that manufactured the reactor for the owning utility. The reactor supplier may be involved in reactor repair and maintenance issues.

- **Generator Supplier.** The name of the company that manufactured the generator for the owning utility. The generator supplier may be involved in generator design and repair issues.

- **Architect or Engineer.** The name of the company that designed the plant complex and its related structures.

- **Constructor.** The name of the company that built the plant complex and its related structures.

⑫ **Key Dates.** Listed in this paragraph are major milestones in the plant's history, including: the year the reactor was ordered; the date construction began (if applicable); the date the plant was cancelled (if applicable); the date criticality was achieved (i.e., the date the plant achieved a self-sustaining nuclear chain reaction), or the date criticality is expected (if applicable); the date the reactor generated its first electrical power or the date first power is expected (if applicable); the date commercial operation began and the plant was connected to a utility's electrical grid, or the date commercial operation is expected (if applicable); the date the plant shut down or is expected to shut down (if applicable); or the year decommissioning began (if applicable).

⑬ **Operating Costs.** Entries in the United States chapter provide information on a plant's operating and maintenance costs, if available. Financial information was obtained directly from the Energy Information Administration's *An Analysis of Nuclear Plant Operating Costs: A 1991 Update*. Operating costs are listed in chronological order, from the first year of operation to 1989. These figures typically include construction costs, fuel costs, capital expenditures, or decommissioning costs.

⑭ **Plant Report.** This section reviews one or more of the following three areas:

- **Plant construction and operating performance.** Includes costs, delays, licensing, awards, honors, accidents, breakdowns, fines, statistics, and more.

- **Plant programs.** Includes safety, staff, technology, radioactive waste, training, environment, public education, and more.

- **Plant news.** Includes social, environmental, governmental, political, international, judicial, regulatory, financial, economic, and health issues surrounding the plant that have made the regional, national, or industry press.

All important names and terms in the plant reports, such as companies, government agencies, treaties, and significant terms, appear in the Alphabetical Index.

⑮ **Sources.** The sources from which the plant report was written are listed in chronological order, beginning with the most recent source.

⑯ **Plant Report Author.** The name of the editor who composed the plant report is provided.

Appendixes

The Country Reports and Plant Profiles section is followed by two appendixes:

Appendix I, **NRC Report Cards**, provides tables of assessments prepared by the U.S. Nuclear Regulatory Commission (NRC) on listed U.S. nuclear power plants. Assessments are presented in chronological order beginning with the most recent data. The dates in the left-hand column indicate the month and year the assessment was published by the NRC. A legend at the bottom of each page explains the NRC coding system used. Each table also includes a See reference to the corresponding descriptive listing on the plant in the Plant Profiles section of NPPW. If necessary, explanatory text regarding the assessment is also provided. Information for this appendix was obtained from *Historical Data Summary of the Systematic Assessment of Licensee Performance* (U.S. Nuclear Regulatory Commission, August 1991).

Appendix II, **Glossary**, includes more than 160 alphabetically arranged terms and acronyms commonly used in the nuclear power industry, with particular emphasis on the terminology used in the country reports and plant profiles. The Glossary also includes See and See Also references as appropriate. Information for the Glossary was obtained from a variety of industry and U.S. federal government sources.

Indexes

NPPW includes two indexes:

The **Technical Problems Index** lists plants that have experienced significant accidents, breakdowns, or technical failures. The terms used to categorize these plants are based on the nine main systems and components of a nuclear power plant, plus one classification for radiation exposure. The terms used to categorize citations are:

Condenser System: Condenser Coolant
Condenser System: Cooling Tower
Condenser System: Tubing

Control Systems: Emergency Water Supply
Control Systems: Fire Safety System
Control Systems: Operations Computer
Control Systems: Safety Control System

Coolant Pipes and Valves

Electrical Engineering Components: Generator
Electrical Engineering Components: Transformer
Electrical Engineering Components: Turbine

Fuel Handling: Cooling Ponds
Fuel Handling: Core Loading Machines and Refueling
Fuel Handling: Waste Handling

Primary Cooling Loop: Coolant Lines
Primary Cooling Loop: Coolant Pumps
Primary Cooling Loop: Heat Exchanger
Primary Cooling Loop: Seals

Radiation Exposure

Reactor: Containment Vessel
Reactor: Control/Dampening Rods
Reactor: Core
Reactor: Fuel Rod Assemblies
Reactor: Moderator System
Reactor: Standpipes and Rod Guide Tubes

Reactor Building: Primary Containment
Reactor Building: Secondary Containment

Secondary Cooling Loop: Coolant Lines and Valves

Secondary Cooling Loop: Coolant Pumps
Secondary Cooling Loop: Seals
Secondary Cooling Loop: Steam Generator

Each citation in the Technical Problems Index includes the name of the nuclear power plant, the country where the plant is located (in parentheses), the entry number of the plant's descriptive listing in the plant profiles, a symbol denoting the plant's current operational status (which is explained in a legend found at the bottom of each index page), and a brief description of the technical problem.

Example:

Reactor: Control/Dampening Rods

Dukovany Nuclear Power Plant (Czech Republic) **30** •
Between 1985 and 1990, safety systems dropped the reactor control rods on 28 occasions, halting the nuclear reaction each time.

For more information about the systems and components of a nuclear power plant, refer to "How a Nuclear Power Plant Works" beginning on page xix.

The **Alphabetical Index** lists all plant names, company names, and other significant details mentioned within the country reports and plant profiles of *NPPW*. This index also includes inversions on selected keywords appearing in the names of the power plants. Bolded numbers following a citation refer to book entry numbers for the plant profiles; numbers preceded by 'p.' are page numbers and refer to names and terms in the country reports. Each plant name citation is followed by a symbol that indicates its current operational status, which is explained in a legend found at the bottom of each page.

Example:

Cleveland Electric Illuminating Co. **242, 267, 331, 332**
Combined Heat and Power Co. (CHP) **160,** p. 211, p. 271
Comissao Nacional de Energia Nuclear (CNEN) p. 19
Enrico Fermi 2 Nuclear Power Plant (United States) **277** •
Fermi 2 Nuclear Power Plant; Enrico (United States) **277** •

Symbols and Abbreviations

Symbols

The following symbols are used to denote the current operational status of the listed power plants. They are used in the plant profiles in both the main section of the directory and in the indexes.

- ● operable
- □ planned
- ■ on order
- ○ under construction
- ☆ cancelled
- △ indefinitely deferred
- ★ shut down
- ▲ decommissioning

Abbreviations

The following abbreviations are used in the main section and indexes of this directory.

&	and	Expy.	Expressway
Admin.	Administrative, Administrator	Fl.	Floor
AK	Alaska	FL	Florida
AL	Alabama	Ft.	Fort
Apt.	Apartment	Fwy.	Freeway
AR	Arkansas	GA	Georgia
Assoc.	Associate	Gen.	General
Asst.	Assistant	GU	Guam
Ave.	Avenue	HI	Hawaii
AZ	Arizona	Hwy.	Highway
Bldg.	Building	IA	Iowa
Blvd.	Boulevard	ID	Idaho
Bus.	Business	IL	Illinois
CA	California	IN	Indiana
CEO	Chief Executive Officer	Inc.	Incorporated
Chm.	Chairman	Intl.	International
Chwn.	Chairwoman	Jr.	Junior
c/o	Care of	KS	Kansas
CO	Colorado	KY	Kentucky
Co.	Company	LA	Louisiana
Coord.	Coordinator	Ln.	Lane
Corp.	Corporation	Ltd	Limited
CT	Connecticut	MA	Massachusetts
Ct.	Court	MD	Maryland
DC	District of Columbia	ME	Maine
DE	Delaware	Mgr.	Manager
Dir.	Director	MI	Michigan
Div.	Division	Mktg.	Marketing
Dr.	Drive	MN	Minnesota
E	East	MO	Missouri
Exec.	Executive	MS	Mississippi

MT	Montana	Rm.	Room
Mt.	Mount	RR	Rural Route
N	North	Rte.	Route
Natl.	National	S	South
NC	North Carolina	SC	South Carolina
ND	North Dakota	SD	South Dakota
NE	Nebraska, Northeast	SE	Southeast
NH	New Hampshire	Sec.	Secretary
NJ	New Jersey	Svc(s).	Service(s)
NM	New Mexico	Sq.	Square
No.	Number	Sr.	Senior
NV	Nevada	St.	Saint, Street
NW	Northwest	Sta.	Station
NY	New York	Ste.	Sainte, Suite
Ofc.	Officer	SW	Southwest
OH	Ohio	Terr.	Terrace, Territory
OK	Oklahoma	TN	Tennessee
OR	Oregon	Tpke.	Turnpike
PA	Pennsylvania	Treas.	Treasurer
Ph.D.	Doctor of Philosophy	TX	Texas
Pkwy.	Parkway	U.S.	United States
Pl.	Place	U.S.A.	United States of America
PO Box	Post Office Box	UT	Utah
PR	Puerto Rico	VA	Virginia
Pres.	President	VI	Virgin Islands
Prof.	Professor	V.Pres.	Vice President
Prog.	Program	VT	Vermont
Rd.	Road	W.	West
RD	Rural Delivery	WA	Washington
Rep.	Representative	WI	Wisconsin
RFD	Rural Free Delivery	WV	West Virginia
RI	Rhode Island	WY	Wyoming

How a Nuclear Power Plant Works

Italicized items indicate terms in the text that are defined in the Glossary on page 525 of this directory.

Commercial nuclear power plants are built to generate electricity. To understand how electricity is produced, it is necessary first to understand atoms, the fission process, nuclear fuel, and nuclear reactors, each of which is described below.

Atoms: The Foundation of Nuclear Energy

The process that produces the heat in nuclear power plants involves *atomic energy*. Atoms consist primarily of three types of particles: protons, neutrons, and electrons.

Inside the *nucleus* at the center of each *atom* are positively charged protons. The number of protons in the nucleus determines to which family or element the atom belongs; for example, all carbon atoms have six protons.

The nucleus also contains uncharged particles called *neutrons* (except for the nucleus of a hydrogen atom, which has no neutrons). Among atoms belonging to the same family or element (i.e., having the same number of protons), it is the number of neutrons that distinguishes one atom from another. For example, the carbon-12 atom contains six protons and six neutrons in the nucleus; the carbon-14 atom has the same number of protons as carbon-12 (six), but has eight neutrons instead of six. Atoms of the same element containing the same number of protons but different numbers of neutrons are called *isotopes*. Isotopes are designated by a number after the element name that indicates the total number of protons and neutrons inside the nucleus.

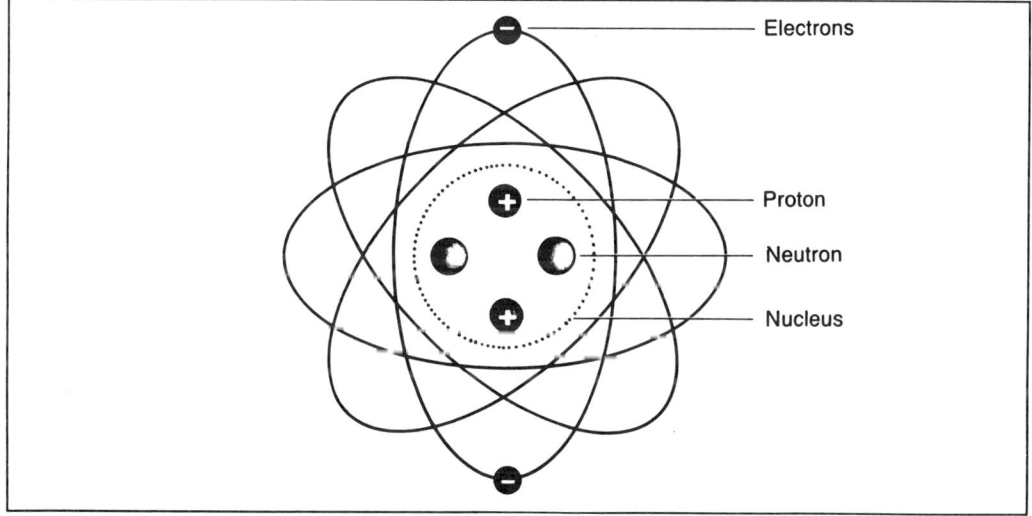

The components of an atom.

Circling the nucleus of each atom at varying distances are tiny, negatively charged *electrons*. In most atoms there are usually the same number of electrons as protons, but if there are more electrons than protons the atom is said to have a negative charge. Conversely, if there are more protons than electrons the atom is said to have a positive charge. If an atom carries either a negative or a positive charge it is called an *ion* or is said to be *ionized*.

Fission: Splitting the Atom

Particles in the nucleus are held together by a force scientists call "nuclear binding energy." It is possible to overcome the binding energy in some large atoms, such as uranium, causing the atoms to split apart or "*fission*." The fission process occurs when a free neutron at a suitable speed from the nucleus of one atom enters the nucleus of a fissionable atom. The nucleus of the fissionable atom immediately becomes unstable, vibrates, and then splits into two fragments that are propelled apart at a high speed. The kinetic energy (energy of motion) of these fragments is turned into heat when the fission fragments collide with surrounding atoms and molecules.

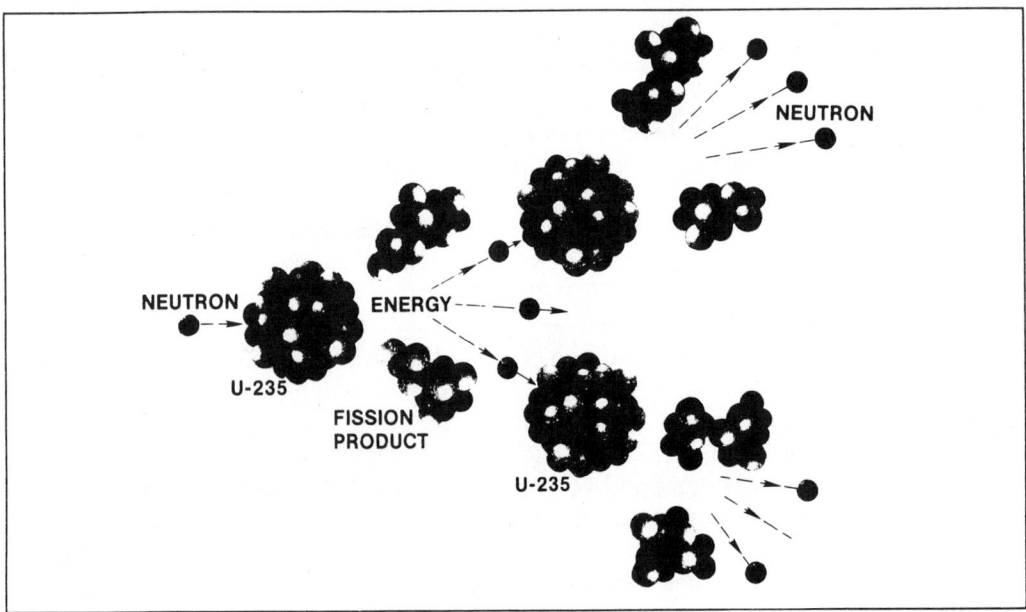

The fission process.

In addition to the fission fragments and heat, a fissioning nucleus also frees two or three additional neutrons. A *nuclear chain reaction* occurs when some of these neutrons strike other fissionable atoms, which release still other neutrons. These neutrons, in turn, hit other fissionable atoms. Controlling the rate at which these "free" neutrons are emitted is the key to sustaining and controlling a nuclear chain reaction.

Reactor Fuel: Uranium and Plutonium

The atom used most often as fuel for nuclear reactors built today is an isotope of *uranium* known as uranium-235 (U-235). When uranium is mined, it contains two isotopes: uranium-238 (U-238), which makes up approximately 99.3 percent of mined uranium, and uranium-235 (U-235), which only makes up approximately 0.7 percent of mined uranium. Although uranium is quite common in nature (about 100 times more common than silver, for example), U-235 is relatively rare. Before uranium can be used as a fuel in a nuclear power plant, the 0.7 percent concentration of U-235 must be *enriched* to around a 3 percent concentration.

Although U-238, the more common uranium, is not fissionable under most conditions, it is fertile, which means that instead of fissioning when it absorbs a neutron it transforms into an atom that is itself fissionable, called *plutonium*-239 (Pu-239). In other words, when neutrons from fissions are absorbed by U-238, the U-238 is converted into Pu-239, which is fissionable and can be used as fuel in some reactors. As reactors burn or consume fuel (U-235), they simultaneously generate fuel by transforming otherwise unusable U-238 into Pu-239.

Nuclear Reactors: Controlling Chain Reactions

The *reactor* is the unique element that distinguishes a nuclear power plant from other electric power plants; the rest of the buildings and equipment are similar from plant to plant. Nuclear reactors are machines that contain and control atomic chain reactions while releasing heat at a controlled rate. The produced heat turns water into steam, which in turn drives *turbine-generators*. The reactor consists of the following four main elements:

- Fuel. The nuclear fuel is the heart of the reactor. In most U.S. reactors, the fuel consists of pellets of enriched uranium dioxide encased in 12-foot-long metal tubes, called *fuel rods*. These fuel rods are bundled to form *fuel assemblies.*

- Control rods. These rods have cross-shaped blades containing materials that absorb neutrons; they are used to regulate the rate of the chain reaction. If they are pulled out of the core, the reaction speeds up. If they are inserted into the core, they capture a larger fraction of the free neutrons and the reaction slows. The *control rods* are interspersed among the fuel assemblies in the core. Boron is a widely used absorber material.

- Coolant. A *coolant*, usually water, is pumped through the reactor to carry away the heat produced by the fissioning of the fuel. This is comparable to the water in the cooling system of a car, which carries away the heat built up in the engine. In large reactors, as much as 330,000 gallons of water per minute flow through the reactor core every minute to carry away the heat. Most U.S. reactors are called *light-water reactors* (LWRs) because they are cooled by ordinary or light water.

- Moderator. Neutrons have a better chance of causing an atom to fission if they move considerably slower than their initial speed after being emitted by a fissioning nucleus. The material used to slow the neutrons is called the *moderator*. Fortunately for reactor designers and owners, water itself is an excellent moderator, so reactors can be moderated by the same water that serves as a coolant. The moderator is essential to maintain a chain reaction; if water is lost from the core, the chain reaction stops (although the residual heat must still be removed).

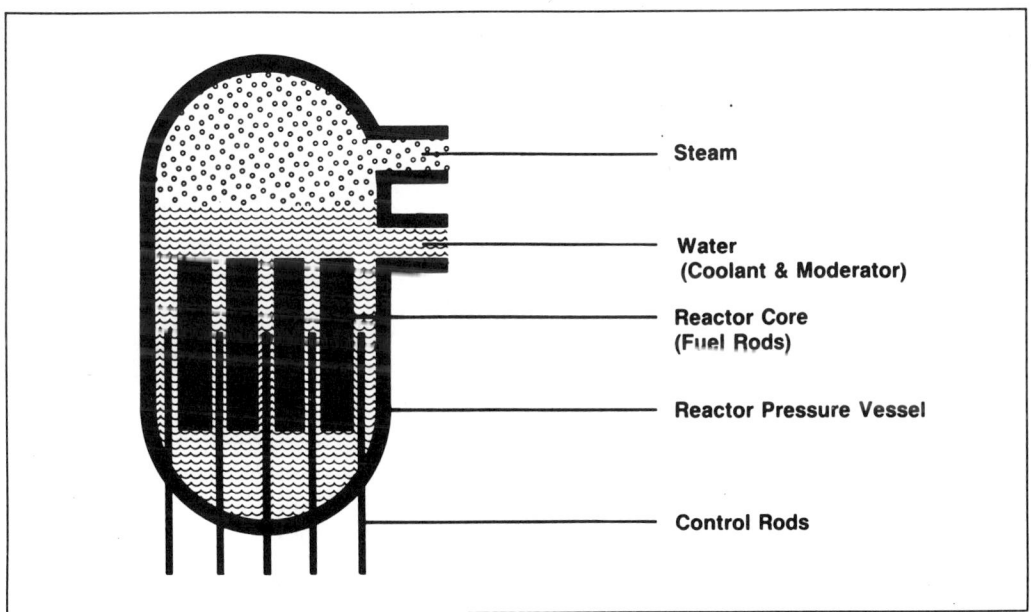

Elements of a water-cooled nuclear reactor.

Although engineering designs are quite complex, these four elements–the fuel, the control rods, the coolant, and the moderator–are the basic components of a nuclear reactor. When the control rods are withdrawn, the uranium fuel begins to fission and release extra neutrons, the neutrons are slowed by the moderator so that they will continue the chain reaction, and the heat is carried away by the coolant.

Below is a summary of the electricity-producing process in a nuclear power plant:

Heat from the fission process turns water into steam;

The steam flows into a turbine and turns a shaft to spin a generator and generate electricity, losing some of its heat and pressure in the process;

The steam then moves to a *condenser*, where water flowing through cooling pipes cools it and condenses it back into water. This water, called "condensate," is preheated to make use of a bit more of the heat in the low-pressure steam and is fed back into the reactor to begin the cycle once again.

The water flowing through the cooling pipes, totally separate from the condensate, is handled differently. Cooling water is necessary for all electric power plants that make steam from a heat source, not just for nuclear plants. For that reason, electric plants of many kinds are typically located near a river, lake, or other body of water. The cooling water for the plant is pumped from the body of water through pipes to the plant, where it cools the steam. In the process of cooling the steam, the temperature of the cooling water itself rises. To dissipate this leftover heat in the cooling water, many electric power plants pump the water through a cooling tower or a specially built pond. Later, the water is fed back into its original source. The cooling water does not come into contact with the nuclear reactor or with radioactive materials.

Types of Nuclear Reactors

Just as there are different approaches to designing and building airplanes and automobiles, engineers have developed different types of nuclear power plants. Several types are used in the United States: boiling water reactors (BWRs), pressurized water reactors (PWRs), and high temperature gas-cooled reactors (HTGRs). PWRs and BWRs are generically called light-water reactors (LWRs) because they use ordinary water as a coolant. The electricity-generating process is essentially the same for all of them; the principal differences lie inside the reactor that produces the heat.

Boiling Water Reactors

In a *boiling water reactor* (BWR), the water that is heated by the core turns directly to steam in the reactor vessel, and the same steam is used to power the turbine-generator.

The water in a BWR is piped around and through the reactor core and is transformed into steam as it flows up between the elements of the nuclear fuel. The steam exits the reactor through a pipe at the top, turns the turbine-generator, is condensed back to water, and is pumped back into the reactor vessel, beginning the process again.

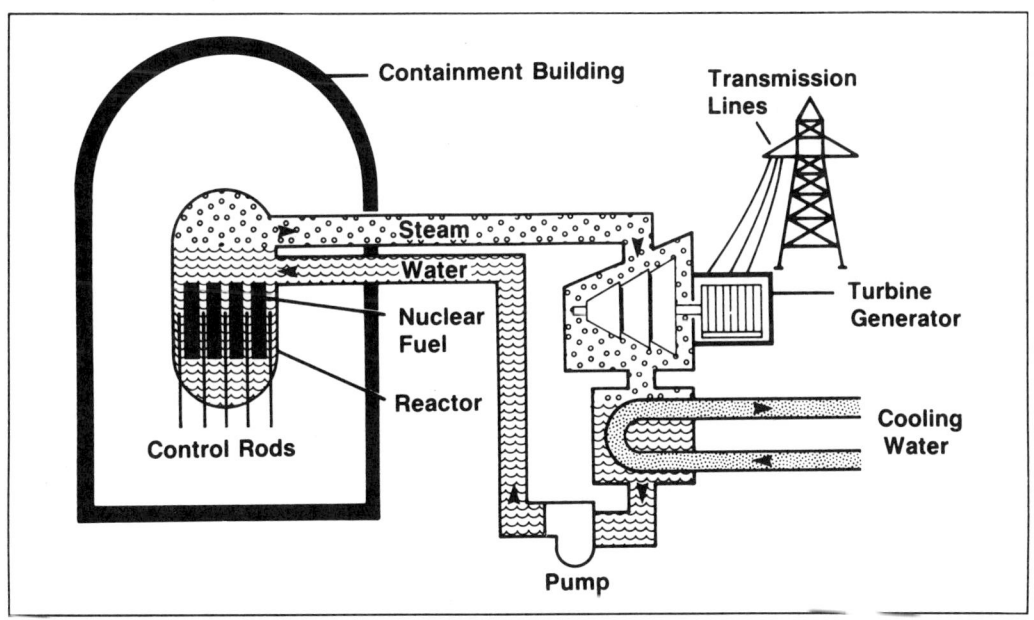

Boiling water reactor.

Normally, water turns to steam at a temperature of 212 degrees Fahrenheit (100 degrees Celsius). But at such a low temperature, steam–like a boiling tea kettle–contains too little energy to be used in a turbine-generator. To raise the temperature and the energy content, the water in a BWR is kept at a pressure of 1,000 pounds per square inch (psi), instead of the normal atmospheric pressure of about 15 psi. Because of this added pressure, the water does not boil and turn to steam until it reaches a temperature of about 545 degrees Fahrenheit (285 degrees Celsius). This higher temperature adds to the energy value of the steam in turning the turbine.

Pressurized Water Reactors

In a *pressurized water reactor* (PWR), the water passing through the core is kept under sufficient pressure so that it does not turn to steam at all–it remains liquid. Steam to drive the turbine is generated in a separate piece of equipment.

The PWR system is known as a double-loop system because it involves two separate circuits of water–or loops–that never physically mix with each other. The water that flows through the part of the reactor known as the *primary loop* is pressurized to about 600 degrees Fahrenheit (315 degrees Celsius) without boiling and leaves the reactor as a liquid. It is pumped through tubes in the steam generator. After transferring its heat to what is called the *secondary loop*, the highly pressurized water in the primary loop is pumped back to the core to be reheated and continue with the process. The secondary loop water circulates around the tubes in the *steam generator*, picking up or "exchanging" heat from the primary loop. This *heat exchange* turns the secondary water into steam, which flows toward the turbine at a temperature of about 500 degrees Fahrenheit (260 degrees Celsius).

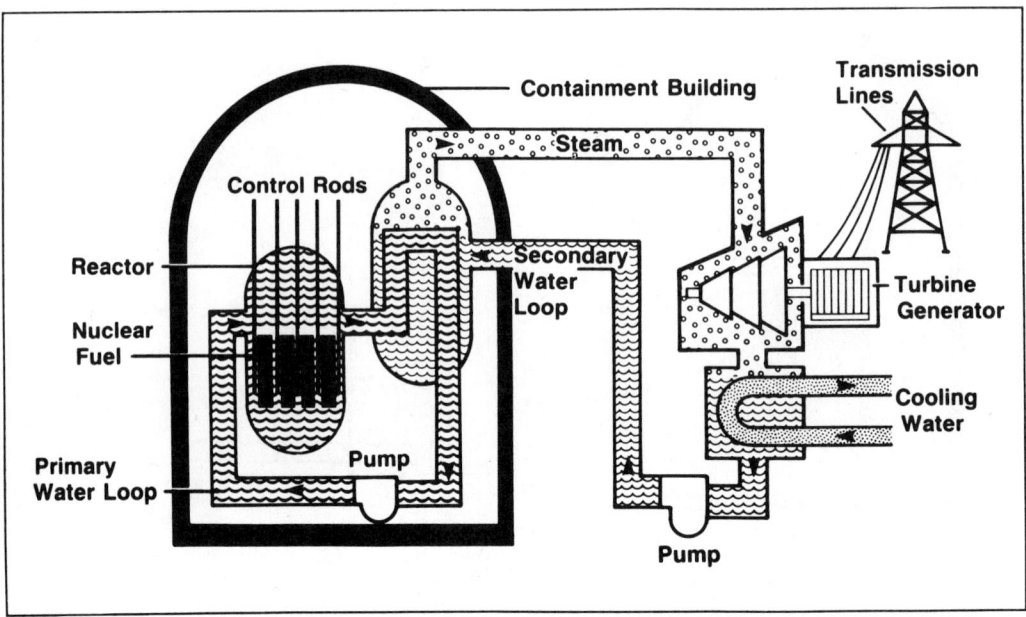

Pressurized water reactor.

High Temperature Gas-Cooled Reactors

High temperature gas-cooled reactors (HTGRs) are also double-loop systems. The principal difference between HTGRs and PWRs is that the coolant in the primary loop–which flows through the core to carry away the heat–is not water, but a gas.

The gas used in HTGRs is helium, which is circulated through pipes in the primary loop by huge blowers. The gas, kept under a pressure of several hundred pounds per square inch, can achieve much higher temperatures than water. In some designs, the gas can be heated to as much as 1,400 degrees Fahrenheit (760 degrees Celsius). As a result, the steam produced from the water in the secondary loop, which powers the turbines, can have temperatures as high as 1,000 degrees Fahrenheit (538 degrees Celsius). This higher temperature leads to improved thermal efficiency; that is, more electric power is generated for the same amount of heat from the fuel.

Another major difference between gas-cooled reactors and water-cooled reactors is the moderator. In water-cooled reactors the water serves as a moderator to slow neutrons and increase the likelihood of atoms fissioning. Gas, however, is not a satisfactory moderator because it is not dense enough, so another material must be included in the core of gas-cooled reactors. The moderator in gas-cooled reactors is graphite, which can withstand the high temperature of these systems. The fuel, uranium carbide particles, is distributed throughout the graphite in the core of an HTGR.

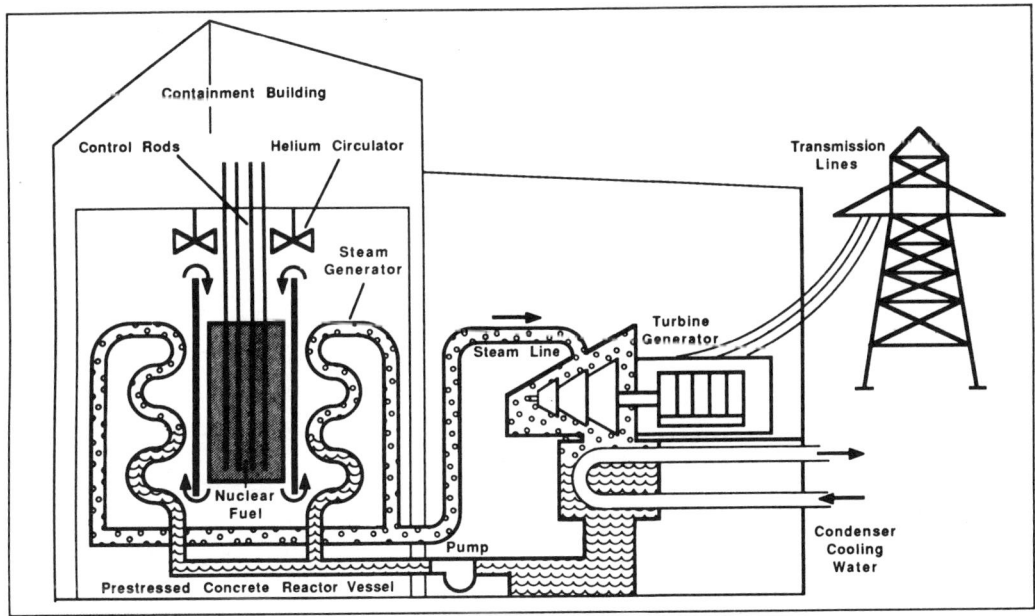

High temperature gas-cooled reactor.

Breeder Reactors

Scientists and engineers have been working for more than three decades on *breeder reactor* technology. Breeder reactors greatly increase the energy extracted from uranium by converting that finite energy resource into a virtually inexhaustible energy supply.

All nuclear power plants produce new fuel material while they are operating–extra neutrons, produced by fission, are absorbed by U-238 atoms, which are then transformed into fissionable plutonium. Some reactors are designed to do this so efficiently that they actually produce more fuel than they consume and are called "breeder" reactors.

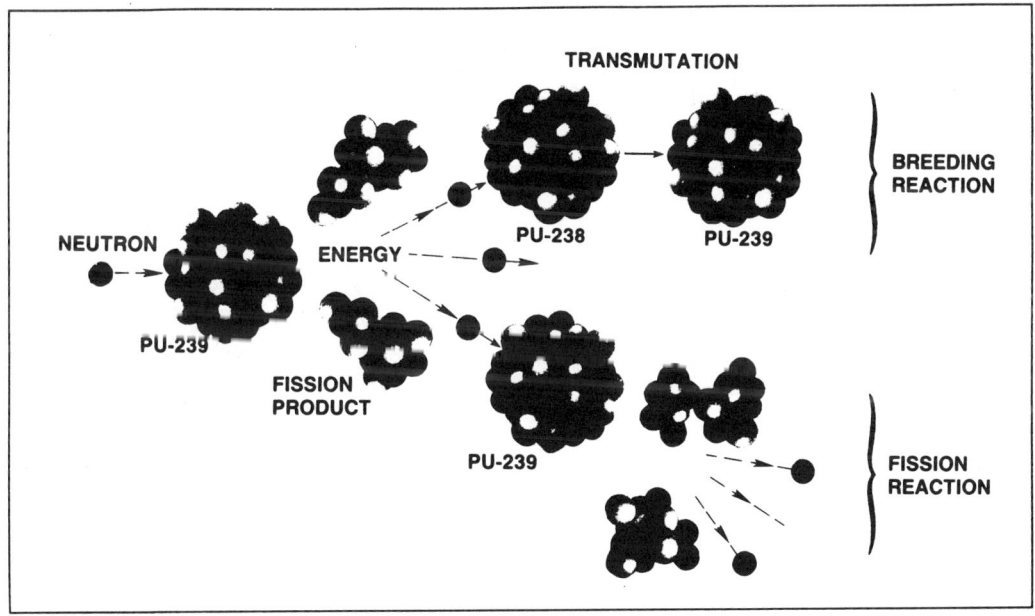

The breeding process.

Breeder reactors are able to multiply the amount of energy available from uranium resources. By using the U-238, which exists in great quantities as an otherwise useless leftover from the uranium enrichment process, a breeder reactor can generate 60 times as much usable energy from natural uranium as today's nuclear power plants.

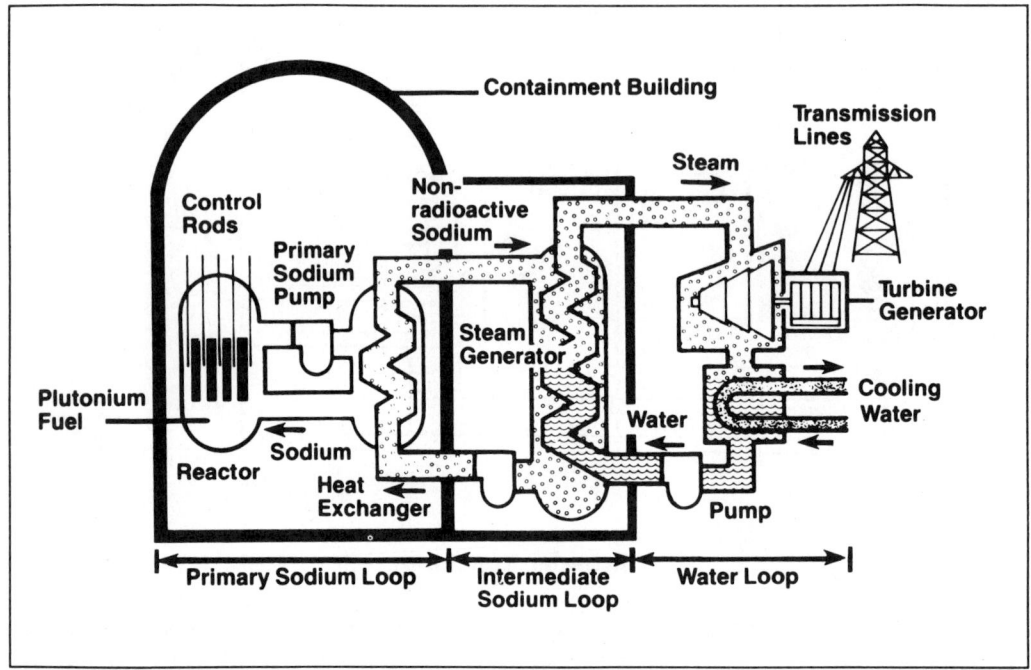

Liquid metal fast breeder reactor.

Several reactor types have the potential for breeding. The one that has been developed most thoroughly through experimental reactors and actual operating plants is cooled by circulating a liquid metal (sodium) through it. It is called the liquid metal fast breeder reactor, or LMFBR.

Advanced Reactors

Research programs are now underway to develop advanced reactor designs that will be suited to future energy needs. Efforts are focused on reactor designs that can be factory-built as small, modular units, according to a standardized design and passive safety system features.

Small, modular reactors would allow utilities to meet an increased need for electricity by installing a single unit, which could be supplemented with additional units as demands for power increased. Factory fabrication would reduce construction costs, while use of standardized designs would reduce the time and costs involved in licensing. Passive safety systems would be integral to the new plant designs; for example, excessive heat would be more easily removed from the reactor core, thus reducing reliance on complex mechanisms and operator control.

Nuclear Power Plant Safety

Safety is of major importance when deciding to license, build, and operate nuclear power plants. In the United States, operators of nuclear power plants must demonstrate to the *U.S. Nuclear Regulatory Commission (NRC)*–the independent federal agency responsible for licensing and regulating civilian nuclear facilities–that each plant is designed and built according to stringent safety requirements. Most of these requirements have one overall objective: to prevent or minimize the accidental release of radioactive material from the plant. The routine operation of nuclear power plants must also meet stringent safety requirements.

Several barriers to trap and contain radioactive material are designed into every nuclear power plant. They include:

- Ceramic fuel pellets. The uranium dioxide fuel material is pressed into pellets to provide a stable form.

- Zircaloy fuel rods. The tubes, or fuel rods, that hold the uranium fuel pellets are made out of a strong alloy of zirconium and tin called "zircaloy." They prevent the solid and gaseous fission products from spreading through the reactor system.

- Containment building. As a further measure of protection, the entire reactor is surrounded by a massive concrete and steel "containment building." It has the single purpose of preventing radioactive materials from reaching the environment in the event that piping systems inside should leak or break. The concrete in the containment building is typically about three feet thick, lined with 3/4-inch steel. The containment building is designed to protect the reactor from being damaged by the direct hit of a large aircraft or tornado winds up to 300 mph.

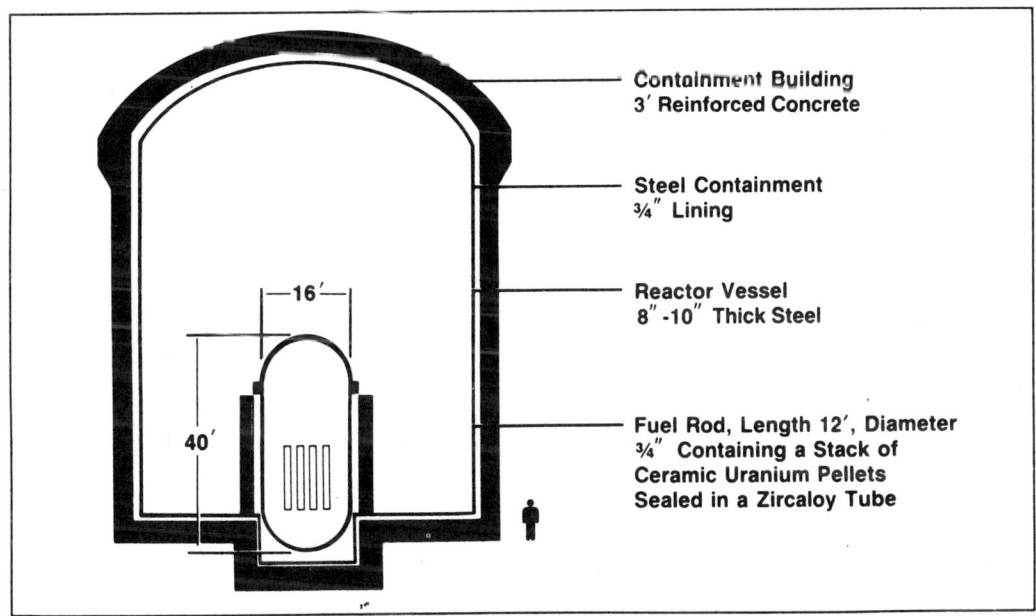

Cross section of a pressurized water reactor (PWR) containment building. Safety features are shown above.

In addition to these physical barriers, nuclear power plants are designed and built with several safety systems and backup safety systems. These systems are designed to protect against malfunctions, mistakes, and potential accidents. For example, the most extensively studied accident is called a *"loss-of-coolant accident"* (LOCA). If the reactor core is not constantly cooled by water, parts of the core can melt, resulting in what's called a *meltdown*. Even after the control rods shut the reactor down, there is still *"decay heat"* that requires cooling. To prevent LOCAs, nuclear plants contain several backup systems that can be called on to cool the core if the primary cooling system should stop functioning.

Source

Atoms to Electricity. U.S. Department of Energy, November 1987.

Maps of Plant Locations

The following nine maps show the locations of all currently operational commercial nuclear reactors worldwide, as well as the locations of commercial plants that are shut down, decommissioning, on order, planned, or indefinitely deferred; cancelled plants are not included. Site locations are numbered on each map, and each map is accompanied by a legend to show the specific names of the plants.

MAP 1

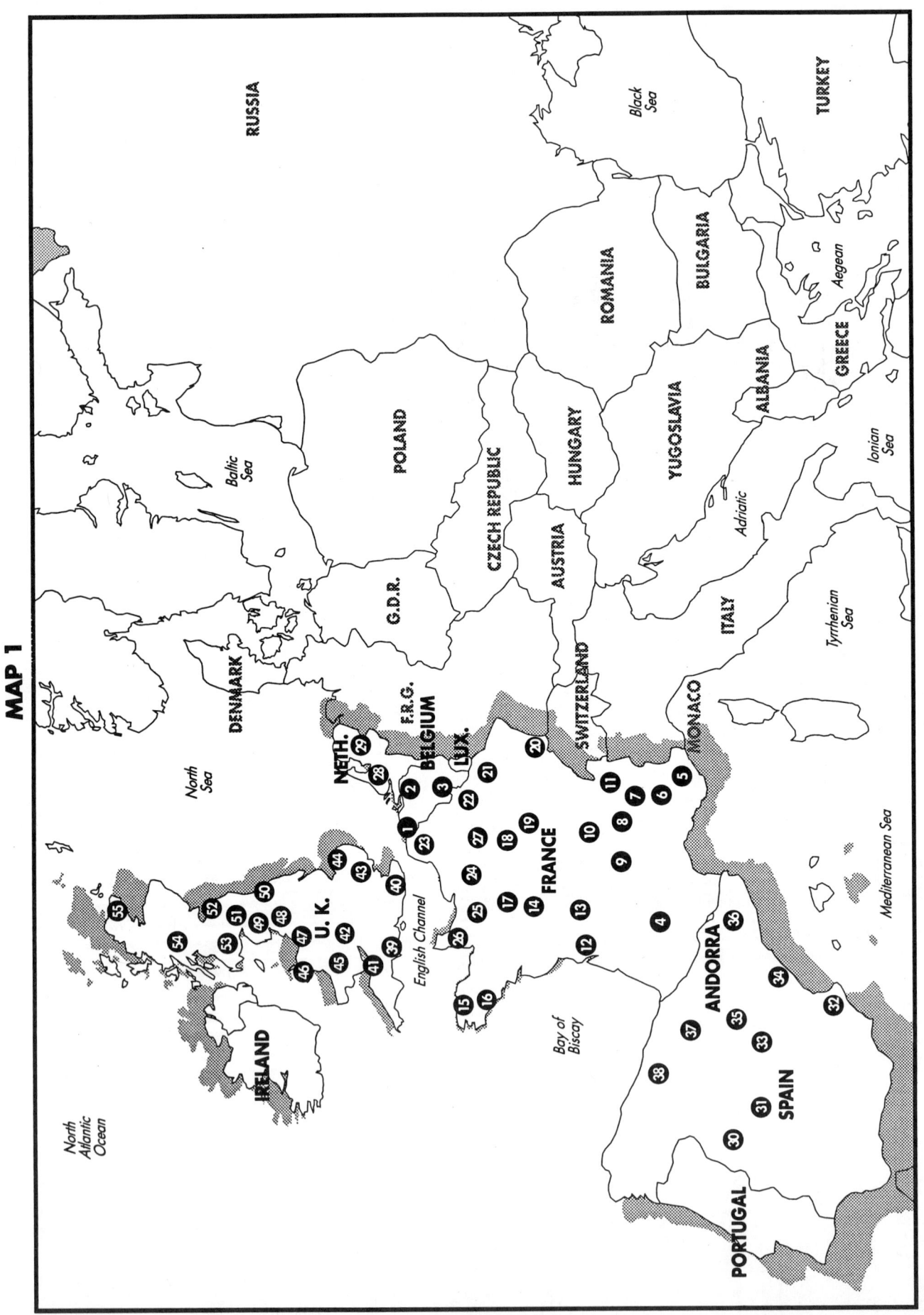

Maps of Plant Locations

BELGIUM
1. Doel Nuclear Power Station
2. Mol BR3
3. Tihange Nuclear Power Plant

FRANCE
4. Golfech Nuclear Power Plant
5. Phenix Prototype Fast Reactor Power Station
6. Marcoule G. Nuclear Power Plant
7. Tricastin Nuclear Power Plant
8. Saint-Alban Nuclear Power Plant
9. Cruas Nuclear Power Plant
10. Bugey Nuclear Power Plant
11. Superphenix Fast Breeder Power Station
12. Blayais Nuclear Power Plant
13. Civaux Nuclear Power Plant
14. Saint-Laurent Nuclear Power Plant
15. Monts d' Arree Nuclear Power Plant
16. Plogoff Nuclear Power Plant
17. Chinon Nuclear Power Plant
18. Dampierre Nuclear Power Plant
19. Belleville Nuclear Power Plant
20. Fessenheim Nuclear Power Plant
21. Cattenom Nuclear Power Plant
22. Chooz Nuclear Power Plant
23. Gravelines Nuclear Power Plant
24. Penly Nuclear Power Plant
38. Palvel Nuclear Power Plant
26. Flamanville Nuclear Power Plant
27. Nogent s/Seine Nuclear Power Plant

NETHERLANDS
28. Borssele Nuclear Power Plant
29. Dodeward Nuclear Power Plant

SPAIN
30. Almaraz Nuclear Center
31. Valdecaballeros Nuclear Center
32. Confrentes Nuclear Power Plant
33. Jose Cabrera Nuclear Power Plant
34. Vandellos Nuclear Center
35. Trillo Nuclear Power Plant
36. Asco Nuclear Center
37. Santa Maria de Garona Nuclear Center
38. Lemoniz Nuclear Power Plant

UNITED KINGDOM
39. Winfrith Steam Generating Heavy Water Reactor
40. Dungeness Power Station
41. Hinkley Point Power Station
42. Berkeley Power Station
43. Bradwell Power Station
44. Sizewell Power Station
45. Oldbury Power Station
46. Wylfa Power Station
47. Transfynydd Power Station
48. Heysham Power Station
49. Calder Hall Nuclear Power Station
50. Hartlepool Power Station
51. Windscale Advanced Gas Cooled Reactor
52. Torness Nuclear Power Station
53. Chaplecross Nuclear Power Station
54. Hunterston Nuclear Power Station
55. Dounreay Prototype Fast Breeder Reactor

MAP 2

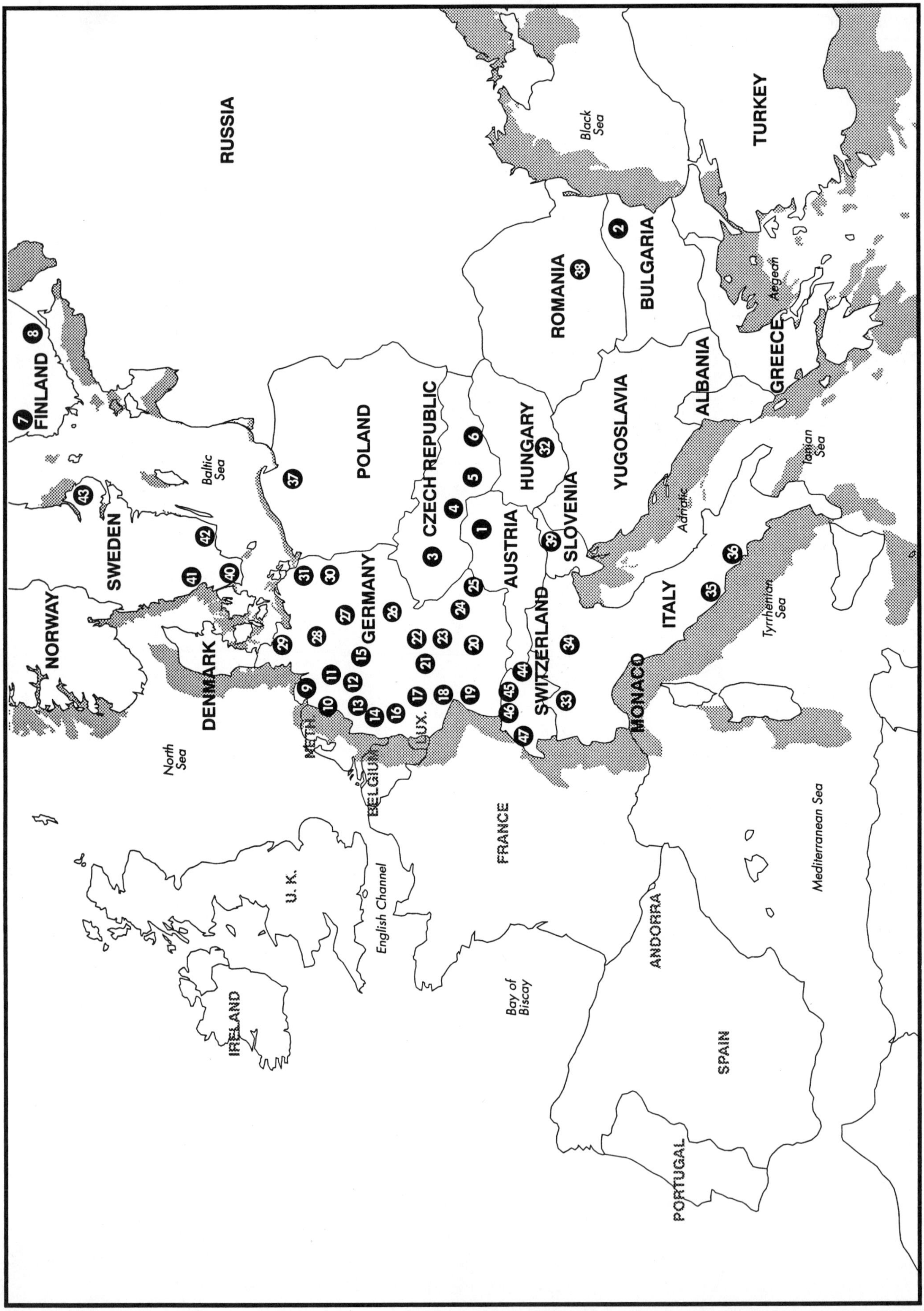

Maps of Plant Locations

AUSTRIA
1 Tullnerfeld Nuclear Power Plant

BULGARIA
2 Kozlodug Nuclear Power Plant

CZECH REPUBLIC
3 Temelin Nuclear Power Plant
4 Dukorany Nuclear Power Plant
5 Bohunice Nuclear Power Plant
6 Mochovce Nuclear Power Plant

FINLAND
7 Olkilouto Nuclear Power Plant
8 Loviisa Power Plant

GERMANY
9 Brunsbuettel Nuclear Power Plant
10 Stade Nuclear Power Plant
11 Brokdorf Nuclear Power Plant
12 Unterweser Nuclear Power Plant
13 Emsland Nuclear Power Plant
14 Lingen Nuclear Power Plant
15 Grohnde Nuclear Power Plant
16 THTR-300 Demonstration Reactor
17 Muelheim-Kaerlich Nuclear Plant
18 Biblis Nuclear Power Plant
19 Karlruhe Nuclear Research Facility
20 Gundremmingen Nuclear Power Plant
21 Obrigheim Nuclear Power Plant
22 Philippsburg Nuclear Power Plant
23 Neckar Nuclear Power Plant
24 Niederaichbach Nuclear Power Plant
25 Isar Nuclear Power Plant
26 Grafenrheinfeld Nuclear Power Plant
27 Stendal Nuclear Power Plant
28 Wuergassen Nuclear Power Plant
29 Kruemmel KKK Nuclear Power Plant
30 Rheinsberg Nuclear Power Plant
31 Greifswald Nuclear PowerPlan

HUNGARY
32 Paks Nuclear Power Plant

ITALY
33 Trino Vercellese Nuclear Power Plant
34 Caorso Nuclear Power Plant
35 Latina Nuclear Power Plant
36 Garigliano Nulcar Power Plant

POLAND
37 Zarnowiec Nuclear Power Plant

ROMANIA
38 Cernavoda Nuclear Power Station

SLOVENIA
39 Krsko Nuclear Power Plant

SWEDEN
40 Barseback Nuclear Power Plant
41 Ringhals Nuclear Power Plant
42 Oskarshamn Nuclear Power Plant
43 Forsmark Nuclear Power Plant

SWITZERLAND
44 Leibstadt Nuclear Power Plant
45 Beznau Nuclear Power Plant
46 Goesgen Nuclear Power Plant
47 Muehleberg Nuclear Power Plant

MAP 3

Maps of Plant Locations

UNITED STATES

Arizona
1. Palo Verde Nuclear Generating Station

California
2. Humboldt Bay Nuclear Power Plant
3. Rancho Seco Nuclear Generating Station
4. Vallecitos Nuclear Power Plant
5. Santa Susana Sodium-Graphite Reactor Experiment
6. Diablo Canyon Nuclear Power Plant
7. San Onofre Nuclear Generating Station

Colorado
8. Fort St. Vrain Nuclear Power Plant, Colorado

Kansas
9. Wolf Creek Nuclear Power Plant, Kansas

Nebraska
10. Fort Calhoun Generating Station
11. Cooper Nuclear Power Plant
12. Hallam Nuclear Power Plant

Oregon
13. Trojan Nuclear Plant

South Dakota
14. Pathfinder Nuclear Power Plant

Texas
15. Comanche Peak Nuclear Power Plant
16. South Texas Project Electric Generating Station

Washington
17. Washington Nuclear Project

MEXICO
18. Laguna Verde Nuclear Power Plant

MAP 4

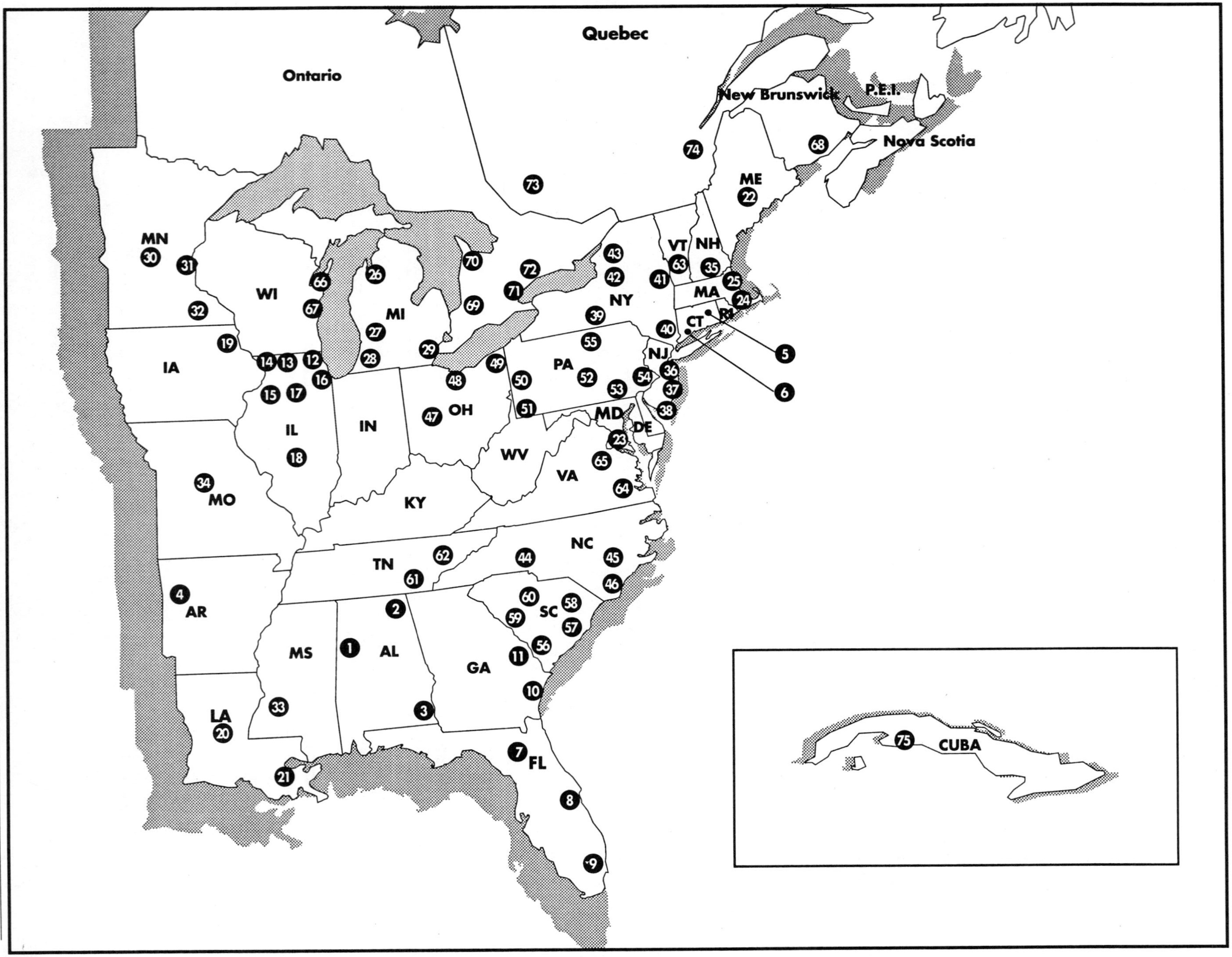

p. xxxvi

Maps of Plant Locations

UNITED STATES

Alabama
1. Browns Ferry Nuclear Plant
2. Bellefonte Nuclear Power Plant
3. Joseph M. Farley Nuclear Power Station

Arkansas
4. Arkansas Nuclear One Steam Electric Station

Connecticut
5. Millstone Nuclear Power Plant
6. Haddam Neck Nuclear Power Plant

Florida
7. Crystal River Energy Complex
8. St. Lucie Nuclear Power Station
9. Turkey Point Nuclear Station

Georgia
10. Edwin I. Hatch Nuclear Power Plant
11. Vogtle Nuclear Power Station

Illinois
12. Zion Station
13. Byron Station
14. Carroll County Nuclear Power Plant
15. Quad Cities Station
16. Dresden Nuclear Station
17. La Salle County Nuclear Station
18. Clinton Nuclear Power Station

Iowa
19. Duane Arnold Energy Center

Louisiana
20. River Bend Station
21. Waterford Power Station

Maine
22. Maine Yankee Nuclear Power Plant

Maryland
23. Calvert Cliffs Nuclear Power Plant

Massachusetts
24. Pilgrim Station
25. Yankee Rowe Nuclear Power Plant

Michigan
26. Big Rock Point Nuclear Power Plant
27. Palisades Nuclear Plant
28. Donald C. Cook Nuclear Plant
29. Enrico Fermi Fast Breeder Reactor Plant

Minnesota
30. Monticello Nuclear Generating Plant
31. Elk River Nuclear Power Plant
32. Prairie Island Nuclear Plant

Mississippi
33. Grand Gulf Nuclear Station

Missouri
34. Callaway Nuclear Plant

New Hampshire
35. Seabrook Station

New Jersey
36. Salem Nuclear Generating Station
37. Oyster Creek Nuclear Generating Station
38. Hope Creek Nuclear Generating Station

New York
39. Robert Emmet Ginna Nuclear Power Plant
40. Indian Point 3 Nuclear Power Plant
41. Shoreham Nuclear Power Plant
42. Nine Mile Point Nuclear Power Plant
43. James A. Fitzpatrick Nuclear Power Plant

North Carolina
44. McGuire Nuclear Station
45. Shearon Harris Nuclear Power Plant
46. Brunswick Nuclear Plant

Ohio
47. Pique Nuclear Power Plant
48. Davis-Besse Nuclear Plant
49. Perry Nuclear Power Plant

Pennsylvania
50. Shippingport Atomic Power Station
51. Beaver Valley Power Station
52. Susquehanna Steam Electric Station
53. Three Mile Island Nuclear Power Plant
54. Peachbottom Nuclear Power Plant
55. Limerick Generating Station

South Carolina
56. Virgil C. Summer Nuclear Station
57. H.B. Robinson Electric Power Plant
58. Catawba Nuclear Station
59. Oconee Nuclear Station
60. Carolinas-Virginia Tube Reactor

Tennessee
61. Sequoyah Nuclear Plant
62. Watts Bar Nuclear Plant

Vermont
63. Vermont Yankee Nuclear Power Plant

Virginia
64. Surry Nuclear Power Station
65. North Anna Nuclear Power Station

Wisconsin
66. Kewaunee Nuclear Plant
67. Point Beach Nuclear Plant

CANADA

New Brunswick
68. Point Lepreau Nuclear Generating Station

Ontario
69. Douglas Point Nuclear Generating Station
70. Bruce Nuclear Power Development
71. Pickering Nuclear Generating Facility
72. Darlington Nuclear Generating Station

Quebec
73. Nuclear Power Demonstration Reactor
74. Gentilly Nuclear Generating Station

CUBA
75. Juragua Nuclear Complex

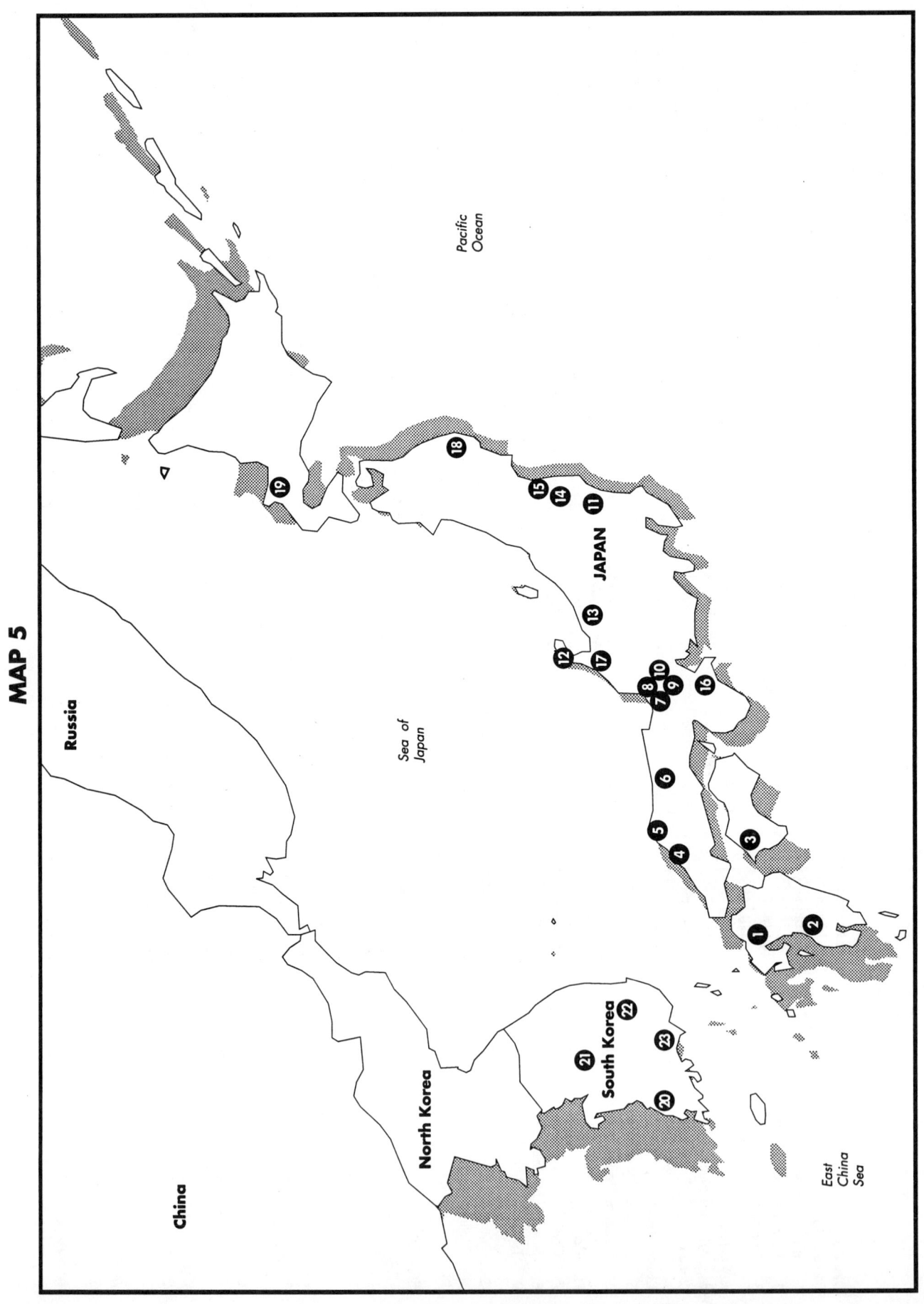

Maps of Plant Locations

JAPAN
1. Genkai Nuclear Power Plant
2. Sendai Nuclear Power Plant
3. Ikata Nuclear Power Plant
4. Shimane Nuclear Power Plant
5. Takahama Nuclear Power Plant
6. Ohi Nuclear Power Station
7. Mihama Power Station
8. Tsuruga Nuclear Power Station
9. Fugen ATR Nuclear Power Plant
10. Monju Nuclear Power Plant
11. Tokai Nuclear Power Station
12. Shika Nuclear Power Plant
13. Maki Nuclear Power Station
14. Fukushima Daiichi Nuclear Power Station
15. Fukushima Daini Nuclear Power Station
16. Hamaoka Nuclear Power Plant
17. Kashiwazaki Kariwa Nuclear Station
18. Onagawa Nuclear Power Station
19. Tomari Nuclear Power Station

REPUBLIC OF KOREA
20. Yonggwang Nuclear Power Plant
21. Kori Nuclear Power Plant
22. Wolsong Nuclear Power Plant
23. Ulchin Nuclear Power Plant

MAP 6

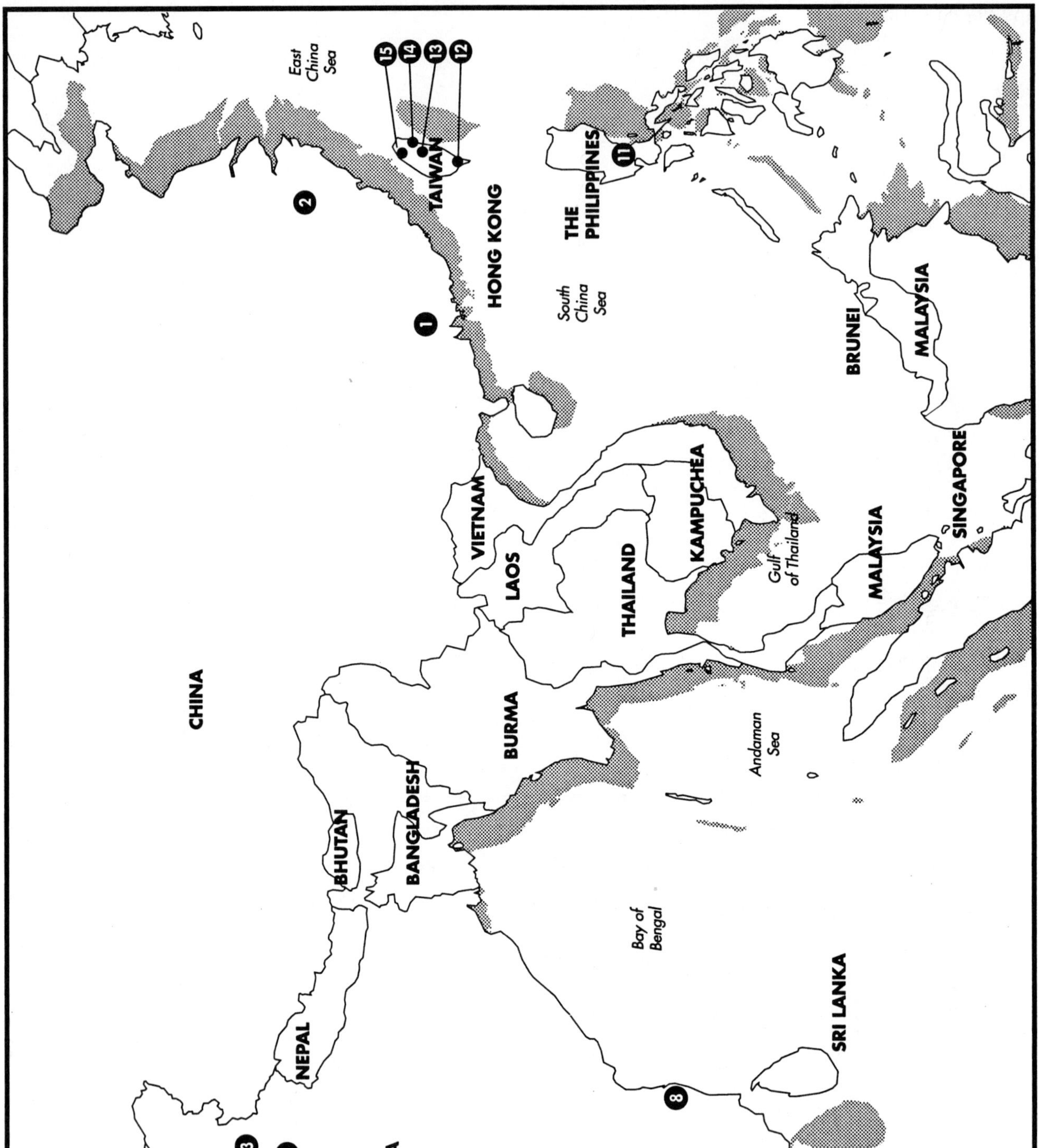

Maps of Plant Locations

CHINA
1. Guang Nuclear Power Plant
2. Qinshan Nuclear Power Plant

INDIA
3. Narora Atomic Power Station
4. Rajasthan Atomic Power Station
5. Kakrapar Atomic Power Station
6. Tarapur Atomic Power Station
7. Kaiga Atomic Power Station
8. Madras Atomic Power Station

PAKISTAN
9. Karachi Nuclear Power Plant
10. Chashma Nuclear Power Plant

PHILIPPINES
11. Philippine Nuclear Power Plant

TAIWAN
12. Maanshan Nuclear Power Station
13. Yenliao Nuclear Power Station
14. Kuosheng Nuclear Power Station
15. Chinshan Nuclear Power Plant

MAP 7

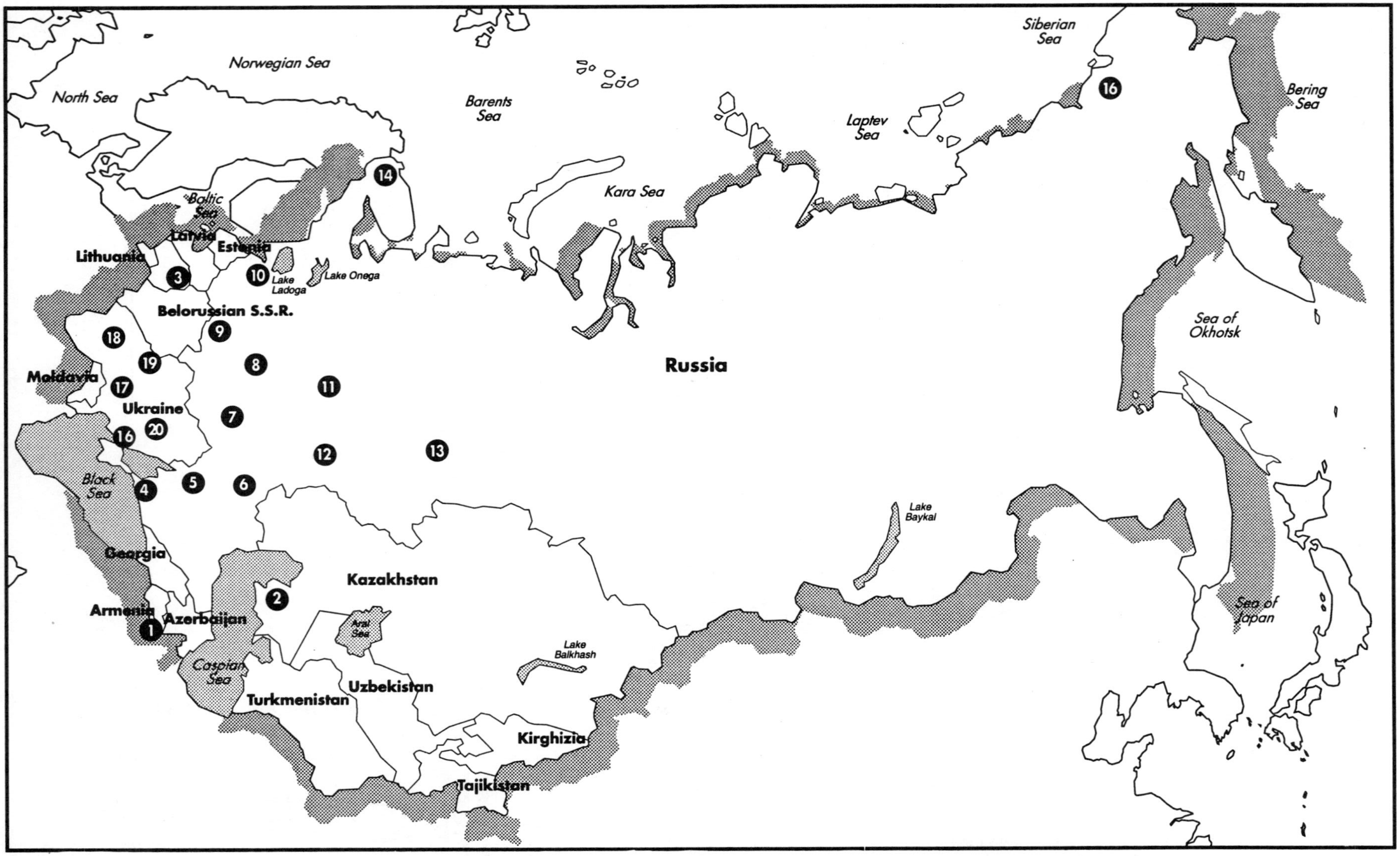

ARMENIA
1 Armenian Nuclear Station

KAZAKHSTAN
2 BN-350 Nuclear Power Plant

LITHUANIA
3 Ignalina Nuclear Power Plant

RUSSIA
4 Kalinin Nuclear Station
5 Kursk Nuclear Station
6 Novovoronezh Nuclear Power Station
7 Rostov Nuclear Power Plant
8 Smolensk Nuclear Power Plant
9 Obninsk Nuclear Power Plant
10 Sosnovy Bor Nuclear Power Plant
11 Gorky Nuclear Facility
12 Balakovo Nuclear Power Plant
13 Kola Nuclear Plant
14 Beloyarsk Nuclear Power Plant
16 Bilibino Nuclear Power Plant

UKRAINE
16 South Ukraine Power Station
17 Khemelinitsty Nuclear Station
18 Rovno Nuclear Power Plant
19 Chernobyl Nuclear Power Station
20 Zaporozhye Nuclear Power Station

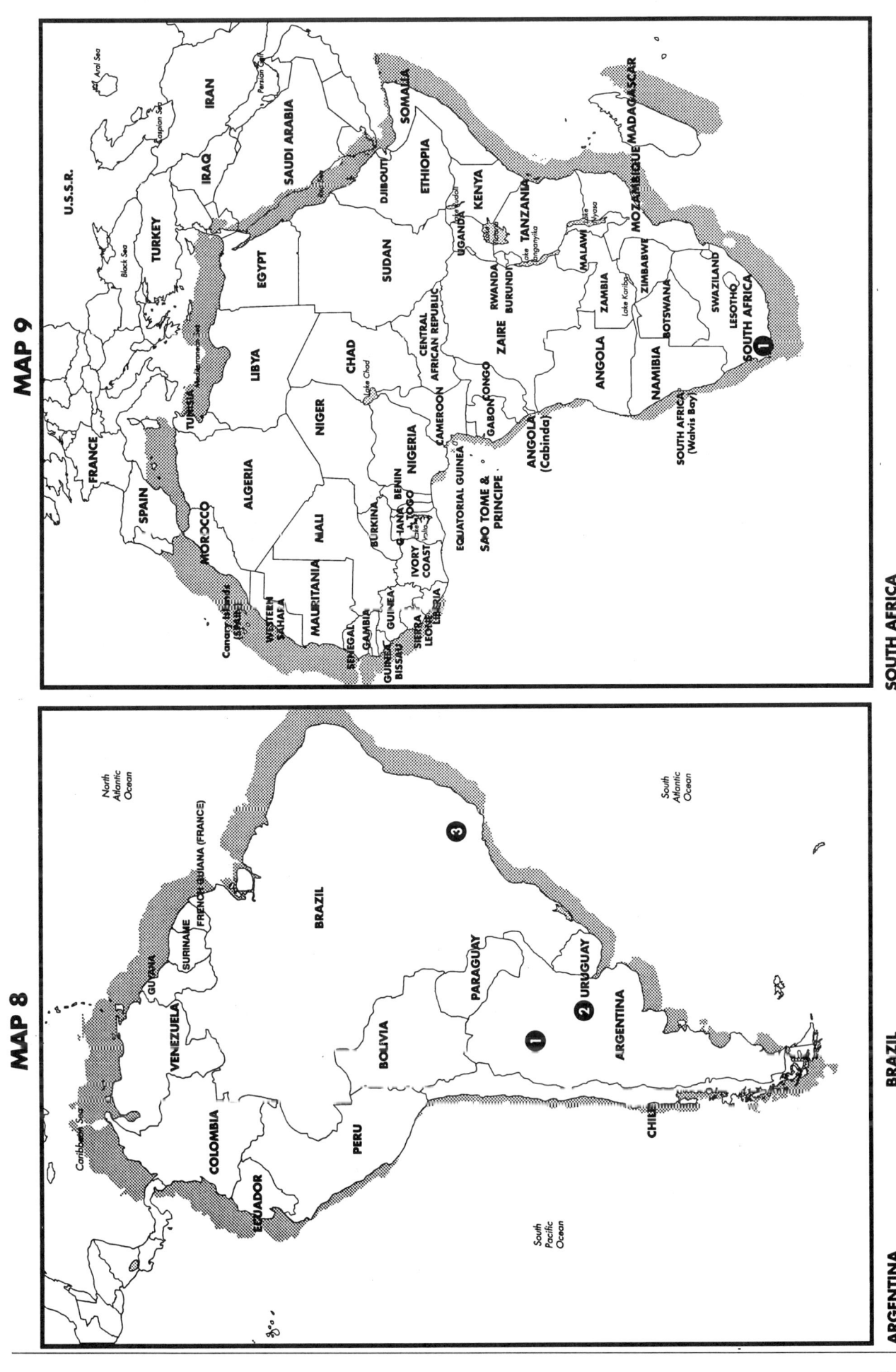

Nuclear Power Plants Worldwide

Argentina

COUNTRY REPORT

Advanced nuclear program in Latin America

Argentina received an Atoms for Peace program grant for a research reactor in 1962 and with India and Pakistan was one of the first Third World adopters of nuclear power for electricity production.

Argentina's nuclear program, the most advanced in Latin America, has been aimed at controlling the complete nuclear cycle—from the mining of uranium to its burning in a nuclear power reactor. Under the Argentine National Atomic Energy Commission (CNEA), formed in 1950, the country has developed the capabilities to prospect for uranium ore at its own deposits, refine and convert the ore into reactor-grade uranium-dioxide, produce zircaloy tubes and accessories required for fuel element fabrication, manufacture fuel elements, enrich uranium through a gaseous diffusion process, recover plutonium produced during nuclear fission, create heavy water used in reactor cooling, and treat and store radioactive waste.

Concern about nuclear weapons

The development of Argentina's nuclear independence has caused concern among some U.S. analysts, for with the technology to enrich uranium comes the capability to make fuel for nuclear explosives. Although former Argentine president Raul Alfonsin said Argentina had no intention of building nuclear weapons, the country has resisted signing the Nuclear Non-Proliferation Treaty supported by over 120 nations. Argentina has also failed to ratify the Tlatelolco Treaty against nuclear weapons in Latin America and has traditionally opposed international inspections and safeguards.

While the Nuclear Non-Proliferation Treaty, in effect since 1970, was supposed to make it harder for nonparticipating nations to obtain nuclear power technology, Argentina has successfully done so. Several countries, including the United States, have transferred key technology to Argentina through the years.

Early in 1982, under the Reagan administration, the U.S. approved the export of an industrial-process computer to Argentina for use in the heavy water production plant under construction at that time.

In August 1982, the U.S. approved the retransfer of 143 tons of U.S.-origin heavy water from West Germany to Argentina.

Canada sold uranium fuel rods to Argentina at the time of the Falkland War.

Both Germany and Canada have assisted Argentina in the construction of its nuclear power plants.

Although these sales had gone toward peaceful uses of nuclear energy, critics argue that nuclear reactor technology can be diverted to develop atomic bombs.

Argentina accepts safeguards

In a joint arrangement with Brazil, Argentina announced late in 1990 that it would accept International Atomic Energy Agency safeguards on all their nuclear facilities. (Historically, Argentina and Brazil have been accused of using their nuclear power programs to compete for ascendancy in Latin American affairs.) Although the militaries in both countries seemed to resist the effort, the safeguard agreement was scheduled to be ratified in December 1991.

Sources

"Atoms in Argentina." *America*, April 12, 1980.

"Power Plant Politics: The Nuclear Care and Feeding of the Third World." *Critical Mass Bulletin*, December 1983.

Nuclear Energy in Argentina. Argentina National Atomic Energy Commission, 1991.

"Brazil's Military May Block Safeguards with Argentina." *Nucleonics Week*, November 28, 1991.

PLANT PROFILES

1 ●
Atucha Nuclear Power Station, Unit 1

[*See Also:* Atucha Nuclear Power Station, Unit 2]

Casilla de Correo 20   Phone: 0328-24671
2806 Lima, Bs. As., Argentina   Telex: CATOM AR 21252
Juan Carlos Duarte, Plant Mgr.

Owning Utility

Comision Nacional de Energia Atomica (CNEA)   Phone: 54-1-544-8370
Av. Del Liberatodor 8250   Fax: 54-1-544-9252
1429 Buenos Aires, Bs. As., Argentina   Telex: 25392 PREAT AR
Manuel A. Mondino, Pres.

Contact

Elida Bustos, Press Office
Comision Nacional de Energia Atomica

Basic Facts

Unit 1 *Status:* Operable. *Type of Reactor:* Pressurized heavy-water reactor. *Megawatts (electric):* 357. *Megawatts (thermal):* 1179. *Reactor System Supplier:* Siemens. *Generator Supplier:* Kraftwerk Union AG (Germany). *Architect Engineer:* Siemens.

Key Dates

Unit 1 *Ordered:* 1967. *Construction began:* 1968. *Criticality:* January 1974. *First Power:* March 1974. *Commercial Operation:* June 1974.

Plant Report

A first for Latin America. The Argentine National Atomic Energy Commission (CNEA) began feasibility studies for Atucha 1 in 1964 and began commercial operation of the plant 10 years later. Located 60 miles north of Buenos Aires on the Parana River, Atucha 1 was the first nuclear power plant in Argentina as well as in Latin America. Together with Argentina's other operating nuclear plant, Embalse, Atucha 1 provides about 15 percent of the country's electricity.

Technical problems. Atucha 1 had several long shutdowns from mid-1987 through 1989. During this time, Argentina experienced major power shortages, and the government ordered rotating electricity cutoffs in Buenos Aires and other areas.

In mid-1987, operations were halted for routine maintenance. When the plant was restarted in October, some pipes broke and a large quantity of water spilled. Although the incident posed no health threat, it caused an additional operating delay.

Late in 1987, plant officials discovered a problem never faced at any nuclear power plant in the world. A probe inside the reactor became unsoldered and broke some of the 253 tubes that hold uranium. To make the repairs, it was necessary to design a tiny robot capable of passing through a tube about five inches in diameter and working in heavy water at a depth of 100 feet.

Power restrictions. Atucha's technical problems, coupled with a drought that dramatically reduced water levels in the rivers that supply hydroelectric plants, caused Argentina's electricity production to fall by 25 percent in early 1989. Power company officials feared a total blackout for Buenos Aires, causing the country to adopt a massive program of restriction on power use.

Sources

"Argentina is Likely to Halt Work on Atom Power Plant." *The New York Times*, April 23, 1988.
"Power Supply Going Down in Argentina." *The New York Times*, January 5, 1989.
Nuclear Energy in Argentina. Argentina National Atomic Energy Commission, 1991.

2 ○
Atucha Nuclear Power Station, Unit 2

[*See Also:* Atucha Nuclear Power Station, Unit 1]

Casilla de Correo 20   Phone: 0328-24671
2806 Lima, Bs. As., Argentina   Telex: CATOM AR 21252
Fernandes De Prado, Mgr. of Plants Under Construction

Owning Utility

Comision Nacional de Energia Atomica (CNEA)   Phone: 54-1-544-8370
Av. Del Liberatodor 8250   Fax: 54-1-544-9252
1429 Buenos Aires, Bs. As., Argentina   Telex: 25392 PREAT AR
Manuel A. Mondino, Pres.

Contact

Luis J. Colangelo, Mgr. of PR Dept.
Comision Nacional de Energia Atomica (CNEA)

Basic Facts

Unit 2 *Status:* Under construction (76 percent complete as of July 1992). *Type of Reactor:* Pressurized heavy-water reactor. *Megawatts (electric):* 745. *Megawatts (thermal):* 2160. *Reactor System Supplier:* Kraftwerk Union AG (Germany). *Generator Supplier:* Kraftwerk Union AG (Germany). *Architect Engineer:* Empresa Nuclear Argentina de Centrales Electricas (ENACE S.A.), a joint arrangement between Kraftwerk Union AG (Germany) and the Comision Nacional de Energia Atomica.

Key Dates

Unit 2 *Ordered:* 1980. *Construction began:* 1981. *Criticality expected* November 1994. *Commercial Operation Expected:* June 1995.

Plant Report

Construction threatened, but not halted. In April 1988, *The New York Times* reported that Argentina was likely to halt construction on Atucha 2, the country's third nuclear power plant. Dr. Emma Perez Ferreira, president of the Argentine National Atomic Energy Commission, said that $750 million was needed by the end of June if work was to continue on the half-finished Atucha 2. Plans were being made for an orderly shutdown of the project, which began construction in 1981.

ARGENTINA

3 Embalse

Burdened by a $54 billion foreign debt, a sharp decline in export income, an opposition-controlled labor movement nervous about falling real wages, and public resistance to higher taxes, the Argentine government seemed in no position to help the faltering project.

However, the Argentine National Atomic Energy Commission reported that construction on Atucha 2 was 71.5 percent complete in early 1991. A 1992 publication of the American Nuclear Society reported the project 76 percent complete.

Sources

"Argentina is Likely to Halt Work on Atom Power Plant." *The New York Times*, April 23, 1988.
Nuclear Energy in Argentina. Argentina National Atomic Energy Commission, 1991.

3 ●

Embalse Nuclear Power Plant

Casilla de Correo 3
5856 Embalse Rio III, Cba., Argentina
Eduardo Diaz, Plant Mgr.

Phone: 0571-22434
Fax: 0571-244577
Telex: 54836 CETOM AR

Owning Utility

Comision Nacional de Energia Atomica (CNEA)
Av. Del Liberatodor 8250
1429 Buenos Aires, Bs. As., Argentina
Manuel A. Mondino, Pres.

Phone: 54-1-544-8370
Fax: 54-1-544-9252
Telex: 25392 PREAT AR

Contact

Luis J. Colangelo, Mgr. of PR Dept.
Comision Nacional de Energia Atomica (CNEA)

Basic Facts

Status: Operable. *Type of Reactor:* Pressurized heavy-water reactor. *Megawatts (electric):* 648. *Megawatts (thermal):* 1179. *Reactor System Supplier:* Atomic Energy of Canada Ltd. *Generator Supplier:* Italimpianti. *Architect Engineer:* Atomic Energy of Canada Ltd; Italimpianti.

Key Dates

Ordered: 1970. *Construction began:* 1973. *Criticality:* March 1983. *First Power:* April 1983. *Commercial Operation:* January 1984.

Armenia

COUNTRY REPORT

Soviet involvement in Armenian nuclear program

Armenian involvement in nuclear power began under the auspices of the former Soviet Union (see Country Report for the former Soviet Union). Government officials blamed a shortage of electrical power for slowing down the development of industry in the republic and thwarting efforts to raise the standard of living among people in rural areas. The former Soviet government decided to meet the increasing demand for electricity with nuclear power. Officials claimed nuclear generation was the best alternative because fossil-fueled and hydroelectric plants were causing environmental damage.

The two-unit Armenia Nuclear Station was built in Yerevan, the Armenian capital, located in the Ararat Valley on the banks of the Razadan River. In 1985, more than a third of the electricity produced in Armenia was generated by its nuclear station. However, demand for power still surpassed the available supply. Before its collapse, the central Soviet government had planned for the construction of additional nuclear units in Armenia.

Effects of earthquake

In 1988, an earthquake caused catastrophic damage in Armenia. Although the Armenia station itself was not damaged, concerns were raised about its ability to survive if a similar earthquake were centered closer to Yerevan. Kremlin officials ordered the permanent closure of both reactors units.

Following the closure, while Armenia was still under Soviet control, the central government ordered neighboring Azerbaijan to supply Armenia with oil and gas in order to meet the increasingly critical demand for electricity. After the demise of the Soviet Union, Azerbaijan's cooperation faltered.

Electrical crisis

The new Armenian government has instituted daily brownouts to conserve energy. As a result, schools have cancelled classes, businesses have been plagued by insufficient power supplies, factory production has been stopped, and some Armenian citizens cook over open fires in the streets.

Because of the severity of the nation's energy crisis, the Armenian government is investigating the possibility of re-opening its nuclear power station. The International Atomic Energy Agency does not have the power to enforce its preference that the plant remain closed. Some Armenian officials feel that the risks of a nuclear accident are less than the risks associated with insufficient power resources, especially during the cold winter months when heat is vital for survival.

Sources

"Introduction." "Improving the Energy Supply in Armenia." *Soviet Life*, June 1986.

Mounfield, Peter. *World Nuclear Power*. Routledge, 1991.

Nightline. ABC News, February 20, 1992.

—Karen Bellenir

PLANT PROFILES

★ Armenia Nuclear Station, Units 1-2

Oktembryan Region
377766 Metsamor, Armenia
Telex: 243563 OMEGA

Basic Facts

Units 1-2 *Status:* Shut down. *Type of Reactor:* Pressurized light-water reactor. *Megawatts (electric):* 408. *Megawatts (thermal):* 1375. *Architect Engineer:* Atomenergyoprojekt (former Soviet Union). *Owner:* Soviet Ministry of Atomic Energy (former Soviet Union). *Operator:* Soviet Ministry of Atomic Energy (former Soviet Union).

Key Dates

Unit 1 *Ordered:* 1968. *Construction began:* 1973. *Criticality:* December 1976. *First Power:* December 1976. *Commercial Operation:* October 1979. *Shut Down:* February 1989.

Unit 2 *Ordered:* 1969. *Construction began:* 1975. *Criticality:* December 1979. *First Power:* December 1979. *Commercial Operation:* May 1980. *Shut Down:* March 1989.

Plant Report

Soviet-styled plants. The Armenia Nuclear Station (also known as the Oktembryan Nuclear Station) consists of two Soviet-styled Model 230, VVER-440 nuclear reactors. Model 230 is a standardized design for a pressurized water reactor similar in many features to western pressurized water reactors but lacking an emergency cooling system and western-styled containment structures.

The power plant is located in Yerevan, the Armenian Capital. The plant began power production in 1976 and played a significant role in providing electricity to Armenia and the Transcaucasia region.

Effect of earthquake. A report published in *Soviet Life*, a publication of the former Soviet embassy, stated that the plants had survived several earthquakes, "attesting to the good job done by the designers." A severe earthquake in December 1988, however, forced a reevaluation of the plant's ability to withstand seismic catastrophes.

As a result of the investigation, both units at the station were shut down after officials discovered they were not capable of withstanding severe earthquakes. Unit 1 was closed on February 25, 1989; Unit 2 closed March 18, 1989.

The future of the Armenian nuclear station remains uncertain. Following the collapse of the central Soviet government, the Armenian government is considering restarting the reactors to help alleviate the nation's acute energy shortage.

Sources

"Introduction." "Improving the Energy Supply in Armenia." *Soviet Life*, June 1986.

"Soviet Nuclear Power Plant Programme Marks Time." *Nature*, January 26, 1989.

Mounfield, Peter. *World Nuclear Power*. Routledge, 1991.

"Soviet Nuclear Energy Plants Raise Concern as Republics Gain Control." *The Wall Street Journal*, August 30, 1991.

Source Book: Soviet-Designed Nuclear Power Plants in the Former Soviet Republics and Czechoslovakia, Hungary and Bulgaria. U.S. Council for Energy Awareness, 1992.

—Karen Bellenir

Austria

COUNTRY REPORT

Austria rejects nuclear option

The Tullnerfeld Nuclear Power Plant, Austria's first nuclear power station, was built by the government-owned utility, Gemeinschaftskraftwerk Tullnerfeld G.m.b.H (GKT). Construction of the plant took six years and was completed in 1978. The plant, however, never went into operation. Austria voted by public referendum to close the facility before it opened, making Tullnerfeld the world's first fully equipped, never-used nuclear power plant to be abandoned.

Neighboring nuclear plants

Austria's opposition to nuclear energy did not stop at Tullnerfeld. The country, which is surrounded by 41 nuclear power plants in neighboring nations, has made special efforts to close down nuclear plants in nearby Czechoslovakia, Hungary, and Yugoslavia. Many of these plants were Soviet designed and are considered by Western experts as extremely unsafe. Austrian officials fear the country is at risk of another nuclear accident like the one at Chernobyl in 1986, when Austria was subjected to the highest per capita radiation dosage of any West European country.

Austrian Chancellor Franz Vranitsky has proposed that the five neighboring states of south-central Europe—Austria, Italy, Hungary, Czechoslovakia, and Yugoslavia—create a nuclear-energy-free zone on their territories. In 1991, Vranitsky offered free electricity to Czechoslovakia in return for the shutdown of nuclear units at the Czech Bohunice site, which is 35 miles from the shared border. The arrangement would cost Austria about $350 million a year.

In 1991, Austria began a nationwide distribution of potassium iodine tablets to combat the effects of thyroid radiation poisoning. It became the first European country to implement this precautionary guideline established by the World Health Organization.

Importing electricity

Although Austria rejected the nuclear energy option, it does import electricity from Germany, Switzerland, and Czechoslovakia, countries dependent on nuclear energy. Due to a lack of primary fuel resources and an inadequate base load generating capacity, Austria has also entered into agreements to import electricity during the winter from Hungary, Poland, and the former Soviet Union. In turn, Austria exports electricity in the summer from its seasonal surplus of hydroelectric capacity.

Sources

"Nuclear Power Plant on Danube's Shores to Become Memory." *The Wall Street Journal,* May 29, 1987.

Mounfield, Peter. *World Nuclear Power.* Routledge, 1991.

"Prague Offered Payoff to Shut Nuclear Plant." *The Washington Post,* January 30, 1991.

PLANT PROFILES

Tullnerfeld Nuclear Power Plant
Zwentendorf, Austria

Owning Utility
Gemeinschaftskraftwerk Tullnerfeld G.m.b.H
A-3435 Zwentendorf, Austria
Dr. Karl Springer, V.Pres., Administration

Phone: 0-22-77-520
Telex: 113347

Basic Facts
Status: Decommissioned. *Type of Reactor:* Boiling water reactor. *Megawatts (electric):* 722. *Megawatts (thermal):* 2100. *Reactor System Supplier:* Telefunken (Germany); Kraftwerk Union (Germany). *Generator Supplier:* Kraftwerk Union (Germany). *Architect Engineer:* Kraftwerk Union (Germany). *Owner:* Gemeinschaftskraftwerk Tullnerfeld G.m.b.H. *Operator:* Gemeinschaftskraftwerk Tullnerfeld G.m.b.H.

Key Dates
Ordered: 1971. *Cancelled:* 1978. *Construction began:* 1972. *Decommissioned:* 1987.

Plant Report

The short history of a never-used nuclear power plant. The Tullnerfeld plant, Austria's first nuclear power station, was built in rich farming country near the town of Zwentendorf, Austria, 30 miles northwest of Vienna on the banks of the Danube River. The government-owned utility, Gemeinschaftskraftwerk Tullnerfeld G.m.b.H (GKT), spent $645 million on the West German-built reactor. Construction of the plant took six years and was completed in 1978. The plant, however, never went into operation. Austria voted by public referendum to cancel the facility before it opened, making Tullnerfeld the world's first fully equipped, never-used nuclear power plant to be abandoned.

The reactor lay dormant for nine years until the dismantling process began in June of 1987. The San Francisco-based Bechtel Group Inc. removed less than half of the plant's equipment for resale to other utilities. The resale of valves, pumps, and other electrical equipment returned some $25 million to GKT—a fraction of the utility's original investment.

U.S. opponents of nuclear energy, who at the time of the Tullnerfeld dismantling were fighting the full start-up of Shoreham and Seabrook nuclear power plants, took heart from the Austrian experience. Nuclear experts around the world looked on in hopes of learning more about the unknown costs and complexities of abandoning nuclear power plants.

Neighboring nuclear plants. Austria's opposition to nuclear energy did not stop at Tullnerfeld. The country, which is surrounded by 41 nuclear power plants in neighboring nations, has made special efforts to close down nuclear plants in nearby Czechoslovakia, Hungary, and Yugoslavia. Many of these plants were Soviet designed and are considered by Western experts to be extremely unsafe. Austrian officials fear that the country is at risk of another nuclear accident like the one at Chernobyl in 1986, when Austria was subjected to the highest per capita radiation dosage of any West European country.

Austrian Chancellor Franz Vranitsky has proposed that the five neighboring states of south-central Europe—Austria, Italy, Hungary, Czechoslovakia, and Yugoslavia—create a nuclear-energy-free zone on their territories. In 1991, the Austrian Chancellor went so far as to offer free electricity to Czechoslovakia in return for the shutdown of nuclear units at the Czech Bohunice site, which is 35 miles from the shared border. The arrangement would cost Austria about $350 million a year.

In 1991, Austria began a nationwide distribution of potassium iodine tablets to combat the effects of thyroid radiation poisoning. It became the first European country to implement this precautionary guideline established by the World Health Organization.

Importing electricity. Although Austria rejected the nuclear energy option, it does import electricity from Germany, Switzerland, and Czechoslovakia, countries that are dependent on nuclear energy. Due to a lack of primary fuel resources and an inadequate base load generating capacity, Austria has also entered into agreements to import electricity from Hungary, Poland, and the former Soviet Union each winter. In turn, Austria exports electricity each summer from its seasonal surplus of hydroelectric capacity.

Sources
"Nuclear Power Plant On Danube's Shores To Become Memory." *The Wall Street Journal*, May 29, 1987.
Mounfield, Peter. *World Nuclear Power*. Routledge, 1991.
"Prague Offered Payoff to Shut Nuclear Plant." *The Washington Post*, January 30, 1991.

Belgium

COUNTRY REPORT

A brief history

Next to France, among the industrialized nations, Belgium relies most heavily on nuclear energy to generate electricity. Over 60 percent of Belgium's electricity comes from the seven reactors on its two sites at Doel and Tihange.

In the early 1950s, Belgium's energy economy was based almost entirely on coal. By the 1970s, the nation had moved to dependence on cheap imported oil for energy. With the construction of nuclear power plants, Belgium's oil dependence was reduced to 6 percent by 1987. All of Belgium's electric utilities have been joined into a single organization, Electrabel, which owns and operates all the nation's power plants, including nuclear plants.

There were four main steps in Belgium's development of nuclear energy applications:

In 1957, a team of Belgian engineers participated in the start-up of the Shippingport Atomic Power Station in the United States, the first commercial pressurized light-water reactor unit in the world.

In 1962, Belgium commissioned the Mol BR3, a 12-megawatt unit and the first pressurized light-water reactor built outside the United States.

In 1967, a joint French-Belgian pressurized light-water reactor unit (310 MWe) was commissioned at Chooz.

Belgium plants Doel 1 and 2 and Tihange 1 were ordered in 1969.

As a result of the country's significant investment in nuclear energy, the Oak Ridge National Laboratory's Carbon Dioxide Analysis Center in Tennessee reports that Belgium's per capita rates of carbon dioxide emissions is among the lowest in the world.

International cooperation

In western Europe, 53 of the 150 power reactors in operation are within 30 miles of a national border, which has often caused conflicts between neighboring countries over safety and environmental issues. Although Belgium's four reactors at Doel are located on the Scheldt estuary within two miles of the Dutch border, Belgium has managed to avoid major controversies over nuclear power plant locations. Instead, Belgium has involved itself in several international cooperative efforts.

For example, Belgium is collaborating with The Netherlands and Germany to fund a commercial fast breeder reactor at Kalkar on the lower Rhine. Belgium's Tihange Unit 1 reactor is a joint venture with France, which consumes 50 percent of the power produced; a quarter of the power produced by the French reactors at the Chooz Nuclear Power Plant will be used by Belgium when contruction at the site is complete.

Sources

"USCEA Survey: 15 New Nuclear Plants Began Commercial Operation in 1990." *Info,* U.S. Council for Energy Awareness, July 1991.

"Heaviest Users of Nuclear Energy have Low Per Capita of CO2 Emissions." *Info,* U.S. Council for Energy Awareness, August 1991.

Mounfield, Peter. *World Nuclear Power.* Routledge, 1991.

1992 World Directory of Nuclear Utility Management. American Nuclear Society, n.d.

Tihange Nuclear Power Plant. Intercom, n.d.

Plant Profiles

PLANT PROFILES

6 ●
Doel Nuclear Power Station, Units 1-2

[*See Also:* Doel Nuclear Power Station, Units 3-4]

Scheldemolenstraat	Phone: 32-3-202-21-11
B-9130 Doel, Belgium	Fax: 32-3-773-37-46
Armand Timmermans, Plant Mgr., Units 1-2	Telex: 71083

Owning Utility

Electrabel	Phone: 32-2-518-61-11
Blvd. du Regent, 8	Fax: 32-2-511-50-20
B-1000 Brussels, Belgium	Telex: 64681 eblbru b
H. Dresse, Responsible for Nuclear Generation	

Contact

Philippe Claes, Image & Commun. Concept	Phone: 32-2-518-65-91
Electrabel	Fax: 32-2-511-65-99

Basic Facts

Units 1-2 *Status:* Operable. *Type of Reactor:* Pressurized light-water reactor. *Megawatts (electric):* 412. *Megawatts (thermal):* 1192. *Reactor System Supplier:* ACEOWEN. *Generator Supplier:* Cockeril Ougree-Providence (Belgium); Franco Toxi SpA (Belgium); Ateliers de Constructions Electriques de Charleroi S.A. (Belgium). *Architect Engineer:* Traction-Electricite (Belgium).

Key Dates

Unit 1 *Ordered:* 1968. *Construction began:* 1969. *Criticality:* 1974. *First Power:* August 1974. *Commercial Operation:* February 1975.

Unit 2 *Ordered:* 1968. *Construction began:* 1971. *Criticality:* 1975. *First Power:* August 1975. *Commercial Operation:* December 1975.

Plant Report

Belgium's first nuclear plant. The Doel Nuclear Power Station occupies a site on the Scheldt estuary near Antwerp, Belgium. Electrabel S.A. owns and operates Doel, which houses four pressurized light-water reactors. Doel 1 and 2 were Belgium's first operating nuclear power units.

Features of Doel 1 and 2

- The reactor core of each unit contains 121 fuel assemblies, with 179 fuel rods per assembly, for a total weight of 31,738 kilograms of uranium oxide.
- Each reactor vessel has an overall inside height of 10,600 millimeters and an inside diameter of 3,327 millimeters, and weighs 160 tons.
- Each unit utilizes two steam generators, which measure 19,505 millimeters high, with overall weights of 284 tons.
- There are 33 control rod clusters, with 16 fingers per cluster, per unit.
- The turbogenerators for each unit have three turbine sections, with turbines spinning at 1,500 revolutions per minute.
- Each reactor has one main and one step-down transformer. There are two start-up transformers and four emergency diesel generator sets for the two units.

Sources

Electricity from Nuclear Energy—Doel Nuclear Power Station. EBES nv, 1986.
Electrabel 1991 Annual Report. N.p., n.d.

—Charles Ceplecha

7 ●
Doel Nuclear Power Station, Units 3-4

[*See Also:* Doel Nuclear Power Station, Units 1-2]

Scheldemolenstraat	Phone: 32-3-730-21-11
B-9130 Doel, Belgium	Fax: 32-3-773-37-33
Alfred Becquaert, Plant Mgr., Units 3-4	Telex: 32878

Owning Utility

Electrabel	Phone: 32-2-518-61-11
Blvd. du Regent, 8	Fax: 32-2-511-50-20
B-1000 Brussels, Belgium	Telex: 64681 eblbru b
H. Dresse, Responsible for Nuclear Generation	

Contact

Philippe Claes, Image & Commun. Concept	Phone: 32-2-518-65-91
Electrabel	Fax: 32-2-511-65-99

Basic Facts

Unit 3 *Status:* Operable. *Type of Reactor:* Pressurized light-water reactor. *Megawatts (electric):* 945. *Megawatts (thermal):* 2775. *Reactor System Supplier:* Framatome (France); Ateliers de Constructions Electriques de Charleroi S.A. (Belgium); Cockeril Ougree-Providence (Belgium). *Generator Supplier:* Alsthom-Atlantique; Ateliers de Constructions Electriques de Charleroi S.A. (Belgium). *Architect Engineer:* Traction-Electricite (Belgium).

Unit 4 *Status:* Operable. *Type of Reactor:* Pressurized light-water reactor. *Megawatts (electric):* 1065. *Megawatts (thermal):* 2988. *Reactor System Supplier:* ACEOWEN: Ateliers de Constructions Electriques de Charleroi S.A. (Belgium); Cockeril Ougree-Providence (Belgium); Westinghouse (Belgium). *Generator Supplier:* Alsthom-Atlantique; Ateliers de Constructions Electriques de Charleroi S.A. (Belgium). *Architect Engineer:* Traction-Electricite (Belgium).

Key Dates

Unit 3 *Ordered:* 1974. *Construction began:* 1974. *Criticality:* June 1982. *First Power:* June 1982. *Commercial Operation:* October 1982.

Unit 4 *Ordered:* 1975. *Construction began:* 1978. *Criticality:* March 1985. *First Power:* April 1985. *Commercial Operation:* July 1985.

Plant Report

Belgium's first nuclear plant. The Doel Nuclear Power Station occupies a site on the Scheldt estuary near Antwerp, Belgium. Electrabel S.A. owns and operates Doel, which houses four pressurized light-water reactors. Doel 1 and 2 were Belgium's first operating nuclear power units.

BELGIUM

8 Doel

New parts for Doel 3 and 4. The third unit of the Doel Nuclear Power Station began operation in 1982. When the tubing in the original steam generators began corroding in 1989, Doel contracted Siemens AG, a global electrical/electronics firm with sales of nearly $40 billion and approximately 360,000 employees, to replace the generators. The value of the supply and installation contracts was more than 100 million deutsche marks. The replacement operation will take approximately 56 days, beginning in mid-1993. The new units are expected to boost the plant's rating by nearly 10 percent.

Other modifications made in 1990 were the addition of a new condenser for Doel 3, and a new water intake for Doel 3 and 4.

Doel 5 cancelled. In 1988, Electrabel planned to put a fifth power station in operation at the Doel site. Doel 5 would have been larger than the four plants presently in operation. However, those plans never materialized, and the project was cancelled or has been indefinitely deferred.

Sources

Electricity from Nuclear Energy—Doel Nuclear Power Station. EBES nv, 1986.
"Siemens to Replace Steam Generators at Belgian Nuclear Power Plant." *Business Wire,* October 18, 1990.
Electrabel 1991 Annual Report. N.p., n.d.

—Charles Ceplecha

8
Doel Nuclear Power Station, Unit 5

Scheldemolenstraat Phone: 32-3-730-21-11
B-9130 Doel, Belgium Fax: 32-3-773-37-33
Alfred Becquaert, Plant Mgr., Units 3-4 Telex: 32878

Owning Utility

Electrabel Phone: 32-2-518-61-11
Blvd. du Regent, 8 Fax: 32-2-511-50-20
B-1000 Brussels, Belgium Telex: 64681 eblbru b
H. Dresse, Responsible for Nuclear Generation

Contact

Philippe Claes, Image & Commun. Concept Phone: 32-2-518-65-91
Electrabel Fax: 32-2-511-65-99

Basic Facts

Unit 5 *Status:* Cancelled. *Type of Reactor:* Pressurized light-water reactor. *Megawatts (electric):* 1450. *Megawatts (thermal):* 4250. *Reactor System Supplier:* Framatome (France).

Key Dates

Unit 5 *Ordered:* 1988. *Cancelled:* Yes. *Construction began:* 1988.

Plant Report

See Plant Report for Doel Nuclear Power Station, Units 3-4.

9 ★
Mol BR3

Mol, Belgium

Basic Facts

Status: Shut down. *Type of Reactor:* Pressurized light-water reactor. *Megawatts (electric):* 12. *Megawatts (thermal):* 40.9. *Reactor System Supplier:* Cockeril Ougree-Providence (Belgium); Westinghouse Electric Corp. (United States). *Generator Supplier:* Westinghouse Electric Corp. (United States). *Architect Engineer:* Gibbs & Hill, Inc. (United States). *Owner:* Studiecentrum Kernenfersie (Belgium); Centre d'Etude de l'Energie Nucleaire (Belgium). *Operator:* Studiecentrum Kernenfersie (Belgium); Centre d'Etude de l'Energie Nucleaire (Belgium).

Key Dates

Ordered: 1955. *Construction began:* 1957. *Criticality:* August 1962. *First Power:* October 1962. *Commercial Operation:* October 1962. *Shut Down:* June 1987.

Plant Report

Belgium's first nuclear reactor. Although a small unit, the Mol BR3 was a big first step in Belgium's nuclear energy program. Belgium's plant was also the first pressurized light-water reactor built outside the United States. Commissioned in 1962, the Mol BR3 operated for 15 years before being shut down.

Sources

Tihange Nuclear Power Plant. Intercom, Information and Public Relations Section, n.d.

10 ●
Tihange Nuclear Power Plant, Units 1-3

avenue de l'Industrie, 1 Phone: 32-2-85-24-30-11
B-4500 Tihange/Huy, Belgium Fax: 32-2-85-24-30-79
Roger Jacquet, Site Mgr. Telex: 59695

Owning Utility

Electrabel Phone: 32-2-518-61-11
Blvd. du Regent, 8 Fax: 32-2-514-50-20
1000 Brussels, Belgium Telex: 64681 eblbru b
H. Dresse, Responsible for Nuclear Generation

Contact

Philippe Claes, Image & Commun. Concept Phone: 32-2-518-65-91
Electrabel Fax: 32-2-511-65-99

Basic Facts

Unit 1 *Status:* Operable. *Type of Reactor:* Pressurized light-water reactor. *Megawatts (electric):* 920. *Megawatts (thermal):* 2660. *Reactor System Supplier:* Ateliers de Constructions Electriques de Charleroi S.A. (Belgium); Cockeril Ougree-Providence (Belgium); Creusot-Loire (France); Framatome (France); Westinghouse (France). *Generator Supplier:* Alsthom; Rateau; LaMeuse; Ateliers de Constructions Electriques de Charleroi S.A. (Belgium); Jeumont-Schneider (France). *Architect Engineer:* Electricite de France; Electrabel.

Unit 2 *Status:* Operable. *Type of Reactor:* Pressurized light-water reactor. *Megawatts (electric):* 940. *Megawatts (thermal):* 2785. *Reactor System Supplier:* FRAMACECO: Framatome (France); Ateliers de Constructions Electriques de Charleroi S.A. (Belgium); Cockeril Ougree-Providence (Belgium). *Generator Supplier:* Alsthom; Rateau; LaMeuse; Ateliers de Constructions Electriques de Charleroi S.A. (Belgium). *Architect Engineer:* Electrabel.

Unit 3 *Status:* Operable. *Type of Reactor:* Pressurized light-water reactor. *Megawatts (electric):* 1070. *Megawatts (thermal):* 3000. *Reactor System Supplier:* ACEOWEN: Ateliers de Constructions Electriques de Charleroi S.A. (Belgium); Cockeril Ougree-Providence (Belgium); Westinghouse (Belgium). *Generator Supplier:* Brown Boveri; Alsthom; Ateliers de Constructions Electriques de Charleroi S.A. (Belgium). *Architect Engineer:* Electrabel.

Key Dates

Unit 1 *Ordered:* 1968. *Construction began:* 1969. *Criticality:* February 1975. *First Power:* March 1975. *Commercial Operation:* October 1975.

Unit 2 *Ordered:* 1974. *Construction began:* 1975. *Criticality:* October 1982. *First Power:* October 1982. *Commercial Operation:* March 1983.

Unit 3 *Ordered:* 1975. *Construction began:* 1977. *Criticality:* June 1985. *First Power:* June 1985. *Commercial Operation:* September 1985.

Plant Report

Background. The Tihange Nuclear Power Plant houses three pressurized light-water reactors and occupies a 173-acre site in the territory of Tihange, incorporated into the commune of Huy. The site is four kilometers from the Meuse River and 25 kilometers southwest of Liege, Belgium.

Operation of all three units requires some 700 permanent employees, with 70 percent living in the Huy-Waremme district. Approximately 1,200 people live within a one-kilometer radius, 18,000 within a three-kilometer radius, and 70,000 within a 10-kilometer radius surrounding the neighboring town of Huy.

Unit 1 is owned jointly with Electricité de France, which consumes 50 percent of the power. The joint venture with France allows Belgium to use the French grid in case of a shutdown or outage for maintenance. Units 2 and 3 are strictly Belgian owned and operated.

Economics and finances. Officials in Belgium claim that the lower price of nuclear fuel compensates the high cost investment in nuclear plants. The resulting savings has been passed along to consumers, making Belgian kilowatt-hours for industrial use the least expensive in Europe.

Tihange's three units were expected to save 4.5 million tons of fuel-oil annually. In 1983, before Tihange 2 and 3 were operational, the aggregate Belgian nuclear program had already saved the country 31 billion francs.

In 1968, Tihange 1 was ordered to curb Belgium's dependency on oil. Plans for Tihange 2 and 3 were made in 1974, during the first oil crisis.

Tihange's three units represent an investment of about 95 billion Belgian francs, and the construction of each unit required a 32 billion hours of labor at factory and onsite activities.

About 50 percent of the construction orders for Tihange 1 were placed with France, and at least 90 percent of the orders for Units 2 and 3 were placed with Belgian companies. Local companies receive orders related to the maintenance and operation of all three units.

Unit 3 honored. Tihange 3 was recognized by *Nuclear Engineering International* as 9th in a ranking of top 10 plants in the world with the best lifetime performance record through June 30, 1991. The unit had maintained an overall capacity factor of 86.5 percent.

Sources

Tihange Nuclear Power Plant. Intercom, Information and Public Relations Section, n.d.

—*Charles Ceplecha*

Brazil

COUNTRY REPORT

Brazil enters field of nuclear power

In 1968, the Government of Brazil decided to enter the field of nuclear power. Because an increase in thermal power had already been planned for the Rio de Janeiro area, Furnas was chosen as the utility to build Brazil's first nuclear power plant—Angra 1. Furnas, a joint government and privately-owned company, is a subsidiary of the holding company Electrobras. Furnas supplies energy for the southeastern and central western region of Brazil and is a member of the Latin American Section of the American Nuclear Society.

Half of the costs for Angra 1 was met by banks in the United States led by the Export-Import Bank. Angra became commercially operational in 1985.

All activities carried out during the design, construction, and operation of nuclear facilities in Brazil are licensed and controlled by Comissao Nacional de Energia Nuclear (CNEN), the regulatory authority of the Brazilian government.

Cooperation with Germany

In 1975 the Brazilian nuclear power program expanded through the Nuclear Cooperation Agreement signed between Brazil and the Federal Republic of Germany. In July 1976, Furnas contracted with German's largest nuclear vendor, Kraftwerk Union Gmbh (KWU), to build Angra 2 and Angra 3, a uranium enrichment plant, and a spent fuel reprocessing plant.

The United States was particularly disturbed by the negotiations between the two countries, as it had forbidden its own industry to export enrichment and reprocessing technologies. The U.S. fear was that Brazil—who had not signed the Nuclear Non-Proliferation Treaty, which had been in effect since 1970—would convert nuclear energy technology into nuclear weapons capabilities. Some felt Brazil, which had originally contracted with Westinghouse to build the new power plants, switched to Kraftwerk Union because Germany agreed not to impose the non-proliferation regulations the U.S. would have imposed.

As a result of delays in licensing procedures, organizational changes, and severe financial problems, Angra 2 and Angra 3 are not scheduled to go online until the turn of the century.

Sources

"Power Plant Politics: The Nuclear Care and Feeding of the Third World." *Critical Mass Bulletin*, December 1983.

Furnas and Nuclear Energy. Furnas Centrais Electricas S.A., Diretoria de Producao Termonuclear, 1990.

Mounfield, Peter. *World Nuclear Power.* Routledge, 1991.

PLANT PROFILES

Angra Nuclear Power Station, Unit 1

[*See Also:* Angra Nuclear Power Station, Units 2-3]

Rodovia Rio-Santos, BR-101-Km132
Praia de Itaorna
23900 Angra dos Reis, RJ, Brazil
Pedro J.D. Figueiredo, Supt.,
 Thermonuclear Generation

Phone: 55-243-42-3355
Fax: 55-243-43-0485

Owning Utility

Furnas-Centrais Eletricas SA
Rua Real Grandeza, 219
22283 Botafogo, RJ, Brazil
Eliseau Resende, Pres.

Phone: 55-21-536-3112
Fax: 55-21-226-7005
Telex: 55-21-21166
FURNBR

Contact

Ayrton Jose Caubit da Silva, Licensing/
 Quality Assurance Coord.
Furnas-Centrais Electricas SA

Basic Facts

Unit 1 *Status:* Operable *Type of Reactor:* Pressurized light-water reactor. *Megawatts (electric):* 657. *Megawatts (thermal):* 1882. *Reactor System Supplier:* Westinghouse Electric Corp. (United States). *Generator Supplier:* Westinghouse Electric Corp. (United States). *Architect Engineer:* Gibbs & Hill, Inc. (United States).

Key Dates

Unit 1 *Ordered:* 1971. *Construction began:* 1971. *Criticality:* 1982. *First Power:* April 1982. *Commercial Operation:* January 1985.

Plant Report

Brazil's first nuclear power plant. In 1968, the government of Brazil decided to enter the field of nuclear power. Because an increase in thermal power had already been planned for the Rio de Janeiro area, Furnas was chosen as the utility to build Brazil's first nuclear power plant. Furnas, a joint government and privately-owned company, is a subsidiary of the holding company Electrobras and supplies energy for the southeastern and central western region of Brazil. Half of the costs for Angra 1 were met by banks in the United States led by the Export-Import Bank.

Delays in construction. After inviting international bids, Furnas awarded the construction contract for Angra 1 to Westinghouse, a U.S. company. The contract included design, procurement, and erection of plant equipment on a turnkey basis. Contracts for provision of the steel containment shell and execution of the civil engineering works were given to Chicago Bridge and Iron and Construtora Norberto Odebrecht, respectively. Also, Westinghouse subcontracted with Gibbs and Hill, Inc. in the U.S. and Promon Engenharia in Brazil for the design engineering work, and they set up a joint venture with Empresa Brasileira de Engenharia for erection activities.

In 1972, the civil engineering work started with a scheduled completion period of five years. However, delays in design, in obtaining both domestic and imported equipment, in managing the different contracts, and the correction of equipment fabrication and erection failures detected during the commissioning phase, detained commercial operations substantially. A decade passed before the the reactor went critical in 1982. Three years later, Angra started commercial operation.

The Firefly. From January 1985 to March 30, 1990, Angra 1 generated over nine billion kilowatt hours of electric energy. During that period, in addition to the scheduled outages for fuel change, plant operation was interrupted to change condenser tubes, to repair the electric generator, and to introduce a number of changes aimed at improving operations. Angra 1 was so frequently turned on and off that it was nicknamed Firefly.

Numbers and statistics. As of 1990, hydroelectric plants accounted for 90 percent of the installed power in Brazil. Of the electricity generated by Furnas, 81 percent is produced by hydroelectric plants, 2 percent by Angra 1, and the remaining power is purchased from other utilities.

Angra is located on Itaorna Beach, Angra dos Reis, State of Rio de Janeiro. It is 130 kilometer from Rio de Janeiro, 220 kilometer from Sao Paulo, and 350 kilometer from Belo Horizonte.

The reactor core of Angra 1 consists of 121 fuel elements with a total load of 51 tons of enriched uranium.

The power produced by the station is integrated into the Furnas system in the southeast region of Brazil through two 500 kv transmission lines: one to Cachoeira Paulista substation and the other to Adrianopolis substation.

Safety. Angra 1 has complied with requirements imposed by Comissao Nacional de Energia Nuclear (CNEN), the regulatory authority of the Brazilian government. The safety requirements were a direct result of the Three Mile Island accident in the United States. Under this safety approach, Furnas and the Federal University of Rio de Janeiro developed the Critical Safety Functions System and the Environment Control System, which with other systems make up the Angra Integrated Computer System. These systems have increased the ability of the operators at Angra to deal effectively with accident conditions.

The plant's Emergency Plan was highly rated by the International Agency of Atomic Energy during inspections in 1985 and 1989.

Environmental monitoring. The following programs are carried out at Angra to assess the radiological and environmental impact of the plant: meteorology and atmospheric dispersion program; marine animal and plant life monitoring program; temperature and chlorine survey in Piraguara de Fora Bay; environmental radiation monitoring program; effluent control; and radiation emergency program.

When all three units at the Angra Station are operating at full power, officials expect a radioactive release of 0.15 millisieverts per year. CNEN has set the maximum limit of radioactive release to the environment at 50 millisieverts per year.

Waste storage. Medium and low-activity wastes resulting from the operation of Angra 1 are placed, after processing, into steel

12 Angra

BRAZIL

and concrete containers and are temporarily being stored on site. A bill is currently under consideration in the Brazilian Congress to provide for permanent storage of wastes.

Spent fuel is being stored in a pool, which is designed to handle 363 fuel elements representing eight to 10 years of operation. Furnas is conducting studies to increase their spent fuel storage capacity until the Brazilian government defines the country's overall fuel cycle policy.

Information available to the public. An information center is located next to the Angra Nuclear Power Station. Through lectures, audiovisual presentations, special displays, and publications, the center answers key questions frequently asked by the public about the station's facilities and operations.

Sources

Furnas and Nuclear Energy. Furnas Centrais Electricas S.A., Diretoria de Producai Termonuclear, 1990.

Mounfield, Peter. *World Nuclear Power.* Routledge, 1991.

—*Sharon Miscavage*

12 ○
Angra Nuclear Power Station, Units 2-3

[See Also: Angra Nuclear Power Station, Unit 1]

Rodovia Rio-Santos, BR-101-Km132
Praia de Itaorna
23900 Angra dos Reis, RJ, Brazil
Moacir F. Gitirana, Head, Construction Dept.

Phone: 55-243-42-3355
Fax: 55-243-43-0485

Owning Utility

Furnas-Centrais Eletricas SA
Rua Real Grandeza, 219
22283 Botafogo, RJ, Brazil
Eliseau Resende, Pres.

Phone: 55-21-536-3112
Fax: 55-21-226-7005
Telex: 55-21-21166
FURNBR

Contact

Ayrton Jose Caubit da Silva, Licensing/Quality Assurance Coord.
Furnas-Centrais Electricas SA

Basic Facts

Units 2-3 *Status:* Under construction (Angra 2, 64 percent complete as of January 1992; Angra 3, 39 percent complete as of January 1992). *Type of Reactor:* Pressurized light-water reactor. *Megawatts (electric):* 1309. *Megawatts (thermal):* 3782. *Reactor System Supplier:* Kraftwerk Union AG (Germany). *Generator Supplier:* Kraftwerk Union AG (Germany). *Architect Engineer:* Kraftwerk Union AG (Germany); Nuclebras Engenaria SA (Brazil).

Key Dates

Unit 2 *Ordered:* 1974. *Construction began:* 1976. *Commercial Operation Expected:* 1998.

Unit 3 *Ordered:* 1974. *Construction began:* 1983. *Commercial Operation Expected:* 2001.

Plant Report

Cooperation with Germany. In 1975, the Brazilian nuclear power program expanded through the Nuclear Cooperation Agreement signed between Brazil and the Federal Republic of Germany. In July 1976, Furnas contracted with German's largest nuclear vendor, Kraftwerk Union GmbH (KWU), to build Angra 2 and Angra 3, a uranium enrichment plant, and a spent fuel reprocessing plant.

The United States was particularly disturbed by the negotiations between the two countries, as it had forbidden its own industry to export enrichment and reprocessing technologies. The U.S. fear was that Brazil—who had not signed the Nuclear Non-Proliferation Treaty, which had been in effect since 1970—would convert nuclear energy technology into nuclear weapons capabilities. Some felt Brazil, which had originally contracted with Westinghouse to build the new power plants, switched to Kraftwerk Union because Germany agreed not to impose the non-proliferation regulations the U.S. would have imposed.

After the purchase of Units 2 and 3, Angra Nuclear Power Station, consisting of three plants and related installations, was subsequently named Almirante Alvaro Alberto Nuclear Power Station under a law enacted by the Brazilian Congress in honor of the Brazilian nuclear energy pioneer.

In 1976, excavation work started for Angra 2. In 1978, the Dutch threatened to halt construction efforts, under an Anglo-Dutch-German treaty governing the production of enriched uranium at the Almelo in The Netherlands. Under the agreement, all three signatories had to approve of uranium exports.

Financial problems, construction delays. Officials at Furnas originally expected Angra 2 to be completed by 1981 and Angra 3 by 1982. However, as a result of delays in licensing procedures, organizational changes, and severe financial problems, the deadlines were extended to 1998 for Angra 2 and 2001 for Angra 3.

By September 1986, the total investment needed to finish Angra 2 and 3 was $2,200 million (U.S.), not including servicing foreign borrowing at $240 million (U.S.). As of 1990, 88 percent of the equipment needed to build the two plants had already been purchased. Of the overall investment, 69 percent of the total figure estimated for Angra 2 and 30 percent of the total amount for Angra 3 have been made.

Sources

"Power Plant Politics: The Nuclear Care and Feeding of the Third World." *Critical Mass Bulletin,* December 1983.

Furnas and Nuclear Energy. Furnas Centrais Electricas S.A., Diretoria de Producao Termonuclear, 1990.

Mounfield, Peter. *World Nuclear Power.* Routledge, 1991.

—*Sharon Miscavage*

Bulgaria

COUNTRY REPORT

Background

Nuclear energy produced at the Kozloduy power plant—Bulgaria's only nuclear energy site—accounts for about 40 percent of all the electricity generated in Bulgaria. Coal-fired plants supply about 48 percent of the country's electricity; oil, 3 percent; natural gas, 1 percent; and self-producers, the remaining amount.

The country struggles to meet the demand for electricity. It often has to impose brownouts and electricity rationing when a power plant is forced off line.

A shortage of electricity, however, is not Bulgaria's only energy problem. Safety has been an additional concern. The physical condition and operating culture of Kozloduy's nuclear units has alarmed the world nuclear community. Public protests over safety have also brought pressure to bear upon nuclear officials. As a result, the government cancelled plans for nuclear expansion in 1991.

Nuclear policy

Before the collapse of the Communist Regime in Eastern Europe, Bulgaria's nuclear program was heavily influenced by the Soviet Union. Since the Communist breakup, however, the country has drawn more on Western expertise. It now has ties with the International Atomic Energy Agency and the World Association of Nuclear Operators. Moreover, cooperation with the West has become increasingly important to Bulgaria as it seeks financial and technical assistance to improve the safety of its Kozloduy plant.

The Bulgarian Energy Committee, which formerly acted as the state-run utility as well as a policy-making body, has been charged by the new government (elected in October 1991) to be solely devoted to energy policy. The responsibility for electricity generation, transmission, and distribution has been transferred to Energetika Association, the state-owned electricity company. As of 1992, there were no plans to privatize the company or the country's power plants.

The Bulgarian Committee on the Uses of Atomic Energy for Peaceful Purposes regulates nuclear power plant operations and promotes the peaceful use of atomic energy.

Sources

"West Urges Bulgarians to Shut Reactors." *The New York Times*, July 10, 1991.

Source Book: Soviet-Designed Nuclear Power Plants in the Former Soviet Republics and Czechoslovakia, Hungary and Bulgaria. U.S. Council for Energy Awareness, 1992.

PLANT PROFILES

13 ☆
Belene Nuclear Power Plant, Units 1-2
Belene, Bulgaria

Owning Utility

Energetika Association
8, Triaditza Str.
1040 Sofia, Bulgaria
Lulin Radoulov, Pres.

Phone: 02-8-61-91
Fax: 02-87-58-26
Telex: 22-707

Basic Facts

Units 1-2 *Status:* Cancelled. *Type of Reactor:* Pressurized light-water reactor. *Megawatts (electric):* 1000. *Reactor System Supplier:* Atomenergoexport (former Soviet Union).

Key Dates

Unit 1 *Ordered:* 1983. *Cancelled:* 1991. *Construction began:* 1984. *Commercial Operation Expected:* indefinite.

Unit 2 *Ordered:* 1985. *Cancelled:* 1991. *Construction began:* 1986. *Commercial Operation Expected:* indefinite.

Plant Report

Construction halted. When the condition of the nuclear units at Bulgaria's Kozloduy plant began to arouse much alarm in the world's nuclear community, public concern also began to grow about the nuclear units under construction at Belene. As a result, the government decided to suspend work at the site until a parliamentary committee could study the safety of the proposed plant. As part of the study, three Western companies were contracted to conduct a seismic analysis.

In light of public protests of the Belene construction, as well as concerns about seismic risks, the Bulgarian government decided in mid-1991 to halt construction of the nuclear units and build a gas-fired power plant at the site instead.

Sources

Source Book: Soviet-Designed Nuclear Power Plants in the Former Soviet Republics and Czechoslovakia, Hungary and Bulgaria. U.S. Council for Energy Awareness, 1992.

Kozloduy Nuclear Power Plant, Units 1-2

Atomna Energeticka Economic
 Organization
3320 Kozloduy, Bulgaria
Kozma Kouzmanov, Plant Mgr.

Phone: 0973-71
Fax: 0973-25-91
Telex: 33416

Owning Utility

Energetika Association
8, Triaditza Str.
1040 Sofia, Bulgaria
Lulin Radoulov, Pres.

Phone: 02-8-61-91
Fax: 02-87-58-26
Telex: 22-707

Basic Facts

Units 1-2 *Status:* Shut down. *Type of Reactor:* Pressurized light-water reactor. *Megawatts (electric):* 440. *Megawatts (thermal):* 1375. *Reactor System Supplier:* Atomenergoexport (former Soviet Union).

Key Dates

Unit 1 *Ordered:* 1967. *Construction began:* 1970. *Criticality:* May 1974. *First Power:* June 1974. *Commercial Operation:* December 1974. *Shut Down:* 1991.

Unit 2 *Ordered:* 1967. *Construction began:* 1970. *Criticality:* August 1975. *First Power:* August 1975. *Commercial Operation:* December 1975. *Shut Down:* 1991.

Plant Report

Soviet-designed reactors. Kozloduy's first nuclear reactor began commercial operation in 1974. Through the next 17 years, Bulgaria built five more reactors at the Kozloduy complex on the Danube river, 130 miles north of Sofia, the Bulgarian capital.

The Kozloduy reactors were all designed by the former Soviet Union, causing Western experts much concern. Nuclear officials in the U.S., France, and Germany have long been worried about the lack of safety of the oldest type of Soviet-built reactors. (Germany closed five such reactors it inherited upon unification with East Germany.) There are about 20 of this type still in operation in the commonwealth republics of the former Soviet Union as well as Eastern Europe—including Units 1-4 at Kozloduy.

Protestors and problems. Concerns about safety at Kozloduy escalated in the early 1990s. At every turn, individuals and organizations were speaking out against Kozloduy.

Specialists from the International Atomic Energy Agency, who inspected the complex for three weeks in June of 1991, reported that four of the six reactors were so full of hazards that "continued operation would be imprudent." They cited numerous oil, steam, and water leaks, including releases of radioactive steam; unsealed and loose electrical cables; valves with missing wheels; severe shortage of staff members; and a leak in the system meant to cool the reactor core in case of an emergency.

A Bulgarian government committee charged with investigating Kozloduy proposed new safety equipment or a permanent shut down. In 1990, the government ordered a checkup of all plant employees' families. After the first 10 children were tested, two proved to be carrying the same radioactive particles as their parents, both of whom worked in the danger zone. The contamination was low, but the news caused such panic that no other parents had their children checked.

Vladimir Benev, president of the Kozloduy Workers Union, told *The New York Times* (October 16, 1991) of his concern about the physical well-being and morale of Kozloduy's 4,600 workers. He reported that 10 workers had received an overdose of radiation; some were on sick leave. He also described how eight to 10 times a month, because of broken pipes or pumps, there was no water for the workers to shower after their shift. By not scrubbing down, the workers risked taking home radioactive particles and contaminating family members. Other union representatives said many employees routinely spent working hours drinking beer.

Georgi Stoilov, president of the Energy Workers' Union, protested low wages. (A reactor operator earns the equivalent of $97 per month, twice the minimum wage.)

More than half of the Soviet specialists responsible for the safety and operation of Kozloduy left the plant in 1991 due to a wage dispute. Skilled Bulgarians left in their wake. The vacuum left a strain on remaining workers. A medical report, noting the dangers of exhaustion, said that one important task was covered by a single dispatcher working a 12-hour shift, instead of the prescribed group of five workers.

Across the Danube in Romania, the ministry of defense was so worried about a potential catastrophe that it planned to conduct a defense drill against a radioactive explosion.

Western pressure, Bulgarian response. Despite Western pressure to shut down Kozloduy, Zhelyu Zhelev, the Bulgarian president, ruled out a permanent closing. He said that over 40 percent of the nation's electricity comes from Kozloduy reactors. The country is often energy deficient; power cuts are common. To close Kozloduy would mean the collapse of the Bulgarian economy. However, Zhelev explained further that the virtually bankrupt country could not afford to bring the plant up to international safety standards.

At a special meeting in July of 1991 in Vienna, officials from six Western nations, including the United States, met with Bulgarian and Soviet specialists on the safety problems at Bulgarian plants. Bulgarian nuclear officials promised that they could temporarily switch off two of the reactors and make improvements on two others at an immediate cost of about $50 million, if they received financial assistance from the West. To completely overhaul the plant, Bulgaria has said it needes up to $1 billion.

In return for vows of international aid for upgrades, Bulgaria closed units 1 and 2 late in 1991. Western nations rallied to shore up the Bulgarian nuclear program:

- Technicians from France arrived in September to fix unsafe electrical cables, oil leaks, and other fire hazards.

- The European Community provided $13.5 million for emergency repairs and technical studies on the plant. The World Association of Nuclear Operators (WANO) is coordinating the European assistance.

- Germany promised to send desperately needed spare parts stripped from reactors of the same model they had closed in 1990.

- Scottish Nuclear Ltd. signed an agreement with the Bulgarian nuclear design and construction organization, Energoproject, to provide technical assistance.

Sources

"Danger in the Balkans." *The Economist*, November 3, 1990.
"Defects at Bulgarian Reactor." *The Wall Street Journal*, July 1, 1991.
"West Urges Bulgarians to Shut Reactors." *The New York Times*, July 10, 1991.
"Bulgaria to Get Nuclear Aid." *The Wall Street Journal*, August 1, 1991.
"Bulgaria Starts Nuclear Closing." *The Wall Street Journal*, September 4, 1991.
"A Peril-Ridden A-Plant: It's a Frightening Legacy." *The New York Times*, October 16, 1991.
"International: Scottish Nuclear to Aid Bulgarian Plant." *Nuclear Industry*, Fourth Quarter 1991.

Kozloduy Nuclear Power Plant, Units 3-6

Atomna Energeticka Economic Organization
3320 Kozloduy, Bulgaria
Kozma Kouzmanov, Plant Mgr.

Phone: 0973-71
Fax: 0973-25-91
Telex: 33416

Owning Utility

Energetika Association
8, Triaditza Str.
1040 Sofia, Bulgaria
Lulin Radoulov, Pres.

Phone: 02-8-61-91
Fax: 02-87-58-26
Telex: 22-707

Basic Facts

Units 3-6 *Status:* Operable. *Type of Reactor:* Pressurized light-water reactor. *Megawatts (electric):* 440 (Units 3-4); 1000 (Units 5-6). *Megawatts (thermal):* 1375 (Units 3-4). *Reactor System Supplier:* Atomenergoexport (former Soviet Union).

Key Dates

Unit 3 *Ordered:* 1972. *Construction began:* 1973. *Criticality:* December 1980. *First Power:* December 1980. *Commercial Operation:* January 1981.

Unit 4 *Ordered:* 1972. *Construction began:* 1973. *Criticality:* 1982. *First Power:* May 1982. *Commercial Operation:* August 1982.

Unit 5 *Ordered:* 1979. *Construction began:* 1980. *Criticality:* November 1987. *First Power:* November 1987. *Commercial Operation:* September 1988.

Unit 6 *Ordered:* 1981. *Construction began:* 1982. *Criticality:* 1989. *First Power:* November 1989. *Commercial Operation:* 1992.

Plant Report

See Plant Report for Kozloduy Nuclear Power Plant, Units 1-2.

Canada

COUNTRY REPORT

Background

By 1993, when the reactors now under construction in the country are completed, Canada's total nuclear system will be 16,700 megawatts and will produce 20 percent of the nation's electricity. Nuclear power accounts for over half of the electricity used in the province of Ontario.

Canadian nuclear power stations are situated in the province of Ontario, which has 20 reactors, and the provinces of Quebec and New Brunswick, which have one reactor each. The provincial government utilities—Ontario Hydro, Hydro Quebec, and New Brunswick Power—own and operate the reactors.

Nuclear energy in Canada is regulated by the Atomic Energy Control Board (AECB), an agency of the federal government. A federal Crown Corporation, Atomic Energy of Canada Limited (AECL) has the mandate to develop and promote the use of nuclear energy in Canada. It designed the world-renowned CANDU (Canada Deuterium Uranium) type reactor, which is fueled with natural uranium and moderated and cooled with heavy water (deuterium oxide). Another federal corporation, Eldorado Nuclear, mines and refines uranium fuel for CANDUs. Journalist Robert Bott wrote in *Saturday Night*, "In the nuclear game, the Canadian federal government is simultaneously player, coach, and referee."

Nuclear history and first accidents

Canada's first reactor was built in 1945 at Chalk River, Ontario, as part of the country's wartime work on the atomic bomb. This reactor, using natural uranium and deuterium to produce plutonium-239 for atomic weapons, became the ancestor of the CANDU and the Canadian nuclear program.

On December 12, 1952, a technician mistakenly lifted at least three of the Chalk River reactor's 12 shut-off rods out of the core, contaminating the interior of the reactor building. The incident lasted only 70 seconds, but it was the first nuclear accident in history and Canada's worst. The reactor was shut down for 14 months. Cleanup from the incident lasted six months. A member of the crew assigned to dismantle the reactor core was Jimmy Carter, later to become the 39th president of the United States.

A second nuclear reactor accident occurred at Chalk River on May 25, 1958. A fuel rod in a precursor to the CANDU reactor broke and caught fire, spreading radioactive dust throughout the 12-story building. More than 600 people helped clean up the contaminated building and 475,000 square yards around it.

In 1952, AECL was created to take over the Chalk River Nuclear Laboratories from the National Research Council with a charge to develop nuclear energy for peaceful purposes. By 1958, the AECL had set up two special design teams instrumental in launching the first nuclear power plants in Canada—NPD and Douglas Point.

The CANDU: Backbone of Canadian nuclear power

CANDU reactors differ significantly from most other nuclear power reactors used elsewhere in the world. They are highly fuel efficient because they use natural rather that enriched uranium. The reactors' fuel rods are mounted horizontally instead of vertically like in other commercial reactors. The rods can also be refuelled while in operation, which avoids the need for costly shutdowns. Some consider the CANDU to be the safest reactor design in the world. CANDUs are consistently rated among the top world leaders in terms of operating performance.

The CANDU is also unique in its use of heavy water as a reactor moderator and coolant. Heavy water is a clear, colorless liquid that looks and tastes like ordinary tap water. It occurs naturally in water in small quantities—about one part heavy water to 7,000 parts ordinary water. An expensive, complex chemical process involving hydrogen sulphide gas and large amounts of steam is used to separate the small proportion of heavy water from ordinary water.

The AECL has sold CANDU reactors to Argentina, Romania, South Korea, Pakistan, and India.

Radioactive waste disposal

In 1978, the governments of Canada and Ontario agreed to cooperate in the development of technologies for the permanent disposal of Canada's nuclear fuel waste. They launched the Nuclear Fuel Waste Management Program. AECL was given the responsibility to assess the concept for the disposal of nuclear fuel wastes in the stable granitic rock formations of the Canadian Shield. Ontario Hydro was asked to develop the technologies for the interim storage and transportation of nuclear fuel.

Public review of the AECL's permanent storage program began in 1991. Disposal is likely to start about 2015. The disposal vault will consist of an engineered excavation 500 to 1,000 meters deep. The fuel waste will be sealed in corrosion-resistant containers and emplaced in the floor of the rooms in the vault. A mixture of sand and clay will surround the containers. When filled, the rooms, access tunnels, and shafts will be filled with a mixture of clay and crushed granite, and sealed.

As part of Canada's Nuclear Fuel Waste Management Program, the AECL is constructing its Underground Research Laboratory in a large body of rock known as the Lac du Bonnet batholith. This test facility will help Canada in developing methodology for geologic characterization of actual disposal sites.

Health concerns

The cancer-causing effects of radiation remain one of the prime concerns of opponents of commercial nuclear energy. To address this concern, the Atomic Energy Control Board of Canada commissioned a major study to be conducted by Drs. Eileen Clarke and John McLaughlin of The Ontario Cancer Treatment and Research Foundation and Dr. Terry Anderson of the University of British Columbia. The three Canadian researchers examined data for 1,894 children aged 14 years or younger who died from leukemia between 1950 and 1987 and who lived within 15 miles of five Canadian nuclear facilities. Results of the study, released in 1991, found no evidence of an increase in childhood leukemia due to living near nuclear facilities.

Future growth, government opposition

Late in 1989, Ontario Hydro released a 25-year, $20 billion energy plan calling for the construction of 10 additional nuclear plants, rated at 880 megawatts each, by the year 2014. It was the first time since the Three Mile Island nuclear power scare in the United States in April 1979 that a North American utility had committed to building new nuclear reactors. Hydro said it would need the plants to meet the growth in electricity demand anticipated in the coming decades.

Almost a year later, in November 1990, the New Democratic Party, Ontario's first socialist government, announced plans to freeze construction on new nuclear power plants. The moratorium, which was expected to last three years, was accompanied by an order to Ontario Hydro to divert $240 million toward energy conservation measures and away from the use that the utility had planned for the money—the design of new nuclear power stations.

In November 1991, a New Democratic Party bill was defeated in the Ontario legislature that would have prohibited the Atomic Energy Control Board from renewing licenses for nuclear power plants and uranium operations.

The provincial utility came under harsh criticism in early 1992 when they revised their 25-year plan, saying new plants would not be needed until 2009. Under the old plan, the first plant was to come online in 2002. The revised plan also called for fewer plants to be built. Ontario Hydro cited conservation efforts and recession as the reasons for the downturn in electricity demand. Critics argued that poor planning and political maneuvering were behind both the switch in long term plans and the utility's recent rate increases—11.8 percent in 1992 and a 44 percent increase over three years.

Despite the reduction in their construction plans, the fact that Ontario Hydro continues to look to nuclear energy sets it apart from many other utilities in the world who dropped plans for nuclear growth throughout the 1980s.

Sources

Nuclear Waste Management: Canada. Organization for Economic Cooperation and Development (OECD) and Atomic Energy of Canada Limited (AECL), n.d.

"The High Cost of Candu." *Maclean's,* May 14, 1984.

"Guarding Against Disaster." *Maclean's,* May 12, 1986.

"Power at Cost." *Saturday Night,* December 1986.

Darlington Generating Station. Ontario Hydro, April 1989.

"Utility Chooses Nuclear Power." *Christian Science Monitor,* January 16, 1990.

"Ontario Liberals Endorse Nuclear Energy." *Nuclear Industry,* Summer 1990.

"Big Shakeup in Tax System." *The Financial Post,* November 21, 1990.

"No Leukemia Rise Near Canadian Nuclear Power Plants." *INFO,* U.S. Council for Energy Awareness, August 1991.

"Canada: Anti-Nuclear Bill Defeated." *Nucleonics Week,* November 28, 1991.

"Hydro's Coin Toss." *The Toronto Sun,* January 18, 1992.

PLANT PROFILES

Bruce Nuclear Power Development, Units 1-4 (BNPD)

PO Box 3000
Tiverton, ON, Canada N0G 2T0
Ken Talbot, Station A Mgr.

Phone: (519)361-2673
Fax: (519)361-4717
Telex: 06-875780

Owning Utility

Ontario Hydro
700 University Ave.
Toronto, ON, Canada M5G 1X6
E.P. Horton, V.Pres., Nuclear Opns. Branch

Phone: (416)592-5111
Fax: (416)592-4087
Telex: 06-217662

Contact

Dave Stevens, Media Employee Commun. Officer
BNPD Information Centre
PO Box 1540
Tiverton, ON, Canada N0G 2T0

Phone: (519)368-7031

Basic Facts

Units 1-4 *Status:* Operable. *Type of Reactor:* Pressurized heavy-water reactor. *Megawatts (electric):* 825. *Megawatts (thermal):* 2606. *Reactor System Supplier:* Atomic Energy of Canada Limited. *Generator Supplier:* NEI-Parsons Limited (Canada). *Architect Engineer:* Atomic Energy of Canada Limited; Ontario Hydro.

Key Dates

Unit 1 *Ordered:* 1968. *Construction began:* 1970. *Criticality:* December 1976. *First Power:* January 1977. *Commercial Operation:* September 1977.

Unit 2 *Ordered:* 1968. *Construction began:* 1970. *Criticality:* July 1976. *First Power:* September 1976. *Commercial Operation:* September 1977.

Unit 3 *Ordered:* 1968. *Construction began:* 1972. *Criticality:* November 1977. *First Power:* December 1977. *Commercial Operation:* January 1978.

Unit 4 *Ordered:* 1968. *Construction began:* 1972. *Criticality:* December 1978. *First Power:* December 1978. *Commercial Operation:* January 1979.

Plant Report

"Nuclear Capital of the World". On the eastern shore of Lake Huron, between Kincardine and Port Elgin, sits the Bruce Nuclear Power Development (BNPD), named after the township in which it's located. BNPD is one of the largest centers of energy production in the world, and is sometimes referred to as the Nuclear Capital of the World. It is capable of meeting the heaviest demands of Metropolitan Toronto, Ottawa, Kingston, Hamilton, London, and Thunder Bay.

Within its boundaries are two nuclear generation stations, a heavy water plant, a bulk steam system, low- and medium-level radioactive waste storage facilities, a nuclear training center, an information center, and various administrative and service buildings. The site also includes the older Douglas Point Generating Station, now retired from service. Adjacent to BNPD is the Bruce Energy Center, an industrial and agricultural park heated by medium pressure steam from BNPD.

Bruce A, Units 1-4. Since its four 825 megawatts electric reactors were completed in the late 1970s, Bruce A has proved to be one of the most successful nuclear generating stations in the world. Bruce A reactors, Units 1-4, have been consistently placed in the top 10 in lifetime operating performance when compared to other large reactors around the world.

Bruce B, Units 5-8. Although essentially a duplicate of Bruce A, the steam generators and turbines at Bruce B are of a different design, giving its four reactors a higher capacity of 915 megawatts electric each. The station's four reactors, Units 5-8, began operation between 1984 and 1987.

Heavy water plants. Heavy water is a clear, colorless liquid that looks and tastes like ordinary tap water. It occurs naturally in water in small quantities—about one part heavy water to 7,000 parts ordinary water. It is used as a moderator and coolant in the CANDU reactors at BNPD. At the Bruce Heavy Water Plant, a complex chemical process involving hydrogen sulphide gas and large amounts of steam is used to separate the small proportion of heavy water from ordinary lake water.

The Bruce Heavy Water Plant A produced its first heavy water in 1973. Heavy Water Plant B, which has the same capacity, went into production in 1981. The Heavy Water Plant C was cancelled in 1976 and construction was stopped on the D plant in 1979, in anticipation of a slowdown in nuclear plant construction. This decreased demand for heavy water also led to a decision to close down Bruce Heavy Water Plant A in 1984. The Bruce Heavy Water Plant provides heavy water for reactors at Darlington as well as BNPD.

CANDU reactors need about 1,000 kilograms of heavy water for each megawatt of electricity they produce. This initial amount of heavy water doesn't have to be replaced during a reactor's lifetime, but a yearly top-up of about 1 percent is necessary.

Electricity backup for the station. Both the nuclear generating stations and the heavy water plant require a constant supply of electricity. Ten large automatic-start combustion turbine electrical generators are available to supply electricity in case of an emergency.

Radioactive waste storage. Low-level wastes such as rags, mops, clothing, and other materials are stored in bins in concrete warehouses or trenches. Intermediate-level wastes, such as filters that remove radioactive materials from liquid systems, are stored in in-ground steel-lined concrete containers and above-ground concrete structures called quadricells. The low/medium level waste site also serves other Canadian plants, specifically the Pickering and Darlington nuclear stations. Used nuclear fuel is kept underwater in used fuel storage pools at each station. All BNPD's radioactive waste facilities are licensed and monitored by the Atomic Energy Control Board of Canada.

Bruce

CANADA

Operations—mishaps and an award. In February 1982, the Bruce Unit 2 was taken out of service because of a pressure tube leak. Another pressure tube leak occurred at Bruce 2 in March of 1986. In light of these tube failures, Energy Probe, Ontario Hydro's principal opponent, called for the temporary closing of BNPD and a metallurgical review of the station.

In 1989, there were 11 incidents at the Bruce Heavy Water Plant that generated odor complaints from local residents.

Early in 1992, a senior official at the Atomic Energy Control Board admitted to the *Sault Daily Star* that minute radioactive particles had been found both outside and on the floor of a reactor building at BNPD. The "hot particle" inside the building was microscopic and picked up by station detectors. The official called these findings minuscule and totally irrelevant.

BNPD Unit 5 was recognized by *Nuclear Engineering International* as 7th in a ranking of top 10 plants in the world with the best lifetime performance record through June 30, 1991. The unit had maintained an overall capacity factor of 86.9 percent.

BNPD Information Centre. BNPD Information Centre is open to the public daily, 9:00 a.m. to 4:00 p.m. mid-April to mid-October, and Monday through Friday the remainder of the year. To arrange group tours, phone (519) 368-8687.

Sources

"Power at Cost." *Saturday Night*, December 1986.
Bruce Nuclear Power Development. Ontario Hydro, September 1989.
Nuclear Generating Division Environmental Summary, 1989. Ontario Hydro, Central Production Services Division, Environmental Protection Department, June 1990.
"Senior AECB official admits routine emission of radioactivity." *The Sault Daily Star*, January 2, 1992.

---

17 ●
Bruce Nuclear Power Development, Units 5-8 (BNPD)

PO Box 4000
Tiverton, ON, Canada N0G 2T0
R.E. Lewis, Station B Mgr.

Phone: (519)361-2673
Fax: (519)361-4998
Telex: 06-875780

Owning Utility

Ontario Hydro
700 University Ave.
Toronto, ON, Canada M5G 1X6
E.P. Horton, V.Pres., Nuclear Opns. Branch

Phone: (416)592-5111
Fax: (416)592-4087
Telex: 06-217662

Contact

Dave Stevens, Media Employee Commun. Officer
BNPD Information Centre
PO Box 1540
Tiverton, ON, Canada N0G 2T0

Phone: (519)368-7031

Basic Facts

Units 5-8 *Status:* Operable. *Type of Reactor:* Pressurized heavy-water reactor. *Megawatts (electric):* 915. *Megawatts (thermal):* 2832. *Reactor System Supplier:* Atomic Energy of Canada Limited. *Generator Supplier:* General Electric Co. (Canada). *Architect Engineer:* Ontario Hydro; Atomic Energy of Canada Limited.

Key Dates

Unit 5 *Ordered:* 1973. *Construction began:* 1978. *Criticality:* November 1984. *First Power:* December 1984. *Commercial Operation:* March 1985.

Unit 6 *Ordered:* 1973. *Construction began:* 1978. *Criticality:* May 1984. *First Power:* June 1984. *Commercial Operation:* September 1984.

Unit 7 *Ordered:* 1973. *Construction began:* 1979. *Criticality:* January 1986. *First Power:* February 1986. *Commercial Operation:* April 1986.

Unit 8 *Ordered:* 1973. *Construction began:* 1979. *Criticality:* February 1987. *First Power:* March 1987. *Commercial Operation:* April 1987.

Plant Report

See Plant Report for Bruce Nuclear Power Development, Units 1-4.

---

18 ●
Darlington Nuclear Generation Station, Units 1-3

Box 100
Bowmanville, ON, Canada L1C 3Z8
A.G. Holt, Stat.Mgr.

Phone: (416)623-6606
Fax: (416)697-7674
Telex: 06-981435

Owning Utility

Ontario Hydro
700 University Ave.
Toronto, ON, Canada M5G 1X6
E.P. Horton, V.Pres., Nuclear Opns. Branch

Phone: (416)592-5111
Fax: (416)592-4087
Telex: 06-217662

Contact

S.E. Stickley, Community Relations Officer
Darlington Nuclear Generation Station

Phone: 800-668-8500

Basic Facts

Units 1-3 *Status:* Operable. *Type of Reactor:* Pressurized heavy-water reactor. *Megawatts (electric):* 935. *Megawatts (thermal):* 2774. *Reactor System Supplier:* Atomic Energy of Canada Limited. *Generator Supplier:* ASEA (Sweden); Brown Boveri (Switzerland). *Architect Engineer:* Ontario Hydro; Atomic Energy of Canada Limited.

Key Dates

Unit 1 *Ordered:* 1973. *Construction began:* 1982. *Criticality:* September 1990. *First Power:* September 1990. *Commercial Operation:* August 1992.

Unit 2 *Ordered:* 1973. *Construction began:* 1981. *Criticality:* September 1989. *First Power:* October 1989. *Commercial Operation:* October 1990.

Unit 3 *Ordered:* 1973. *Construction began:* 1985. *Commercial Operation:* July 1992.

Plant Profiles

Darlington

Plant Report

Canada's newest reactors. Darlington Nuclear Generating Station, one of North America's largest energy projects and Canada's newest, is located on the shores of Lake Ontario in the town of Newcastle, 70 kilometers east of Toronto. When completed, Darlington's four reactors will generate 3,524,000 kilowatts of electricity, enough to service a city of two million people. Together with the province's other nuclear power stations, Darlington will provide some two thirds of the electricity used in Ontario, Canada's most populous province.

Site selection, construction process. Selected because of its proximity to the residential, industrial, and commercial energy markets of Ontario, the 1,200-acre site has good transportation access, an abundant supply of cooling water, relative isolation, and excellent bedrock for station foundations. Electricity from the four units will be fed into Ontario Hydro's system 500,000 volt transmission lines that already cross the site.

The Darlington project was approved by the provincial government in July 1977. Site preparation, which began in 1978, involved the removal of 45- to 95-foot cliffs along the lakeshore so the station could be built on bedrock. By 1981, the first concrete was poured for the foundations. Construction activity peaked in 1986 when over 7,000 workers were on the site. When completed, Darlington will provide jobs for about 1,100 people.

Community impact. Following an extensive series of environmental, engineering, and community studies, a number of agreements were signed by Ontario Hydro and the local municipalities to ensure Darlington's construction would not unduly disrupt the social and economic life of the area. These included a March 1977 agreement with the town of Newcastle, and a similar agreement in June of the same year between Hydro, Newcastle, and the Durham Region. Ontario Hydro and the town of Newcastle meet regularly to administer the 1977 construction impact agreements.

Physical features. When finished, Darlington will contain nearly 20 times the concrete of Toronto's CN Tower. The site is made up of six basic buildings or systems:

The powerhouse is the primary structure housing the four reactor buildings and the turbine hall; it measures six football fields long and 12 stories high.

The four reactor buildings are made of heavily reinforced concrete (six feet thick) to enclose the reactors and related equipment. Each reactor consists of a large, heavily-shielded vessel or calandria that holds 6,240 bundles of uranium fuel. The calandrias are placed in separate steel tanks that contain water and steel balls for shielding. The control rooms, surrounded by steel and cement, are built to withstand an airplane crash or a tornado, and each reactor has a complete set of parallel computer and control systems in place. Darlington has a very sophisticated version of the automatic fuel-handling machinery that allows its reactors to be refuelled without shutting down.

The turbine hall—approximately 1,255 feet long, 179 feet wide, and 148 feet high—houses four turbine generators. Each generator has a single shaft rotating at 1,800 revolutions per minute.

The vacuum building is connected to the reactor building by a pressure relief duct. Maintained at negative atmospheric pressure, the vacuum building is intended to suck any release of radioactivity from the pressurized systems and prevent its escape outside the station. The top of the vacuum tower holds 10-million liters of water, which would shower down to quench steam rushing into the structure.

The irradiated fuel bays contain two water-filled pools that will store the irradiated fuel when it's removed by robots from the reactors. The pools, which cool the fuel and shield station workers from radiation, have the capacity to store 17 years worth of irradiated fuel.

The cooling water system uses lake water to condense steam after it has passed through thee turbines. When operating at capacity, the station will require 34,100 gallons of lake water per second for cooling purposes. The system's diffuser system is designed to limit lake temperature differences caused by the return of warmer discharge water and protect potential whitefish spawning beds in the near-shore area. The intake and discharge tubes are the size of subways and extend more than a kilometer into the lake.

Controversial, cryogenic feature. Installed at Darlington is a $109 million tritium recovery system, cryogenic-distillation vessels that use chilling to separate tritium from deuterium. Tritium, a key radioactive ingredient in nuclear fusion reactions, builds up in the heavy water of Darlington's reactor cores. It can enter the human body through skin pores or be ingested or inhaled, and it has a half-life of 12.5 years.

In addition to its own reactors, Darlington treats heavy water from Bruce and Pickering nuclear power plants. This has sparked some concern for people who live along the route where tanks of radioactive liquid are being transported.

Darlington economics. Darlington's first two reactors became operational in 1990 and 1992 and cost Ontario Hydro about $8.7 billion. When the remaining two reactors, scheduled for operation in 1992 and 1993, are completed, the total bill is expected to reach over $13.5 billion—significantly higher than the original estimate in 1975 of $3.5 billion. Much of Hydro's 11.8 percent rate increase in 1992 was due directly to costs of the utility's nuclear program and the Darlington Nuclear Generating Station.

Critics of Darlington argue that the energy it produces is not needed, and that the utility could save the province a lot of money by meeting electricity needs through conservation efforts and better management of peak demand. The utility's rationale for building Darlington was based on expectation of electricity demand to grow in the province by about 2.6 percent annually. Growth in the early 1980s was low. By the mid 1980s, growth rates were over 4 percent annually. However, by the early 1990s, growth had slowed dramatically. By 1982, it was clear to the utility that Darlington's power would be in surplus of Ontario's needs. Hydro then delayed the construction of Darlington units three and four by several years and began looking for ways to expand their markets. In 1986, Hydro had 39 percent more generating capacity than it needed to meet peak demand.

Darlington politics. In the early 1980s, both Liberals and the New Democrats urged cancellation of Darlington. However, Ontario Premier William Davis had made Darlington the centerpiece of a 1981 election campaign promising job creation and high technology. Even though the need was questionable, Darlington went forward as an engine of growth for the provincial economy.

The Liberal Party in Ontario made an about face and endorsed nuclear energy at the party's April 1990 convention. The resolution, adopted after open floor debate called for plants to be strategically situated throughout Ontario.

In November 1990, the New Democratic Party, Ontario's first socialist government, announced plans to freeze construction on new nuclear power plants. The moratorium, which was expected to last three years, was accompanied by an order to Ontario Hydro to divert $240 million toward energy conservation measures and away from the design of new nuclear power stations. However, the government authorized Hydro to finish the Darlington nuclear station, which was 80 percent complete at the time.

Darlington Information Centre. Copies of the construction impact agreements signed by Ontario Hydro and the surrounding municipalities are available at the Information Centre. Special group tours of Darlington can also be arranged by contacting the Centre.

Sources

"Power Failure." *Canadian Business*, January 1984.
"Power at Cost." *Saturday Night*, December 1986.
Darlington Generating Station, Ontario Hydro, April 1989.
"Ontario Liberals Endorse Nuclear Energy." *Nuclear Industry*, Summer 1990.
"Big Shakeup in Tax System." *The Financial Post*, November 21, 1990.
"Firms Welcome Efforts to Hold Hydro Costs." *Toronto Daily Star*, January 11, 1992.
"Hydro Modifies Plan to Expand." *Toronto Globe and Mail*, January 15, 1992.

19 ○
Darlington Nuclear Generation Station, Unit 4

Box 100 Phone: (416)623-6606
Bowmanville, ON, Canada L1C 3Z8 Fax: (416)697-7674
A.G. Holt, Stat.Mgr. Telex: 06-981435

Owning Utility

Ontario Hydro Phone: (416)592-5111
700 University Ave. Fax: (416)592-4087
Toronto, ON, Canada M5G 1X6 Telex: 06-217662
E.P. Horton, V.Pres., Nuclear Opns. Branch

Contact

S.E. Stickley, Community Relations Officer Phone: 800-668-8500
Darlington Nuclear Generation Station

Basic Facts

Unit 4 *Status:* Under construction (80% complete as of January 1992). *Type of Reactor:* Pressurized heavy-water reactor. *Megawatts (electric):* 935. *Megawatts (thermal):* 2774. *Reactor System Supplier:* Atomic Energy of Canada Limited. *Generator Supplier:* ASEA (Sweden); Brown Boveri (Switzerland). *Architect Engineer:* Ontario Hydro; Atomic Energy of Canada Limited.

Key Dates

Unit 4 *Ordered:* 1973. *Construction began:* 1986. *Commercial Operation Expected:* March 1993.

Plant Report

See Plant Report for Darlington Nuclear Generation Station, Units 1-3.

20 ★
Douglas Point Nuclear Generating Station

Tiverton, ON, Canada

Owning Utility

Ontario Hydro Phone: (416)592-5111
700 University Ave. Fax: (416)592-4087
Toronto, ON, Canada M5G 1X6 Telex: 06-217662
E.P. Horton, V.Pres., Nuclear Opns. Branch

Contact

Dave Stevens, Media Employee Commun. Officer Phone: (519)368-7031
BNPD Information Centre
PO Box 1540
Tiverton, ON, Canada N0G 2T0

Basic Facts

Status: Shut down. *Type of Reactor:* Pressurized heavy-water reactor. *Megawatts (electric):* 218. *Megawatts (thermal):* 703.8. *Reactor System Supplier:* Atomic Energy of Canada Limited. *Operator:* Ontario Hydro.

Key Dates

Ordered: 1960. *Construction began:* 1961. *Criticality:* November 1966. *First Power:* January 1967. *Commercial Operation:* September 1968. *Shut Down:* May 1984.

Plant Report

Start-up in Canada. Canada's first reactor began operation in 1945 at Chalk River, Ontario, as part of the country's wartime work on the atomic bomb. This reactor, using natural uranium and deuterium, produced plutonium-239 for atomic weapons.

In 1952, Atomic Energy of Canada Limited (AECL) was created to take over the Chalk River Nuclear Laboratories from the National Research Council with a mandate to develop nuclear energy for peaceful purposes. In 1954, a Nuclear Power Group was set up to explore such possibilities. Out of the deliberations came the idea to build a pilot reactor, Nuclear Power Demonstration (NPD), as a three-part undertaking of Canadian General Electric, Ontario Hydro, and AECL.

Before construction was finished on NPD, strictly a pilot project, the Nuclear Power Group had begun pushing for a full-scale power reactor. As a result, another team called the Nuclear Power Plant Division was set up in 1958 in Toronto. The new project team was based in Ontario Hydro's A.W. Manby Center and began design studies for a 200 megawatts electric power reactor.

After securing full approval from Canada's highest levels of government, the Douglas Point project was set in full motion by 1959. The project placed Canada into the world nuclear power scene along with the United Kingdom, which was building two 300 megawatts electric Magnox reactors, and the United States, which had the 60 megawatts electric Shippingport pressurized light-water reactor in operation and was building a 200 megawatts electric boiling water reactor at Dresden.

Site selection. Douglas Point was chosen as the site for Canada's first full-scale power reactor after a long process of elimination. Under original consideration were sites along the shoreline north of Manitoulin Island, and from Tobermory on the Bruce Peninsula to Goderich. It was gradually narrowed down to the stretch between Goderich and Southampton.

Douglas Point, 90 miles straight across Lake Huron from Port Huron, Michigan, was finally chosen because the bed rock is close to the surface—so close that trees near the station had to be planted using a jackhammer.

Construction, the community, and costs. The Douglas Point site is located in Bruce Township. When clearing of the site began in 1960, the region was experiencing an economic slump. People in the area welcomed the construction, which went forward with no antinuclear sentiment. The 2,300-acre site went for $50 to $70 an acre, and a building permit was never obtained from the Township.

Construction of Douglas Point changed the way of life for much of the surrounding farming community. Most farmers ran mixed operations and kept hired hands, which they could no longer afford once union wages came in. As a result, many farmers changed over to beef farming, because it wasn't as labor-intensive. Also, many farmers went to work on the site themselves.

The target date for completion of Douglas Point was 1964, five years after the decision to build it. Cost estimate in 1959 was $81.5 million. Because of construction delays and an 11 percent federal tax imposed on materials and equipment, the plant cost $91 million by the time it began operating commercially in 1968.

Douglas Point was originally scheduled to be a two-unit undertaking. But in light of the successful, larger reactors just built at Pickering, a second Douglas Point reactor was considered too small and was never built.

Operational experience helps other plants. Douglas Point encountered equipment problems from the start, including trouble with the primary circuit pumps, the fueling machines, and many valves. As the engineers and technicians straightened out thousands of snags at Douglas Point, they paved the way for smoother installations and operations of other Canadian reactors. Plants at Pickering and Bruce, which were constructed shortly after Douglas Point, especially benefited.

Douglas Point experienced many firsts in the Canadian nuclear industry. The reactor was the first in which a pressure tube and a calandria tube were changed. Development of the "mouse," a remotely controlled tool that crawled into the core of the reactor and repaired a leak in 1976, was a major advancement. The CAN-DECON decontamination process was also a first for Douglas Point. Finally, the station was the only CANDU plant in the world with an oil-filled window, which offered visitors a view of the reactor face—even while operating.

Douglas Point's overall capacity factor during its years of operation was 54 percent. However, toward the end of it's life, the plant improved significantly. In 1982, its capacity factor was 75 percent, in 1983; it was 77 percent; and in 1984, it was 84 percent.

Two incidents. In 1980-81 the Douglas Point reactor ran for six months without a proper filter to contain radioactive gases. In January 1986, it was discovered that every hour 60 liters of water were leaking out of the used-fuel storage bay. Opponents of nuclear energy in Canada pointed to the accumulation of small incidents like these as the main danger of nuclear power.

Economics lead to closing. Although its operations were improving through the years, the harsh economic realities of Douglas Point convinced AECL to retire the plant on May 5, 1984. The plant was owned by AECL, which also paid for all of its operating expenses, including the salaries of its 264 workers. Douglas Point was operated by Ontario Hydro and was the facility's sole customer; Ontario Hydro reimbursed the federal corporation for all the power generated by Douglas Point. However, it never showed a profit; in the three fiscal years before its closing, the plant lost $16 million. To close the plant permanently cost AECL $100 million.

Legacy of Douglas Point. In 1968, Ontario Hydro began building the Bruce Nuclear Power Development (BNPD) site around the Douglas Point Plant. The Douglas Point reactor now sits inactive among BNPD's eight power plants except for the cooling and ventilation circuits which are kept operating to cool the plant and equipment.

The Douglas Point project set the pattern for many Canadian nuclear projects to come. It provided the essential learning experience in design, development, manufacturing, and project organization as well as in public acceptance. It was Canada's first large-scale CANDU nuclear station, pioneering the use of heavy water for reactor cooling and for the control of the nuclear reaction.

Gentilly

Sources

"The High Cost of CANDU." *Macleans*, May 14, 1984.
Atomic Energy of Canada Limited, CANDU Operations. "The Douglas Point Story." *Power Projections Special Edition*, June 1984.
"Power at Cost." *Saturday Night*, December 1986.

21 ▲
Gentilly Nuclear Generating Station, Unit 1

[*See Also:* Gentilly Nuclear Generating Station, Unit 2]

PO Box 360
Gentilly, PQ, Canada G0X 1G0
Phone: (819)298-2943
Fax: (819)298-5203

Owning Utility

Hydro-Quebec
75 W. Rene Levesque Blvd.
Montreal, PQ, Canada H2Z 1A4
Richard Drouin, Chairman
Phone: (514)289-2211
Fax: (514)843-3163
Telex: 055-60708

Contact

Gilles Lafontaine, Pub.Info.
Gentilly 2 Nuclear Power Plant

Basic Facts

Unit 1 *Status:* Decommissioned. *Type of Reactor:* Boiling water reactor. *Megawatts (electric):* 266. *Megawatts (thermal):* 840. *Architect Engineer:* Atomic Energy of Canada Limited; Hydro-Quebec. *Owner:* Atomic Energy of Canada Limited. *Operator:* Hydro-Quebec.

Key Dates

Unit 1 *Ordered:* 1965. *Construction began:* 1966. *Criticality:* 1970. *First Power:* April 1971. *Commercial Operation:* May 1972. *Shut Down:* June 1977. *Decommissioned:* June 1986.

Plant Report

Twenty-eight weeks of operation. In 1970 Atomic Energy of Canada Limited (AECL) finished building Gentilly 1, an $88 million CANDU experimental reactor on the south shore of the St. Lawrence River opposite Trois-Rivieres. After it began operating, the plant was in service intermittently for a total of only 200 days before the federal energy control board shut it down for safety reasons in 1977. Intended to be a temporary closure, the plant never reopened. The board demanded $120 million worth of safety equipment. Hydro-Quebec, the provincial utility operating the unit, decided not to buy the plant.

The plant was decommissioned between 1984 and 1986. In June 1986, the AECL sealed the doors with concrete. Some considered the experiment a failure. One AECL official said the plant used more electricity than it produced.

Alternative reactor type. At the time Gentilly 1 was being designed, the Canadian government was also planning for a nuclear power plant in Bruce Township, Ontario. At this Douglas Point nuclear power plant, the AECL was experimenting with the use of heavy water as a reactor coolant and moderator. Because there were certain unknown factors in the use of heavy water, the government decided to use light water as a coolant in the Gentilly reactor. The plant was designed to be an alternative to Douglas Point, so the AECL could analyze the performance of both types of reactors in the early stages of the country's nuclear industry.

Sources

Atomic Energy of Canada Limited, CANDU Operations. "The Douglas Point Story." *Power Projections Special Edition*, June 1984.
"Guarding Against Disaster." *Macleans*, May 12, 1986.
"Power at Cost." *Saturday Night*, December 1986.
Mounfield, Peter. *World Nuclear Power*. Routledge, 1991.

22 ●
Gentilly Nuclear Generating Station, Unit 2

[*See Also:* Gentilly Nuclear Generating Station, Unit 1]

PO Box 360
Gentilly, PQ, Canada G0X 1G0
Roger Emard, Dir., Nuclear Operations
Phone: (819)298-2943
Fax: (819)298-5203

Owning Utility

Hydro-Quebec
75 W. Rene Levesque Blvd.
Montreal, PQ, Canada H2Z 1A4
Richard Drouin, Chairman
Phone: (514)289-2211
Fax: (514)843-3163
Telex: 055-60708

Contact

Gilles Lafontaine, Pub.Info.
Gentilly 2 Nuclear Power Plant

Basic Facts

Unit 2 *Status:* Operable. *Type of Reactor:* Pressurized heavy-water reactor. *Megawatts (electric):* 685. *Megawatts (thermal):* 2184. *Reactor System Supplier:* Atomic Energy of Canada Ltd. *Generator Supplier:* GE Canada. *Architect Engineer:* Atomic Energy of Canada Ltd; Canatom Ltd. (Canada); Hydro-Quebec.

Key Dates

Unit 2 *Ordered:* 1972. *Construction began:* 1974. *Criticality:* September 1982. *First Power:* September 1982. *Commercial Operation:* October 1983.

Plant Report

Minor role in the province. Construction on Gentilly 2, Quebec's only working nuclear plant, began in 1974. The Quebec plant experienced dramatic cost escalations—from a first estimate of $385 million to a projected price of $1.36 billion at the end of 1982. Although construction on Gentilly 2 began a year before the Point Lepreau Nuclear Generation Station, it was almost a year behind the New Brunswick plant in starting commercial operations in October 1983.

In Quebec's overall power scene, Gentilly 2 plays a minor role. The province fills its domestic and export needs easily with water-powered turbines at the mammoth James Bay project and other hydroelectric plants. A 1986 report in the Canadian journal *Saturday Night* said that Gentilly 2 seldom operated because of a shortage of skilled staff and a lack of demand for its expensive power. In that same year, Hydro Quebec, the provincial utility that owns and runs Gentilly 2 and one of Canada's largest corporations by assets, had a generating surplus of 5,000 megawatts of electricity. It isn't likely that Quebec will invest in further nuclear projects.

Sources

"An Uneasy Nuclear Debut." *Maclean's*, September 25, 1982.
"The High Cost of CANDU." *Maclean's*, May 14, 1984.
"Power at Cost." *Saturday Night*, December 1986.

23 ★
Nuclear Power Demonstration Reactor (NPD)

Rolphton, ON, Canada

Owning Utility

Ontario Hydro
700 University Ave.
Toronto, ON, Canada M5G 1X6
E.P. Horton, V.Pres., Nuclear Opns. Branch

Phone: (416)592-5111
Fax: (416)592-4087
Telex: 06-217662

Basic Facts

Status: Shut down. *Type of Reactor:* Pressurized heavy-water reactor. *Megawatts (electric):* 25. *Megawatts (thermal):* 96.2. *Reactor System Supplier:* General Electric Canada. *Architect Engineer:* Ontario Hydro; General Electric Canada. *Owner:* Ontario Hydro; Atomic Energy of Canada Limited. *Operator:* Ontario Hydro.

Key Dates

Ordered: 1957. *Construction began:* 1958. *Criticality:* April 1962. *First Power:* June 1962. *Commercial Operation:* October 1962. *Shut Down:* August 1987.

Plant Report

Peaceful nuclear energy. Canada's first reactor began operation up in 1945 at Chalk River, Ontario, as part of the country's wartime work on the atomic bomb. This reactor, using natural uranium and deuterium, produced plutonium-239 for atomic weapons.

In 1952, Atomic Energy of Canada Limited (AECL) was created to take over the Chalk River Nuclear Laboratories from the National Research Council with a mandate to develop nuclear energy for peaceful purposes. In 1954, a Nuclear Power Group was set up to explore such possibilities. Out of the deliberations came the idea to build the Nuclear Power Demonstration (NPD) reactor, a pilot project, as a three-part undertaking of Canadian General Electric, Ontario Hydro, and AECL. The primary purpose of NPD was to establish the feasibility of electricity production using heavy water technology.

In 1955, the Nuclear Power Group set up a design time at Canadian General Electric in Peterborough, Ontario. Several people from electrical utilities and related industries were brought to the site to work with AECL designers.

The prototype reactor, one of the most conservative reactors in the world, began operation in 1962 at Rolphton, Ontario, on the Ottawa River. Its licensed net electrical output was 22 megawatts. It had a successful operating history with no abnormal shutdowns before finally closing in 1987. The reactor had a vertical pressure vessel, unlike the unique horizontal tube design of the CANDU reactor adopted by the AECL in 1957.

Before construction was finished on NPD, strictly a pilot project, the Nuclear Power Group had begun pushing for a full-scale power reactor. As a result, another team called the Nuclear Power Plant Division was set up in 1958 in Toronto. The Division's efforts resulted in the construction of the 200 megawatt electric Douglas Point reactor, which launched Canada into the world nuclear power scene.

Sources

Atomic Energy of Canada Limited, CANDU Operations. "The Douglas Point Story." *Power Projections Special Edition*, June 1984.
"Power at Cost." *Saturday Night*, December 1986.

24 ●
Pickering Nuclear Generating Station, Units 1-8

Box 160
Pickering, ON, Canada L1V 2R5
R.O. Schuelke, Stat.Mgr.

Phone: (416)839-1151
Fax: (416)839-7994

Owning Utility

Ontario Hydro
700 University Ave.
Toronto, ON, Canada M5G 1X6
Gary Lem, Corp. Relations Advisor

Phone: (416)592-5111
Fax: (416)592-4087
Telex: 06-217662

Contact

J.W. Muir, Corp. Relations Officer
Pickering Energy Information Centre

Phone: (416)839-0465

Basic Facts

Units 1-8 *Status:* Operable. *Type of Reactor:* Pressurized heavy-water reactor (CANDU). *Megawatts (electric):* Units 1-4, 542; Units 5-8, 540. *Megawatts (thermal):* Units 1-4, 1744; Units 5-8, 1754. *Reactor System Supplier:* Atomic Energy of Canada Limited. *Generator Supplier:* NEI-Parsons Ltd. (Canada). *Architect Engineer:* Ontario Hydro; Atomic Energy of Canada Limited.

Key Dates

Unit 1 *Ordered:* 1964. *Construction began:* 1965. *Criticality:* February 1971. *First Power:* April 1971. *Commercial Operation:* July 1971.

Unit 2 *Ordered:* 1964. *Construction began:* 1965. *Criticality:* September 1971. *First Power:* October 1971. *Commercial Operation:* December 1971.

Unit 3 *Ordered:* 1965. *Construction began:* 1966. *Criticality:* April 1972. *First Power:* May 1972. *Commercial Operation:* June 1972.

Unit 4 *Ordered:* 1965. *Construction began:* 1966. *Criticality:* May 1973. *First Power:* May 1973. *Commercial Operation:* June 1973.

Unit 5 *Ordered:* 1968. *Construction began:* 1974. *Criticality:* October 1982. *First Power:* December 1982. *Commercial Operation:* May 1983.

Unit 6 *Ordered:* 1968. *Construction began:* 1974. *Criticality:* October 1983. *First Power:* November 1983. *Commercial Operation:* February 1984.

Unit 7 *Ordered:* 1973. *Construction began:* 1974. *Criticality:* October 1984. *First Power:* November 1984. *Commercial Operation:* January 1985.

Unit 8 *Ordered:* 1973. *Construction began:* 1974. *Criticality:* December 1985. *First Power:* January 1986. *Commercial Operation:* February 1986.

Plant Report

About the station. Located about 30 kilometers east of Toronto, the Pickering Nuclear Generating Station has one of the best performance records of any station in the world, according to Ontario Hydro, the station's owner and operator. Since it first began producing electricity in 1971, the utility claims that the station has consistently ranked at or near the top in terms of reliability and performance. A 1978 Ontario royal commission report on electric power planning predicted that there was likely to be only one meltdown in 10,000 years at Pickering.

The station's first four nuclear reactors, Units 1-4, went into service between February 1971 and May 1973. The second four reactors, Units 5-8, began operating between May 1983 and February 1986. Each of the reactors can generate 540 megawatts of electricity. The station's total capacity is over 4,300 megawatts, enough electricity to meet the needs of a city the size of metropolitan Toronto. It generates power at a cost 40 percent lower than coal-fired power.

The station's most noticeable features are the eight domed buildings that house the reactors and boilers. The largest structure in the station is a 51 meter high vacuum building. All the reactor buildings are connected to this silo-like structure, which is maintained at near vacuum. In the event of a major break in any of the main components of a reactor, the steam and radioactive material would be sucked automatically into the vacuum building. Then the steam would be condensed by water sprayed from a 9.2 million liter tank housed under the roof of the vacuum building.

The station site occupies 670 acres and includes a 100-acre wildlife sanctuary and recreation area. The station itself is built on shale rock, a strong foundation for the heavy building.

Problems with Unit 2. In the 1970s, Pickering Unit 2 ran for a year-and-a-half with an unnoticed hole, more than a hundred square centimeters, in its containment building.

In August 1983, while units 7 and 8 were still under construction, the rupture of a pressure tube in Pickering's Unit 2 resulted in a spill of contaminated coolant within the plant. It was caused by an improperly designed spring that allowed highly radioactive fuel bundles to shift out of position.

Although the unit was shut down normally without requiring the multiple emergency backup systems, and although there was no hazardous leak from the plant and no one was injured, the incident is generally considered one of the most serious in Canada since the first Canadian reactor began operating at Chalk River.

The accident led Ontario Hydro to completely retube Units 1 and 2, a $720 million repair that took three years and resulted in one- to two-percent rate increases in each of the three years the plant was closed. The immediate cost of the shutdown of Pickering Unit 2 was $250,000 a day to supply power from coal-fired stations.

The breakdown at Pickering Unit 2 after only 12 years of operation raised some accounting questions in the industry. In 1982, Ontario Hydro lengthened the amortization period for nuclear plants from 30 years to 40. However, at the time, no CANDU plant had operated that long. Some wondered if the Pickering incident was a sign that other Canadian plants would become less reliable with age.

Leaky transformer in Unit 4. In April 1984, Pickering Unit 4 was closed for about 10 days after one of its transformers leaked oil. The shutdown was costly—$2 million to replace the lost nuclear power with coal-generated electricity.

An exceptional Unit 7. Pickering Unit 7 was recognized by *Nuclear Engineering International* as fifth in a ranking of top 10 plants in the world with the best lifetime performance record through June 30, 1991. The unit had maintained an overall capacity factor of 87 percent.

Radioactive waste. Of the nuclear waste that is produced at the station, over 99 percent of the long-lived radioactive material is confined to fuel bundles that are stored after use in specially designed pools at the station. The bundles lose 90 percent of their heat into the water in the first five years, but are not considered safe to handle for at least another 300 years. Low-level and medium-level radioactive materials are transported to the Bruce Nuclear Power Development. There, Ontario Hydro operates a carefully monitored waste storage facility.

In some 20 years of operation, Pickering has never had a single employee lose a day of work due to radiation exposure. More than 1,500 people work at the station. Some 100,000 people live within 10 kilometers of the site.

Benefits beyond electricity. Coolwater Farms Ltd. owes its existence to the Pickering station. This company uses heated water discharged from the station to support a fish farm that produces about one million pounds of rainbow trout each year.

Pickering is a major source of Cobalt-60, a key element in the treatment of cancer and used also in diagnosing disease, analyzing blood and tissue samples, and sterilizing medical supplies. Ontario Hydro provides about 85 percent of the world's Cobalt-60, which is easily produced in the CANDU reactor.

Energy Information Centre. Each year, about 20,000 visitors tour the Energy Information Centre at the Pickering station. The Centre uses a variety of audiovisual presentations, a large

Plant Profiles

CANDU reactor model, and many interactive displays. It is open to the public seven days a week from 9 a.m. to 4 p.m. Special group presentations or in-plant tours can be arranged by calling (416) 839-0465.

Sources

"Power Failure." *Canadian Business*, January 1984.
"The High Cost of CANDU." *Macleans*, May 14, 1984.
"Guarding Against Disaster." *Macleans*, May 12, 1986.
"Power at Cost." *Saturday Night*, December 1986.
Pickering Generating Station. Ontario Hydro, Public Relations Division, September 1989.

25 •
Point Lepreau Nuclear Generating Station, Unit 1

PO Box 10
Lepreau, NB, Canada E0G 2H0
A. Rex Johnson, Plant Mgr.

Phone: (506)659-2000
Fax: (506)659-2703

Owning Utility

New Brunswick Electric Power
  Commission
PO Box 2000
Fredericton, NB, Canada E3B 4X1
Lin Titus, Pres.

Phone: (506)458-4444
Fax: (506)458-4390
Telex: 014-46285

Contact

Christine Nassrallah, Dir. Pub.Aff.
New Brunswick Electric Power
  Commission

Basic Facts

Unit 1 *Status:* Operable. *Type of Reactor:* Pressurized heavy-water reactor. *Megawatts (electric):* 680. *Megawatts (thermal):* 2156. *Reactor System Supplier:* Atomic Energy of Canada Ltd. *Generator Supplier:* NEI-Parsons Ltd. (Canada). *Architect Engineer:* Atomic Energy of Canada Ltd; Canatom Ltd. (Canada); New Brunswick Electric Power Commission.

Key Dates

Unit 1 *Ordered:* 1974. *Construction began:* 1975. *Criticality:* July 1982. *First Power:* September 1982. *Commercial Operation:* February 1983.

Plant Report

Sole nuclear plant in Atlantic Canada. The Point Lepreau Nuclear Generating Station consists of a single CANDU-6 nuclear reactor with a net capacity of 630,000 kilowatts. The station is located on the Lepreau peninsula, overlooking the Bay of Fundy, 40 kilometers southwest of Saint John. It is operated by a staff of 425.

The only nuclear installation in Atlantic Canada, Point Lepreau supplies about 30 percent of New Brunswick's electrical energy. It is owned and operated by the provincial utility New Brunswick Electric Power Commission and produces 26 percent of the company's total output from hydro, oil/coal, and nuclear plants.

The most noticeable feature at Lepreau is the domed building that houses the nuclear reactor system. The building is 42 meters high with reinforced concrete walls 1.2 meters thick. In August 1991, a new building housing a simulated nuclear control room and a public affairs information center was opened adjacent to the reactor building.

Construction highlights. Construction of Point Lepreau began in May 1975 and was completed in late 1981. At the peak of construction in 1979, 3,300 workers were employed on the project—a boon for the New Brunswick economy which had a 14 percent unemployment rate at the time. Point Lepreau was built with provision for an additional 600 megawatt unit on the site and is essentially the same as CANDU-6 reactors in Quebec, Argentina, and Korea. Although started third, Lepreau was the first of these CANDUs to be licensed for operation, the first to achieve criticality, and the first to begin commercial operation.

Primary technical data

- Gross station output—680,000 kilowatts
- Reactor containment building—low pressure containment, pre-stressed concrete vessel
- Fuel channels—380
- Fuel—natural uranium (UO_2)
- Fuel inventory—4,560 bundles, each with 37 elements and weighing 22 kilograms and measuring 50 centimeters long
- Refuelling method—on-power
- Heavy water inventory—494 tons
- Primary coolant pumps—4
- Boilers—4
- Generator voltage output—26,000 volts
- Turbine—one double-flow high pressure cylinder and three, double-flow low pressure cylinders, all arranged on the same shaft
- Turbine speed—1,800 rpm
- Steam temperature—500 °F
- Cooling water flow—seawater from the Bay of Fundy at 341,000 Igpm; water is later returned to the bay 11 Celsius degrees warmer through specially designed submerged diffuser nozzles
- Reactor regulation system—direct digital control dual computer system

Liberal opposition. Point Lepreau was not built without opposition. While it was championed by the provincial premier Richard Hatfield and his Progressive Conservative government, the construction of Point Lepreau was intensely opposed by the Liberal party within the New Brunswick legislature. The Liberals called the project a "debacle" and charged that political patronage had been involved in awarding some of the contracts for Point Lepreau. They also cited the results of a consultants' study that showed labor productivity at the construction site in 1977 to be very low. On average, pipefitters, electricians, masons, and other workers were actually busy only two hours a day. Another study, done in April 1980, showed productivity had dropped to 18.6 percent or 1.5 hours of productive work a day.

Liberals were also concerned about Lepreau's construction costs. The original estimate in 1974 was $466 million. A year later, estimates were coming in at $684 million. The plant's final tab was approximately $1.3 billion. New Brunswick Power, a government-owned utility, claimed that of 32 North American nuclear plants coming on line around the same time as Point Lepreau, the cost overrun at Lepreau was the sixth lowest.

In 1982, Point Lepreau became a campaign issue for Liberal candidates running for provincial premier in the October 12th election. Liberal leader Douglas Young called Point Lepreau "the most ridiculous undertaking in New Brunswick history."

Personal resistance. Resistance took on a more personal dimension by protestors such as Jim and Kay Bedell, who went on a nine-day, water-only fast early in 1980. During the fast, the couple moved into a hotel in Fredericton and passed the time with media interviews, daily visits to the New Brunswick legislature, and talks with visitors to their room. Among their visitors were the New Brunswick Premier Richard Hatfield and provincial Liberal leader Joseph Daigle.

The fast was not the Bedell's first antinuclear protest. They were among a group that in 1977 sailed out to sea in rubber dinghies to try and block the arrival of Point Lepreau's calandria, the sealed cooling vessel for the reactor. The couple was forced up on a dock by the wash of the freighter as it backed toward the shore.

Energy exports. In 1982, New Brunswick Power became the first Canadian utility to successfully contract the sale of power from a nuclear power station to the United States. The utility planned to sell more than half of the plant's generating capacity through 1990. Its first contracts were signed with the Massachusetts Municipal Wholesale Electric Company (for 100 megawatts), Boston Edison (for 100 megawatts), and the small Eastern Maine Electric Co-op of Calais, Maine (for 5 megawatts).

The exports caused some concern among Canadian citizens, but New Brunswick Power saw the contracts as a way of sharing risk: U.S. customers would pay more than half of Lepreau's annual carrying costs of $250 million during the early years of operation, regardless of plant performance.

Four spills, but still a world leader in performance. In the initial testing of Lepreau before full operation, the plant experienced some difficulties that made New Brunswick headlines. A series of four spills of radioactive heavy water occurred from July to September 1982. Three of the spills were caused by equipment problems; the fourth occurred when a worker prematurely opened a valve. The New Brunswick Power Electric Power Commission asserted that all the spills were minor and that they were routinely mopped up by maintenance personnel.

Since its start-up, Point Lepreau has consistently outperformed all other nuclear reactors in its range throughout the world. It was ranked by *Nuclear Engineering International* as first among 343 large reactors worldwide for best lifetime performance through the end of June 1991. The plant has maintained an overall capacity factor over 90 percent. Its capacity factor for fiscal year 1990-91 was 98.5 percent.

Unit 2 still in the planning stages. The Point Lepreau site was built with provision for two nuclear reactors. Construction on Unit 1 began in May 1975 and, by the time it was completed in late 1981, momentum was building for a second Lepreau nuclear unit. At the time New Brunswick Power's annual growth was projected at 3.7 percent. At that rate, the utility claimed it would need additional generating capacity by 1995. The plans were to build Unit 2 and sell its surplus power to U.S. customers until the power was needed within the province.

New Brunswick Power never commenced construction on Unit 2, but as of 1992, the utility had not permanently cancelled the project.

Sources

"A Terrible Thing Worth Fasting For." *Maclean's*, April 7, 1980.
"Slimming Down a Durable Tory." *Maclean's*, July 7, 1980.
"An Uneasy Nuclear Debut." *Maclean's*, September 25, 1982.
The New Brunswick Electric Power Commission Annual Report 1990/1991. New Brunswick Electric Power Commission, n.d.
Point Lepreau Generating Station. New Brunswick Electric Power Commission, July 1991.
Interview with Christine Nasarrallah, Director of Public Affairs, New Brunswick Power. April 22, 1992.

Point Lepreau Nuclear Generating Station, Unit 2

PO Box 10
Lepreau, NB, Canada E0G 2H0
A. Rex Johnson, Plant Mgr.

Phone: (506)659-2000
Fax: (506)659-2703

Owning Utility

New Bruswick Electric Power Commission
PO Box 2000
Fredericton, NB, Canada E3B 4X1
Lin Titus, Pres.

Phone: (506)458-4444
Fax: (506)458-4390
Telex: 014-46285

Contact

Christine Nassrallah, Dir. Pub.Aff.
New Brunswick Electric Power Commission

Basic Facts

Unit 2 *Status:* Indefinitely deferred. *Type of Reactor:* Pressurized heavy-water reactor. *Megawatts (electric):* 450. *Megawatts (thermal):* 1290.

Plant Report

See Plant Report for Point Lepreau Nuclear Generating Station, Unit 1.

Cuba

COUNTRY REPORT

Economic problems

Construction on the Juragua nuclear complex, Cuba's first attempt at nuclear energy, began with the hopes that the finished plant would help solve the country's crippling energy crisis. However, the joint Cuban-Soviet project, which started in 1983, was suspended in 1992 in the face of mounting economic problems.

Cuba's sugar-based economy was severely affected by the fall of socialism in the Soviet Union and Eastern Europe. The country's purchasing power was slashed from the 1989 level of $8.1 billion to about $2.2 billion in 1992.

When Cuba's Russian partners wanted $200 million in hard currency to continue working on the Juragua nuclear complex, Cuba was forced to abandon construction efforts on the needed plant. Not having enough money for oil, Cuba had to impose daily blackouts, which darkened cities and cut back factory work hours.

Sources

"Cuba Halts Work on Nuclear Plant, Blaming Russians." *The Wall Street Journal,* September 8, 1992.

PLANT PROFILES

27 △
Juragua Nuclear Complex, Units 1-2

UPICEN Juragua
Aptdo. 409
Cienfuegos, Cuba
Rafael Soler Deschapells, Stat.Mgr.

Owning Utility

Cuban Ministry of Basic Industries
Carlos III No. 666/7 Piso
Havana, Cuba
Miguel Angel Serradet Acosta, Dir.

Basic Facts

Units 1-2 *Status:* Indefinitely deferred. *Type of Reactor:* Pressurized light-water reactor. *Megawatts (electric):* 440. *Reactor System Supplier:* Atomenergoexport (former USSR). *Generator Supplier:* Atomenergoexport (former USSR). *Architect Engineer:* Filial Leningradense de Atomenergoprojekt (former USSR/Cuba).

Key Dates

Unit 1 *Ordered:* 1983. *Construction began:* 1983. *Commercial Operation Expected:* indefinite.

Unit 2 *Ordered:* 1983. *Construction began:* 1985. *Commercial Operation Expected:* indefinite.

Plant Report

Construction suspended. The Juragua nuclear complex is located near Cienfuegos, Cuba, about 150 miles from the coast of Florida. Construction on the joint Cuban-Soviet project began in 1983, and Unit 1 of the two-reactor complex was expected to start operating commercially in 1995, at least five years later than the initial deadline. However, the financially-strapped government suspended construction on the plant in September of 1992. The future of the plant, in which Cuba invested more than $1 billion, is uncertain. Fidel Castro's son, Fidelito, was in charge of the building program.

Safety concerns. The plant generated safety concerns among federal experts in the U.S., Florida lawmakers, as well as Cuban scientists. On May 30, 1991, the U.S. U.S. State Department told Cuba it expected the island country to follow high safety standards in building Juragua—the same type of standards observed by the United States. State department spokesman Richard Boucher told the Associated Press that "Cuba and the Soviets do have a responsibility to the countries of the region to ensure that the facility is constructed and operated safely, and in accordance with international standards." Representatives from the United States had been following the construction of Juragua and had visited the site in 1989.

Jorge Oro, a Cuban scientist who defected, said that the likelihood of a nuclear accident at Juragua was possible as only a few managers at the plant were adequately trained.

Sources

"Farewell Fidel? Not So Fast, South Florida." *Florida Trend*, July 1990.
"U.S. Warns Cuba on Safety at New Nuclear Power Plant." *The Washington Post*, May 31, 1991.
"Cuba Halts Work on Nuclear Plant, Blaming Russians." *The Wall Street Journal*, September 8, 1992.

Czech Republic

COUNTRY REPORT

A firm commitment to nuclear energy

In the early 1980s, the Czech Republic aimed to produce 30 percent of it's electricity from the atom by the mid 1990s. In the country's 1981-1985 plan, their commercial nuclear energy program received almost a third of the total national investment.

The Czech Republic's dedication to nuclear energy stems from a desperate need for electricity and a growing limitation on the country's non-nuclear energy sources. It relies on the countries of the former Soviet Union for 93 percent of its oil and struggles to maintain natural gas supplies. A heavy reliance on coal for electricity (second only to the U.S.) has depleted the country's coal reserves and has produced a severe pollution problem from sulphur dioxide emissions—especially in older industrial districts.

By July 1991, the Czech Republic was close to meeting their goal and generated 28.4 percent of their electricity from nuclear energy. And, despite pressure from neighboring Austria, the Czechs seemed committed to growing their nuclear energy capabilities even further. That same year, the country had eight operable commercial nuclear reactors, six under construction, and plans for six more.

Cooperation with Western companies

The Czech Republic is one of the few Eastern European countries with a strong nuclear-power construction capacity: the Czech company Skoda manufactures forgings, components for turbines, and other reactor equipment; the firm Vitkovice produces special steels; and Sigma specializes in pipes and pumps. Nevertheless, the country has encouraged Western companies to participate in its nuclear energy program.

In 1990, Westinghouse Electric Corp. (United States) announced that it had been selected by Skoda to help complete two partially built nuclear power plants and construct two others. The deal was expected to bring the U.S. company $1 billion in revenues over 10 years. Westinghouse was hoping the two companies could also bid on other joint ventures in Eastern and Central Europe.

Also in 1990, the Czech Republic was considering a contract with West Germany's Siemens AG to construct a nuclear power plant to provide heat for the city of Pilsen. The facility, a water cooled nuclear reactor, would supply only heat, not electricity, and thus would operate at lower temperatures than full-scale power reactors.

In 1991, a consortium of Czech Republic utilities, the national fuel company, and an architect-engineering firm invited eight Western suppliers to bid on the construction of two large light-water reactors. Although the Czechs had not yet chosen a site for the nuclear plant, it was evidence of their continued commitment to generate electricity by nuclear energy. Outside of the former Soviet Union, the Czech Republic has the largest commitment to nuclear power in Eastern Europe.

In February 1991, the French Commissariat a l'Energie Atomique concluded an agreement with Czech Republic to help of modernize the country's operable reactors as well as help develope future reactors.

Austrian opposition

Czech nuclear power plants have not suffered a great deal of internal opposition. However, the country's nuclear energy program has been severely criticized by its national neighbor Austria. Austria officials have proposed that the five neighboring states of south-central Europe—Austria, Italy, Hungary, Czech Republic, and Yugoslavia—create a nuclear-energy-free zone on their territories. At a meeting with Czechoslovak Premier Marian Calfa in January of 1991, Austrian Chancellor Franz Vranitsky went so far as to offer free electricity to the Czech Republic in return for the shutdown of nuclear units at the Czech Bohunice site, which is 35 miles from the shared border. The arrangement would cost Austria about $350 million a year.

In fear of a release of radioactivity from Czech plants, in early 1991 Austria began a nationwide distribution of potassium iodine tablets to combat the effects of thyroid radiation poisoning. It became the first European country to implement this precautionary guideline established by the World Health Organization. Czech officials quickly followed suit and began distributing similar pills.

Sources

"Nuclear Power Plant for Pilsen?" *The Wall Street Journal,* July 5, 1990.

"WE to Help Construct Czech Nuclear Plants." *Pittsburgh Post-Gazette,* December 13, 1990.

"Prague Offered Payoff to Shut Nuclear Plant." *The Washington Post,* January 30, 1991.

"Czechoslovakia Will Not Close Bohunice-1 and -2." *Nuclear News,* February 1991.

"15 New Nuclear Plants Began Commercial Operation in 1990." *INFO,* U.S. Council for Energy Awareness, July 1991.

"International." *INFO,* U.S. Council for Energy Awareness, July 1991.

Mounfield, Peter. *World Nuclear Power.* Routledge, 1991.

PLANT PROFILES

28 ★
Bohunice A1 Nuclear Power Plant

Atomove Electrame Bohunice
919 31 Jaslovske Bohunice, Czech Republic
Phone: 42-805-213-01-05
Fax: 42-805-244-67
Telex: 936 31

Owning Utility

Slovak Power Board
Prazska 25
Bratislava, Czech Republic

Basic Facts

Status: Shut down. *Type of Reactor:* Gas-cooled reactor. *Megawatts (electric):* 143. *Megawatts (thermal):* 540. *Reactor System Supplier:* Skoda (Czech Republic). *Generator Supplier:* Skoda (Czech Republic). *Architect Engineer:* Energoprojeckt Skoda Lotep (Czech Republic).

Key Dates

Ordered: 1958. *Construction began:* 1958. *Criticality:* October 1972. *First Power:* December 1972. *Commercial Operation:* December 1972. *Shut Down:* May 1979.

29 ●
Bohunice Nuclear Power Plant, Units 1-4

Atomove Electrame Bohunice
919 31 Jaslovske Bohunice, Czech Republic
Juraj Kmosena, Plant Mgr.
Phone: 42-805-213-01-05
Fax: 42-805-244-67
Telex: 936 31

Owning Utility

Slovak Power Board
Prazska 25
Bratislava, Czech Republic

Basic Facts

Units 1-4 *Status:* Operable. *Type of Reactor:* Pressurized light-water reactor. *Megawatts (electric):* 440. *Megawatts (thermal):* 1375. *Reactor System Supplier:* Units 1-2, Atomenergoexport (former USSR); Units 3-4, Skoda (Czech Republic). *Generator Supplier:* Units 1-2, Atomenergoexport (former USSR); Units 3-4, Skoda (Czech Republic). *Architect Engineer:* Energoprojeckt Skoda Lotep (Czech Republic).

Unit 2 *Ordered:* . *Criticality:* March 1980. *Commercial Operation:* January 1981.

Unit 3 *Ordered:* . *Criticality:* August 1984. *Commercial Operation:* May 1985.

Unit 4 *Ordered:* . *Criticality:* August 1985. *Commercial Operation:* March 1986.

Plant Report

Target of Austrian antinuclear campaign. The Bohunice Nuclear Power Plant has been the center of an international conflict between Austria and the Czech Republic. Austria is surrounded by 41 nuclear power plants in neighboring nations. Many of these plants were Soviet designed and are considered by many as extremely unsafe. Fearing another nuclear accident like the one at Chernobyl in 1986 when Austria was subjected to the highest per capita radiation dosage of any West European country, Austrian officials have made efforts to close down nuclear plants in nearby Czech Republic, Hungary, and Yugoslavia. Bohunice has been a special target of Austria's antinuclear campaign.

An Austria report on conditions at Bohunice said that two of the site's four reactors—in operation for more than a decade—lack emergency cooling systems and protective enclosures to prevent the accidental escape of radiation. The report also cited welded joints that were brittle and inadequate fire protection. A fire at the plant on January 15, 1991, alarmed the Austrians, but Czech officials said there were no injuries and no escape of radiation in the incident.

At a meeting with Czech Premier Marian Calfa in January of 1991, Austrian Chancellor Franz Vranitsky offered free electricity to the Czech Republic in return for the shutdown of nuclear units at the Bohunice site, which is 35 miles from the shared border. The arrangement would cost Austria about $350 million a year. The Czech government said it would not close the plant unless various teams of international and Czech experts agreed that the plant was unsafe.

Bohunice inspected, improved, kept open. A January 12, 1991, a statement from Jozef Vavrousek, chairman of the Czechoslovak Federal Environment Committee, said that safety inspections had been carried out by reputable foreign experts as well as by a team from the International Atomic Energy Agency. All the commissions, except experts from Austria, agreed that there was no immediate need to close the plant. Vavrousek accused Austrian officials of acting unprofessionally in their efforts to get the plant shut down.

In a previous press conference, Bohunice station director, Juraj Kmosena, said that foreign experts made about 80 reecommendations for improvements at the site. Kmosena committed to implement the improvements at Unit 1 by 1992. This would ensure operation until 1995, and further backfitting could extend the operating life to 2003. With a similar program, Unit 2 could be operated until 2005. The units entered service in 1979 and 1981, respectively.

Preventive medicine. In fear of a release of radioactivity from Czech plants and from Bohunice in particular, Austria at the beginning of 1991 began a nationwide distribution of potassium iodine tablets to combat the effects of thyroid radiation poisoning. It became the first European country to implement this precautionary guideline established by the World Health Organization. The pills are available free of charge to children under 16 and pregnant and breastfeeding women.

To allay public fears that were aroused by the Austrian campaign, Czech officials quickly followed suit and began distributing similar pills.

Dukovany

Sources

"Prague Offered Payoff to Shut Nuclear Plant." *The Washington Post*, January 30, 1991.
"Czechoslovakia Will Not Close Bohunice-1 and -2." *Nuclear News*, February 1991.
"Austria Distributing KI Tablets Near Bohunice." *Nuclear News*, February 1991.

Dukovany Nuclear Power Plant, Units 1-4

Jaderna Electrarna Dukovany
675 50 Dukovany, Czech Republic
Jan Krenk, Plant Mgr.

Phone: 00-42-9231-32
Fax: 00-42-509-922495

Owning Utility

Czech Power Board
Sbalena 1
110 00 Prague, Czech Republic

Contact

Frantisek Rynes, Personnel Mgr. & Pub.Info.
Dukovany Nuclear Power Plant

Basic Facts

Units 1-4 *Status:* Operable. *Type of Reactor:* Pressurized light-water reactor. *Megawatts (electric):* 432. *Megawatts (thermal):* 1375. *Reactor System Supplier:* Skoda (Czech Republic). *Generator Supplier:* Skoda (Czech Republic). *Architect Engineer:* Energoprojeckt Skoda Lotep (Czech Republic).

Key Dates

Unit 1 *Ordered:* 1971. *Construction began:* 1974. *Criticality:* February 1985. *First Power:* February 1985. *Commercial Operation:* March 1985.
Unit 2 *Ordered:* . *Criticality:* January 1986. *First Power:* January 1986. *Commercial Operation:* September 1986.
Unit 3 *Ordered:* . *Criticality:* October 1986. *First Power:* December 1986. *Commercial Operation:* June 1987.
Unit 4 *Ordered:* . *Criticality:* June 1987. *First Power:* June 1987. *Commercial Operation:* January 1988.

Plant Report

The site, surroundings, and statistics. The four-reactor Dukovany nuclear power plant complex is situated approximately 60 kilometers southwest of Brno on the left bank of the river Jihlava and the artificial dam Dalesice. The Dukovany site lies on the territories of three districts—Trebic, Znojmo, and Brno. The majority of the plant's employees live in the nearby towns of Maoravsky Krumlov and Trevic.

At the start of construction, two villages and one settlement were moved outside the three-kilometer wide safety zone encircling Dukovany. Czech Power Company. The Company's nuclear-, coal-, and hydro-powered plants supply electricity for 5,326,289 customers. In 1990, the utility earned total revenues of 31.6 thousand million Czech Republic crowns and realized a profit of 5.5 thousand million Czech Republic crowns.

Construction history. In 1970, the Soviet Union and the Czech Republic signed an intergovernmental agreement to build two pressurized light water reactors (V 230 model). Design preparations were launched in 1971, and in 1974 preparation of the site began. The same year it was decided that a different model of pressurized light-water reactor would be used at Dukovany, V 213. Construction was interrupted from 1975 to 1978. In 1978, work resumed on Dukovany, which was now scheduled to be a four-reactor site. It cost 21 thousand million Czech Republic crowns to build Dukovany.

Operational history. In 1985, when only Unit 1 was operating, the site generated 2.97 percent of the Czech Republic's electricity. By 1990, when all four reactors were operating, Dukovany produced 14.53 percent of the nation's electricity. The complex produces 22.92 percent of the Czech Power Company's total output.

The site's load factor—the ratio of units actually generated to those which could have been if the plant had worked continuously at full power—has hovered around 80 percent each year from 1985 through 1990.

From 1985 through 1990, the four units at Dukovany experienced a total of 28 reactor scrams—aa shutdown of the reactor by rapid insertion of control rods, either automatically or manually. Unit 1 had a rough start in 1985 and 1986 with 8 and 4 scrams respectively.

A fire at Dukovany on January 21, 1991 alarmed residents of Austria, which shares a border with the Czech Republic. However, Czech officials said there were no injuries and no escape of radiation in the incident.

Technical details

- 215 tons of steel were used in building each reactor.
- 42 tons of 3.6 percent enriched uranium is held in each reactor core.
- Each reactor contains 349 fuel assemblies and 37 control assemblies.
- Each reactor unit has two turbine sets with an output of 220 megawatts each.
- Steam entering the turbines is 256° C.

Training. Personnel training at Dukovany is conducted in accordance with the guidelines of the Czechoslovakia Atomic Energy Commission as well as the Ministry of the Economy. Training is conducted on site as well as at the Nuclear Power Plants Research Institute in Trnava and at the Czech Power Board Syndicate Training Centre.

Radiation monitoring. Radiation is monitored at three levels at Dukovany, including operational, personnel, and environmental monitoring. SEYVAL, an installed monitoring, warning, and information system, monitors the radiation levels in the technological systems and working areas of Dukovany. It is supplemented by backup monitors that check the amount of radioactivity in the vent-air stack and the waste-water channel.

The personnel dosimetry check monitors personnel surface contamination. It systematically measures and records dose equivalents for all personnel entering the controlled areas of the plant.

Outside the plant, the environmental monitoring laboratory is responsible for monitoring the radioactive material transport released into the environment in small quantities with gaseous and liquid effluents. Checks of the following are taken: air (aerosols, rainfalls, radioiodine), water (surface, underground, drinking), sediments, soil, agricultural products, milk, and fish.

If radiation measurements exceed established limits, the state hygienic authority is informed and an investigation is undertaken to identify and remove the cause.

Radioactive waste. Dukovany did not have a permanent storage plan for radioactive waste when construction began in 1978. The plant expected to store wastes on site temporarily until the government established a permanent repository.

In 1981, the Czech government began the process of choosing a storage site and in 1985 decided to make Dukovany the central radioactive waste depository for the entire republic. Construction of the Czech Disposal Centre of Radioactive Waste was completed in 1991.

Facing the past and planning for the future. Most of the equipment used in building Dukovany originated in the former Soviet Union. With the breakup of the socialist governments in the USSR and Eastern Europe, officials at Dukovany began to acknowledge that the plant was started under unfavorable conditions. Although the plant is considered one of the best of its kind in the Eastern Bloc countries, its technology and operations fall short of Western standards in many areas.

In 1991, the plant began a program of technical improvement to bring the plant up to par with nuclear power plants in Germany and Switzerland. They planned to replace pumps, valves, measuring transducers, measuring instruments, and industrial computers, as well as maintenance materials such as lubricants, glues, and tools. Many of these technologies were not available to the plant when under socialist rule.

Officials at the plant also planned to construct or add new equipment and systems—which were not included in or were eliminated from Dukovany's original plans for mostly political reasons—including such investments as:

- temporary deposit of burnt-up fuel
- incinerator for nonradioactive wastes
- incinerator for radioactive wastes
- security system
- information system
- system for early warning and advising the inhabitants living near the power plant in the case of an emergency
- health center for plant staff
- substitution of copper materials with titanium ones in the secondary circuit
- reconstruction or substitution of control and measuring systems of the entire power plant.

The improvements would be costly for the utility and were expected to be implemented through 1993 and beyond.

Joining the club. As part of its technology and operations improvement program, Dukovany joined the Club VVER 440-V 213 within the World Association of Nuclear Operators. The Club was formed for representatives of power plants having Soviet V 213 reactors to exchange information concerning the operation and maintenance of these unique nuclear units. Club members also commonly purchase equipment to reduce unit costs.

To facilitate stronger cooperation and interaction with Western utilities, officials at Dukovany introduced a comprehensive program of English lessons for select employees. Most of the staff were Russian-speaking, because the construction and commissioning of the plant took place in the presence of Soviet specialists.

Public relations. Dukovany welcomes visitors to the plant and issues 12,000 monthly bulletins to residents in the vicinity of the plant.

Sources

Nuclear Power Plant Duökovany. Czech Power Company, 1991.
CEZ 1990. Czech Power Company, April 1991.
"Prague Offered Payoff to Shut Nuclear Plant." *The Washington Post,* January 30, 1991.

Mochovce Nuclear Power Plant, Units 1-4

Atomove Elektrarne
Koncernovy Podnik
Mochovce, Czech Republic
Martin Spirko, Plant Mgr.

Phone: 972 19
Telex: 98557

Owning Utility

Slovak Power Board
Prazska 25
Bratislava, Czech Republic

Contact

Jozef Valach, Pub.Info.
Mochovce Nuclear Power Plant

Basic Facts

Units 1-4 *Status:* Under construction. *Type of Reactor:* Pressurized light-water reactor. *Megawatts (electric):* 440. *Megawatts (thermal):* 1375. *Reactor System Supplier:* Skoda (Czech Republic). *Generator Supplier:* Skoda (Czech Republic). *Architect Engineer:* Energoprojekt Skoda Lotep (Czech Republic).

Key Dates

Unit 1 *Ordered:* 1982. *Construction began:* 1983. *Commercial Operation Expected:* December 1993.

Unit 2 *Ordered:* 1982. *Construction began:* 1983. *Commercial Operation Expected:* 1994.

Unit 3 *Ordered:* 1982. *Construction began:* 1985. *Commercial Operation Expected:* 1995.

Unit 4 *Ordered:* 1982. *Construction began:* 1985. *Commercial Operation Expected:* April 1996.

Plant Report

Antipollution efforts. To reduce its reliance on highly polluting brown coal for electricity, the Czech Republic is pressing ahead with the completion of the nuclear station under construction at Mochovce in Slovakia. The site's four units are expected to come online between 1993 and 1996.

Sources

Source Book: Soviet-Designed Nuclear Power Plants in the Former Soviet Republics and Czechoslovakia, Hungary and Bulgaria. U.S. Council for Energy Awareness, 1992.

32 ○
Temelin Nuclear Power Plant, Units 1-2

Jaderna Elektrarna Temelin
373 05 Temelin-Electrarna, Czech Republic
Frantisek Poukar

Phone: 0342-905-360
Telex: 144 682

Owning Utility

Czech Power Board
Sbalena 1
110 00 Prague, Czech Republic

Contact

Augustin Hlava, Pub.Info.
Temelin Nuclear Power Plant

Basic Facts

Units 1-2 *Status:* Under construction. *Type of Reactor:* Pressurized light-water reactor. *Megawatts (electric):* 1014. *Megawatts (thermal):* 3000. *Reactor System Supplier:* Atomenergoexport (former USSR); Skoda (Czech Republic). *Generator Supplier:* Atomenergoexport (former USSR); Skoda (Czech Republic).

Key Dates

Unit 1 *Ordered:* 1982. *Construction began:* 1984. *Commercial Operation Expected:* May 1994.

Unit 2 *Ordered:* 1982. *Construction began:* 1985. *Commercial Operation Expected:* November 1995.

Plant Report

Help from Westinghouse. The breakup of the socialist governments in the former Soviet Union and Eastern Europe in the early 1990s had a significant effect on the Czech nuclear power program. Czech authorities began an intensive program to bring operating plants up to Western safety standards and to build new plants with Western technology and in cooperation with Western companies. Work at Temelin, under construction at the time, reflected the changes.

In 1990, Westinghouse Electric Corp. (United States) announced that it had been selected by Skoda, a manufacturer of pressure vessels and project engineer for Czech nuclear reactor projects, to help complete the partially built Units 1 and 2 at the Temelin Nuclear Power Plant. The U.S. company would also join with Skoda in the design, manufacture, and construction of the next two nuclear units in the country—Temelin, Units 3 and 4. Temelin 3 and 4, projects originally scheduled to use Soviet hardware, had been indefinitely deferred while the Czech Republic reoriented its nuclear program toward Western standards.

The deal was expected to bring Westinghouse $1 billion in revenues over 10 years. For Skoda, the arrangement would augment the company's technology base and management practices, as well as provide access to international power systems markets. Westinghouse was hoping the two companies could also bid on other joint ventures in Eastern and Central Europe.

Inspections. At the request of the Czech Republic, the International Atomic Energy Agency (IAEA) sent a Pre-Operational Safety Team (Pre-OSART) mission to the two units under construction at Temelin. In its three-part review that began in 1990, the IAEA found the following:

Evaluations

The second mission evaluated construction quality and start-up preparations and found that quality assurance met international standards, but recommended an overhaul of the operating organization.

On its third visit, the team reviewed reactor-core safety and systems design, and subsequently called for improved documentation and supported a staff proposal to improve reactor design.

U.S. Contracts. Late in 1991, the Czech Power Board awarded General Physics International Engineering & Simulation Inc. (GPI) in cooperation with the Czech firm Orgrez a $10 million, three-year contract to provide a full scope plant-referenced simulator for Temelin. GPI is based in Maryland. Orgrez has manufactured simulators in the Czech Republic since the late 1970s, but has been operating with out-of-date equipment. Simulators assist in training reactor operators.

Sources

"WE to Help Construct Czech Nuclear Plants." *Pittsburgh Post-Gazette*, December 13, 1990.

"Skoda Picks Westinghouse." *Nuclear News*, February 1991.

"U.S. Firm Wins Czech Contract for Temelin Simulator Project." *Nucleonics Week*, November 28, 1991.

Source Book: Soviet-Designed Nuclear Power Plants in the Former Soviet Republics and Czechoslovakia, Hungary and Bulgaria. U.S. Council for Energy Awareness, 1992.

Temelin Nuclear Power Plant, Units 3-4

Jaderna Elektrarna Temelin
373 05 Temelin-Electrarna, Czech Republic
Frantisek Poukar
Phone: 0342-905-360
Telex: 144 682

Owning Utility
Czech Power Board
Sbalena 1
110 00 Prague, Czech Republic

Contact
Augustin Hlava, Pub.Info.
Temelin Nuclear Power Plant

Basic Facts
Units 3-4 *Status:* Indefinitely deferred. *Type of Reactor:* Pressurized light-water reactor.

Plant Report
See Plant Report for Temelin Nuclear Power Plant, Units 1-2.

Finland

COUNTRY REPORT

Nuclear power in Finland

Finland generates 35 percent of its electricity from four nuclear reactors on two sites. The Loviisa power plant is operated by Imatran Voima Oy (IVO) and the Olkiluoto plant by Teollisuuden Voima Oy (TVO). In addition, a research reactor has been in operation since 1962 at the Technical Research Centre of Finland (VTT) at Otaniemi, Espoo. The first commercial nuclear reactor was commissioned in 1977.

In 1983, the Finnish government established the national policy on waste management and the Ministry of Trade and Industry was given responsibility to supervise the program. The Finnish waste management system includes no centralized facilities: Both IVO and TVO have their own waste management plans with different time schedules for implementation. Each utility is also responsible for the costs involved in the safe management of wastes and for other research and development work. The utilities have established a joint Nuclear Waste Commission to coordinate R & D work. Finland's nuclear facilities accumulate some 70 tons of spent fuel annually in addition to low- and medium-level radioactive waste.

The new Nuclear Energy Act and Decree passed in 1988 established the regulatory basis for the country by defining responsibilities, licensing procedures, and practical financing principles.

The Finnish Centre for Radiation and Nuclear Safety (STUK) supervises the safety of the country's nuclear facilities. In 1990, the U.S. Nuclear Regulatory Commission signed the second renewal of an information exchange arrangement with STUK. The renewal continues safety cooperation between the NRC and STUK for five years through 1995.

New nuclear plant by 1998?

Growth in Finland's total energy consumption slackened substantially in the late 1980s, but the consumption of electricity continued to increase. Although electricity consumption was projected to slow down in the 1990s, it was forecast that every second year Finland would need 445 megawatts more electricity—a capacity equal to one nuclear unit of the Loviisa plant. As of December 1991, Finland imported 15 percent of its electricity from the former Soviet Union.

Almost completed plans for a new reactor in Finland were stalled due to the Chernobyl accident in 1986. The coalition government formed in 1987 agreed that no decision would be made on a fifth reactor before the next election in 1991.

After the 1991 elections, Finland's utilities asked the new government for a "decision in principle" that would allow them to build the country's fifth nuclear unit at either the Loviisa or Olkiluoto sites. Besides a new plant unit, the application included new nuclear fuel storage, interim storage for spent nuclear fuel, and storage and final repository for reactor and decommissioning waste.

In response to the utility request, public hearings were organized by the Ministry of Trade and Industry for local residents to voice their opinions. The government also requested assessments from Loviisa city and the Eurajoki municipal councils, as well as other municipalities close to the possible sites.

If the government agrees to the unit, the question will be submitted to Parliament, which will have the final word. The Finnish utilities were looking for a decision by the Spring of 1992 and hoped to bring the power plant into operation by 1998. French, German, Swedish, and Russian companies had already submitted bids for the reactor in anticipation of a favorable response from the government.

Sources

Loviisa Power Plant. Imattran Voima Oy, 1990.

Nuclear Waste Management: Finland. Organization for Economic Cooperation and Development and the Nuclear Waste Commission of Finnish Power Companies, 1990.

U.S. Nuclear Regulatory Commission. *U.S. Nuclear Regulatory Commission 1990 Annual Report.* Washington, D.C., n.d.

"Bids Being Submitted for Fifth Reactor." *Nuclear News,* February 1991.

"15 New Nuclear Plants Began Commercial Operation in '90." *INFO,* U.S. Council for Energy Awareness, July 1991.

So Why Nuclear Power? Imatran Voima Oy and Teollisuuden Voima Oy, August 1991.

"International: New Orders for Nuclear Power Plants in Four Countries." *Nuclear Industry,* Third Quarter, 1991.

"Nuclear Energy Forum." *INFO,* U.S. Council for Energy Awareness, November/December 1991.

PLANT PROFILES

Loviisa Power Plant, Units 1-2

PO Box 23
SF-07900 Loviisa, Finland
Jussi Helske, Plant Mgr.

Phone: 358-15-5501
Fax: 358-15-5504435
Telex: 1819 ivolo sf

Owning Utility

Imatran Voima Oy
PO Box 138
SF-00101 Helsinki, Finland
Anders Palmgren, V.Pres., Nuclear

Phone: 358-0-60901
Fax: 358-0-6940896
Telex: 124608 voima sf

Contact

Antti Ruuskanen, Mgr. of Pub.Info.
Imatran Voima Oy

Basic Facts

Units 1-2 *Status:* Operable. *Type of Reactor:* Pressurized light-water reactor. *Megawatts (electric):* 465. *Megawatts (thermal):* 1375. *Reactor System Supplier:* Atomenergoexport (former USSR). *Generator Supplier:* Atomenergoexport (former USSR). *Architect Engineer:* Imatran Voima Oy.

Key Dates

Unit 1 *Ordered:* 1969. *Construction began:* 1971. *Criticality:* 1977. *First Power:* February 1977. *Commercial Operation:* May 1977.

Unit 2 *Ordered:* 1971. *Construction began:* 1972. *Criticality:* 1980. *First Power:* November 1980. *Commercial Operation:* January 1981.

Plant Report

Background. Finland's first nuclear power plant is situated on the island of Hastholmen in Loviisa, about 15 kilometers southeast of the town center. The foundations for the plant were blasted out of the island's rapakivi granite bedrock.

The Loviisa Power Plant is operated by the state-owned Imatran Voima Oy (IVO). IVO generates nearly half of the electricity consumed in Finland and operates the national high voltage grid. The company owns power generation and transmission systems worth $7,650 million and invests over $34 million per year in research and development. Founded in 1932, IVO now employs over 5,000 people had has two main lines of business: energy production and distribution and comprehensive, international engineering services in the energy field.

Loviisa is comprised of two 465 Soviet-built pressurized light water reactors of the VVER-44 type and generates almost half of the owning utility's electricity output. It has a permanent staff of about 600. A temporary staff of nearly 1,000 is needed during annual refueling periods.

Radioactive waste and release. Nuclear fuel used at Loviisa comes from the former Soviet Union. The fuel is transported to the plant as complete fuel assemblies. About one third of the spent fuel assemblies are removed from the reactor every year and replaced with fresh fuel. Spent fuel is cooled under water for five years on site and then is returned by rail to Chelyabinsk in the former Soviet Union.

The plant plans to build a nuclear waste repository in the bedrock of the island of Hastholmen in the late 1990s. Currently, low and intermediate-level nuclear waste is processed at the power plant and solidified in concrete, if necessary. The wastes will eventually be stored in the final repository. When the plant is decommissioned, the caverns of the bedrock repository of Hastholmen will be enlarged to store the radioactive decommissioning wastes.

In 1990, construction of a cesium separation facility for treatment of liquid reactor waste was started at Loviisa. After separation of cesium, the capacity of the liquid waste store will be sufficient for all liquid waste that accumulates during the expected operational life of the plant. The facility was scheduled for completion in 1991.

Radioactive releases from Loviisa to the sea are reported by the utility as 1-10 percent of the authorized release limits. Radioactive releases to the atmosphere usually fall below 1/1000 of the release limit value.

Operating experience. In May of 1990, Loviisa 1 experienced a failure in a feed water pipe, which caused neither a radiation risk nor human injury. The plant was out of performance for two weeks for repairs and inspections. During this time, Loviisa 2 was also inspected. The long-term outage of Loviisa 2—done every four years—took five weeks.

In 1980, three years after its startup, the load factor of Loviisa 1 was only 36.7 percent. (Load factor is the ratio of electricity units actually generated to those which could have been if the plant had worked continuously at full power.) The poor performance was due to an extensive welding inspection of the steam generators. Since 1981, the load factors of both Loviisa 1 and 2 have been above the international average. In 1990, the load factor for Loviisa 1 was 84.9 percent, and for Loviisa 2 it was 84.5 percent. Loviisa 2 placed eighth in *Nuclear Engineering International's* rating of the world's top 10 reactors with the best lifetime performance as of June 30, 1991.

From 1988-1990, Loviisa undertook a costly process of replacing worn-out components and upgrading the plant to meet tightened safety requirements. At the end of 1990, the State of Finland invited the International Atomic Energy Agency (IAEA) to perform an operational safety inspection of Loviisa. According to IVO, IAEA considered the plant to be safe, well functioning, and managed with skill.

Sources

Loviisa Power Plant. Imatran Voima Oy, 1990.
Imatran Voima 1990 Annual Report. Imatran Voima, n.d.
"Introduction." *IVO International News,* May 1991.

Olkiluoto Nuclear Power Plant, Units 1-2

SF-27160 Olkiluoto, Finland
Rauno Mokka, Plant Mgr.
Phone: 358-38-3811
Fax: 358-38-3812109
Telex: 65154 tvo sf

Owning Utility

Teollisuuden Voima Oy (TVO)
Fredrikinkatu 51-53 B
SF-00100 Helsinki, Finland
Magnus Von Bonsdorff, Managing Dir.
Phone: 358-0-605-022
Fax: 358-0-605-135
Telex: 122065

Contact

Ahti Rastas, Mgr. of Info. Office
TVO

Basic Facts

Units 1-2 *Status:* Operable. *Type of Reactor:* Boiling water reactor. *Megawatts (electric):* 735. *Megawatts (thermal):* 2160. *Reactor System Supplier:* ASEA-Atom. *Generator Supplier:* Stal-Laval. *Architect Engineer:* ASEA-Atom.

Key Dates

Unit 1 *Ordered:* 1972. *Construction began:* 1974. *Criticality:* July 1978. *First Power:* September 1978. *Commercial Operation:* October 1979.

Unit 2 *Ordered:* 1974. *Construction began:* 1975. *Criticality:* October 1979. *First Power:* February 1980. *Commercial Operation:* July 1982.

Plant Report

TVO: The Industrial Power Company. On the west coast of Finland, in Eurajoki, Teollisuuden Voima Oy (TVO) operates two boiling water reactors—Olkiluoto 1 and 2, which are also known as TVO 1 and 2. The Industrial Power Company (Teollisuuden Voima Oy) was founded in 1969 by a number of Finnish industrial companies with the purpose of building and operating large power plants. The company supplies electricity to its shareholders at cost. In 1990, it generated 18.6 percent of the total supply of electricity in Finland.

The plant employs approximately 500 employees, and uses an additional outside workforce of up to 900 during refueling.

Technical data

- Reactor steam flow—1,080 kilograms/second
- Feedwater temperature—180 °C
- Number of fuel assemblies—500
- Fuel rods per assembly—63-100
- Total fuel weight—86-90 tons of Uranium
- Number of control rods—121
- Weight of pressure vessel—524 tons
- Inner diameter of pressure vessel—5,540 millimeters
- Inner height of pressure vessel—20,593 millimeters
- Thickness of wall—134 millimeters
- Live steam temperature in turbine—283 °C
- Rated speed of turbine—3,000 rpm

One outstanding feature of the TVO units is the division of the safety-related systems into four redundant parts, separated from each other. Systematic separation provides optimum built-in protection against hazards associated with fires, flooding, crashing aircraft, and other external impacts.

Operational experience. In 1990 TVO set a new record for annual output of electricity from its Olkiluoto Nuclear Power Plant. The load factor was 94.4 percent for TVO 1 and 92.7 percent for TVO 2. *Nucleonics Week* listed TVO 1 10th and TVO 2 11th in its review comparing capacity factors of all western reactors—an excellent achievement considering that the review included plants at which no refueling was carried out during the year. TVO 1 and 2 were shut down for a total of 31 days in 1990 for refueling.

In 1990, a number of enhancements took place at the plant. A simulator at the company's Training Centre was commissioned. An Information Centre, which is open to the public, was renovated and reopened. And the replacement of the reactor core grid—scheduled for completion at both plants by 1994—and an operator's support system to by installed in the control room by the end of 1991 were both initiated in 1990.

In 1990, the plants had no incidents that according to the Finnish Centre for Radiation and Nuclear Safety would have impaired safety levels. Gaseous radioactive emissions from the plant were .05 percent of permitted annual rates, and liquid emissions were 10 percent of the allowable rates. Average work dose to radiation was 1.47 millisieverts (mSv), a low rate when compared to permissible doses.

Fuel management and radioactive waste storage. TVO purchases uranium from Canada and China under long-term delivery contracts. Enrichment services are purchased from the former Soviet Union as well as the Netherlands and Germany.

Construction of the VLJ Repository for Operating Wastes (low- and medium-level nuclear waste) at Olkiluoto began in 1988 and was scheduled to be in operation in 1992. The repository is built in the crystalline bedrock at a depth of 70-100 meters. The repository will be enlarged to house the radioactive wastes resulting from the eventual decommissioning of the plant.

In accordance with a policy decision of the Finnish government in 1983, TVO is responsible for the development of a repository system for its spent fuel (high-level radioactive waste). Milestones in TVO's plans to develop a repository for spent fuel include the following:

- by 1985—select possible areas for site investigations and update technical plans for disposal;
- by 1992—perform preliminary investigations at several sites—including airborne survey, deep and shallow drillings, as well as surface and borehole measurements and sampling—and choose sites for detailed studies;
- by 2000—carry out detailed investigations and select a site for a repository meeting the requirements of safety and environmental protection;
- by 2010—present detailed technical plans and safety reports for construction permit application;

- by 2020—construct and commission the repository.

Interim plans for spent fuel include on site storage as well as foreign services.

Sources

TVO 1990 Annual Report. Teollisuuden Voima Oy, n.d.
Nuclear Waste Management: Finland. Organization for Economic Cooperation and Development and the Nuclear Waste Commission of Finnish Power Companies, 1990.
TVO Nuclear Power Plant Units I and II. Teollisuuden Voima Oy, 1991.

France

COUNTRY REPORT

Nuclear revolution

In the early 1970s, France imported more than 75 percent of its energy needs in the form of oil from the Middle East. Its own domestic supplies of coal, oil, and lignite could not support its growing industrialization. When the OPEC oil embargoes began in 1973, France saw a threat to its security. Determined to re-assert energy independence, it launched a massive expansion of its fledgling nuclear program in 1975. By 1990, France led the world in nuclear-energy generating-capacity, receiving 74.5 percent of its electricity from 54 operating nuclear plants. At approximately 30 percent of French energy consumption, uranium displaces the equivalent of 59 million tons of oil.

As a result of its heavy reliance on nuclear energy, France has one of the lowest per capita rates of carbon dioxide emissions among industrialized nations and boasts the lowest electricity prices in Europe. In 1992, nuclear-generated electricity cost 4.3 cents per kilowatts-hours (KWh) compared to 6.3 cents per KWh for coal and 8.3 cents per KWh for oil.

History

The French nuclear program began even before World War II. In 1939, France became the first country to patent a nuclear power station design, a concept revived in 1952 as part of a national energy strategy. Remembering the humiliation of Germany's easy initial victories, the nation enthusiastically supported General Charles de Gaulle when he announced, in the late 1950s, that France was to build its own nuclear weapons. The support widened in the 1960s to include nuclear power.

The country built its first nuclear reactor, a gas-graphite model, at Marcoule in 1954 as part of that military effort. The plutonium-producing Marcoule G1 generated only 2.5 megawatts electric (MWe). Later versions of the gas-graphite reactor produced considerably more electricity, reaching a high with the largest unit, the 515-MWe St. Laurent des Eaux A2 in 1966.

The national electricity utility, Electricité de France (EDF) doubted the viability of nuclear generated power. It called its collaborations at Chinon with the new government body, the Atomic Energy Commission (CEA), "experiments on a commercial scale."

Ideal conditions for nuclear industry

France's tradition of a centralized administration encouraged the growth of its nuclear industry. It simplified the structure by creating a single dominant energy utility and building standardized reactors under a streamlined regulatory system. EDF designs, constructs, and operates all nuclear power plants as well as the nation's conventional fossil fuel and hydro plants. It employs 25,000 people.

Similarly, large single suppliers, many owned 51 percent by the French government, dominate the nuclear component industry. All reactor vessels come from Framatome. Alsthom produces all of France's turbine generators. Generale d'Electricité (CGE) manufactures much of the industry's electric power equipment along with telephones and high-speed trains.

In all, the industry supports more than 160,000 jobs spread over 600 companies.

That unified approached extends into the government itself. No major political party professes an antinuclear stance. Both the political right and left have demonstrated a long-term commitment to the development of nuclear energy over all other energy sources. Even the Socialist Party, which placed a freeze on planning and construction of nuclear power plants after Francois Mitterrand came to power in 1981, changed its mind once it realized how many jobs would disappear if the nuclear construction program halted. In 1990, acknowledging the importance of the nuclear energy industry to the nation, the Socialist government adopted a policy of 51 percent government ownership of major nuclear corporations.

Technology—standardization and independence

Much of the success of the French nuclear effort resulted from the decision to use a standardized reactor rather than develop a series of prototypes. That reduced construction costs overall and increased construction efficiency and quality. It simplified and streamlined safety inspections, licensing, and maintenance. However, flaws discovered in one reactor could result in an enormous overhaul of the entire system—an extremely costly, and potentially crippling process.

After abandoning the early gas-graphite technology in 1969, France turned to the American pressurized water reactor (PWR). It chose to devellop a Westinghouse design into its own 900 MWe model. Between 1977 and 1984, EDF built 34 of these reactors on nine different sites.

As antinuclear opposition began to mount in the early 1980s, EDF switched to a 1,300 MWe design, also based on the Westinghouse PWR. The larger unit allowed the utility to produce more power at each site, reducing the number of new locations needed. By 1994, the utility will build 20 of the larger reactors.

The next stage of development calls for an even larger model. The first 1,450 MWe reactor at Chooz Nuclear Power Plant was scheduled to begin operations in July, 1991, but construction difficulties delayed that several years. EDF developed this reactor as an all-French design, not based on the Westinghouse PWR.

Parallel to the PWR, EDF has developed a Fast Breeder (FBR) technology in concert with other European nations. It has two operating FBR plants with more planned.

France views nuclear power as an unbroken circle

The completion of the FBR program will allow France to close the fuel cycle, making it energy self-sufficient. It can draw from a domestic uranium reserve of more than 95,000 tons, using advanced mining techniques it developed. It leads the world in uranium enrichment with plants providing more that one third of the world's processing capacity. It has developed its own uranium and plutonium fuel fabrication facilities. With nuclear opposition growing in other countries, France has become a leader in reprocessing spent fuel for other nations. Its waste vitrification process allows it to stabilize nuclear waste for disposal in its country of origin. It not only builds more nuclear power plants than any other country except the United States, it exports components and designs to other nations. In concert with Kraftwerk Union AG (Germany), Framatome exports technology globally. Together, they are exploiting the opening Eastern European market.

Radioactive waste

When France embarked on its nuclear construction program in 1975, it understood the coming problem of waste management. To protect the environment it passed an act in July 1975 that required the nuclear industry to arrange for the disposal of waste at its own cost, using an agent approved by the government.

In 1979, the government created the National Agency for Radioactive Waste Management (ANDRA) as a department of the CEA. That body developed waste disposal technology, chose specific storage sites and will operate long-term disposal depositories.

Individual nuclear plants and research reactors perform their own short-term storage, usually in cooling ponds within the reactor containment buildings.

Major spent-fuel reprocessing centers at Marcoule and La Hague use a vitrification technique developed in 1978 to safely stabilize nuclear waste for long-term storage. By 1988, those centers processed 560 tons of vitrified waste.

The glass-like material cools in special air-conditioned concrete bunkers. After about 30 years, it can be permanently disposed of in a deep geological repository yet to be built.

During the 1980s, medical researchers observed a higher-than-expected rate of leukemia around reprocessing plants in the United Kingdom and at sites in the United States. French researchers conducted a study in 1989 to check for similar problems there. They concluded that proximity to nuclear facilities had no effect on leukemia rates in France.

Opposition to nuclear power

EDF president Marcel Boiteux described the popular French view of the state as a "protective mother." For the most part, they remain confident in the ability and intention of their government to protect their interests safely. While nuclear protest exploded in other European nations throughout the 1980s, the movement found little success in France. An attack on nuclear power was seen as an attack on the integrity of the state.

In many countries, anti-nuclear groups had used the courts to stop nuclear development. In France, the courts refused to rule on anything but procedural matters; licensing and regulation of the industry was a governmental concern.

The initial nuclear building program called for 200 reactors by the year 2000. To minimize opposition, the government decided to build four reactor nuclear parks, inviting little local input into site decisions. At Golfech, a regional referendum gave an 80 percent "NO" to a proposed nuclear project. A public utility inquest overturned the decision, and the government issued a building permit. That caused some protest among opposition parties about the undemocratic method of locating reactors, but little opposition to the overall plan.

Some protest spilled over from neighboring countries, since France often built its reactors close to its bborders. In 1977, 30,000 protestors fought 4,000 police at Creys-Malville. In 1982, at the same site, a protest group fired an anti-tank rocket at the buildings. The Breton-Liberation Front exploded a bomb at Brennilis, taking their lead from a German group's attack on Fassenheim.

In 1980, the 2,360 villagers of Plogoff in western Brittany, astonished the authorities when 20,000 marchers and 15 sheep descended on a proposed nuclear construction site in

January, 1980. Riot police battled the protestors for six weeks. The incident gained wide coverage in the European press, but the government approved construction.

Socialist government slows nuclear rush

The political upheaval of 1981 swept the Socialists into power and froze nuclear development for a while. Plogoff was cancelled permanently. However, the process quickly reasserted itself, even adding an additional reactor at Golftech.

International concern about French safety provisions is increasing, particularly among its neighbors. Pan-European environmental standards may force a stiffening of French regulations, raising reactor costs as other nations pressure France to be a good neighbor.

That could be bad news for EDF. In 1990, the company owed $45.2 billion as a result of its ambitious building program. At 1.4 times the annual turnover of the company, EDF rejects suggestions of financial insecurity. It claims the ratio is normal for capital-intensive-electric utilities. It intends to reduce the price of its electricity 1.5 percent each year until the end of the century, but increasing construction costs and declining electricity demand projections may claim that revenue.

Sources

"Where the Atom is Admired." *Time*, February 18, 1980:

"Off and Running in a Nuclear Race." *Maclean's*, October 20, 1980.

"France: The Nuclear Dream Maintained." *Technology Review*, November 1982.

"France Weighs Benefits, Risks of Nuclear Gamble." *Science*, August 29, 1986.

"Commonwealth North Learns the Virtues of Nuclear From French." *Alaska Journal of Commerce*, February 23, 1987.

Nuclear Waste Management: France. Nuclear Energy Agency of the Organization for Economic Cooperation and Development (OECD/NEA) and the French National Agency for Radioactive Waste Management (ANDRA), 1989.

"Nuclear Plant Venture in Europe." *The New York Times*, April 14, 989.

"France's Nuclear Energy Boom." *World Press Review*, March 1990.

"CGE's Effort to Control Framatome May be Fading." *The Wall Street Journal*, May 30, 1990.

"Overall Mortality and Cancer Mortality Around French Nuclear Sites." *Nature*, October 25, 1990.

"CGE, French Government Agree on Terms of Control." *The Wall Street Journal*, October 31, 1990.

Mounfield, Peter. *World Nuclear Power*. Routledge, 1991.

"15 New Nuclear Plant Began Commercial Operation in 1990." *INFO*, U.S. Council for Energy Awareness, July 1991.

"Greenpeace Notwithstanding, EDF Says Nuclear Doing Well." *INFO*, U.S. Council for Energy Awareness, July 1991.

"Heaviest Users of Nuclear Energy have Low Per Capita CO_2 Emissions." *INFO,* U.S. Council for Energy Awareness, August 1991.

"A Good Year For Nuclear Energy Abroad." *INFO*, U.S. Council for Energy Awareness, April 1992.

—Al Cook

Plant Profiles	Blayais 37

PLANT PROFILES

36 ●
Belleville Nuclear Power Plant, Units 1-2

BP 11
F-18240 Lere, France
Jean-Pierre Delcasso, Plant Mgr.
Phone: 48-54-50-50
Fax: 48-54-24-39

Owning Utility

Electricite de France
2 rue Louis Murat
F-75005 Paris, France
Christian Nadal, Dir., Commun.
Phone: 1-42-56-94-00
Telex: 280041

Contact

Nicolas Petsitis, Pub.Info.
Belleville Nuclear Power Plant

Basic Facts

Units 1-2 Status: Operable. *Type of Reactor:* Pressurized light-water reactor. *Megawatts (electric):* 1363. *Megawatts (thermal):* 3817. *Reactor System Supplier:* Framatome (France). *Generator Supplier:* Alsthom. *Architect Engineer:* Electricite de France.

Key Dates

Unit 1 Ordered: 1978. *Construction began:* 1981. *Criticality:* May 1987. *First Power:* October 1987. *Commercial Operation:* June 1988.

Unit 2 Ordered: 1978. *Construction began:* 1981. *Criticality:* May 1988. *First Power:* July 1988. *Commercial Operation:* September 1988.

Plant Report

Top producer. According to data compiled by the industry publication *Nucleonics Week*, in December of 1990, Belleville Unit 2 ranked eighth among the top 20 producers of electricity in the world. That month, Belleville generated 995,973 megawatt-hours.

Sources

"Palo Verde Units Top Producers in U.S. for December." *Business Wire*, February 15, 1991.

37 ●
Blayais Nuclear Power Plant, Units 1-4

BP 27
F-33820 St. Ciers sur Gironde, France
Jean-Pierre Abraham, Site Mgr.
Phone: 57-33-33-33
Fax: 57-33-32-89

Owning Utility

Electricite de France
2 rue Louis Murat
F-75005 Paris, France
Christian Nadal, Dir., Commun.
Phone: 1-42-56-94-00
Telex: 280041

Contact

Maud Sterlingots, Pub.Info.
Blayais Nuclear Power Plant

Basic Facts

Units 1-4 Status: Operable. *Type of Reactor:* Pressurized light-water reactor. *Megawatts (electric):* 951. *Megawatts (thermal):* 2785. *Reactor System Supplier:* Framatome (France). *Generator Supplier:* Alsthom. *Architect Engineer:* Electricite de France.

Key Dates

Unit 1 Ordered: 1975. *Construction began:* 1976. *Criticality:* May 1981. *First Power:* June 1981. *Commercial Operation:* December 1981.

Unit 2 Ordered: 1975. *Construction began:* 1976. *Criticality:* June 1982. *First Power:* July 1982. *Commercial Operation:* February 1983.

Unit 3 Ordered: 1977. *Construction began:* 1977. *Criticality:* July 1983. *First Power:* August 1983. *Commercial Operation:* November 1983.

Unit 4 Ordered: 1977. *Construction began:* 1977. *Criticality:* May 1983. *First Power:* May 1983. *Commercial Operation:* October 1983.

Plant Report

Production facts. The Blayais Nuclear Power Plant occupies a site along the Gironde Estuary near Bordeaux, France and houses four nuclear reactors. Yearly production figures for Blayais, in billion kilowatts, were as follows:

- 1.6 in 1981
- 7.8 in 1982
- 13.7 in 1983
- 25 in 1984
- 25.6 in 1985
- 25.3 in 1986
- 22 in 1987
- 19.5 in 1988
- 23.7 in 1989
- 22.2 in 1990

Blayais' 1990 production of 22.2 billion kilowatts was achieved with a 78.2 percent availability. This represents 5.5 percent of national production, and is 1.4 times the consumption of the Aquitaine region. The plant's production was equal to 7,391,000 tons of coal or 4,928,000 tons of oil.

There were several factors which contributed to the decrease in production at Blayais from 1989 to 1990. During the summer of 1990, the water level of the Gironde reached exceptionally high temperatures; production had to be decreased to keep the temperature of the release of cooling water within authorized limits. Also, there were extended outages for modification to bring the plant into line with the later standardized plants, and for maintenance of steam pipes of the secondary system. The availability during normal operations was satisfactory, but the average availability was affected by the duration of the outages.

Bugey

1990 safety facts. During the year, there were 18 accidents at Blayais. EDF considered most of these accidents to have been of minor seriousness.

1990 environmental facts. During the year, radioactive effluent release from Blayais was markedly lower than authorized limits. The liquid effluent, for radioelements other than tribium, potassium 40 and radium, did not exceed 4.9 percent of the limits. Tririum remained at 52.5 percent of the limits. Gaseous releases were lower than 7.8 percent of their authorized limits, with gaseous halogens and aerosols being under a hundredth of their limits.

There were 55 shipments of waste removed from Blayais, and sent to a storage site of the National Agency for the Management of Radioactive Waste.

The ecological monitoring of the Gironde Estuary and the measurements taken for environmental monitoring represented an annual expenditure of over five million Francs.

1990 maintenance facts. As a consequence of incidents discovered at a national level, which were caused by anomalies in maintenance operations, EDF's Nuclear and Fossil Fuel Generation Division implemented a general procedure aimed at improving the safety standards of nuclear installations. Blayais participated in the analysis, and created operational groups that studied themes such as surveillance, outage structure, and equipment monitoring. The teams were advised, assisted, and checked on by the Nuclear Safety and Quality Assurance body.

During the year, Blayais was the subject of 14 inspections by the National Safety Authorities. Also, periodical exercises that simulated incidents were carried out.

1990 employee facts. The workforce at Blayais rose to 1,053 employees by the end of 1990. More than 100,000 hours of training, which represents approximately 100 hours per person and six percent of the plant's budget, were given to the workers.

Public access. There is a reception and information center called the "Belvedere," which is located at the entrance to the Blayais site. During 1990, more than 17,000 visitors were received there, and some access was granted to the nuclear buildings during shutdown.

The Belvedere is open all year from Monday to Saturday. Information can be obtained by calling: 57333030 or 573333203.

Sources

1990 Blayais in Brief. Electricité de France, n.d.

—*Charles Ceplecha*

---

38 ●
Bugey Nuclear Power Plant, Units 1-5

BP 14
F-01366 Camp de la Valbonne Cedex, France
Jean Fluchere, Site Mgr.

Phone: 74-34-33-33
Fax: 74-34-17-32

Owning Utility

Electricite de France
2 rue Louis Murat
F-75005 Paris, France
Christian Nadal, Dir., Commun.

Phone: 1-42-56-94-00
Telex: 280041

Contact

Philippe Richez, Pub.Info.
Bugey Nuclear Power Plant

Basic Facts

Unit 1 *Status:* Operable. *Type of Reactor:* Gas-cooled reactor. *Megawatts (electric):* 555. *Megawatts (thermal):* 2000. *Reactor System Supplier:* Various. *Generator Supplier:* Rateau; Jeumont-Schneider (France). *Architect Engineer:* Electricite de France.

Units 2-5 *Status:* Operable. *Type of Reactor:* Pressurized light-water reactor. *Megawatts (electric):* Units 2-3, 955; Units 4-5, 937. *Megawatts (thermal):* Units 2-3, 2785; Units 4-5, 2785. *Reactor System Supplier:* Framatome (France). *Generator Supplier:* Alsthom. *Architect Engineer:* Electricite de France.

Key Dates

Unit 1 *Ordered:* 1965. *Construction began:* 1966. *Criticality:* 1972. *First Power:* April 1972. *Commercial Operation:* July 1972.

Unit 2 *Ordered:* 1971. *Construction began:* 1972. *Criticality:* April 1978. *First Power:* May 1978. *Commercial Operation:* March 1979.

Unit 3 *Ordered:* 1972. *Construction began:* 1974. *Criticality:* August 1978. *First Power:* September 1978. *Commercial Operation:* March 1979.

Unit 4 *Ordered:* 1973. *Construction began:* 1974. *Criticality:* February 1979. *First Power:* May 1979. *Commercial Operation:* July 1979.

Unit 5 *Ordered:* 1974. *Construction began:* 1974. *Criticality:* July 1979. *First Power:* November 1979. *Commercial Operation:* January 1980.

Plant Report

Experiments with nuclear technology. During the 1960s, the industrial region of Rhône-Alpes outgrew the available hydroelectric power generated in the region. One answer to the problem was the Bugey nuclear plant, built on the Rhône River in 1965 at a site 35 kilometers east of Lyon and only 11 kilometers from Geneva.

That early plant used a 560-megawatts electric (MWe) graphite-moderated reactor. As with the British Magnox reactor, Bugey-1 used carbon dioxide gas as the coolant and wrapped its natural uranium fuel in a magnesium casing.

Old site becomes home for new reactors When the French government embarked on its 1975 nuclear expansion program, the success of that first reactor made the Bugey site a prime candidate for one of the new multi-reactor industrial parks to be built along the Rhône.

Between May 1978 and July 1979, four 900-MWe pressurized water (PWR) units went on-line at Bugey.

The first three units drew their cooling water from the Rhône in an open circuit, dumping the somewhat warmer water back into the river. The last two incorporated cooling towers into their design to reduce the thermal load on the river.

Plant Profiles

French regulations for PWR reactors require extensive re-inspections after 10 years of operation. For Bugey, the teardowns began in November of 1989 and lasted until December 1991. Each unit went out of service for about six months.

Technicians at the plant took the opportunity to install modern safety systems and make technical improvements designed to bring the plants up to the technological level of Chinon B4, the latest 900 MWe PWR.

Corrosion cracks. During the high-pressure hydro-leak testing of Bugey-3's reactor vessel, workers discovered cracks in the Inconel-600 standpipes that pass through the lid of the vessel.

The cracking of the high nickel-content stainless steel is a common form of corrosion, according to James Lang, of the Electric Power Research Institute in the United States. The phenomenon happens at a specific ideal combination of temperature, pressure, and primary water environment.

French researchers termed the problem a generic defect of both the 900- and 1300-MWe PWR reactors. Although, the utility does not consider the cracks an immediate safety hazard, the correction will involve considerable technical expertise and imagination, according to Jean Fluchere of EDF.

To counteract a similar problem involving Inconel taps on pressurizers and deforming steam generator tubes, EDF lowered the temperature in all of its 1,300-MWe PWRs by 4.2 degrees Celsius. French designers based both their 900-MWe and the 1,300-MWe PWR on American Westinghouse reactors. Westinghouse Electric Corp. (United States) notes it has not received reports of similar problems with U.S. reactors. However, American standards for hydro-testing of reactor vessels call for a much lower test pressure than French standards do.

Sources

Centre de Production Nucléaire du Bugey. Electricité de France, Paris, 1986.
Bugey 1989: En bref. Electricité de France Paris, 1989.
Bugey 1990: En bref. Electricité de France, Paris, 1990.
Mounfield, Peter. *World Nuclear Power.* Routledge, 1991.
Bugey 1991: En bref. Electricité de France, Paris, 1991.
EDF 1990 Annual Report. Generation and TransmissionGroup, Paris, April 1991.
"Cracking of French PWRs Raises Curiosity, But Little Concern in U.S," *Inside N.R.C.*, December 2, 1991.

—Al Cook

Cattenom Nuclear Power Plant, Units 1-4

BP 41
F-57570 Cattenom, France
Bernard Dupraz, Site Mgr.
Phone: 82-51-70-00
Fax: 82-55-30-83

Owning Utility

Electricite de France
2 rue Louis Murat
F-75005 Paris, France
Christian Nadal, Dir., Commun.
Phone: 1-42-56-94-00
Telex: 280041

Contact

Roland Ropars, Pub.Info.
Cattenom Nuclear Power Plant

Basic Facts

Units 1-4 *Status:* Operable. *Type of Reactor:* Pressurized light-water reactor. *Megawatts (electric):* 1362. *Megawatts (thermal):* 3817. *Reactor System Supplier:* Framatome (France). *Generator Supplier:* Alsthom. *Architect Engineer:* Electricite de France.

Key Dates

Unit 1 *Ordered:* 1979. *Construction began:* 1979. *Criticality:* October 1986. *First Power:* November 1986. *Commercial Operation:* April 1987.
Unit 2 *Ordered:* 1980. *Construction began:* 1980. *Criticality:* August 1987. *First Power:* September 1987. *Commercial Operation:* February 1988.
Unit 3 *Ordered:* 1979. *Construction began:* 1982. *Criticality:* 1990. *First Power:* 1990. *Commercial Operation:* February 1991.
Unit 4 *Ordered:* 1984. *Construction began:* 1984. *Criticality:* 1991. *First Power:* 1991. *Commercial Operation:* December 1991.

Plant Report

France's 55th reactor. During the summer of 1990, Cattenom nuclear power plant, unit number three, began producing electricity for the French grid. By January 1991, the new reactor would be in full commercial service.

Similar plants opened at Golfech and Penly. That made three new 1,300-megawatts electric (MWe) plants beginning operation in time for the most severe winter for that country in four years.

All three locations have one more unit scheduled for construction within the next three years. Once in service, they will mark the end of the 1,300 MWe building program. In all, France will have built 20 of that reactor model.

However, the choice of site for the four-reactor set did not make for good relations with its neighbor, Luxembourg. The plant sits near the Moselle River only 8 kilometers from the Luxembourg border.

France started building Cattenom in 1978, shortly after Luxembourg decided not to build a nuclear plant on its side of the Moselle at Remersham. Even though Luxembourg imports 95 percent of its electricity, it rejected nuclear generation for ecological reasons.

Neighboring countries watch construction. Two thirds of the population of Luxembourg live within 40 kilometers of Cattenom, but Luxembourg has no control over the environmental standards for emissions from the plant.

In July 1986, fierce protests were staged by German antinuclear groups, backed by officials from Luxembourg and the state of Saarland, about the safety norms at Cattenom. Both Luxembourg and Germany attempted to block commissioning of the first reactor, but a Strasbourg court rejected the petitions.

Luxembourg did win some control of day-to-day activities in a 1986 Franco-German convention, but still questions environmental standards.

Germany restricts similar reactors to a liquid discharge limit of 3 curies per year. France arbitrarily set the Cattenom limit at 15 curies without consultation with its neighbors.

After construction of the first two reactor blocks, the future of Cattenom came into question during the early 1980s. Unexpectedly low electricity demand left France with an excess in generating capacity, and the antinuclear lobby was gaining strength and scoring victories.

Temporary halt for Units 3 and 4. In 1981, the Socialist government of French President Francois Mitterrand came into power with a promise to review nuclear power. One of its first acts was to freeze further construction starts for nine new reactors, including two at Cattenom. However, on November 25, 1982, it approved construction of six, giving the go-ahead for Cattenom 3 and 4.

The actual commissioning dates slipped for the two reactors as technical problems with the 1,300 MWe reactor suurfaced in 1989. Technicians performed special inspections of steam generators looking for signs of corrosion cracking and tube deformation that occurred in earlier plants. Researchers expect a change in operating temperatures they instituted for all 1,300-MWe reactors should avoid the problem at Cattenom.

Sources

"France: The Nuclear Dream Maintained." *Technology Review*, November/December, 1982.
Mounfield, Peter. *World Nuclear Power*. Routledge, 1991.
"International: Three New French Plants Enter Service." *Nuclear News*, February 1991.
EDF 1990 Annual Report. Generation and Transmission Group, Paris, April 1991.
"Palo Verde Units 1 & 2 Top American Producers in May; Unit 2 Continues Record Run." *Business Wire*, July 12, 1991.
"Cracking at French PWRs Raises Curiosity, But Little Concern in U.S." *Inside N.R.C.*, December 2, 1991.

—Al Cook

---

40 ★

Chinon A Nuclear Power Plant, Units 1-3

BP 80
F-37240 Avoine, France
Christian LaSeure, Plant Mgr., Unit A3
Phone: 47-98-60-60
Fax: 47-98-77-09

Owning Utility

Electricite de France
2 rue Louis Murat
F-75005 Paris, France
Christian Nadal, Dir., Commun.
Phone: 1-42-56-94-00
Telex: 280041

Contact

Edouard Poulain, Pub.Info.
Chinon Nuclear Power Plant

Basic Facts

Units 1-3 *Status:* Shut down. *Type of Reactor:* Gas-cooled reactor. *Megawatts (electric):* Unit A1, 83; Unit A2, 210; Unit A3, 375. *Megawatts (thermal):* Unit A1, 300; Unit A2, 800; Unit A3, 1300. *Reactor System Supplier:* Various companies. *Generator Supplier:* Units A1 and A2, Alsthom; Unit A3, Alsthom and Jeumont Schneider Ste de Constructions Electromecaniques (France). *Architect Engineer:* Electricite de France; Commissariat al'Energie Atomique (France).

Key Dates

Unit 1 *Ordered:* 1956. *Construction began:* 1957. *Criticality:* September 1962. *First Power:* September 1962. *Commercial Operation:* February 1964. *Shut Down:* April 1973.

Unit 2 *Ordered:* 1957. *Construction began:* 1958. *Criticality:* August 1964. *First Power:* February 1965. *Commercial Operation:* March 1965. *Shut Down:* June 1985.

Unit 3 *Ordered:* 1960. *Construction began:* 1961. *Criticality:* 1965. *First Power:* 1965. *Commercial Operation:* 1967. *Shut Down:* June 1990.

Plant Report

A first for France. Chinon A1 was France's first commercial reactor built in the late 1950s and early 1960s along the Loire at Chinon. It generated electricity for 10 years before being shut down. Two more reactors were also constructed on the Chinon-A site and had life spans of 20 and 25 years, respectively.

Sources

"France's Nuclear Energy Boom." *World Press Review*, March 1980.

---

41 ●

Chinon B Nuclear Power Plant, Units 1-4

BP 80
F-37240 Avoine, France
Pierre Decaix, Site Mgr.
Phone: 47-98-60-60
Fax: 47-98-77-09

Owning Utility

Electricite de France
2 rue Louis Murat
F-75005 Paris, France
Christian Nadal, Dir., Commun.
Phone: 1-42-56-94-00
Telex: 280041

Contact

Edouard Poulain, Pub.Info.
Chinon Nuclear Power Plant

Basic Facts

Units 1-4 *Status:* Operable. *Type of Reactor:* Pressurized light-water reactor. *Megawatts (electric):* Units B1 and B2, 919; Units B3 and B4, 970. *Megawatts (thermal):* Units B1 and B2, 2785; Units B3 and B4, 2905. *Reactor System Supplier:* Framatome (France). *Generator Supplier:* Alsthom. *Architect Engineer:* Electricite de France.

Plant Profiles — Civaux 44 — FRANCE

Key Dates

Unit 1 *Ordered:* 1976. *Construction began:* 1977. *Criticality:* October 1982. *First Power:* November 1982. *Commercial Operation:* February 1984.

Unit 2 *Ordered:* 1976. *Construction began:* 1977. *Criticality:* September 1983. *First Power:* November 1983. *Commercial Operation:* August 1984.

Unit 3 *Ordered:* 1979. *Construction began:* 1981. *Criticality:* September 1986. *First Power:* October 1986. *Commercial Operation:* March 1987.

Unit 4 *Ordered:* 1979. *Construction began:* 1982. *Criticality:* October 1987. *First Power:* November 1987. *Commercial Operation:* March 1988.

---

42 ★

Chooz A Nuclear Power Plant

Centrale Nucleaire des Ardennes
BP 160
F-80600 Givet, France
Andre Senne, Head

Phone: 24-42-05-26
Fax: 24-42-03-37
Telex: 840-304-F

Owning Utility

Societe d'Energie Nucleaire Franco-Belge des Ardennes (SENA)
12, place les Etats Unis
F-75116 Paris, France
Remy Carle, Pres. & Dir. Gen.

Phone: 16-1-47-64-78-66

Basic Facts

Status: Shut down. *Type of Reactor:* Pressurized light-water reactor. *Megawatts (electric):* 320. *Megawatts (thermal):* 1000. *Reactor System Supplier:* Ateliers de Constructions Electriques de Charleroi S.A. (Belgium); Cockeril Ougree-Providence (Belgium); Westinghouse; Framatome (France). *Generator Supplier:* Rateau; Creusot-Loire (France). *Architect Engineer:* Gibbs & Hill Inc. (United States); Spie Batignolles SA (France).

Key Dates

Ordered: 1961. *Construction began:* 1962. *Criticality:* 1966. *First Power:* April 1967. *Commercial Operation:* April 1967. *Shut Down:* 1991.

---

43 ○

Chooz B Nuclear Power Plant, Units 1-2

BP 174
F-08600 Givet, France
Joseph Kandel, Plant Mgr.

Phone: 24-42-60-00
Fax: 24-42-61-80

Owning Utility

Electricite de France
2 rue Louis Murat
F-75005 Paris, France
Christian Nadal, Dir., Commun.

Phone: 1-42-56-94-00
Telex: 280041

Contact

Bruno Richard, Pub.Info.
Chooz B Nuclear Power Plant

Basic Facts

Units 1-2 *Status:* Under construction (Unit 1, 85 percent complete as of January 1992; Units 2, 60 percent complete as of January 1992). *Type of Reactor:* Pressurized light-water reactor. *Megawatts (electric):* 1516. *Megawatts (thermal):* 4270. *Reactor System Supplier:* Framatome (France). *Generator Supplier:* Alsthom. *Architect Engineer:* Electricite de France.

Key Dates

Unit 1 *Ordered:* 1984. *Construction began:* 1984. *Commercial Operation Expected:* May 1995.

Unit 2 *Construction began:* 1986. *Commercial Operation Expected:* July 1995.

Plant Report

Belgium to share output from Chooz. At a point just south of Dinant, Belgium, the French border extends into Belgium with a strip of land 15 kilometers long and 3 kilometers wide. The two reactors under construction at Chooz are located at the tip of this point. Belgium will take 25 percent of the power produced at the Chooz site, once it's completed.

The Chooz units are the first of a new series of 1450-megawatts electric (MWe) (net) reactors to be built by Electricité de France. EDF finished constructing the last of 20 1300 MWe (net) units at the beginning of 1991.

Sources

Mounfield, Peter. *World Nuclear Power.* Routledge, 1991.
"International: Three New French Plants Enter Service." *Nuclear News,* February 1991.

---

44 ○

Civaux Nuclear Power Plant, Unit 1

Centrale de Civaux
BP 1
F-86320 Civaux, France
Mazet, Plant Mgr.

Phone: 49-91-40-00
Fax: 49-91-40-06

Owning Utility

Electricite de France
2 rue Louis Murat
F-75005 Paris, France
Christian Nadal, Dir., Commun.

Phone: 1-42-56-94-00
Telex: 280041

Basic Facts

Unit 1 *Status:* Under construction (10% complete as of January 1992). *Type of Reactor:* Pressurized light-water reactor. *Megawatts (electric):* 1516. *Megawatts (thermal):* 4250. *Reactor System Supplier:* Framatome (France). *Generator Supplier:* Alsthom. *Architect Engineer:* Electricite de France.

Key Dates

Unit 1 *Ordered:* 1991. *Construction began:* 1991. *Commercial Operation Expected:* April 1997.

● Operable ○ Under Construction △ Indefinitely Deferred ▲ Decommissioning □ Planned ■ On Order ☆ Cancelled ★ Shut Down

45 Civaux

Plant Report

France's first new reactor in four years. Civaux, Unit 1, ordered in the summer of 1991, was the first new reactor ordered by Electricité de France (EDF), the national electric company, since 1987. EDF is also planning a second unit at Civaux, pending approval by the French Ministry of Industry. The order for the second unit, which should take seven years to build, is expected in late 1992 or 1993.

Sources

"International: French Planning Two New Units." *Nuclear Industry,* Third Quarter 1991.

45 □
Civaux Nuclear Power Plant, Unit 2

Centrale de Civaux
BP 1
F-86320 Civaux, France
Mazet, Plant Mgr.
Phone: 49-91-40-00
Fax: 49-91-40-06

Owning Utility

Electricite de France
2 rue Louis Murat
F-75005 Paris, France
Christian Nadal, Dir., Commun.
Phone: 1-42-56-94-00
Telex: 280041

Basic Facts

Unit 2 *Status:* Planned. *Type of Reactor:* Pressurized light-water reactor. *Megawatts (electric):* 1516. *Megawatts (thermal):* 4250. *Reactor System Supplier:* Framatome (France). *Generator Supplier:* Alsthom. *Architect Engineer:* Electricite de France.

Key Dates

Unit 2 *Ordered:* 1993. *Commercial Operation Expected:* 2000.

Plant Report

See Plant Report for Civaux Nuclear Power Plant, Unit 1.

46 ●
Cruas Nuclear Power Plant, Units 1-4

BP 31
F-38550 Le Peage du Roussillon, France
Max Morel, Site Mgr.
Phone: 74-29-32-32
Fax: 74-29-69-81

Owning Utility

Electricite de France
2 rue Louis Murat
F-75005 Paris, France
Christian Nadal, Dir., Commun.
Phone: 1-42-56-94-00
Telex: 280041

Contact

Jean-Louis Di Mayo, Pub.Info.
Cruas Nuclear Power Plant

Basic Facts

Units 1-4 *Status:* Operable. *Type of Reactor:* Pressurized light-water reactor. *Megawatts (electric):* Units 1, 3, and 4, 921; Unit 2, 956. *Megawatts (thermal):* 2785. *Reactor System Supplier:* Framatome (France). *Generator Supplier:* Alsthom. *Architect Engineer:* Electricite de France.

Key Dates

Unit 1 *Ordered:* 1977. *Construction began:* 1978. *Criticality:* April 1983. *First Power:* April 1983. *Commercial Operation:* April 1984.

Unit 2 *Ordered:* 1977. *Construction began:* 1978. *Criticality:* August 1984. *First Power:* September 1984. *Commercial Operation:* April 1985.

Unit 3 *Ordered:* 1977. *Construction began:* 1978. *Criticality:* April 1984. *First Power:* May 1984. *Commercial Operation:* September 1984.

Unit 4 *Ordered:* 1977. *Construction began:* 1978. *Criticality:* October 1984. *First Power:* October 1984. *Commercial Operation:* February 1985.

47 ●
Dampierre Nuclear Power Plant, Units 1-4

BP 18
F-45570 Ouzouer sur Loire, France
Jean Vedrinne, Site Mgr.
Phone: 38-29-70-70
Fax: 38-67-68-02

Owning Utility

Electricite de France
2 rue Louis Murat
F-75005 Paris, France
Christian Nadal, Dir., Commun.
Phone: 1-42-56-94-00
Telex: 280041

Contact

Jean-Paul Ladaigue, Pub.Info.
Dampierre Nuclear Power Plant

Basic Facts

Units 1-4 *Status:* Operable. *Type of Reactor:* Pressurized light-water reactor. *Megawatts (electric):* 937. *Megawatts (thermal):* 2785. *Reactor System Supplier:* Framatome (France). *Generator Supplier:* Alsthom. *Architect Engineer:* Electricite de France.

Key Dates

Unit 1 *Ordered:* 1973. *Construction began:* 1975. *Criticality:* March 1980. *First Power:* March 1980. *Commercial Operation:* September 1980.

Unit 2 *Ordered:* 1974. *Construction began:* 1974. *Criticality:* December 1980. *First Power:* December 1980. *Commercial Operation:* February 1981.

Unit 3 *Ordered:* 1974. *Construction began:* 1975. *Criticality:* January 1981. *First Power:* January 1981. *Commercial Operation:* May 1981.

Unit 4 *Ordered:* 1974. *Construction began:* 1976. *Criticality:* August 1981. *First Power:* August 1981. *Commercial Operation:* November 1981.

Plant Profiles

48 ●
Fessenheim Nuclear Power Plant, Units 1-2

BP 15
F-68740 Fessenheim, France
Philippe Gustin, Site Mgr.
Phone: 89-26-51-26
Fax: 89-48-64-08

Owning Utility
Electricite de France
2 rue Louis Murat
F-75005 Paris, France
Christian Nadal, Dir., Commun.
Phone: 1-42-56-94-00
Telex: 280041

Contact
Christian Chapus, Pub.Info.
Fessenheim Nuclear Power Plant

Basic Facts
Units 1-2 *Status:* Operable. *Type of Reactor:* Pressurized light-water reactor. *Megawatts (electric):* 920. *Megawatts (thermal):* 2660. *Reactor System Supplier:* Framatome (France). *Generator Supplier:* Alsthom. *Architect Engineer:* Electricite de France.

Key Dates
Unit 1 *Ordered:* 1970. *Construction began:* 1971. *Criticality:* March 1977. *First Power:* April 1977. *Commercial Operation:* December 1977.

Unit 2 *Ordered:* 1970. *Construction began:* 1971. *Criticality:* June 1977. *First Power:* October 1977. *Commercial Operation:* March 1978.

49 ●
Flamanville Nuclear Power Plant, Units 1-2

BP 4
F-50340 Les Pieux, France
Francois Buffet, Plant Mgr.
Phone: 33-08-95-95
Fax: 33-04-13-00

Owning Utility
Electricite de France
2 rue Louis Murat
F-75005 Paris, France
Christian Nadal, Dir., Commun.
Phone: 1-42-56-94-00
Telex: 280041

Contact
Jacques Ropars, Pub.Info.
Flamanville Nuclear Power Plant

Basic Facts
Units 1-2 *Status:* Operable. *Type of Reactor:* Pressurized light-water reactor. *Megawatts (electric):* 1382. *Megawatts (thermal):* 3817. *Reactor System Supplier:* Framatome (France). *Generator Supplier:* Alsthom. *Architect Engineer:* Electricite de France.

Key Dates
Unit 1 *Ordered:* 1975. *Construction began:* 1979. *Criticality:* September 1985. *First Power:* December 1985. *Commercial Operation:* December 1986.

Unit 2 *Ordered:* 1975. *Construction began:* 1979. *Criticality:* June 1986. *First Power:* July 1986. *Commercial Operation:* March 1987.

Plant Report
Top producer. According to data compiled by the industry publication *Nucleonics Week*, in December of 1990, Flamanville Units 2 and 1 ranked 19th and 20th respectively among the top twenty producers of electricity in the world. That month, the two units at Flamanville generated a total of 1,822,308 megawatt-hours.

Sources
"Palo Verde Units Top Producers in U.S. for December." *Business Wire,* February 15, 1991.

50 ●
Golfech Nuclear Power Plant, Unit 1

BP 24 Golfech
F-82400 Valence d'Agen, France
Jacques Regaldo, Site Mgr.
Phone: 63-29-39-49
Fax: 63-29-39-50

Owning Utility
Electricite de France
2 rue Louis Murat
F-75005 Paris, France
Christian Nadal, Dir., Commun.
Phone: 1-42-56-94-00
Telex: 280041

Contact
Phillippe Denis, Pub.Info.
Golfech Nuclear Power Plant

Basic Facts
Unit 1 *Status:* Operable. *Type of Reactor:* Pressurized light-water reactor. *Megawatts (electric):* 1363. *Megawatts (thermal):* 3817. *Reactor System Supplier:* Framatome (France). *Generator Supplier:* Alsthom. *Architect Engineer:* Electricite de France.

Key Dates
Unit 1 *Ordered:* 1978. *Construction began:* 1983. *Criticality:* 1990. *First Power:* 1990. *Commercial Operation:* January 1991.

Plant Report
Last of the 1300-MWe series. Electricité de France (EDF) had three 1300-MWe (net) pressurized water reactors enter commercial service at the beginning of 1991—Penly-1, Golfech-1, and Cattenom-3. The plants were just in time to help meet demand in what was a more severe winter that the previous four years. These were the last of 20 1300-MWe reactors ordered by EDF.

Sources
"International: Three New French Plants Enter Service." *Nuclear News,* February 1991.

FRANCE

| 51 | ○ |

Golfech Nuclear Power Plant, Unit 2

BP 24 Golfech　　　　　　　　　　Phone: 63-29-39-49
F-82400 Valence d'Agen, France　　Fax: 63-29-39-50
Jacques Regaldo, Site Mgr.

Owning Utility

Electricite de France　　　　　Phone: 1-42-56-94-00
2 rue Louis Murat　　　　　　Telex: 280041
F-75005 Paris, France
Christian Nadal, Dir., Commun.

Contact

Phillippe Denis, Pub.Info.
Golfech Nuclear Power Plant

Basic Facts

Unit 2 Status: Under construction (95% complete as of January 1992). *Type of Reactor:* Pressurized light-water reactor. *Megawatts (electric):* 1363. *Megawatts (thermal):* 3817. *Reactor System Supplier:* Framatome (France). *Generator Supplier:* Alsthom. *Architect Engineer:* Electricite de France.

Key Dates

Unit 2 *Ordered:* 1978. *Construction began:* 1984. *Criticality:* November 1992. *First Power:* January 1993. *Commercial Operation Expected:* May 1993.

Plant Report

See Plant Report for Golfech Nuclear Power Plant, Unit 1.

| 52 | ● |

Gravelines Nuclear Power Plant, Units 1-6

BP 149　　　　　　　　　　　Phone: 28-68-40-00
F-59820 Gravelines, France　　Fax: 28-68-42-08
Georges Bouchard, Site Mgr.

Owning Utility

Electricite de France　　　　　Phone: 1-42-56-94-00
2 rue Louis Murat　　　　　　Telex: 280041
F-75005 Paris, France
Christian Nadal, Dir., Commun.

Contact

Jacques Decriem, Pub.Info.
Gravelines Nuclear Power Plant

Basic Facts

Units 1-6 Status: Operable. *Type of Reactor:* Pressurized light-water reactor. *Megawatts (electric):* 951. *Megawatts (thermal):* 2785. *Reactor System Supplier:* Framatome (France). *Generator Supplier:* Alsthom. *Architect Engineer:* Electricite de France.

Key Dates

Unit 1 *Ordered:* 1973. *Construction began:* 1974. *Criticality:* February 1980. *First Power:* March 1980. *Commercial Operation:* December 1980.

Unit 2 *Ordered:* 1973. *Construction began:* 1975. *Criticality:* August 1980. *First Power:* August 1980. *Commercial Operation:* December 1980.

Unit 3 *Ordered:* 1973. *Construction began:* 1975. *Criticality:* November 1980. *First Power:* December 1980. *Commercial Operation:* June 1981.

Unit 4 *Ordered:* 1973. *Construction began:* 1976. *Criticality:* May 1981. *First Power:* June 1981. *Commercial Operation:* October 1981.

Unit 5 *Ordered:* 1979. *Construction began:* 1979. *Criticality:* August 1984. *First Power:* August 1984. *Commercial Operation:* January 1985.

Unit 6 *Ordered:* 1979. *Construction began:* 1979. *Criticality:* July 1985. *First Power:* August 1985. *Commercial Operation:* October 1985.

| 53 | ★ |

Marcoule G Nuclear Power Plant, Units 1-3

Bagnois sur Ceze, France

Owning Utility

Commissariat a l'Energie Atomique　　Phone: 40-56-10-00
　(CEA)　　　　　　　　　　　　　　Telex: 200-671 ENAT
31-33 rue de la Federation
F-75015 Paris Cedix 15, France
Jacques Bouchard, Dir., Nuclear
　Reactors

Basic Facts

Units 1-3 Status: Shut down. *Type of Reactor:* Gas-cooled reactor. *Megawatts (electric):* Unit 1, 2; Units 2-3, 42. *Megawatts (thermal):* Unit 1, 38; Units 2-3, 255. *Reactor System Supplier:* Units 2-3, Societe Alsacienne de Constructions Mecaniques (France). *Generator Supplier:* Rateau. *Architect Engineer:* Unit 1, Commissariat a l'Energie Atomique (CEA); Unit 2-3, Societe Alsacienne de Constructions Mecaniques (France).

Key Dates

Unit 1 *Ordered:* 1954. *Construction began:* 1954. *Criticality:* September 1956. *First Power:* September 1956. *Commercial Operation:* September 1956. *Shut Down:* October 1968.

Unit 2 *Ordered:* 1955. *Construction began:* 1956. *Criticality:* July 1958. *First Power:* July 1958. *Commercial Operation:* April 1959. *Shut Down:* February 1980.

Unit 3 *Ordered:* 1955. *Construction began:* 1956. *Criticality:* June 1959. *First Power:* April 1960. *Commercial Operation:* May 1960. *Shut Down:* July 1984.

| 54 | ★ |

Monts d'Arree Nuclear Power Plant

Brennilis, France

Plant Profiles

Owning Utility

Commissariat a l'Energie Atomique (CEA)
31-33 rue de la Federation
F-75015 Paris Cedix 15, France
Jacques Bouchard, Dir., Nuclear Reactors

Phone: 40-56-10-00
Telex: 200-671 ENAT

Basic Facts

Status: Shut down. *Type of Reactor:* Gas-cooled reactor. *Megawatts (electric):* 75. *Reactor System Supplier:* Commissariat a l'Energie Atomique; Indatom. *Generator Supplier:* Compagnie Electro-Mechanique (France). *Architect Engineer:* Indatom. *Owner:* Commissariat a l'Energie Atomique. *Operator:* Electricite de France.

Key Dates

Ordered: 1962. *Construction began:* 1962. *Criticality:* December 1966. *First Power:* July 1967. *Commercial Operation:* March 1968. *Shut Down:* July 1985.

---

55 •

Nogent s/Seine Nuclear Power Plant, Units 1-2

BP 62
F-10400 Nogent sur Seine, France
Claude Jeandron, Site Mgr.

Phone: 25-39-30-00
Fax: 25-39-32-40

Owning Utility

Electricite de France
2 rue Louis Murat
F-75005 Paris, France
Christian Nadal, Dir., Commun.

Phone: 1-42-56-94-00
Telex: 280041

Contact

Brigitte Bosch, Pub.Info.
Nogent s/Seine Nuclear Power Plant

Basic Facts

Units 1-2 *Status:* Operable. *Type of Reactor:* Pressurized light-water reactor. *Megawatts (electric):* 1363. *Megawatts (thermal):* 3817. *Reactor System Supplier:* Framatome (France). *Generator Supplier:* Alsthom. *Architect Engineer:* Electricite de France.

Key Dates

Unit 1 *Ordered:* 1978. *Construction began:* 1981. *Criticality:* August 1987. *First Power:* October 1987. *Commercial Operation:* February 1988.

Unit 2 *Ordered:* 1979. *Construction began:* 1982. *Criticality:* October 1988. *First Power:* December 1988. *Commercial Operation:* May 1989.

---

56 •

Paluel Nuclear Power Plant, Units 1-4

BP 48
F-76450 Cany Barville, France
Michel Leroy, Site Mgr.

Phone: 35-57-57-57
Fax: 35-57-58-88

Owning Utility

Electricite de France
2 rue Louis Murat
F-75005 Paris, France
Christian Nadal, Dir., Commun.

Phone: 1-42-56-94-00
Telex: 280041

Contact

Yannick Couasnon, Pub.Info.
Paluel Nuclear Power Plant

Basic Facts

Units 1-4 *Status:* Operable. *Type of Reactor:* Pressurized light-water reactor. *Megawatts (electric):* 1382. *Megawatts (thermal):* 3817. *Reactor System Supplier:* Framatome (France). *Generator Supplier:* Alsthom. *Architect Engineer:* Electricite de France.

Key Dates

Unit 1 *Ordered:* 1975. *Construction began:* 1977. *Criticality:* May 1984. *First Power:* June 1984. *Commercial Operation:* December 1985.

Unit 2 *Ordered:* 1977. *Construction began:* 1977. *Criticality:* August 1984. *First Power:* September 1984. *Commercial Operation:* December 1985.

Unit 3 *Ordered:* 1977. *Construction began:* 1977. *Criticality:* August 1985. *First Power:* September 1985. *Commercial Operation:* February 1986.

Unit 4 *Ordered:* 1977. *Construction began:* 1977. *Criticality:* March 1986. *First Power:* April 1986. *Commercial Operation:* June 1986.

Plant Report

Massive French nuclear program. The nuclear initiative launched by the French government in 1975 called for the construction of 200 atomic reactors by the year 2000. The massiveness of that scheme can be seen in the tremendous scale of the Paluel nuclear power facility.

Electricité de France already had several 900-megawatts electric (MWe) pressurized water reactors (PWR) either operating, under construction, or planned, but for Paluel, the utility chose a larger version of the technology, the 1,300-MWe PWR.

With the completion of its fourth block in August 1986, Paluel became the largest nuclear facility in the world, producing enough electricity to power the city of Tokyo. At 35 million kilowatt-hour (KWh), Paluel generates 1.5 times the entire electrical production of France in 1950.

Engineers move mountains. However, before it could do that, French engineers had to prepare a place for the plant on the cliffs of Normandy's coast, between Fecamp and Dieppe.

That meant gouging a 10 million cubic meter bite out of the stubborn rock, to create a gentle berm rising from the sea to the farm land above.

Engineers used the resulting boulders to create a pair of smoothly curving breakwaters that protect the facility from the channel waves, while funneling 180 cubic meters per second of sea water into the plant's coolant water intakes.

French workers dug four concrete-lined tunnels 70 meters below the channel's bed to exhaust heated water from the plant's boilers. Each gallery measures 4.3 meters. The 7.3 meter diam-

57 Penly

eter Eurotunnel, connecting England and France used a similar technology in its construction.

Four mammoth turbine halls now replace that missing section of natural white cliffs, but from the fields above, they remain invisible. Only the four round tops of the reactor containment buildings, standing neatly in a row, mark the starting point for the high-tension towers that seem to march into the horizon.

Corrosion cracks limits plant's output. During 1989, EDF experienced disappointing availability levels for its 1,300 MWe PWRs. Investigators discovered design and manufacturing faults in welds involving Inconel-600 as well as deformation of steam generator tubes.

The cracking of the high nickel-content stainless steel is a common form of corrosion, according to James Lang, of the Electric Power Research Institute in the United States. The phenomenon happens at a specific ideal combination of temperature, pressure, and primary water environment.

To counteract the problem, EDF lowered the temperature in all of its 1,300-MWe PWRs by 4.2 degrees Celsius.

French designers based both their 900-MWe and the 1,300-MWe pressurized light-water reactor on American Westinghouse reactors. Westinghouse in the United States notes it has not received reports of similar problems with U.S. reactors. However, American standards for hydro-testing of reactor vessels call for a much lower test pressure than French standards do.

For December 1990, two Paluel reactor units earned a spot in *Nucleonic Week*'s list of the 20 top-producing nuclear power plants. Unit four rated ninth, while the first Paluel block claimed the 14th spot.

Sources

Paluel 1976-1986, EDF. Region d'Equipement Clamart, Paris, 1986.
Mounfield, Peter. *World Nuclear Power*. Routledge, 1991.
EDF 1990 Annual Report. Generation and Transmission Group, Paris, April 1991.
"Palo Verde Units 1 & 2 Top American Producers in May; Unit 2 Continues Record Run." *Business Wire*, July 12, 1991.
"Cracking at French PWRs Raises Curiosity, But Little Concern in U.S." *Inside N.R.C.*, December 2, 1991.

—Al Cook

57

Penly Nuclear Power Plant, Units 1-2

BP 854 Phone: 35-40-60-00
F-76370 Neuville Les Dieppe, France Fax: 35-40-60-99
Bernard Payen, Site Mgr.

Owning Utility

Electricite de France Phone: 1-42-56-94-00
2 rue Louis Murat Telex: 280041
F-75005 Paris, France
Christian Nadal, Dir., Commun.

Contact

Andre Michel, Pub.Info.
Penly Nuclear Power Plant

Basic Facts

Units 1-2 *Status:* Operable. *Type of Reactor:* Pressurized light-water reactor. *Reactor System Supplier:* Framatome (France). *Generator Supplier:* Alsthom. *Architect Engineer:* Electricite de France.

Key Dates

Unit 1 *Ordered:* 1978. *Construction began:* 1983. *Criticality:* 1990. *First Power:* 1990. *Commercial Operation:* December 1990.
Unit 2 *Ordered:* 1978. *Construction began:* 1984. *Commercial Operation:* 1992.

Plant Report

Last of the 1300-MWe series. Electricité de France had three 1300-MWe (net) pressurized water reactors enter commercial service at the beginning of 1991—Penly-1, Golfech-1, and Cattenom-3. The plants were just in time to help meet demand in what was France's most severe winter in four years. These were the last of 20 1300-MWe reactors ordered by EDF. Each site was due to have an additional unit enter service within three years.

According to data compiled by the industry publication *Nucleonics Week*, in December of 1990, Penly -1 ranked sixth among the top 20 producers of electricty in the world. That month the plant generated 1,007,444 megawatt-hours.

Sources

"International: Three New French Plants Enter Service." *Nuclear News*, February 1991.
"Palo Verde Units Top Producers in U.S. for December." *Business Wire*, February 15, 1991.

58

Phenix Prototype Fast Reactor Power Station

Centrale Phenix Phone: 66-79-50-00
BP 171
F-30205 Bagnois sur Ceze, France
Michel Gelee, Dir., Centrale Phenix

Owning Utility

Commissariat a l'Energie Atomique Phone: 40-56-10-00
(CEA) Telex: 200-671 ENAT
31-33 rue de la Federation
F-75015 Paris Cedix 15, France
Jacques Bouchard, Dir., Nuclear Reactors

Basic Facts

Status: Operable. *Type of Reactor:* Liquid metal fast breeder reactor. *Megawatts (electric):* 250. *Megawatts (thermal):* 568. *Reactor System Supplier:* Commissariat a l'Energie Atomique; Electricite de France; Novatome. *Generator Supplier:* Compagnie Electro-Mechanique

Plant Profiles — Phenix 58

(France). *Architect Engineer:* Commissariat a l'Energie Atomique; Novatome; Electricite de France.

Key Dates
Ordered: 1967. *Construction began:* 1968. *Criticality:* August 1973. *First Power:* November 1973. *Commercial Operation:* July 1974.

Plant Report

New breed of reactor. For atomic scientists, the thermal nuclear reactors in common use today represent a crude beginning. They remain inefficient, burning only a small portion of the available uranium-235. However, they provided the first steps towards the modern machines of the future that would "close the fuel cycle," reducing the need to mine uranium by increasing the burn-efficiency by as much as 600 percent.

The Fast Breeder Reactor (FBR) uses both plutonium and enriched uranium, U-238 in an unmoderated core. The bombarded U-238 in the core becomes plutonium and can be recycled. In fact, the FBR produces more plutonium than it uses, earning it the name breeder.

Because the core is not moderated by some substance like light or heavy water or graphite, the reactor reaches extremely high temperatures. That increases the thermal efficiency of the system.

Those two characteristics named France's first prototype FBR. Like the legendary bird that rises, reborn from its own ashes, the Phenix creates the fuel that sustains it, within the fiery inferno of its own core.

The Phenix is born. The 250-megawatts electric (MWe) prototype plant, located at Marcoule in the South of France, went critical for the first time in August, 1973, beating the similar British project at Dounreay by almost a year.

Only the Soviet Union seems to have had an FBR in operation before France. The BNR-350 at Schevchenko started up in 1972.

For France, the project began with the experimental reactor, Rapsodie, at the Cadarache Center in Provence in 1967. That initial investigation into the use of sodium as a coolant prompted the construction of Phenix in 1968.

After 10 years of operation, French scientists labelled the prototype a success. In 1980, they fabricated a fuel element using plutonium extracted from the core, closing the fuel cycle for the first time.

In the meantime, work on a commercial-scale FBR proceeded. The Superphenix 1,200 MWe FBR went critical in 1985.

Concept may be stillborn. However, the FBR concept was beginning to fall out of favor, as development costs escalated and time tables lengthened. By 1991, United States, Britain, and Germany had all abandoned FBR programs as too expensive and unnecessary given the ready-availability of cheap natural uranium.

During its first 10 years of operation, Phenix experienced two serious incidents that called for major overhaul and component redesign.

The first incident involved the intermediate heat exchanger that passes heat from the radioactive sodium pool surrounding the core to the reccirculating non-radioactive sodium in the cooling system.

On two separate occasions in 1976, thermal expansion and contraction of a rigid metal plate caused a weld failure at the floor level of the exchanger. Liquid sodium leaked through the failed welds to combust spontaneously with the air of the reactor hall. Emergency fire-protection equipment contained the fires. Since the leaked sodium came from the secondary side of the exchanger, no radiation escaped.

Researchers replaced the rigid plate with a flexible closure. To do so, they had first to remove the six heat exchangers from the core. They used shielded flasks to move the components to a washing pit where decontamination took place. Normal machining operations completed the repair.

Total renovation time for all six major components was 18 months. Previously, many scientists had expressed doubts that such reconstruction of irradiated components was possible.

Phenix survives greatest technical challenge. Between April 29, 1982 and August 12, 1983, Phenix experienced its most dreaded problem, leaks that resulted in the mixing of sodium and water. When sodium contacts water, the material effectively burns, generating hydrogen and, in a closed system, increasing pressure quickly.

In 1973, a sodium-water explosion devastated the Soviet FBR at Shevchenko, creating a fire detected by an American surveillance satellite.

Researchers at Cadarache anticipated the eventuality. Since the steam generators presented the most likely place for such a leak to happen, they designed a special safety system for them. A hydrogen detector senses a build-up early, and initiates a safe shut-down. The system immediately depressurizes the water-steam side of the boilers, allowing the sodium to leak harmlessly.

On four occasions, the safety system worked as intended. As a result of the incidents, scientists redesigned the boilers, which were replaced without incident.

The Phenix proved the technology worked and that the system could operate safely. Next, the researchers had to prove the economics of FBR technology. When nuclear scientists conceived the Phenix, they pointed to the high core temperatures as a significant advantage over earlier reactors. Much reactor development work focused on increasing that core temperature in order to improve the burn-efficiency of the uranium fuel. That increased efficiency should translate into greater electricity generation at a lower cost. To prove that, researchers needed a commercial-scale reactor.

On January 10, 1984, Great Britain, Germany, and France agreed to pool their efforts in perfecting the European Fast Breeder Reactor.

Under the terms of the agreement, each country would build a demonstration plant. France already had the Superphenix nearing completion, and intended to build six replicas in France and a seventh in Germany.

Germany had a research reactor operating at Kalkar, but the growing strength of the antinuclear lobby had already stalled that country's thermal nuclear construction program. Nuclear opponents had effectively blocked the issuance of building permits for all proposed reactors between 1978 and 1982. Only by adopting a standardized pressurized light-water reactor design had the German nuclear industry been able to forge ahead. A new design like the FBR would be certain to attract opposition.

Britain had a similar problem. Its FBR research at Dounreay continued to fall behind schedule, pushing the possibility of a demonstration plant into the next century. Meanwhile, the industry was trying to switch from the unsuccessful British-designed Advanced Gas Cooled reactor to the more common American PWR.

Eventually, both Britain and Germany abandoned plans for demonstration plants, pushing instead for the construction of the next generation at Marcoule. However, that depended on the success, or failure, of Superphenix.

Sources

Prototype Fast Reactor Power Station Phenix. Commissariat à l'Energie Atomique and Electricité de France, Paris, September 1974.
The Phenix Nuclear Plant After 10 Years of Operation. Centrale Phenix, Bagnols-Sur-Ceze, July 1984.
"Breeder Reactor Politics in Europe." *Bulletin of the Atomic Scientists*, May 1986.
La Centrale Nucléaire Phenix Experience D'Exploitation. Centrale Phenix, Bagnols-Sur-Ceze, 1988.
Phenix. Commissariat l'Energie Atomique and Electricité de France, Avignon, October 1991.

—Al Cook

59 ☆
Plogoff Nuclear Power Plant

Plogoff, France

Basic Facts

Status: Cancelled.

Plant Report

Public protest and government construction freeze. In the 1970s, the French government had strongly supported nuclear energy development. Antinuclear protesters had a difficult time making an impact on the decision-making process. However, with the election of Socialist leader Francois Mitterrand in 1981, France experienced a political upheaval and the antinuclear protest movement was revitalized. The Socialist Party was divided on the issue of nuclear power and had a minority wing favoring a nuclear-energy freeze.

When legislative elections were held after Mitterrand took office, many Socialist candidates faced with public opposition to planned-nuclear plants in their localities promised to stop construction. The Socialist Party won a majority and voted on July 30, 1981 to freeze planning and construction of nuclear power plants at five sites. The proposed plant in Plogoff on the Brittany coast, where local opposition had been the most radical, was abandoned.

In January 1980, the 2,360 citizens of the Breton fishing port, Plogoff, had rebelled against the prospect of four 1,300 megawatt nuclear reactors. Paris called in the riot police and army, turning the town into a six-week battlefield.

Although the government later lifted the construction freeze off three of the five sites, plans for Plogoff were never reinstated.

Sources

"Off and Running in a Nuclear Race." *Maclean's*, October 20, 1980.
"France: The Nuclear Dream Maintained." *Technology Review*, November/December, 1982.

60 ●
Saint-Alban Nuclear Power Plant, Units 1-2

BP 31
F-38550 Le Peage du Roussillon, France
Yves Corre, Plant Mgr.

Phone: 74-29-32-32
Fax: 74-29-69-81

Owning Utility

Electricite de France
2 rue Louis Murat
F-75005 Paris, France
Christian Nadal, Dir., Commun.

Phone: 1-42-56-94-00
Telex: 280041

Contact

Jean-Paul Lancon, Pub.Info.
Saint-Alban Nuclear Power Plant

Basic Facts

Units 1-2 *Status:* Operable. *Type of Reactor:* Pressurized light-water reactor. *Megawatts (electric):* 1381. *Megawatts (thermal):* 3817. *Reactor System Supplier:* Framatome (France). *Generator Supplier:* Alsthom. *Architect Engineer:* Electricite de France.

Key Dates

Unit 1 *Ordered:* 1978. *Construction began:* 1978. *Criticality:* August 1985. *First Power:* August 1985. *Commercial Operation:* May 1986.
Unit 2 *Ordered:* 1978. *Construction began:* 1979. *Criticality:* June 1986. *First Power:* July 1986. *Commercial Operation:* March 1987.

Plant Report

Grand vision for river region. During the 1960s, the picturesque mountain valleys and rivers of the Rhône-Alps region contributed to France's energy needs with large scale hydroelectric projects. When Electricité de France (EDF) began its ambitious nuclear building program in 1975, the company looked once again to that region as a prime source of generating capacity. It envisioned 34 reactors on nine separate energy parks along the 500-kilometer Rhône River between Marseille and Switzerland.

One of the first sites chosen was Saint-Alban/Saint-Maurice. The utility planned to build four 1,300-megawatts electric

(MWe) pressurized water reactors (PWR) using the standardized French design.

In August 1985 and July 1986, the first two units went on-line as France's 34th and 38th operating commercial reactors.

Reassessment may cut project in half. However, the fate of the other two units remains uncertain. In pursuing its nuclear expansion program, the French government wished to decrease the country's 75 percent dependance on imported oil by substituting the use of locally mined uranium. To do so, France began building nuclear power plants at a rate of six per year.

However, as in other countries, early estimates of increasing electrical demand proved wildly inaccurate. Conservation and slowing economic growth forced a reassessment.

By the time Saint-Alban's first two units came on line, France had excess generating capacity. New reactor construction dropped to one per year, pushing units three and four well into the future, at best.

Plant ranks as high producer. In terms of production, Saint-Alban generates 13 percent of the electricity used by the Rhône-Alps region. On the national scale, it generates 3.5 percent of EDF's total output. It supplies enough energy to power the city of Lyon nine times over. Internationally, in December 1990, *Nucleonics Week* ranked Saint Alban, unit 1, as the seventh most productive plant in the world.

In 1988, the International Atomic Energy Agency, (IAEA) sent an international team of experts to examine Saint-Alban. The Operational Safety Review Team (OSART) identified 54 "local actions" to improve plant safety and efficiency.

During 1989, many of the 1,300-MWe PWR units like Saint-Alban, experienced steam-generator-tube deformation and corrosion cracking in Inconel-600 taps on the pressurizers. French designers termed the problem a generic fault of the PWR design, including both the 1,300-MWe model and the 900-MWe model.

To counteract the failure, EdF lowered the average reactor-coolant temperature on all its 1,300 MWe units by 4.2 degrees Celsius. Researchers calculate that move will eliminate the phenomenon by avoiding the ideal temperature and pressure conditions that cause it.

High-pressure hydro-leak testing of the pressure vessel at Bugey 3 revealed a similar problem in the 900-MWe PWR. Inconel-600 stand-pipes, passing through the reacotr vessel lid developed longitundinal cracks from corrosion.

French designers based both their 900-MWe and the 1,300-MWe PWR on American Westinghouse reactors. Westinghouse in the United States notes it has not received reports of similar problems with U.S. reactors. However, American standards for hydro-testing of reactor vessels call for a much lower test pressure than French standards do.

In 1990, plant operators cut back production at Saint-Alban because of unusually low levels of water in the Rhône caused by a severe summer-long drought.

Sources

Le programme electronucleaire francais. Electricité de France, Paris, November 1987.
1990 In Brief: Nuclear and Fossil Generation. Electricité de France, Paris, February 1991.
1990 en bref: Saint-Alban Saint-Maurice. Electricité de France, Paris, February 1991.
EDF 1990 Annual Report. Generation and Transmission Group, Paris, April 1991.
Saint-Alban/Saint-Maurice. Electricité de France, Paris, July 1991.
"Palo Verde Units 1 & 2 Top American Producers in May; Unit 2 Continues Record Run." *Business Wire,* July 12, 1991.
"Cracking at French PWRs Raises Curiosity, But Little Concern in U.S." *Inside N.R.C.,* December 2, 1991.
Mounfield, Peter. *World Nuclear Power.* Routledge, 1991.

—Al Cook

61
Saint-Laurent A Nuclear Power Plant, Unit A-1

BP 42
F-41220 La Ferte St. Cyr, France
Jean-Claude Piednoir, Plant Mgr., Units A1 & A2

Phone: 54-44-84-84
Fax: 54-87-22-45

Owning Utility

Electricite de France
2 rue Louis Murat
F-75005 Paris, France
Christian Nadal, Dir., Commun.

Phone: 1-42-56-94-00
Telex: 280041

Contact

Pierre Germain, Pub.Info.
Saint-Laurent Nuclear Power Plant

Basic Facts

Unit A-1 *Status:* Shut down. *Type of Reactor:* Gas-cooled reactor. *Megawatts (electric):* 405. *Megawatts (thermal):* 1570. *Reactor System Supplier:* various. *Generator Supplier:* Alsthom. *Architect Engineer:* Electricite de France.

Key Dates

Unit A-1 *Ordered:* 1963. *Construction began:* 1963. *Criticality:* 1969. *First Power:* March 1969. *Commercial Operation:* June 1969. *Shut Down:* April 1990.

62
Saint-Laurent A Nuclear Power Plant, Unit A-2

BP 42
F-41220 La Ferte St. Cyr, France
Jean-Claude Piednoir, Plant Mgr., Units A1 & A2

Phone: 54-44-84-84
Fax: 54-87-22-45

Owning Utility

Electricite de France
2 rue Louis Murat
F-75005 Paris, France
Christian Nadal, Dir., Commun.

Phone: 1-42-56-94-00
Telex: 280041

| 63 | **Saint-Laurent** | Nuclear Power Plants Worldwide, First Edition |

FRANCE

Contact
Pierre Germain, Pub.Info.
Saint-Laurent Nuclear Power Plant

Basic Facts
Unit A-2 *Status:* Operable. *Type of Reactor:* Gas-cooled reactor. *Megawatts (electric):* 465. *Megawatts (thermal):* 1690. *Reactor System Supplier:* various. *Generator Supplier:* Alsthom. *Architect Engineer:* Electricite de France.

Key Dates
Unit A-2 *Ordered:* 1965. *Construction began:* 1966. *Criticality:* 1971. *First Power:* August 1971. *Commercial Operation:* November 1971.

| 63 | ● |

Saint-Laurent B Nuclear Power Plant, Units B1-B2

BP 42
F-41220 La Ferte St. Cyr, France
Marc Boursier, Plant Mgr., Units B1 & B2
Phone: 54-44-84-84
Fax: 54-87-22-45

Owning Utility
Electricite de France
2 rue Louis Murat
F-75005 Paris, France
Christian Nadal, Dir., Commun.
Phone: 1-42-56-94-00
Telex: 280041

Contact
Pierre Germain, Pub.Info.
Saint-Laurent Nuclear Power Plant

Basic Facts
Units B1-B2 *Status:* Operable. *Type of Reactor:* Pressurized light-water reactor. *Megawatts (electric):* 921. *Megawatts (thermal):* 2785. *Reactor System Supplier:* Framatome (France). *Generator Supplier:* Alstom. *Architect Engineer:* Electricite de France.

Key Dates
Unit B-1 *Ordered:* 1975. *Construction began:* 1976. *Criticality:* 1981. *First Power:* January 1981. *Commercial Operation:* August 1983.
Unit B-2 *Ordered:* 1975. *Construction began:* 1976. *Criticality:* May 1981. *First Power:* June 1981. *Commercial Operation:* August 1983.

| 64 |

Superphenix Fast Breeder Power Station

Centrale de Creys-Malville
BP 63
F-38510 Morestel, France
Andre Lacroix, Plant Mgr.
Phone: 74-33-34-35
Fax: 74-33-34-37
Telex: Sphenix 380 693

Owning Utility
Centrale Nucleaire Europeenne a Neutrons Rapides S.A. (NERSA)
177 rue Garibaldi
F-69003 Lyon, France

Contact
Brigitte d'Heilly, Pub.Info.

Basic Facts
Status: Construction 100% complete, commercial operation postponed pending safety studies. *Type of Reactor:* Liquid metal fast breeder reactor. *Megawatts (electric):* 1242. *Megawatts (thermal):* 3000. *Reactor System Supplier:* Novatome; Nucleare Italiana Reattori Avanzati (Italy). *Generator Supplier:* Ansaldo. *Architect Engineer:* Centrale Nucleaire Europeenne a Neutrons Rapides S.A. (NERSA).

Key Dates
Ordered: 1972. *Construction began:* 1977. *Criticality:* September 1985. *First Power:* January 1986. *Commercial Operation Expected:* indefinite.

Plant Report

Theory becomes a reality. After three decades of preparation, Superphenix, the world's first commercial-scale Fast Breeder Reactor (FBR) prepared to begin operation under the auspices of the Centrale Nucléaire à Neutrons Rapides S.A. (NERSA).

As a nod of respect to the man who created and sustained the French FBR research and development program, NERSA officials advanced the opening by 48 hours to coincide with George Vendryés' 65th birthday.

The reactor is located at Creys-Malville, population 50, along the Rhône River near the Swiss border. The power station is also known as the Creys-Malville Nuclear Power Plant.

Challenge after challenge. The FBR program had survived difficult technology, mastering the handling of liquid sodium coolant. An early Soviet FBR self-destructed when sodium mixed with water, but French designers crafted a detection system that efficiently safeguarded against the problem.

The FBR survived political ambivalence when Vendryés gained the support of General de Gaulle for a cooperative European effort despite de Gaulle's reluctance to involve France with other European countries.

It survived antinuclear protests that included often-violent mass demonstrations. In 1977, a confrontation between 4,000 heavily-armed police and 30,000 demonstrators at the Superphenix construction site, resulted in one man dead and another hundred injured. In 1982, during construction of the plant, nuclear-opponents fired anti-tank rockets at the buildings.

However, with the opening of Superphenix at Creys-Malville, the FBR technology faced its biggest challenge.

Closing the fuel cycle. According to its proponents, the FBR would generate electricity more efficiently than conventional nuclear reactors. Its higher core temperature would mean greater thermal efficiency with more-even fuel burn. By recycling the fuel through the core several times, the reactor would utilize a charge of uranium 600 times more efficiently than the conventional reactor. In doing so, it would create plutonium which could be extracted and recycled into fuel, closing the fuel cycle and making France self-sufficient in energy.

President Valery Giscard d'Estaing bragged on the fast-breeder as a "vein of energy comparable to Saudi Arabia's oil reserves." He and other officials claimed a fast-breeder reactor program would make the country energy-independent because of their ability to feed on plutonium—the waste produced by France's conventional pressurized water reactors—and because it "fast-breeds" more plutonium than it consumes. Socialist deputy Paul Quiles and other government officials called the boast a "gigantic hoax."

However, in its first few years of operation, Superphenix generated electricity at a cost 2.2 times that of a pressurized water reactor (PWR).

In part, that was because France operated its PWRs very efficiently, at a cost 35-to-40 percent less than similar reactors in the United States. The use of standardized designs in bulk orders pushed construction costs down for the PWR, but Superphenix was essentially a prototype. The next FBR French designers have in mind should cost 30 percent less, using a less stringent containment structure.

However, first they have to get the cost of operating Superphenix down, says Pierre Delaporte, chairman of Electricité de France (EDF). In 1987, he insisted the reactor would have to operate competitively with French PWRs within the next three-to-five years, if the program is to continue, with or without international support.

A British radioprotection report predicted that an accident which resulted in a 5 percent leak from the core of the Superphenix would cause a 100 percent death rate within a three-kilometer radius, 60 percent within five kilometers, and that within 10 kilometers 16 percent of the population would eventually die of cancer.

Early experiments. For France, the project began with the experimental reactor, Rapsodie, at the Cadarache Center in Provence in 1967. That initial investigation into the use of sodium as a coolant prompted the construction of the 250-megawatts electric prototype, Phenix in 1968. That plant, located at Marcoule in the South of France, went critical for the first time in August, 1973, beating the similar British project at Dounreay by almost a year.

The Dounreay project and the German effort at Kalkar paralleled the French development. On January 10, 1984, Great Britain, Germany and France agreed to pool their efforts in perfecting the European Fast Breeder Reactor.

Under the terms of the agreement, each country would build a demonstration plant. France already had the Superphenix nearing completion, and intended to build six replicas in France and a seventh in Germany. However, the growing strength of the antinuclear forces in Germany and Britain and the escalating cost of reactor development prompted both countries to abandon their FBR programs.

Germany still wanted the next generation FBR, the 1,800-megawatts electric Hyperphenix, but France had the advantage of a more nuclear-friendly political climate. If built, the most likely place for the next FBR would be at Marcoule. Britain opted for developing the reprocessing technology needed to close the fuel cycle.

Uses of Plutonium. Not closing the fuel cycle would please many scientists. They fear that breeder reactors will destroy the effectiveness of nuclear non-proliferation treaties by increasing the availability of weapons-grade plutonium.

The question of the ultimate use of the plutonium produced in the Superphenix arose shortly before the reactor began operation. The actual core contains both enriched uranium-238 and 5,500 kilograms of plutonium. A blanket of uranium surrounds that mass to catch the fast-moving neutrons as they leave the core. The impact of the neutrons changes the uranium into plutonium. Over one year of operation, Superphenix produces 200 kilograms of plutonium-239. That plutonium could be used to make 60 atomic bombs if the French military chose to do so.

When questioned about their intentions, French officials indicated they had no plans to make weapons out of the material, but reserved the right to do so if they wished. Member countries of NERSA, which provided plutonium to the original core, are entitled to a portion of the material produced in the core, proportional to their original contribution.

For EDF, the argument revolves around efficiency. The longer it leaves the plutonium in the core, the more efficiently the reactor operates. Long burn-times mean less fuel changes and less cost. They also mean a deterioration of the weapons-quality plutonium into plutonium-240, making it less valuable for the military. Unless over-ridden by the military, EDF will maximize the reactor's electricity production.

Fine tuning technology. The plant still experiences technical problems. A major sodium leak began draining 800 liters of coolant through a 6-inch crack daily in April 1987. Eventually, the plant went into a maintenance shutdown to rectify the problem. That may prove difficult and expensive. EDF officials estimate a major redesign of the faulty storage tank could cost $15 million.

In fact, commercial operation of the Superphenix has been indefinitely postponed pending completion of studies on the safety of operation without the availability of the fuel storage vessel.

Electricity production in 1989 was 2 billion kilowatt-hours; 1990 production was projected at 4.5 billion kilowatt-hours.

Sources

"France's Nuclear Energy Boom." *World Press Review*, March 1980.
"Off and Running in a Nuclear Race." *Maclean's*, October 20, 1980.
"French Military Plans for Superphenix?" *Bulletin of the Atomic Scientists*, November 1984.
"Breeder Reactor Politics in Europe." *Bulletin of the Atomic Scientists*, May 1986.
"France Weighs Benefits, Risks of Nuclear Gamble." *Science*, August 29, 1986.
"Despite Superphenix Startup, Outlook for Breeders is Poor." *Physics Today*, September 1986.
"Vendryés, pioneer in breeders, regrets their cost." *Physics Today*, September 1986.
"Superphenix Springs a Leak." *Science*, April 17, 1987.
"Slowdown for French Breeders?" *Science*, October 1987.
Mounfield, Peter. *World Nuclear Power*. Routledge, 1991.

65 Tricastin

"Vendryés, pioneer in breeders, regrets their cost." *Physics Today*, September 1986.
"Superphenix Springs a Leak." *Science*, April 17, 1987.

—*Al Cook*

65 ●
Tricastin Nuclear Power Plant, Units 1-4

BP 9
F-26130 St. Paul Trois Chateaux, France
Michel Andrieux, Site Mgr.

Phone: 75-50-39-99
Fax: 75-96-84-20

Owning Utility

Electricite de France
2 rue Louis Murat
F-75005 Paris, France
Christian Nadal, Dir., Commun.

Phone: 1-42-56-94-00
Telex: 280041

Contact

Jacques Michard, Pub.Info.
Tricastin Nuclear Power Plant

Basic Facts

Units 1-4 *Status:* Operable. *Type of Reactor:* Pressurized light-water reactor. *Megawatts (electric):* 955. *Megawatts (thermal):* 2785. *Reactor System Supplier:* Framatome (France). *Generator Supplier:* Alsthom. *Architect Engineer:* Electricite de France.

Key Dates

Unit 1 *Ordered:* 1974. *Construction began:* 1974. *Criticality:* February 1980. *First Power:* May 1980. *Commercial Operation:* December 1980.

Unit 2 *Ordered:* 1974. *Construction began:* 1974. *Criticality:* July 1980. *First Power:* August 1980. *Commercial Operation:* December 1980.

Unit 3 *Ordered:* 1973. *Construction began:* 1974. *Criticality:* November 1980. *First Power:* February 1981. *Commercial Operation:* May 1981.

Unit 4 *Ordered:* 1973. *Construction began:* 1974. *Criticality:* May 1981. *First Power:* June 1981. *Commercial Operation:* November 1981.

Germany

COUNTRY REPORT

Germany's entry into nuclear world delayed by Allies

The terms of the peace treaty which ended World War II restricted Germany's use of nuclear power until 1956. After a decade-long hiatus, German scientists resumed their wartime research into heavy-water-moderated nuclear energy. However, the first commercial stations used American light water designs, a trend which continued throughout Germany's nuclear construction program.

Commercial utilization of the atom began in 1968 with two demonstration plants, the 254-MWe Lingen boiling water reactor and the 357-MWe Obrigheim pressurized water reactor. With these early successes, the German nuclear industry moved on to two commercial 600-MWe plants which started operations in 1972.

In the early seventies, the German government saw nuclear power as the energy of the future. It predicted a one percent growth in market share for the technology over the next fifteen years giving nuclear power a 45-percent share of installed electrical generating capacity. To meet that expected demand, the government proposed building a fleet of new reactors with a capacity of 45,000-to-50,000 MWe by 1985. However, diminishing demand over that time period reduced the optimistic proposals first to 30,000 MWe and then to 24,000 MWe or 29 percent of installed capacity. As 1985 approached, Germany found itself with fifteen operating reactors with a capacity of 10,358 MWe. On the drawing boards, electrical utilities had another 12 plants with an additional capacity of 13,155 MWe. The industry peaked in 1989, with 22 commercial plants producing 23,919 MWe.

East Germany uses radical new Soviet technology

In East Germany, the state-run industry concentrated on Soviet-designed and built pressurized water reactors, beginning with the 70 MWe Rheinsberg facility in 1966. Between 1971 and 1979, the Soviets built four reactors at Lubmin near Greifswald. A second building program called for four more PWRs at Lubmin and another at Stendal, but German reunification in 1990 imposed Western safety standards on the East German facilities. By the end of the year, all ten reactors were moth-balled, leaving East Germany with no nuclear power generation.

German industry demands new source of cheap fuel

German industrial growth in the first half of the twentieth century outstripped the local supply of hydro-electric and lignite-fueled plants. In the 1960s, Germany thought it had found a source of cheap, plentiful energy in the abundant supplies of oil from the Middle East. However, the OPEC oil embargoes of the 1970s made clear the need for energy self-sufficiency. To that end, the German government promoted the use of nuclear power using imported uranium. It intended to close the fuel-cycle with a commercially viable Fast Breeder Reactor by the end of the century.

That justification vanished in the late 1980s as advanced safety designs pushed the cost of nuclear energy steadily up, while the price of fossil fuels, especially lignite, fell.

However, in the 1990s, the nuclear industry discovered new life with growing public concern over pollution and global warming. In 1991, the use of nuclear plants reduced carbon dioxide (CO_2) emissions by 147 million tons according the industry's Nuclear Forum. Since 1961, the industry has prevented the release of 1.6 billion tons of CO_2. Responding to public concern, the German government announced in 1990 that it would seek an over-all CO_2-reduction of 25-30 percent by the year 2005. That would require a continued commitment to nuclear energy. In 1990, nuclear energy supplied 38 percent of West Germany's electricity demand.

East Germany Europe's worst polluter

By comparison, before re-unification, East Germany generated only 10 percent of its electricity with nuclear power. That has now dropped to zero with the closing of all reactors in the new federal states. Instead lignite, (brown coal), provided 80 percent of all electrical generation. The result was an inefficient energy sector which made East German air, water, and soil the most polluted in Europe. In 1989, East Germany emitted 15 times more CO_2 than West Germany. With the abandonment of its nuclear program, East Germany is mining even more lignite. In 1988, before closing its reactors, it took 318 million tons, or one third of the total international usage of lignite.

German city-states create fractured utility structure

The structure of Germany's energy distribution and generation industry can best be described as a series of interconnected islands. Rather than one national governing body, the country has close to 700 utilities serving cities, towns or even individual factories and railways. Between these islands stretch the 380-kV, bridges of the "Nine first-level utilities." Most generating capacity and all nuclear plants lie under the control of the Nine.

The electrical agreements of August 20, 1990, created a tenth first-level utility in East Germany. Bayernwerk, RWE Energie, and PreussenElektra split 75 percent of Vereinigte Energiewerke AG (VEAG). As in the west, the new distribution company will supply electricity to local utilities to supplement their own production. Until VEAG can build sufficient generating capacity in the East, it will supply power to the eastern grid by means of an immense extension cord-like transmission link from the western grid. The utility-owned 380-kV transmission grid forms a part of the pan-European grid system.

Nuclear opposition promotes industry standardization

Mirroring the fractured nature of Germany's energy industry, is its extremely complex licensing procedure for new nuclear plants. Each of the eleven individual states approves specific site-choices, which must then be sanctioned by the federal Ministry of the Interior. The Ministry also reviews each application for technical and environmental merit. The resulting overlapping regulations and administrative guidelines provide ample opportunities for environmentalists and citizens' groups to use the court system to stall targeted nuclear projects. Between 1977 and 1982, the strategy worked so well it effectively created a moratorium on nuclear power construction.

In response, the federal government launched what it called the convoy system. Under the scheme, the Ministry now approves drawings for a standardized plant which can be used without re-examination on any suitable site. That has forced anti-nuclear activists to seek other judicial impediments and to continue with massive, and often violent, demonstrations.

For the industry, the success of the strategy has limited the development of reactor technology. Despite success with new, safer and more efficient designs, the utilities are reluctant to depart from the standardized 1300-MWe PWR.

German utilities improve efficiency of American design

The design for that basic convoy-class PWR came from the United States, but German reactors operate significantly more efficiently than their American cousins. German PWRs consistently deliver 80 percent of their rated output, while similar American plants lag around 57 percent. Experts cite the more-spacious German containment structures as one reason for the difference. In order to reduce building costs, American utilities construct cramped reactor buildings, adding to maintenance costs and down-time later because of the lack of working-room.

By German law, no nuclear plant can operate without a viable method of dealing with its nuclear waste. Each plant can store up to ten years worth of spent fuel rods in cooling ponds located inside its containment dome. After that, the utilities use special steel flasks to transport the rods to reprocessing centers in France and Great Britain.

Originally, Germany intended to open its own reprocessing center at Wackersdorf in Bavaria. However, massive public protest, the falling value of uranium and plutonium, and the rising cost of reprocessing technology conspired to scuttle the project in 1988, after an investment of 2,000 million DM. With it went the Fast Breeder concept.

Now, the industry will opt for longer dwell times for fuel rods. Instead of leaving the uranium-235 inside the core for three years, new technology will allow it to remain safely for five years. As a result a higher percentage of the uranium burns, making reprocessing unnecessary. Instead, a new facility will condition waste material to make it more compact and then pack it into specially designated direct storage sites.

In 1993, the Konrad Iron Mine may become the first direct-disposal site. In preparing Konrad, researchers used the former-salt-mine, Asse, near Remlingen, as a test project, storing low- and medium-activity waste in barrels in its tunnels between 1965 and 1978.

Ultimate waste disposal site

Disposal experts see the salt-mines of Gorleben as a perfect place for their conditioning plant and ultimate storage site. However, nuclear opponents do not agree. Protestors battled police periodically with massive, violent demonstrations in 1977 and 1984, without convincing the government to abandon Gorleben.

In 1988, a bribery and illegal-practices scandal rocked the nuclear waste-disposal industry. Hints of $540,000-worth of bribes by employees of Transnuklear, Germany's only nuclear waste transport company, lead federal prosecutors to even more serious allegations. Eventually, they located 2,438 drums of nuclear waste containing traces of highly toxic plutonium-239 and cobalt-60. One informant testified that bomb-grade nuclear fuel had been shipped to Pakistan and Libya. The government suspended the company's operating license pending the outcome of the investigation, halting all shipments of fuel and waste for Germany's nuclear plants.

Sources

"West Germany: Can a Nuclear Convoy' Run Over Its Opposition?" *Business Week*, August 9, 1982.

"The German Reactor Advantage." *Reporter*, April 1985.

"Nuclear Industry Under Fire in Bonn." *The New York Times*, February 7, 1988.

"Anti-Nuclear Emotions Raised by Bavarian Reprocessing Plant." *Nature*, May 5, 1988.

GKN Annual Report 1988. Gemeinschaftskernkraftwerk Neckar GmbH, Neckarwestheim, 1989.

Nuclear Waste Management. Physikalisch-Technische Bundesandstalt, Berlin, 1989.

Mounfield, Peter. *World Nuclear Power.* Routledge, 1991.

"German Nuclear Plants cut CO_2 Emissions 15 Percent." *INFO.* U.S. Council for Energy Awareness, February 1992.

—Al Cook

PLANT PROFILES

Biblis Nuclear Power Plant, Units A-B

Kernkraftwerke Biblie
Postfach 11 40
W-6843 Biblis, Germany
K. Distler, Stat.Mgr.

Phone: 0-62-45-2-11
Fax: 0-62-45-21-3180

Owning Utility

RWE Energie AG
Kruppstrasse
Postfach 103165
W-4300 Essen 1, Germany
K. Kuhnt, Chairman of the Mgmt. Board

Phone: 02-01-185-1
Fax: 02-01-185-4313
Telex: 857851 Energ

Contact

H. Kirschke, Mgr. for Business Administration
Biblis Nuclear Power Plant

Basic Facts

Units A-B *Status:* Operable. *Type of Reactor:* Pressurized light-water reactor. *Megawatts (electric):* Unit A, 1204; Unit B, 1300. *Megawatts (thermal):* Unit A, 3517; Unit B, 3733. *Reactor System Supplier:* Kraftwerk Union AG (Germany). *Generator Supplier:* Kraftwerk Union AG (Germany). *Architect Engineer:* Kraftwerk Union AG (Germany); Hochtief AG (Germany).

Key Dates

Unit A *Ordered:* 1969. *Construction began:* 1970. *Criticality:* July 1974. *First Power:* August 1974. *Commercial Operation:* February 1975.

Unit B *Ordered:* 1971. *Construction began:* 1972. *Criticality:* March 1976. *First Power:* March 1976. *Commercial Operation:* January 1977.

Plant Report

German reactors get bigger. In 1974, RWE Energie commissioned Germany's first nuclear power station in the over-1000 megawatts electric class, Biblis A. Two years later, it added the slightly larger Biblis B. The two stations, located in the state of Hesse near Mannheim, produce a net output of 2,386 megawatts electric between them. The site sits on the Rhine River about 80 km from the French border.

In the period between the commissioning of Biblis A and the addition of Biblis B, Germany's official attitude towards the utilization of nuclear power took a severe turn. In November, 1975, the Free Democratic faction of the ruling Social Democratic Government convinced Chancellor Helmut Schmidt to refuse new building permits for nuclear reactors until the government devised a final disposition for nuclear waste. Schmidt also required future calculations of the price of electricity to include the full cost of the entire fuel cycle, including decommissioning of shutdown plants. Safety first became the new watchwords, taking precedence over economic development.

Antinuclear groups. Mounting opposition to nuclear power and the growth of the German Green Movement became apparent at Biblis on May 25, 1986. More than 20,000 gathered outside the station to protest nuclear power in the wake of the Chernobyl accident. An opinion poll taken on that day showed that 83 percent of Germans were opposed to the expansion of Germany's nuclear system.

Nuclear accident. One year later, in December 1987, the plant experienced a serious accident apparently caused by operator error. Attempts to hush up the incident failed when unnamed sources in the United States Nuclear Regulatory Commission leaked details to *Nucleonics Week* in December 1988. Those sources indicated that the accident would have triggered immediate response as a "top-level event" if it had occurred at an American plant. West German officials became aware of it only after a technical review in March 1988.

In the fifteen hours before the accident, operators failed to notice a warning light indicating an open valve between the primary and secondary cooling systems. Once they realized the problem, they attempted to close the valve by opening another, thus reducing pressure, which would allow the stuck valve to move. The result was the release of a small amount of radioactive steam at high pressure into piping designed for low pressure.

The valve did close without further incident, but the pipes could have burst at the point where they emerge from the containment structure. That could have drained the Pressurized Water Reactor of its critical coolant, causing a meltdown of the core.

Federal minister for the environment, Klause Toepfer, acknowledged the incident as serious, but denied there was ever any danger of meltdown. He indicated that German reactors experienced 305 safety-related incidents that year, 11 of which deserved "rapid attention."

The Hessen environment minister, Karlheinz Weimar, later accused RWE Energie of making a false report of the accident. The company logged the incident with the Organization for Economic Co-operation and Development (OECD) in Paris, but requested that it be kept secret.

Utility opposes government-ordered time-table. In March 1991, the Technical Control Association (TV) performed an extensive safety analysis of the two plants as ordered by the licensing authority. Section 17 of the German Atomic Energy Law requires RWE Energie to comply with all the recommendations of the report by the plant's 1993/94 inspection outage. However, the company has called the short time-frame unrealistic and has requested clarification of certain items. Additionally, it has brought a legal challenge against the directives.

RWE has commenced the installation of a new emergency safety system which should prevent operator-error incidents. The company expects to complete that system along with a new containment building filtration system by 1995 5**

In the meantime, it has applied for a change in the operating licenses of the plants to allow the use of Mixed Oxide fuel elements. The new fuel should reduce operating costs by allowing the company to use reprocessed material from COGEMA in

67 Brokdorf

France and British Nuclear Fuels Limited in Britain. RWE has reprocessing contracts with those companies until 2015.

In autumn 1991, Gorleben Interim Storage Facility will accept the first irradiated fuel elements from Biblis.

Sources

"West Germany Admits to 'Top-Level' Nuclear Accident." *New Scientist*, December 17, 1988.
Mounfield, Peter. *World Nuclear Power*. Routledge, 1991.
"Betriebsergebnisse der westdeutschen Kernkraftwerke 1990." *atomwirtschaft*, May 1991.
RWE Energie Annual Report 1990/91. Essen, September 6, 1991.

—Al Cook

67

Brokdorf Nuclear Power Plant

Postface
W-2211 Brokdorf, Germany
Helmut Verfuerth, Plant Mgr., Technical

Phone: 4829-750
Telex: 17482921

Owning Utility

Kernkraftwerk Brokdorf GmbH (KBR)
Hoerstener Str. 49
W-2000 Hamburg 50, Germany
Peter Hartmann, Utility Mgmt., Commercial

Contact

Wolfgang Kurtze, Plant Mgmt., Commercial
Brokdorf Nuclear Power Plant

Basic Facts

Status: Operable. *Type of Reactor:* Pressurized light-water reactor. *Megawatts (electric):* 1365. *Megawatts (thermal):* 3765. *Reactor System Supplier:* Kraftwerk Union AG (Germany). *Generator Supplier:* Kraftwerk Union AG (Germany). *Architect Engineer:* Kraftwerk Union AG (Germany). *Owner:* PreussenElektra AG; Hamburgische Electricitaets-Werke AG (HEW). *Operator:* Hamburgische Electricitaets-Werke AG (HEW).

Key Dates

Ordered: 1974. *Construction began:* 1981. *Criticality:* 1986. *First Power:* July 1986. *Commercial Operation:* December 1986.

Plant Report

Hamburg's newest reactor shared by utilities. PreussenElektra AG (PE) operates seven nuclear reactors. It shares the output from its newest and most modern plant, KBR Brokdorf, with Hamburgische Electricitaets-Werke AG (HEW) as part of a four-reactor operating agreement. Under that agreement, PE retains 80 percent of the electrical output of the 1,395 megawatts electric station, along with operating control. HEW feeds the other 20 percent into its Hamburg grid as part of the city's base load electricity requirement.

The four reactors, Brokdorf, Brunsbüttel, Stade, and Krümmel, contribute 10,243 million kWh to HEW's integrated power and district heat distribution system. That represents about 82 percent of the city's electrical demand. The utility also employs a pumped-hydro electric station and thermal plants burning oil, gas, coal, and garbage. Many of the generating stations also heat water, which HEW pumps to over 270,000 apartments and 120,000 private homes, eliminating the need for individual furnaces. The 380-kV Hamburg grid forms a small part of the pan-European grid system stretching from North Cape to Sicily.

Plant returns Elbe River water cleaner than it found it. Each of the four reactors draws its cooling water from the Elbe River. However, HEW sends that water back to the river, cleaner than it found it. Its treatment plants extract most of the heavy metals dumped into the Elbe by industries upstream. As a final step, the utility injects oxygen before returning the water to the river.

PE considers Brokdorf's design as the model for modern safe nuclear systems. Completed in 1986, the pressurized-water reactor uses a series of containment devices. The core sits within a steel pressure vessel. A partitioned concrete dome forms a biological shield around the containment vessel and the plant's cooling ponds. Spent fuel rods remain immersed in water for one full year, reducing their radioactivity by 99 percent. Once "cooled" the rods travel in special containers to reprocessing plants in France and England.

The concrete dome is surrounded by a spherical steel shell. In case of a serious accident, resulting in a pipe rupture, the shell can hold the entire volume of the coolant system. Finally, another concrete dome encloses the entire structure to protect against external impacts like earthquakes and aircraft crashes.

Antinuclear protest stalls construction. The Brokdorf plant came very near to never being built. In November 1976, 2,000 militant antinuclear activists battled police for possession of the construction site. Police used water cannon and tear gas grenades to restore control. Twenty-six police and more than one hundred protesters suffered injuries during the fray. As a result of the intense media coverage of the incident, the Schleswig-Holstein Administrative Court ordered a halt to construction until the government came up with an effective radwaste disposal program, which took them until May 1983 to accomplish.

During 1990, PE shut Brokdorf down for safety up-grading to meet increasingly stringent government regulations for the nuclear industry. At that time, the utility installed new fire prevention systems. Even so, nuclear generation provided 57.9 percent of PE's output, and *Nucleonics Week* rated Brokdorf as the third highest producing plant in the world for December 1990.

Sources

PreussenElektra Strom. PreussenElektra Aktiengesellschaft, Hanover, 1989.
"Palo Verde Units Top Producers in U.S. for December." *Business Wire*, February 15, 1991.
PreussenElektra Annual Report 1990. Hanover, April 1991.
HEW in Focus. Hamburger Electricitätswerke AG, Hamburg, April 6, 1991.
HEW: Excerpt from Annual Report 1990. Hamburger Electricitätswerke AG, Hamburg, April 6, 1991.
"Betriebsergebnisse der westdeutschen Kernkraftwerke 1990." *atomwirtschaft*, May 1991.

Plant Profiles

Mounfield, Peter. *World Nuclear Power*. Routledge, 1991.

—Al Cook

68 •
Brunsbuettel Nuclear Power Plant

Otto Hahn-Strasse
W-2212 Brunsbuettel, Germany
Volker Brodale, Plant Mgr.

Phone: 04852-89-0
Fax: 04852-89-2012
Telex: 48-52-10

Owning Utility

Hamburgische Electricitaets-Werke AG (HEW)
Ueberseering 12
W-2000 Hamburg 60, Germany
Werner Hartel, Mgr. of Nuclear Div.

Phone: 040-6396-1
Fax: 040-6396-2770
Telex: 40-30-68

Contact

Robert Schulte, Pub.Info.
Hamburgische Electricitaets-Werke AG (HEW)

Basic Facts

Status: Operable. *Type of Reactor:* Boiling water reactor. *Megawatts (electric):* 806. *Megawatts (thermal):* 2292. *Reactor System Supplier:* Kraftwerk Union AG (Germany). *Generator Supplier:* Kraftwerk Union AG (Germany). *Architect Engineer:* Kraftwerk Union AG (Germany). *Owner:* PreussenElektra AG; Hamburgische Electricitaets-Werke AG (HEW). *Operator:* Hamburgische Electricitaets-Werke AG (HEW).

Key Dates

Ordered: 1969. *Construction began:* 1970. *Criticality:* June 1976. *First Power:* July 1976. *Commercial Operation:* February 1977.

Plant Report

Hamburg expands nuclear generating capacity. During the first half of this century Hamburgische Electricitaets-Werke AG (HEW) met the electricity needs of the city of Hamburg by burning coal. In 1972, Stade power station introduced a new technology utilizing uranium, which would quickly displace all fossil fuels. The experiment proved so successful, HEW decided to build a plant of its own. In 1976, KKB Brunsbüttel went on line. By 1990, nuclear energy produced 82.7 percent of Hamburg's electricity.

A four-reactor operating agreement gives PreussenElektra (PE) 34 percent of Brunsbüttel's 806-MWe output. HEW operates the plant and feeds the remaining output into its Hamburg grid as part of the city's base load electricity requirement.

Other power options. The four reactors, Brokdorf, Brunsbüttel, Stade, and Krümmel, contribute 10,243 million kWh to HEW's integrated power and district heat distribution system. The utility also employs a pumped-hydro electric station and thermal plants burning oil, gas, coal, and garbage. Many of the generating stations also heat water, which HEW pumps to over 270,000 apartments and 120,000 private homes, eliminating the need for individual furnaces.

The 380-kV Hamburg grid forms a small part of the pan-European grid system stretching from North Cape to Sicily.

Each of the four reactors draws its cooling water from the Elbe River. However, HEW sends that water back to the river cleaner than it found it. Its treatment plants extract most of the heavy metals dumped into the Elbe by industries upstream. As a final step, the utility injects oxygen before returning the water to the river.

Reactor uses recycled uranium. The boiling-water reactor uses a Mixed Oxide Fuel (MOX Fuel) made from natural uranium from Canada, Australia, and the United States enriched to 3 percent uranium-238. British Nuclear Fuels Limited in Britain recycles spent fuel rods to make the oxide second-generation fuel.

Brunsbüttel has sufficient storage capacity for all its nuclear waste until 2000. By then, the utility hopes the nuclear depository under construction at Gorleben will be operational.

Sources

Schroeder, Friedrich. *PreussenElektra Strom*. PreussenElektra Aktiengesellschaft, Hanover, 1989.
PreussenElektra Annual Report 1990. Hanover, April 1991.
HEW in Focus. Hamburger Elektrizitätswerke AG. Hamburg, April 6, 1991.
HEW: Excerpt from Annual Report 1990. Hamburger Elektrizitätswerke AG, Hamburg, April 6, 1991.
"Betriebsergebnisse der westdeutschen Kernkraftwerke 1990." *atomwirtschaft*, May 1991.

—Al Cook

69 •
Emsland Nuclear Power Plant

Kernkraftwerk Emsland
Am Hilgenberg
W-4450 Lingen, Germany
Wilke, Plant Mgr.

Phone: 0591-8060
Fax: 0591-806-25-49
Telex: 0591-98-897

Owning Utility

Kernkraftwerke Lippe-Ems GmbH
Rheinlanddamm 24
W-4600 Dortmund 1, Germany
Dieter Junge, Dir.

Phone: 0231-438-44-32
Fax: 0231-438-21-47
Telex: 0231-8-22-121

Basic Facts

Status: Operable. *Type of Reactor:* Pressurized light-water reactor. *Megawatts (electric):* 1328. *Megawatts (thermal):* 3850. *Reactor System Supplier:* Kraftwerk Union AG (Germany). *Generator Supplier:* Kraftwerk Union AG (Germany). *Architect Engineer:* Kraftwerk Union AG (Germany).

Key Dates

Ordered: 1982. *Construction began:* 1982. *Criticality:* April 1988. *First Power:* April 1988. *Commercial Operation:* June 1988.

Plant Report

Modern reactor aging plant. In 1977, Vereinigte Elektrizitätswerke Westfalen AG (VEW) faced a difficult decision. Its nine-year-old boiling water reactor, Emsland, faced excessively expensive upgrading costs to meet new safety regulations. Rather than trying to get the old plant past the new licensing requirements, the firm decided to build a new, more modern facility. However, they first had to get around the antinuclear lobby. In Germany, nuclear protest often took the form of public and sometimes violent demonstrations involving thousands of protestors and well-armed police.

Nuclear opponents also discovered they could hold up individual construction projects by using the German court system. Between 1977 and 1982, the antinuclear lobby effectively blocked all new reactor building-permits.

VEW's new Lippe-Ems (KLE) project, near the town of Lingen in Lower Saxony, became enmeshed in the legal process with little prospect of ever passing the planning stage.

Promoting nuclear construction. After five years of frustration, the federal government introduced a new strategy, the convoy system, to combat the stalling campaign. It involved designs for standardized plants that would only need to be examined once. When approved, they could be used for any new project on any suitable site without re-examination. Kraftwerk Union AG (KWU), the major German nuclear power plant builder, backs the concept fully, maintaining it reduces paperwork by 30 percent and repetitive engineering work by 25 percent.

Nuclear power opponents recognized the danger posed by the convoy system to their cause. "If we don't win on the first challenge, we are finished," said Ludwig Trautman, nuclear power critic actively opposed to Isar-2. They did not win on their first challenge, and the Isar-2 project proceeded, quickly followed by Lippe-Ems and Neckar GKN-2.

Each plant used an identical 1,300-MWe pressurized water design built by KWU.

To help defray the construction costs, VEW teamed up with Kommunales Elektrizitätswerk Mark AG (Elektromark), giving the local utility a 25 percent share of KLE.

Emission standards provide new weapon for nuclear opponents. However, the antinuclear lobby has not given up. At both Isar-2 and Lippe-Ems, they have issued new legal challenges, citing harmful emissions from the cooling towers, excessive water temperature increases in the plant's river-source of coolant water, and excessive emissions of chemicals and radiation into the rivers.

Lippe-Ems faced its first set of challenges on August 22, 1989, but the Administrative Court of Oldenburg dismissed the complaints in January. The complainant has asked leave to appeal. On February 9, 1990, KLE received official permission to increase the thermal output of its reactor by 2 percent to 3,859 MW. Electrical output and operating efficiency increased proportionately. According to *Nucleonics Week*, Lippe-Ems ranks as the world's most productive nuclear power plant, generating 1,086,447 MWH in December 1990.

With the abandonment of the Wackersdorf nuclear fuel recycling plant, KLE had to find alternative ways to deal with its waste products. German law requires all nuclear facilities to have firm feasible methods in place or cease operation. To that end, KLE entered into a long-term agreement with British Nuclear Fuels Limited in England. Under the contract, KLE will begin shipping spent fuel rods in 1994, and will continue to do so until 2006 with an option to extend the pact to 2015. By that time, the direct storage facility at Gorleben should be in full operation.

Sources

"West Germany: Can a Nuclear Convoy' Run Over Its Opposition?" *Business Week*, August 9, 1982.
Das Kernkraftwerk Emsland. Kernkraftwerke Lippe-Ems GmbH, Lingen, July 1990.
Mounfield, Peter. *World/Nuclear Power*. Routledge, 1991.
Bericht über das Geschäftsjahr 1990 KLE. Kernkraftwerke Lippe-Ems GmbH, Lingen, March 13, 1991.
"Palo Verde Units 1 & 2 Top American Producers in May; Unit 2 Continues Record Run." *Business Wire*, July 12, 1991.

—Al Cook

70 ●

Grafenrheinfeld Nuclear Power Plant

Postfach 7
W-8722 Grafenrheinfeld, Germany
Phone: 0-97-23-62-1
Telex: 622998

Owning Utility

Bayernwerk AG
Strom fur Bayern
Nymphenburger Strasse 39
Postfach 20 03 40
W-8000 Munchen 2, Germany
Phone: 089-12-54-1
Fax: 089-12-54-3906
Telex: 523-172-bagh-d

Basic Facts

Status: Operable. *Type of Reactor:* Pressurized light-water reactor. *Megawatts (electric):* 1300. *Megawatts (thermal):* 3765. *Reactor System Supplier:* Kraftwerk Union AG (Germany). *Generator Supplier:* Kraftwerk Union AG (Germany). *Architect Engineer:* Kraftwerk Union AG (Germany).

Key Dates

Ordered: 1974. *Construction began:* 1975. *Criticality:* December 1981. *First Power:* April 1982. *Commercial Operation:* June 1982.

Plant Report

Plant still generates opposition. In December 1991, the International Atomic Energy Agency, (IAEA), gave the 1300-megawatt (MW) Grafenrheinfeld nuclear power facility a birthday present. A little later that month, the environmental group, David Gegen Goliath (David versus Goliath) tried to hold a party in the plant's information center.

On the 10th anniversary of the start-up of the plant's reactor, the IAEA's Operational Safety Review Team (OSART) completed an inspection of the facility and awarded it high marks.

The nuclear experts from 10 member nations that made up the team pronounced the plant well-built and well-run. They noted that for the last five years the reactor's operating efficiency stayed above 85 percent. Apart from regularly scheduled maintenance, the station operated without interruption for the last three years. All personnel exhibited an honest concern for safety that pervaded their work ethic.

In its first 10 years, Grafenrheinfeld generated more than 90 million kilowatt hours of electricity, 10 million in 1991. Bayernwerk officials proudly claim the reactor's output has safely displaced more than 90 million metric tons of CO_2 emissions. In addition, the cost of that electricity has remained stable for six years. The last price increase came in 1983.

Despite these successes, the company denies the suggestion by environmentalists that it plans to construct a second reactor on the site before the end of this century.

The anti-nuclear lobby would like to see the original plant turned off immediately. They claim the plant threatens the neighboring cities of Schweinfurt and Wurzburg.

To make their point clear, the David Gegen Goliath group invaded the company's information center, waving placards and chanting slogans. Plant security removed the protestors, and the Bayernwerk management banned group members from plant property. A company spokesman defended the reaction, noting the center provides information for people legitimately interested in nuclear power generation and its ecological impact.

Ecology Institute paints fearful scenario. During Schweinfurt's 1991 review of disaster preparedness planning, Lothar Hahn of the Ecology Institute warned that Grafenrheinfeld could potentially create a catastrophe greater than Chernobyl. He claimed an explosion from a hydrogen bubble or a build-up of steam pressure could breach the containment structure releasing clouds of radioactive vapor over the city.

A spokesman for Bayernwerk dismissed the analysis, noting the advanced safety design and operation of the plant. He claimed modern safety standards made such explosions impossible and pointed to a risk study performed by Gesellschaft für Reaktorsicherheit (GRS) (The Society for Reactor Safety) as evidence.

German safety standards have challenged traditional reactor construction methods. Massive containment structures constitute the German nuclear industry's "technically and economically optimum solution" to those increasingly stringent safety regulations.

The familiar cylindrical concrete containment structure, lined with steel, sits within a second concrete reactor building. This provides an air-lock between the shields from which the operators can extract any contaminated gases or fluids.

Reactor can withstand impact of jet aircraft. The building itself improves the reactors ability to survive external impacts like earthquakes, bomb blasts, and aircraft crashes. In recent years, crashes of military jets near Greifswald and Karlsruhe raised concerns of possible disasters. In 1990, member countries of the North Atlantic Treaty Organization (NATO) based about 1,150 warplanes in Germany. They flew more than 68,000 hours of low-level training flights yearly.

In case of an accident, Kraftwerke Union AG—now Siemens AG—the plant's designers, provided two separate and automatic reactor trip mechanisms activated by a computerized protection system.

Boron-control rods, held by electromagnets above the core, can drop into position with the slightest break in control power.

As a back-up to the boron control rods, an emergency water-feed system can inject neutron-absorbing boric acid into the core coolant.

That system also protects against the Pressurized Water Reactor's most common accident, the loss of coolant incident. Low and high pressure injection pumps remove decay heat from the core until shutdown completes.

Meanwhile, the containment isolation system double-seals the structure from the outside environment with multiple sets of redundant valves in each pipe that penetrates the containment wall.

During normal operation, technicians exchange 20 percent of the reactors enriched uranium fuel rods each year. The reactor shuts down during the procedure, providing an opportunity for maintenance on other systems.

Storage of spent fuel rods. Spent fuel remains within the containment structure, immersed in cooling ponds for at least seven years. The pools can store up to 10 years worth of used rods. After that, Bayernwerk loads the material into special forged-steel flasks for shipment to reprocessing plants in France and Great Britain. With the abandonment of Germany's own reprocessing project at Wackersdorf, the industry may now move to direct storage in salt mines at Gorleben.

Outside the containment structure, two gigantic cooling towers process 9,200 cubic meters of filtered river water hourly in a closed loop system. Water drawn from the Main River replaces that lost by evaporation in the towers. With this system, 97 percent of the coolant recycles, reducing the demand on the Main River.

Nuclear power provides 66 percent of all electricity generated for the public grid in Bavaria. Grafenrheinfeld feeds its 380-kilovolt output into Bayernwerk's own grid system, providing power to industry throughout Bavaria. In addition, as part of the pan-European grid stretching from North Cape to Sicily, Grafenrheinfeld exchanges generating capacity with plants in other European countries. With the completion of the Nordring Grafenrheinfeld-Oberhaid-Redwitz extension of the high-voltage line, Bayernwerk has positioned itself to sell electricity into the former German Democratic Republic state, Thuringen.

Sources

Pressurized Water Reactor. Kraftwerk Union Aktiengesellschaft, Mulheim an der Ruhr, 1982.
Kernkraftwerk Grafenrheinfeld. Bayernwerk AG, Munich, November, 1988.
Geschäftsbericht 1989/90 Bayernwerk. Bayernwerk AG, Munich, January 31, 1991.
Muhlberger, Egon. "Pressemitteilung 35/91." Bayernwerk, Munich, July 1, 1991.
Muhlberger, Egon. "Pressemitteilung 51/91." Bayernwerk, Munich, September 27, 1991.

Greifswald

Geib, Armin. "Pressemitteilung 52/91." Bayernwerk, Munich, September 30, 1991.

Muhlberger, Egon. "Pressemitteilung 64/91." Bayernwerk, Munich, November 19, 1991.

Muhlberger, Egon. "Pressemitteilung 65/91." Bayenwerk, Munich, November 26, 1991.

Erwin, R. Ashley, and Brian M. Moore. "Pressemitteilung OSART-Untersuchung des Kernkraftwerks Grafenrhienfeld." International Atomic Energy Agency, Vienna, December 13, 1991.

"Reaktor Grafenrheinfeld läuft seit zehn Jahren." *Suddeutsche Zeitung*, December 19, 1991.

"Atomkraftwerk auf Prufstand." *Frankfurter Rundschau*, November 26, 1991.

"öko-Institut warnt vor Reaktor Grafenrheinfeld." *Suddeutsche Zeitung*, September 30, 1991.

Muhlberger, Egon. "Pressemitteilung 70/91." Bayernwerk, Munich, December 2, 1991.

"Reaktor Grafenrheinfeld läuft seit zehn Jahren." *Suddeutsche Zeitung*, December 19, 1991.

"Atomkraftwerk auf Prufstand." *Frankfurter Rundschau*, November 26, 1991.

"öko-Institut warnt vor Reaktor Grafenrheinfeld." *Suddeutsche Zeitung*, September 30, 1991.

—Al Cook

Greifswald Nuclear Power Plant, Units 1-4

O-2200 Greifswald, Germany

Owning Utility

Energie Werke Norde GmbH (EWN)
Vorstand
Postfach 35
O-2205 Lubmin, Germany
Ulrich Loeschhorn, Managing Dir.

Phone: 0082297-40
Fax: 0082297-2-24-71
Telex: 3981883z-ewn-d

Basic Facts

Units 1-4 *Status:* Shut down. *Type of Reactor:* Pressurized light-water reactor. *Megawatts (electric):* 440. *Megawatts (thermal):* 1375. *Reactor System Supplier:* Atomenergoexport (former USSR). *Generator Supplier:* Atomenergoexport (former USSR).

Key Dates

Unit 1 *Ordered:* 1966. *Construction began:* 1967. *Criticality:* December 1973. *First Power:* December 1973. *Commercial Operation:* July 1974. *Shut Down:* February 1990.

Unit 2 *Ordered:* 1967. *Construction began:* 1969. *Criticality:* December 1974. *First Power:* December 1974. *Commercial Operation:* April 1975. *Shut Down:* February 1990.

Unit 3 *Ordered:* 1972. *Construction began:* 1972. *Criticality:* October 1977. *First Power:* November 1977. *Commercial Operation:* May 1978. *Shut Down:* June 1990.

Unit 4 *Ordered:* 1972. *Construction began:* 1972. *Criticality:* July 1979. *First Power:* August 1979. *Commercial Operation:* October 1979. *Shut Down:* December 1990.

Plant Report

"Dangerous" reactors provide heat for thousands. Before reunification, the four 440-megawatts electric (MWe) reactors at Lubmin provided 10 percent of East Germany's electricity needs. They also supplied zone-heating water to 65,000 people in the Baltic coast city of Greifswald. However, Western experts placed the facility on a "blacklist of the world's especially dangerous nuclear installations."

The Soviet-designed VVER-440 model 230 reactor differs considerably from the Chernobyl style graphite-moderated system. For the VVER, Soviet researchers experimented with pressurized water (PWR).

Between 1971 and 1981, the Soviets built 14 VVER-440s in East Germany, Czechoslovakiaa, Bulgaria, and the Soviet Union.

They used an earlier, smaller pressurized light-water reactor design for Rheinsberg, East Germany in 1966.

Soviet reactors fail to meet Western standards. Western concerns over the VVER center on four main areas:

- No blast-hardened containment building. West German reactors use multiple-layered containment structures designed to withstand the impact of an aircraft crash and to hold the entire volume of the cooling system in case of a catastrophic leak.

- The use of a thermal annealing process to repair brittle reactor vessels. Every 25 years, technicians heat the vessel to 480 degrees Celsius for 120 hours to restore tensile strength to the steel. Western experts doubt the system works.

- Lack of redundant safety controls. Both main and auxiliary power cables run through the same duct work, making the entire system vulnerable to fire. No emergency water feed system exists, meaning a break in even the smallest cooling-water line could result in a core melt-down.

- Sloppy maintenance and poor management. In February 1990, a team from the Atomic Energy Agency in Vienna inspected Greifswald and reviewed its last two years of operation. Noting a total of 1,191 minor operating problems over that time period, the team assessed the operating staff as technically qualified and blamed management for the difficulties.

Reactor accident kept secret for 15 years. Earlier that year, the East German government revealed it had kept a potentially serious accident at Greifswald secret for more than 15 years.

In 1975, a duct-work fire destroyed both normal and emergency control cables for one of the reactors. That stopped the water feed pumps, threatening a loss-of-coolant incident, and a probable melt-down. Fortuitously, workers had previously connected a single water pump to an outside power source. That one pump supplied coolant for the reactor for six hours allowing repair crews to install temporary replacement lines. As reunification approached in 1990, the East Berlin government began to reluctantly agree to West Germany's demands to close some of the Greifswald reactors and abandon construction of four more at the site.

The first two reactor blocks closed for repairs in February, but never re-opened. A third closed in June and the fourth in December, once planners devised an alternate method of providing heat to Greifswald.

Under the German unity treaty, the Greifswald reactors were licensed to operate until June 30, 1995. With such a short time remaining, the Bonn Environment Ministry decided up-grading the VVERs to Western safety standards was economically unfeasible.

The other four reactors under construction at the site are a more modern pressurized light-water reactor design. The No. 5 block began its power build-up in 1989, but failed part way through the process. Construction already lagged three years behind schedule because of safety modifications.

In December 1990, Bonn halted construction of all four reactors at Greifswald and another at Stendahl, 75 miles west of Berlin.

In order to offset the loss of generating capacity, the newly-created East German energy authority hastily ordered construction of a $30 million oil-fired power plant and began running an enormous extension cord to link the East and West German grid systems.

On August 22, 1990, three major West German utilities joined forces to form Vereinigte Energiewerke AG (VEAG). Between them, PreussenElektra, Bayernwerk, and RWE Energie control 75 percent of the new grid company.

In early 1990, PreussenElektra began to supply electricity to the new Federal states through its just-completed 380-kilovolt Helmstedt-Wolmirstedt transmission line.

Greifswald explores new technology. Future use of nuclear power in the Eastern states remains uncertain with the growing influence of antinuclear sentiment in Germany.

However, officials at Greifswald continue to push ahead with nuclear programs. They hope to sell the four more-modern reactors to one of the utilities and gain the blessing of the German government to complete construction.

If they cannot achieve that goal, they have other plans as well. The site could become the home of a new experimental Fusion reactor.

Energie Werke Nord (EWN), the operators of the site—which is also known as Nord Nuclear Power Plant—entered into discussions with Euratom, proposing Greifswald as a possible location for a research fusion plant. International researchers built an early version at Culham, England.

EWN foresees employment for more than 2,000 workers in the project with construction stretching from 1994 to 2000. Another 1,000 would find work operating the reactor. That would help to offset the loss of 5,000 power station jobs that disappeared with the reactor closings and 5,000 construction jobs that vanished when work stopped on the four new reactors.

At Greifswald, the populous responded to the West German decree closing their plant with East Germany's first large scale public protest against West German policy.

Sources

"East Germany Discloses Serious Accident at Nuclear Plant in 1976." *The New York Times*, January 23, 1990.
"East Berlin Heeds Bonn and Shuts Nuclear Facility." *The Wall Street Journal*, February 16, 1990.
"East German Plants Face Western Tests On Nuclear Safety." *The Wall Street Journal*, February 20, 1990.
"Campaign to Shut Plant." *Nature*, March 22, 1990.
"East German Nuclear Shutdown." *The Wall Street Journal*, June 30, 1990.
"East Germany's Nuclear Plants." *The Wall Street Journal*, September 12, 1990.
"Germans to Shut 5 Atom Plants Built by Soviets." *The New York Times*, October 21, 1990.
"Last Soviet Reactor in Eastern Germany Shut." *The New York Times*, December 16, 1990.
Geschäftsbericht 1989/90 Bayernwerk. Bayernwerk AG, Munich, January 30, 1991.
PreussenElektra Annual Report 1990. PreussenElektra AG, Hanover, April 1991.
RWE Energie Annual Report 1990/91. RWE Energie AG, Essen: September 6, 1991.
Forshungsfusionsreakto am Standort des Kernkraftwerkes Greifswald. Energie Werke Nord, Greifswald, January 1991.
Betriebserfahrungen mit den Reaktoranlagen WWER-440/W-230. Energie Werke Nord, Greifswald, May 1991.
PreussenElektra Annual Report 1990. PreussenElektra AG, Hanover, April 1991.
RWE Energie Annual Report 1990/91. RWE Energie AG, Essen: September 6, 1991.
Forshungsfusionsreakto am Standort des Kernkraftwerkes Greifswald. Energie Werke Nord, Greifswald, January 1991.

—Al Cook

Greifswald Nuclear Power Plant, Units 5-8

O-2200 Greifswald, Germany

Owning Utility

Energie Werke Norde GmbH (EWN)
Vorstand
Postfach 35
O-2205 Lubmin, Germany
Ulrich Loeschhorn, Managing Dir.

Phone: 0082297-40
Fax: 0082297-2-24-71
Telex: 3981883z-ewn-d

Basic Facts

Units 5-8 *Status:* Indefinitely deferred. *Type of Reactor:* Pressurized light-water reactor. *Megawatts (electric):* 440. *Megawatts (thermal):* 1375. *Reactor System Supplier:* Atomenergoexport (former USSR). *Generator Supplier:* Atomenergoexport (former USSR).

Key Dates

Unit 5 *Ordered:* 1978. *Construction began:* 1980. *Criticality:* March 1989. *First Power:* April 1989. *Shut Down:* December 1990.

Unit 6 *Ordered:* 1978. *Construction began:* 1980. *Shut Down:* December 1990.

Unit 7 *Ordered:* 1978. *Construction began:* 1981. *Shut Down:* December 1990.

Unit 8 *Ordered:* 1978. *Construction began:* 1981. *Shut Down:* December 1990.

Plant Report

See Plant Report for Greifswald Nuclear Power Plant, Units 1-4.

| 73 | **Grohnde**

Grohnde Nuclear Power Plant

Postfach 12 30
W-3254 Emmerthal 1, Germany
Herbert Dittmar, Plant Supt.

Phone: 0-51-55-67-1
Fax: 0-51-55-67-23-80
Telex: 92891 KWG

Owning Utility

Gemeinschaftskernkraftwerk Grohnde GmbH
Soenke Albrecht, Utility Mgmt., Technical

Contact

Gerhard Alternberend, Utility Mgmt., Commercial
Grohnde Nuclear Power Plant

Basic Facts

Status: Operable. *Type of Reactor:* Pressurized light-water reactor. *Megawatts (electric):* 1365. *Megawatts (thermal):* 3765. *Reactor System Supplier:* Kraftwerk Union AG (Germany). *Generator Supplier:* Kraftwerk Union AG (Germany). *Architect Engineer:* Kraftwerk Union AG (Germany).

Key Dates

Ordered: 1975. *Construction began:* 1976. *Criticality:* August 1984. *First Power:* September 1984. *Commercial Operation:* February 1985.

Plant Report

Nuclear protestors. Grohnde KWG nuclear power plant, located near Hamien in Lower Saxony, began its existence amidst a storm of controversy and violence, although most of the confrontation centered on the planned Gorleben nuclear waste disposal site nearby.

In March 1977, militant protest groups, equipped with wire- and bolt-cutters, ropes, grappling irons, acetylene cutting torches, and aluminum foil to interfere with police radio-transmissions, invaded Grohnde's construction site. The police responded with gas grenades, water cannon, barbed wire, and attack-dogs. When the dust settled, over 300 people were injured, and Saxony Prime Minister Herr Ernest Albrecht announced he would allow Gorleben to proceed.

The violent confrontations continued, with more than 2,000 demonstrators blocking the access road to Gorleben in April 1984.

Meanwhile, construction proceeded on the 1300-megawatts electric (MWe) Pressurized Water Reactor at Grohnde. It began producing electricity for the public grid early in 1985.

Exceptional operating record. By 1990, the station had established itself as a reliable source of base-load electricity with an availability rate of about 98 percent. Early in the year, the plants operators, Gemeinschaftskernkraftwerk Grohnde, increased the thermal rating of the reactor by 29 megawatts. That brought the net electrical output to 1,325 MWe, with a gross output of 1394 MWe.

Internationally, the station rates as the second most productive plant in the world. Of the over 400 operating nuclear power reactors operating world wide, only Emsland outproduces Grohnde.

Sources

"Palo Verde Units Top Producers in U.S. for December." *Business Wire,* February 15, 1991.
"Betriebsergebnisse der westdeutschen Kernkraftwerke 1990." *atomwirtschaft,* May 1991.
Mounfield, Peter. *World Nuclear Power.* Routledge, 1991.

—Al Cook

| 74 | ▲

Gundremmingen Nuclear Power Plant, Unit A

Postfach 300
W-8871 Gundremmingen, Germany
Walter Reim, Plant Mgr.

Phone: 08224-78-1
Fax: 08224-78-2900
Telex: 531143

Owning Utility

Kernkraftwerke Gundremmingen Betriebsgesellschaft mbH (KGB)

Contact

Albrecht Schonder, Supt., Commercial Tasks
Gundremmingen Nuclear Power Plant

Basic Facts

Unit A *Status:* Decommissioned. *Type of Reactor:* Boiling water reactor. *Megawatts (electric):* 250. *Megawatts (thermal):* 801. *Reactor System Supplier:* General Electric Co. (United States); Ruhrstahl. *Generator Supplier:* Telefunken.

Key Dates

Unit A *Ordered:* 1962. *Construction began:* 1962. *Criticality:* August 1966. *First Power:* December 1966. *Commercial Operation:* April 1967. *Shut Down:* January 1980.

Plant Report

See Plant Report for Gundremmingen Nuclear Power Plant, Units B-C.

| 75 | ●

Gundremmingen Nuclear Power Plant, Units B-C

Postfach 300
W-8871 Gundremmingen, Germany
Walter Reim, Plant Mgr.

Phone: 08224-78-1
Fax: 08224-78-2900
Telex: 531143

Owning Utility

Kernkraftwerke Gundremmingen Betriebsgesellschaft mbH (KGB)

Plant Profiles

Contact

Albrecht Schonder, Supt., Commercial Tasks
Gundremmingen Nuclear Power Plant

Basic Facts

Units B-C *Status:* Operable. *Type of Reactor:* Boiling water reactor. *Megawatts (electric):* Unit B, 1300; Unit C, 1308. *Megawatts (thermal):* 3840. *Reactor System Supplier:* Kraftwerk Union AG (Germany). *Generator Supplier:* Kraftwerk Union AG (Germany). *Architect Engineer:* Kraftwerk Union AG (Germany); Hochtief AG (Germany).

Key Dates

Unit B *Ordered:* 1974. *Construction began:* 1976. *Criticality:* March 1984. *First Power:* March 1984. *Commercial Operation:* July 1984.
Unit C *Ordered:* 1974. *Construction began:* 1976. *Criticality:* October 1984. *First Power:* November 1984. *Commercial Operation:* January 1985.

Plant Report

Sylvan setting provides siting for nuclear series. The river Danube flows through the woodlands and farms of Guenzburg. In that region, at a point roughly equidistant from Munich, Nurnburg, and Stuttgard, the town of Gundremmingen hosts three nuclear reactors and gives them their names.

The first reactor built on the site began operation in 1966. The 250-megawatts electric (MWe) boiling water reactor began decommissioning in 1980, as Kraftwerk Union AG (now Siemens AG) prepared to bring units B and C on line. After eight years of construction, both plants started generating electricity in 1984.

The new Gundremmingens dwarf the A-station, both physically and in terms of electrical production. Generating 1300 megawatts electric each, Blocks B and C fill the 35-hectare site with a new style of nuclear power station construction. The new buildings demonstrate the advancement of the boiling water-reactor design. Gundremmingen A was Germany's second such reactor. The station's new containment buildings protect the ninth and 10th.

Nuclear industry responds to stiffening regulations. Those containment structures constitute the German nuclear industry's "technically and economically optimum solution" to increasingly stringent safety regulations.

The familiar cylindrical concrete containment structure, lined with steel, sits within a second concrete reactor building. This provides an air-lock between the shields from which the operators can extract any contaminated gases or fluids.

The building itself improves the reactor's ability to survive external impacts like earthquakes, bomb blasts, and aircraft crashes. In recent years, crashes of military jets near Greifswald and Karlsruhe raised concerns of possible disasters. In 1990, member countries of the North Atlantic Treaty Organization (NATO) based about 1,150 warplanes in Germany. They flew more than 68,000 hours of low-level training flights yearly.

A new addition to the architecture features two gigantic cooling towers processing 9,200 cubic meters of filtered river water hourly. Previously, the plant used an open system, extracting water from the Danube by means of a one-kilometer long canal. The heated water returned to the river without any attempt at cooling. Now the plant cools both new reactors and the support buildings with a closed loop system. The water drawn from the river replaces that lost by evaporation in the towers.

Two utilities share output and up-grading of nuclear facility. RWE Energie operates Gundremmingen in conjunction with its minority partner, Bayenwerk AG. RWE holds a 75 percent interest in the plant.

In 1990, the two companies engaged in extensive back-fitting and up-grading of safety systems to meet current regulations in Germany's Atomic Energy Law.

The companies improved the filtering system for the containment structure pressure relief system. In the future, they will add an additional decay-heat removal system.

Consequently, nuclear power's share of electrical generation dropped that year. Even so, Gundremmingen-B maintained an availability ratio of 84 percent, earning it 13th spot in *Nucleonics Week*'s list of the world's top producers with 968,937 megawatt hours generated.

Sources

Gundremmingen Nuclear Power Plant Units B and C. Kraftwerk Union AG, Mulheim an der Ruhr, 1982.
"Stillegung und Beseitigung von Kernkraftwerken - Betrieb auch ohne Stromerzeugung." *Atomkern-Energie Kerntechnik,* April 1990.
"Jet Accident Revives Controversy Over Flights." *The Washington Post,* April 19, 1990.
Kernkraftwerk Gundremmingen Block B und C. Siemens AG, January 1991.
Kernkraftwerke Gundremmingen RWE - Bayernwerk. Kernkraftwerke Gundremmingen Betriebsgesellschaft mbH, February 10, 1991.
"Palo Verde Units Top Producers in U.S. for December." *Business Wire,* February 15, 1991.
Geschaftsbericht 1989/90 Bayernwerk. Bayernwerk AG, Munchen, March 18, 1991.
RWE Energie Annual Report 1990/91. Essen, September 6, 1991.

—Al Cook

Isar Nuclear Power Plant, Units 1-2

Kernkraftwerk Isar
W-8307 Essembach, Germany

Phone: 08702-99-0
Fax: 08702-992461
Telex: 870282 KKIESS

Owning Utility

Kernkraftwerk Isar GmbH (KKI)

Basic Facts

Units 1-2 *Status:* Operable. *Type of Reactor:* Unit 1, Boiling water reactor; Unit 2, Pressurized light-water reactor. *Megawatts (electric):* Unit 1, 907; Unit 2, 1390. *Megawatts (thermal):* Unit 1, 2575; Unit 2, 3765. *Reactor System Supplier:* Kraftwerk Union AG (Germany). *Generator Supplier:* Kraftwerk Union AG (Germany). *Architect Engineer:* Kraftwerk Union AG (Germany).

Isar

Key Dates

Unit 1 *Ordered:* 1971. *Construction began:* 1972. *Criticality:* November 1977. *First Power:* December 1977. *Commercial Operation:* March 1979.

Unit 2 *Ordered:* 1980. *Construction began:* 1982. *Criticality:* January 1988. *First Power:* February 1988. *Commercial Operation:* April 1988.

Plant Report

Bavaria develops nuclear power. Bavaria, the industrialized southeastern region of Germany, has long had a need for energy. However, its own local sources of hydro-electric sites and lignite coal quickly proved inadequate. By the 60s, Bavaria imported vast quantities of oil, but the OPEC oil embargoes of the early 70s forced a reassessment.

With the experience of Germany's first nuclear plant, the 15-megawatt (MW) Kahl experimental station built in 1961, Bavaria saw a future in atomic power.

That success led to the 250-MW Gundremmingen pilot plant in 1966, followed in 1977 by the 900-MW Isar-1 near the village of Ohu. Eventually, these early boiling-water reactors evolved into the standard 1300-MW design used by Kraftwerk Union AG (now Siemens AG) for Gundremmingen B and C. For Isar 2, Siemens adopted the American Pressurized Water Reactor (PWR) and added the extensive German multiple-containment system. The 1,390-MW plant began generating electricity in 1988.

Experimenting with original German technology. In 1972, Germany's nuclear industry experimented with a heavy-water gas-cooled design. The 100-MW Niederaichbach prototype, built on the Isar site, operated for only two years, closing in 1974. Once fully decommissioned, Niederaichbach will be the first nuclear reactor in Germany to be completely dismantled.

By 1989, nuclear power sources provided 66 percent of all electricity generated for the public grid in Bavaria. By reducing dependence on oil-fired and gas-fired plants, Bavaria's major utilities provided the region's industries with the lowest-priced electricity in the Federal Republic of Germany.

Cooperative effort creates nuclear enterprise. In 1971, two of Germany's oldest public utilities, Bayernwerk AG and Isar-Ampwerke AG, joined forces as equal partners to create Kernkraftwerk Isar Gmbh. That company commissioned and operates Isar-1. owns, builds, and operates large power plants using both conventional and nuclear power sources. In addition, it supplies electricity to regional utilities through its high-voltage grid system. The 380-kilovolt line forms a portion of the pan-European grid stretching from North Cape to Sicily.

Isar-Ampwerke AG dates back even further. The regional utility grew out of the 1955-union of Isarwerke AG, founded in 1894, and Amperwerke Elektrizitaets-AG, founded in 1908.

For Isar-2, the original partners teamed up with Stadwerke Muenchen and Energieversorgung Ostbayern (OBAG). Bayernwerk AG maintained a 40 percent interest in the new plant and functions as general coordinator.

The two new partners operate as regional utilities. OBAG serves most of upper Bavaria while Stadwerke Muenchen supplies the city of Munich.

Consortium breaks through antinuclear wall. Granting the consortium a construction licence broke a five-year-long moratorium on nuclear power plants created by antinuclear lobbies. Citizens' groups and environmentalists had used the courts to tie up applications for years, but a new convoy system made that more difficult.

Now a series of projects using identical reactors gained approval at one time instead of individual sites.

Both Isar 1 and 2 draw cooling water from the Isar River, but utilize it differently.

Isar-1's design calls for direct cooling by the river. Only on exceptionally hot days, when oxygen levels drop in the water, do the plant's operators call on the reactor's cooling towers for assistance. Then powerful fans kick in to disperse the waste reactor heat through a bank of 12 towers. Under its 1967 license, granted under the German Law of Water Rights, Isar-1 can increase the water temperature in the river by 5 degrees Celsius. With the cooling towers, even including the hottest days, the actual heat rise averages about 2.5 degrees Celsius.

For Isar-2, the designers built a single large cooling tower, to operate continuously, rather than relying on the river. This system uses natural-draft to create the distinctive white plume of condensing water. A closed loop recirculates the coolant through the reactor's boilers. Fresh water drawn from the Isar replaces what is lost as steam from the tower.

Not everyone believes the system works, or that it does not create other problems than temperature rise.

Nuclear opponents try new tactic. Throughout 1991, the antinuclear group, Burgerforum gegen Atomkraftwerken, campaigned to convince the Landshut city council to deny the renewal of Isar-2's water rights. The group gained support from the cities of Dingolfing and Landau, the town of Wallersdorf and sport-fishing association, Angelsportverein Landshut/Bayern.

They argued that the plant exceeded the allowable temperature increase by 320 percent in August 1990. Company spokesmen responded by noting the water-rights license allows the plant to exceed the restricted temperature rise for as long as 30 days during emergency or regularly scheduled repairs. During normal operations, only Isar-1 affects the temperature of the river since Isar-2 does not return coolant to the Isar.

The anti-nuclear group also accused the power facility of dumping ammonia and ammonium nitrogen (NH4N), in excess of the allowable limit of 10 mg/liter. In April 1990, the chemical discharge reached 15 mg/liter according to Burgerforum. Resulting changes in the PH balance of the river water could damage fish stocks.

However, power plant officials deny excessive dumping has occurred. They note that the company monitors the PH level of the river and, each year, injects 50 kilograms of sulfuric acid and 2 kilograms of caustic soda to maintain a proper balance.

Isar-2 rates as one of the world's top producing nuclear power plants. In its July 11, 1991 list, *Nucleonics Week* placed it second in the world behind another German plant, Brokdorf.

Sources

"West Germany: Can a Nuclear 'Convoy' Run Over Its Opposition?" *Business Week*, August 9, 1982.
Kernkraftwerk Isar. Kernkraftwerk Isar GmbH, Landshut, n.d.
Vollradt, J., W. Harbecke, G. Lukacs, and W. Stang. "Stillegung und Beseitigung von Kernkraftwerken - Betrieb auch ohne Stromerzeugung." *Atomkern-Energie Kerntechnik*, April 1990.
Kreuzer and Dr. Brosche. "Letter to Landratsamt, July 2, 1991." Gemeinshaftskernkraftwerk Isar2 GmbH, July 2, 1991.
Von Taeuffenbach, Thomas. "Presseerklaerung v. 7/7/91." Burgerforum gegen Atomkraftwerke Landshut u. Umgebung, July 7, 1991.
Geiss, Armin. "Pressemitteilung 38/91." Bayernwerk, Munich, July 8, 1991.
"Landratsamt Und Burgerforum streiten sich." *Landshut Zeitung*, Landshut, July 9, 1991.
"Palo Verde Units 1 & 2 Top American Producers in May; Unit 2 Continues Record Run." *Business Wire*, July 12, 1991.
Mühlberger, Egon. "Pressemitteilung 43/91." Bayernwerk, Munich, August 9, 1991.
Mounfield, Peter. *World Nuclear Power*. Routledge, 1991.
"Landratsamt Und Burgerforum streiten sich." *Landshut Zeitung*, Landshut, July 9, 1991.
"Palo Verde Units 1 & 2 Top American Producers in May; Unit 2 Continues Record Run." *Business Wire*, July 12, 1991.

—Al Cook

77

Karlsruhe Nuclear Research Facility

[*See Also:* Civaux Nuclear Power Plant, Unit 2]

Lepoldshafen, Germany

Plant Report

German wartime research program resumes with FR2. The modern era of nuclear energy started for Germany at Karlsruhe, about 50 miles west of Stuttgard near the French border.

The country's first power plants had to rely on foreign technology because Germany did not have a nuclear research facility until 1956, two years after it renounced all nuclear weapons research. German researchers quickly restarted the World War II nuclear program, building a heavy-water moderated pile they called Otto Hahn or FR2. The reactor went critical in 1962.

By 1968 German scientists were working on light water designs similar to the current American technology. Several different small research reactors developed over the years, including the 20-megawatts electric (MWe) plutonium-burning KNK and KNK-II.

In 1984, the multipurpose research reactor, MZFR, began decommissioning. That reactor operated for 18 years, starting in 1965. The other research reactors continue to provide power to the German grid.

German research fails to produce commercial design. Even though the center has produced several workable prototypes, no Karlsruhe design has been built in Germany.

Argentina purchased a 200-MWe heavy-water reactor in 1970, and the Soviet Union entered into an as-yet uncompleted contract for a high temperature reactor in 1988.

Sources

Vollradt, T., W. Harbecke, G. Lukacs, and W. Stang. "Stillegung and Beseitigung von Kernkraftwerken - Betrief auch ohne Stromerzeugung." *Kerntechnik*, Munich, April 1990.
Mounfield, Peter. *World Nuclear Power*. Routledge, 1991.

—Al Cook

78 ●

Kruemmel KKK Nuclear Power Plant

Elbuferstrasse 82 Phone: 04152-15-0
W-2054 Geesthacht, Germany Fax: 04152-15-2008
Peter Gerdes, Plant Mgr. Telex: 41-52-10

Owning Utility

Hamburgische Electricitaets-Werke AG (HEW) Phone: 040-6396-1
Ueberseering 12 Fax: 040-6396-2770
W-2000 Hamburg 60, Germany Telex: 40-30-68
Werner Hartel, Mgr. of Nuclear Div.

Contact

Robert Schulte, Pub.Info.
Hamburgische Electricitaets-Werke AG (HEW)

Basic Facts

Status: Operable. *Type of Reactor:* Boiling water reactor. *Megawatts (electric):* 1316. *Megawatts (thermal):* 3690. *Reactor System Supplier:* Kraftwerk Union AG (Germany). *Generator Supplier:* Kraftwerk Union AG (Germany). *Architect Engineer:* Kraftwerk Union AG (Germany).

Key Dates

Ordered: 1972. *Construction began:* 1974. *Criticality:* September 1983. *First Power:* September 1983. *Commercial Operation:* March 1984.

Plant Report

Hamburg forges ahead with nuclear development. With the successful commissioning of Brunsbuettel power station in 1976, Hamburgische Electricitaets-Werke AG (HEW) decided to build a third nuclear generation plant. KKK Kruemmel went on line in 1983. By 1990, the utility used nuclear energy to produce 82.7 percent of city of Hamburg's electricity.

A four-reactor operating agreement gives PreussenElektra (PE) 50 percent of Kruemmel's 1,316-megawatts electric (MWe) output. HEW operates the plant and feeds the other 50 percent into its Hamburg grid as part of the city's base load electricity requirement.

The four reactors, Brokdorf, Brunsbuettel, Stade, and Kruemmel, contribute 10,243 million kilowatt-hours (kWh) to HEW's integrated power and district heat distribution system. The utility also employs a pumped-hydro electric station and thermal plants burning oil, gas, coal and garbage. Many of the generating stations also heat water, which HEW pumps to over 270,000 apartments and 120,000 private homes, eliminating the need for individual furnaces.

New transmission lines for East Germany. The 380-kilovolt Hamburg grid forms a small part of the pan-European grid system stretching from North Cape to Sicily. In 1990, PE completed a new link between Kruemmel and Grohnde power plant. This line, and future northerly links, will plug the old East German states into the western grid, providing a ready route for excess West-German power.

Each of the four reactors draws its cooling water from the Elbe River. However, HEW sends that water back to the river, cleaner than it found it. Its treatment plants extract most of the heavy metals dumped into the Elbe by industries upstream. As a final step, the utility injects oxygen before returning the water to the river.

The boiling-water reactor uses a MOX fuel of natural uranium from Canada, Australia, and USA enriched to 3 percent uranium-235.

Kruemmel has sufficient storage capacity for all its nuclear waste until 2000. By then, the utility hopes the nuclear depository under construction at Gorleben will be operational.

Sources

Schroeder, Friedrich. *PreussenElektra Strom*. PreussenElektra Aktiengesellschaft, Hanover, 1989.
PruessenElektra Annual Report 1990. Hanover, April 1991.
HEW in Focus. Hamburger Electricitéaets-Werke AG, Hamburg, April 6, 1991.
HEW: Excerpt from Annual Report 1990. Hamburger Electricitaets-Werke AG, Hamburg, April 6, 1991.
"Betriebsergebnisse der westdeutschen Kernkraftwerke 1990." *atomwirtschaft*, May 1991.

—Al Cook

Lingen Nuclear Power Plant

Lingen, Dame, Germany

Basic Facts

Status: Decommissioned. *Type of Reactor:* Boiling water reactor. *Megawatts (electric):* 268. *Megawatts (thermal):* 520. *Reactor System Supplier:* Telefunken. *Generator Supplier:* Telefunken. *Architect Engineer:* Telefunken; Hochtief.

Key Dates

Ordered: 1964. *Construction began:* 1964. *Criticality:* January 1968. *First Power:* July 1968. *Commercial Operation:* October 1968. *Shut Down:* January 1977. *Decommissioned:* March 1979.

Plant Report

Prototype plant fails to meet new standards. In 1977, Lingen power station fell victim to Germany's changing attitude towards nuclear power.

The 254-megawatts electric (MWe) boiling water reactor began operations in 1968. Only nine years later, the plant's management decided to take Lingen out of service.

Reactor technicians had stopped the reactor for a scheduled maintenance inspection and refueling, but new, tighter safety regulations also called for substantial up-grading of the facility.

The political climate of the time also suggested the operators would find it difficult to get official approval for the refitting or a license to restart once work was completed.

Decommissioning will take half a century to complete. In March 1979, Vereinigte Elektrizitätswerke Westfalen (VEW) decided to decommission the plant.

By the end of 1983, the company removed all nuclear fuels from the site and sent spent fuel rods to England for reprocessing.

Non-nuclear buildings and equipment were removed from the site and footings prepared for the installation of a safety barrier.

Work on the construction of the barrier began in mid-1983, finishing in March 1988. The barrier will remain in place for 30 years to allow radioactivity to disperse safely.

When safe to do so, the company will dismantle the entire reactor and dispose of the waste material in a depository like Gorleben.

Five technicians remain on site to monitor the radiation decay in the core.

In 1989, Euratom designated the decommissioning project as a European Joint Venture. Observations and special studies on the progress of the work will be made until December 1999.

Sources

"Stillegung and Beseitigung von Kernkraftwerken - Betrief auch ohne Stromerzeugung." *Kerntechnik*, Munich, April 1990.
Bericht Uber das Geschäftsjahr 1990 KWL. Kerkraft Lingen GmbH, Lingen, March 7, 1991.
"Das steillgelegte Kernkraftwerk Lingen." *Atomwirtschaft*, December 1991.

—Al Cook

Muelheim-Kaerlich Nuclear Power Plant

Kernkraftwerke Muelheim-Kaerlich Phone: 0-26-37-6-41
W-5403 Muelheim-Kaerlich, Germany Fax: 0-26-37-64-2260
H.B. Gutmann, Stat.Mgr.

Owning Utility

RWE Energie AG Phone: 02-01-185-1
Kruppstrasse Fax: 02-01-185-4313
Postfach 103165 Telex: 857851 Energ
W-4300 Essen 1, Germany
K. Kuhnt, Chairman of the Mgmt. Board

Muelheim-Kaerlich

Contact
K.H. Klawunn, Mgr. for Business Administration

Basic Facts
Status: Operable. *Type of Reactor:* Pressurized light-water reactor. *Megawatts (electric):* 1308. *Megawatts (thermal):* 3760. *Reactor System Supplier:* Babcock-Brown Boveri Reaktor, GmbH (Germany). *Generator Supplier:* ASEA (Sweden); Brown Boveri (Switzerland). *Architect Engineer:* ASEA; Brown Boveri; Babcock-Brown Boveri Reaktor, GmbH (Germany).

Key Dates
Ordered: 1973. *Construction began:* January 1975. *Criticality:* March 1986. *First Power:* March 1986. *Commercial Operation:* October 1987.

Plant Report

Completed reactor not allowed to operate. The idling nuclear power plant, Muelheim-Kaerlich, near Koblenz on the Rhine River, provides a graphic representation of the battle between the German nuclear industry and its opponents.

Planning and construction of the 1300-megawatts electric (MWe) pressurized water reactor (PWR) began in 1971. Rheinisch-Westfälisches Elektrizitätswerk AG (RWE Energie) chose to site the plant in Rhineland-Palatinate because the increasing electrical demand of industry in that region was forcing the utility to import power from other areas.

Drawing on its experience with Gundremmingen power station, a pilot reactor at Kahl, and the continuing construction of Biblis, RWE decided to use the proven American pressurized light-water reactor design.

Brown Boveri Reaktor GmbH (BBR) of Mannheim built the nuclear portion of the plant using components licenced from Babcock & Wilcox.

Despite growing opposition, the reactor progressed through the planning stages. The first structural work began in January 1975.

Court order slows construction. On February 2, 1977, the antinuclear lobby scored its first victory. The Koblenz Administrative Court issued a stop-work order. Construction ceased for three months before RWE could untangle that legal knot.

However, construction did continue. On March 1, 1986, Muelheim-Kaerlich's reactor went critical for the first time as part of the necessary series of tests to prove the system ready for commercial operation.

In September of that year, the project again stalled. This time new emission standards for the cooling tower required more court action and public hearings. That resulted in another eight-month delay.

With the last apparent technical hurdle out of the way, RWE again synchronized the plant's generator with the German electrical grid on August 18, 1987.

However, on September 9, 1988, the Supreme Administrative Court in Berlin revoked the plant's operating licence, ruling in favor of nuclear opponents who claimed the original construction permits were technically invalid.

One step forward, two steps back. That forced the utility to repeat the entire licensing procedure, a process which took the next two years.

In January 1991, RWE completed the re-licensing, but Alfred Beth, environment minister for the state of Rhineland-Palatinate, refused to allow the plant to restart until the court again reviewed the matter later that summer.

As a result of the further delay, RWE suspended an electricity exchange agreement with Switzerland and Italy, and purchased 500 megawatts electric from Electricité de France because of a lack of generating capacity to meet base-load requirements during the winter.

On May 24, 1991, the Koblenz Higher Administrative Court again revoked the operating licennse and refused the company leave to appeal. RWE has appealed to a higher court.

Nuclear industry learns the lesson of Muelheim-Kaerlich. The experience of Muelheim-Kaerlich has shown the German Nuclear industry the strength of the antinuclear lobby in that country. Public demonstrations and violent confrontations have given way to quiet but effective use of the courts to block individual building projects. Between 1977 and 1982, nuclear opponents effectively stopped all reactor licensing.

In response, the federal government devised a counter-strategy called the convoy system.

The 1982 federal plan eliminates the need to individually examine the plans for standard plant designs. Once approved, a design can be used on any acceptable site without repeating the technical review.

Three plants quickly burst through the antinuclear road-block. The first was Isar KKI-2 followed by Emsland KKE and Neckar GKN-2.

Kraftwerk Union AG, the major German nuclear power plant builder, backs the concept fully, saying it reduces paper-work by 30 percent and repetitive engineering work by 25 percent.

Nuclear opponents cry foul, search for new weapons. Nuclear power opponents see the danger to their cause. "If we don't win on the first challenge, we are finished," said Ludwig Trautman, a nuclear power critic actively opposed to Isar-2.

However, the concept does not help Muelheim-Kaerlich. That plant did not use the Isar-convoy design. For now at least, RWE is on its own, fighting a determined antinuclear lobby.

Sources
"West Germany: Can a Nuclear Convoy' Run Over Its Opposition?" *Business Week*, August 9, 1982.
Mülheim-Kärlich. Rheinisch-Westfälisches Elektrizitätswerke AG, Essen, April 1987.
"Mülheim-Kärlich Was Still Blocked from Restart." *Nuclear News*, February 1991.
RWE Energie Annual Report 1990/91. Essen, September 6, 1991.
Mounfield, Peter. *World Nuclear Power.* Routledge, 1991.

Kerkraftwerk Mülheim-Kärlich, ABB Kraftwerke AG, Mannheim, n.d.

—Al Cook

81 •

Neckar Nuclear Power Plant, Units 1-2

W-7129 Neckarwestheim, Germany Phone: 07133-13-0
Kolditz, Plant Mgr. Fax: 07133-13-2825
 Telex: 7-28-029 gkn

Owning Utility

Gemeinschaftskernkraftwerk Neckar Phone: 07133-13-0
  GmbH Fax: 07133-13-2825
Postfach Telex: 7-28-029 gkn
W-7129 Neckarwestheim, Germany

Contact

Lessow, Pub.Info. Supt.
Gemeinschaftskernkraftwerk Neckar
  GmbH

Basic Facts

Units 1-2 *Status:* Operable. *Type of Reactor:* Pressurized light-water reactor. *Megawatts (electric):* Unit 1, 855; Unit 2, 1316. *Megawatts (thermal):* Unit 1, 2497; Unit 2, 3765. *Reactor System Supplier:* Kraftwerk Union AG (Germany). *Generator Supplier:* Kraftwerk Union AG (Germany). *Architect Engineer:* Kraftwerk Union AG (Germany).

Key Dates

Unit 1 *Ordered:* 1971. *Construction began:* 1972. *Criticality:* May 1976. *First Power:* June 1976. *Commercial Operation:* December 1976.

Unit 2 *Ordered:* 1975. *Construction began:* 1982. *Criticality:* December 1988. *First Power:* January 1989. *Commercial Operation:* April 1989.

Plant Report

Long term investment to protect. The operators of Neckar nuclear power stations, GKN-1 and GKN-2, expect their plants to operate productively for 50 to 60 years. To ensure that they do so, Gemeinschaftskernkraftwerk Neckar GmbH continually up-grades the facilities and puts its technicians through extensive ongoing retraining.

In 1988, the average worker received 234 hours of specialized instruction, emphasizing preventative measures against radiation contamination and fire.

During the annual maintenance shutdown in the same year, the company replaced 364 core-baffle screws as a precautionary measure. Quality assurance workers had found a better austenitic material.

Located on the Neckar River, near the towns of Neckarwestheim, Gemmrigheim, and Kirchheim, GKN, along with the Obrigheim nuclear power plant and some hydroelectric plants, has provided the base-load electricity supply for the region of Baden-Wurtemberg since GKN-1 first went into operation in 1976.

New regulations modify construction plans. Originally, when construction of the 800-megawatt (MW) unit 1 began in 1972, the utility shareholders planned to build a second identical plant, a few years later. However, tightening government regulations and improvements in reactor design prompted them to try a different configuration,

Kraftwerk Union AG (KWU) proposed a 1300 MW convoy class Pressurized Water Reactor, which it also planned to use at Lanushut (Isar-2) and Lingen (Emsland). The standardized design reduced cost and streamlined the approval process. Before the convoy system, antinuclear groups succeeded in stoping all new reactor projects between 1978 and 1982.

Even so, the resulting public inquiry proved the longest to date for a nuclear facility in the Federal Republic. After the 19-day hearing, construction began in November 1982. With the granting of the final nuclear operating permit and water rights, the utilities commissioned the plant in December 1988.

Nuclear protest ineffective. However, the new plant did not go into service without some difficulty. Antinuclear lobbyists launched a last-minute campaign to block its opening, claiming the site's geology was unsound. Protestors blockaded a shipment of unirradiated fuel assemblies. Even so, construction and testing finished ahead of schedule.

With both units operating, the station provides about 30 percent of the regions domestic and industrial electricity and supplies power to one out of every three electric locomotives operated by the Deutsche Bundesbahn (German Federal Railways) between Flensburn and Freilassing.

Internationally, the station ranks third out of over 400 operating nuclear power stations. Only the German plants Grohnde and Emsland outprooduce GKN.

Safety measures built into new reactor designs. German safety standards have challenged traditional reactor construction methods. Massive containment structures constitute the German nuclear industry's "technically and economically optimum solution" to those increasingly stringent safety regulations.

The familiar cylindrical concrete containment structure, lined with steel, sits within a second concrete reactor building. This provides an air-lock between the shields from which the operators can extract any contaminated gases or fluids.

The building itself improves the reactors ability to survive external impacts like earthquakes, bomb blasts and aircraft crashes. In recent years, crashes of military jets near Greifswald and Karlsruhe raised concerns of possible disasters. In 1990, member countries of the North Atlantic Treaty Organization (NATO) based about 1,150 warplanes in Germany. They flew more than 68,000 hours of low-level training flights yearly.

In case of an accident, KWU provided two separate and automatic reactor trip mechanisms activated by a computerized protection system.

Boron-control rods, held by electromagnets above the core, can drop into position with the slightest break in control power.

Plant Profiles

As a back-up to the boron control rods, an emergency water-feed system can inject neutron-absorbing boric acid into the core coolant.

That system also protects against the Pressurized Water Reactor's most common accident, the loss of coolant incident. Low and high pressure injection pumps remove decay heat from the core until shutdown completes.

Meanwhile, the containment isolation system double-seals the structure from the outside environment with multiple-sets of redundant valves in each pipe that penetrates the containment wall.

Yearly fuel change sends spent rods to Britain. During normal operation, technicians exchange 20 percent of the reactors' enriched uranium fuel rods each year. The reactor shuts down during the procedure, providing an opportunity for maintenance on other systems.

Spent fuel remains within the containment structure, immersed in cooling ponds for at least seven years. The pools can store up to 10 years worth of used rods. After that, GKN loads the material into special forged-steel flasks for shipment to reprocessing plants in France and Great Britain. With the abandonment of Germany's own reprocessing project at Wackersdorf, the industry may now move to direct storage in salt mines at Gorleben.

In 1990, 20 to 30 antinuclear demonstrators tried to block the movement of spent fuel assemblies from the plant.

New cooling tower technology protects river water. One obvious way in which the two Neckar reactors differ is in their cooling method. The original plant relies heavily on the Neckar River. However, when the river temperature rises, two banks of 17 cooling towers each release the plant's waste heat into the atmosphere instead.

GKN-2 uses a single huge cooling tower. The special hybrid device operates as a normal tower, dropping the coolant water through an air stream created by large fans mounted in the tower's side walls. However, to avoid the distinctive white plume of rising steam, the tower's design allows the water to flow through pipes mounted on the intake side of the fans. The effect is to preheat and dry the air entering the tower, reducing the relative humidity of the air stream.

With the commissioning of GKN-2, the utility completed its nuclear expansion program. However, in 1990, a court order closed KWO Obrigheim until it could prove the safety of its turbines and reactor containment system. Upgrading of the 357-megawatts electric (MWe) plant, built in 1968, began immediately, but re-licensing may prove difficult and expensive. In the meantime, GKN will have to provide as much of the unexpected shortfall as possible.

Sources

GKN Block II. Gemeinschaftskernkraftwerk Neckar GmbH, Neckarwestheim, 1981.

"West Germany: Can a Nuclear Convoy Run Over Its Opposition?" *Business Week*, August 9, 1982.

GKN. Gemeinschaftskernkraftwerk Neckar GmbH, Neckarwestheim, 1989.

Obrigheim | 83

GKN Annual Report 1988. Gemeinschaftskernkraftwerk Neckar GmbH, Neckarwestheim, 1989.

GKN Annual Report 1989. Gemeinschaftskernkraftwerk Neckar GmbH, Neckarwestheim, 1990.

"Jet Accident Revives Controversy Over Flights." *The Washington Post*, April 19, 1990.

GKN Annual Report 1990. Gemeinschaftskernkraftwerk Neckar GmbH, Neckarwestheim, 1991.

"Betriebsergebnisse der westdeutschen Kernkraftwerke 1990." *atomwirtschaft*, May 1991.

—Al Cook

---

82 ▲
Niederaichbach Nuclear Power Plant

Landshut, Bavaria, Germany

Basic Facts

Status: Decommissioned. *Type of Reactor:* Gas-cooled reactor. *Megawatts (electric):* 106. *Megawatts (thermal):* 320. *Reactor System Supplier:* Siemens. *Generator Supplier:* Siemens. *Owner:* Kernforschungzentrum Karlsruhe; Kernkraftwerk Niederaichbach. *Operator:* Kernkraftwerk Niederaichbach.

Key Dates

Ordered: 1964. *Construction began:* 1966. *Criticality:* December 1972. *First Power:* January 1973. *Commercial Operation:* January 1973. *Shut Down:* August 1974.

Plant Report

Experimenting with original German technology. In 1972, Germany's nuclear industry experimented with a heavy-water gas-cooled design. The 100-megawatt Niederaichbach prototype, built on the same site as the Isar Nuclear Power Plant, operated for only two years, closing in 1974. Once fully decommissioned, Niederaichbach will be the first nuclear reactor in Germany to be completely dismantled.

—Al Cook

---

83 ●
Obrigheim Nuclear Power Plant

Kraftwerkstrasse 1 Phone: 06261-651
W-6952 Obrigheim, Germany Fax: 06261-390
Ernst Pickel, Plant Mgr. Telex: 0466121 kwo d

Owning Utility

Kernkraftwerk Obrigheim GmbH (KWO) Phone: 06261-651
Kraftwerkstrasse 1 Fax: 06261-390
W-6952 Obrigheim, Germany Telex: 0466121 kwo d
Karl Stabler, Chairman of the Board

Contact

Karlfried Theilig, Dir. of Pub.Info.
Kernkraftwerk Obrigheim GmbH (KWO)

Basic Facts

Status: Operable. *Type of Reactor:* Pressurized light-water reactor. *Megawatts (electric):* 357. *Megawatts (thermal):* 1050. *Reactor System*

Supplier: Siemens. *Generator Supplier:* Siemens. *Architect Engineer:* Siemens.

Key Dates

Ordered: 1965. *Construction began:* 1965. *Criticality:* September 1968. *First Power:* October 1968. *Commercial Operation:* March 1969.

Plant Report

Euratom project. For Germany, the age of large-scale electricity production from atomic energy began in 1968 with KWO Obrigheim in Baden-Wurtemberg.

The 357-megawatts electric (MWe) plant, built by Siemens-Schuckertwerke, explored early pressurized-water technology as part of Euratom. Data from the experimental plant provided valuable reactor experience for the modern designs that followed.

When first constructed, it pioneered the technology in Europe as the largest operating light-water reactor. Over the next 20 years, it provided reliable base electric generation, supplying 50 billion kilowatt-hours (KWh) with an average availability rate of 84 percent.

Over its operating life, the plant has contributed a significant portion of the electricity consumed by the Bavarian region. In order to continue to do so, the plant's owners, Kernkraftwerk Obrigheim GmbH, have maintained a policy of constant upgrading and back-fitting.

First major nuclear-component-change. In 1983, they replaced both steam generators, something never before attempted at a European nuclear power plant.

A 1988 safety analysis of KWO by the Reactor Safety Commission rated the plant as secure as any new plant. Two other inspection teams made similar evaluations over the next two years.

Even so, on May 25, 1990, the Administrative Court of Baden-Wurtemberg, ordered Obrigheim to cease operations because of safety concerns with its high pressure steam turbines and the reactor containment vessel.

KWO immediately began an accelerated program of upgrading and testing to regain its operating license.

Over the following 14 months, it spent 50 million DM and earmarked another 100 million DM for the project.

Economic projections indicated the modernization would be feasible if the plant operated an additional 10 years. When the plant halted operations in 1989, it had been producing electricity for about 5.87 pfennigs per KWh, about two-thirds the cost of a coal-fired plant. After the renovations, the 1990 cost would rise to 9.62 pfennigs per KWh, comparable to coal-fired.

Supreme court gives plant five years. However, the Supreme Administrative Court in Berlin was not convinced. It refused to lift the stop-work order, but gave KWO five years to complete its scheduled repairs. The court insisted on independent verification of the plant's safety before it would consider returning the operating license.

In the meantime, other plants will have to pick up the unexpected short-fall of electric generation.

Sources

Das Kernkraftwerk Obrigheim. Obrigheim am Neckar, 1968.
"Aout Us." *The Kernkraftwerk Obrigheim GmbH,* Obrigheim am Neckar, October 1989.
"Informationen Uber das Kernkraftwerk Obrigheim." *Obrigheim GmbH,* Obrigheim am Neckar, April 9, 1990.
"Betriebsergebnisse der westdeutschen Kernkraftwerke 1990." *atomwirtschaft,* May 1991.
"Pressemitteilung NR. 15/1991." *Bundesverwaltungsgericht,* Berlin, June 7, 1991.

—*Al Cook*

Philippsburg Nuclear Power Station, Units 1-2

Postfach 1140
W-7522 Philippsburg 1, Germany
H. Schenk, Technical Dir.

Phone: 07256-95-1
Fax: 07256-6998
Telex: 7822357 KKPD

Owning Utility

Kernkraftwerk Philippsburg GmbH (KKP)

Contact

H. Blaeske, Pub.Info.
Philippsburg Nuclear Power Station

Basic Facts

Units 1-2 *Status:* Operable. *Type of Reactor:* Unit 1, Boiling water reactor; Unit 2, Pressurized light-water reactor. *Megawatts (electric):* Unit 1, 900; Unit 2, 1349. *Megawatts (thermal):* Unit 1, 2575; Unit 2, 3765. *Reactor System Supplier:* Kraftwerk Union AG (Germany). *Generator Supplier:* Kraftwerk Union AG (Germany). *Architect Engineer:* Kraftwerk Union AG (Germany).

Key Dates

Unit 1 *Ordered:* 1970. *Construction began:* 1971. *Criticality:* March 1979. *First Power:* May 1979. *Commercial Operation:* March 1980.
Unit 2 *Ordered:* 1975. *Construction began:* 1977. *Criticality:* December 1984. *First Power:* December 1984. *Commercial Operation:* April 1985.

Plant Report

Island forms nuclear park on Rhine. Rheinschanz Island, on the Rhine River, about 30 kilometers south of Manheim, hosts the two reactors of Philippsburg Nuclear Power Station, KKP.

As equal partners, the two utilities, Badenwerk AG and Energie-Versorgung Schwaben AG (EVS), commissioned KKP-1 as a 900-megawatts electric (MWe) boiling water reactor from Kraftwerk Union AG (KWU) in 1970. KKP-1 began operation in November 1978, three years after planning had begun in earnest on KKP-2.

The second plant used the American pressurized water reactor design to produce 1,350 MWe. It first fed electricity into the south-west German 380-kilovolt (kV) grid system on December 17, 1984.

As with all of Germany's nuclear power stations, increasingly stringent government regulations mean constant up-grading of safety and operating systems, particularly to ensure the safety of the containment system.

Reactor shuts for major repairs. In 1990, KKP-1 shut down from June 8 to September 21, a period of 105 days, for safety inspections, back-fitting of safety features, and repairs to its generator-turbine system and steam piping.

The Technical Control Association (TÜV) supervised testing of the reactor containment vessel.

As a result of that up-grading, the Baden-Wurttemberg environment ministry improved the reactor's safety classification from "E" (corrective action needed within one year) to "N" (normal operation).

KKP-2 suffered a generator malfunction from excessive cooling water temperature in 1990. The company expects to correct that situation as well as make repairs to steam piping and other safety up-grades during its scheduled shutdown in August.

During December of 1990, KKP-2 delivered 1,017,250 megawatts electric to the German grid, making it the fifth highest producing nuclear power plant in the world, according to *Nucleonics Week*.

Sources

Philippsburg Nuclear Power Station. Kernkraftwerk Philippsburg GmbH, November 1988.
Aktuelle Informationen: Betriebsbericht 1990. Kernkraftwerk Philippsburg GmbH, May 1991.
"Betriebsergebnisse der westdeutschen Kernkraftwerke 1990." *atomwirtschaft*, May 1991.
"Palo Verde Units 1 & 2 Top American Producers in May; Unit 2 Continues Record Run." *Business Wire*, July 12, 1991.

—Al Cook

Rheinsberg Nuclear Power Plant

Kernkraftwerk Rheinsberg
O-1955 Rheinsberg, Germany

Owning Utility

Energie Werke Norde GmbH (EWN) Phone: 0082297-40
Vorstand Fax: 0082297-2-24-71
Postfach 35 Telex: 3981883z-ewn-d
O-2205 Lubmin, Germany
Ulrich Loeschhorn, Managing Dir.

Basic Facts

Status: Shut down. *Type of Reactor:* Pressurized light-water reactor. *Megawatts (electric):* 440. *Megawatts (thermal):* 1375. *Reactor System Supplier:* Atomenergoexport (former USSR). *Generator Supplier:* Atomenergoexport (former USSR).

Key Dates

Ordered: 1956. *Construction began:* 1960. *Criticality:* March 1966. *First Power:* May 1966. *Commercial Operation:* October 1966. *Shut Down:* October 1990. *Decommissioned:* 1994.

Plant Report

New technology used for first time outside Soviet Union. In 1966, the 70-megawatts electric (MWe) Rheinsberg station became the first nuclear power station built in the Eastern block outside of the Soviet Union.

It was also only the second Pressurized Water Reactor built by the Soviets. The first—the 210 MWe Novovoronezh plant in the Soviet Union—went into operation in 1964.

After the Chernobyl incident, Soviet-designed reactors became suspect. That included Rheinsberg even though its pressurized light-water reactor design bore no resemblance to the graphite moderated unit at Chernobyl.

As German reunification approached in 1990, West German experts became alarmed with the safety features of Rheinsberg, located 75 miles west of Berlin, and other East German reactors.

They objected to the lack of a blast-hardened containment building. West German reactors use multiple-layered containment structures designed to withstand the impact of an aircraft crash and to hold the entire volume of the cooling system in case of a catastrophic leak.

Safety concerns. They also questioned the lack of redundant safety controls. Both main and auxiliary power cables run through the same duct work, making the entire system vulnerable to fire. No emergency water feed system exists, meaning a break in even the smallest cooling-water line could result in a core melt-down.

In Western reactors, designers shield the reactor vessel from direct bombardment by radioactive particles. This helps prevent the vessel material becoming brittle. However, Soviet designers devised a thermal annealing process to repair brittle reactor vessels. Every 25 years, technicians heat the vessel to 480 degrees Celsius for 120 hours to restore tensile strength to the steel. Western experts doubt the system works.

Under the German unity treaty, the operating license for the Rheinsberg plant expired in 1992. Given the questionable safety features of the plant and the short time left to recoup any cost of up-grading, German authorities decided to close the plant in 1990, after 23 years of operation. Decommissioning will commence in 1994.

Sources

"East Berlin Heeds Bonn and Shuts Nuclear Facility." *The Wall Street Journal*, February 16, 1990.
"East Germany's Nuclear Plants." *The Wall Street Journal*, September 12, 1990.
"Germans to Shut 5 Atom Plants Built by Soviets." *The New York Times*, October 21, 1990.
Kernkraftwerk Rheinsberg: Ruckblick auf 23 Jahre Betrieb. Energie Werke Nord, June 1991.

—Al Cook

SNR 300 Kalkar Nuclear Power Plant

[*See Also:* Karlsruhe Nuclear Research Facility]

Kalkar, Germany

Basic Facts

Status: Cancelled. *Type of Reactor:* Liquid metal fast breeder reactor. *Megawatts (electric):* 327. *Megawatts (thermal):* 762. *Reactor System Supplier:* Internationale Natrium Brutreaktro Baiges; Interatom. *Generator Supplier:* Kraftwerk Union AG (Germany). *Architect Engineer:* Internationale Natrium Brutreaktro Baiges; Interatom; Kraftwerk Union AG (Germany). *Owner:* Schnell-Brueter-Kernkraftwerks-Gesellschaft. *Operator:* Schnell-Brueter-Kernkraftwerks-Gesellschaft.

Key Dates

Ordered: 1972. *Cancelled:* April 1991. *Construction began:* 1973.

Plant Report

Kalkar develops final link in nuclear chain. From the very earliest days, nuclear researchers viewed the Fast Breeder Reactor (FBR) as a necessary component of the uranium fuel cycle.

Early thermal reactors, using natural uranium, gave way to higher-temperature producing designs which used enriched uranium and eventually mixed-oxide (MOX). Several reprocessing plants in various countries now produce Mixed Oxide Fuel (MOX Fuel) with recycled fuel from thermal reactors.

However, those facilities could not recycle one by-product from the nuclear reaction, plutonium, into anything except nuclear weapons, or fuel for a high temperature fast reactor.

Even as the first thermal reactors began operations in the early '50s and '60s, researchers began development of a system that would efficiently burn that by-product. They believed that since the fast reactor would derive 60 times as much energy from a load of uranium as a thermal system could, the FBR would eventually be the cheapest and most efficient way to produce electricity.

In addition, the reactor would produce more plutonium than it used making uranium-poor countries self-sufficient in energy.

Germany works to advance FBR technology. Germany entered the field of FBR technology well behind Britain and France. Its first prototype, the 20 megawatts electric (MWe) KNK at the Karlsruhe nuclear research facility, did not become operational until 1974. Both Britain and France had second-generation FBRs by then.

However, Germany's FBR program was moving into high gear. The Federal Ministry of Research and Technology (BMFT) teamed up with RWE Energie to build a 300 megawatts electric commercial prototype, SNR-300 Kalkar, located 70 kilometers west of Dusseldorf in North Rhine Westphalia.

Since the site lay close to the Dutch border, its promise of future secure electricity attracted investment from Dutch and Belgian utilities.

As with other FBR development projects, researchers discovered the high temperatures of the reactor core required specialized materials and fuels delaying completion of SNR-300 for decades.

Major FBR players announce compact. On January 10, 1984, Germany, Great Britain, and France announced they would pool their FBR efforts. Under the Centrale Nucléaire Européenne à Neutrons Rapides SA (NERSA) agreement, each country would build a commercial demonstration plant and share the reactor experience gained.

France's Superphenix became the first, on September 7, 1985, despite public protest and demonstrations at the construction site. Planners optimistically foresaw an additional six replicas in France and one more in Germany.

However, by 1988, Kalkar was in serious trouble. After spending 7,000 million DM over 16 years, and obtaining 18 different planning permits because of design changes, the project seemed stalled by North Rhine-Westphalia's Social Democratic government.

Support for Kalkar had been slipping since Chernobyl in 1986, especially in the coal-and-lignite rich land of North Rhine Westphalia. Even though that state contained most of the largest electric demand centers, no nuclear plant operated within its borders.

An increasingly vocal and effective German Green Party called for an abandonment of nuclear energy. Abandoning Kalkar would mean paying the Dutch and Belgian investors 1,300 million DM.

In the meantime, Japanese researches embarked on an FBR development program of their own, primarily to ensure energy self-sufficiency.

Economics threaten breeder concept. The experience of Superphenix showed the expected cost savings of the FBR did not yet exist. Power from the French plant cost 2.2 times as much as from thermal reactors. New discoveries of rich veins of uranium and increasingly efficient European thermal reactors made the cost differential difficult to overcome.

Meanwhile, continuing delays at Kalkar pushed construction costs up by 10 million DM each month.

In April 1991, RWE Energie announced it would abandon the project because of continued stalling by North Rhine-Westphalia on licensing, evaporating financial support, and the prospect of many more years of development before the plant began producing power.

As with Great Britain, Germany will now concentrate FBR efforts on the Superphenix project.

RWE has retained all 200 workers from the project and will investigate an alternative use for the site and facilities.

Sources

"Breeder Reactor Politics in Europe." *Bulletin of the Atomic Scientists*, May 1986.
"Fast Breeder Generates Controversy." *Nature*, July 1988.
RWE Energie Annual Report 1990/91. Essen, September 6, 1991.
Mounfield, Peter. *World Nuclear Power.* Routledge, 1991.

Plant Profiles

87 ●
Stade Nuclear Power Plant

Postfach 17 80
W-2160 Stade, Germany
Horst Salcher, Plant Mgmt., Technical

Phone: 4141-15-1
Telex: 218140

Owning Utility

Kernkraftwerk Stade GmbH (KKS)
Hoerstener Str. 49
W-2000 Hamburg 90, Germany
Bernhard Broecker, Utility Mgmt., Technical

Contact

Peter Hartmann, Utility Managment, Commercial
Kernkraftwerk Stade GmbH (KKS)

Basic Facts

Status: Operable. *Type of Reactor:* Pressurized light-water reactor. *Megawatts (electric):* 672. *Megawatts (thermal):* 1892. *Reactor System Supplier:* Siemens. *Generator Supplier:* Siemens. *Architect Engineer:* Siemens.

Key Dates

Ordered: 1967. *Construction began:* 1967. *Criticality:* 1972. *First Power:* January 1972. *Commercial Operation:* May 1972.

Plant Report

Old reactor gets facelift. With the completion of extensive refitting in 1990, PreussenElektra (PE) brought its Stade nuclear power plant up to current safety and technical standards. The oldest of PE's seven reactors, KKS Stade began operations in 1972. As part of a four-reactor operating agreement, PE shares its output with Hamburgische Electricitaets-Werke AG (HEW).

Under that agreement, PE retains 64 percent of the electrical output of the 672-megawatts electric (MWe) station, along with operating control. HEW feeds the other 33 percent into its Hamburg grid as part of the city's base load electricity requirement.

Four reactor set provides base-load requirements of city. The four reactors, Brokdorf, Brunsbüttel, Stade and Krümmel, contribute 10,243 million kilowatt-hours to HEW's integrated power and district heat distribution system. That represents about 82 percent of the city's electrical demand. The utility also employs a pumped-hydro electric station and thermal plants burning oil, gas, coal and garbage. Many of the generating stations also heat water, which HEW pumps to over 270,000 apartments and 120,000 private homes, eliminating the need for individual furnaces.

The 380-kilovolt (kV) Hamburg grid forms a small part of the pan-European grid system stretching from North Cape to Sicily.

Each of the four reactors draws its cooling water from the Elbe River. However, HEW sends that water back to the river, cleaner than it found it. Its treatment plants extract most of the heavy metals dumped into the Elbe by industries upstream. As a final step, the utility injects oxygen before returning the water to the river.

As part of a planned schedule of maintenance and retro-fitting, PE replaced steam piping and installed new safety and blocking valves into the cooling system over several years.

Sources

PreussenElektra Strom. PreussenElektra Aktiengesellschaft, Hanover, 1989.
PruessenElektra Annual Report 1990. Hanover, April 1991.
HEW in Focus. Hamburger Electricitätswerke AG, Hamburg, April 6, 1991.
HEW: Excerpt from Annual Report 1990. Hamburger Electricitaets-Werke AG, Hamburg, April 6, 1991.
"Betriebsergebnisse der westdeutschen Kernkraftwerke 1990." *atomwirtschaft*, May 1991.

—*Al Cook*

88 △
Stendal Nuclear Power Plant, Units 1-2

Stendal, Germany

Basic Facts

Units 1-2 *Status:* Indefinitely deferred. *Type of Reactor:* Pressurized light-water reactor. *Megawatts (electric):* 1000. *Megawatts (thermal):* 3000. *Reactor System Supplier:* Atomenergoexport (former USSR).

Key Dates

Units 1-2 *Ordered:* 1983. *Cancelled:* December 1990. *Construction began:* 1984.

Plant Report

Reactor design may be reworked. In 1989, East Germany emitted 15 times more sulphur dioxide than West Germany. With the closing of the last Soviet-made nuclear power plant in the new Federal states, that area will have to rely even more heavily on electricity generated by burning lignite and coal, pushing sulphur dioxide emissions even higher.

That one consideration may salvage Stendal nuclear power plant, located 50 miles west of Berlin on the Elbe River, along with four similar reactors at Greifswald.

In December 1990, the German government ordered construction stopped on the station, because it used a Soviet pressurized water reactor.

The modern version of the Soviet VVER-440, 440 megawatts electric (MWe) design lacks many common Western safety features.

The containment system does not meet Germany's criteria for resistance to impact from aircraft. Scientists who have examined the partially completed plant complain of poor maintenance, low-quality building materials, inadequate fire protection and reactor cooling systems, and the lack of sufficiently redundant safety controls.

Future depends on cost. Studies continue, weighing all the factors to determine whether it would be better to modify the plant to meet German safety standards or to dismantle it.

Currently, the German Finance Ministry owns the plant, as a result of the German unity treaty that went into effect in 1990.

Sources

"Germans to Shut 5 Atom Plants Built by Soviets." *The New York Times*, October 21, 1990.

"Last Soviet Reactor in Eastern Germany Shut." *The New York Times*, December 16, 1990.

PreussenElektra Annual Report 1990. PreussenElektra AG, Hanover, April 1991.

—Al Cook

THTR-300 Demonstration Reactor

W-4700 Hamm-Uentrop, Germany
K. Maull, Radiological Protection/Waste Mgmt.

Owning Utility

Hochtemperatur-Kernkraftwerk GmbH
Sigenbeckstrasse 10
W-4700 Hamm 1, Germany

Phone: 0-23-88-32-0
Fax: 0-23-88-7-22-18
Telex: 08-28-884

Basic Facts

Status: Decommissioned. *Type of Reactor:* High temperature gas reactor. *Megawatts (electric):* 308. *Megawatts (thermal):* 750. *Reactor System Supplier:* Hochtemperatur Reaktorbau (HRB). *Generator Supplier:* Brown Boveri et Cie. *Architect Engineer:* Brown Boveri et Cie; Hochtemperatur Reaktorbau (HRB). *Owner:* Hochtemperatur Kernkraftwerk GmbH. *Operator:* Hochtemperatur Kernkraftwerk GmbH.

Key Dates

Ordered: 1970. *Construction began:* 1971. *Criticality:* 1983. *First Power:* September 1985. *Commercial Operation:* June 1987. *Shut Down:* September 1988. *Decommissioned:* 1992.

Plant Report

German researchers develop better reactor. For the nuclear researchers and engineers at Hochtemperatur Reaktorbau (HRB), 1985 was a very exciting year. They were ready to demonstrate the commercial feasibility of a powerful new, and inherently safer, reactor technology.

Early work on the High Temperature Gas Reactor (HTR) technology began in 1964, when the Organization for Economic Co-operation and Development (OECD) sponsored research projects at Julich, Germany, and Winfrith, Great Britain.

From 1967 until the end of 1988, the 15-megawatts electric (MWe) experimental facility—Julich AVR—consistently supplied electricity to the German grid.

On June 4, 1974, the OECD announced a European Joint Venture under the auspices of Euratom. German researchers officially titled it the THTR-300.

Gulf General Atomic (now GA Technologies Inc.) used the early reactor design for its Fort St. Vrain plant in Colorado with unsatisfactory results.

However, THTR-300 was the world's first Pebble-Bed Reactor, abandoning the block fuel-component configuration of Fort St. Vrain in favor of 675,000 60-mm thorium and carbon spheres.

Thorium kernels keep rector core popping. Unlike conventional reactors, which use metal cladding to contain the uranium fuel, the THTR uses a hard-shelled graphite ball embedded with 30,000 fuel kernels. Each kernel consists of about 10 percent uranium-235 and 90 percent thorium-232.

When burned, the thorium converts to uranium-233 which can be enriched to create more uranium-235.

Fuel and core designers opted for the spherical design to facilitate the reactors unique refueling process. The funnel-shaped graphite floor of the core allows spheres to discharge by gravity from the bottom of the reactor during normal operation. A computer sorts damaged elements to a storage area and recycles good ones back into service. The continuous refueling ensures greater efficiency by promoting uniformly high burn rates.

Helium gas cools the system, emerging from the core base at a temperature of 750 degrees Celsius. Researchers envision outlet gas temperatures of 950 degrees Celsius but that would mean new heat exchanger component materials that could withstand the heat.

Reactor design has process capability. The high-temperature capability of the THTR allows it to make more than just steam for electric turbines. The process steam could also be used by the chemical industry to produce ammonia, methane and methanol, or by the steel industry for direct reduction of iron ore.

The demonstration reactor, built at Schmehausen near Dortmund, shares its site with the Westfalen Coal Works, making future coal-gasification projects likely.

Even though the system produces extremely hot gas and steam, researchers firmly believe it to be inherently safer than the more common light water reactors (LWR).

In the case of a loss-of-coolant accident, like Three Mile Island, a LWR quickly heats up threatening a core melt-down if emergency systems do not activate in time. However, the HTGR has a core power density one-tenth that of an LWR. For the HTGR, the core temperature rises slowly even without the flow of Helium gas to cool it.

Even if the loss-of-coolant incident were allowed to continue uncorrected for an extended period of time, the core structure would not melt or catch fire. Graphite sublimes at about 3,650 degrees Celsius. In so-called graphite fires at Chernobyl and Windscale, impurities in the graphite cores caught fire, but the graphite itself did not.

Commercial version on the order book. After THTR-300, HRB plans to build an HTR-500 for a group of 16 West German utilities. That system would use uranium-238 rather than thorium as a fuel. By 1985, the consortium had completed the projects initial conceptual phase.

A 500 MWe plant would cost about the same as a 1200 MWe LWR and operate more economically and safely, according to HRB.

Plant Profiles

The consortium foresaw an operational commercial plant by the early 90s, if all went well with the licensing procedures.

Nuclear protest kills safer reactor. However, all did not go well. THTR-300 did start operations in 1985 and continued until a scheduled maintenance shutdown on September 29, 1988. Inspections then revealed considerable damage to bolts and insulation materials in the high temperature gas ducting.

With antinuclear sentiment increasing in Germany, coupled with an over-supply of generating capacity and the unexpected repair and redesign cost, Hochtemperatur-Kernkraftwerk GmbH decided to end the project.

Using the experience of Niederaichbach in Germany and Shippingport in the United States, researchers will unload the core during 1992, and enclose the entire structure within a safety barrier for 30 years to allow excess radioactivity to disperse. After that they will completely dismantle the reactor.

Sources

"Pebble-Bed Nuclear Reactor Readied For Power Generation." *Chemical & Engineering News*, March 5, 1984.
Der Thorium-Hochtemperaturereaktor THTR 300. Hochtemperatur-Kernkraftwerk GmbH (HGK), Hamm-Uentrop, September 1985.
"Nuclear Safety: Some Like it Hot." *New Scientist*, November 4, 1989.
"Stillegung and Beseitigung von Kernkraftwerken - Betrieb auch ohne Stromerzeugung." *Kerntechnik*, Munich, April 1990.
HKG Geschäftsbericht 1990. Hochtemperatur-Kernkraftwerk Gmbh (HGK), Hamm-Uentrop, April 1988.
"Stand der Arbeiten zur Stillegung des THTR-300." *atomwirtschaft*, December 1991.

—Al Cook

90 ●
Unterweser Nuclear Power Plant

Postfach 1 40
W-2883 Rodenkirchen-Stadland 1, Germany
Gerhard Guether, Plant Mgmt., Technical

Phone: 4732-80-1
Fax: 4732-8659
Telex: 238303

Owning Utility

Kernkraftwerk Unterweser GmbH (KKU)
Tresckowstrasse 5
W-3000 Hannover 1, Germany
Peter Hartmann, Utility Mgmt., Commercial

Phone: 511-439-0
Fax: 511-439-23-75
Telex: 922756

Basic Facts

Status: Operable. *Type of Reactor:* Pressurized light-water reactor. *Megawatts (electric):* 1300. *Megawatts (thermal):* 3733. *Reactor System Supplier:* Kraftwerk Union AG (Germany). *Generator Supplier:* Kraftwerk Union AG (Germany). *Architect Engineer:* Kraftwerk Union AG (Germany).

Key Dates

Ordered: 1971. *Construction began:* 1971. *Criticality:* 1978. *First Power:* October 1978. *Commercial Operation:* September 1979.

Plant Report

Utility supplies one-fifth of Germany from seven reactors. In 1978, PreussenElektra completed construction of Unterweser KKU nuclear power plant at Nordenham on the North Sea near Bremen. Along with Unterweser, the company has interests in seven nuclear power facilities.

PreussenElektra, founded in Berlin in 1927, supplies electricity to over 15 million people in Schleswig-Holstein, Lower Saxony, and parts of North Rhine-Westphalia and Hesse. In 1990, it supplied more than 57 billion kilowatt-hours (kwh) of electricity to consumers from the Danish border to the River Main, one-fifth of Germany's territory.

The company sells its power to regional utilities by means of its own 380- kilovolt transmission lines. Those lines make up part of the pan-European grid, allowing PreussenElektra to exchange generating capacity with utilities in other countries.

East Germany plugs into Western grid. In 1990, East Germany received electricity from the Western grid for the first time. PreussenElektra used a new line from Helmstedt to Madgeburg as the first link between east and west. The company hopes to complete the integration of the East German Grid by 1992.

During Unterweser's 1990 scheduled maintenance shut-down, PreussenElektra replaced one of three turbine wheels. Eventually, all three wheels will be replaced with a newer, more efficient design.

As part of its drive to publicize the benefits of nuclear power, the company opened a tourist information center on the site in the summer of 1991.

According to *Nucleonics Week*, Unterweser ranked as the 17th highest producing nuclear power plant in the world in December 1990, with a gross output of 953,585 megawatt hours.

Sources

"Palo Verde Units Top Producers in U.S. for December." *Business Wire*, February 15, 1991.
PreussenElektra Annual Report 1990. Hanover, April 1991.
"Betriebsergebnisse der westdeutschen Kernkraftwerke 1990." *atomwirtschaft*, May 1991.
VEBA Group 1991/92. N.p., July 1991.

—Al Cook

91 ●
Wuergassen Nuclear Power Plant

Postfach 13 61
W-3472 Beverungen 1, Germany
Jorg-Dieter Peters, Plant Managment, Technical

Phone: 5273-91-1
Fax: 5273-91-23-50
Telex: 931727

Owning Utility

PreussenElektra Aktiengesellschaft
Postfach 48 49
W-3000 Hannover 1, Germany
Hermann Kraemer, Chairman

Phone: 511-439-0
Fax: 511-439-23-75
Telex: 922756

● Operable ○ Under Construction △ Indefinitely Deferred ▲ Decommissioning □ Planned ■ On Order ☆ Cancelled ★ Shut Down

Wuergassen

Contact

Ulrich Straske, Plant Mgmt., Commercial
Wuergassen Nuclear Power Plant

Basic Facts

Status: Operable. *Type of Reactor:* Boiling water reactor. *Megawatts (electric):* 670. *Megawatts (thermal):* 1912. *Reactor System Supplier:* Allgemeine Elektricitaets-Gessellschaft; AEG Telefunken (Germany). *Generator Supplier:* Allgemeine Elektricitaets-Gessellschaft; AEG Telefunken (Germany); Kraftwerk Union AG (Germany). *Architect Engineer:* Allgemeine Elektricitaets-Gessellschaft; AEG Telefunken (Germany); Kraftwerk Union AG (germany).

Key Dates

Ordered: 1967. *Construction began:* 1968. *Criticality:* 1971. *First Power:* December 1971. *Commercial Operation:* November 1975.

Plant Report

Utility builds first of seven nuclear reactors. In 1972, PreussenElektra entered the arena of nuclear power generation with the opening of the 670-megawatts electric (MWe) Wuerrgassen generating station, near Güttingen, not far from the East German border. Eventually, the company would have interests in seven nuclear power facilities.

PreussenElektra, founded in Berlin in 1927, supplies electricity to over 15 million people in Schleswig-Holstein, lower Saxony and parts of north Rhine-Westphalia and Hesse. In 1990, it supplied more than 57 billion kilowatt-hour (kwh) of electricity to consumers from the Danish border to the River Main, one-fifth of Germany's territory.

The company sells its power to regional utilities by means of its own 380-kilovolt transmission lines. Those lines make up part of the pan-European grid, allowing PreussenElektra to exchange generating capacity with utilities in other countries.

In 1990, East Germany received electricity from the western grid for the first time. PreussenElektra used a new line from Helmstedt to Madgeburg as the first link between east and west. The company hopes to complete the integration of the East German grid by 1992.

New plant doubles output of earlier reactors. When commissioned, Wuergassen was the largest nuclear power plant in Germany, more than doubling the earlier 250 MWe-Gundremmingen-A and the 268-MWe-Lingen.

The plant provided badly-needed power for PruessenElektra's north/south grid system as the country began to move away from a dependence on imported foreign oil. In 1990, nuclear plants supplied 62.4 percent of PreussenElektra's electricity.

In May 1982, safety concerns from the Reactor Safety Commission closed the boiling water reactor for 16 months. Extensive back-fitting and up-grading brought the plant back on-line in July 1984.

New fire regulations issued in 1989 resulted in a further closure from December 1989 to October 1990 to refit the plant's fire-protection system. This closure, along with a similar problem at Brokdorf, reduced the amount of power produced by nuclear reactors to a five-year low.

Even so, PreussenElektra credits its use of atomic stations to meet its base load requirements for keeping the price of its electricity constant since 1983.

Sources

PreussenElektra Kernkraftwerk Wuergassen. PreussenElektra Aktiengesellschaft, Hanover, 1991.
PreussenElektra Annual Report 1990. Hanover, April 1991.
"Betriebsergebnisse der westdeutschen Kernkraftwerke 1990." *atomwirtschaft,* May 1991.
VEBA Group 1991/92. N.p., July 1991.

—Al Cook

Hungary

COUNTRY REPORT

Getting involved in nuclear power production

Hungary began considering nuclear power during the 1960's because the country lacked sufficient domestic energy sources to reach its goals for economic development. The site for Paks, the nation's only nuclear station, was selected in 1967, but in 1969 the project was postponed. Following oil price increases in the 1970's, plans to build the nuclear facility were resumed.

Hungary has one uranium mine. During its years under communist rule, all the output from the mine was delivered to the former Soviet Union. The USSR supplied all the fuel used at Hungary's nuclear units. Spent fuel, after "resting" in a fuel pit at Paks for two years, was returned to the Soviet Union.

Before the fall of Hungarian communism, officials had planned to construct two additional nuclear units at the Paks site. In 1989 the Hungarian Electricity Trust (MVMT), the national utility that owns and operates Paks, cancelled its order for two VVER-1000's and began negotiating with western suppliers for additional nuclear construction.

Although negotiations with French and Canadian suppliers continue, the utility is prohibited from entering into a contract until the new government and the country's parliament finalize an energy policy. Decisions on whether to increase generating capacity with nuclear or coal-burning power plants are expected to be made by the end of 1992. Currently Paks supplies approximately half of Hungary's electricity. The remaining electricity is generated from coal-fired plants (23 percent), gas (21 percent), oil (4 percent) and self-producers (3 percent).

A new proposal submitted by Atomic Energy of Canada Limited (AECL) offered the construction of a 450-megawatt Candu heavy water reactor which would use natural Hungarian uranium and have the ability to reuse spent fuel already stored at the Paks site. The AECL offer also contained the option of two more additional units in the future. Costs were estimated at $1 billion per unit. A decision is expected in 1993.

Political changes affect structure of utility

During 1992 MVMT will begin being restructured as Hungary moves toward a market economy. The utility will become a shareholder company but initially the government will hold all shares. Long-term privatization is expected after 1995.

Along with the changing economic climate in Hungary, plans are underway to connect the nation to the Western European power grid.

International cooperation

Hungary was the first of the Eastern European nations to turn to the west for assistance in overseeing reactor safety following the explosion at Chernobyl in 1986. The Hungarian government officially adopted West German safety practices in 1986. The nation was also the first Eastern European country to request an International Atomic Energy Agency (IAEA) inspection of its nuclear reactors.

In 1990, Hungary entered into an agreement with the U.S. Nuclear Regulatory Commission to exchange information and cooperate in the supervision of nuclear facilities. The accord called for mutual assistance to analyze safety methods, share operating experience, research plant life extension, study next-generation reactors and address concerns regarding radioactive waste.

Hungary also entered into agreements with the French Atomic Energy Commission to research nuclear safety, radiation protection, radioecology, and waste-management issues and with Tecnatom S.A. (Spain) to establish a joint venture for reactor pressure-vessel inspection services in Eastern Europe.

In 1990, Hungary formed a nuclear society and expressed an interest in joining Formatom, the European Atomic Forum. In 1991 The Hungarian AEC's Nuclear Safety Inspectorate and Formatom's Working Group on Quality Assurance co-sponsored a workshop in Budapest on nuclear quality assurance.

Hungary is a member of the World Association of Nuclear Operators (WANO). The Paks Nuclear Power Plant has participated in WANO plant exchange programs and was scheduled for a peer review of plant operations in February 1992.

Nuclear regulatory responsibilities

In 1987, the country's Presidential Council amended the Hungarian Nuclear Power Act of 1980 to provide for public and environmental protection from radiation originating from any source, not just domestic, and gave responsibility for enforcement to the Council of Ministers.

In 1991, the Hungarian government reorganized its regulation of nuclear power. The Atomic Energy Commission (AEC) was made an independent regulatory agency and it's Nuclear Safety Inspectorate was given responsibility for nuclear safety and operator licensing.

Preparing for the future

With an eye to the future, Hungary established a secondary school specifically to train technicians, mechanics, and electrical specialists for the nation's nuclear industry. Most of the vocational instruction is done by Paks's engineers, physicists, computer engineers, and economists.

Sources

"Eastern Europe's Nuclear Power Plans." *World Press Review*, December 1986.

Nuclear Power Plant of Paks. Paksi Atomeromu Vallalat, n.d.

"Hungary: New Reactor Bid from AECL." *Nucleonics Week*, November 28, 1991.

Mounfield, Peter. *World Nuclear Power*. Routledge, 1991.

Source Book: Soviet-Designed Nuclear Power Plants in the Former Soviet Republics and Czechoslovakia, Hungary and Bulgaria. U.S. Council for Energy Awareness, April 1992.

—*Karen Bellenir*

PLANT PROFILES

92 •
Paks Nuclear Power Plant, Units 1-4

PO Box 71
Erno Petz, Plant Mgr.
Phone: 36-75-11-222
Fax: 36-11-551-332
Telex: 14402 paea h

Owning Utility

Hungarian Power Companies Ltd.
Vam. U. 5-7
H-1011 Budapest 1, Hungary

Contact

Geza Rosa, PR
Paks Nuclear Power Plant

Basic Facts

Units 1-4 *Status:* Operable. *Type of Reactor:* Pressurized light-water reactor. *Megawatts (electric):* 440. *Megawatts (thermal):* 1375. *Reactor System Supplier:* Atomenergoexport (former USSR); Skoda. *Generator Supplier:* Atomenergoexport (former USSR); Ganz Electrical Works (Hungary). *Architect Engineer:* Power Station and Network Engineering Co. (Hungary).

Key Dates

Unit 1 *Ordered:* 1971. *Construction began:* 1974. *Criticality:* October 1982. *First Power:* December 1982. *Commercial Operation:* August 1983.

Unit 2 *Ordered:* 1971. *Construction began:* 1974. *Criticality:* August 1984. *First Power:* September 1984. *Commercial Operation:* November 1984.

Unit 3 *Ordered:* 1977. *Construction began:* 1979. *Criticality:* August 1986. *First Power:* September 1986. *Commercial Operation:* December 1986.

Unit 4 *Ordered:* 1977. *Construction began:* 1979. *Criticality:* July 1987. *First Power:* August 1987. *Commercial Operation:* November 1987.

Plant Report

Paks role in Hungarian energy picture. The Paks Nuclear Power Plant, located on the Danube River, plays an important role in Hungarian power production. It is owned and operated by the Hungarian Electricity Trust (MVMT), the national utility and supplies approximately half of the nation's electricity.

The four operating units. The four operating units at Paks are Soviet designed VVER-440 Model V213 reactors. The Model V213's are second-generation VVER plants which were developed at the same time the Soviets instituted uniform safety requirements. Improvements over older VVER reactors include an upgraded accident localization system, the addition of emergency core-cooling and auxiliary feedwater systems, and a stainless steel lining in the reactor vessel to reduce the problem of embrittlement.

Construction on the first of four units began in 1974 and commercial operation of Unit 4 began in November 1987. Completion of the four units marked the end of the first phase of the project.

Plans to construct two additional units at Paks, both Soviet-designed VVER-1000 reactors, were cancelled by the Hungarian government in 1989 following the collapse of the communist government in Hungary. Negotiations continue with French and Canadian organizations for possible future units at the Paks station.

Governmental regulation. Paks is regulated by the Hungarian Atomic Energy Commission (AEC), which was reorganized in 1991. The AEC's Nuclear Safety Inspectorate, which is responsible for nuclear safety and operator licensing, employs 10 resident inspectors at Paks.

International cooperation. Hungary was the first Eastern European country to request an International Atomic Energy Agency (IAEA) inspection of its nuclear reactors following the 1986 explosion at a Soviet-designed unit in the Chernobyl complex. An OSART (Operational Safety Review Team) inspection conducted at Paks 3 commended the unit for achieving an 86 percent availability factor. The team also noted that during the unit's 14 years of operation it had experienced a low number of unplanned outages and had experienced no serious accidents. In a follow-up visit, the team stated that Paks management was committed to safe operation and that 91 percent of the OSART recommendations had been carried out or were in process.

Further IAEA assistance is expected. An ASSET mission was scheduled to go to Paks in the fall of 1992 to assess corrective actions and exchange information regarding incident prevention.

In addition to its involvement with IAEA, Hungary is a member of the World Association of Nuclear Operators (WANO). Personnel at Paks participated in plant exchanges with the Three Mile Island Nuclear Power Plant and Limerick Nuclear Power Plant in Pennsylvania. Also, under a new WANO program, a peer review of plant operations is scheduled to be conducted in February 1992.

Upgrades at Paks. When the Paks facility first began operation, most of the staff training was done in the Soviet Union and at Eastern Europe Soviet-styled VVER reactors. Hungarian officials focused on developing domestic resources for training. The country established its own training facilities, which began to be used in 1983. In 1986, a separate Training Center was completed which included models, mock-ups, closed circuit television, a video laboratory and a special simulator. A full-scale simulator was completed in 1988 to help train staff and to test emergency procedures.

Other upgrades at the Paks Nuclear Power Plant include:

- A contract with the Finnish company Imatra Voima Oy for inspection services and quality-control support.
- An IAEA contract with Tecnatom, S.A., a Spanish firm, to help upgrade Paks's in-service inspection abilities.
- The development of an anti-terrorist guard unit, which is expected to be operational in 1992.

Sources

Nuclear Power Plant of Paks. Paksi Atomeromu Vallalat, n.d.
"Operational Statistics." Nuclear Power Plant Paks Operation Management, n.d.
Source Book: Soviet-Designed Nuclear Power Plants in the Former Soviet Republics and Czechoslovakia, Hungary and Bulgaria. U.S. Council for Energy Awareness, April 1992.

—Karen Bellenir

India

COUNTRY REPORT

Third World power production

India, where traditional sources of energy included firewood, charcoal, and dung, was the first Third World nation to operate a nuclear power station. The country's interest in nuclear generation began in 1945 with the creation of the Tata Institute of Fundamental Research in Bombay. The India Atomic Energy Commission was established in 1948. India's efforts received a needed impetus in 1953 when Dwight D. Eisenhower initiated an international program to develop peaceful uses of atomic power. India was one of thirteen Third World countries to receive personnel, training, and fissile materials.

A Department of Atomic Energy was set up in 1954 and the nation's first experimental reactor, the Apsara reactor, began operating in 1956. Although Apsara was built by Indian scientists, its fuel was supplied by the UK Atomic Energy Authority. During the 1960's India continued constructing experimental reactors to test various technologies including heavy-water moderated and plutonium fueled units. The nation's first commercial reactor, the Tarapur Atomic Power Station (TAPS), was built by the General Electric Co. (US). Its two boiling water units began production in 1969.

Reliance on heavy water technology

India turned to Canada for assistance in developing its nuclear industry. As a result, many of the nation's nuclear stations employ Candu-styled reactors, which use natural uranium for fuel and heavy water as a moderator. Canadian support, however, was terminated in 1974 when India exploded a nuclear device. Following Canadian withdrawal, India continued to develop domestic technology and to pursue independence from international supervision.

International safeguards refused

Although a few of India's early plants are monitored by the International Atomic Energy Agency (IAEA) to ensure safe operation and the peaceful use of fissile materials, many of the nation's domestically constructed stations operate without such safeguards. India refused to join 120 other nations in signing the Nuclear Non-Proliferation Treaty.

In 1983, the United States entered into an agreement to supply spare parts necessary for safe opeeration at TAPS. The agreement placed the station under international safeguards for ten years; however, when it expires India retains the right to remove TAPS from IAEA supervision. The Indian government has also asserted its right to reprocess spent fuel at TAPS and generate weapons-grade plutonium.

In 1984 and 1986, two units at the Madras Atomic Power Plant (MAPP) entered commercial production. They began operation behind schedule because the Indian government refused to permit IAEA monitoring which was necessary for the country to legally import the heavy water needed as a moderator. Although Indian officials insist their country produces sufficient quantities of heavy water to supply its own needs, some commentators have accused the country of illegally importing heavy water.

India's plans for future development

Controversy has not slowed India's ambitious nuclear construction program. The country plans to increase its nuclear power production through the construction of an additional 10,000 megawatts of generating capacity by the end of the century. By the year 2000, officials estimate nuclear production will account for ten percent of the country's total generating capacity.

In order to achieve nuclear independence, Indian researchers are constructing a waste management facility at Trombay. The country is also pursuing Fast Breeder Reactor technology. A small fast-breeder reactor at the Indira Gandhi Center for Atomic Research achieved criticality in 1985 and scientists continue working on a prototype for a 500-megawatt breeder.

Sources

"Power Plant Politics." *Critical Mass Bulletin*, December 1983.

"Heavy-Water Drought." *Far Eastern Economic Review*, August 31, 1989.

Mounfield, Peter. *World Nuclear Power*. Routledge, 1991.

Rahman, A., ed. *Science and Technology in India, Pakistan, Bangladesh and Sri Lanka*. Longman Guide to World Science and Technology, n.d.

—Karen Bellenir

PLANT PROFILES

93 ○
Kaiga Atomic Power Station, Units 1-2

Kaiga, Uttara Kannada, India

Owning Utility

India Department of Atomic Energy,   Phone: 4952811
    Nuclear Power Board   Telex: 011-2510
Dr. Homi Bhabha Rd.
Colaba
Bombay 40 00 05, India
S.L. Kati, Chairman

Contact

K.V.M. Rao, Secretary to the Board
India Department of Atomic Energy,
    Nuclear Power Board

Basic Facts

Units 1-2 *Status:* Under construction (40% complete as of January 1992). *Type of Reactor:* Pressurized heavy-water reactor. *Megawatts (electric):* 235. *Megawatts (thermal):* 801. *Reactor System Supplier:* Nuclear Power Corporation of India Ltd. *Architect Engineer:* India Department of Atomic Energy, Nuclear Power Board.

Key Dates

Unit 1 *Ordered:* 1987. *Construction began:* 1988. *Commercial Operation Expected:* December 1996.
Unit 2 *Ordered:* 1987. *Construction began:* 1988. *Commercial Operation Expected:* June 1997.

94 ○
Kakrapar Atomic Power Plant, Units 1-2 (KAPP)

PO: Anumala   Phone: 026264-245
Surat, Gujarat, India   Telex: 188 396 KAPP IN
J.P. Kalaiya, Station Supt.

Owning Utility

India Department of Atomic Energy,   Phone: 4952811
    Nuclear Power Board   Telex: 011-2510
Dr. Homi Bhabha Rd.
Colaba
Bombay 40 00 05, India
S.L. Kati, Chairman

Contact

K.V.M. Rao, Secretary to the Board
India Department of Atomic Energy,
    Nuclear Power Board

Basic Facts

Units 1-2 *Status:* Under construction (Unit 1, 93% complete as of January 1992; Unit 2, 75% complete as of January 1992). *Type of Reactor:* Pressurized heavy-water reactor. *Megawatts (electric):* 235. *Megawatts (thermal):* 801. *Reactor System Supplier:* Larson & Toubro (India); Walchandnagar Industries Ltd. (India). *Generator Supplier:* Bharat Heavy Electrical Ltd. (India). *Architect Engineer:* India Department of Atomic Energy, Nuclear Power Board.

Key Dates

Unit 1 *Ordered:* 1981. *Construction began:* 1983. *Commercial Operation Expected:* 1992.
Unit 2 *Ordered:* 1981. *Construction began:* 1983. *Commercial Operation Expected:* 1993.

95 ●
Madras Atomic Power Station, Units 1-2 (MAPS)

PO: Kalpakkam   Phone: Kalpakkam 331
Tamil Nadu, India   Fax: 04117-316
V. Rangarajan, Chief Supt.

Owning Utility

India Department of Atomic Energy,   Phone: 4952811
    Nuclear Power Board   Telex: 011-2510
Dr. Homi Bhabha Rd.
Colaba
Bombay 40 00 05, India
S.L. Kati, Chairman

Contact

K.V.M. Rao, Secretary to the Board
India Department of Atomic Energy,
    Nuclear Power Board

Basic Facts

Units 1-2 *Status:* Operable. *Type of Reactor:* Pressurized heavy-water reactor. *Megawatts (electric):* 235. *Megawatts (thermal):* 801. *Reactor System Supplier:* Larson & Toubro (India). *Generator Supplier:* Bharat Heavy Electrical Ltd. (India). *Architect Engineer:* India Department of Atomic Energy, Nuclear Power Board.

Key Dates

Unit 1 *Ordered:* 1967. *Construction began:* 1970. *Criticality:* July 1983. *First Power:* July 1983. *Commercial Operation:* January 1984.
Unit 2 *Ordered:* 1971. *Construction began:* 1971. *Criticality:* August 1985. *First Power:* September 1985. *Commercial Operation:* March 1986.

96 ●
Narora Atomic Power Plant, Units 1-2 (NAPP)

Narora   Phone: 05734-24660
Bulandshar UP, India   Fax: 05734-22177
C.M. Kothari, Station Supt.

Owning Utility

India Department of Atomic Energy,   Phone: 4952811
    Nuclear Power Board   Telex: 011-2510
Dr. Homi Bhabha Rd.
Colaba
Bombay 40 00 05, India
S.L. Kati, Chairman

| 97 | **Rajasthan**

Contact

K.V.M. Rao, Secretary to the Board
India Department of Atomic Energy,
   Nuclear Power Board

Basic Facts

Units 1-2 *Status:* Operable. *Type of Reactor:* Pressurized heavy-water reactor. *Megawatts (electric):* 235. *Megawatts (thermal):* 801. *Reactor System Supplier:* Unit 1, Walchandnager Industries Ltd. (India); Unit 2, Richardson & Cruddas (India). *Generator Supplier:* Bharat Heavy Electrical Ltd. (India). *Architect Engineer:* India Department of Atomic Energy, Nuclear Power Board.

Key Dates

Unit 1 *Ordered:* 1974. *Construction began:* 1976. *Criticality:* March 1989. *First Power:* July 1989. *Commercial Operation:* January 1991.
Unit 2 *Ordered:* 1974. *Construction began:* 1976. *Criticality:* August 1991. *First Power:* November 1991. *Commercial Operation:* April 1992.

| 97 | ●

Rajasthan Atomic Power Station, Units 1-2 (RAPS)

PO: Anushakti-Rawatbhata   Phone: KOTA 24412
Via Kota   Telex: 0305 240
Rajasthan, India
T.S.V. Ramanan, Chief Supt.

Owning Utility

India Department of Atomic Energy,   Phone: 4952811
   Nuclear Power Board   Telex: 011-2510
Dr. Homi Bhabha Rd.
Colaba
Bombay 40 00 05, India
S.L. Kati, Chairman

Contact

K.V.M. Rao, Secretary to the Board
India Department of Atomic Energy,
   Nuclear Power Board

Basic Facts

Units 1-2 *Status:* Operable. *Type of Reactor:* Pressurized heavy-water reactor. *Megawatts (electric):* 220. *Megawatts (thermal):* 693.5. *Reactor System Supplier:* Unit 1, GE Canada; Unit 2, Larson & Toubro (India). *Generator Supplier:* English Electric Co., Ltd. (Canada). *Architect Engineer:* Atomic Energy of Canada Ltd; Montreal Engineering Co. (Canada).

Key Dates

Unit 1 *Ordered:* 1964. *Construction began:* 1964. *Criticality:* August 1972. *First Power:* November 1972. *Commercial Operation:* December 1973.
Unit 2 *Ordered:* 1967. *Construction began:* 1968. *Criticality:* October 1980. *First Power:* November 1980. *Commercial Operation:* April 1981.

| 98 | ○

Rajasthan Atomic Power Station, Units 3-4 (RAPS)

PO: Anushakti-Rawatbhata   Phone: KOTA 24412
Via Kota   Telex: 0305 240
Rajasthan, India
T.S.V. Ramanan, Chief Supt.

Owning Utility

India Department of Atomic Energy,   Phone: 4952811
   Nuclear Power Board   Telex: 011-2510
Dr. Homi Bhabha Rd.
Colaba
Bombay 40 00 05, India
S.L. Kati, Chairman

Contact

K.V.M. Rao, Secretary to the Board
India Department of Atomic Energy,
   Nuclear Power Board

Basic Facts

Units 3-4 *Status:* Under construction (31% complete as of January 1992). *Type of Reactor:* Pressurized heavy-water reactor. *Megawatts (electric):* 235. *Megawatts (thermal):* 801. *Reactor System Supplier:* Nuclear Power Coproration of India Ltd. *Generator Supplier:* Bharat Heavy Electrical Ltd. (India). *Architect Engineer:* India Department of Atomic Energy, Nuclear Power Board.

Key Dates

Unit 3 *Ordered:* 1986. *Construction began:* 1988. *Commercial Operation Expected:* May 1995.
Unit 4 *Ordered:* 1986. *Construction began:* 1988. *Commercial Operation Expected:* November 1997.

| 99 | ●

Tarapur Atomic Power Plant, Units 1-2 (TAPP)

PO: TAPP   Phone: TARAPUR 2221
DIST: THANE   Telex: 132 209 TAPS IN
Maharashtra 401 504, India
B.K. Bhasin, Chief Supt.

Owning Utility

India Department of Atomic Energy,   Phone: 4952811
   Nuclear Power Board   Telex: 011-2510
Dr. Homi Bhabha Rd.
Colaba
Bombay 40 00 05, India
S.L. Kati, Chairman

Contact

K.V.M. Rao, Secretary to the Board
India Department of Atomic Energy,
   Nuclear Power Board

Basic Facts

Units 1-2 *Status:* Operable. *Type of Reactor:* Boiling water reactor. *Megawatts (electric):* 160. *Megawatts (thermal):* 535. *Reactor System*

Supplier: General Electric Co. (United States). *Generator Supplier:* General Electric Co. (United States). *Architect Engineer:* Bechtel.

Key Dates

Unit 1 *Ordered:* 1964. *Construction began:* 1964. *Criticality:* February 1969. *First Power:* April 1969. *Commercial Operation:* October 1969.

Unit 2 *Ordered:* 1964. *Construction began:* 1964. *Criticality:* February 1969. *First Power:* May 1969. *Commercial Operation:* October 1969.

Tarapur Atomic Power Plant, Units 3-4 (TAPP)

PO: TAPP
DIST: THANE
Maharashtra 401 504, India
B.K. Bhasin, Chief Supt.

Phone: TARAPUR 2221
Telex: 132 209 TAPS IN

Owning Utility

India Department of Atomic Energy,
   Nuclear Power Board
Dr. Homi Bhabha Rd.
Colaba
Bombay 40 00 05, India
S.L. Kati, Chairman

Phone: 4952811
Telex: 011-2510

Contact

K.V.M. Rao, Secretary to the Board
India Department of Atomic Energy,
   Nuclear Power Board

Basic Facts

Units 3-4 *Status:* On order. *Type of Reactor:* Pressurized heavy-water reactor. *Reactor System Supplier:* Nuclear Power Corporation of India, Ltd. *Architect Engineer:* India Department of Atomic Energy, Nuclear Power Board.

Unit 4 *Ordered:* . *Commercial Operation Expected:* May 2001.

Indonesia

COUNTRY REPORT

Indonesia launches nuclear power program

Early in the twenty-first century, Indonesia will begin operating its first nuclear power facilities. Over a twelve year period beginning in 2003, twelve nuclear power plants are scheduled to start operations. Power units, which will range from 600 to 1000 megawatts, will come on line at one- to two-year intervals according to Djali Ahimsa, director general of Batan, the Indonesian national atomic energy agency. The plants will help generate power to meet Indonesia's ever-growing demand for electricity—a need which will increase by 27,000 megawatts by the year 2015.

Although coal, hydropower, and geothermal energy will provide a substantial amount of electricity, Indonesia says nuclear power is necessary to provide service to its 180 million people as well as the rapidly expanding manufacturing sector which are currently plagued by serious power shortages. The first plant will be located on Java where demand increases about fifteen percent each year.

Competition for lucrative Indonesian contract

In the past decade, global construction of nuclear reactors has dramatically slowed; thus, a contract for the construction of Indondesia's twelve new reactors is highly attractive to companies such as Westinghouse and General Electric.

Westinghouse, a dominant force in the nuclear construction industry, is working to develop a new type of nuclear reactor, the AP600, and hopes to sell Indonesia on the idea of trying them out. The new reactors are not radically different from existing pressurized water reactors and boiling water reactors, but they have been streamlined for ease in construction and are cheaper and safer to operate. Among the most attractive features of the AP600 are its simplified controls and an emergency cooling system that makes use of natural forces such as gravity rather than more complex and vulnerable pump systems. Currently, the AP600 design is scheduled to be ready for commercial construction at the end of the 1990s—the same time that construction in Indonesia should begin.

Westinghouse's competition for the 600-megawatt reactors is coming from Mitsubishi, a company that works very closely with Westinghouse. By offering the traditional pressurized water reactors, the Mitsubishi proposal appeals to those who would rather not be part of an experimental nuclear program, as well as those who are concerned about construction costs. Other companies competing for the contract have concentrated on higher wattage reactors, arguing that Indonesia's power needs are increasing rapidly enough to justify the construction of the more powerful reactors.

Funding a problem

Regardless of which company is awarded the construction contract, implementation of the program cannot begin if Indonesia is unable to pay for the reactors. This is a serious question in light of the nation's $45 billion debt which would only increase with the construction of twelve nuclear power plants. Unfortunately for Indonesia, restrictions placed on members of the Organization for Economic Cooperation and Development (which includes the United States and Japan) prohibit loans for the export of nuclear technology. Thus, international funding is unlikely. Furthermore, some experts argue that Indonesia does not need nuclear power but should rely on its vast supply of natural gas to fuel conventional plants. All of these negative factors notwithstanding, the Indonesian President Suharto and Bucharuddin J. Habibie, Minister of Research and Technology, support the development of nuclear power and are pursuing that course.

Sources

"If You Can't Build Nukes at Home." *Business Week*, November 4, 1991.

"Indonesia Launches Nuclear Program." *Nuclear Industry*, First Quarter 1992.

—Linda M. Ross

Italy

COUNTRY REPORT

An uncertain start

Nuclear power plant construction in Italy began during the 1960's. Latina, a 160-megawatt MAGNOX reactor and Trino, a 270-megawatt pressurized water reactor, entered commercial operation in 1964 and 1965.

Environmental and political opposition, however, impeded further development of Italy's nuclear power industry. Seismic considerations complicated siting decisions (although many parts of the nation are considered geologically stable). Water, necessary for vital plant cooling functions, exists in appropriate supplies only in coastal areas. Throughout the decades of rapid nuclear expansion in the United States and other countries, Italy's shifting national governments lacked the stability to override local opposition to siting decisions. In March 1983, when a consensus of the political parties supported nuclear power expansion, the central government won for itself the ability to dictate location decisions.

Italy abandons nuclear power

The 1986 Chernobyl accident in the former Soviet Union increased public opposition to nuclear power in Italy. Following the incident, an Italian poll revealed that 71 percent of the participants were against nuclear power. This public opposition led to further political division.

In November 1987, Italy's citizens approved a referendum to abandon nuclear power. Despite the election results, Prime Minister Giovanni Goria asked the Cabinet to permit continued construction at Montalto di Castro Nuclear Power Plant near Rome. Montalto, with two 1,000 megawatt BWR units, was two-thirds complete. Mr. Goria claimed that the high costs of suspending work at the plant warranted the decision. His political opponents accused him of trying to overrule the referendum and he was forced to resign on March 11, 1988.

In August 1988, Italy's Counsel of Ministers adopted a national Energy Plan which called for a five-year suspension of all nuclear plant construction and a study of alternatives including the possibility of converting nuclear stations to fossil- generating plants. On December 10, 1988 all construction efforts at Montalto were officially abandoned.

In addition to Montalto, 13 other planned projects were cancelled or indefinitely postponed, including:

— Two units at Termoli nuclearr power station

- Cirene nuclear power plant
- Two units at Trino, Piedmont nuclear power station
- Two units at Apulia nuclear power station
- Units 13-16 at the Italy Power Station
- Two units at Lombardy nuclear power station

Two operating plants, Trino 1 and Caorso, were temporarily shut down in 1988. In 1990, their closure was affirmed and made permanent. The closure mandate also affected the completed Latina and Gargliano stations.

Possible future development

Italy's ENEL (Ente Nazionale per l'Energia Elettrica) continues to cooperate with on-going industry and international research and development. Intrinsically and passively safe reactors may be incorporated into a new nuclear power effort in the future.

Sources

World Nuclear Industry Handbook 1991. Reed Business Publishing Group, 1990.

Ente Nazionale Per L'Energia Elettrica (ENEL). Direzione Relazioni Publiche E Comunicazione, November 12, 1991.

Mounfield, Peter. *World Nuclear Power.* Routledge, 1991.

—*Karen Bellenir*

PLANT PROFILES

101 ★
Caorso Nuclear Power Plant

ENEL-DPT SEDE Distaccata Milano
Centrale Caorso
Casella Postale 7
29012 Caorso, Italy
Giuseppe Bolla, Plant Mgr.

Phone: 0523-821741
Telex: 530137

Owning Utility

Ente Nazionale per l'Energia Elettrica (ENEL)
Via G.B. Martini, 3
3-00198 Rome, Italy
Alberto Negroni, Gen.Mgr.

Phone: 06-85091
Fax: 06-85093-278
Telex: 610518

Contact

Pub. Relations & Commun. Dir.
Ente Nazionale per l'Energia Elettrica (ENEL)

Basic Facts

Status: Shut down. *Type of Reactor:* Boiling water reactor. *Megawatts (electric):* 882. *Megawatts (thermal):* 2651. *Reactor System Supplier:* Ansaldo Meccanico Nucleare SpA (Italy); General Electric Technical Services Co. (United States). *Generator Supplier:* Ansaldo Meccanico Nucleare SpA (Italy); Asgen. *Architect Engineer:* Gibbs & Hill, Inc. (United States).

Key Dates

Ordered: 1969. *Construction began:* 1970. *Criticality:* December 1977. *First Power:* May 1978. *Commercial Operation:* December 1981. *Shut Down:* June 1990.

102 ★
Garigliano Nuclear Power Plant

Sessa Aurunca, Italy

Owning Utility

Ente Nazionale per l'Energia Elettrica
Via G.B. Martini
3-00198 Rome, Italy
Alberto Negroni, Gen.Mgr.

Phone: 06-85091
Fax: 06-85093-278
Telex: 610518

Contact

Pub. Relations & Commun. Dir.
Ente Nazionale per l'Energia Elettrica (ENEL)

Basic Facts

Status: Shut down. *Type of Reactor:* Boiling water reactor. *Megawatts (electric):* 160. *Megawatts (thermal):* 506. *Reactor System Supplier:* General Electric Co. (United States). *Generator Supplier:* Ansaldo Meccanico Nucleare SpA (Italy). *Architect Engineer:* Ebasco.

Key Dates

Ordered: 1958. *Construction began:* 1959. *Criticality:* June 1963. *First Power:* January 1964. *Commercial Operation:* June 1964. *Shut Down:* March 1982.

103 ★
Latina Nuclear Power Plant

ENEL-DPT SEDE Distaccata Roma
Centrale Latina
Casella Postale 118
04100 Latina, Italy
Massiamo Sturvi, Plant Mgr.

Phone: 0773-28016

Owning Utility

Ente Nazionale per l'Energia Elettrica
Via G.B. Martini
3-00198 Rome, Italy
Alberto Negroni, Gen.Mgr.

Phone: 06-85091
Fax: 06-85093-278
Telex: 610518

Contact

Pub. Relations & Commun. Dir.
Ente Nazionale per l'Energia Elettrica (ENEL)

Basic Facts

Status: Shut down. *Type of Reactor:* Gas-cooled reactor. *Megawatts (electric):* 160. *Megawatts (thermal):* 650. *Reactor System Supplier:* The Nuclear Power Group (United Kingdom). *Generator Supplier:* Parsons; Ansaldo Meccanico Nucleare SpA (Italy). *Architect Engineer:* The Nuclear Power Group (United Kingdom).

Key Dates

Ordered: 1958. *Construction began:* 1958. *Criticality:* December 1962. *First Power:* May 1963. *Commercial Operation:* January 1964. *Shut Down:* December 1987.

104 ☆
Montalto di Castro Nuclear Power Plant, Units 1-2

Montalto di Castro, Italy

Owning Utility

Ente Nazionale per l'Energia Elettrica (ENEL)
Via G.B. Martini, 3
3-00198 Rome, Italy
Alberto Negroni, Gen.Mgr.

Phone: 06-85091
Fax: 06-85093-278
Telex: 610518

Contact

Pub. Relations & Commun. Dir.
Ente Nazionale per l'Energia Elettrica (ENEL)

Basic Facts

Units 1-2 *Status:* Cancelled (Construction 66.6% complete). *Type of Reactor:* Boiling water reactor. *Megawatts (electric):* 1009. *Megawatts (thermal):* 2894.

105 | Trino

Key Dates
Units 1-2 *Ordered:* 1974. *Cancelled:* December 1988. *Construction began:* 1977.

105 | ★
Trino Vercellese Nuclear Power Plant

ENEL-DPT SEDE Distaccata Torino Phone: 0161-8271
Centrale Trino
Strada Statale, 31 Bis
13039 Trino Vercellese, Italy
Gian Luigi Sacco, Plant Mgr.

Owning Utility
Ente Nazionale per l'Energia Elettrica Phone: 06-85091
Via G.B. Martini Fax: 06-85093-278
3-00198 Rome, Italy Telex: 610518
Alberto Negroni, Gen.Mgr.

Contact
Pub. Relations & Commun. Dir.
Ente Nazionale per l'Energia Elettrica
 (ENEL)

Basic Facts
Status: Shut down. *Type of Reactor:* Pressurized light-water reactor. *Megawatts (electric):* 270. *Megawatts (thermal):* 870. *Reactor System Supplier:* Westinghouse Electric Corp. (United States). *Generator Supplier:* Franco Tosi. *Architect Engineer:* Gibbs & Hill, Inc. (United States).

Key Dates
Ordered: 1956. *Construction began:* 1961. *Criticality:* June 1964. *First Power:* October 1964. *Commercial Operation:* January 1965. *Shut Down:* June 1990.

Japan

COUNTRY REPORT

Japan's generating mix

Japan is a nation with few domestic power sources. More than 80 percent of the raw material used to produce energy must be imported. Before World War II, Japan met most of its energy demand with coal, charcoal, firewood, and hydroelectric production. During the 1950's, Japan's nine electric utilities increased their generating ability by building more hydroelectric stations. During the 1960's they turned increasingly to oil-burning generating plants. By 1973, 71 percent of all the power generated in Japan came from oil, 99 percent of which was imported.

The Arab oil embargo of 1973-74 and subsequent price increases had a profound effect on Japanese interest in nuclear production. Demand for electricity in Japan had tripled between 1960 and 1975. The Japanese economy was vulnerable because of its dependence on foreign sources for energy. Policy makers insisted that the future prosperity of Japan depended on an available supply of reasonably priced energy. They turned to nuclear technology for solutions to the country's power problems.

The history of nuclear power in Japan

The Japanese nuclear power program began in 1952 when U.S. occupation ended and a ban on nuclear research was lifted. President Eisenhower's "Atoms for Peace" speech in 1953 intensified the country's interest in nuclear development.

The Japaneses government passed the Atomic Energy Basic Law in 1955 which required that the development of nuclear energy be a civilian undertaking. The Japan Atomic Energy Commission (AEC) was established to supervise the promotion of nuclear power. In 1956 the AEC set its primary goal: to achieve self-reliance through the development of breeder reactors and a complete domestic fuel cycle. This early goal has been consistently maintained and reaffirmed. Modern policy makers, anticipating shortages of fossil fuels in the 21st century, still embrace it.

Japan's first attempts at nuclear production began between 1957 and 1963 with the completion of six experimental reactors ranging in capacity from 10 to 50 megawatts. The first commercial reactor, a 166-megawatt, gas-cooled British unit was completed in 1966 at Tokai-mura 65 miles northeast of Tokyo.

Japan's next reactors were built by U. S. corporations who licensed Japanese companies to construct plants using U. S. technology. General Electric Co. (US) licensed Hitachi and

Toshiba to produce boiling water reactors. Westinghouse licensed Mitsubishi Heavy Industries to build pressurized water reactors. Japanese scientists modified the designs, and Japan now has its own version of a standardized light water reactor (LWR).

Nuclear power in the 1990s

Japan entered the 1990's with plans for aggressive growth in nuclear capacity. An official with the Tokyo Electric Power Co. stated that the utility would be unable to supply sufficient power to the region unless it added 2,000 megawatts of generating capacity per year.

Projects underway by Japanese nuclear interests include:

Hitachi and Toshiba are involved in a project with GE to perfect a simplified boiling water reactor.

Researchers are developing an Advanced Thermal Reactor (ATR), a heavy-water reactor which uses plutonium.

A Very High Temperature Reactor (VHTR) is under development. The VHTR would help alleviate nuclear waste problems because of its high efficiency. It also offers the possibility of direct use in the steel and heavy chemical industries.

The Japanese hope work on an integral fast reactor, which combines breeder and reprocessing facilities, will help overcome problems inherent in shipping plutonium.

The development of fusion is seen as the ultimate end of Japan's nuclear energy programs.

In Japan the 1990's will also see increased efforts in decommissioning research. Two experimental decommissioning procedures are being tested by the Japan Atomic Energy Research Institute (JAERI), in Tokai, site of Japan's first reactor.

Japan currently has more than 40 reactors in operation. Combined, they supply more than 25 percent of the nation's need for electricity. The Japanese government plans to increase nuclear generation to produce 60 percent of the demand by 2030.

Partnership of government and industry

Cooperation between the Japanese government and the nation's nuclear industry has been credited for rapid advances in technology. Advocates of the country's programs claim that the efficiency of Japan's centralized system enabled the Japanese to catch up with the rest of the world. Opponents of nuclear power, however, charge that progress has been made at the expense of public discussion about nuclear safety issues.

The Japanese Ministry of International Trade & Industry (MITI), a policy-setting agency with extensive responsibilities, has continued Japan's push toward achieving self-sufficiency in energy production. MITI regulates utility rates, licenses new reactor construction, monitors safety at nuclear plants, and is supposed to act as an impartial arbiter between government and industry. Critics claim MITI has conflicting interests because of its wide scope of authority and they accuse the agency of acting as an industry representative rather than in the public interest. For example, in the early 1980's, in order to achieve a ten percent cost reduction in nuclear power plant construction, MITI agreed to relax safety precautions, including some earthquake defense systems.

Siting policy

One of the most controversial parts of the Japanese nuclear program is plant siting. The country is a little smaller than California and has a population of 117 million. Japan has ten percent of the world's active volcanoes and the country is subject to frequent earthquakes, about 500 per year. In some areas, earthquakes registering 7 and 8 on the Richter scale are not uncommon. Seismic activity sometimes causes giant tidal waves called tsunamis. Most of the population lives on only 30 percent of the land.

The Japanese geography has resulted in plant siting decisions which have placed many nuclear plants in small, densely populated areas. Coastal regions are preferred because they offer a steady supply of water for plant cooling systems. Japanese rivers are frequently unsuited for nuclear power plants because they tend to be small with extreme seasonal fluctuations.

In 1974, the Japanese government passed a series of laws to make nuclear siting attractive to rural sea-side communities. Nuclear plants were proposed as a means to spur local economic growth. Tax benefits and construction projects were promised to towns accepting nuclear power projects. Discounted electric bills were offered to people who lived in areas near the plants.

With the coordinating abilities of MITI, utility companies were able to acquire land and licenses quickly. Secrecy kept protestors away. By the middle 1980's, the best sites had been taken. Critics claim that in subsequent construction, safety has been compromised by placing nuclear plants in locations prone to earthquakes. The government has considered the possibility of developing offshore sites.

Uranium reprocessing and enrichment

In order to fuel its reactors, Japan imports uranium from Canada, Australia, the U.S., Britain, France, and Africa. China is a potential future supplier. Enrichment and reprocessing are both necessary if Japan is to achieve self-sufficiency. In 1980, Japan made an important step towards energy independence when the Fugen plant was re-fueled with domestically recovered plutonium.

A reprocessing plant began operation at the Tokaimura site on January 17, 1981, but because of numerous mishaps it produced only about half of its anticipated output. Plans for two additional reprocessing plants are on the drawing board. Some analysts, however, are fearful about the development of reprocessing technology. They see it as a dangerous precedent for other countries who may wish to produce plutonium for weapons.

A pilot enrichment plant began operation in central Japan in March 1982. The facility is able to produce 12 tons per year (a six-month supply for a 1,000 megawatt station). A larger demonstration enrichment facility was completed at the same site in 1988.

A project to complete Japan's domestic nuclear fuel cycle is underway in the Aomori Prefecture. The facility, called Rokkasho, will include uranium enrichment, reprocessing, and low-level waste disposal. Originally slated to begin operation in the 1990's, Rokkasho is now scheduled to open in 2000. Until the Japanese are able to handle their own spent fuel, it is sent to the UK and France for reprocessing.

Plutonium shipments

Japan's original enriched uranium fuel came from the United States. Under the agreement, Japan needed U.S. permission for each planned shipment of used fuel intended for reprocessing. In 1988, the U.S. and Japan signed a new agreement giving Japan a 30-year approval to extract plutonium from its spent fuel.

In 1984, a plutonium shipment made between France and Japan was escorted by both French and U.S. naval vessels because of the risks of terrorist activities and theft. Japan has been charged with the responsibility of developing its own escort vessel to protect future shipments. The plans have drawn international attention because the amount of plutonium involved is enough to produce 20,000 bombs. Environmental groups have also expressed concern about the potential hazards posed if a ship were sunk, either as a result of terrorist activity or an accident.

Japan planned to use plutonium in its commercial breeder reactors, but because of delays in developing breeder technology and because of the current abundance in uranium supplies for conventional reactors, the nation may experience a plutonium glut. The AEC estimates that Japan's inventory of plutonium will be 84 tons by 2010; and, therefore, officials have recommended using it in light water reactors by combining it with regular uranium fuel.

Development of breeder technology

Japanese research in breeder technology began in 1967. The Breeder Reactor was deemed vital to Japan's nuclear program because supplies of uranium were expected to last only 50 years. This projection could be extended with breeder capability because breeder reactors produce more fuel than they consume. A small experimental breeder station, called Joyo, began operation in 1977.

Monju is a larger breeder-reactor prototype with a capacity of 280-megawatts. It is expected to be 70 times more efficient in its use of uranium than traditional light water reactors. Although original estimates called for completion in 1989 at a price of $1.6 billion, the station was not finished until 1991 and final costs totaled $4.3 billion. Monju is scheduled to achieve criticality in October 1992. Japanese planners do not expect to have a full-scale commercial breeder reactor operational until 2030.

Growing public opposition

The early years of nuclear power development in Japan met little public opposition. Prior to 1970 the country had no method for conducting public hearings. Environmental groups began impacting siting decisions during the 1970's but did not gain momentum until after the former Soviet Union's Chernobyl accident in 1986. Following Chernobyl, for the first time in Japan's nuclear history, more of its citizens opposed nuclear power than supported it. Three million Japanese petitioned for a nuclear moratorium. A government poll in 1990 indicated that 90 percent of the respondents were "uneasy" about nuclear power, 46 percent thought it was unsafe, but 65 percent thought it was necessary.

To allay fears, Japanese officials cite favorable safety statistics, the low number of unplanned shutdowns (0.4 per unit per year in 1988), and better than average capacity factors. Critics charge that the statistics only reflect a tendency to keep plants operating even when they are in trouble. An accident at the Mihama station in 1991 prompted MITI to set up guidelines on when to shut down plants. Under the new rules, an abnormal increase in radioactivity of 20 percent or more results in a shutdown.

Recent years have seen an increase in the number of anti-nuclear protests in Japan. Some examples include:

In January 1991, a provincial meeting regarding a proposed plant in the Shimane Prefecture was protested by thousands of demonstrators. Twelve hundred riot police were called in to guard the meeting.

Four hundred anti-nuclear protestors built "solidarity" huts at the Kashiwazaki-Kariwa nuclear plant. The huts were destroyed in February 1991.

Protestors accused the government of bribing families in Ikata to win their approval of a nuclear construction project.

Officials blame anti-nuclear forces for complicating nuclear plant siting, lengthening the time required to start building and for causing Japan to fall behind its projected construction timetable. Despite these problems, the Japanese government approved an official policy of continued nuclear development as recently as 1991.

Problems of nuclear waste management

Nuclear waste generated in Japan is stored in temporary locations until a long-term storage facility and nuclear waste policy can be developed. The possibility of ocean-dumping was considered during the 1970's and early 1980's. Ocean-dumping proved unworkable because it was protested by Japan's domestic fishing industry and international organizations. The idea was officially abandoned in 1984. This decision led to the policy of storing waste at reactor sites which incited increased opposition from local communities.

MITI hopes to reduce the amount of waste through reprocessing efforts, but significant amounts of waste will remain. The Rokkasho facility, if it opens on schedule, will be able to process 800 tons per year, but plans for permanent storage at the site were thwarted in 1990 when local officials voted against it.

Sources

"Japan: The Nuclear Victor?" *Forbes*, December 6, 1982.

"Nuclear Power in Japan." *The Bulletin of the Atomic Scientists*, May 1983.

"Japan's Nuclear Juggernaut." *Management Today*, October 1984.

"Development of Nuclear Energy in Public Utility — Present Status and Perspectives of PWR." *Business JAPAN*, February 1989.

"GE Nuclear Wins $50 Million Program for Simplified Reactor." *Business Wire*, September 7, 1989.

"Public Understanding Needed in Building Nuclear Power Plants." *Business JAPAN*, February 1990.

"Nuclear Power Development Program of Kansai Electric Power Company." *Business JAPAN*, February 1990.

"Nuclear Power Generation in Japan." *Business JAPAN*, February 1990.

"Plutonium Terror on the High Seas?" *The New York Times*, April 28, 1990.

"Tokyo's Nuclear-Power Plans." *The Wall Street Journal*, June 6, 1990.

"Japan Pushes Nuclear Decommissioning Work." *ENR*, September 13, 1990.

"Japan Nuclear Plants Get Rules in Wake of Mishap." *The Wall Street Journal*, February 19, 1991.

"15 New Nuclear Plants Began Commercial Operation in '90." *INFO*, U.S. Council for Energy Awareness, July 1991.

"Japan Debates Plutonium." *Nature*, July 4, 1991.

"Monju Fast-Breeder Reactor Completed." *Nature*, July 4, 1991.

Mounfield, Peter. *World Nuclear Power*. Routledge, 1991.

"Japan's Nuclear Dilemma." *Technology Review*, October 1991.

"Japanese Cabinet: More Nuclear Energy." *Nuclear Industry*, First Quarter 1992.

—*Karen Bellenir*

PLANT PROFILES

106 ●
Fugen ATR Prototype Nuclear Power Plant

3 Myoujin-cho
Tsuruga-city
Fukui-ken, Japan
Tanemichi Kitahara, Dir.

Phone: 0770-26-1221

Owning Utility

Power Reactor & Nuclear Fuel
   Development Corp.
1-9-13 Akasaka
Minato-ku
Tokyo 107, Japan
Michia Akebi, Exec. Dir., Constr. & Opr.
   Fugen & Monju

Phone: 03-586-3311
Telex: J26462

Contact

Hideaki Ashiwara, Dir., PR Office
Power Reactor & Nuclear Fuel
   Development Corp.

Basic Facts

Status: Operable. *Type of Reactor:* Heavy-water/light-water reactor. *Megawatts (electric):* 165. *Megawatts (thermal):* 557. *Reactor System Supplier:* Hitachi; Mitsubishi Heavy Industries, Ltd. (Japan); Sumitomo Heavy Industries, Ltd. (Japan); Fuji. *Generator Supplier:* Toshiba. *Architect Engineer:* Power Reactor & Nuclear Fuel Development Corp. (PNC); Electric Power Development Co., Ltd. (Japan).

Key Dates

Ordered: 1970. *Construction began:* 1971. *Criticality:* May 1978. *First Power:* July 1978. *Commercial Operation:* March 1979.

Plant Report

ATR technology. Japan's ultimate nuclear goal is the development of a reliable domestic power source. In 1966, the Japan Atomic Energy Commission named Fast Breeder Reactor (FBR) technology a national project. An Advanced Thermal Reactor (ATR) was seen as an intermediate step necessary to FBR development. The ATR, intended for use until breeder technology could become commercially feasible, features a heavy water reactor design which was developed by Japan's Power Reactor and Nuclear Fuel Development Corporation (PNC). It can use uranium and plutonium for fuel. Japanese officials expect ATR reactors to help use the stockpiles of plutonium produced from reprocessing the spent fuel used in Japan's light water reactors.

The Fugen facility. The Fugen Nuclear Power Station, a 165-megawatt prototype ATR plant, was built in Tsuruga, in the Fukui Prefecture. It began operation in 1979 and has provided power to three utility companies. The station has achieved a life-time capacity factor of 63 percent, considered high for a research facility. A milestone was reached in 1988 when Fugen was refueled with fuel reprocessed from its own spent fuel.

Other advances made at the Fugen station include developing a refueling machine, designing a heavy water upgrader, and improving inservice inspection techniques. Continued research to develop high-performance fuels and study the use of artificial intelligence in operations and maintenance is underway.

Next steps. The goal of the Fugen project is to transfer its technological advances to commercial application. The Electric Power Development Company plans to build a 606-megawatt ATR demonstration plant based on the Fugen experience. The new ATR station will be sited in Ohma in the Aomori Prefecture, which is in northern Japan. Construction is expected to begin in 1993 and the plant is slated for initial operation by 2000.

Sources

PNC: *Power Reactor and Nuclear Fuel Development Corporation.* N.p., September 1991.
"Japan's Nuclear Dilemma." *Technology Review,* October 1991.

—Karen Bellenir

107 ●
Fukushima Daiichi Nuclear Power Station, Units 1-6

[*See Also:* Fukushima Daini Nuclear Power Station, Units 1-4]

No. 22 Kitahara, Ottozawa
Ookumi-Machi, Futaba-Gun
Fukushima Prefecture 979-13, Japan
Tachimori Ooba, Supt.

Phone: 0240-32-2101

Owning Utility

Tokyo Electric Power Co.
No. 1-3, 1-Chome
Uchisaiwai-cho
Chiyoda-ku
Tokyo 100, Japan
Shiro Sasaki, Managing Dir., Nuclear
   Power

Phone: 03-3501-8111
Fax: 03-3596-8544
Telex: TODEN-J 2224045

Contact

Hiroshi Kanagawa, Gen.Mgr., Nuclear
   Power Info. Center
Tokyo Electric Power Co.

Basic Facts

Units 1-6 *Status:* Operable. *Type of Reactor:* Boiling water reactor. *Megawatts (electric):* Unit 1, 460; Units 2-5, 784; Unit 6, 1100. *Megawatts (thermal):* Unit 1, 1380; Units 2-5, 2381; Unit 6, 3293. *Reactor System Supplier:* Units 1, 2, and 6, General Electric (United States); Units 3 and 5, Toshiba; Unit 4, Hitachi. *Generator Supplier:* Units 1, 2, and 6, General Electric (United States); Units 3 and 5, Toshiba; Unit 4, Hitachi. *Architect Engineer:* Units 1, 2, and 6, Ebasco; Units 3 and 5, Toshiba; Unit 4, Hitachi.

Key Dates

Unit 1 *Ordered:* 1966. *Construction began:* 1967. *Criticality:* October 1970. *First Power:* November 1970. *Commercial Operation:* March 1971.

Fukushima

Unit 2 *Ordered:* 1968. *Construction began:* 1969. *Criticality:* May 1973. *First Power:* December 1973. *Commercial Operation:* July 1974.

Unit 3 *Ordered:* 1970. *Construction began:* 1970. *Criticality:* September 1974. *First Power:* October 1974. *Commercial Operation:* March 1976.

Unit 4 *Ordered:* 1972. *Construction began:* 1972. *Criticality:* January 1978. *First Power:* February 1978. *Commercial Operation:* October 1978.

Unit 5 *Ordered:* 1971. *Construction began:* 1971. *Criticality:* August 1977. *First Power:* September 1977. *Commercial Operation:* April 1978.

Unit 6 *Ordered:* 1972. *Construction began:* 1973. *Criticality:* March 1979. *First Power:* May 1979. *Commercial Operation:* October 1979.

Plant Report

Ten reactors. The Fukushima nuclear complex has 10 reactors on two sites, the largest collection of reactors in Japan. It is owned by the Tokyo Electric Power Company (TEPCO) and is located on the Pacific coast approximately 120 miles north of Tokyo. Fukushima I's first reactor is a 460-megawatt boiling water reactor. Four other units at Fukushima I are 784-megawatt boiling water reactors. The largest unit at Fukushima I is a 1,100-megawatt boiling water reactor. Fukushima II has four units, all of which are 1,100 megawatt BWR's. At Fukushima II the Japanese method of standardization enabled builders to achieve an average construction period of 65 months.

Initial opposition. Construction at Fukushima began in 1967, but not without local protesting. Farmers in the area challenged the siting and caused delays. Protestors accused the Ministry of International Trade and Industry (MITI) of holding hearings only after decisions were already made. In a 1973 decision, the Fukushima District Court rejected a suit against the government for approving the first two reactors at Fukushima. The judge ruled that the government had the right to make siting decisions.

In a campaign to win local approval and settle disputes, the Japanese government offered incentives and TEPCO, the largest of Japan's nine electric utilities, gave gifts to area residents. The gifts came in the form of direct presents and loans for farming equipment. TEPCO also subsidized festivals for the community and sponsored cultural events.

Community relations and operating troubles. Although the project brought initial benefits to the surrounding community, it also resulted in increased unemployment. Workers were laid off when construction was completed and the new farming equipment eliminated many traditional jobs. Critics claimed that fishing grounds had been contaminated and that plant workers were being exposed to excessive radiation.

Fukushima's attempts to win the approval of the community were further thwarted by an incident at the plant in 1989. Operators were forced to shut down one of the 1,100-megawatt reactors because of excessive vibration in a recirculation pump. Investigators discovered that the pump had been partially destroyed and that pieces had entered the reactor vessel, clogging the circulation system and threatening to cut off coolant. Improper welding and operator error were blamed. Clean-up efforts, safety inspections, and negotiations with local officials kept the reactor shut down for two years.

Sources

"Japan's Nuclear Juggernaut." *Management Today*, October 1984.
The Present Status of Nuclear Power Development. Kyushu Electric Power Co., June 1991.
"Japan's Nuclear Dilemma." *Technology Review*, October 1991.
Mounfield, Peter. *World Nuclear Power.* Routledge, 1991.

—Karen Bellenir

Fukushima Daini Nuclear Power Station, Units 1-4

[*See Also:* Fukushima Daiichi Nuclear Power Station, Units 1-6]

No. 12 Kohamasaku, Namikura
Naraha-Machi, Futaba-Gun
Fukushima Prefecture 979-06, Japan
Takaya Seko, Supt.

Phone: 0240-25-4111

Owning Utility

Tokyo Electric Power Co.
No. 1-3, 1-Chome
Uchisaiwai-cho
Chiyoda-ku
Tokyo 100, Japan
Shiro Sasaki, Managing Dir., Nuclear Power

Phone: 03-3501-8111
Fax: 03-3596-8544
Telex: TODEN-J 2224045

Contact

Hiroshi Kanagawa, Gen.Mgr., Nuclear Power Info. Center
Tokyo Electric Power Co.

Basic Facts

Units 1-4 *Status:* Operable. *Type of Reactor:* Boiling water reactor. *Megawatts (electric):* 1100. *Megawatts (thermal):* 3293. *Reactor System Supplier:* Units 1 and 3, Toshiba; Units 2 and 4, Hitachi. *Generator Supplier:* Units 1 and 3, Toshiba; Units 2 and 4, Hitachi. *Architect Engineer:* Units 1 and 3, Toshiba; Units 2 and 4, Hitachi.

Key Dates

Unit 1 *Ordered:* 1975. *Construction began:* 1975. *Criticality:* June 1981. *First Power:* July 1981. *Commercial Operation:* April 1982.

Unit 2 *Ordered:* 1978. *Construction began:* 1979. *Criticality:* April 1983. *First Power:* June 1983. *Commercial Operation:* February 1984.

Unit 3 *Ordered:* 1980. *Construction began:* 1980. *Criticality:* October 1984. *First Power:* December 1984. *Commercial Operation:* June 1985.

Unit 4 *Ordered:* 1980. *Construction began:* 1980. *Criticality:* October 1986. *First Power:* December 1986. *Commercial Operation:* September 1987.

Plant Report

Ten reactors. The Fukushima nuclear complex has 10 reactors on two sites, the largest collection of reactors in Japan. It is owned by the Tokyo Electric Power Company (TEPCO) and is located on the Pacific coast approximately 120 miles north of Tokyo. Fukushima I's first reactor is a 460-megawatt boiling water reactor. Four other units at Fukushima I are 784-megawatt boiling water reactors. The largest unit at Fukushima I is a 1,100-megawatt boiling water reactor. Fukushima II has four units, all of which are 1,100 megawatt BWR's. At Fukushima II the Japanese method of standardization enabled builders to achieve an average construction period of 65 months.

Initial opposition. Construction at Fukushima began in 1967, but not without local protesting. Farmers in the area challenged the siting and caused delays. Protestors accused the Ministry of International Trade and Industry (MITI) of holding hearings only after decisions were already made. In a 1973 decision, the Fukushima District Court rejected a suit against the government for approving the first two reactors at Fukushima. The judge ruled that the government had the right to make siting decisions.

In a campaign to win local approval and settle disputes, the Japanese government offered incentives and TEPCO, the largest of Japan's nine electric utilities, gave gifts to area residents. The gifts came in the form of direct presents and loans for farming equipment. TEPCO also subsidized festivals for the community and sponsored cultural events.

Community relations and operating troubles. Although the project brought initial benefits to the surrounding community, it also resulted in increased unemployment. Workers were laid off when construction was completed and the new farming equipment eliminated many traditional jobs. Critics claimed that fishing grounds had been contaminated and that plant workers were being exposed to excessive radiation.

Fukushima's attempts to win the approval of the community were further thwarted by an incident at the plant in 1989. Operators were forced to shut down one of the 1,100-megawatt reactors because of excessive vibration in a recirculation pump. Investigators discovered that the pump had been partially destroyed and that pieces had entered the reactor vessel, clogging the circulation system and threatening to cut off coolant. Improper welding and operator error were blamed. Clean-up efforts, safety inspections, and negotiations with local officials kept the reactor shut down for two years.

Sources

"Japan's Nuclear Juggernaut." *Management Today*, October 1984.
The Present Status of Nuclear Power Development. Kyushu Electric Power Co., June 1991.
"Japan's Nuclear Dilemma." *Technology Review*, October 1991.
Mounfield, Peter. *World Nuclear Power*. Routledge, 1991.

—Karen Bellenir

---

109 ●
Genkai Nuclear Power Plant, Units 1-2

Oaza-Imamura, Genkai-cho　　　　Phone: 0955-52-6821
Higashimatsuura-gun
Saga-ken 847-14, Japan
Hiroto Kitamura, Gen.Mgr.

Owning Utility

Kyushu Electric Power Co., Inc.　　　Phone: 092-761-3031
1-82 Watanabe-Dori　　　　　　　　Fax: 092-761-4622
2-Chrome, Chuo-ku　　　　　　　　Telex: 725497 KYUDEN J
Fukuoka 810, Japan
Hideo Kodama, Gen.Mgr., Nuclear
　Operations

Contact

Yusufumi Yarimizu, Gen.Mgr., PR
Kyushu Electric Power Co., Inc.

Basic Facts

Units 1-2 *Status:* Operable. *Type of Reactor:* Pressurized light-water reactor. *Megawatts (electric):* 559. *Megawatts (thermal):* 1650. *Reactor System Supplier:* Mitsubishi Heavy Industries, Ltd. (Japan). *Generator Supplier:* Mitsubishi Heavy Industries, Ltd. (Japan); Mitsubishi Electric Corp. (Japan). *Architect Engineer:* Mitsubishi Atomic Power Industries, Inc. (Japan).

Key Dates

Unit 1 *Ordered:* 1968. *Construction began:* 1969. *Criticality:* 1974. *First Power:* February 1975. *Commercial Operation:* October 1975.
Unit 2 *Ordered:* 1976. *Construction began:* 1976. *Criticality:* 1980. *First Power:* June 1980. *Commercial Operation:* March 1981.

Plant Report

Two operating units. The Genkai Nuclear Power Station occupies a 210-acre site in Imamura, Genkai Town in the Saga Prefecture on the island of Kyushu. Kyushu is the most southern of Japan's four main islands. Genkai is one of two nuclear stations owned and operated by Kyushu Electric Power Company; the other is Sendai. Nuclear generation provides approximately one third of the electricity supplied by Kyushu Electric. The remaining power is supplied from oil, gas, coal, and hydroelectric sources, with a small amount from geothermal facilities.

The Genkai station has two operating reactors and two additional units under construction. Genkai-1, the first nuclear unit on Kyushu, was also Japan's first domestically manufactured pressurized water reactor and it played a role in the development and construction of other PWR's in the nation.

Additional units under construction. In December 1978, in order to meet anticipated future demand for power, Kyushu Electric requested government approval for the addition of two new units at Genkai. Construction on Genkai 3 and 4 began in 1985. Both will be 1,180 megawatt units. Unit 3 is scheduled to begin operation in March 1994 and Unit 4 in July 1997. They feature an improved design to achieve higher reliability. The new units will also be housed in Prestressed Concrete Containment Vessels (PCCV's). The PCCV, an innovative containment vessel, is smaller than standard containment structures and is designed to withstand greater seismic dangers.

A High Temperature Incinerator (HTI), which will be used to reduce the volume of radioactive wastes, is also being developed at the site. Burned radioactive waste will be converted into granules which are stable and leach-resistant.

Sources

Kyushu Electric Power Company, Incorporated: Annual Report 1991. Kyushu Electric Power Co., n.d.

Present Status of Nuclear Power Development. Kyushu Electric Power Co., June 1991.

Genkai Nuclear Power Station Units 3 & 4. Kyushu Electric Power Co., n.d.

—Karen Bellenir

Genkai Nuclear Power Plant, Units 3-4

Oaza-Imamura, Genkai-cho Phone: 0955-52-6821
Higashimatsuura-gun
Saga-ken 847-14, Japan
Hiroto Kitamura, Gen.Mgr.

Owning Utility

Kyushu Electric Power Co., Inc. Phone: 092-761-3031
1-82 Watanabe-Dori Fax: 092-761-4622
2-Chrome, Chuo-ku Telex: 725497 KYUDEN J
Fukuoka 810, Japan
Hideo Kodama, Gen.Mgr., Nuclear Operations

Contact

Yusufumi Yarimizu, Gen.Mgr., PR
Kyushu Electric Power Co., Inc.

Basic Facts

Units 3-4 *Status:* Under construction (Unit 3, 83% complete as of January 1992; Unit 4, 29% complete as of January 1992). *Type of Reactor:* Pressurized light-water reactor. *Megawatts (electric):* 1180. *Megawatts (thermal):* 3423. *Reactor System Supplier:* Mitsubishi Heavy Industries, Ltd. (Japan). *Generator Supplier:* Mitsubishi Heavy Industries, Ltd. (Japan); Mitsubishi Electric Corp. (Japan). *Architect Engineer:* Mitsubishi Atomic Power Industries, Inc. (Japan).

Key Dates

Unit 3 *Ordered:* 1985. *Construction began:* 1985. *Commercial Operation Expected:* March 1994.

Unit 4 *Ordered:* 1985. *Construction began:* 1985. *Commercial Operation Expected:* July 1997.

Plant Report

See Plant Report for Genkai Nuclear Power Plant, Units 1-2.

Hamaoka Nuclear Power Plant, Units 1-3

5561 Sakura, Hamaoka-cho Phone: 0537-86-3481
Ogasa-gun Fax: 0537-86-4393
Shizuoka Prefecture, 437-16, Japan
Yoshikazu Ando, Gen.Mgr.

Owning Utility

Chubu Electric Power Co. Phone: 052-951-8211
No. 1 Toshin-cho Fax: 052-962-4624
Higashi-ku
Nagoya, 461-91, Japan
Jun Aoki, Gen.Mgr., Info. & Pub.Aff.

Contact

Hiroshi Inaba, Mgr. of PR Sect.
Hamaoka Nuclear Power Plant

Basic Facts

Units 1-3 *Status:* Operable. *Type of Reactor:* Boiling water reactor. *Megawatts (electric):* Unit 1, 540; Unit 2, 840; Unit 3, 1100. *Megawatts (thermal):* Unit 1, 1593; Unit 2, 2436; Unit 3, 3293. *Reactor System Supplier:* Toshiba. *Generator Supplier:* Hitachi. *Architect Engineer:* Unit 1, Toshiba; Units 2-3, Toshiba, Hitachi.

Key Dates

Unit 1 *Ordered:* 1969. *Construction began:* 1971. *Criticality:* June 1974. *First Power:* August 1974. *Commercial Operation:* March 1976.

Unit 2 *Ordered:* 1972. *Construction began:* 1974. *Criticality:* March 1978. *First Power:* May 1978. *Commercial Operation:* November 1978.

Unit 3 *Ordered:* 1982. *Construction began:* 1982. *Criticality:* November 1986. *First Power:* January 1987. *Commercial Operation:* September 1987.

Plant Report

Site called "Nest of Earthquakes". The Hamaoka Nuclear Power Station, owned by the Chubu Electric Power Company, is situated on Japan's Pacific coast, south of Tokyo. The power plant has three operating reactors, and a fourth unit is under construction.

Antinuclear activists have criticized Japan's Ministry of International Trade and Industry (MITI) and Chubu Electric for choosing the Hamaoka site which they term a "Nest of Earthquakes." Jinzaburo Takagi, a nuclear chemist and industry opponent, claimed that the fact that Hamaoka-3 was built with much more stringent seismic standards than the first two units indicates that the earlier standards were too lenient.

Cooling system failure. Hamaoka-3 was temporarily shut down in April 1991 following a pump failure. Although no radiation was released during the incident, it added to the growing antinuclear sentiment in Japan.

Sources

"Japan's Nuclear Juggernaut." Management Today, October 1984.
"Another Nuclear Plant is Shut." The Wall Street Journal, April 5, 1991.

Plant Profiles

Present Status of Nuclear Power Development. Kyushu Electric Power Co., June 1991.

—Karen Bellenir

112 ○
Hamoaka Nuclear Power Plant, Unit 4

5561 Sakura, Hamaoka-cho Phone: 0537-86-3481
Ogasa-gun Fax: 0537-86-4393
Shizuoka Prefecture, 437-16, Japan
Yoshikazu Ando, Gen.Mgr.

Owning Utility

Chubu Electric Power Co. Phone: 052-951-8211
No. 1 Toshin-cho Fax: 052-962-4624
Higashi-ku
Nagoya, 461-91, Japan
Jun Aoki, Gen.Mgr., Info. & Pub.Aff.

Contact

Hiroshi Inaba, Mgr. of PR Sect.
Hamoaka Nuclear Power Plant

Basic Facts

Unit 4 *Status:* Under construction (88% complete as of January 1992). *Type of Reactor:* Boiling water reactor. *Megawatts (electric):* 1137. *Megawatts (thermal):* 3293. *Reactor System Supplier:* Toshiba. *Generator Supplier:* Hitachi. *Architect Engineer:* Toshiba, Hitachi.

Plant Report

See Plant Report for Hamoaka Nuclear Power Plant, Units 1-3.

113 ●
Ikata Nuclear Power Station, Units 1-2

Kuchogoshi, Ikata-cho Phone: 0894-39-0221
Nishi Uwa-gun
Ehime Prefecture, 796-04, Japan
Kazunori Ishii, Supt.

Owning Utility

Shikoku Electric Power Co. Phone: 0878-21-5061
2-5, Marunouchi Fax: 0878-25-3012
Takamatsu City Telex: 5822115
Kagawa Prefecture, Japan
Hiroshi Nonaka, Managing Dir.,
  Thermal & Nuclear Power

Contact

Tetsuya Nagano, Gen.Mgr., Pub.Info.
Shikoku Electric Power Co.

Basic Facts

Units 1-2 *Status:* Operable. *Type of Reactor:* Pressurized light-water reactor. *Megawatts (electric):* 556. *Megawatts (thermal):* 1650. *Reactor System Supplier:* Mitsubishi Heavy Industries, Ltd. (Japan). *Generator Supplier:* Mitsubishi Heavy Industries, Ltd. (Japan); Mitsubishi Electric Corp. (Japan). *Architect Engineer:* Mitsubishi Atomic Power Industries, Inc. (Japan).

Key Dates

Unit 1 *Ordered:* 1973. *Construction began:* 1973. *Criticality:* January 1977. *Commercial Operation:* September 1977.
Unit 2 *Ordered:* 1977. *Construction began:* 1978. *Criticality:* 1981. *First Power:* 1981. *Commercial Operation:* March 1982.

Plant Report

Controversial site. The Ikata nuclear power station, located in the Ehime Prefecture on the smallest of Japan's four main islands, is owned by the Shikoku Electric Power Company. The station has two operating 566-megawatt pressurized water reactors. A third unit, an 890-megawatt pressurized light-water reactor is under construction.

Controversy surrounded the siting of the Ikata station. In 1981, Japan's Ministry of International Trade and Industry authorized increases in the subsidies available to local governments willing to accept nuclear plants. After an official suggested that Shikoku Electric give each family a 5,000 yen monthly subsidy, protestors accused the government of bribing residents. According to antinuclear activists subsidies would be unnecessary if the plants were safe.

Sources

"Nuclear Power in Japan." *The Bulletin of the Atomic Scientists,* May 1983.
Present Status of Nuclear Power Development. Kyushu Electric Power Co., June 1991.

—Karen Bellenir

114 ○
Ikata Nuclear Power Station, Unit 3

Kuchogoshi, Ikata-cho Phone: 0894-39-0221
Nishi Uwa-gun
Ehime Prefecture, 796-04, Japan
Kazunori Ishii, Supt.

Owning Utility

Shikoku Electric Power Co. Phone: 0878-21-5061
2-5, Marunouchi Fax: 0878-25-3012
Takamatsu City Telex: 5822115
Kagawa Prefecture, Japan
Hiroshi Nonaka, Managing Dir.,
  Thermal & Nuclear Power

Contact

Tetsuya Nagano, Gen.Mgr., Pub.Info.
Shikoku Electric Power Co.

Basic Facts

Unit 3 *Status:* Under construction (48% complete as of January 1992). *Type of Reactor:* Pressurized light-water reactor. *Megawatts (electric):* 890. *Megawatts (thermal):* 2652. *Reactor System Supplier:* Mitsubishi Heavy Industries, Ltd. (Japan). *Generator Supplier:* Mitsubishi Heavy Industries, Ltd. (Japan); Mitsubishi Electric Corp. (Japan). *Architect Engineer:* Mitsubishi Atomic Power Industries, Inc. (Japan).

Kashiwazaki

Key Dates
Unit 3 *Ordered:* 1985. *Construction began:* 1986. *Criticality expected* 1994. *Commercial Operation Expected:* March 1995.

Plant Report
See Plant Report for Ikata Nuclear Power Station, Units 1-2.

Kashiwazaki Kariwa Nuclear Station, Units 1, 2, & 5

No. 16-46 Aoyama-Machi
Kashiwazaki-Shi
Niigata Prefecture 945, Japan
Masao Takuma, Supt. (Power Station)
Phone: 0257-45-3131

Owning Utility
Tokyo Electric Power Co.
No. 1-3, 1-Chome
Uchisaiwai-cho
Chiyoda-ku
Tokyo 100, Japan
Shiro Sasaki, Managing Dir., Nuclear Power
Phone: 03-3501-8111
Fax: 03-3596-8544
Telex: TODEN-J 2224045

Contact
Hiroshi Kanagawa, Gen.Mgr., Nuclear Power Info. Center
Tokyo Electric Power Co.

Basic Facts
Units 1-2 *Status:* Operable. *Type of Reactor:* Boiling water reactor. *Megawatts (electric):* 1100. *Megawatts (thermal):* 3293. *Reactor System Supplier:* Toshiba. *Generator Supplier:* Toshiba. *Architect Engineer:* Toshiba.

Unit 5 *Status:* Operable. *Type of Reactor:* Boiling water reactor. *Megawatts (electric):* 1100. *Megawatts (thermal):* 3293. *Reactor System Supplier:* Hitachi. *Generator Supplier:* Hitachi. *Architect Engineer:* Hitachi.

Key Dates
Unit 1 *Ordered:* 1978. *Construction began:* 1978. *Criticality:* December 1984. *First Power:* February 1985. *Commercial Operation:* September 1985.

Unit 2 *Ordered:* 1983. *Construction began:* 1983. *Criticality:* November 1989. *First Power:* February 1990. *Commercial Operation:* October 1990.

Unit 5 *Ordered:* 1983. *Construction began:* 1983. *Criticality:* July 1989. *First Power:* September 1989. *Commercial Operation:* April 1990.

Plant Report
World's largest nuclear station. The Kashiwazaki Kariwa Nuclear Station, being constructed by the Tokyo Electric Power Company (TEPCO), is situated on the coast of the Sea of Japan in the Niigata Prefecture approximately 140 miles northwest of Tokyo. The station has three operating 1,100-megawatt boiling water reactors and four units under construction. Two of the units currently being built are 1,100-megawatt boiling water reactors and two are 1,356-megawatt Advanced Boiling Water Reactors (ABWR). When completed, the station will be the world's largest nuclear power plant and will be capable of generating 8.2 gigawatts of power.

Public opposition. TEPCO, Japan's largest electric utility, announced its plans for the Kashiwazaki Kariwa plant in 1969 and was immediately met with local protesting. Because of intense public opposition, it took 16 years to generate the first kilowatt of electricity, but after initial delays, construction proceeded quickly.

TEPCO began a major public relations campaign to quell public opposition and instituted a plant-visit program. Despite promotional efforts, 400 antinuclear protestors built "solidarity" huts on the construction site in February 1991. The huts were torn down following clashes between the protesters and riot police.

Advanced Boiling Water Reactors. The Japanese Ministry of International Trade and Industry (MITI) granted construction approval of Kashiwazaki Kariwa's newest units in May 1991. The process involved a review conducted by Japan's Atomic Energy Commission and Nuclear Safety Commission and included extensive safety analyses and public hearings. Units 6 and 7 will be the world's first nuclear units to use the 1,356-megawatt Advanced Boiling Water Reactor (ABWR) design. The ABWR design, developed by General Electric and its Japanese partners Hitachi and Toshiba, incorporates the best features from traditional boiling water reactors and takes advantage of technological advances made during the past decade. The design is said to be safer and more efficient with lower construction costs and reduced operating expenses.

Groundbreaking for Units 6 and 7 took place in September 1991. GE will supply the reactors, nuclear fuel, and turbine generator. The Japanese firms involved in the venture will be responsible for the rest of the plant and for overseeing the civil construction work. TEPCO expects the construction period to last five years and anticipates commercial operation at the two units to begin in 1996 and 1997.

Once the ABWR reactors are added to TEPCO's production mix, nuclear power will account for thirty percent of the utility's generating capacity.

Sources
"Nuclear Power in Japan." *The Bulletin of the Atomic Scientists*, May 1983.
"Japan Not Retreating from Nuclear Power." *Chicago Tribune*, November 4, 1990.
"Construction Approval Given for World's First Advanced Nuclear Power Plants." *Business Wire*, May 20, 1991.
Present Status of Nuclear Power Development. Kyushu Electric Power Co., June 1991.
"Groundbreaking Japanese-Style Marks Construction Start for Two GE Nuclear Power Plants Using Advanced Design." *INFO*, U.S. Council for Energy Awareness, October 1991.

—*Karen Bellenir*

Plant Profiles

116 ○
Kashiwazaki Kariwa Nuclear Station, Units 3, 4, 6, & 7

No. 16-46 Aoyama-Machi Phone: 0257-45-3131
Kashiwazaki-Shi
Niigata Prefecture 945, Japan
Masao Takuma, Supt. (Power Station)

Owning Utility

Tokyo Electric Power Co. Phone: 03-3501-8111
No. 1-3, 1-Chome Fax: 03-3596-8544
Uchisaiwai-cho Telex: TODEN-J 2224045
Chiyoda-ku
Tokyo 100, Japan
Shiro Sasaki, Managing Dir., Nuclear Power

Contact

Hiroshi Kanagawa, Gen.Mgr., Nuclear Power Info. Center
Tokyo Electric Power Co.

Basic Facts

Unit 3 *Status:* Under construction (86% complete as of January 1992). *Type of Reactor:* Boiling water reactor. *Megawatts (electric):* 1100. *Megawatts (thermal):* 3293. *Reactor System Supplier:* Toshiba. *Generator Supplier:* Toshiba. *Architect Engineer:* Toshiba.

Units 4, 6, & 7 *Status:* Under construction (65%, 7%, and 0% complete respectively as of January 1992). *Type of Reactor:* Boiling water reactor. *Megawatts (electric):* Unit 4, 1100; Units 6-7, 1356. *Megawatts (thermal):* Unit 4, 3293; Units 6-7, 3926. *Reactor System Supplier:* Hitachi. *Generator Supplier:* Hitachi. *Architect Engineer:* Hitachi.

Key Dates

Unit 3 *Ordered:* 1987. *Construction began:* 1987. *Commercial Operation Expected:* July 1993.
Unit 4 *Ordered:* 1987. *Construction began:* 1988. *Commercial Operation Expected:* July 1994.
Unit 6 *Ordered:* . *Construction began:* 1991. *Commercial Operation Expected:* July 1996.
Unit 7 *Ordered:* . *Construction began:* 1993. *Commercial Operation Expected:* July 1997.

Plant Report

See Plant Report for Kashiwazaki Kariwa Nuclear Station, units 1-2.

117 ■
Maki Nuclear Power Station

Maki-cho, Japan

Owning Utility

Tohoku Electric Power Co., Inc. Phone: 022-225-2111
7-1, Ichibancho 3-Chrome Fax: 022-263-3659
Aoba-ku
Sendai, Miyagi 980, Japan
Yasuo Matsumoto, Gen.Mgr., Nuclear Power Dept.

Mihama 118

Basic Facts

Status: On order. *Type of Reactor:* Boiling water reactor. *Megawatts (electric):* 825. *Megawatts (thermal):* 2436.

Plant Report

Delayed construction. The Maki Nuclear Power Station is expected to have a single 825-megawatt boiling water reactor. The power plant, still in the planning phase, is a project of the Tohoku Electric Power Company. Tohoku Electric, one of Japan's nine electric utilities, also operates the Onagawa nuclear station.

Maki's site, on the Sea of Japan in central Japan, has been the object of increased local opposition. Following the Three Mile Island accident in 1979 and the Chernobyl accident in 1986, siting negotiations conducted among municipalities, local utilities, and Japan's Electric Power Coordinating Council, the agency responsible for construction authorizations, have become more difficult.

Sources

"Japan's Nuclear Dilemma." *Technology Review*, October 1991.
Present Status of Nuclear Power Development. Kyushu Electric Power Co., n.d.

—Karen Bellenir

118 ●
Mihama Power Station, Units 1-3

5-3 Kawasakayama, 66 Nyu Phone: 0770-39-1111
Mihama-cho, Mikata-gun
Fukui 919-22, Japan
Y. Okuda, Supt.

Owning Utility

Kansai Electric Power Co., Inc. Phone: 06-441-8821
3-22 Nakanoshima 3-Chrome Fax: 06-443-2659
Kita-ku Telex: 5248320 KEPCO J
Osaka 530-70, Japan
M. Mori, Dir., Nuclear Power Operations

Basic Facts

Unit 1 *Status:* Operable. *Type of Reactor:* Pressurized light-water reactor. *Megawatts (electric):* 340. *Megawatts (thermal):* 1031. *Reactor System Supplier:* Westinghouse Electric Corp. (United States). *Generator Supplier:* Mitsubishi Heavy Industries Ltd. (Japan); Mitsubishi Electric Corp. (Japan). *Architect Engineer:* Kansai Electric Power Co., Inc; Gilbert.

Unit 2 *Status:* Operable. *Type of Reactor:* Pressurized light-water reactor. *Megawatts (electric):* 500. *Megawatts (thermal):* 1456. *Reactor System Supplier:* Mitsubishi Heavy Industries Ltd. (Japan). *Generator Supplier:* Mitsubishi Heavy Industries Ltd. (Japan); Mitsubishi Electric Corp. (Japan). *Architect Engineer:* Kansai Electric Power Co., Inc; Mitsubishi Atomic Power Industries, Inc. (Japan).

Unit 3 *Status:* Operable. *Type of Reactor:* Pressurized light-water reactor. *Megawatts (electric):* 826. *Megawatts (thermal):* 2440. *Reactor System Supplier:* Mitsubishi Heavy Industries Ltd. (Japan). *Generator Supplier:* Mitsubishi Heavy Industries Ltd. (Japan); Mitsubishi Electric

Corp. (Japan). *Architect Engineer:* Kansai Electric Power Co.,Inc; Mitsubishi Atomic Power Industries, Inc. (Japan).

Key Dates

Unit 1 *Ordered:* 1966. *Construction began:* 1967. *Criticality:* July 1970. *First Power:* August 1970. *Commercial Operation:* November 1970.

Unit 2 *Ordered:* 1967. *Construction began:* 1968. *Criticality:* April 1972. *First Power:* April 1972. *Commercial Operation:* July 1972.

Unit 3 *Ordered:* 1971. *Construction began:* 1971. *Criticality:* January 1976. *First Power:* February 1976. *Commercial Operation:* December 1976.

Plant Report

Part of Kansai Electric's power mix. The Mihama Power Station, in the Niu-district of Mihama-cho, is 220 miles west of Tokyo. The plant site occupies a coastal area on the Sea of Japan inside the Wakasa Bay National Park. Local industry consists primarily of agriculture and fishing.

Mihama is owned by the Kansai Electric Power Company, one of Japan's nine electric utilities. Kansai serves approximately 11 million customers in the central region of Japan's main island. The company established a Nuclear Power Department in 1957 and in 1970 became the first Japanese utility to use a pressurized water reactor (PWR). Currently, approximately one-quarter of the power supplied by Kansai is nuclear-generated. The region is undergoing economic expansion and several large-scale projects, including the construction of the Kansai International Airport, are underway. The company predicts increases in power demands and plans to meet customer needs through additional nuclear capacity. Kansai expects nuclear-generated power to make up 60 percent of its output by 2030.

Construction of three units. The Mihama Power Station has three PWR's. Unit 1, Japan's first pressurized water reactor, was imported from Westinghouse Electric (USA). Its construction began in August 1967. The unit was completed in time to supply electric power to EXPO '70, held in Osaka in August 1970. It entered commercial production the following November. In a landmark refueling conducted in 1988, Mihama-1 was loaded with Mixed Oxide Fuel (MOX Fuel) which is made from recycled plutonium.

The second and third units at Mihama were manufactured by Mitsubishi Heavy Industries. Unit 2, a 500-megawatt pressurized light-water reactor began its operating life in 1972. Unit 3, an 826-megawatt PWR, began operation in 1976. It was noted for using the first dome-shaped design in Japan, a design intended to reduce radiation leakage.

Steam generator accidents. All three units at Mihama have experienced difficulties with steam generator tubes. Unit 1 was closed from 1974 to 1983 for repairs and improvements. Unit 2 closed following an accident in February 1991; its steam generator is scheduled to be replaced in 1994. Following the accident at Unit 2, an inspection of Unit 3 revealed similar pipe damage.

The 1991 accident at Mihama-2 marked the first time in the history of Japanese nuclear power that an Emergency Core Cooling System was activated. At 12:40 p.m. on February 9, a worker noticed that one gauge indicated a sharp increase in the generator's level of radioactivity. A second gauge did not show the increase. Workers left the reactor running at full capacity for 50 minutes while they attempted to determine which gauge was correct. By the time they established that an accident had taken place, the radiation levels had increased further and pressure inside the reactor was dropping. If the pressure had dropped far enough, water inside the reactor would have boiled. According to the Japanese press, water inside the reactor was within a few degrees of boiling. This could have exposed the core, which could have resulted in a melt down. The Emergency Core Cooling System activated automatically and flooded the core. Although the cooling system worked, two valves that should have opened to release the resulting excess pressure failed to open, forcing additional radioactive water to leak out of the primary water loop.

The accident was attributed to a break in a steam generator tube which permitted radioactive water from the reactor to leak into the steam-generating water supply. One of Japan's newspapers reported that approximately 20 tons of water escaped into the secondary water system before the plant was shut down. Authorities initially denied that any radiation had been released, but an investigation revealed that some had escaped into the atmosphere. Estimates of actual harm done were small, but the accident fueled Japan's growing antinuclear movement which threatens to hinder the government's plans to double its nuclear capacity over the next twenty years.

A follow-up investigation conducted by Japan's Ministry of International Trade and Industry (MITI) discovered that improperly installed antivibration bars had caused the initial break. The error had not been previously discovered because the bars were not subject to annual inspection. Following this discovery, MITI ordered an inspection of antivibration bars on all similar reactors. MITI also issued new regulations requiring that plants be shut down whenever radioactivity increases by 20 percent or more.

Efforts to improve safety. Kansai Electric operates two training centers. Through the use of computers to simulate accidents, plant employees keep their skills upgraded. Kansai also participates in international efforts to promote nuclear safety. The International Atomic Energy Agency (IAEA) sent an Operational Safety Review Team (OSART) mission to review Kansai's nuclear program. In addition, the utility is a member of the World Association of Nuclear Operators (WANO). As part of the WANO exchange program, Kansai hosted a team from the Zaporozhe Nuclear Power Station in the former Soviet Union.

Kansai Electric also takes prides in its efforts to protect and enhance the environment surrounding Mihama. In Niu Bay, for example, the plant's pumping system increased natural circulation. The company planted oysters and scallops and developed a breeding program for red snappers and yellow tail.

Visitors. The Mihama visitors' center originally opened in 1967 and was enlarged and modernized in 1983. Approximately 130,000 people visit annually to learn about nuclear power generation. The center offers plant tours and displays, including a life-size model of a reactor.

Sources

"Japan Now Tells of Radiation Release." *The New York Times*, February 12, 1991.
"Japan Steps Up Nuclear Plant Safety." *Chicago Tribune*, February 19, 1991.
"Japanese Are Hardly Reassured on Nuclear Peril." *The New York Times*, March 5, 1991.
"More Damage at Nuclear Plants." *The Wall Street Journal*, April 1, 1991.
Introduction to Kansai Electric's Nuclear Power Generation, 1991. Kansai Electric Power Company, n.d.
Current Information: 1991. Kansai Electric Power Co., n.d.
Mihama Power Station. Kansai Electric Power Co., n.d.
"Japan's Nuclear Dilemma." *Technology Review*, October 1991.

—Karen Bellenir

119 ○
Monju Nuclear Power Plant

2-1 Shiraki Phone: 0770-39-1031
Tsuruga-city
Fukui-ken, Japan
Tadao Takahashi, Dir.

Owning Utility

Power Reactor & Nuclear Fuel   Phone: 03-586-3311
   Development Corp.   Telex: J26462
1-9-13 Akasaka
Minato-ku
Tokyo 107, Japan
Michia Akebi, Exec. Dir., Constr. & Opr.
   Fugen & Monju

Contact

Hideaki Ashiwara, Dir., PR Office
Power Reactor & Nuclear Fuel
   Development Corp.

Basic Facts

Status: Under construction (97% complete as of January 1992). *Type of Reactor:* Liquid metal fast breeder reactor. *Megawatts (electric):* 280. *Megawatts (thermal):* 714. *Reactor System Supplier:* Toshiba; Hitachi; Mitsubishi Heavy Industries, Ltd. (Japan); Fuji. *Generator Supplier:* Toshiba. *Architect Engineer:* Power Reactor & Nuclear Fuel Development Corp. (PNC); FBR Engineering Co., Ltd. (Japan).

Key Dates

Ordered: 1984. *Construction began:* 1985. *Criticality expected* October 1992. *Commercial Operation Expected:* 1993.

Plant Report

A prototype reactor. The Monju nuclear plant is a prototype Fast Breeder Reactor (FBR) with a capacity of 280 megawatts. It is named for Monju Bodhisattva, a spiritual symbol of power controlled by wisdom. The unit, currently in its final stages of construction, is located on the coast of the Sea of Japan in Tsuruga City in the Fukui Prefecture. The site is inside the Wakasa Bay park and is surrounded by mountains. Japanese officials expect it to be used as the foundation for nuclear power development in Japan during the 21st century.

The basis for the Monju project was established in 1967 when the Power Reactor and Nuclear Fuel Development Corporation Law created the Power Reactor and Nuclear Fuel Development Corporation (PNC). PNC was given the responsibility of developing technologies for using plutonium, including reprocessing spent fuel, fabricating uranium-plutonium Mixed Oxide Fuel (MOX Fuel), and developing FBR and Advanced Thermal Reactor (ATR) designs. PNC's study of large-scale breeder reactors began in 1975. The development of FBR's was named as a top governmental priority.

The PNC is responsible for managing the Monju project. Construction is being supervised by thhe Japan Atomic Power Company (JAPCO), an organization representing Japan's nine electric utilities. Through this arrangement, the technological lessons learned at Monju will be transferred to the utilities and nuclear industries for future commercialization.

How FBRs differ from LWRs. Traditional light water reactors use a fuel made of uranium, U235, which can undergo atomic fission. Natural uranium is composed of 99.3 percent U238, which does not undergo atomic fission. A FBR converts U238 into plutonium 239, a substance which does not occur naturally. The plutonium 239 fuels the reactor. A breeder reactor produces more fuel than it consumes. Japan is particularly interested in breeder technology because it anticipates that the supply of uranium will dwindle in the next century and that breeder reactors will provide a significant source of power.

A FBR produces power by a method similar to a LWR: heat is used to produce steam which turns a turbine to generate electricity. In addition to using a different fuel, however, a FBR also employs a different cooling system. At Monju, liquid sodium will be used to remove excess heat.

Experimental predecessor. Monju was based on an experimental reactor, Joyo, built in Oarai-machi in the Ibaraki Prefecture. The Joyo reactor achieved initial criticality in April 1977. It was first designed with a capacity of 50 megawatts. A new core installed in 1983 brought the capacity up to 100 megawatts. A major milestone was achieved in 1984 when Joyo was refueled with fuel fabricated from its own spent fuel.

Construction history. The experiences gained at Joyo were used to construct Monju. Site preparation started in 1983 and construction began in 1985. Monju's main components were manufactured by Toshiba, Hitachi, Fuji Electric, and Mitsubishi Heavy Industries. The reactor containment vessel was completed in April 1987 and the reactor vessel was installed in October 1988. Functional testing began in May 1991 and fuel loading began during the summer of 1992. Costs through the end of the construction period totaled $4.3 billion.

Initial criticality at Monju is expected to be achieved in October 1992. Planners have targeted September 1993 for the first power generation. Full capacity and production is anticipated during 1994.

After Monju. After Monju becomes operational, the Japan Atomic Power Co. (JAPC) expects to construct a demonstration reactor beginning in the latter part of the 1990's. JAPC anticipates using FBR technology on a serviceable level sometime between 2020 and 2030. After 2030 FBR's are expected to become Japan's main source of nuclear power.

In addition to contributing to nuclear power technology, researchers at Monju expect to apply their findings to many diverse areas, such as maintaining life in space, changing natural materials into new materials and eliminating radioactivity from radioactive substances

Sources

"Monju Fast-Breeder Reactor Completed." *Nature*, July 4, 1991.
"Nuclear Power Reactor "Monju" Begins Operation Tests." *Business JAPAN*, August 1991.
PNC: Power Reactor and Nuclear Fuel Development Corporation. N.p., September 1991.

—Karen Bellenir

---

120 ●

Ohi Nuclear Power Station, Units 1-3

1-1 Aza Yoshimi, 1 Ohshima Phone: 0770-77-1331
Ohi-cho, Ohi-gun
Fukui 919-22, Japan
Y. Fujita, Supt.

Owning Utility

Kansai Electric Power Co., Inc.   Phone: 06-441-8821
3-22 Nakanoshima 3-Chrome   Fax: 06-443-2659
Kita-ku   Telex: 5248320 KEPCO J
Osaka 530-70, Japan
M. Mori, Dir., Nuclear Power Operations

Basic Facts

Units 1-2 *Status:* Operable. *Type of Reactor:* Pressurized light-water reactor. *Megawatts (electric):* 1175. *Megawatts (thermal):* 3423. *Reactor System Supplier:* Westinghouse Electric Corp. (United States). *Generator Supplier:* Mitsubishi Heavy Industries, Ltd. (Japan); Mitsubishi Electric Corp. (Japan). *Architect Engineer:* Kansai Electric Power Co., Inc; Gilbert.

Unit 3 *Status:* Operable. *Type of Reactor:* Pressurized light-water reactor. *Megawatts (electric):* 1180. *Megawatts (thermal):* 3423. *Reactor System Supplier:* Mitsubishi Heavy Industries, Ltd. (Japan). *Generator Supplier:* Mitsubishi Heavy Industries, Ltd. (Japan) Mitsubishi Electric Corp. (Japan). *Architect Engineer:* Kansai Electric Power Co., Inc; Mitsubishi Atomic Power Industries, Inc. (Japan).

Key Dates

Unit 1 *Ordered:* 1970. *Construction began:* 1972. *Criticality:* December 1977. *First Power:* December 1977. *Commercial Operation:* March 1979.
Unit 2 *Ordered:* 1970. *Construction began:* 1972. *Criticality:* September 1978. *First Power:* October 1978. *Commercial Operation:* December 1979.
Unit 3 *Ordered:* 1986. *Construction began:* 1987. *Criticality:* 1991. *First Power:* 1991. *Commercial Operation:* December 1991.

Plant Report

Additional units nearing completion. The Ohi Power Station is one of three nuclear power plants operated by the Kansai Electric Power Company. Units 1-3 at Ohi are operable. Unit 4, still under construction, is expected to begin operation in early 1993.

The two newer units at Ohi will both have a Prestressed Concrete Containment Vessel (PCCV). The PCCV design is smaller and lighter than traditional containment structures for the purpose of greater seismic stability. Unit 4 will also feature a special room where visitors can watch control room operations through a glass partition and view actual spent fuel pits, turbines and generators.

Continued construction program. Kansai Electric, however, will not be finished with nuclear power plant construction when Ohi's third and fourth units are complete. Anticipating greater increases in demand for electricity, the utility is considering new sites for future nuclear stations. Two are Suzu on the Noto peninsula which is northeast of the Ohi plant, and Oura in Wakayama, which is southeast of Osaka.

Steam generator repairs. The PWR's operated by Kansai Electric have been plagued with steam generator difficulties. An accident, called the worst in Japanese nuclear power history, occurred at the utility's Mihama station. Because the reactors were of similar design, Kansai Electric extended a 1991 inspection outage at Ohi-1 to examine and repair its steam generator.

Sources

Introduction to Kansai Electric's Nuclear Power Generation, 1991. Kansai Electric Power Company, n.d.
Current Information: 1991. Kansai Electric Power Co., n.d.
Profile of Nos. 3 & 4 Plant Extension Work of Ohi Power Station. Ohi Nuclear Power Station Construction Office, n.d.

—Karen Bellenir

---

121 ○

Ohi Nuclear Power Station, Unit 4

1-1 Aza Yoshimi, 1 Ohshima   Phone: 0770-77-1331
Ohi-cho, Ohi-gun
Fukui 919-22, Japan
Y. Fujita, Supt.

Owning Utility

Kansai Electric Power Co., Inc.   Phone: 06-441-8821
3-22 Nakanoshima 3-Chrome   Fax: 06-443-2659
Kita-ku   Telex: 5248320 KEPCO J
Osaka 530-70, Japan
M. Mori, Dir., Nuclear Power Operations

Basic Facts

Unit 4 *Status:* Under construction (90% complete as of January 1992). *Type of Reactor:* Pressurized light-water reactor. *Megawatts (electric):* 1180. *Megawatts (thermal):* 3423. *Reactor System Supplier:* Mitsubishi Heavy Industries, Ltd. (Japan). *Generator Supplier:* Mitsubishi Heavy

Industries, Ltd. (Japan) Mitsubishi Electric Corp. (Japan). *Architect Engineer:* Kansai Electric Power Co., Inc; Mitsubishi Atomic Power Industries, Inc. (Japan).

Key Dates

Unit 4 *Ordered:* 1986. *Construction began:* 1987. *Criticality:* 1991. *First Power:* 1992. *Commercial Operation Expected:* February 1993.

Plant Report

See Plant Report for Ohi Nuclear Power Station, Units 1-3.

---

122 •

Onagawa Nuclear Power Station, Unit 1

1 Maeda Tsukahama Onagawacho
Oshikagun
Miyagi Prefecture 986-22, Japan
Tsuneji Hashimoto, Supt.

Phone: 0225-53-3111
Fax: 0225-54-4379

Owning Utility

Tohoku Electric Power Co., Inc.
7-1, Ichibancho 3-Chrome
Aoba-ku
Sendai, Miyagi 980, Japan
Yasuo Matsumoto, Gen.Mgr., Nuclear Power Dept.

Phone: 022-225-2111
Fax: 022-263-3659

Contact

Katsuo Hino, Mgr. of PR
Onagawa Nuclear Power Station

Basic Facts

Unit 1 *Status:* Operable. *Type of Reactor:* Boiling water reactor. *Megawatts (electric):* 825. *Megawatts (thermal):* 2436. *Reactor System Supplier:* Toshiba. *Generator Supplier:* Toshiba. *Architect Engineer:* Toshiba.

Key Dates

Unit 1 *Ordered:* 1969. *Construction began:* 1979. *Criticality:* 1983. *First Power:* November 1983. *Commercial Operation:* June 1984.

Plant Report

Tohoku Electric looks to nuclear power. The Onagawa Nuclear Power Station is located on a site that stretches between Onagawa and Oshika, Oshika-gun in the Miyagi Prefecture. The power plant is operated by the Tohoku Electric Power Company. Tohoku Electric, one of Japan's nine regional electric utilities, was founded on May 1, 1951, when the national electric power industry was reorganized. The company supplies power to more than a million customers in a service area covering the seven northernmost prefectures on Honshu, Japan's main island.

Tohoku Electric selected the Onagawa site on the Pacific coast in January 1968 and the Japanese government issued a nuclear construction permit in December 1970. Following the oil crises in the 1970's, Tohoku began making efforts to reduce its dependance on foreign oil. The company switched to other fuels such as liquefied natural gas and intensified efforts to build a nuclear power plant.

Lengthy site negotiations. Construction at Onagawa, however, was delayed by local opposition. Agreements between the utility and local residents concerning safety and fishermen's rights were finalized in 1978 and construction began in December 1979. Although local residents were paid 12.2 billion yen in compensation, opposition to the siting continued. In 1981, when the plant was 55 percent complete, a civil suit was filed in the Sendai District Court seeking to halt construction. Tohoku Electric called the action "regrettable" and said that the company had taken every possible safety precaution.

Commercial operation at Unit 1, a 524-megawatt boiling water reactor manufactured by Toshiba, began in June 1984. The unit consists of four buildings: the reactor building, the turbine building, the control building and a radioactive waste building. Earthquake design specifications are three times as strict as normal architectural standards. A harbor was constructed in front of the plant to facilitate shipping and provide sea water to cool the main condensers.

Construction on a second unit, which will be an 825-megawatt BWR, started in August 1989. Unit 2 is expected to begin operation in July 1995. A third unit is also in the planning stages.

In addition to the generating units, Tohoku Electric opened the Nuclear Power Operation and Maintenance Training Center at the Onagawa station in October 1990.

Environmental issues. Radioactive wastes in three forms, gaseous, liquid and solid, are produced and treated at Onagawa. Gaseous waste is released after treatment; liquid waste is treated and reused. Solid waste is stored on-site. The Environmental Radioactivity Monitoring Center samples sea and agricultural products in the area surrounding the power plant. The environmental impact of discharging warmed seawater, 7° C higher than natural water temperatures, is also being studied.

Sources

"Nuclear Power in Japan." *The Bulletin of the Atomic Scientists,* May 1983.
Onagawa Unit No. 1 Nuclear Power Station. Tohoku Electric Power Co., n.d.
Tohoku Electric Power Co., Inc.: Annual Report 1991. Tohoku Electric Power Co., n.d.

—Karen Bellenir

---

123 ○

Onagawa Nuclear Power Station, Unit 2

1 Maeda Tsukahama Onagawacho
Oshikagun
Miyagi Prefecture 986-22, Japan
Tsuneji Hashimoto, Supt.

Phone: 0225-53-3111
Fax: 0225-54-4379

124 Sendai

Owning Utility

Tohoku Electric Power Co., Inc.
7-1, Ichibancho 3-Chrome
Aoba-ku
Sendai, Miyagi 980, Japan
Yasuo Matsumoto, Gen.Mgr., Nuclear
  Power Dept.

Phone: 022-225-2111
Fax: 022-263-3659

Contact

Katsuo Hino, Mgr. of PR
Onagawa Nuclear Power Station

Basic Facts

Unit 2 *Status:* Under construction (36% complete as of January 1992). *Type of Reactor:* Boiling water reactor. *Megawatts (electric):* 825. *Megawatts (thermal):* 2436. *Reactor System Supplier:* Toshiba. *Generator Supplier:* Toshiba. *Architect Engineer:* Toshiba.

Plant Report

See Plant Report for Onagawa Nuclear Power Station, Unit 1.

124 ●
Sendai Nuclear Power Station, Units 1-2

1765-3, Aza-Katahirayama
Kumizaki-cho
Sendai-shi, Kagoshima-ken 895-01,
  Japan
Shigeharu Ozono, Gen.Mgr.

Owning Utility

Kyushu Electric Power Co., Inc.
1-82 Watanabe-Dori
2-Chrome, Chuo-ku
Fukuoka 810, Japan
Hideo Kodama, Gen.Mgr., Nuclear
  Operations

Phone: 092-761-3031
Fax: 092-761-4622
Telex: 725497 KYUDEN J

Contact

Yusufumi Yarimizu, Gen.Mgr., PR
Kyushu Electric Power Co., Inc.

Basic Facts

Units 1-2 *Status:* Operable. *Type of Reactor:* Pressurized light-water reactor. *Megawatts (electric):* 890. *Megawatts (thermal):* 2660. *Reactor System Supplier:* Mitsubishi Heavy Industries, Ltd. (Japan). *Generator Supplier:* Mitsubishi Heavy Industries, Ltd. (Japan); Mitsubishi Electric Corp. (Japan). *Architect Engineer:* Mitsubishi Atomic Power Industries, Inc. (Japan).

Key Dates

Unit 1 *Ordered:* 1979. *Construction began:* 1979. *Criticality:* August 1983. *First Power:* September 1983. *Commercial Operation:* July 1984.

Unit 2 *Ordered:* 1981. *Construction began:* 1981. *Criticality:* March 1985. *First Power:* April 1985. *Commercial Operation:* November 1985.

Plant Report

Two operating reactors. The Sendai Nuclear Power Station occupies a 358-acre sea-side site in Gumizaki-cho, which is located in the Kagoshima Prefecture on the island of Kyushu. Kyushu is the southern-most of Japan's four main islands. Sendai is one of two nuclear stations operated by Kyushu Electric Power Company. Together with the Gendai station it contributes approximately one third of the power supplied by Kyushu Electric. The remaining electricity is generated from oil, gas, coal, and hydroelectric facilities, with a small percentage originating from geothermal sources.

Both of Sendai's units are 890-megawatt pressurized water reactors. Unit 1 was the first pressurized light-water reactor produced under the Japanese "Standardization and Improvement of LWR (light water reactor) Program."

Safety features at Sendai. The Sendai units were designed with safety features approved by the Ministry of International Trade and Industry (MITI) and the Atomic Energy Safety Committee. MITI performed a safety check during the construction phase and checks safety features during refueling outages. The central focus of Sendai's safety program is to prevent accidents from occurring and to protect the public if an accident does occur. Both reactor buildings are made of steel reinforced concrete. The requirements for seismic defense are three times stronger than for conventional buildings. To prevent radiation damage to the environment, Kyushu Electric Power Co., and the Kagoshima Prefecture government monitor the surrounding area.

Visitors welcome. The Kyushu Electric Power Company welcomes visitors at the Sendai Nuclear Power Station Exhibition Hall. For more information telephone (0996) 27-3506.

Sources

Present Status of Nuclear Power Development. Kyushu Electric Power Co., June 1991.
Sendai Nuclear Power Station. Kyushu Electric Power Co., n.d.

—Karen Bellenir

125 ○
Shika Nuclear Power Plant

1-Akasume, Shika-Machi
Hakui-gun
Ishikawa 925-01, Japan
Minoru Ishihara, Dir. & Gen.Mgr.

Phone: 0767-32-2666
Fax: 0767-32-3820

Owning Utility

Hokuriku Electric Power Co.
15-1 Ushijima-cho
Toyama City 930, Japan
Yuzuru Hieda, Gen.Mgr., Nuclear
  Power Dept.

Phone: 0764-41-2511
Fax: 0764-33-9963
Telex: 5152-915
  HOKDEN J

Contact

Toshihiko Maeda, Deputy Gen.Mgr., PR
Shika Nuclear Power Plant

Basic Facts

Status: Under construction (91% complete as of January 1992). *Type of Reactor:* Boiling water reactor. *Megawatts (electric):* 540. *Megawatts (thermal):* 1593. *Reactor System Supplier:* Hitachi. *Generator Supplier:* Hitachi. *Architect Engineer:* Hitachi.

Key Dates

Ordered: 1988. *Construction began:* 1988. *Criticality expected* 1992. *Commercial Operation Expected:* July 1993.

Plant Report

Thwarted construction efforts. The Shika Nuclear Power Plant, under construction on Japan's west coast, is expected to have a single 540-megawatt boiling water reactor. Hokuriku Electric Power Company, the utility building Shika, has no other nuclear stations.

Plans to build a nuclear plant at Shika were complicated following the 1979 accident at Three Mile Island. Officials, uneasy about safety issues, negotiated more than 10 years before the Electric Power Coordinating Council authorized construction to begin.

Sources

"Japan's Nuclear Dilemma." *Technology Review,* October 1991.
Present Status of Nuclear Power Development. Kyushu Electric Power Co., n.d.

—*Karen Bellenir*

126 •

Shimane Nuclear Power Station, Units 1-2

654-1 Kataku, Oaza Phone: 0852-82-2220
Kashima-cho, Yatsuka-gun Fax: 0852-82-3017
Shimane 690-03, Japan
Yasuhiko Inoue, Supt.

Owning Utility

Chugoku Electric Power Co, Inc. Phone: 082-241-0211
4-33 Komachi Naka-ku Fax: 082-244-1741
Hiroshima City 730-91, Japan
Toshiaki Doi, Gen.Mgr., Nuclear Power Dept.

Contact

Yasuyuki Matsuo, Managing Dir. of PR
Chugoku Electric Power Co, Inc.

Basic Facts

Units 1-2 *Status:* Operable. *Type of Reactor:* Boiling water reactor. *Megawatts (electric):* Unit 1, 460; Unit 2, 820. *Megawatts (thermal):* Unti 1, 1380; Unit 2, 2436. *Reactor System Supplier:* Hitachi. *Generator Supplier:* Hitachi. *Architect Engineer:* Unit 1, Hitachi; Unit 2, Hitachi and Kajima.

Key Dates

Unit 1 *Ordered:* 1970. *Construction began:* 1970. *Criticality:* 1973. *First Power:* December 1973. *Commercial Operation:* March 1974.
Unit 2 *Ordered:* 1984. *Construction began:* 1984. *Criticality:* May 1988. *First Power:* July 1988. *Commercial Operation:* February 1989.

Plant Report

Location. The Shimane Nuclear Power Station is located on the Sea of Japan near Matsue, north of Hiroshima.

Sources

Present Status of Nuclear Power Development. Kyushu Electric Power Co., June 1991.

127 •

Takahama Nuclear Power Station, Units 1-4

1 Tanoura Phone: 0770-76-1221
Takahama-cho, Ohi-gun
Fukui 919-23, Japan
A Kishida, Supt.

Owning Utility

Kansai Electric Power Co., Inc. Phone: 06-441-8821
3-22 Nakanoshima 3-Chrome Fax: 06-443-2659
Kita-ku Telex: 5248320 KEPCO J
Osaka 530-70, Japan
M. Mori, Dir., Nuclear Power Operations

Basic Facts

Units 1-2 *Status:* Operable. *Type of Reactor:* Pressurized light-water reactor. *Megawatts (electric):* 826. *Megawatts (thermal):* 2440. *Reactor System Supplier:* Unit 1, Westinghouse Electric Corp. (United States); Unit 2, Mitsubishi Heavy Industries Ltd. (Japan). *Generator Supplier:* Mitsubishi Heavy Industries Ltd. (Japan); Mitsubishi Electric Corp. (Japan). *Architect Engineer:* Unit 1, Kansai Electric Power Co., Inc., Gilbert; Unit 2, Kansai Electric Power Co., Inc., Mitsubishi Atomic Power Industries, Inc. (Japan).

Units 3-4 *Status:* Operable. *Type of Reactor:* Pressurized light-water reactor. *Megawatts (electric):* 870. *Megawatts (thermal):* 2660. *Reactor System Supplier:* Mitsubishi Heavy Industries, Ltd. (Japan). *Generator Supplier:* Mitsubishi Heavy Industries, Ltd. (Japan); Mitsubishi Electric Corp. (Japan). *Architect Engineer:* Kansai Electric Power Co., Inc; Mitsubishi Atomic Power Industries, Inc (Japan)

Key Dates

Unit 1 *Ordered:* 1969. *Construction began:* 1970. *Criticality:* March 1974. *First Power:* March 1974. *Commercial Operation:* November 1974.
Unit 2 *Ordered:* 1970. *Construction began:* 1971. *Criticality:* December 1974. *First Power:* Janaury 1975. *Commercial Operation:* November 1975.
Unit 3 *Ordered:* 1981. *Construction began:* 1981. *Criticality:* April 1984. *First Power:* May 1984. *Commercial Operation:* January 1985.
Unit 4 *Ordered:* 1981. *Construction began:* 1981. *Criticality:* October 1984. *First Power:* November 1984. *Commercial Operation:* June 1985.

Tokai

Plant Report

A four-unit complex. The Takahama Power station is the largest of three nuclear plants operated by Kansai Electric Power Company. Kansai Electric, formed in 1951 when Japan's electric utility system was reorganized, supplies power to more than 11 million customers in the central region of Japan's main island.

Takahama has four nuclear units. Units 3 and 4 entered commercial production in 1985, a decade after Units 1-2 started. Many modifications were incorporated into Takahama's third and fourth units. These included the use of robots to reduce workers' exposure to radiation and design improvements in plant systems such as steam generators, reactor vessels, turbines, and coolant pumps. Unit 3 played an important role in Japan's nuclear progress when it became the first Japanese pressurized light-water reactor to use a high burn-up fuel which is expected to lower the fuel cycle cost.

Nuclear Power Plant Maintenance Training Center. The Kansai Electric Company opened its Nuclear Power Plant Maintenance Training Center in Takahama-cho in 1983. The center uses simulation replicas and actual instruments to train employees. A see-through model was added in April 1989. It gives students a chance to see the internal status of a reactor under normal, transient, and accident conditions.

International efforts. In an effort to contribute to international efforts toward increased safety, Kansai Electric welcomed an inspection team from the International Atomic Energy Agency (IAEA) to the Takahama facility. An Operational Safety Review Team (OSART) reviewed Units 2 and 3 in 1988 and praised the utility for its achievements in maintenance technology and safety attitudes.

Affected by Mihama accident. The Takahama plant was adversely affected by a 1991 accident at the Mihama station, also operated by Kansai Electric. At Mihama, a steam generator tube burst and leaked radioactive water into the secondary water circulation system. Following the incident Japan's Ministry of International Trade and Industry (MITI) ordered an inspection of other Japanese PWR's including the Takahama complex. During the examination, inspectors found an improperly installed antivibration bar on Unit 2, the same kind of error blamed for the Mihama accident. Officials ordered the unit to close until the problem could be corrected.

Sources

Introduction to Kansai Electric's Nuclear Power Generation, 1991. Kansai Electric Power Company, n.d.

Current Information: 1991. Kansai Electric Power Co., n.d.

"Japan Shuts Down 2d Nuclear Plant." *The New York Times*, March 21, 1991.

—*Karen Bellenir*

Tokai Nuclear Power Station, Units 1-2

1-1, Shirakata, Tokai-mura
Naka-gun
Ibaragi Prefecture 319-11, Japan
Yohichi Iwasaki, Supt.

Phone: 0292-82-1211
Fax: 0292-82-1211

Owning Utility

Japan Atomic Power Company (JAPCO)
Ohtemachi Bldg. No. 6-1, 1-Chrome
Ohtemachi, Chiyoda-ku
Tokyo 100, Japan
Hiroshi Shimokawa, Managing Dir., Plant Mgmt.

Phone: 03-3201-6631
Fax: 03-3285-0541
Telex: J24592 (JATOPOW)

Contact

Yukio Sugano, Gen.Mgr., PR Dept.
Japan Atomic Power Company (JAPCO)

Basic Facts

Unit 1 *Status:* Operable. *Type of Reactor:* Gas-cooled reactor. *Megawatts (electric):* 166. *Megawatts (thermal):* 587. *Reactor System Supplier:* General Electric Co. (United Kingdom). *Generator Supplier:* General Electric Co. (United Kingdom). *Architect Engineer:* General Electric Co. (United Kingdom).

Unit 2 *Status:* Operable. *Type of Reactor:* Boiling water reactor. *Megawatts (electric):* 1100. *Megawatts (thermal):* 3293. *Reactor System Supplier:* General Electric Co. (United States). *Generator Supplier:* General Electric Co. (United States). *Architect Engineer:* Ebasco.

Key Dates

Unit 1 *Ordered:* 1959. *Construction began:* 1960. *Criticality:* 1965. *First Power:* 1966. *Commercial Operation:* July 1966.

Unit 2 *Ordered:* 1972. *Construction began:* 1972. *Criticality:* 1978. *First Power:* 1978. *Commercial Operation:* November 1978.

Plant Report

Japan's first commercial reactor. Tokai-mura, Maka-gun, located in the Ibaraki Prefecture is the site of Japan's first commercial nuclear generating station. The village of Tokai, located approximately 80 miles northeast of Tokyo on Japan's Pacific shore, has been called "the very birthplace of peaceful use of nuclear energy in Japan."

Tokai-1 is a UK-designed 166-megawatt graphite moderated, carbon dioxide gas cooled reactor. It uses natural uranium fuel. Some design modifications were necessary to achieve the seismic protections necessary in Japan. Tokai-mura was selected as a potential site in 1957 and application for construction was made in March 1959 and commercial operation began on July 25, 1966. During it's first 25 years of production, Tokai-1 achieved an average availability factor of 79 percent and an average capacity factor of 63.8 percent.

A second unit. Application for the construction of a second unit at Tokai was made on December 21, 1971 and approved two days later. Tokai-2, a 1,100 megawatt boiling water reactor, was the first boiling water reactor in its capacity range to be built in Japan. The unit uses enriched uranium fuel. Tokai-2 has produced more electricity than any other single reactor in Japan.

Tokai's operator. The Tokai station is operated by Japan Atomic Power Company (JAPCO). JAPCO was established by the Prime Minister's Office on November 1, 1957. Its two-fold objectives are to develop nuclear power generation for commercial purposes and to offer technological assistance in operations related to nuclear power plants. More than 97 percent of JAPCO's shares are held by electric power companies and nuclear power industry groups. JAPCO established a training center in May 1968. The center was reorganized at Tokai-mura in January 1989.

Next to the Tokai-mura station, the Japan Atomic Energy Research Institute is using two small retired test reactors to research methods of decommissioning. One reactor was moved into an underground chamber for permanent storage and a new 20-megawatt research reactor was installed on top of it. The other is being disassembled using a remote-controlled arc saw.

Sources

"Japan Pushes Nuclear Decommissioning Work." *ENR*, September 13, 1990.
Mounfield, Peter. *World Nuclear Power.* Routledge, 1991.
Outline of Tokai, Tokai II Power Stations. Japan Atomic Power Company, 1991.
The Japan Atomic Power Company 1991. N.p., n.d.

—*Karen Bellenir*

129 •
Tomari Nuclear Power Station, Units 1-2

Ohaza-horikappumura 726
Tomarimura, Furuu-gun 045-02,
  Hokkaido, Japan
Takeshi Endo, Dir./Gen.Mgr., Nuclear
  Office

Phone: 0135-75-3331
Fax: 0135-75-3705

Owning Utility

Hokkaido Electric Power Co., Inc.
2 Higashi, 1-Chome, Ohdori
Chuoku, Sapporo
Hokkaido 060-91, Japan
Makoto Tanimura, Managing Dir.,
  Nuclear

Phone: 011-251-1111
Fax: 011-221-1864
Telex: 922-303 HEPCO J

Contact

Keigo Tuteoku, Head of PR Center,
  Nuclear Office
Tomari Nuclear Power Station

Basic Facts

Units 1-2 *Status:* Operable. *Type of Reactor:* Pressurized light-water reactor. *Megawatts (electric):* 579. *Megawatts (thermal):* 1645. *Reactor System Supplier:* Mitsubishi Heavy Industries Ltd. (Japan). *Generator Supplier:* Mitsubishi Heavy Industries Ltd. (Japan); Mitsubishi Electric Corp. (Japan). *Architect Engineer:* Mitsubishi Atomic Power Industries Inc. (Japan).

Key Dates

Unit 1 *Ordered:* 1986. *Construction began:* 1984. *Criticality:* November 1988. *First Power:* December 1988. *Commercial Operation:* June 1989.

Unit 2 *Ordered:* 1986. *Construction began:* 1984. *Criticality:* July 1990. *First Power:* August 1990. *Commercial Operation:* April 1991.

Plant Report

Controversial siting. The Tomari Nuclear Power Station is the only nuclear power plant on Hokkaido, the northern-most of Japan's four main islands. Plans to construct a nuclear station on Hokkaido took twenty years to complete. Only five years were required for actual construction; 15 years of planning were consumed by lengthy negotiations with the community. Agreement was reached only after the Japanese government promised Hokkaido's residents to bolster the local economy and provide enough jobs to stop emigration.

The official announcement of Tomari as the site for Hokkaido's nuclear station was made on March 27, 1982. The following day two small earthquakes hit the region. In 1983 an intense earthquake injured more than 100 people and destroyed roads.

Sources

"Nuclear Power in Japan." *The Bulletin of the Atomic Scientists*, May 1983.
"Japan: The Nuclear Victor?" *Forbes*, December 6, 1982.
"Public Understanding Needed in Building Nuclear Power Plants." *Business JAPAN*, February 1990.
Present Status of Nuclear Power Development. Kyushu Electric Power Co., June 1991.

—*Karen Bellenir*

130 •
Tsuruga Nuclear Power Station, Units 1-2

1 Myojin-cho
Tsuruga-shi
Fukui Prefecture 914, Japan
Takamichi Miike, Mgr., Plant Mgmt. &
  Coordination

Phone: 0770-26-1111
Fax: 0770-26-1111

Owning Utility

Japan Atomic Power Company (JAPCO)
Ohtemachi Bldg. No. 6-1, 1-Chrome
Ohtemachi, Chiyoda-ku
Tokyo 100, Japan
Hiroshi Shimokawa, Managing Dir.,
  Plant Mgmt.

Phone: 03-3201-6631
Fax: 03-3285-0541
Telex: J24592
  (JATOPOW)

Contact

Yukio Sugano, Gen.Mgr., PR Dept.
Japan Atomic Power Company (JAPCO)

Basic Facts

Unit 1 *Status:* Operable. *Type of Reactor:* Boiling water reactor. *Megawatts (electric):* 357. *Megawatts (thermal):* 1064. *Reactor System Sup-*

plier: General Electric Co. (United States). *Generator Supplier:* General Electric Co. (United States). *Architect Engineer:* Ebasco.
Unit 2 *Status:* Operable. *Type of Reactor:* Pressurized light-water reactor. *Megawatts (electric):* 1160. *Megawatts (thermal):* 3423. *Reactor System Supplier:* Mitsubishi Heavy Industries Ltd. (Japan). *Generator Supplier:* Mitsunishi Heavy Industries Ltd. (Japan); Mitsubishi Electric Corp. (Japan). *Architect Engineer:* Mitsubishi Heavy Industries Ltd. (Japan); Mitsubishi Atomic Power Industries Inc. (Japan).

Key Dates

Unit 1 *Ordered:* 1965. *Construction began:* 1966. *Criticality:* 1969. *First Power:* 1969. *Commercial Operation:* March 1970.
Unit 2 *Ordered:* 1981. *Construction began:* 1982. *Criticality:* May 1986. *First Power:* June 1986. *Commercial Operation:* February 1987.

Plant Report

A unique station. The Tsuruga Power Station is operated by the Japan Atomic Power Company (JAPCO). The Japanese Prime Minister's office established JAPCO on November 1, 1957 to oversee the development of commercial nuclear power generation. The station supplies electricity to three separate utility companies, the Kansai Electric Power Co., the Chubu Electric Power Co., and the Hokuriku Electric Power Co.

Tsuruga's 1,327-acre site faces the Sea of Japan on the northern side of the Tsuruga Peninsula in the Fukui Prefecture. The plant is unique in Japan because its units represent two different types of light water reactors. Unit 1 is a boiling water reactor; Unit 2 is a pressurized water reactor.

The site was selected in November 1962 and construction began in April 1966. Unit 1, a 357-megawatt boiling water reactor using enriched uranium fuel, was Japan's first commercial light water reactor. It began operation in March 1970 after a construction period of only 48 months. Construction on Unit 2 started in 1982. The 1,160 megawatt pressurized light-water reactor was Japan's first domestically improved and standardized pressurized light-water reactor plant. It began commercial operation on February 17, 1987.

Design improvements. Several design improvements were implemented at Tsuruga. The Prestressed Concrete Containment Vessel (PCCV), made of compressed reinforced concrete, was developed to improve seismic stability. The PCCV design enabled containment vessels to be smaller and lighter and, therefore, better suited to earthquake prone regions.

The Tsuruga facility is designed to withstand earthquakes, typhoons, and floods. Earthquake meters are set to automatically shut down the reactor if vibration exceeds a predefined level. To further reduce potential earthquake damage, Unit 2 was designed with only three buildings: a reactor building, an auxiliary building, and a turbine building. A conventional unit layout normally included six or seven buildings.

Waste disposal programs. Researchers at Tsuruga have pioneered the development of several nuclear waste disposal programs. The "Off-Gas Hold-up System," which holds noble gases to reduce the amount of radioactivity discharged into the atmosphere, was first used at Tsuruga. The system is now employed at all of Japan's domestic boiling water reactors.

A radioactive waste treatment facility was constructed at the site. Liquid radioactive wastes are filtered and treated. Some are reused and others are discharged into the sea if radioactive levels are low enough. Other liquid waste is solidified. Solid wastes are burned to reduce their volume and low-level waste is stored in an on-site storage building. High-level waste is also stored on-site in specially prepared bunkers.

Cover-up and environmental concerns. Thermal effluent released from the station raises the temperature in Urasoko Bay by as much as 6° C. The Fukui Prefectural Fishery Laboratory is researching the possibility of exploiting the higher than normal temperatures to achieve faster fish growth.

The Tsuruga Power Station is surrounded by the Wakasa Bay National Park. Radiation levels are continuously monitored by the Fukui Environmental Radiation Monitoring Council.

In April 1981, mud and seaweed taken from Tsuruga Bay were found to have 10 to 100 times more radioactivity than normal. Upon investigation, JAPCO officials admitted that a leak of radioactive materials in March 1981 had resulted in the dumping of radioactive sludge directly into the bay. A warning signal had been turned off and the event had not been reported. At the time, the incident was termed the worst accident in Japan's nuclear program.

Continued investigation at Tsuruga revealed 21 additional unreported accidents at Tsuruga, including two coolant leaks in which radioactive materials had been washed directly into the bay in January 1980. During questioning by the Japanese Parliament, company officials admitted that it had been a common practice to release contaminated waste water into the bay.

The cover-up surrounding the incident caused public confidence in nuclear power to waiver. The Japanese Government threatened the possibility of lawsuits against company officials, but no indictment was possible because the prosecutor ruled that the law did not clearly define "excessive irradiation" or "excessive contamination."

Sources

"Chilling Reminder of Things Past." *Macleans*, May 4, 1981.
"Nuclear Power in Japan." *The Bulletin of the Atomic Scientists*, May 1983.
Tsuruga Power Station. Japan Atomic Power Company, n.d.
The Japan Atomic Power Company 1991. N.p., n.d.

—Karen Bellenir

Kazakhstan

COUNTRY REPORT

No information channels

As of January 1992, no formal contacts had been established between the government or electrical authorities in Kazakhstan and international nuclear organizations, such as the International Atomic Energy Agency and the World Association of Nuclear Operators. Little is known about the country's nuclear power plant.

Sources

1993 World Directory of Nuclear Utility Management. American Nuclear Society, July 1992.

PLANT PROFILES

131 ●
BN-350 Nuclear Plant
Shevchenko
Mangyshlak, Kazakhstan

Basic Facts

Status: Operable. *Type of Reactor:* Liquid metal fast breeder reactor. *Megawatts (electric):* 150. *Megawatts (thermal):* 1000.

Key Dates

Ordered: 1963. *Construction began:* 1964. *Criticality:* November 1972. *First Power:* July 1973. *Commercial Operation:* July 1973.

Plant Report

Dual-purpose plant. The BN-350 Nuclear Plant—also known as the Shevchenko Nuclear Power Plant—is a dual-purpose facility. Beyond its power output, it also produces 200 MWe-equivalent for desalination of water. The unit desalinates about 80,000 tons of water for the city of Shevchenko, which is on the Caspian Sea. The plant was troubled by a sodium/water reaction in 1975 that resulted in a two-hour fire. international nuclear organizations, such as the International Atomic Energy Agency and the World Association of Nuclear Operators.

Sources

1993 World Directory of Nuclear Utility Management. American Nuclear Society, July 1992.

Source Book: Soviet-Designed Nuclear Power Plants in the Former Soviet Republics and Czechoslovakia, Hungary and Bulgaria. U.S. Council for Energy Awareness, 1992.

Lithuania

COUNTRY REPORT

Nuclear power in Lithuania under Soviet control

Lithuania relied on nuclear power more heavily than any of the other former Soviet republics (see the former Soviet Union). The country had practically no domestic sources of energy and also depended on the former Soviet Union for coal, gas, and oil. Its only nuclear generating station, Ignalina, was built near the Lithuanian capital, Vilnius. Ignalina produced approximately two-thirds of the electricity generated in Lithuania. Traditionally, Lithuania used about half of its power production and exported the rest.

The two units at Ignalina are both RBMK reactors, similar in design to the reactor that exploded at Chernobyl in April 1986. Following that accident, opposition to nuclear power surfaced in Lithuania. Because Kremlin officials planned to construct additional nuclear units at the Ignalina site, the Lithuanian government established a commission to investigate the project. The commission raised questions about the safety of a third unit and determined that a fourth unit at the site would be an environmental hazard. Lithuanian officials and the central Soviet government in Moscow also disagreed about funding for the project. Moscow ultimately yielded to public protests and cancelled the third and fourth units.

Although still under Soviet control, Lithuanian officials were successful in pioneering international cooperation. In 1989 Ignalina became the first Soviet nuclear station to be visited by an inspection team from the International Atomic Energy Agency. The team reported that plant operation was acceptable and that the number of unplanned automatic shutdowns experienced at Ignalina was among the lowest in the world. Plant operators also participated in a personnel exchange program conducted by the World Organization of Nuclear Operators (WANO).

Nuclear power in Lithuania after independence

Lithuania, along with the other Baltic states of Estonia and Latvia, were the first republics to separate from the former Soviet Union. The Soviet Atomic Ministry officially transferred ownership of the Ignalina nuclear power station in August 1991. Responsibility for near-term operations, however, remained with the Soviet Ministry of Atomic Power and Industry (MAPI).

Following the break-up of the Soviet Union, responsibility for security and safety at Ignalina were transferred to the Lithuanian government and the Lithuanian Institute for Physical and Engineering Problems of Energy Research. The need to improve plant security was emphasized in early 1992 when an employee was arrested for suspected sabotage. Jurgis

Vilemas, director of the Institute, requested Western aid claiming that Lithuania did not have the funds to increase security and improve safety.

In other efforts to improve conditions at the Ignalina plant, the Lithuanian government entered into an agreement with Swedish experts to evaluate and compare the Ignalina station with the Barsebaeck plant in Sweden. In addition, the Swedish Nuclear Power Inspectorate is working with the Lithuanian government to help create a Lithuanian nuclear safety authority.

Lithuania's energy minister, Leonas Asmantas, predicted that the demand for electricity would increase as a result of anticipated growth in the nation's the free market economy. If his forecasts prove true, the Lithuanian government may consider expanding the nation's nuclear generation program by the year 2000.

Sources

"Lithuania Gets Control of A-Plant." *Los Angeles Times*, August 31, 1991.

"Borrowed Time: The Next Chernobyl." *Nightline*, ABC News, February 20, 1992.

"Many Chernobyls Just Waiting to Happen." *Business Week*, March 16, 1992.

Source Book: Soviet-Designed Nuclear Power Plants in the Former Soviet Republics and Czechoslovakia, Hungary and Bulgaria. U.S. Council for Energy Awareness, 1992.

—Karen Bèllenir

PLANT PROFILES

132 ●
Ignalina Nuclear Power Plant, Units 1-2

Snieckus 4761, Lithuania
Victor Shevaldin, Plant Mgr.
Phone: 012-66-31-31
Fax: 012-66-29-350
Telex: 303084 Briz

Owning Utility

Lithuania Ministry of Energy
234761 Lietuva
Snieckus 4761, Lithuania
Phone: 012-66-31-31
Fax: 012-66-29-350
Telex: 303084 Briz

Basic Facts

Units 1-2 *Status:* Operable. *Type of Reactor:* Light-water-cooled, graphite-moderated reactor. *Megawatts (electric):* 1500. *Megawatts (thermal):* 4800. *Generator Supplier:* Elektrosila.

Key Dates

Unit 1 *Ordered:* 1974. *Construction began:* 1977. *Criticality:* October 1983. *First Power:* December 1983. *Commercial Operation:* May 1985.

Unit 2 *Ordered:* 1974. *Construction began:* 1978. *Criticality:* December 1986. *Commercial Operation:* August 1987.

Plant Report

Two Soviet-styled RBMK reactors. The Ignalina Nuclear Power Station is a two-unit power plant located in Snieckus near the Lithuanian capital of Vilnius. Each of the two nuclear units contains a 1,500 megawatt Soviet-styled RBMK reactor, the largest RBMK's built by the Soviets. Ignalina's units have been operated at less than full capacity since the Chernobyl accident in 1986. Because of the power reduction, the plant's total generating capacity is 2,500 megawatts. The station produces two-thirds of all the power generated in Lithuania.

The central Soviet government intended to bring Ignalina's capacity up to 6,000 megawatts through the addition of two more units. After Chernobyl, however, public protests erupted and thwarted the construction program. Environmental studies ordered by the Lithuanian government suggested that the planned increases in generating capacity would have an adverse effect on the local environment. The report cited a potential rise in the temperature of Lake Druksaia and the creation of a possible subsidence hazard at the site. It also expressed concern about increased public exposure to radiation. Ignalina-3 and -4 were officially cancelled in 1988.

Operating history and events. 1988 also proved to be a difficult year for the existing reactors at Ignalina. Ignalina-1 closed for major repairs on August 9, 1988. Ignalina-2 suffered a minor fire on July 20. Although its reactor was not damaged, station management received heavy criticism for not making an official announcement for five days. Ignalina-2 suffered a second fire on September 5, 1988 and was forced out of service for minor repairs. The fire started in a control cable and was extinguished by automatic safety systems. According to a *Tass* report, no injuries were suffered and no radiation was released.

International cooperation. In 1989 Lithuanian officials invited the International Atomic Energy Agency to direct an ASSET (Assessment of Safety Significant Events Team) review at the Ignalina station. The purpose of an ASSET mission is to study a plant's operating history and incident-prevention programs, to determine the root cause of safety-significant events, and to make recommendations for corrective actions. The investigation marked the first time an ASSET study was conducted in the former Soviet Union. The team reported that Ignalina's operation was acceptable by international standards. The plant was also commended for its low number of unplanned automatic shutdowns and for achieving a drop in safety-significant events. Such incidents were down from 40 in 1987 to four in 1989. A second ASSET mission is planned for 1993.

In another effort toward increased international cooperation, personnel from Ignalina participated in a plant exchange program sponsored by the World Association of Nuclear Operators (WANO). In 1991 plant officials visited the Duane Arnold Energy Center in Iowa. Duane Arnold officials also visited Ignalina.

Political unrest threatens operation. In addition to coping with technological problems, the Ignalina station has had to cope with political unrest. Russian plant workers staged a two-hour strike during the winter of 1990 to protest Lithuanian efforts toward independence. The central Soviet government also stopped oil shipments to Lithuania which threatened the plant's oil-dependent cooling system.

Lithuania achieved political sovereignty, and the Soviets transferred ownership of Ignalina to the newly-autonomous government in August 1991. Although the Soviet Ministry of Atomic Power and Industry (MAPI) continued short-term operation of the plant, their jurisdiction ended when the former USSR dissolved the following December.

Political conflict, however, continues. A sabotage attempt at Ignalina-1 was blamed on nationalism. In 1992, a senior programmer of Russian descent was accused of trying to plant a virus in the computer responsible for overseeing the plant's cooling system.

Sources

"Soviet Worries About Nuclear Safety After Chernobyl." *Nature*, August 1988.
"No Escape of Radiation Seen In Fire at a Lithuanian Plant." *The New York Times*, September 6, 1988.
"Lithuanians Protest at Soviet Nuclear Plans." *New Scientist*, October 8, 1988.
"Soviet Nuclear Energy Plants Raise Concern as Republics Gain Control." *The Wall Street Journal*, August 30, 1991.
"Lithuania Gets Control of A-Plant." *Los Angeles Times*, August 31, 1991.
"Many Chernobyls Just Waiting to Happen." *Business Week*, March 16, 1992.
Source Book: Soviet-Designed Nuclear Power Plants in the Former Soviet Republics and Czechoslovakia, Hungary and Bulgaria. U.S. Council for Energy Awareness, 1992.

—Karen Bellenir

133 Ignalina

133 ☆
Ignalina Nuclear Power Plant, Units 3-4

Snieckus 4761, Lithuania
Victor Shevaldin, Plant Mgr.

Phone: 012-66-31-31
Fax: 012-66-29-350
Telex: 303084 Briz

Owning Utility

Lithuania Ministry of Energy
234761 Lietuva
Snieckus 4761, Lithuania

Phone: 012-66-31-31
Fax: 012-66-29-350
Telex: 303084 Briz

Basic Facts

Units 3-4 *Status:* Cancelled. *Type of Reactor:* Light-water-cooled, graphite-moderated reactor. *Megawatts (electric):* 1500. *Megawatts (thermal):* 4800. *Generator Supplier:* Elektrosila.

Plant Report

See Plant Report for Ignalina Nuclear Power Plant, Units 1-2.

Luxembourg

COUNTRY REPORT

Luxembourg rejects nuclear option

When France decided to build a four-reactor nuclear complex at Cattenom, it did not make for good relations with its neighbor, Luxembourg. The plant sits near the Moselle River only 8km from the Luxembourg border.

France started building Cattenom in 1978, shortly after Luxembourg decided not to build a nuclear plant on its side of the Moselle at Remerschem. Even though Luxembourg imports 95 percent of its electricity, a strong anti-nuclear sentiment and the environmental pressure group Mouvement Ecologique caused it to reject nuclear generation.

Two thirds of the population of Luxembourg live within 40 km of Cattenom, but Luxembourg has no control over the environmental standards for emissions from the plant.

In July 1986, fierce protests were staged by German antinuclear groups, backed by officials from Luxembourg and the state of Saarland, about the safety norms at Cattenom. Both Luxembourg and Germany attempted to block commissioning of the first reactor, but a Strasbourg court rejected the petitions.

Luxembourg did win some control of day-to-day activities in a 1986 Franco-German convention, but still questions environmental standards.

Germany restricts similar reactors to a liquid discharge limit of 3 curies per year. France arbitrarily set the Cattenom limit at 15 curies without consultation with its neighbors.

Sources

"France: The Nuclear Dream Maintained." *Technology Review*, November/December, 1982.

"International: Three New French Plants Enter Service." *Nuclear News*, February 1991.

EDF 1990 Annual Report. Generation and Transmission Group, Paris, April 1991.

"Palo Verde Units 1 & 2 Top American Producers in May: Unit 2 Continues Record Run." *Business Wire*, July 12, 1991.

"Cracking at French PWRs Raises Curiosity, But Little Concern in U.S." *Inside N.R.C.*, December 2, 1991.

Mounfield, Peter. *World Nuclear Power*. Routledge, 1991.

—Al Cook

PLANT PROFILES

134 ☆
Remerschem Nuclear Power Plant
Remerschem, Luxembourg

Basic Facts

Status: Cancelled. *Type of Reactor:* Pressurized light-water reactor. *Megawatts (electric):* 1330. *Megawatts (thermal):* 3760. *Reactor System Supplier:* The Babcock & Wilcox Co. (United States). *Generator Supplier:* Brown Boveri et Cie. (Switzerland). *Owner:* Societe Luxembourgoise d'Energie Nucleaire. *Operator:* Societe Luxembourgoise d'Energie Nucleaire.

Key Dates

Ordered: 1976. *Cancelled:* 1977.

Mexico

COUNTRY REPORT

U.S. Atoms for Peace spurs Mexican nuclear program

Encouraged by the U.S. Atoms for Peace program, in the late 1960s many countries, including Mexico, designed ambitious programs for nuclear power plants. The Mexican government sent a number of scientists and technicians for foreign training. They planned to build at least 20 nuclear power plants by the end of the century, which would produce a total of 21,000 megawatts, estimated to represent 16.6 percent of total electricity production in Mexico by the year 2000. But cost overruns, a decade of delays, and heavy interest payments on foreign debt led to the shelving of plans for all nuclear plants except Laguna Verde. (Laguna Verde Unit 2 is still under construction with an expected commercial operation date of mid-1994.)

Sources

"Wooing Mexico to Nuclear Power." *Bulletin of the Atomic Scientists*, July/August 1989.

PLANT PROFILES

135 ●
Laguna Verde Nuclear Power Plant, Unit 1

Apartado Postal No. 53
91680 Cd. Cardel, Veracruz, Mexico
Rafael Fernandez de la Garza, Nuclear Power Plants Mgr.

Phone: (29) 37-45-11
Fax: (29) 37-47-76
Telex: 151190-CFLVME

Owning Utility

Comision Federal de Electricidad (CFEM)
Rio Rodano No. 14
7 Piso
06500 Colonia Cuauhtemoc, D.F., Mexico

Phone: (5) 5-53-64-00
Fax: (5) 5-53-64-24
Telex: 01772551

Contact

Vinicio Serment Cabrero, Pub.Info. Dept. Head
Laguna Verde Nuclear Power Plant

Basic Facts

Unit 1 *Status:* Operable. *Type of Reactor:* Boiling water reactor. *Megawatts (electric):* 675. *Megawatts (thermal):* 1931. *Reactor System Supplier:* General Electric Co. (United States). *Generator Supplier:* Mitsubishi. *Architect Engineer:* Ebasco.

Key Dates

Unit 1 *Ordered:* 1969. *Construction began:* 1975. *Criticality:* 1989. *First Power:* April 1989. *Commercial Operation:* July 1990.

Plant Report

Unprecedented opposition. The construction of Laguna Verde 1, the nation's first nuclear power plant, did not go forward without opposition. Many critics, grouped around the Coordinadora Nacional Contra Laguna Verda (CON-CLAVE)—a national coalition of some 80 antinuclear and environmental groups, housekeepers, farmers, cattle ranchers, shrimpers, and intellectuals—felt that the plants' construction was shoddy, the design faulty, and its performance poor.

Critics questioned the need for a nuclear power plant in a country so rich in non-nuclear resources, asserting that electricity generated by Laguna Verde would cost double that produced by plants using oil. Critics also claimed the plant was in an earthquake and volcano zone and accused authorities for their radioactive waste handling policies. (The government, as of 1989, had not yet announced long-range plans for disposing of the plant's wastes, which were stockpiled in the reactor's pools.)

Among the groups that fought the construction of Laguna Verde were the Group of 100, an organization of prominent intellectuals and artists active on ecological issues; the Veracruz Ecological Group, headed by Dr. Thomas Berlin Schaller, professor of urban planning at Veracruz University in Xalapa; and the Veracruz Mothers' Committee.

Other critics included past presidential candidates from the political opposition, past top government energy officials such as Jacinto Miqueira and Jorge Young, and prominent scientists. Several major demonstrations in Mexico City and vigils outside the reactor site in Veracruz suggested that a sizeable portion of the general public resisted the project.

The attack against Laguna Verde was unusual for Mexico where there is little tradition of this type of lobbying or civic involvement. At one point Mexican Army troops were deployed to the plant and surrounding towns and villages fearing protestor activities would impede functioning of the plant.

Design flaws, accidents, and a cover-up. Public concern about the safety of Laguna Verde was not unfounded. Design flaws at Laguna Verde were first identified in a U.S. Atomic Energy Commission memoranda written between 1971 and 1972, and released in 1978 by the Union of Concerned Scientists. The Reed Report, a General Electric internal safety assessment study made in 1975 but not released until 1987, also noted reactor defects.

The flaws of the boiling water reactor may have been compounded during construction, when the steel reactor vessel was dropped from a crane and fell to the ground. Also, during operation tests, sea water accidentally entered the reactor containment building, and this may have led to corrosion.

Two months after the plant was loaded for pre-operational testing in October 1988, two accidents occurred, producing ruptured pipes. Had the plant been online, the accidents would have been severe. (Critics of the plants emergency plan argue that there are inadequate medical facilities, roads, communications, and economic resources to handle a major radioactive release.)

In one case, plant officials and local authorities tried to hide the accident, claiming that military, police, and medical personnel were mobilized as part of a safety exercise. Eventually the authorities were forced to admit the accident, and a commission of the Mexican House of Representatives who inquired into the incidents recommended that a number of officials be dismissed for having misled its investigators and the public. The accused officials were J. Eibenschutz, deputy director of CFEM; Rafael Fernandez de la Garza, director of the plant, and Miguel Medina, director of the Comision Nacional de Seguridad Nuclear y Salvaguardas (CNSNS), Mexico's safety agency.

Worries about safety oversight. Safety oversight was another area of concern to Laguna Verde critics. The Comision Nacional de Seguridad Nuclear y Salvaguardas (CNSNS) is required by law to watch over safety and, as an independent agency, grant an operating license to the Comision Federal de Electricidad, the state-owned utility for power production which runs the power plant. However, as of 1989, there were no documents showing that CNSNS conducted any safety assessment.

CNSNS has been criticized for negligence in the past due to its poor management of radiation accidents. In 1984 in Ciudad Juarez, a frontier city, one the most severe radiation accidents in North America occurred when cobalt 60 pellets from a cancer therapy machine were inadvertently smelted in an iron foundry. Some of the metal pieces produced at the foundry were exported to the U.S., where the radioactivity was first detected. In the meantime, several houses were built with radioactive iron rods, and their occupants received unknown levels of radiation. CNSNS did not release a study of this accident or make public any plan to institute preventive measures. U.S. officials expressed concern about CNSNS' handling of the accident.

Opposition countered. In the face of public protest, the government mounted an expensive publicity campaign: "The French love life too" said one of the many publicity slogans. (France generates 80 percent of its energy by nuclear power plants.) Plant authorities released a video tape that was shown to peasants and fisherman in nearby villages and sponsored plant tours for thousands of school children and delegations of local mayors, layers, and doctors.

In response to the concerns of critics, government officials did not dispute Laguna Verde's financial inefficiency as compared to the cost of electricity produced by oil, but argued that with so much money spent, it would be foolish to abandon the plant. In countering other arguments, authorities said the Laguna Verde site was chosen for its seismic stability and had capabilities to store radioactive waste on site for 50-100 years.

Moreover, Mexican nuclear power officials reported that the plant was constructed following international norms and quality standards, arguing that critics were misinformed. And International Atomic Energy Agency officials Noramyl Bin Muslim and Morris Rossen noted that "Laguna Verde guarantees the maximum safety."

Construction overdue, cost overrun, more opposition. In July of 1990, Laguna Verde 1 began commercial operation— 14 years overdue. The cost of the plant mushroomed from initial estimates of $128 million to $3.5 billion.

In November 1990 a federal judge in Mexico agreed to hear a motion by environmentalists to halt commercial operation of Laguna Verde 1. The motion was the first of about 20,000 legal complaints filed against the plant by environmentalists since 1987 to receive a hearing. Agency officials Noramyl Bin Muslim and Morris Rossen noted that "Laguna Verde guarantees safety."

Sources

"Storm Gathers Over Mexico A-Plant." *The New York Times*, May 2, 1987.
"Mexico to Start First Atom Plant." *The New York Times*, October 16, 1988.
"Wooing Mexico to Nuclear Power." *Bulletin of the Atomic Scientists*, July/August 1989.
"World Wire: Environmentalists' Victory." *The Wall Street Journal*, November 2, 1990.

Laguna Verde Nuclear Power Plant, Unit 2

Apartado Postal No. 53
91680 Cd. Cardel, Veracruz, Mexico
Rafael Fernandez de la Garza, Nuclear Power Plants Mgr.

Phone: (29) 37-45-11
Fax: (29) 37-47-76
Telex: 151190-CFLVME

Owning Utility

Comision Federal de Electricidad (CFEM)
Rio Rodano No. 14
7 Piso
06500 Colonia Cuauhtemoc, D.F., Mexico

Phone: (5) 5-53-64-00
Fax: (5) 5-53-64-24
Telex: 01772551

Contact

Vinicio Serment Cabrero, Pub.Info. Dept. Head
Laguna Verde Nuclear Power Plant

Basic Facts

Unit 2 *Status:* Under construction (73% complete as of January 1992). *Type of Reactor:* Boiling water reactor. *Megawatts (electric):* 675. *Megawatts (thermal):* 1931. *Reactor System Supplier:* General Electric Co. (United States). *Generator Supplier:* Mitsubishi. *Architect Engineer:* Ebasco; Comision Federal de Electricidad.

Key Dates

Unit 2 *Ordered:* 1969. *Construction began:* 1975. *Commercial Operation Expected:* June 1994.

Plant Report

See Plant Report for Laguna Verde Nuclear Power Plant, Unit 1.

Netherlands

COUNTRY REPORT

A limited nuclear industry

Up to 1900, only two commercial nuclear power plants resided in the Netherlands: Dodewaard, in operation since 1969, and Borssele, in operation since 1973. These plants produce 6 percent of the country's electricity.

Two nuclear research reactors also exist in the Netherlands, one at Delft University, and one at the Netherlands Energy Research Foundation ECN at Petten. Petten also hosts a storage facility for low- and medium-level radioactive wastes.

Possibility for expansion

The possibility of expanding the share of nuclear-generated electricity in the Netherlands to at least 30 percent of the total electricity production has been discussed several times in the last 15 years. Since the accident at Chernobyl, the majority of the people in the Netherlands feel quite uncomfortable about the safety risks concerning certain reactor types. New research will have to show that these risks can be avoided or made sufficiently small. No decision was expected before 1991.

Nuclear energy history

In 1927, the N.V. tot Keuring van Elektrotechnische Materialen (KEMA) was founded to do electricity research and to advise electric companies in the Netherlands. In 1946, KEMA started a study program on nuclear energy and its possible contribution to the country.

In 1949, the N.V. Samenwerkende Elektriciteitis Produktiebedrijven (SEP) was founded to coordinate the electrical supply in the Netherlands. In the 1950s, SEP recommended the building of a 150-megawatt nuclear power plant, but further research proved the plant would be too costly. A final decision recommended the building of a 50-megawatt plant.

On February 2, 1965, the N.V. Gemeenschappelijke Kernenergiecentrale Nederland (GKN) was established to build and run Dodewaard, which became the Netherlands' first commercial nuclear power plant.

Nuclear waste management

The Netherlands' limited nuclear industry produces a relatively small amount of nuclear waste. In 1982, the government decided to build a long-term, above-ground interim storage facility. A three-part strategy was worked out in the Radioactive Waste Policy Paper, and accepted by the parliament in 1984: 1) build one facility (at Petten), 2) found a waste management organization, the Centrale Organisatie Voor Radioactief Afval (COVRA), and 3) fund continuing research. COVRA, a privately owned, nonprofit organization, is responsible for the managing, conditioning, packaging, transportation, and interim storage of radioactive waste in the Netherlands.

The Petten site is scheduled to operate until 1994, and a new site at Borssele is planned, with wastes being transferred there also in 1994. Up to 1990, the Netherlands sent all its spent fuel to France or the United Kingdom for reprocessing, with return to the country by 2000.

Sources

The Netherlands: Nuclear Waste Management. The Nuclear Energy Agency of the Organization for Economic Cooperation and Development (NEA/OECD), and the Centrale Organisatie Voor Radioactief Afval (COVRA), n.d.

The Dodewaard Nuclear Power Plant. Gemeenschappelijke Kernenergiecentrale Nederland, May 1990.

—Charles Ceplecha

PLANT PROFILES

137 • Borssele Nuclear Power Plant

Zeedijk 32
4454 PM Borssele, Netherlands
J. Den Boer, Site Mgr.

Phone: (0) 1100-21000
Fax: (0) 1105-2385
Telex: 55399

Owning Utility

N.V. Elektriciteits-Produktiemaatschappij
   Zuid-Nederland (EPZ)
Zeedijk 32
Postbus 130
4380 AC Vlissingen, Netherlands
J.J. Th. Beckers, Social Aff. &
   Organization Dir.

Phone: 01100-21000
Fax: 01105-2550
Telex: 55489

Contact

Mr. G. Villerius, Pub.Aff.
N.V. Elektriciteits-Produktiemaatschappij
   Zuid-Nederland (EPZ)

Basic Facts

Status: Operable. *Type of Reactor:* Pressurized light-water reactor. *Megawatts (electric):* 481. *Megawatts (thermal):* 1366. *Reactor System Supplier:* Kraftwork Union AG (Germany); Rotterdamse Drookdok Madtdschappij (Netherlands). *Generator Supplier:* Kraftwork Union AG (Germany); Stork. *Architect Engineer:* Kraftwork Union AG (Germany).

Key Dates

Ordered: 1969. *Construction began:* 1969. *Criticality:* June 1973. *First Power:* July 1973. *Commercial Operation:* October 1973.

138 • Dodewaard Nuclear Power Plant

Waalbandijk 112A
6669 MG Dodewaard, Netherlands
J. Hoekstra, Plant Mgr.

Phone: (08885) 88-11
Fax: (08885) 21-28

Owning Utility

N.V. Gemeenschappelijke
   Kernenergiecentrale Nederland (GKN)
Postbus 40
6669 ZG Dodewaard, Netherlands
H. Arnold, Pres.

Phone: (08885) 88-11
Fax: (08885) 21-28

Contact

I.W. Carsouw Van Gulijk, Pub.Info.
N.V. Gemeenschappelijke
   Kernenergiecentrale Nederland (GKN)

Basic Facts

Status: Operable. *Type of Reactor:* Boiling water reactor. *Megawatts (electric):* 58. *Megawatts (thermal):* 183. *Reactor System Supplier:* General Electric Co. (United States); Rotterdamse Drookdok Madtdschappij (Netherlands). *Generator Supplier:* Stork. *Architect Engineer:* N.V. Gemeenschappelijke Kernenergiecentrale Nederland (GKN).

Key Dates

Ordered: 1963. *Construction began:* 1965. *Criticality:* June 1968. *First Power:* September 1968. *Commercial Operation:* January 1969. *Shut Down Expected:* January 2004.

Plant Report

A first for The Netherlands. The Dodewaard Nuclear Power Plant occupies a site along the Waal River, near Arnem. It was the country's first commercial nuclear power plant.

N.V. Gemeenschappelijke Kernenergiecentrale Nederland (GKN), the owner of Dodewaard, employs 138 people at Dodewaard, and plans to operate the plant until January 1, 2004.

Construction history. Dodewaard's site was chosen for three reasons: its closeness to the Waal River, its closeness to Arntem (the seat of KEMA), and its closeness to the national power grid.

The government established GKN specifically to run Dodewaard. An agreement with the General Electric Corporation (GE), in which GE became plant advisor, allowed as many parts as possible to be produced in the Netherlands and let experts from Dutch utilities participate in the project. Dutch firms did more than 95 percent of the work on the plant.

The Dutch government provided financial support for the construction and used its financess to stimulate industrial development in nuclear techniques. Euratom also provided financial support for the project.

Small, but educational. Dodewaard houses one 60-megawatt boiling water reactor—a comparatively lower output than most commercial nuclear power plants. The primary reason for building the plant was to provide the Dutch electric companies with insight on nuclear energy. The plant has additional instruments for use in detailed studies concerning different aspects of nuclear power, providing the country with knowledge while attaining experience.

The N.V. tot Keuring van Elektrotechnische Materialen (KEMA) conducts research on the plant, and advises other Dutch utilities. Dodewaard also plays a part in the Dutch government's plans concerning new legislature and the establishment of efficient and effective controlling agencies.

A provider of research opportunities. Many different kinds of research have taken place at Dodewaard. Research on optimizing fuel use resulted in a higher energy supply per fuel element.

Researchers have checked the fluctuations of different variables, such as temperature and pressure, in the reactor to understand its dynamic behavior as an early warning sign of potential fault. The Interuniversitair Reaktor Instituut at Delft collaborated with GKN in this research.

For materials research, scientists have put different materials in the vessel to study their behavior when exposed to radiation. This also included the study of corrosive products, some of which are released from the turbine into the water, which can build up and cause undesired radiation fields.

Research into limiting the volume of radioactive waste produced two possible procedures: acid digestion and incineration. The incineration procedure has already been tested at Dodewaard.

Sources

The Dodewaard Nuclear Power Plant. Gemeenschappelijke Kernenergiecentrale Nederland, May 1990.

—Charles Ceplecha

Pakistan

COUNTRY REPORT

The beginning of Pakistan's nuclear program

Pakistan established an Atomic Energy Commission in 1955 and received its first research reactor under an Atoms for Peace grant in 1960. Construction on the country's sole operating commercial nuclear power station, the Karachi Nuclear Power Plant (KANUPP) began in 1966. KANUPP, a 137-megawatt pressurized heavy water reactor, was built by Canadian General Electric on a "turnkey" basis, and it began providing power in 1972. KANUPP is a single unit station located near Karachi, a southern Pakistani port city. Under the terms of the construction agreement, the facility is subject to international inspection and monitoring.

Increasing international concern

In the middle 1970's France and Pakistan entered into an agreement to construct a nuclear waste treatment plant. The deal aroused international concern about potential diversion of spent fuel into a weapons program. Under pressure from the United States, France backed away from the project.

In the early 1980's France and Pakistan again began nuclear negotiations, this time over the proposed sale of a 900-megawatt pressurized water reactor. The proposed plant, called the Chashma Nuclear Power Project (CHASNUPP) was planned for construction 150 miles south of Islamabad. Although Pakistani officials indicated their willingness to place the light water reactor under the supervision of the International Atomic Energy Agency (IAEA), they refused to include Pakistani uranium-enrichment facilities in the inspection agreement. Again, France backed away from the deal because of western pressure.

The Pakistani government consistently denies diverting peaceful power-producing technology into weapons programs; however, many western commentators feel the nation has a nuclear weapons program capable of constructing nuclear bombs. During the 1980's on two occasions, Pakistani nationals were convicted of attempting to smuggle nuclear-explosive technology out of the United States.

Resumption of power plant construction program

Pakistan resumed negotiations with nuclear suppliers in 1990 for the construction of the CHASNUPP plant. In 1990, French President Francois Mitterand and Pakistani Prime Minister Benazir Bhutto entered into an agreement for the construction of a 900-megawatt nuclear station. Bhutto vowed to submit the plant to IAEA safeguards and monitoring.

Ms. Bhutto's government collapsed in August 1990. Following her ouster, Pakistani Prime Minister Nawaz Sharif continued nuclear discussions with both France and China. The Chinese proposal called for the construction of a 300-megawatt reactor. Pakistani officials expect to finalize their plans for the CHASNUPP station and begin construction during the early 1990's. Commercial production is tentatively slated for 1998 or 1999.

Pakistan's nuclear future

Because Pakistan has few domestic fossil fuel supplies and an inadequate number of sites suitable for hydroelectric generation, nuclear power promises to be an important source of electricity for the nation's economic development. The nation has suffered from electricity shortages and the cost of imported oil is blamed for thwarting growth. Pakistani refusal to sign the Nuclear Non-Proliferation Treaty, however, has created difficulties in its relationship with potential suppliers. Although Prime Minister Sharif called nuclear development essential to the nation's progress, he remains insistent that Pakistan's nuclear program remain independent and removed from foreign oversight.

Sources

"France to Sell Nuclear Power Plant to Pakistan." *The Washington Post*, February 20, 1990.

"Pakistani Vows to Step Up Nuclear Program." *The New York Times*, November 8, 1990.

Shafique, Muhammad. Pakistan Atomic Energy Commission. December 19, 1991.

Mounfield, Peter. *World Nuclear Power*. Routledge, 1991.

"Pakistan Nuclear Deal for China." *The Wall Street Journal*, January 2, 1992.

"Power Station Deal." *Toronto Globe and Mail*, January 18, 1992.

—Karen Bellenir

PLANT PROFILES

139 ☐
Chashma Nuclear Power Plant (CHASNUPP)

PO Box 1133 Phone: 92-21-820463
Islamabad, Pakistan
Mirza Azfar Beg, Gen.Mgr.

Owning Utility

Pakistan Atomic Energy Commission Phone: 92-51-819030-9
PO Box 1114
Islamabad, Pakistan
Ishfaq Ahmad, Chairman

Contact

Mohammad Shafique, Member, Power
Pakistan Atomic Energy Commission

Basic Facts

Status: Planned. *Type of Reactor:* Pressurized light-water reactor. *Megawatts (electric):* 900.

140 ●
Karachi Nuclear Power Plant (KANUPP)

PO Box 3183 Phone: 92-21-7737445
Paradise Point
Karachi-29, Pakistan
Jamshed A. Hashmi, Gen.Mgr.

Owning Utility

Pakistan Atomic Energy Commission Phone: 92-51-819030-9
PO Box 1114
Islamabad, Pakistan
Ishfaq Ahmad, Chairman

Contact

Mohammad Shafique, Member, Power
Pakistan Atomic Energy Commission

Basic Facts

Status: Operable. *Type of Reactor:* Pressurized heavy-water reactor. *Megawatts (electric):* 137. *Megawatts (thermal):* 433. *Reactor System Supplier:* General Electric Canada. *Generator Supplier:* Hitachi. *Architect Engineer:* General Electric Canada.

Key Dates

Ordered: 1965. *Construction began:* 1966. *Criticality:* 1971. *First Power:* October 1971. *Commercial Operation:* October 1972.

People's Republic of China

COUNTRY REPORT

China adopts nuclear power

China depends on coal for 70 percent of its energy needs. Many of China's provinces, especially those in the north, have large reserves of coal from which they generate electricity. However, a small rail system makes it difficult to transport surplus coal to those areas of the country lacking in energy resources. In the industrialized and heavily populated southeastern coastal regions, shortages of coal and electricity have held back economic development. To expand its energy capabilities in these growing areas, China has turned to nuclear energy options.

Because China has used massive amounts of coal to generate electricity, it ranks among the world's most inefficient countries in carbon dioxide emissions. In 1989, China emitted 6.01 metric tons of carbon dioxide per $1,000 of gross national production—more than any other nation, according to the National Academy of Sciences' Commission on Policy Options for Global Warming. As the country's economy grows, some experts are concerned that its carbon dioxide emissions will increase even more. An active nuclear energy program would help the country meet its growing need for electricity without adding greenhouse gases to the atmosphere.

China has encountered limitations with other fossil fuels as well as hydroelectric power. Offshore oil developments have fallen short of expectations. The country's natural gas reserves are very small. Hydroelectric dams are not favored for further development as they flood valuable arable land needed to feed the nation's population of over a billion people. Moreover, existing hydroelectric stations are too far inland for power to be transmitted economically to large population centers in the coastal areas.

For these reasons—excessive carbon dioxide emissions from coal, poor distribution channels, and the limitations of hydroelectricity and oil—China began considering nuclear energy in the late 1970s.

12 plants, 20 years, $20 billion

In 1980, China announced to the world that they intended to build 12 nuclear power plants over the next 20 years. The country also said it would open its doors to foreign contractors to help in the $20 billion construction process.

Companies in the United Stataes had much to gain from open trade arrangements with China. Westinghouse Electric Corp. (United States) would be a prime contender for selling nuclear reactors. Control Data envisioned doing hundreds of thousands of dollars worth of business in software and consulting services. The Tennessee Valley Authority wanted to sell parts from the two Yellow Creek nuclear units that the utility had discontinued building.

U.S.-China cooperation

In 1981, U.S. President Ronald Reagan began negotiations with China to clear the way for U.S. companies to compete for a role in China's new nuclear boom. Negotiations culminated in 1984 with the signing of an agreement for nuclear cooperation by President Reagan and China's Premier Zhao Ziyang. China promised not to carry out the enrichment or reprocessing of nuclear fuel from U.S.-built reactors, or to store nuclear materials capable of use in nuclear weapons without U.S. consent.

Critics in the U.S. Congress had problems with the agreement because China had refused to sign the Nuclear Non-Proliferation Act (NNPA), which is designed to prevent the spread of nuclear weapons, and had failed to place its nuclear facilities under international safeguards. Although Premier Zhao Ziyang had said at a White House dinner that "we do not engage in nuclear proliferation ourselves, nor do we help other countries develop nuclear weapons," the country would not commit itself on paper to a nonproliferation policy. Although China had not developed commercial nuclear energy capabilities, the country has had an active nuclear weapons program. Antinuclear groups feared China would use plutonium produced in civilian nuclear reactors to make additional warheads.

A joint resolution of Congress placed three basic restrictions on the Presidential agreement: the president had to certify that China is not helping any other country build nuclear weapons; effective safeguards must be in place to guarantee the peaceful use of any technology transferred under the agreement; and the United States can disapprove future Chinese requests.

In September 1985, China's deputy nuclear minister said the country would open civilian nuclear sites for inspection by the International Atomic Energy Agency (IAEA). Although it wasn't clear which sites the country would place under IAEA safeguards (at the time, China had no commercial nuclear power plants), and despite the pledge's vagueness about timing, U.S. officials felt it was a significant step. China seemed eager to trade with U.S. suppliers, as well as companies in Europe and Japan.

Chinese independence—on again, off again

In 1987, China's safety chief at the Ministry of Nuclear Industry announced that the country would revert to self-reliance. The Guongdong nuclear plant at Daya Bay, a project being built with Western technology, would be China's first and last joint venture to set up a foreign-designed nuclear plant. Qinshan Nuclear Power Plant, also under construction, would go forward as an indigenous project.

The announcement had political overtones. Zhao Ziyang, the former prime minister who favored Western investment in China and signed the nuclear cooperation agreement with U.S. President Reagan, had recently been succeeded by Li Peng, a Soviet trained leader who favored nuclear independence.

China's self-sufficient attitude didn't last long. In 1989, the country attempted to purchase equipment from a cancelled U.S. nuclear power plant in the Washington Public Power Supply System. However, Congressional pressure deferred the $500 million deal on the basis of restrictions placed on the 1985 U.S.-China agreement. The president had not been able to certify Chinese nonproliferation. In fact, U.S. intelligence sources were reporting deals with Pakistan and a number of Middle Eastern states.

China also maintained ties with Germany to assist in the planning and construction of addition reactors at Qinshan.

U.S. companies involved in the nuclear industry have yet to benefit from President Reagan's negotiations with China in the early 1980s.

victory for Chinese autonomy

The start-up of Qinshan Nuclear Power Plant in December 1991 is seen by the Chinese as a victory in national self-reliance. Although Western companies provided some components, over 600 plants in China participated in the design and construction process. With Qinshan's inauguration, China became the 26th nation in the world to generate commercial electricity with nuclear energy. The country is now one of six in the world that can design and build nuclear reactors.

Future plans

Although the country was still in desperate need of increased generating capacity in the late 1980s—evidenced by blackouts and industrial shutdowns that were common in much of the country—skepticism about the economic benefits of nuclear power was growing. China has reduced the number of planned nuclear plants from 12 to 10, and it is likely the number might be further reduced. Some speculate that plans for expanding China's nuclear program beyond the Qinshan and Guongdong projects will be shelved for many years.

Sources

"Critics Reserve Judgment on China Nuclear Accord." *Science,* May 11, 1984.

"Nuclear Cooperation: China and the United States." *America,* July 14, 1984.

"China to Permit Inspection of Some Nuclear Facilities." *Science,* October 11, 1985.

"Shock Action; Chinese Nuclear Power." *The Economist,* December 12, 1987.

"White House Defers Deal for Nuclear Equipment." *The Washington Post,* June 12, 1989.

"China Begins Trial Operation of First Atomic Power Plant." *The New York Times,* December 19, 1991.

Mounfield, Peter. *World Nuclear Power.* Routledge, 1991.

Qinshan Nuclear Power Plant Under Construction. Qinshan Nuclear Power Company, n.d.

"China Becomes 26th Nation to Generate Commercial Nuclear Electricity." *INFO,* U.S. Council for Energy Awareness, January 1992.

PLANT PROFILES

Guangdong Nuclear Power Plant, Units 1-2

Shenzhen City, Guangdong, People's Republic of China

Owning Utility

Guangdong Nuclear Power Joint Venture Company, Ltd.
Nuclear Power Bldg.
Central Chennan Rd
Shenzhen City, Guangdong, People's Republic of China
Zan Yun Long, Gen.Mgr.

Phone: 366566

Contact

Qian Zhi Min, Head of PR Branch
Guangdong Nuclear Power Joint Venture Company, Ltd.

Basic Facts

Units 1-2 *Status:* Under construction (Unit 1, 80 percent complete as of January 1992; Unit 2, 73 percent complete as of January 1992). *Type of Reactor:* Pressurized light-water reactor. *Megawatts (electric):* 936. *Megawatts (thermal):* 2785. *Reactor System Supplier:* Framatome (France). *Generator Supplier:* General Electric Co. (United Kingdom); Alsthom. *Architect Engineer:* Electricite de France; General Electric Co. (United Kingdom); Alstthom; Others.

Key Dates

Unit 1 *Ordered:* 1986. *Construction began:* 1986. *Criticality expected* 1993. *Commercial Operation Expected:* 1993.

Unit 2 *Ordered:* 1986. *Construction began:* 1987. *Criticality expected* 1994. *Commercial Operation Expected:* 1994.

Plant Report

Location. The Guangdong Nuclear Power Plant is located in the Guangdong province of China. Also known as Daya Bay Nuclear Power Plant, the Guangdong plant is just less than 50 kilometers northeast of Hong Kong's 5.5 million inhabitants.

With the construction of two power plants at the Guangdong site, Chinese officials hope to increase the generating capacity of the fast-growing province from 8,500 megawatts to 15,000 megawatts. When complete, over 70 percent of the electricity generated will be sold to Hong Kong. The remainder of Guangdong's output will flow into the Guangdong electrical grid to not only relieve the area's energy shortage, but to alleviate pollution problems.

Guangdong's proximity to the Hong Kong territory border has caused great concern among Hong Kong citizens. Lingering doubts about the plant's safety in the wake of Chernobyl have not been remedied by the Chinese government.

Political stand against Guangdong. Hong Kong residents have repeatedly protested the building of the Guangdong plant through public outcry and petitions. In August 1986, a petition reportedly signed by one million Hong Kong citizens protesting construction of the Guangdong plant was rejected by China's Minister of Nuclear Industry, Jiang Xinxiong. This marked an important political stand for Hong Kong, whose residents are known to be quite reticent to sign petitions or advocate controversial positions.

In his September 1986 speech, Premier Zhao Ziyang stated that China should pay greater attention to the safety of the plant, but that the government's attitude toward developing a nuclear power industry in the country would not change. This same sentiment was reiterated on March 8, 1991 by the governor of Guangdong, Ye Xuanping. To calm Hong Kong, the Chinese government in the summer of 1986 announced that a new safety committee would oversee the course of the project. This committee would not have any legal authority, but would serve as a consulting group.

Radioactive fallout and earthquakes. One of the concerns of Hong Kong residents was their close proximity to the plant in the event of a disaster like the one at Chernobyl. In response, the Chinese claimed that they had adhered to International Atomic Energy Agency (IAEA) site selection standards by choosing an area that has a 1.2 kilometer uninhabited zone around the plant, a lightly populated zone that extends 5 kilometers from the first zone, and a population center that does not exist within a 20-kilometer radius.

However, Hong Kong residents were not pacified. To ease their fears, the Chinese agreed to install a sand filtration system designed to control leakage of radioactive hydrogen should an accident occur. The new system was expected to cost an additional $45 million.

Another concern of the residents of Hong Kong is Guangdong's ability to withstand an earthquake. In 1987, the Shuitou-Xichong fault was discovered to be located less than five miles from Guangdong. Somehow this fault went unnoticed by surveyors who had examined the area to a depth of 30 kilometers.

Chinese seismologists are predicting an earthquake along the Shuitou-Xichong fault that would exceed 6 on the Richter scale during the 1990s. Although the plant has been built according to French and Japanese seismic safety standards, and it is designed to withstand an earthquake measuring 8 on the Richter scale, Hong Kong residents are not reassured.

Construction difficulties. Originally, the 1,800 megawatt plant was supposed to cost $3.5 billion, site excavation was scheduled to begin in 1984, and it was to be operational sometime in 1991. Delays have pushed back the operational date for Unit 1 to mid-1993.

In September 1987, construction was temporarily halted when it was discovered that workers laying the concrete foundation for Unit 1 had misread the blueprints and left out 316 of 576 reinforcing rods. Hong Kong investors were not informed of the problem for 12 days, and plant officials did not admit construction had stopped until October 9, 1987. After a two-month delay, China gave the go-ahead to begin full-scale construction on the containment vessel. To account for the missing rods, the second layer of the foundation was to include an extra 10 tons of steel bars. However, the Chinese government would not allow an independent inspection of its proposed solution.

Sources

"A Great Leap into Foreign Borrowing." *Business Week*, April 2, 1984.
"Nuclear Alarm Stirs Hong Kong Activism." *The New York Times*, January 4, 1987.
"Chinese Nuclear Power Shock Action." *The Economist*, December 12, 1987.
"Hong Kong Fears Chinese Chernobyl." *The Bulletin of the Atomic Scientists*, October 1991.
"China Begins Trial Operation of First Atomic Power Plant." *The New York Times*, December 19, 1991.

—Yvette M. Costa

142 ○

Qinshan Nuclear Power Plant, Unit 1

Haiyan County 314300, Zhejiang,
People's Republic of China

Owning Utility

Qinshan Nuclear Power Company
China National Nuclear Corp.
Haiyan County 314300, Zhejiang,
  People's Republic of China
Zhang Huai Lin, Exec. Gen.Mgr.

Phone: 0571-551547
Telex: 37215 HYNPP CN

Contact

Zhu Wei He, Head of Pub.Info. Center
Qinshan Nuclear Power Company

Basic Facts

Unit 1 *Status:* Under construction (95% complete as of January 1992). *Type of Reactor:* Pressurized light-water reactor. *Megawatts (electric):* 300. *Reactor System Supplier:* China National Nuclear Corporation. *Generator Supplier:* China National Nuclear Corporation. *Architect Engineer:* China National Nuclear Corporation.

Key Dates

Unit 1 *Ordered:* 1983. *Construction began:* 1983. *Criticality:* October 31, 1991. *First Power:* December 15, 1991. *Commercial Operation Expected:* June 1992.

Plant Report

China's first nuclear power plant. Qinshan Nuclear Power Plant, the first nuclear power plant designed and built by China, is located by the Hangzhou Bay at the foot of Qinshan Mountain. Although China has built smaller nuclear research reactors, Qinshan is its first to generate commercial electricity. When Qinshan's 300 megawatt reactor became operational on December 15, 1991, China became the 26th nation in the world to generate commercial electricity with nuclear energy. The country is one of six in the world that can design and build nuclear reactors.

The plant feeds electricity into a grid serving Shanghai and three eastern Chinese provinces. China has sought to expand energy capabilties in industrialized coastal regions where shortages of coal and electricity impede economic development.

Victory for Chinese independence. The start-up of Qinshan is seen by the Chinese as a victory in national self-reliance. Although Western companies provided some components, over 600 plants in China participated in the construction process, including the Shanghai Turbine Factory that made the steam turbine rotors and Shanghai Boiler Factory that made the steam generator.

The plant enjoys the support of the Communist Party and was inspected several times during construction by state leaders, including General Secretary Comrade Jiang Zemin and Premier Li Peng.

Although many of the Qinshan workers were trained in France, Germany, Spain, and Yugoslavia, the country hopes to do more training at home in the future. Operating licenses were issued by Spain and Yugoslavia to 36 main control room operators.

Revolution and earthquakes. China first announced its plan to develop a civilian nuclear energy program in February 1978. Qinshan was approved around the same time, but work on the plant was delayed by the chaos of the Cultural Revolution. Actual construction began at the site in 1985.

Construction experienced a hiatus in 1990 due to the implementation of increased quality standards and the need to further study Qinshan's resistance to earthquakes.

An International Atomic Energy Agency inspector who visited the plant in January 1991 expressed confidence in the Qinshan construction program. Inspector Jean-Paul Bember told *The New York Times*, "I think they are very careful about what they are doing."

Although the plant became operational at the end of 1991, full operation was not expected until June 1992 after the plant had been visited by international inspectors. The completed concrete reactor containment building is 62.5 meters high and 38 meters in diameter and is the third protective screen of the nuclear power station.

Qinshan 2 and 3. China planned to build two additional reactor units at the Qinshan site as part of a 20-year plan announced in 1980 to increase the country's electricity generating capacity. In 1987, it was agreed that Kraftwerk Union of Germany would help with the design and construction of Qinshan 2 and 3. As of 1992, construction had not started on either of the sister plants.

Sources

Qinshan Nuclear Power Plant Under Construction. Qinshan Nuclear Power Company, n.d.
Mounfield, Peter. *World Nuclear Power.* Routledge, 1991.
"Hong Kong Fears Chinese Chernobyl." *The Bulletin of the Atomic Scientists*, October 1991.
"China Begins Trial Operation of First Atomic Power Plant." *The New York Times*, December 19, 1991.
"China Becomes 26th Nation to Generate Commercial Nuclear Electricity." *INFO*, U.S. Council for Energy Awareness, January 1992.

Plant Profiles

143 □
Qinshan Nuclear Power Plant, Units 2-3

Haiyan County 314300, Zhejiang,
People's Republic of China

Owning Utility

Qinshan Nuclear Power Company
China National Nuclear Corp.
Haiyan County 314300, Zhejiang,
  People's Republic of China
Zhang Huai Lin, Exec. Gen.Mgr.

Phone: 0571-551547
Telex: 37215 HYNPP CN

Contact

Zhu Wei He, Head of Pub.Info. Center
Qinshan Nuclear Power Company

Basic Facts

Units 2-3 *Status:* Planned. *Type of Reactor:* Pressurized light-water reactor. *Megawatts (electric):* 300. *Reactor System Supplier:* China National Nuclear Corporation. *Generator Supplier:* China National Nuclear Corporation. *Architect Engineer:* China National Nuclear Corporation.

Plant Report

See Plant Report for Qinshan Nuclear Power Plant, Unit 1.

Philippines

COUNTRY REPORT

Philippines' introduction into nuclear energy

The Philippines' introduction into nuclear power was a slow and well-researched process. Beginning in 1955, the country signed the Atoms for Peace program agreement with the United States. By signing the agreement, the Philippine government committed itself to the peaceful use of atomic energy.

On October 26, 1956, the Philippine government was one of 82 nations present at a United Nations meeting that established the International Atomic Energy Agency (IAEA). This agency was established to encourage the development of atomic energy for peaceful purposes for the needs of the under-developed areas of the world. The statute was entered into force between the United Nations and the Philippines on September 2, 1958.

The Philippine Science Act of 1958 created the Philippine Atomic Energy Commission (PAEC) in order to conduct or cause the research and development of materials and devices used in the production of atomic energy.

Considering its first nuclear plant

In the late 1950's, Meralco, a subsidiary of Gilbert Associates, a United States consulting firm, was commissioned to make a preliminary study on the feasibility of a nuclear power plant. The study concluded that the time wasn't right to undertake the project.

Then in 1960, the Philippine government asked the IAEA for assistance in conducting another survey for the use of nuclear power in the country over the next decade. The IAEA began the survey in October 1960, and in August 1961, submitted its report entitled "Prospects of Nuclear Power in the Philippines." The report concluded that toward the end of the 1960's, due to a lack of fossil fuels in the country and limited hydro potential on Luzon Island, a relatively large nuclear plant might compete favorably with thermal stations using imported fuel oil.

In 1967, Meralco invited bids for a 300 to 500 megawatt nuclear power plant to be completed in early 1975. However, it was concluded that a nuclear power plant could not be supported by the Manila market until the late 1970's or early 1980's.

On June 15, 1968, the Atomic Energy Regulatory and Liability Act of 1968 empowered the PAEC to issue licenses for the construction, possession, or operation of any atomic energy facility. This is the law by which the Philippine Nuclear Power Plant (PNPP-1) was licensed.

Site selection

In 1972, a second feasibility study was conducted with the assistance of the UN Development Programme. The survey proposed five possible sites for the location of a nuclear power plant which were then narrowed to two sites: Bagac and Limay, Bataan. The final site selection was completed in January 1976.

The site selection was a time-consuming process because the region was so susceptible to earthquakes, volcanic eruptions, and tsunamis. Originally, the site for the plant was to be located at Kabayo Point at Bagac, but after additional tests were performed, the site was moved to Napot Point, Morong on the Bataan peninsula, even though it is still relatively close to an earthquake fault and next to a volcano that has been dormant for 70,000 years.

Oil crisis revives nuclear option

In late 1973, the first international oil crisis erupted. Not only did the price of crude oil rise from $2.55 per barrel in April 1973 to $10.84 by December 1974, but OPEC also imposed a partial embargo on oil exports to the Philippines. Consequently, the oil import bill of the Philippines jumped from $230 million in 1973 to $680 million in 1974; at the same time, the total volume of crude oil imports dropped 12 percent.

The partial embargo created doubts as to the continuing supply of imported oil which accounted for 92 percent of the country's energy conssumption in 1973. This fact revived interest in the construction of a nuclear power plant in order to reduce the country's dependence on imported oil.

Construction begins

In early 1974, discussions with General Electric Co. (US) and Westinghouse Electric Corp. (US) began and concluded with the decision to enter into a contract with Westinghouse because of the overwhelming preference for it's Pressurized Water Reactor (PWR) design. Negotiations with Westinghouse were conducted from June 1974 to January 1976.

In February 1976, the contract for a 620-megawatt nuclear power plant was signed with Westinghouse at a cost of $562 million. However, by the time financing and cost escalation were figured in, the price had increased to $1.1 billion. On October 21, 1976, the National Power Corporation (NPC) filed an application with the PAEC to start groundwork. By October 1977, the PAEC had granted the NPC its first limited work authority for the plant.

PNPP entangled in politics

Although construction on the first nuclear reactor in the Philippines was finished by its scheduled date of January 1985, it never went into commercial operation. Unfortunately, the plant became entangled in politics when President Corazon Aquino took office in February 1986, after the fall of former President Ferdinand Marcos. The plant became a symbol of the Marcos regime so the new government would not have anything to do with it.

It was also believed by Aquino, that Westinghouse had bribed Marcos through an associate to obtain the contract, and had also built an inferior and unsafe plant. After years of litigation, a settlement was finally reached in early 1992, when the International Chamber of Commerce of the United Nations could not find evidence to support Aquino's claims.

Power outages until PNPP goes online

Because PNPP-1 had never gone on-line, the Philippines began experiencing severe power outages that started in the summer of 1990. By 1992, 12-hour brownouts daily and four-day work weeks became common place. Even though an agreement has been reached, because PNPP-1 will not be operational until 1995 at a cost of $2.6 billion, the Philippines is likely to experience greater shortages of power.

Sources

"Brief on the Philippine Nuclear Power Plant." *National Power Corporation,* n.d.

"Philippines Nuclear Power Plant." *Nuclear Engineering International,* January 1982.

"The $2.2 Billion Nuclear Fiasco." *Fortune,* September 1, 1986.

"Dark Days in the Philippines." *Nuclear Industry,* Summer 1990.

"Philippines' Nuclear Plant Slated to Go on Line." *Info,* U.S. Council for Energy Awareness, March 1992.

—Yvette M. Costa

PLANT PROFILES

Philippine Nuclear Power Plant (PNPP)

Napot Point, Morong
Bataan, Philippines
Antonio T. Carpuz, Plant Mgr.

Phone: 632-922-40-31
Fax: 632-922-40-44
Telex: 40120

Owning Utility

Philippine National Power Corporation
Quezon Ave.
cor. Agham Rd.
Diliman, Quezon City 10183,
   Philippines
Francisco T. Delgado, Sr.V.Pres.,
   Engineering & Nuclear

Phone: 632-921-35-41
Fax: 632-922-84-52
Telex: 40120

Contact

Patricia C. Gutierrez, Mgr., Pub.Aff.
   Dept.
Philippine National Power Corporation

Basic Facts

Status: Under construction (98% complete as of January 1992). *Type of Reactor:* Pressurized light-water reactor. *Megawatts (electric):* 651. *Megawatts (thermal):* 1876. *Reactor System Supplier:* Westinghouse Electric Corp. (United States). *Generator Supplier:* Westinghouse Electric Corp. (United States). *Architect Engineer:* Burns & Roe Inc. (United States).

Key Dates

Ordered: 1974. *Construction began:* 1976. *Commercial Operation Expected:* 1995.

Plant Report

Philippines' first attempt at nuclear power. The Philippine Nuclear Power Plant (PNPP-1) was the first and only attempt to date by the Philippine government to construct and operate a nuclear facility. Since early inception of the plant, the plan has been fraught with heated controversy.

In a deal that has been debated as to its legality for years, Westinghouse Electric Corporation of the United States signed a contract with the Philippine government in February 1976 for $562 million. After the costs of additional project components, financing, and cost escalation for continuing services and equipment provided by Westinghouse, the price tag of PNPP-1 had risen to $1.1 billion. Although the original plans called for two 620-megawatt reactors to be built, by the time a formal contract was signed, just one 626-megawatt (net) plant was planned because of financing restraints of the Philippine government.

Environmental concerns. In choosing a site to build the Philippines first nuclear reactor, the National Power Corporation (NPC), a Philippine government-owned electric utility, had to consider several environmental factors. During eleven months of investigation, researchers studied topography, geology faulting, hydrography, and land use. When the results were tabulated, the site for the plant was moved from Kabayo Point at Bagac to Napot Point, Morong on the Bataan peninsula.

Final site selection made by the NPC received some criticism regarding safety from a potential earthly disaster. Built on the side of a volcano, beside an earthquake fault, environmentalists voiced their concern over the possibility of a volcanic eruption. Experts worried about the ash fall and lava flow should an eruption occur from neighboring Mount Natib, even though the volcano had not erupted for an estimated 70,000 years.

Since the volcano had not erupted for such a long time, it was concluded by the consultant firm hired by the NPC, Ebasco Services, a subsidiary of Enserch Corporation of Dallas, that the volcano would not pose a serious threat. However, PNPP-1 is located just 143 km southeast of Mount Banahao, a volcano that has previously erupted in the year 1739 and as recently as 1909.

The controversy surrounding the plant's susceptibility to an earthquake has also been addressed by NPC engineers. The plant itself was designed to withstand four times the level of shaking a typical Metro-Manila building may incur. Also, the Napot Point site was chosen because ground elevation is 18 meters above sea level on solid rock. The ability for the plant to withstand a tsunami or tidal wave was also addressed by the increase in ground elevation. Since the highest recorded tsunami in the Philippines measured 17 meters, the move to the site at Napot Point was expected to be safe by experts.

Plant design. Westinghouse was chosen to build the reactor after an investigation by the NPC revealed that there was an overwhelming preference for the Pressurized Water Reactor (PWR) design produced by the company and not the Boiling Water Reactor (BWR) design produced by the General Electric company. PNPP-1 was an adaption of the Krsko Nuclear Power Plant in Yugoslavia that included some modifications for the Philippine terrain. However, one item that the Philippine government did insist upon was that the plant be built with the same codes and standards established by the Nuclear Regulatory Commission (NRC) of the United States.

Safety of the Napot Point site. Although site tests were not complete, and the NPC had not obtained a construction permit, Westinghouse began clearing the area in March 1976. On October 21, 1976, the NPC filed an application with the Philippine Atomic Energy Commission (PAEC) for a permit to begin groundwork preparation; the license was granted on December 20, 1976. On July 12, 1977, the NPC applied for its construction permit with the PAEC and it obtained a license with limited work authority on October 3, 1977.

Although the final construction permit was issued in April 1979, construction was halted by President Ferdinand Marcos following the accident at Three Mile Island which resulted in an investigation on the safety features of PNPP-1. The Puno Commission, as it was called, concluded that the plant design was unsafe. Although this claim was disputed by Westinghouse, the company did incorporate safety upgrades in the plant design in order to return to work at the beginning of 1981.

When renegotiations of the contract were finished, the price of PNPP-1 had increased to $1.8 billion. Once Westinghouse resumed construction, it was in a hurry to have PNPP-1 operational as originally promised in January 1985. Therefore, by September 1981, 20 percent of overall construction was complete, with 70 percent of the engineering work done, and over 60 percent of civil construction complete. By 1982, the mechanical and electrical installation work in lower elevations of several buildings had begun.

Radioactive waste disposal. Since PNPP-1 is the Philippine's first nuclear power plant, much consideration has been given to how to handle the low- and high-level radioactive wastes produced at the plant. By the year 2000, PNPP-1 is expected to accumulate 240 metric tons of spent fuel. Currently, the plant will store spent fuel rods on-site in underwater storage pools. This facility will have a 10 year storage capacity that could be expanded to twenty years. However, the Philippines does not have any plans to develop a permanent repository. Since the technology is so new to them, they are going to wait and see what the western countries develop.

The low-level radioactive waste (LLRW) of resins, filters, clothing, rags, and paper will be solidified in concrete and placed in 55-gallon drums to be stored on-site in warehouses. Currently, the plant will be able to accomodate thirty years of storage. However, the NPC is planning a volume reduction system in order to reduce all LLRW at the plant.

Training of plant operators. Because workers who would be employed at PNPP-1 would not have had prior experience in nuclear energy, part of the contract between Westinghouse and the NPC called for a comprehensive training program. The NPC itself requires that its plant operators have engineering degrees and licenses. These operators were subjected to over 160 weeks of structured curricula at the Westinghouse training facilities in the United States, and they spent one year on-site in Morong, Battan.

Westinghouse suspected of bribery. On December 1, 1988, the Philippine government, under the guidance of President Corazon Aquino, filed suit against Westinghouse, claiming that the company paid bribes to former President Ferdinand Marcos in order to win the contract to build a nuclear power plant. The suit also claimed that the plant was built with inferior grade materials and was unsafe to operate. The Philippine government was seeking to rescind the huge contract and force Westinghouse to repay some of what had been paid in kickbacks. Also named in the suit was Burns & Roe Enterprises, Inc. of Oradell, New Jersey, and a subsidiary of Westinghouse, Westinghouse International Projects Company.

The suit was filed under the Racketeer Influenced and Corrupt Organizations statute (RICO) which states that the plaintiff is entitled to triple the amount of damages. The suit demanded an unspecified amount of punitive damages and repayment of the money spent by the Philippine government on the plant.

Mysterious payments. By the time the Marcos regime had been overthrown, PNPP-1 was complete. However, the plant did not go into operation due to the decision by President Corazon Aquino to scrap the plans for the plant. It was believed that the Aquino government made a political rather than a technical decision against operating the plant because it was viewed as a symbol of the Marcos regime and no matter how safe or unsafe it was, Aquino would never allow it to go into operation.

This reluctance stems from the belief that Marcos was bribed by Westinghouse, through an associate, Herminio Disini, for the contract to build a nuclear power plant. Through Jesus Vergara, the former head of Westinghouse's Philippine sales affiliate, Aquino investigators discovered that Westinghouse paid Disini a commission of $50 million of which Marcos received about $30 million. However in a contradictory statement released by a Westinghouse attorney, Vergara said that he did not have any personal knowledge of payments made to Marcos through Disini by Westinghouse and that his original statement was made under pressure by the Philippine government.

When Marcos and his wife fled the Philippines in February 1986, the United States Customs Service seized documents of which one showed a payment of $11.2 million to a Disini firm, Herdi's Management and Investment Corp., by Westinghouse over the course of six years. Another Herdi's document showed two further payments described as "W commission" of $19.5 million.

Bribes disguised as work. Another issue raised by the suit was that three companies, in which Disini had a controlling interest, obtained contracts from Westinghouse for plant-related work. First, Power Contractors Inc. was awarded the bulk of the construction work on the plants. Second, the Summa Insurance Company wrote a $688 million policy on the reactor which was the largest policy ever written by a Philippine insurance company. And last, Ecco-Asia, an electric company was paid $27 million for work that was never performed.

Philippine government officials believed that these companies were actually owned by Marcos, whereby he received large, unspecified amounts of money that had been funneled through the companies disguised as work on the reactor. The suit also claimed that these firms were either overpaid or paid outright for work that was not performed, and that the NPC was double-billed by the defendants for work performed on the reactor.

Reaction by Westinghouse. Westinghouse responded to the allegations of bribery by denying that any money reached Marcos. However, officials did acknowledge that $17.3 million was paid to two companies owned by Disini. Westinghouse explained that Disini was viewed as a special sales representative by which it could gain access to Marcos, and that these expenditures conformed with the company's international policies and were similar to payments made by other companies.

Westinghouse said that Disini's help was enlisted so that he could represent its technology and assist in the details of complying with Philippine regulations. Westinghouse also professed its innocence by reiterating that it was cleared of any wrongdoing by the Securities and Exchange Commission (SEC), the United States Justice Department, and the Philippine government in the late 1970's and early 1980's, even though the

Justice Department was still conducting a criminal investigation of payments made to Disini by Westinghouse.

All during the suit action, Westinghouse asserted that the plant was safe and ready to be loaded with fuel for operation, and that it was proud of the plant that it built. Westinghouse said that the allegations were a re-hash of old ones that had already been investigated, and that no evidence of wrongdoing had been found.

Arbitration versus a lawsuit. In May 1989, the suit was divided into two parts. First, a majority of the claims, which included breach of contract, fraud, negligence, and the antitrust matter were sent to the International Chamber of Commerce of the United Nations for arbitration along with $30 million in fees that Westinghouse claimed was still owed to them by the Philippine government. Second, in July 1989, a federal judge ruled that the Philippine government could proceed with the suit because only a court could decide if bribery was involved.

All along, Westinghouse had wanted to settle its dispute with the Philippine government through arbitration at the International Chamber of Commerce. However, President Aquino did not. Even though the contract called for arbitration to settle a dispute, since bribery may have been involved, the Philippine government claimed the contract was not enforceable. Aquino also stated that they made every attempt to settle the dispute with Westinghouse before filing the suit.

Westinghouse and Aquino reach an agreement. Before the trial was scheduled to begin on February 6, 1992, a settlement was reached between Westinghouse and President Aquino. Aquino was persuaded to settle for two reasons. First, the International Chamber of Commerce did not find any evidence that Westinghouse had paid bribes to Marcos for obtaining the contract. And second, the Philippines has been experiencing severe power outages which has caused 12-hour brownouts daily and has forced businesses into a four-day work week.

As outlined in the agreement, Westinghouse will operate PNPP-1 and will use some of the revenue to help the Philippines pay for the cost of bringing the plant up to current NRC requirements. Westinghouse will also pay the Philippine government $100 million in cash and services. Westinghouse considered the agreement to be in the best interests of all involved.

The plant is now scheduled to go on line in 1995 at a cost of over $2.6 billion.

Sources

"Brief on the Philippine Nuclear Power Plant." *National Power Corporation*. N.p, n.d.

"14 Questions About Nuclear Power, With Answers." *National Power Corporation*. N.p., n.d.

"Philippines Nuclear Power Plant." *Nuclear Engineering International*, January 1982.

"The $2.2 Billion Nuclear Fiasco." *Fortune*, September 1, 1986.

"Philippines May Sue Westinghouse Over Nuclear Plant." *The Wall Street Journal* , November 30, 1988.

"Philippines Expected to File Suit Against Westinghouse." *The New York Times*, December 1, 1988.

"Westinghouse Seeks Arbitrator in Dispute with Philippines Over $2.4 Billion Plant." *The Wall Street Journal*, December 2, 1988.

"Judge Says Philippines May Proceed With Bribery Suit Against Westinghouse." *The Wall Street Journal*, September 15, 1989.

"Dark Days in the Philippines." *Nuclear Industry*, Summer, 1990.

"Westinghouse Says It Did Not Bribe Ferdinand Marcos." *The Wall Street Journal* , November 25, 1991.

"Westinghouse Documents Released; Damages Limited to Philippine Case." *Nucleonics Week*, November 28, 1991.

"Philippines' Nuclear Plant Slated to Go on Line." *INFO*, U.S. Council for Energy Awareness, March 1992.

—Yvette M. Costa

Poland

COUNTRY REPORT

Poland changes energy policy

Poland has the lowest energy output per head in eastern Europe. The production of electricity in total production costs is five times higher than in western industrial countries.

The crisis led the government to change their energy policy in 1990. As part of the policy shift, Poland cancelled their only nuclear power plant, still under construction in Zarnowiec near the Baltic port of Gdansk. The government envisioned nuclear power in the country's energy balance only after the year 2000. (Plans drawn up under the former communist government provided for construction of nuclear power plants supplying over 7,000 megawatts of electricity by the year 2000 and 28,000 megawatts by the year 2020.)

The Polish government was not alone it its questioning of nuclear power:

The country's environmental movement had fought hard against the Zarnowiec plant arguing that the plant was unsafe, too expensive, and would not make that significant of a contribution to Poland's energy mix.

Public opposition to nuclear power had already caused the government to cancel plans for a second nuclear power plant in western Poland.

Other new democratic governments across Eastern and Central Europe were reassessing ambitious nuclear-power construction programs begun under communist regimes.

Although Poland's new energy policy stressed rational use of available resources and more effective environmental protection, there is some question of how Poland will meet its burgeoning energy needs and break its dependence on heavily polluting brown coal. Energy prices were expected to rise as a result of the government plan—by as much as five or six times their 1990 level. Natural gas prices were forecasted to sore by between eight and ten times 1990 levels.

Sources

"World Wire: Poland's Nuclear-Energy Plans." *The Wall Street Journal*, September 6, 1990.

"Poles Halt Construction on Nuclear Power Plant." *The Washington Post*, September 6, 1990.

"World Digest: Nuclear Plant Abandoned." *Modern Power Systems*, October 1990.

PLANT PROFILES

145 ★
Zarnowiec Nuclear Power Plant, Units 1-4

Kartoszyno, Poland

Basic Facts

Units 1-4 *Status:* Shut down. *Type of Reactor:* Pressurized light-water reactor. *Megawatts (electric):* 465. *Megawatts (thermal):* 1375. *Reactor System Supplier:* Skodaexport; Praha. *Generator Supplier:* Zamech-Elblag; Dolmel.

Key Dates

Units 1-2 *Ordered:* 1974. *Construction began:* 1982. *Shut Down:* September 1990.

Units 3-4 *Ordered:* 1974. *Construction began:* 1988. *Shut Down:* September 1990.

Plant Report

Potential electricity for Poland. The Zarnowiec Nuclear Power Plant occupies a site in Zarnowiec, Poland, near the Baltic port of Gdansk. Construction at Zarnowiec began in 1982, and the station's first unit was originally slated to open in 1990.

Government cancellation. In September 1990, the Polish government cancelled the construction of Zarnowiec, and recommended delaying further development of nuclear energy until after the year 2000.

The decision was made in light of a change in Poland's energy policy. Officials had been reviewing the future of Zarnowiec for months before the decision, and work was suspended in 1989.

The Polish Cabinet noted that no foreign company had been willing to complete Zarnowiec and provide top-quality safety controls for the plant. The cabinet also found that it would have cost more to complete the plant than it would to build a conventional coal-fired station. According to the government newspaper, the plant had already cost $1 billion.

Some 40 percent of the total investment in Zarnowiec's construction had already been committed. The plant was to have been linked to the national grid, with two blocks operating in 1992. Eventually, it was to have four blocks operating, with a total capacity of 1860 megawatts.

Opposition. Public opposition to Zarnowiec had been building for several years, and intensified after the 1986 accident at the Chernobyl nuclear power plant in the Soviet Ukraine. The public opposition to nuclear power had already scuttled plans for a second nuclear plant in western Poland.

Critics contended that if Zarnowiec were to be completed, its life span would not have been more than 30 years. Poland's fledgling environmental movement had also argued that the plant was unsafe, highly expensive, and would not have made a significant contribution to Poland's energy needs.

Sources

"World Wire: Poland's Nuclear-Energy Plans." *The Wall Street Journal*, September 6, 1990.

"Poles Halt Construction on Nuclear Power Plant." *The Washington Post*, September 6, 1990.

"World Digest: Nuclear Plant Abandoned." *Modern Power Systems*, October 1990.

Republic of Korea

COUNTRY REPORT

A growing nuclear power industry

As of 1992, South Korea was home to four sites, housing nine operating nuclear reactors, with three more under construction. The Kori Nuclear Power Plant had four operating units, the Ulchin Nuclear Power Plant had two with two more planned, the Wolsong Nuclear Power Plant had one with another under construction, and the Yonggwang Nuclear Power Plant had two with two under construction. All nuclear power plants in Korea are owned and operated by the Korea Electric Power Corporation (KEPCO).

The first electric business in Korea was the Seoul Electric Company, established in 1898. In 1961, Seoul Electric, along with two other companies, Korea Electric and South Korea Electric, merged to form the Korea Electric Company Limited (KECO). In 1982, KECO converted to a public corporation, which became KEPCO.

KEPCO opened its first nuclear plant, Kori, in 1982. As of 1992, the utility's nine operating plants had a total generating capacity of 7,616 megawatts, and provided roughly 50 percent of Korea's electricity. The annual average capacity factors of all the plants has been over 70 percent since 1984, and, as of 1992, no accidents or single radioactive releases to the public have occurred.

A plan for expansion

In 1989, a government-sponsored academic study recommended the building of 55 nuclear power plants in South Korea by 2031. At the time, the country's electricity demand had risen 13 percent, and was projected to rise 6.3 percent annually for the next 15 years.

The study took on more weight in 1990, when South Korea suffered from an electricity generating capacity shortage, causing routine cut-offs in some facets of the power supply. According to data from the Korea Atomic Industrial Forum, nuclear power generated cheaper electricity than fossil fuels during the year, which also added significance to the study.

A 15-year draft plan, to be launched in 1992, was drawn up, which called for the addition of 18 nuclear plants by 2006. This Long-term Power Development Plan was revised in 1991, and called for the building of nine units (8,100 total megawatts) from 1992 to 2001, and nine more units (8,100 total megawatts) from 2002 to 2006. KEPCO hoped to provide Korea with a total of 23,229 megawatts of nuclear-generated electricity by the end of 2006.

International cooperation

KEPCO has a technical cooperation agreement with 12 utilities from nine foreign countries for the promotion of technology development, exchange of information, training and other necessary cooperation concerning Korea's nuclear industry. The countries involved in the agreement are: Argentina, Belgium, Canada, England, France, Italy, Japan, the Republic of China, and the United States.

A closer relationship with Canada

Korea's nuclear relationship with Canada began around 1982 when Atomic Energy of Canada Limited (AECL) sold KEPCO one of its CANDU reactors for the utility's Wolsong plant. In 1990, KEPCO signed a contract with AECL for Wolsong's second CANDU reactor, expected to be operable in 1997.

Up to 1990, South Korea and Canada were each others fifth-largest trading partners, with trade expected to reach $4 billion in 1990.

Aside from technology, Canada also provides South Korea with uranium. Up to December 1990, one-third of the country's uranium came from Canada, especially Saskatchewan, where KEPCO had a minority interest in a uranium mine.

Moving toward a self-reliant nuclear industry

Korea emulated Japan by increasing indigenous commercial and technological input into its nuclear plants. As part of its Technology Self Reliance in Nuclear Power Plant Construction Program, KEPCO planned to have as many Korean suppliers, designers, and manufacturers as possible take part in its plants. The ratio of plant localization has been steadily increasing since the utility shifted its project philosophy from turnkey to non-turnkey contracts.

The entities participating in the program are KEPCO (overall project management), the Korea Atomic Energy Research Institute (nuclear steam system and fuel design), the Korea Heavy Industries and Construction Company (nuclear steam system and turbine/generator design and fabrication), the Korea Nuclear Fuel Company (nuclear fuel fabrication), and the Korea Power Engineering Company (plant design and engineering). By 1995, KEPCO expected to achieve more than 95 percent independence in all areas.

Fuel fabrication

In the 1970s, KEPCO created a plan to study nuclear fuel fabrication technology. As of 1992, all nuclear plants in Korea used fuel fabricated from domestic facilities.

Regulatory facts

The Korean Atomic Energy Commission (AEC), which belongs to the prime minister, discusses and approves important issues concerning nuclear energy and safety. The Korean Ministry of Science and Technology (MOST) does nuclear regulatory and licensing work, which includes the issuing of construction permits and operating licenses. MOST also inspects and reviews nuclear safety issues, including radiation protection. The Korea Institute of Nuclear Safety provides technical support for MOST.

To get a nuclear plant construction permit in Korea, an Environmental Impact Report and a Preliminary Safety Analysis Report must be approved by MOST, with a final review by the

AEC. For an operating license, a Final Safety Analysis Report and an Operational Technical Specification and Radiation Plan must be approved by MOST.

The Korean Nuclear Act has safety regulatory requirements which are applied to all nuclear plants during design, fabrication, construction, and operation. Also, the U.S. NRC's Post Three Mile Island Action Plan is utilized for construction and operation.

Sources

"Cracking Up." *Far Eastern Economic Review*, December 14, 1989.

"Canada-Korea A-Plant Pact." *The New York Times*, December 31, 1990.

Korea Electric Power Corporation '91 Nuclear Power Program. Korea Electric Power Corporation, n.d.

Mounfield, Peter. *World Nuclear Power*. Routledge, 1991.

"Perini Affiliate Part of Consortium, Canatom, Awarded Portion of Contract." *Business Wire*, February 1, 1991.

"U.S. Manufacturer, ABB/CE, Wins Contract for Two South Korean Nuclear Plants." *INFO*, U.S. Council for Energy Awareness, August 1991.

Korea Electric Power Corporation 1992 Annual Review. Korea Electric Power Corporation, n.d.

"Nuclear Power Cheaper Than Fossil Fuel in South Korea." *INFO*, U.S. Council for Energy Awareness, March 1992.

—Charles Ceplecha

PLANT PROFILES

146 •

Kori Nuclear Power Plant, Units 1-4

216, Ko-Ri, Changan-Up
Yangsan-Gun
Kyongsangnam-Do 626-950, Republic of Korea
Chang Tong Choi, Dir., Ko-Ri Nuclear Power Div.

Phone: (Yangsan) 376-2114
Fax: KECKORY K53807
Telex: (Yangsan) 376-2214

Owning Utility

Korea Electric Power Corporation
167, Samsong-Dong
Kangnam-Gu
Seoul 135-791, Republic of Korea
Jong Sok Kim, Gen.Mgr., Nuc. Power Generation Dept.

Phone: (Seoul) 553-3114
Fax: (Seoul) 550-5981
Telex: KELECCO K23717

Basic Facts

Units 1-4 *Status:* Operable. *Type of Reactor:* Pressurized light-water reactor. *Megawatts (electric):* Unit 1, 595; Unit 2, 650; Units 3-4, 950. *Megawatts (thermal):* Unit 1, 1729; Unit 2, 1876; Units 3-4, 2775. *Reactor System Supplier:* Westinghouse Electric Corp. (United States). *Generator Supplier:* General Electric Co. (United Kingdom). *Architect Engineer:* Units 1-2, Gilbert; Units 3-4, Bechtel.

Key Dates

Unit 1 *Ordered:* 1969. *Construction began:* 1970. *Criticality:* 1977. *First Power:* June 1977. *Commercial Operation:* April 1978.

Unit 2 *Ordered:* 1974. *Construction began:* 1977. *Criticality:* 1983. *First Power:* April 1983. *Commercial Operation:* July 1983.

Unit 3 *Ordered:* 1978. *Construction began:* 1978. *Criticality:* 1985. *First Power:* January 1985. *Commercial Operation:* September 1985.

Unit 4 *Ordered:* 1978. *Construction began:* 1980. *Criticality:* 1985. *First Power:* 1985. *Commercial Operation:* March 1986.

Plant Report

Korea's first nuclear power plant. The Kori Nuclear Power Plant occupies a 447-acre site in Kori, Korea, on the country's southeast coast. The site sits 32 kilometers northeast of Pusan.

Kori was the first and largest nuclear power plant in Korea. The Korea Electric Power Corporation (KEPCO) owns and operates the plant, which houses four pressurized water reactors. Units 1 and 2 were turnkey projects, which means they were essentially handed over to KEPCO as finished units. Units 3 and 4 were non-turnkey projects.

The combined capacity of all four reactors totals 3,137 megawatts. From 1980 to 1991, Kori accounted for almost 18 percent of KEPCO's total power generation.

Features of Unit 1. Unit 1 has an output of 587 gross megawatts. The core stands 3.66 meters high, and measures 2.46 meters in diameter.

The reactor contains 121 fuel assemblies, with 179 fuel rods per assembly, and 29 control rods.

The containment vessel is a cylindrical, concrete and steel structure, standing 11.88 meters high. It has an inner diameter of 3.35 meters, with walls measuring 165.1 millimeters thick.

The unit's turbine has a 587-megawatt rating, and spins at 1,800 revolutions per minute.

Features of Unit 2. Unit 2 has an output of 650 gross megawatts. The core stands 3.65 meters high, and measures 2.46 meters in diameter.

The reactor contains 121 fuel assemblies, with 235 fuel rods per assembly, and 33 control rods.

The containment vessel is a cylindrical, concrete and steel structure, standing 11.9 meters high. It has an inner diameter of 3.35 meters, with walls measuring 168.4 millimeters thick.

The unit's turbine has a 650-megawatt rating, and spins at 1,800 revolutions per minute.

Costs of Units 1 and 2. The total construction cost of Unit 1 totaled approximately $299 million (U.S. dollars). About $174 million of this total was obtained by loans from foreign countries.

The total construction cost of Unit 2 was approximately $936 million (U.S. dollars). About $541 million of this total was obtained by loans from foreign countries.

Features of Units 3 and 4. Both units have outputs of 950 gross megawatts. Their cores contain 157 fuel assemblies and 52 control rods each.

Containment vessels for the units are made of concrete, reinforced and lined with steel.

Plant localization. As part of its Technology Self Reliance in Nuclear Power Plant Construction Program, KEPCO planned to have as many Korean suppliers, designers, and manufacturers as possible take part in its nuclear plants. The ratio of plant localization has been steadily increasing since the utility shifted its project philosophy from turnkey to non-turnkey contracts, starting with Units 3 and 4. The domestic participation ratio for the units was 37 percent in design engineering, 10 percent in nuclear steam system fabrication, 11 percent in turbine/generator fabrication, 33 percent in fuel fabrication, and 40 percent in equipment manufacturing. By 1995, KEPCO hopes to achieve more than 95 percent indepedence in all areas.

Kori performance. Annual power generation for Units 1 and 2 has fluctuated since the units first went into commercial operation. Unit 1's lowest generation was 1611 gigawatt-hours in 1978, and its highest was 4622 gigawatt-hours in 1991. Unit 2's lowest generation was 2005 gigawatt-hours in 1983, and its highest was 5375 gigawatt-hours in 1989.

Capacity factors for the units have also fluctuated, ranging from 45.8 percent in 1988 to 94.0 percent in 1987 for Unit 1, and 70.1 percent in 1985 to 94.4 percent in 1989 for Unit 2. Unit 1's annual capacity measured higher than the world average eight times from 1979 to 1991, and Unit 2's factor surpassed the world average in every year of its operation, up to 1991.

Capacity factors for Unit 3 ranged from 71.7 percent in 1986 to 89.7 percent in 1985. Unit 4's factors ranged from 73.0 percent in 1987 to 94.2 percent in 1986.

In 1990, Unit 1 had an availability factor of 74.6 percent and a station consumption of 5.6 percent. Unit 2 had an availability factor of 84.3 percent and a station consumption of 5.8 percent.

In 1990, Unit 3 produced 7,144,871 megawatt-hours of electricity, with an availability factor of 90.4 percent and a station consumption of 4.3 percent. Unit 4 produced 6,501,336 megawatt-hours of electricity, with an availability factor of 81.5 percent and a station consumption of 4.2 percent.

Kori outage history. From 1978 to 1991, Unit 1 amassed 13 planned outages, usually one per year. The unit had 98 unplanned outages, with the most, 17, in its first year of operation.

From 1983 to 1991, Unit 2 amassed seven planned outages, usually one per year. The unit had 39 unplanned outages, with the most, 15, in 1985.

During 1990, Units 3 and 4 had one planned and three unplanned outages each.

Radiation exposure at Kori. According to the Atomic Energy Law, the permissible exposure limit for rad-workers in Korea is five rem per year. The occupational radiation exposure rate, total collective in man rem, for Kori averaged 627.6 from 1977 to 1990, with a high rate of 1,495.6 in 1988. The rate, based on average individual dose in rem per man, averaged 0.361 for the same period, with a high rate of 0.622 in 1981.

Fuel fabrication. In the 1970s, KEPCO created a plan to study nuclear fuel fabrication technology. As of 1992, all nuclear plants in Korea used fuel fabricated from domestic facilities.

Spent fuel storage and transshipment. As of 1992, Unit 1 used 562 storage racks, for an 11-year storage capacity, and Unit 2 used 920 racks, for a 20-year storage capacity. Units 3 and 4 used 756 storage racks each, which gave each unit a 14-year capacity.

KEPCO plans to increase the spent fuel storage capacity of all its nuclear plants by transshipping spent fuel between neighboring units.

Long-term Fuel Utilization Plan. To increase plant capacity factors, KEPCO is moving toward extended cycle operation of its nuclear plants. Since 1987, Unit 2 has adopted a 12-month cycle, and since 1989, Unit 1 has loaded with a 15-month cycle. Both units load approximately 18 tons of fuel per year. Units 3 and 4 are scheduled to gradually adopt 18-month cycles from 1992 on.

Environmental monitoring and surveillance. From the beginning stages of Kori's construction, KEPCO conducted extensive surveillance and evaluation of environmental impacts around the plant. Within 30 kilometers from the site, environmental samples, such as airborne particulates, soil and water, are collected and analyzed periodically at the plant's environmental laboratory. Some samples are sent to the Korea Atomic Energy Research Institute to confirm the reliability of the data.

To monitor air radiation dose, thermoluminescent dosimeters stand at about 40 locations around the site. All results of radiological surveillance go to the Ministry of Science and Technology, quarterly and annually. As of 1991, no significant environmental impacts existed.

To check the health of residents around Kori, KEPCO calculates off-site dose periodically, using computer programs approved by the U.S. NRC or Canada AECB. The results are also sent to the ministry, and have shown that the maximum individual dose to residents in a year measures less than one millirem. This is far less than the average national background level of 249 millirem per year.

KEPCO also performs non-radiological surveillance which consists of biota and abiota surveys, including surveys on warm discharged water. This surveillance report also goes to the ministry, quarterly and annually, and no evidence of environmental damage has been detected.

Emergency planning and response. Kori has a radiation emergency plan in cooperation with local government and relevant agencies.

Employees and training. Unit 1's site organization has 341 employees. The Korea Electric Power Operating Service Company Limited, a KEPCO affiliate, has 307 employees, who are responsible for maintenance of mechanical and electrical systems, at the site.

All new employees at Kori must take three weeks of general training and seven weeks of power general basic training at KEPCO's Seoul Training Center. Employees are also required to take 36 weeks of nuclear training at the KEPCO Nuclear Training Center, which contains a plant model (1/24 scale) of Units 3 and 4, as well as two plant simulators, which are applicable to all four units.

In August 1990, KEPCO completed work on a new nuclear training center, located on a 55-acre site at Kori. The site includes a maintenance training center, and chemistry and health physics laboratories, designed to improve the technical capability of operation and maintenance personnel through intensive exercises and practices.

Cleaning up Unit 1. In June 1990, KEPCO awarded a contract to the PN Services Group (PNS), and ABB Reaktor of Mannheim, Germany, to perform the chemical cleaning of secondary side of two steam generators of Unit 1. PNS, a part of Pacific Nuclear Systems Incorporated, had an approximate share of $2 million in the contract, which included laboratory testing, engineering, equipment and supplies, as well as actual chemical cleaning services. Work was to begin in fall 1990.

Achievements of Kori. In 1988, Unit 1 had one continuous run of over 150 days before it was shut down for overhauling. From 1985 to 1991, Unit 2 had four continuous runs of over 150 days, including a Korean record run of 387 days, form April 1990 to March 1991. During this period, Unit 2's 99.4-percent capacity factor ranked it first among 202 operating pressurized water reactors in the world.

Plant Profiles

Sources

"Pacific Nuclear Systems Inc. Receives $2-Million Korean Chemical Cleaning Contract." *Business Wire*, June 20, 1990.

Korea Electric Power Corporation '91 Nuclear Power Program. Korea Electric Power Corporation, n.d.

Korea Electric Power Corporation 1992 Annual Review. Korea Electric Power Corporation, n.d.

Kori Nuclear Power Plant 1 Status: May 1992. Kori Nuclear Power Division, Korea Electric Power Corporation, n.d.

Information Of Kori NPP. KEPCO Kori NPP 1&2 Technical Staff Section, May 1, 1992.

—Charles Ceplecha

Ulchin Nuclear Power Plant, Units 1-2

84-4, Pugu-Ri, Puk-Myon
Ulchi-Gun
Kyongsangbuk-Do 767-890, Republic of Korea
Joo Bo Hong, Dir., Ulchin Nuclear Power Div.

Phone: (Ulchin) 80-0404
Fax: (Ulchin) 80-2214
Telex: KECULJN K54430

Owning Utility

Korea Electric Power Corporation
167, Samsong-Dong
Kangnam-Gu
Seoul 135-791, Republic of Korea
Jong Sok Kim, Gen.Mgr., Nuclear Power Generatio

Phone: (Seoul) 553-3114
Fax: (Seoul) 550-5981
Telex: KELECCO K23717

Basic Facts

Units 1-2 *Status:* Operable. *Type of Reactor:* Pressurized light-water reactor. *Megawatts (electric):* 950. *Megawatts (thermal):* 2785. *Reactor System Supplier:* Framatome (France). *Generator Supplier:* Alsthom. *Architect Engineer:* Framatome (France); Alsthom (France).

Key Dates

Unit 1 *Ordered:* 1980. *Construction began:* 1982. *Criticality:* February 1988. *First Power:* 1988. *Commercial Operation:* September 1988.

Unit 2 *Ordered:* 1980. *Construction began:* 1982. *First Power:* 1989. *Commercial Operation:* August 1989.

Plant Report

A Korean power center. The Ulchin Nuclear Power Plant occupies a site near Ulchin, Korea, on the country's western coast. The site is about 138 miles southwest of Seoul.

Ulchin is owned and operated by the Korea Electric Power Corporation (KEPCO). As of 1991, the plant housed two 950-megawatt pressurized water reactors, manufactured by Framatome of France. At the time, these two units were the only French-supplied reactors in Korea. Estimated cost for the two units was about $3 billion. The addition of the units boosted Korea's nuclear power generating capacity to 7.6 million kilowatts.

Adding two more reactors. In 1990, KEPCO set up a long-term power development plan to have nuclear power share with base load, the coal-fired with intermediate load, and the hydro, including pumped storage, with peak load. According to the plan, five units, including Units 3 and 4 at Ulchin, were to be put into operation every five years, from 1995 to 1999. The additional Ulchin units were to be Korea's 13th and 14th nuclear plants.

On July 19, 1990, the Atomic Energy Commission authorized KEPCO's construction plan for Unit's 3 and 4. On July 22, 1991, KEPCO ordered the nuclear steam systems for the new units. The contract was estimated at about $200 million, and was for the compound system 80 design, a standardized design used at the Palo Verde Nuclear Plant in Arizona.

Plant localization. As part of its Technology Self Reliance in Nuclear Power Plant Construction Program, KEPCO planned to have as many Korean suppliers, designers and manufacturers as possible take part in its nuclear plants. The ratio of plant localization has been steadily increasing since the utility shifted its project philosophy from turnkey to non-turnkey contracts. The domestic participation ratio for Ulchin's Units 1 and 2 is 46 percent in design engineering and 40 percent in equipment manufacturing. By 1995, KEPCO expected to achieve more than 95 percent independence in all areas of nuclear plant construction and maintenance.

Fuel fabrication and operation cycles. In the 1970s, KEPCO created a plan to study nuclear fuel fabrication technology. As of 1992, all nuclear plants in Korea used fuel fabricated from domestic facilities.

In order to increase its plant capacity figure, KEPCO is moving toward longer fuel cycle operation. For Ulchin's Units 1 and 2, an 18-month cycle operation is to be gradually adopted from 1992 on.

Radiation exposure. According to the Atomic Energy Law, the permissible exposure limit for rad-workers in Korea is five rem per year. Ulchin's occupational radiation exposure rate (total collective, in man rem) was 7.6 in 1988, 100.4 in 1989, and 162.9 in 1990. The exposure rate (average individual dose, in rem per man) was 0.005 in 1988, 0.073 in 1989, and 0.138 in 1990.

Environmental monitoring and surveillance. From the beginning stages of Ulchin's construction, KEPCO conducted extensive surveillance and evaluation of environmental impacts around the plant. Within 30 kilometers from the site, environmental samples, such as airborne particulates, soil and water, are collected and analyzed periodically at the plant's environmental laboratory. Some samples are sent to KAERI to confirm the reliability of the data.

To monitor air radiation dose, thermoluminescent dosimeters stand at about 40 locations around the site. All results of radiological surveillance go to the Ministry of Science and Technology, quarterly and annually. As of 1991, no significant environmental impacts existed.

To check the health of residents around Ulchin, KEPCO calculates off-site dose periodically, using computer programs approved by the U.S. NRC or Canada AECB. The results are also sent to the ministry, and have shown that the maximum individual dose to residents in a year is less than one millirem. This is far less than the average national background level of 249 millirem per year.

KEPCO also performs non-radiological surveillance which consists of biota and abiota surveys, including surveys on warm discharged water. This surveillance report is also sent to the ministry, quarterly and annually, and no evidence of environmental damage has been detected.

Emergency plan. Ulchin has a radiation emergency plan in cooperation with local government and relevant agencies.

Employee training. All new employees at Ulchin must take three weeks of general training and seven weeks of power general basic training at KEPCO's Seoul Training Center. Employees are also required to take 36 weeks of nuclear training at the KEPCO Nuclear Training Center.

To reinforce the training of Ulchin operators, KEPCO added an additional plant specific simulator of Units 1 and 2 at the plant in January 1990. The simulator was fabricated in France by Thomson.

Problems with the turbine blades. In October 1989, Unit 1 was shut down after the discovery of loose and cracked turbine blades. Similar problems were found in Unit 2, which was shut down in November. The shutdowns forced KEPCO to turn to higher-cost power sources, such as coal, oil, and hydro power to replace the 1,900-megawatt output of the two units.

The problems tested relations between KEPCO, Alsthom, and the subcontractor, KHIC. The subcontract was part of South Korea's program of developing local industry. At the time, KEPCO officials said that Alsthom conceded that the turbines had design flaws, but Alsthom declined to comment. The company did send a team of engineers to Ulchin to try and resolve the issue, and the utility hoped to restart the reactors in January 1990.

Earlier in 1990, KEPCO asked Alsthom for damages after problems in a generator unit forced the shutdown of Units 1 and 2. The utility's contract called for penalties, to be paid by Alsthom, after a 60-day shutdown involving a failure of their components. KEPCO officials said the company, and KHIC, agreed to pay a daily penalty of Sfr 75,000 for 47 of the 107 days the plant was closed.

Performance. Capacity factors for both of Ulchin's units have risen since the start of commercial operation. The factors for Unit 1 were 40.9 percent in 1988, 65.2 percent in 1989, and 78.5 percent in 1990. Unit 2's factors were 45.8 percent in 1989 and 70.3 percent in 1990.

In 1990, Unit 1 produced 6,529,450 megawatt-hours of electricity, while Unit 2 produced 5,849,342.

In 1990, plant availability factors were 81.7 percent for Unit 1 and 73.0 percent for Unit 2.

In 1990, station consumption was 5.6 percent for Unit 1 and 5.2 percent for Unit 2.

Sources

"Cracking Up." *Far Eastern Economic Review*, December 14, 1989.
"South Korea's Nuclear Power." *The Wall Street Journal*, February 28, 1990.
Korea Electric Power Corporation '91 Nuclear Power Program. Korea Electric Power Corporation, n.d.
"ABB Will Build 2 Nuclear Plants." *The New York Times*, July 23, 1991.
"U.S. Manufacturer, ABB/CE, Wins Contract for Two South Korean Nuclear Plants." *INFO*, August 1991.
Korea Electric Power Corporation 1992 Annual Review. Korea Electric Power Corporation, n.d.
Kori Nuclear Power Plant 1 Status: May 1992. Kori Nuclear Power Division, Korea Electric Power Corporation, n.d.

—Charles Ceplecha

148 ○
Ulchin Nuclear Power Plant, Units 3-4

84-4, Pugu-Ri, Puk-Myon
Ulchi-Gun
Kyongsangbuk-Do 767-890, Republic of Korea
Joo Bo Hong, Dir., Ulchin Nuclear Power Div.

Phone: (Ulchin) 80-0404
Fax: (Ulchin) 80-2214
Telex: KECULJN K54430

Owning Utility

Korea Electric Power Corporation
167, Samsong-Dong
Kangnam-Gu
Seoul 135-791, Republic of Korea
Jong Sok Kim, Gen.Mgr., Nuclear Power Generatio

Phone: (Seoul) 553-3114
Fax: (Seoul) 550-5981
Telex: KELECCO K23717

Basic Facts

Units 3-4 *Status:* Under construction (6% complete as of January 1992). *Type of Reactor:* Pressurized light-water reactor. *Megawatts (electric):* 950. *Megawatts (thermal):* 2785. *Reactor System Supplier:* Korea Heavy Industries and Construction Co; Combustion Engineering, Inc. (United States). *Generator Supplier:* Korea Heavy Industries and Construction Co; Combustion Engineering, Inc. (United States). *Architect Engineer:* Sargent & Lundy Engineers (United States).

Key Dates

Unit 3 *Ordered:* 1991. *Construction began:* 1992. *Commercial Operation Expected:* June 1998.

Unit 4 *Ordered:* 1991. *Construction began:* 1992. *Commercial Operation Expected:* June 1999.

Plant Report

See Plant Report for Ulchin Nuclear Power Plant, Units 1-2.

149 ●
Wolsong Nuclear Power Plant, Unit 1

260, Naa-Ri Yangnam-Myon
Kyongju-Gun
Kyongsangbuk-Do 778-840, Republic of Korea
Byong Koo Yoon, Dir., Wolsong Nuclear Power Div.

Phone: (Yangnam) 40-0101
Fax: (Yangnam) 40-3228
Telex: KECWS K54360

Plant Profiles

Wolsong | 149

Owning Utility

Korea Electric Power Corporation
167, Samsong-Dong
Kangnam-Gu
Seoul 135-791, Republic of Korea
Jong Sok Kim, Gen.Mgr., Nuc. Power Generation Dept.

Phone: (Seoul) 553-3114
Fax: (Seoul) 550-5981
Telex: KELECCO K23717

Basic Facts

Unit 1 *Status:* Operable. *Type of Reactor:* Pressurized heavy-water reactor. *Megawatts (electric):* 679. *Megawatts (thermal):* 2180. *Reactor System Supplier:* Atomic Energy of Canada Ltd. *Generator Supplier:* NEI-Parsons Ltd. (Canada). *Architect Engineer:* Canatom.

Key Dates

Unit 1 *Ordered:* 1973. *Construction began:* 1977. *Criticality:* 1982. *First Power:* December 1982. *Commercial Operation:* April 1983.

Plant Report

Electricity for Korea. The Wolsong Nuclear Power Plant occupies a site near Ulsan, Korea, on the Sea of Japan. The site sits about 160 miles southwest of Seoul.

CANDU reactor. Wolsong houses one 687.7-megawatt pressurized heavy water reactor, the only one of its kind in South Korea up to 1992. Wolsong's Unit 1 was a turnkey project, in which AECL and its contractors basically handed over a finished plant to KEPCO. South Korea and Canada are each others fifth-largest trading partners.

Radiation exposure at Wolsong. According to the Atomic Energy Law, the permissible exposure limit for rad-workers in Korea is five rem per year. Wolsong's occupational radiation exposure rate (total collective, in man rem) was 26.3 in 1983, 103.3 in 1984, 86.6 in 1985, 184.0 in 1986, 56.0 in 1987, 168.8 in 1988, 70.6 in 1989, and 117.1 in 1990. The exposure rate (average individual dose, in rem per man) was 0.043 in 1983, 0.122 in 1984, 0.148 in 1985, 0.184 in 1986, 0.094 in 1987, 0.195 in 1988, 0.071 in 1989, and 0.119 in 1990.

Environmental monitoring and surveillance. From the beginning stages of Wolsong's construction, KEPCO conducted extensive surveillance and evaluation of environmental impacts around the plant. Within 30 kilometers from the site, environmental samples, such as airborne particulates, soil and water, are collected and analyzed periodically at the plant's environmental laboratory. Some samples are sent to the Korea Atomic Energy Research Institute to confirm the reliability of the data.

To monitor air radiation dose, thermoluminescent dosimeters stand at about 40 locations around the site. All results of radiological surveillance go to the Ministry of Science and Technology, quarterly and annually. As of 1991, no significant environmental impacts existed.

To check the health of residents around Wolsong, KEPCO calculates off-site doses periodically, using computer programs approved by the U.S. NRC or Canada AECB. The results are also sent to the ministry, and have shown that the maximum individual dose to residents in a year measures less than one millirem. This is far less than the average national background level of 249 millirem per year.

KEPCO also performs non-radiological surveillance which consists of biota and abiota surveys, including surveys on warm discharged water. This surveillance report also goes to the ministry, quarterly and annually, and no evidence of environmental damage has been detected.

Emergency planning and response. Wolsong has a radiation emergency plan which is based on the Basic Civil Defense Program and operated in cooperation with local government and relevant agencies.

Employee training. All new employees at Wolsong must take three weeks of general training and seven weeks of power general basic training at KEPCO's Seoul Training Center. Employees are also required to take 36 weeks of nuclear training at the KEPCO Nuclear Training Center.

Fuel fabrication. In the 1970s, KEPCO created a plan to study nuclear fuel fabrication technology. As of 1992, all nuclear plants in Korea used fuel fabricated from domestic facilities.

Spent fuel storage and transshipment. In 1990, KEPCO planned to increase the spent fuel storage capacity of all its plants by transshipping spent fuel between neighboring units. The utility also planned to expand storage capacity at Wolsong by building a silo-like, dry concrete storage facility.

Wolsong's overall performance. Capacity factors for Wolsong have fluctuated since the start of commercial operation. Capacity factors measured 61.9 percent in 1983, 66.8 percent in 1984, 94.4 percent in 1985, 79.7 percent percent in 1986, 92.9 percent in 1987, 79.4 percent in 1988, 91.0 percent in 1989 and 85.9 percent in 1990. The plant's 99.1 percent capacity factor from April 1, 1989, to March 31, 1990, ranked it second among 341 operating plants in the world.

In 1990, Wolsong produced 5,106,228 megawatt-hours of electricity, with an 86.0-percent availability factor. Station consumption measured 6.7 percent.

Plans for another CANDU reactor. In 1990, KEPCO announced plans to build Unit 2 at Wolsong. The unit was to be a 700-megawatt CANDU pressurized heavy water reactor.

In December 1990, KEPCO signed the main contract, worth $535 million (U.S. dollars), with AECL. The company was to take charge of the plant's architecture, engineering, and nuclear steam system supply. AECL and the Korea Heavy Industry Construction Company (KHIC) were to manufacture the reactor, with KHIC and General Electric supplying the turbine/generator. The estimated cost of the turbine/generator totaled about $1 billion.

Relying on domestic technology. KEPCO made plans for Unit 2 under its Technology Self Reliance in Nuclear Power Plant Construction Program. The program was designed to develop Korea's indigenous nuclear industry, and involved using Korean firms at several project levels.

While AECL would fabricate the major components of Unit 2's nuclear steam supply system, the vendor estimated that when

● Operable   ○ Under Construction   △ Indefinitely Deferred   ▲ Decommissioning   □ Planned   ■ On Order   ☆ Cancelled   ★ Shut Down   p. **197**

other equipment and overall assembly were taken into account, the steam system would be 60 percent AECL and 40 percent Korean. On the balance of the plant, the company estimated that Korean firms would do 70 percent of the work, and non-Korean firms, 30 percent.

Contracts awarded for Unit 2. In August 1990, Nucon International, a consultant and equipment provider for the nuclear energy industry, received a contract to provide KEPCO with a gaseous rad-waste system for Unit 2. The system delays the release of radioactive gases until they are no longer dangerous. In February 1992, Nucon secured finances for another contract, worth $1 million, to provide advice and equipment for the plant.

In January 1991, Canatom Incorporated, a three-member Canadian consortium, was awarded a major portion of the engineering design and project management service contract for Unit 2. Approximately two-thirds of the $517-million contract awarded to the Canadian government was to go to Canadian engineering companies and manufacturers of reactor parts.

Also in February 1992, KHIC awarded a contract to Babcock & Wilcox Industries Limited for the supply of steam generators and heat exchanger coomponents for Unit 2. The contract, estimated at about $28 million, called for the supply of four lower steam generator sections, component design and tubing for all heat exchangers, primary hydrostatic test and pre-service non-destructive examination for the four steam generators. Delivery to the plant was scheduled for mid-1993.

Work begins on Unit 2. On October 23, 1991, excavation began for Unit 2. A construction permit was expected in 1992, with concrete pouring soon afterward. The unit was scheduled for completion by June 1997.

Plans to add Units 3 and 4. In late 1991, invitations for bids were issued for two more 700-megawatt pressurized heavy water reactors at Wolsong . The schedule for these plants included the awarding of prime contracts in December 1992, and completion by June 1998 and June 1999.

Sources

"Korea To Add New Nuclear Plant As Electric Demand Soars." *Nuclear Industry*, First Quarter 1990.
"Canada-Korea A-Plant Pact." *The New York Times*, December 31, 1990.
Korea Electric Power Corporation '91 Nuclear Power Program. Korea Electric Power Corporation, n.d.
"KEPCO buys second CANDU for Wolsong plant site." *Nuclear News*, February 1991.
"Perini Affiliate Part of Consortium, Canatom, Awarded Portion of Contract." *Business Wire*, February 1, 1991.
"Nuclear Consultant Lands $1 Million Korean Contract." *Business First-Columbus*, February 18, 1991.
"B&W Unit Receives Contract Valued at $28 Million for Korean Nuclear Power Plant Components." *Business Wire*, February 19, 1991.
Korea Electric Power Corporation 1992 Annual Review. Korea Electric Power Corporation, n.d.

—*Charles Ceplecha*

150 ○
Wolsong Nuclear Power Plant, Unit 2

260, Naa-Ri Yangnam-Myon
Kyongju-Gun
Kyongsangbuk-Do 778-840, Republic of Korea
Byong Koo Yoon, Dir., Wolsong Nuclear Power Div.

Phone: (Yangnam) 40-0101
Fax: (Yangnam) 40-3228
Telex: KECWS K54360

Owning Utility

Korea Electric Power Corporation
167, Samsong-Dong
Kangnam-Gu
Seoul 135-791, Republic of Korea
Jong Sok Kim, Gen.Mgr., Nuc. Power Generation Dept.

Phone: (Seoul) 553-3114
Fax: (Seoul) 550-5981
Telex: KELECCO K23717

Basic Facts

Unit 2 *Status:* Under construction (12% complete as of January 1992). *Type of Reactor:* Pressurized heavy-water reactor. *Megawatts (electric):* 679. *Megawatts (thermal):* 2180. *Reactor System Supplier:* Atomic Energy of Canada Ltd; Korea Heavy Industries and Construction Co. *Generator Supplier:* Korea Heavy Industries and Construction Co; General Electric Co. (United States). *Architect Engineer:* Atomic Energy of Canada Ltd.

Plant Report

See Plant Report for Wolsong Nuclear Power Plant, Unit 1.

151 ■
Wolsong Nuclear Power Plant, Units 3-4

260, Naa-Ri Yangnam-Myon
Kyongju-Gun
Kyongsangbuk-Do 778-840, Republic of Korea
Byong Koo Yoon, Dir., Wolsong Nuclear Power Div.

Phone: (Yangnam) 40-0101
Fax: (Yangnam) 40-3228
Telex: KECWS K54360

Owning Utility

Korea Electric Power Corporation
167, Samsong-Dong
Kangnam-Gu
Seoul 135-791, Republic of Korea
Jong Sok Kim, Gen.Mgr., Nuc. Power Generation Dept.

Phone: (Seoul) 553-3114
Fax: (Seoul) 550-5981
Telex: KELECCO K23717

Basic Facts

Units 3-4 *Status:* On order. *Type of Reactor:* Pressurized heavy-water reactor.

Key Dates

Unit 3 *Ordered:* 1992. *Commercial Operation Expected:* June 1998.
Unit 4 *Ordered:* 1992. *Commercial Operation Expected:* June 1999.

Plant Report

See Plant Report for Wolsong Nuclear Power Plant, Unit 1.

Yonggwang Nuclear Power Plant, Units 1-2

517, Kyema-Ri, Hongnong-UP
Yonggwang-Gun
Chonlanam-Do 513-880, Republic of Korea
Eun Rae Roh, Dir., Yonggwang Nuclear Power Div.

Phone: (Yonggwang) 356-3111
Fax: (Yonggwang) 356-2214
Telex: KECYGK K66781

Owning Utility

Korea Electric Power Corporation
167, Samsong-Dong
Kangnam-Gu
Seoul 135-791, Republic of Korea
Jong Sok Kim, Gen.Mgr., Nuc. Power Generation Dept.

Phone: (Seoul) 553-3114
Fax: (Seoul) 550-5981
Telex: KELECCO K23717

Basic Facts

Units 1-2 *Status:* Operable. *Type of Reactor:* Pressurized light-water reactor. *Megawatts (electric):* 996. *Megawatts (thermal):* 2775. *Reactor System Supplier:* Westinghouse Electric Corp. (United States). *Generator Supplier:* Westinghouse Electric Corp. (United States). *Architect Engineer:* Bechtel.

Key Dates

Unit 1 *Ordered:* 1978. *Construction began:* 1979. *Criticality:* January 1986. *First Power:* March 1986. *Commercial Operation:* August 1986.

Unit 2 *Ordered:* 1978. *Construction began:* 1979. *Criticality:* October 1986. *First Power:* November 1986. *Commercial Operation:* July 1987.

Plant Report

Adding units. The Yonggwang Nuclear Power Plant occupies a site near Mokpo, Korea, on the country's southwest coast. The site is about 145 miles west of Pusan.

As of 1992, Yonggwang housed two operable 950-megawatt pressurized water reactors and two more were under construction. In April 1987, the utility had awarded a $200-million contract for the units' nuclear steam supply systems and fuel components to Asea Brown Boveri Group/Combustion Engineering Nuclear Power of Windsor, Connecticut. The steam systems were to be the company's system 80 design, and a South Korean construction company was to build the components. At the time, the order was the first to the United States in a decade.

KEPCO planned commercial operation in 1995, for Unit 3, and 1996, for Unit 4. Up to December 1989, the utility had not received a construction license from the Ministry of Science and Technology, concrete pouring had not yet began, and the project was six months behind schedule. Analysts said that the delay resulted from antinuclear groups' requests for safety checks on reactor design.

In late December 1989, the construction permit was issued, and concrete work began on December 23.

Law suit concerning Units 1 and 2. In July 1987, the Hyundai Construction & Engineering Company threatened to sue KEPCO after the South Korean government canceled a $353 million contract the utility awarded the company. The contract called for Hyundai to provide engineering work and other services for Units 1 and 2.

Government officials overturned the selection of Hyundai, citing the lack of competitive bidding, but company officials maintained that the contract was legally sound. KEPCO's president, Park June Kee, resigned to assume responsiblity for the controversy, but denied any wrongdoing.

Plant standardization. In 1990, KEPCO planned to develop a standardized nuclear power plant design to ensure technology self-reliance, to promote plant localization, and to reduce construction time and cost. The proposed design utilized a 1,000-megawatt pressurized light water reactor, developed by incorporating Unit 3 and 4's design and new technology of advanced light water reactors.

Plant localization. As part of its Technology Self Reliance in Nuclear Power Plant Construction Program, KEPCO planned to have as many Korean suppliers, designers and manufacturers as possible take part in its nuclear plants. The ratio of plant localization has been steadily increasing since the utility shifted its project philosophy from turnkey to non-turnkey contracts. By 1995, KEPCO expected to achieve more than 95 percent independence in all areas of nuclear plant construction and operations.

The domestic participation ratios for Units 1 and 2 were 44 percent in design engineering, 16 percent in nuclear steam system manufacturing, 28 percent in turbine/generator manufacturing, and 45 percent in fuel fabrication.

In 1990, the ratios for Units 3 and 4 were 75 percent in design engineering, 50 percent in nuclear steam system design, 50 percent in fuel design, 63 percent in nuclear steam system manufacturing, 94 percent in turbine/generator manufacturing, and 100 percent in fuel fabrication.

Resident study. Beginning in April 1990, the Seoul National University Hospital performed a one-year epidemiological study for residents adjacent to Yonggwang, with results to be utilized as important basic data when controversy on the radiological effects from Korean nuclear plants occurs.

Radiation exposure. According to the Atomic Energy Law, the permissible exposure limit for rad-workers in Korea is five rem per year. At Yonggwang, the occupational radiation exposure rate (total collective in man rem) was 43.6 in 1986, 281.9 in 1987, 408.9 in 1988, 315.5 in 1989 and 373.6 in 1990. The average individual dose in rem per man was 0.032 in 1986, 0.169 in 1987, 0.227 in 1988, 0.216 in 1989 and 0.241 in 1990.

Performance. Capacity factors for Units 1 and 2 have fluctuated since the start of commercial operation. Capacity factors for Unit 1 were 88.2 percent in 1986, 75.2 percent in 1987, 77.6 percent in 1988, 81.0 percent in 1989, and 86.5 percent in 1990. Unit 2's capacity factors were 95.9 percent in 1987, 78.6 percent in 1988, 71.6 percent in 1989, and 74.9 percent in 1990.

Yonggwang

REPUBLIC OF KOREA

In 1990, Unit 1 produced 7,200,582 megawatt-hours of electricity, with an 85.7 percent availability factor. Station consumption was 4.3 percent.

In 1990, Unit 2 produced 6,235,192 megawatt-hours of electricity, with a 77.1 percent availability factor. Station consumption was 4.5 percent.

Construction continues on Units 3 and 4. In December 1991, Unit 3's 400-ton pressure vessel and four 616-ton steam generators were lifted into place. This stage of construction was completed a week ahead of schedule, only two years and two days after work first started.

The lifting process was unique. Normally, pressure vessels and steam generators are lowered into place before the containment dome is finished, but Unit 3's dome was done first, with components put in through a hatch on the side of the containment. This process insured that no construction delays existed while builders waited for the delivery of the vessel and generators.

At the time, Units 3 and 4 were on a five-year schedule. Unit 2 was expected to attain 71.7 percent completion during 1992, with installation of the turbine/generators and start-up testing.

Fuel fabrication and operation cycles. In the 1970s, KEPCO created a plan to study nuclear fuel fabrication technology. As of 1992, all nuclear power plants in Korea used fabricated fuel from domestic facilities.

In order to increase its plant capacity figure, KEPCO is moving toward longer fuel cycle operation. For Yonggwang's Units 3 and 4, an 18-month cycle operation is planned.

Environmental monitoring and surveillance. From the beginning stages of Yonggwang's construction, KEPCO conducted extensive surveillance and evaluation of environmental impacts around the plant. Within 30 kilometers from the site, environmental samples, such as airborne particulates, soil and water, are collected and analyzed periodically at the plant's environmental laboratory. Some samples are sent to KAERI to confirm the reliability of the data.

To monitor air radiation dose, thermoluminescent dosimeters stand at about 40 locations around the site. All results of radiological surveillance go to the Ministry of Science and Technology, quarterly and annually. As of 1991, no significant environmental impacts existed.

To check the health of residents around Yonggwang, KEPCO calculates off-site dose periodically, using computer programs approved by the U.S. NRC or Canada AECB. The results are also sent to the ministry, and have shown that the maximum individual dose to residents in a year is less than one millirem. This is far less than the average national background level of 249 millirem per year.

KEPCO also performs non-radiological surveillance which consists of biota and abiota surveys, including surveys on warm discharged water. This surveillance report is also sent to the ministry, quarterly and annually, and no evidence of environmental damage has been detected.

Emergency plan. Yonggwang has a radiation emergency plan in cooperation with local government and relevant agencies.

Employee training. All new employees at Yonggwang must take three weeks of general training and seven weeks of power general basic training at KEPCO's Seoul Training Center. Employees are also required to take 36 weeks of nuclear training at the KEPCO Nuclear Training Center.

Potential plant operators utilize the Unit 2 Simulator at the KEPCO Training Center. The Westinghouse-built simulator is applicable to Yonggwang's Units 1 and 2, and was completed in December 1986.

Plans for Units 5 and 6. In 1992, KEPCO made plans for the construction of Units 5 and 6 at Yonggwang. The plans were expected to be finalized by mid-year, with completion of the plants by June 2000 and 2001.

Sources

"Combustion Engineering Gets $200 Million Contract." *The Wall Street Journal*, April 10, 1987.
"Hyundai Construction Says It May Sue Utility in Korea." *The Wall Street Journal*, July 21, 1987.
"Cracking Up." *Far Eastern Economic Review*, December 14, 1989.
Korea Electric Power Corporation '91 Nuclear Power Program. Korea Electric Power Corporation, n.d.
"Pacific Nuclear Systems Inc. Awarded Korean Technical Assistance Contract." *Business Wire*, April 15, 1991.
"U.S. Manufacturer, ABB/CE, Wins Contract for Two South Korean Nuclear Plants." *INFO*, U.S. Council for Energy Awareness, August 1991.
Korea Electric Power Corporation 1992 Annual Review. Korea Electric Power Corporation, n.d.
Kori Nuclear Power Plant 1 Status: May 1992. Kori Nuclear Power Division, Korea Electric Power Corporation, n.d.

—*Charles Ceplecha*

Yonggwang Nuclear Power Plant, Units 3-4

517, Kyema-Ri, Hongnong-UP
Yonggwang-Gun
Chonlanam-Do 513-880, Republic of Korea
Eun Rae Roh, Dir., Yonggwang Nuclear Power Div.

Phone: (Yonggwang) 356-3111
Fax: (Yonggwang) 356-2214
Telex: KECYGK K66781

Owning Utility

Korea Electric Power Corporation
167, Samsong-Dong
Kangnam-Gu
Seoul 135-791, Republic of Korea
Jong Sok Kim, Gen.Mgr., Nuc. Power Generation Dept.

Phone: (Seoul) 553-3114
Fax: (Seoul) 550-5981
Telex: KELECCO K23717

Basic Facts

Units 3-4 *Status:* Under construction (58% complete as of January 1992). *Type of Reactor:* Pressurized light-water reactor. *Megawatts (electric):* 1000. *Megawatts (thermal):* 2825. *Reactor System Supplier:* Korea Heavy Industries and Construction Co; Combustion Engineering, Inc. (United States). *Generator Supplier:* Korea Heavy Industries and

Construction Co; General Electric Co. (United States). *Architect Engineer:* Korea Power Engineering Co., Ltd; Sargent & Lundy Engineers (United States).

Key Dates

Unit 3 *Ordered:* 1987. *Construction began:* 1989. *Criticality expected* September 1994. *Commercial Operation Expected:* March 1995.
Unit 4 *Ordered:* 1987. *Construction began:* 1989. *Criticality expected* September 1995. *Commercial Operation Expected:* March 1996.

Plant Report

See Plant Report for Yonggwang Nuclear Power Plant, Units 1-2.

154 ■ Yonggwang Nuclear Power Plant, Units 5-6

517, Kyema-Ri, Hongnong-UP
Yonggwang-Gun
Chonlanam-Do 513-880, Republic of Korea
Eun Rae Roh, Dir., Yonggwang Nuclear Power Div.

Phone: (Yonggwang) 356-3111
Fax: (Yonggwang) 356-2214
Telex: KECYGK K66781

Owning Utility

Korea Electric Power Corporation
167, Samsong-Dong
Kangnam-Gu
Seoul 135-791, Republic of Korea
Jong Sok Kim, Gen.Mgr., Nuc. Power Generation Dept.

Phone: (Seoul) 553-3114
Fax: (Seoul) 550-5981
Telex: KELECCO K23717

Basic Facts

Units 5-6 *Status:* On order.

Key Dates

Unit 5 *Ordered:* 1992. *Commercial Operation Expected:* 2000.
Unit 6 *Ordered:* 1992. *Commercial Operation Expected:* 2001.

Plant Report

See Plant Report for Yonggwang Nuclear Power Plant, Units 1-2.

Republic of South Africa

COUNTRY REPORT

Why nuclear power?

In South Africa, water resources are unreliable, and the coal reserves are only expected to last until the middle of the next century. Also, South Africa has large resources of uranium and is one of the top uranium producing nations of the world with the third largest reserves. Uranium occurs in the gold bearing conglomerates of the Witwatersrand and the copper bearing ores of the Phalaborwa complex. Minor deposits of uranium are found at Pilanesberg, in the Karoo and North Western Cape.

Add to these factors the health and environmental problems associated with the burning of fossil fuels, and it is easy to see why nuclear energy became a viable alternative for South Africa in the early 1970s.

A nuclear power/nuclear weapons connection?

Construction on the Koeberg Nuclear Power Station, South Africa's only commercial nuclear power plant, began in 1977, and the U.S. played a controversial role in the construction process.

South Africa has resisted signing the Nuclear Non-Proliferation Treaty. In 1983, Over 120 nations had signed the Treaty, thereby agreeing to peaceful uses of nuclear technology and accepting international inspections at their nuclear facilities. While the Nuclear Non-Proliferation Treaty, in effect since 1970, was supposed to make it harder for nonparticipating nations to obtain nuclear power technology, South Africa was able to do so.

Although the U.S. was barred by law from exporting nuclear reactors and nuclear fuel to South Africa, nothing prohibited technology transfer that provided training and services for reactors built and fueled by other countries for South Africa. The loophole allowed the following transactions to take place between U.S. companies and the French-built Koeberg plant:

The Reagan administration approved seven licenses for nuclear exports to South Africa from June 1981 to June 1982.

In October 1983, the Department of Energy granted several American companies, including Westinghouse Electric Corp. (United States), the right to sign a 10-year service maintenance contract with the Koeberg plant.

Even though the technology transfer arrangements were for commercial nuclear technology, opponents of the deals maintain that the technology would help South Africa develop the capacity to make the fuel used in nuclear explosives. Although by 1989, Koeberg had joined the Association of Nuclear Operators (WANO) and was in contact with the International Atomic Energy Agency (IAEA), South Africa may have developed nuclear weapons capabilities in the period when its nuclear installations were not under inspection by the IAEA.

Waste disposal

The licensing of nuclear installations in South Africa is the statutory responsibility of the South African Council for Nuclear Safety (CNS). Waste disposal falls under the country's Atomic Energy Commission. South Africa's site for the disposal of nuclear waste is situated at Vaalputs—600 kilometers north of Cape Town—where the annual evaporation exceeds the annual rainfall. South African nuclear officials say even if radioactivity should escape at Vaalputs, it could not contaminate the ground water which may find its way to the surface.

The site—known as Vaalputs Radioactive Waste Disposal Facility—was acquired in 1983; the delivery of the first nuclear waste was scheduled for 1986. The area allocated for burial of metal drums and concrete containers containing low-level and intermediate-level waste is sufficient for storing the nuclear waste of three power stations the size of Koeberg for 40 years. Although officials expect no health hazard, the area is fenced off and monitored.

Sources

"Power Plant Politics: The Nuclear Care and Feeding of the Third World." *Critical Mass Bulletin*, December 1983.

"Comment." *Critical Mass Bulletin*, August 1984.

Koeberg General Information Brochure. Eskom, November 1989.

Koeberg—Energy for the Future: Technical Information. Eskom, January 1990.

PLANT PROFILES

155 •
Koeberg Nuclear Power Station, Units 1-2

Private Bag X10
Kernkrag 7440, Republic of South Africa
B.E. Oaten, Power Stat.Mgr.

Phone: 27-21-0224-2133
Fax: 27-21-0224-3421
Telex: 558178 KPSPRO

Owning Utility

ESKOM
PO Box 1091
Johannesburg 2000, Republic of South Africa
J.H. Henderson, Div.al Mgr. (for Koeberg)

Phone: 27-11-800-8111
Fax: 27-11-800-5771
Telex: 424481-SA

Contact

Donne B. Murray, Head, Commun.
Koeberg Nuclear Power Station

Basic Facts

Units 1-2 *Status:* Operable. *Type of Reactor:* Pressurized light-water reactor. *Megawatts (electric):* 965. *Megawatts (thermal):* 2775. *Reactor System Supplier:* Framatome (France). *Generator Supplier:* Alsthom. *Architect Engineer:* Framatome (France); Alsthom; Framateg.

Key Dates

Unit 1 *Ordered:* 1976. *Construction began:* 1977. *Criticality:* March 1984. *First Power:* April 1984. *Commercial Operation:* July 1984.

Unit 2 *Ordered:* 1976. *Construction began:* 1977. *Criticality:* July 1985. *First Power:* July 1985. *Commercial Operation:* November 1985.

Plant Report

Site selection. Koeberg was chosen as a site for South Africa's first nuclear power plant because:

- the economic and industrial development of the surrounding area caused an increased demand for electricity
- the Duynefontein area is geologically stable
- the cold water of the Atlantic Ocean is ideal for cooling
- Eskom's national grid is easily accessible
- Koeberg supplies the Western Cape with electricity making it less dependent on Transvaal power stations 1,500 kilometers away.

Safety. Koeberg has affiliated with several international organizations in an effort to keep the plant up to par with world safety standards. In 1989 Eskom became a member of the World Association of Nuclear Operators (WANO), an organization supplying international support and immediate assistance in the case of a nuclear accident. Officials at Koeberg are also in contact with other international bodies such as Nuclear Utility Maintenance Experience Exchange (NUMEX), Electricité de France (EDF), and the International Atomic Energy Agency (IAEA).

Safety was also considered in the design of the plant. The 25-centimeter thick stainless steel reactor vessel rests in a two-meter thick reinforced concrete biological shield. A massive one-meter thick, steel-lined post-stressed concrete structure built to withstand incredible forces, including the impact of falling aircraft, houses the entire nuclear steam supply system. The nuclear section of the power station is built on aseismic bearings to prevent damage from even major earthquakes. All critical systems are fully duplicated and the backup systems are on permanent stand-by should the original ones fail.

A safety injection system ensures that the core is cooled in the event of an accident. In case of a reactor coolant pipe break, this system injects cooling water as quickly as possible into the reactor core. In case of a steam pipe break, the system injects as quickly as possible concentrated boric acid solution into the reactor core to stop the reaction. Finally, in the event of an accident, all radioactive material would be trapped in the containment building, which is tested regularly at a pressure higher than would occur in this eventuality.

Radiation. The International Commission on Radiological Protection (ICRP) has set the international maximum limit of radiation which an individual member of the public can receive at 5,000 microsievert per year. (The Sievert is a unit of radiation dosage equal to 100 joules of energy absorbed per kilogram of matter.) The South African Council for Nuclear Safety license for Koeberg stipulates a maximum dose of 250 microsievert. Eskom decided to set their limit at 100 microsievert. In 1988 the maxiumum exposure calculated at Koeberg for a hypothetical unclothed individual permanently subjected to the maximum exposure in the station was 10 microsievert.

Samples of fish, meat, vegetables, milk, water, and grass within a 30-kilometer radius of Koeberg—as well as the marine life and indigenous fauna—are monitored regularly for the possible affects of radiation.

Nuclear waste. Koeberg produces three types of nuclear waste: low-level waste (LLW), paper, gloves, glassware, disposable overalls, etc; intermediate-level waste (ILW), radioactive resins and sludges, spent fuel cartridges, scrap metal, etc; and and high-level waste (HLW), spent fuel assemblies.

At Koeberg, LLW is compressed into steel drums, sealed, and transported to the Vaalputs Radioactive Waste Disposal Facility. The site generates 500 drums of LLW per year. ILW is solidified by combining it with a sand/cement mix which is poured into concrete drums. Koeberg produces 300 drums of ILW waste per year which are transported to Vaalputs.

Spent fuel assemblies are stored at the plant under water for 10 years. Then, the spent fuel is placed in dry storage flasks until uranium recovery becomes economically viable. At that point, the spent fuel is sent to a reprocessing plant. After reprocessing, HLW in synthetic rock form will be finally sealed into stainless steel flasks to be buried about one kilometer deep at a waste repository.

Koeberg

Visitor's center. Koeberg has a visitor's center which is open to the public free of charge. Hours are Monday-Friday, 8:30-12:30 and 2:00-4:30; and the second and last Sunday of every month, 2:00-4:30. The center is closed on Saturdays and public holidays. For further information, telephone 02224-2133.

Sources

Koeberg Visitors Center Brochure. Eskom, n.d.
Koeberg General Information Brochure. Eskom, November 1989.
Koeberg—Energy for the Future: Technical Information. Eskom, January 1990.

Romania

COUNTRY REPORT

Central government impacts nuclear project

Romania's nuclear power construction program began in 1979 with plans to construct a multi-reactor facility at Cernavoda. Because the country is rich in uranium deposits, officials decided to implement CANDU technology, which uses a natural uranium fuel. This decision enabled Romania to avoid becoming dependent on foreign sources for fuel enrichment. Romanian officials claim nuclear generating capacity is needed because the country faces power shortages and is suffering from pollution associated with fossil-fuel generation.

The Ceausescu regime's central economy, however, thwarted construction efforts and hindered progress at Cernavoda. The plant remained unfinished in 1989 when the Ceausescu government was overthrown. Romania's new government invited the International Atomic Energy Agency to inspect the progress at Cernavoda and assist in efforts to plan for the plant's completion.

In 1991, the Canadian government approved a $227 million loan to help complete the first of five units at Cernavoda. Unit 1 is scheduled to begin commercial operation in 1995. Although four additional units are planned, Romanian officials doubt that domestic demand will require all the output. They hope power exports to other countries will justify the added generating capacity.

Sources

Mounfield, Peter. *World Nuclear Power*. Routledge, 1991.

"Romania to get First Nuclear Plant." *Nuclear Industry*, Fourth Quarter 1991.

—*Karen Bellenir*

PLANT PROFILES

156 ○
Cernavoda Nuclear Power Station, Units 1-5

Filiala Centrala Nuclearoelectrica Cernavoda
Romania, Jud. Constanta
PO Box 18
8625 Cernavoda, Romania
Viorel Marculescu, Stat.Mgr.

Phone: 912-38-337
Fax: 400-124-407
Telex: INCVR 14523

Owning Utility

Romanian Electricity Authority (RENEL)
Bd. Magheru 33
Bucharest 1, Romania
Tiberiu Campureanu, Gen. Dir. of Nuclear Power Group

Phone: 90-59-6000
Fax: 400-120-800
Telex: 11279

Contact

I. Rotaru, Deputy Gen. Dir.
Romanian Electricity Authority, Nuclear Power Group

Basic Facts

Units 1-2 *Status:* Under construction (Unit 1, 62% complete as of January 1992; Unit 2, 46% complete as of January 1992). *Type of Reactor:* Pressurized heavy-water reactor. *Megawatts (electric):* 700. *Megawatts (thermal):* 2180. *Reactor System Supplier:* Atomic Energy of Canada Ltd. *Generator Supplier:* Ansaldo Meccanico Nucleare SpA (Italy); General Electric Co. (United States). *Architect Engineer:* Romanian Institute for Power Studies and Design.

Units 3-5 *Status:* Under construction (Unit 3, 23% complete as of January 1992; Unit 4, 12% complete as of January 1992; Unit 5, 8% complete as of January 1992). *Type of Reactor:* Pressurized heavy-water reactor. *Megawatts (electric):* 700. *Megawatts (thermal):* 2180. *Reactor System Supplier:* Romanian Ministry of of Machine Building Industry. *Generator Supplier:* Romanian Ministry of of Machine Building Industry. *Architect Engineer:* Romanian Institute for Power Studies and Design.

Key Dates

Unit 1 *Ordered:* 1979. *Construction began:* 1980. *Commercial Operation Expected:* 1995.

Unit 2 *Ordered:* 1981. *Construction began:* 1982. *Commercial Operation Expected:* 1996.

Unit 3 *Ordered:* 1983. *Construction began:* 1984. *Commercial Operation Expected:* 1997.

Unit 4 *Ordered:* 1984. *Construction began:* 1985. *Commercial Operation Expected:* 1998.

Unit 5 *Ordered:* 1985. *Construction began:* 1986. *Commercial Operation Expected:* 1999.

Plant Report

Political system hinders construction. The Romanian Electricity Authority (RENEL) began on-site construction work on Romania's first nuclear power plant, Cernavoda in 1980. The site is located in southeast Romania, about 160 km east of Bucharest on the Danube River.

Although the initial proposal called for four units at the station, plans for a fifth were later added. Under the Ceausescu regime and its controlled central economy, the project suffered many deficiencies including inferior workmanship, poor materials, and delays. In October 1989, Romania's new government ordered all work halted until problems could be corrected. Construction resumed in the early 1990's.

New government makes commitment to plant. In 1990, the Romanian government invited the International Atomic Energy Agency (IAEA) to conduct a Pre-OSART mission at the plant. The mission's goal was to give advice on how to improve management at the project. The pre-OSART report made the following recommendations:

- Reorganize Romania's electricity-supply industry.
- Involve experienced nuclear construction companies in the project's management.
- Improve quality control.
- Improve social conditions near the site.
- Develop training programs.
- Improve corporate and individual attitudes about quality and safety.
- Improve site cleanliness.
- Improve equipment and record storage.

In 1991, an IAEA follow-up team reported that all the Pre-OSART proposals had been addressed and that "solutions to more than three-quarters of the issues have been fully implemented or are progressing satisfactorily."

Work continues. Also in 1991, Romanian officials entered into a contract with Atomic Energy of Canada Limited (AECL) and the Italian Ansaldo Consortium (AAC) to finish work on the first unit at Cernavoda, a 700-megawatt CANDU reactor. The Canadian government approved a $277 million loan and AAC agreed to manage the construction, provide commissioning, training, and up to 18 months of operation. The unit is scheduled to enter commercial production in 1995.

Work on Units 2-5 at Cernavoda is in various stages of completion ranging from 8 percent to 46 percent. Continued work on the additional four units is dependant on the availability of funding.

Sources

IAEA *Pre Operational Safety of Nuclear Installations: Cernavoda Nuclear Power Plant, 24 September - 12 October 1990.* Report to the Government of Romania.
RENEL 1991 Information Bulletin.
"IAEA Follow-up Review of the 1990 Pre-OSART Mission to Cernavoda Nuclear Power Plant." Press release. IAEA, September 13, 1991.

—*Karen Bellenir*

Russia

COUNTRY REPORT

The Soviet legacy

When the Soviet Union dissolved (see the former Soviet Union), the Russian Federation inherited more nuclear generating stations than any other former republic. Nine separate power-producing stations with a combined total of 28 nuclear units were left within Russian borders. These units consisted of 12 VVER reactors, 11 RBMK-1000 reactors, four small RBMK reactors and one 560-megawatt liquid metal fast breeder reactor (LMFBR). In addition to these power-generating units, the Russian Federation also gained ownership of several Combined Heat and Power (CHP) plants and some small experimental breeder reactors.

Energy situation in the Russian Federation

Nuclear generation plays an important part in the Russian Federation's power mix. Dependance on nuclear sources, however, varies regionally. In the northwestern districts surrounding St. Petersburg, nuclear power contributes 33 percent of the available electricity. In the Moscow area, nuclear generation accounts for 22 percent. Nationwide, approximately 11 percent of the power in Russia is from nuclear sources. Fossil-fueled plants generate 74 percent of the nation's electricity, and 15 percent is from hydroelectric facilities.

During 1991 on the eve of the central government's break-up, Soviet officials stated that new nuclear power facilities were needed to support the standard lifestyle in several Russian areas. Seven regional parliaments had formally requested the construction of more nuclear generating units. With the exception of Siberia, every regional power system expected major shortages by 1995.

As the former Soviet republics began achieving their independence, the Russian Federation faced an energy crisis. In the southern area of Chelyabinsk Oblast, power was being rationed. The Vovoronezh city council opted to complete a nuclear powered district heating plant. The Kursk, Murmansk, and Chelyabinsk regional parliaments voted to continue plans for nuclear construction. Power shortages compelled Russian officials to operate four VVER-440 Model V230 reactor units even though the International Atomic Energy Agency (IAEA) had recommended phasing them out. The Russians imposed no deadlines for completing safety backfits or for shutting down unsafe plants. Instead, plans called for the continued operation of marginal units for another eight to 10 years.

In October 1991, the Russian nuclear plants began forming a consortium to allow each station financial autonomy. On January 1, 1992, each plant became an independent unit and was permitted to export electricity for hard currency in order to purchase equipment. Under the Soviet regime, nuclear plants had produced electricity at a loss, recovering only one-third of their operating expenses. Boris Yeltsin did not include the price of electricity in his decree to float all prices in Russia, so the price of electricity in the Russian Federation remains fixed.

Changes in regulatory agencies and research facilities

When the central Soviet government dissolved, several nuclear regulatory agencies were restructured. The Soviet Ministry of Atomic Power and Industry (MAPI) was split in January 1992. Gospromatomnadzor (GPAN), a Soviet nuclear safety agency created in response to the Chernobyl accident, was dismantled and replaced by new nuclear regulatory organizations, one in Russia and one in Ukraine.

Boris Yeltsin signed a decree on January 29, 1992 to create a new Russian Federation Ministry for Atomic Energy (Minatom RF). Minatom RF's responsibilities included taking over several MAPI functions, such as the administration of nuclear safety, research and design. The new agency was also charged with modernizing the country's nuclear program and converting former military facilities to civilian power generation. A Russian State Committee received the responsibility of researching, designing, and planning international programs and defense policies. A separate State Committee for the Supervision of Nuclear and Radiation Safety was formed to oversee reactor safety.

Critics of the Russian Federation's ability to manage its nuclear industry have suggested that the regulators' effectiveness has been compromised. One of the reasons cited is the problem of nuclear experts leaving government work to take jobs in private corporations for higher salaries. Another difficulty is the uncertain legal standing of the newly organized regulatory agencies.

The former I.V. Kurchatov Institute of Atomic Energy, an important research organization, separated from MAPI in December 1991 to become the Russian Kurchatov Institute. Another research facility affected by the unfolding political events was the Soviet National Academy of Sciences which became the Russian Academy of Sciences and the Institute of Nuclear Safety. The All-Union Institute of Nuclear Power Operation remained intact.

Nuclear by-products and waste issues

After the break-up of the Soviet Union more complete information became available about accidents involving radiation and about nuclear waste hazards in the Russian Federation. Under Soviet rule, public disclosure was forbidden. This policy led to delayed containment of nuclear accidents and hampered cleanup efforts.

Alexander Penyagin, former chairman of the Soviet Subcommittee on Atomic Energy and Nuclear Ecology prepared a map documenting nuclear accidents and inappropriate radioactive waste disposal sites. According to Penyagin, radioactive waste materials were improperly stored in tanks in the Chelyabinsk region. One tank exploded in 1957. If a second explosion occurs at the site, it could release 20 times the amount of radiation released during the Chernobyl accident. The storage facility has also been blamed for contaminating the Techa River and Lake Karachay, which some claim is so polluted with radioactive materials that standing on the shore for an hour would be fatal.

A secret underground Siberian city, Atomgrad, has been accused of dumping nuclear waste by-products into the Yenisey River from 1950 until 1990. The city, with three small RBMK breeder reactors, was built under Stalin's orders. The plutonium producing reactors are scheduled to be decommissioned beginning in 1992, and the site may be converted to a spent-fuel reprocessing facility. According to the newspaper *Investia*, local officials support the proposition.

Improper storage of nuclear material has also been documented at a nuclear power plant 40 miles west of St. Petersburg. Radiation leaks from the storage facility because of its inadequate design and poor seals. Levels of strontium 90 in the groundwater near the plant are 350 times above normal.

The current Russian government does not have enough money to effectively clean up the nuclear mess it inherited. International attention to nuclear disarmament and nonproliferation has forced the government to spend its limited funds on restructuring former military activities. Some activists fear this policy may cause the nation's serious environmental issues to be overlooked.

Security at civilian power plants, which were formerly guarded by the KBG, has also been questioned. Russian officials said that in order to prevent the danger of possible thefts of radioactive material, civilian nuclear facilities must be made more secure. One official suggested that all U.S. and Russian nuclear power generating facilities should be put under international control.

A related concern regarding radioactive materials management in the Russian Federation is what to do with the plutonium produced by the nation's breeder reactors. Evgeny Mikerin, head of science and technology in Russia's Atomic Energy Ministry, opposed using the plutonium as a fuel in civilian power plants or diluting it. Instead, he favored conserving the plutonium "as a national wealth."

Continuing international cooperation

Following the disastrous accident at Chernobyl, the former Soviet government began increased efforts at cooperating with other nations to improve reactor safety. The World Atomic Nuclear Operators (WANO) was formed at a meeting in Moscow before the Soviet collapse. WANO established four regional centers, located in Moscow, Atlanta, Paris, and Tokyo. Following the break-up of the former USSR, the Moscow Center, headed by Anotoly Kontisevoy, continues to operate and to coordinate plant personnel exchanges between eastern and western nuclear stations.

In 1988 the former Soviet Union and the United States entered into an agreement to exchange information about plant operation and operator proficiency. As a result the U.S./Soviet Joint Coordinating Committee on Civilian Nuclear Reactor Safety (JCCNRS) established twelve separate working groups. Russian plants and Russian research organizations were involved in the project. The U.S. government expressed hope that political changes would not disrupt the ongoing work of the groups.

After the collapse of the USSR, The Russian Federation also inherited all of the Soviet nuclear training institutes and production facilities for plant equipment. Minatom RF's head, Vicktor Mikhailov was given the responsibility to make contractual arrangements with Ukraine and Lithuania for continued operation of their nuclear stations.

Continuing controversy surrounding Soviet-styled reactors

Some nuclear experts claim that RBMK graphite-moderated reactors cannot be modified to meet western safety standards. Although newer Soviet VVER designs can be upgraded, the Russian economy cannot support the needed safety backfits, which are estimated to require 10 years and $15 billion to complete. The Russian Federation has expressed its desire to comply with international safety standards, but official action has been hampered because of the evolving regulatory climate and political restructuring taking place within the newly autonomous nation.

Several of the Russian Federation's European neighbors have demanded that all RBMK reactors be shut down. The State Committee for the Supervision of Nuclear and Radiation

Safety acknowledged that RBMKs should be closed, but the agency had no authority to enforce such a decision. The cancellation of nuclear construction projects underway at the time of the Chernobyl accident left older, more dangerous reactors unchanged and failed to provide substitute means for power generation. Russian officials now claim that the country cannot afford to shut down its unsafe reactors.

Sources

"Siberian Production Reactors to be Shut, Newspaper Says." *Nucleonics Week,* November 28, 1991.

"Moscow's Dirty Nuclear Secrets." *U.S. News & World Report,* February 10, 1992.

"Many Chernobyls Just Waiting to Happen." *Business Week,* March 16, 1992.

"Experts Fault Efforts to Shut Down Old Reactors." *The Detroit News,* March 25, 1992.

Source Book: Soviet-Designed Nuclear Power Plants in the Former Soviet Republics and Czechoslovakia, Hungary and Bulgaria. U.S. Council for Energy Awareness, 1992.

"Chernobyl: The Final Irony." *Info,* May 1992.

"Nuclear Black Market Scares Experts." *The Detroit News,* July 20, 1992.

—Karen Bellenir

PLANT PROFILES

157 ●
Balakovo Nuclear Power Plant, Units 1-3

413800
Saratovskaya obl
g. Balakovo, ul. Titova, 2a, Russia
Pavel Ipatov, Plant Mgr.

Owning Utility

Russian Ministry of Atomic Energy (MINATOM)
24/26 B. Ordynka
Moscow, Russia
Victor Mikhailov, Minister

Phone: 095-230-24-20

Contact

Vitaliy F. Konovalov, Deputy Minister
Ministry of Atomic Energy (MINATOM)

Basic Facts

Units 1-3 *Status:* Operable. *Type of Reactor:* Pressurized light-water reactor. *Megawatts (electric):* 1000. *Megawatts (thermal):* 3000. *Generator Supplier:* Elektrosila.

Key Dates

Unit 1 Ordered: 1978. Construction began: 1980. Criticality: December 1985. First Power: December 1985. Commercial Operation: May 1986.

Unit 2 Ordered: 1980. Construction began: 1981. Criticality: October 1987. First Power: October 1987. Commercial Operation: January 1988.

Unit 3 Ordered: 1980. Construction began: 1982. Criticality: December 1988. First Power: December 1988. Commercial Operation: April 1989.

Plant Report

Three-unit station. The Balakovo Nuclear Power Plant, located in Balakovo, Saratov, in the Russian Federation is a three unit nuclear generating station. All three units are Soviet-styled third-generation VVER-1000 reactors. The design incorporated some international safety features absent in previous Soviet nuclear plants, such as the use of a steel-lined containment building. In February 1992, Russian officials announced their intention to complete a fourth unit at the site.

Fire forces shutdown. A fire on March 4, 1992 in the electrical equipment at Balakovo caused the station to shut down automatically. According to reports, plant personnel brought the fire under control in 40 minutes.

International cooperation. Russian officials asked the International Atomic Energy Agency (IAEA) to perform an ASSET (Assessment of Safety Significant Events Team) review of the Balakovo station. The purpose of an ASSET mission is to study a plant's operating history and incident-prevention programs, to determine the root cause of safety-significant events, and to make recommendations for corrective actions.

Sources

Source Book: Soviet-Designed Nuclear Power Plants in the Former Soviet Republics and Czechoslovakia, Hungary and Bulgaria. U.S. Council for Energy Awareness, 1992.

—Karen Bellenir

158 □
Balakovo Nuclear Power Plant, Unit 4

413800
Saratovskaya obl
g. Balakovo, ul. Titova, 2a, Russia
Pavel Ipatov, Plant Mgr.

Owning Utility

Russian Ministry of Atomic Energy (MINATOM)
24/26 B. Ordynka
Moscow, Russia
Victor Mikhailov, Minister

Phone: 095-230-24-20

Contact

Vitaliy F. Konovalov, Deputy Minister
Ministry of Atomic Energy (MINATOM)

Basic Facts

Unit 4 *Status:* Planned. *Type of Reactor:* Pressurized light-water reactor. *Megawatts (electric):* 1000. *Megawatts (thermal):* 3000. *Generator Supplier:* Elektrosila.

Plant Report

See Plant Report for Balakovo Nuclear Power Plant, Units 1-3.

159 ★
Beloyarsk Nuclear Power Plant, Units 1-2

624051
Sverdlovskaya obl
pos. Zarechniy, Russia
Oleg Saraev, Plant Mgr.

Fax: 34377-31070
Telex: 721543 ATOM SU

Owning Utility

Russian Ministry of Atomic Energy (MINATOM)
24/26 B. Ordynka
Moscow, Russia
Victor Mikhailov, Minister

Phone: 095-230-24-20

Contact

Nikolai Oshkanov, Deputy Plant Mgr.
Beloyarsk Nuclear Power Plant

Basic Facts

Unit 1 *Status:* Shut down. *Type of Reactor:* Light-water-cooled, graphite-moderated reactor. *Megawatts (electric):* 108. *Megawatts (thermal):* 286.

● Operable ○ Under Construction △ Indefinitely Deferred ▲ Decommissioning □ Planned ■ On Order ☆ Cancelled ★ Shut Down

Beloyarsk

Unit 2 *Status:* Shut down. *Type of Reactor:* Light-water-cooled, graphite-moderated reactor. *Megawatts (electric):* 160. *Megawatts (thermal):* 530. *Generator Supplier:* Elektrosila.

Key Dates
Unit 1 *Ordered:* 1958. *Construction began:* 1958. *Criticality:* September 1963. *First Power:* April 1964. *Commercial Operation:* April 1964. *Shut Down:* 1983.

Unit 2 *Ordered:* 1959. *Construction began:* 1959. *Criticality:* October 1967. *First Power:* December 1967. *Commercial Operation:* December 1969. *Shut Down:* October 1989.

Plant Report
See Plant Report for Beloyarsk Nuclear Power Plant, Unit 3.

Beloyarsk Nuclear Power Plant, Unit 3

624051
Sverdlovskaya obl
pos. Zarechniy, Russia
Oleg Saraev, Plant Mgr.

Fax: 34377-31070
Telex: 721543 ATOM SU

Owning Utility
Russian Ministry of Atomic Energy (MINATOM)
24/26 B. Ordynka
Moscow, Russia
Victor Mikhailov, Minister

Phone: 095-230-24-20

Contact
Nikolai Oshkanov, Deputy Plant Mgr.
Beloyarsk Nuclear Power Plant

Basic Facts
Unit 3 *Status:* Operable. *Type of Reactor:* Liquid metal fast breeder reactor. *Megawatts (electric):* 600. *Megawatts (thermal):* 1470. *Generator Supplier:* Elektrotyazhmash.

Key Dates
Unit 3 *Ordered:* 1966. *Construction began:* 1968. *Criticality:* February 1980. *First Power:* April 1980. *Commercial Operation:* November 1981.

Plant Report

Pioneering site. Officials in the former Soviet Union described the Beloyarsk nuclear power station as "a large-scale test ground for various types of nuclear power reactors, a large industrial laboratory." The facility, located in Zarechny, Sverdolosvsk, in the Russian Federation, has indeed been a pioneering site for the development of nuclear technology.

Construction on the first unit, a 100-megawatt boiling water, graphite moderated reactor began in 1958. The unit operated from April 26, 1964 until it was closed in 1981. Unit 2, a 200-megawatt boiling water, graphite moderated reactor, began power production in 1967 and was taken out of service in 1989. Both units were early prototype RBMK reactors.

Breeder technology. In 1968, construction began on Beloyarsk-3, a 560-megawatt sodium-cooled fast-breeder reactor intended to be the prototype for future commercial breeder reactors. The Soviet government wanted to develop breeder reactors to reduce the amount of natural uranium consumed and to produce plutonium. Beloyarsk-3 is a BN-600 Liquid-Metal Fast Breeder Reactor (LMFBR) designed to generate new fuel during its operation. The main components are submerged in a pool of liquid sodium. Its modular design enables repairs to be made on the steam generators without taking the plant off line. The unit began operation in November 1981. It is the second-largest Fast Breeder Reactor in the world after the French Super Phenix.

Beloyarsk-3 also became the prototype for Soviet Combined Heat and Power (CHP) plants. Hot water from the reactor is used to provide heat to Zarechny's 27,000 residents.

The BN-600 design is different from western reactors. Most significantly, it has no containment building. The reactor is housed in a standard industrial building. A protective shell was installed to prevent sodium leakage in the event of a rupture within the reactor vessel, and special barrier shrouds prevent radiation from leaking into the environment.

A second breeder unit (Beloyarsk-4), a larger BN-800, was under construction during the later years of the Soviet regime. The BN-800 had an increased core size and a greater number of fuel assemblies. Other improvements included seismic safeguards. Work ground to a near halt in 1990. Continuation was contingent on the results of an Environment Impact Inquiry.

Operating concerns and emergency preparedness. Beloyarsk-3 experienced early problems with leaking fuel and steam generator tube breaks. This resulted in volatile sodium/water interactions. In 1991, officials planned to backfit the sodium-to-air heat removal system and to use plutonium instead of enriched uranium for fuel. The current status of these proposals in the post-Soviet era is uncertain.

The computerized control system, called "Kompleks-Uran," was developed to monitor reactor functions and indicate potential problems in plant equipment. Fire suppression techniques necessary to put out a sodium fire were also developed. The system employed at Beloyarsk includes a powder compound kept in special trays and mechanisms to drain the sodium to emergency tanks. The tanks keep the sodium in a relatively air-tight environment which helps to suppress fire. In addition, the station maintains an emergency back-up power supply.

Officials keep a watchful eye on radiation levels within the facility and in the surrounding community. According to reports, the level of radioactive materials released into the atmosphere is 10 times below the permitted level. The dosage of radiation received by plant personnel is also below established limits.

Sources
"Summary of the Beloyarsk NPP History." Reference Information Group. Beloyarsk Nuclear Power Plant, n.d.

Beloyarskaya Nuclear Power Plant Named After I. V. Kurchatov. Atomenergoexport. Beloyarsk Nuclear Power Plant, n.d.

Mounfield, Peter. *World Nuclear Power.* Routledge, 1991.
Source Book: Soviet-Designed Nuclear Power Plants in the Former Soviet Republics and Czechoslovakia, Hungary and Bulgaria. U.S. Council for Energy Awareness, 1992.

—Karen Bellenir

161 ○
Beloyarsk Nuclear Power Plant, Unit 4

624051
Sverdlovskaya obl
pos. Zarechniy, Russia
Oleg Saraev, Plant Mgr.

Fax: 34377-31070
Telex: 721543 ATOM SU

Owning Utility
Russian Ministry of Atomic Energy (MINATOM)
24/26 B. Ordynka
Moscow, Russia
Victor Mikhailov, Minister

Phone: 095-230-24-20

Contact
Nikolai Oshkanov, Deputy Plant Mgr.
Beloyarsk Nuclear Power Plant

Basic Facts
Unit 4 *Status:* Under construction. *Type of Reactor:* Liquid metal fast breeder reactor. *Megawatts (electric):* 800. *Generator Supplier:* Elektrotyazhmash.

Plant Report
See Plant Report for Beloyarsk Nuclear Power Plant, Unit 3.

162 ●
Bilibino Nuclear Power Plant, Units 1-4

Bilibino, Chukotka, Russia

Owning Utility
Russian Ministry of Atomic Energy (MINATOM)
24/26 B. Ordynka
Moscow, Russia
Victor Mikhailov, Minister

Phone: 095-230-24-20

Contact
Vitaliy F. Konovalov, Deputy Minister
Ministry of Atomic Energy (MINATOM)

Basic Facts
Units 1-4 *Status:* Operable. *Type of Reactor:* Light-water-cooled, graphite-moderated reactor. *Megawatts (electric):* 12. *Megawatts (thermal):* 62.

Key Dates
Unit 1 *Ordered:* 1965. *Construction began:* 1970. *Criticality:* December 1973. *First Power:* January 1974. *Commercial Operation:* April 1974.

Unit 2 *Ordered:* 1965. *Construction began:* 1970. *Criticality:* December 1973. *First Power:* January 1974. *Commercial Operation:* February 1975.

Unit 3 *Ordered:* 1965. *Construction began:* 1970. *Criticality:* December 1975. *First Power:* December 1975. *Commercial Operation:* February 1976.

Unit 4 *Ordered:* 1965. *Construction began:* 1970. *Criticality:* December 1976. *First Power:* December 1976. *Commercial Operation:* January 1977.

Plant Report
Heat and electricity. Bilibino is a mining town in the northeastern region of the Russian Federation. Bilibino is also the home of the smallest Russian nuclear power plant. The station consists of four 12-megawatt reactors, similar in design to the reactor that exploded at Chernobyl in 1986. It supplies electricity to a gold mine and two Arctic port towns. In addition to electrical power, the station produces hot water. The water flows through special conduits to provide heat to 14,000 people.

Constructing a nuclear station inside the Arctic Circle posed unique engineering challenges to Soviet builders. At Bilibino, the reactors were built on monolithic slabs which float on the permafrost. Soviet researchers intended to use the technique to build additional nuclear plants in Siberia.

Effects of Chernobyl accident. Prior to the disastrous accident at the Chernobyl nuclear station, the Soviet government planned to build three new reactors at Bilibino. As a result of that explosion, the planners were forced to incorporate new safety features in the proposed units which increased costs by more than forty percent. Officials at the station also established emergency evacuation plans for residents.

After the collapse of the Soviet system, further construction at Bilibino and other Siberian sites is uncertain, but the four RBMK's continue to supply electricity and steam heat.

Sources
"What Price Nuclear Power? In Siberia, It's High." *The New York Times,* April 20, 1987.
Mounfield, Peter. *World Nuclear Power.* Routledge, 1991.

—Karen Bellenir

163 ●
Gorky Nuclear Facility, Units 1-2

Gorky, Russia

Owning Utility
Russian Ministry of Atomic Energy (MINATOM)
24/26 B. Ordynka
Moscow, Russia
Victor Mikhailov, Minister

Phone: 095-230-24-20

Gorky

RUSSIA

Contact
Vitaliy F. Konovalov, Deputy Minister
Ministry of Atomic Energy (MINATOM)

Basic Facts
Units 1-2 *Status:* Operable. *Type of Reactor:* Boiling water reactor. *Megawatts (electric):* 0. *Megawatts (thermal):* 500.

Plant Report

Originally a heat-supplying plant. The Gorky nuclear facility, consisting of two 500-megawatt heat-supplying, boiling water reactors, was the object of public opposition following the 1986 Chernobyl accident. Citizens raised questions about safety at the station because it was constructed near the center of the city.

Nuclear planners in the former Soviet Union agreed to allow the International Atomic Energy Agency (IAEA) to review their plans. The IAEA made an Independent Safety Review of the site and also a Pre-OSART (Operational Safety Review Team) mission. The purpose of a pre-OSART review is to examine construction quality, arrangements for commissioning, and other preparations that will impact future operational safety.

Uncertain future. Plans to construct a power-producing unit at the Gorky station were disclosed by the former Soviet government in the fall of 1991. According to the announcement, a 630-megawatt pressurized water reactor was being designed. The Soviet regime collapsed shortly afterwards and future construction at Gorky under the Russian Federation's autonomy is uncertain.

Sources
"Soviet Nuclear Power Plant Programme Marks Time." *Nature*, January 26, 1989.
"Soviet Citizens Not Impressed by IAEA." *Nature*, March 23, 1989.
Mounfield, Peter. *World Nuclear Power*. Routledge, 1991.
Source Book: Soviet-Designed Nuclear Power Plants in the Former Soviet Republics and Czechoslovakia, Hungary and Bulgaria. U.S. Council for Energy Awareness, 1992.

—Karen Bellenir

---

164 ☐

Gorky Nuclear Facility, Unit 3
Gorky, Russia

Owning Utility
Russian Ministry of Atomic Energy (MINATOM)   Phone: 095-230-24-20
24/26 B. Ordynka
Moscow, Russia
Victor Mikhailov, Minister

Contact
Vitaliy F. Konovalov, Deputy Minister
Ministry of Atomic Energy (MINATOM)

Basic Facts
Unit 3 *Status:* Planned. *Type of Reactor:* Pressurized light-water reactor. *Megawatts (electric):* 630.

Plant Report
See Plant Report for Gorky Nuclear Facility, Units 1-2.

---

165 ●

Kalinin Nuclear Station, Units 1-2
171850   Telex: 171354 Raduga
Tverskaya obl
pos. Udomlya, Russia
Gennadi Shchapov, Plant Mgr.

Owning Utility
Russian Ministry of Atomic Energy (MINATOM)   Phone: 095-230-24-20
24/26 B. Ordynka
Moscow, Russia
Victor Mikhailov, Minister

Contact
Vitaliy F. Konovalov, Deputy Minister
Ministry of Atomic Energy (MINATOM)

Basic Facts
Units 1-2 *Status:* Operable. *Type of Reactor:* Pressurized light-water reactor. *Megawatts (electric):* 1000. *Megawatts (thermal):* 3000. *Generator Supplier:* Elektrosila.

Key Dates
Unit 1 *Ordered:* 1971. *Construction began:* 1977. *Criticality:* April 1984. *First Power:* May 1984. *Commercial Operation:* June 1985.
Unit 2 *Ordered:* 1971. *Construction began:* 1982. *Criticality:* November 1986. *First Power:* December 1986. *Commercial Operation:* March 1987.

Plant Report

Background. The Kalinin Nuclear Station is located in Tver, Volga in the Russian Federation. The station consists of two VVER-1000 units. The VVER-1000 is a third-generation Soviet-styled pressurized water reactor which incorporates some western safety features, such as a containment structure. Units 1 and 2 began commercial operation June 1985 and March 1987. A third unit at the site is under construction, and a fourth unit is planned.

International cooperation. The Kalinin station has taken part in several international efforts aimed at improving nuclear safety. It participated in plant exchanges sponsored by the World Association of Nuclear Operators (WANO). Personnel visits were arranged with the Susquehanna nuclear power plant in Pennsylvania and the Shearon Harris plant in North Carolina. A team from the United States visited Kalinin in September 1991 to examine the diagnostic systems used to detect problems in Soviet plants. The study was part of the U.S./Soviet Joint Coordi-

Plant Profiles

nating Committee on Civilian Nuclear Reactor Safety (JCCNRS) and done by members of the committee's Working Group 9.

Steam generators. Officials at Kalinin plan to replace the plant's steam generators. The station announced its plans to obtain a steam-generator tube inspection/repair manipulator and machinery to remove bolts from steam-generator manhole covers from the French company Framatome.

Economic effect of the Soviet breakup. In October 1991, the Kalinin plant manager announced that all Russian nuclear plants were forming a consortium which would enable the individual plants to achieve financial independence. Under the agreement, the plants would be permitted to export electricity for hard currency. The sales would enable operators to raise funds to purchase plant equipment.

Sources

Source Book: *Soviet-Designed Nuclear Power Plants in the Former Soviet Republics and Czechoslovakia, Hungary and Bulgaria*. U.S. Council for Energy Awareness, 1992.

—Karen Bellenir

166 ○
Kalinin Nuclear Station, Unit 3

171850 Telex: 171354 Raduga
Tverskaya obl
pos. Udomlya, Russia
Gennadi Shchapov, Plant Mgr.

Owning Utility

Russian Ministry of Atomic Energy Phone: 095-230-24-20
 (MINATOM)
24/26 B. Ordynka
Moscow, Russia
Victor Mikhailov, Minister

Contact

Vitaliy F. Konovalov, Deputy Minister
Ministry of Atomic Energy (MINATOM)

Basic Facts

Unit 3 *Status:* Under construction. *Type of Reactor:* Pressurized light-water reactor. *Megawatts (electric):* 1000. *Megawatts (thermal):* 3000. *Generator Supplier:* Elektrosila.

Plant Report

See Plant Report for Kalinin Nuclear Station, Units 1-2.

167 □
Kalinin Nuclear Station, Unit 4

171850 Telex: 171354 Raduga
Tverskaya obl
pos. Udomlya, Russia
Gennadi Shchapov, Plant Mgr.

Owning Utility

Russian Ministry of Atomic Energy Phone: 095-230-24-20
 (MINATOM)
24/26 B. Ordynka
Moscow, Russia
Victor Mikhailov, Minister

Contact

Vitaliy F. Konovalov, Deputy Minister
Ministry of Atomic Energy (MINATOM)

Basic Facts

Unit 4 *Status:* Planned. *Type of Reactor:* Pressurized light-water reactor. *Megawatts (electric):* 1000. *Megawatts (thermal):* 3000. *Generator Supplier:* Elektrosila.

Plant Report

See Plant Report for Kalinin Nuclear Station, Units 1-2.

168 ●
Kola Nuclear Plant, Units 1-4

184151 Telex: 126716 Salma
Murmanskaya obl
pos. Polyaniye Zori, Russia
Vladmir Schmidt, Plant Mgr.

Owning Utility

Russian Ministry of Atomic Energy Phone: 095-230-24-20
 (MINATOM)
24/26 B. Ordynka
Moscow, Russia
Victor Mikhailov, Minister

Contact

Vitaliy F. Konovalov, Deputy Minister
Ministry of Atomic Energy (MINATOM)

Basic Facts

Units 1-4 *Status:* Operable. *Type of Reactor:* Pressurized light-water reactor. *Megawatts (electric):* 440. *Megawatts (thermal):* 1375. *Generator Supplier:* Elektrosila.

Key Dates

Unit 1 *Ordered:* 1966. *Construction began:* 1970. *Criticality:* June 1973. *First Power:* June 1973. *Commercial Operation:* December 1973.

Unit 2 *Ordered:* 1966. *Construction began:* 1970. *Criticality:* November 1974. *First Power:* December 1974. *Commercial Operation:* February 1975.

Unit 3 *Ordered:* 1974. *Construction began:* 1977. *Criticality:* February 1981. *First Power:* March 1981. *Commercial Operation:* December 1982.

Unit 4 *Ordered:* 1974. *Construction began:* 1976. *Criticality:* October 1984. *First Power:* October 1984. *Commercial Operation:* December 1984.

Kola

Plant Report

Four units, two designs. The Kola Nuclear Plant, located in Murmansk, in the Russian Federation, consists of four Soviet-styled, pressurized water reactors. Although Kola's units are all VVER-440's, they incorporate two different phases in the development of nuclear technology. Units 1 and 2 represent the earliest VVER design, Model V230. These first-generation VVER reactors lack a containment structure and do not have emergency core-cooling systems or auxiliary feedwater systems. They are also prone to reactor vessel embrittlement.

Kola's second two units are VVER-440 Model V213's. Kola's third and fourth units are second-generation VVER's. The Model V213 incorporated some safety features such as an upgraded Accident Localization System, emergency core-cooling and auxiliary feedwater systems, and a stainless steel lining in the reactor vessel to address embrittlement concerns.

In the fall of 1991, Soviet officials announced plans to construct a fifth and sixth unit at Kola. According to the proposal, the two new reactors would be VVER-1000's, a third-generation design which incorporated many western-styled safety features. The status of any new construction at Kola, however, is uncertain following the collapse of the central Soviet government.

International cooperation. The Kola plant has been the subject of several international research efforts. The International Atomic Energy Agency (IAEA) conducted an on-site safety review at Kola in 1991. The visit was part of an extensive IAEA study of generic and plant-specific safety issues related to the VVER-440 Model V230. Although the investigation confirmed design deficiencies and substandard safety features, the team commended Kola's staff for demonstrating "good management and high professionalism."

In addition to IAEA sponsored work, the Kola station agreed to participate in plant exchanges sponsored by the World Association of Nuclear Operators (WANO). Personnel were scheduled to visit the North Anna plant in Virginia in 1992.

Efforts at improving safety and operation. In 1987, Kola's management began efforts to improve operator training programs. They initiated new programs and created a department to select and train personnel. Programs were upgraded to include psychological and physical testing as well as working on a training simulator.

Plant equipment upgrades were begun in 1989. Improvements included modifications to the boron-injection system, annealation of welds in the reactor vessel to correct embrittlement problems, and renovations to the fire protection and suppression systems.

Although the IAEA recommended phasing out Kola's two older units, in February 1992 Russian officials announced their intention to continue operating the reactors. No deadlines have been set to complete safety backfits.

Sources

Source Book: Soviet-Designed Nuclear Power Plants in the Former Soviet Republics and Czechoslovakia, Hungary and Bulgaria. U.S. Council for Energy Awareness, 1992.

—Karen Bellenir

Kola Nuclear Plant, Units 5-6

184151 Telex: 126716 Salma
Murmanskaya obl
pos. Polyaniye Zori, Russia
Vladmir Schmidt, Plant Mgr.

Owning Utility

Russian Ministry of Atomic Energy (MINATOM) Phone: 095-230-24-20
24/26 B. Ordynka
Moscow, Russia
Victor Mikhailov, Minister

Contact

Vitaliy F. Konovalov, Deputy Minister
Ministry of Atomic Energy (MINATOM)

Basic Facts

Units 5-6 *Status:* Planned. *Type of Reactor:* Pressurized light-water reactor. *Megawatts (electric):* 440. *Megawatts (thermal):* 1375. *Generator Supplier:* Elektrosila.

Plant Report

See Plant Report for Kola Nuclear Plant, Units 1-4.

Krasnodar Nuclear Power Plant

Krasnodar, Russia

Owning Utility

Russian Ministry of Atomic Energy (MINATOM) Phone: 095-230-24-20
24/26 B. Ordynka
Moscow, Russia
Victor Mikhailov, Minister

Contact

Vitaliy F. Konovalov, Deputy Minister
Ministry of Atomic Energy (MINATOM)

Basic Facts

Status: Cancelled. *Type of Reactor:* Pressurized light-water reactor. *Megawatts (electric):* 1000. *Megawatts (thermal):* 3200.

Plant Report

Requested plant. In the middle 1980's, Krasnodar, a Russian town located near the Black Sea, expected an energy crisis. Officials in the region projected that by the year 2000, power supplies in the northern Caucasus Mountains would fall 8 million kilowatts short of demand. Faced with fears that additional coal-burning power plants would pollute the Kuban Valley, leaders approached the central Soviet government in Moscow and asked for a nuclear plant.

Then, an accident at the Chernobyl plant in 1986 changed the public's perception about the safety of nuclear power.

Cancelled plant. Following the Chernobyl accident, public opposition to nuclear power began to be felt throughout the former Soviet Union. The intensity of the controversy forced the Soviet government to halt construction at Krasnodar. Although the Council of Ministers officially declared that the plant was being cancelled because the region is an earthquake zone, many analysts cite the public outcry against nuclear power as the primary reason. The Soviets abandoned the plant after spending more than $40 million.

Sources

"Soviet Scraps a New Atomic Plant in Face of Protest Over Chernobyl." *The New York Times*, January 28, 1988.
"Ukrainian Scientists Protest Nuclear Plans." *Science*, March 4, 1988.

—Karen Bellenir

---

171 ●

Kursk Nuclear Power Station, Units 1-4

307239   Telex: 137185 Alfa
Kurskaya obl.
g. Kurchatov, Russia
Vladimir Gusarov, Plant Mgr.

Owning Utility

Russian Ministry of Atomic Energy   Phone: 095-230-24-20
  (MINATOM)
24/26 B. Ordynka
Moscow, Russia
Victor Mikhailov, Minister

Contact

Vitaliy F. Konovalov, Deputy Minister
Ministry of Atomic Energy (MINATOM)

Basic Facts

Units 1-4 *Status:* Operable. *Type of Reactor:* Light-water-cooled, graphite-moderated reactor. *Megawatts (electric):* 1000. *Megawatts (thermal):* 3200. *Generator Supplier:* Elektrosila.

Key Dates

Unit 1 *Ordered:* 1968. *Construction began:* 1972. *Criticality:* October 1976. *First Power:* December 1976. *Commercial Operation:* October 1977.
Unit 2 *Ordered:* 1968. *Construction began:* 1973. *Criticality:* December 1978. *First Power:* January 1979. *Commercial Operation:* August 1979.
Unit 3 *Ordered:* 1974. *Construction began:* 1978. *Criticality:* August 1983. *First Power:* October 1983. *Commercial Operation:* March 1984.

Unit 4 *Ordered:* 1974. *Construction began:* 1981. *Criticality:* October 1985. *First Power:* December 1985. *Commercial Operation:* February 1986.

Plant Report

Four operating units. The Kursk Nuclear Power Station, located in Kursk in the Russian Federation, has four operating RBMK reactors, similar in design to the reactor that exploded at Chernobyl in 1986. Combined, they give the station a total capacity of 3,700 megawatts.

A fifth unit was announced in the fall of 1991 by Soviet officials following regional parliamentary support for more nuclear capacity in the area. The proposed unit would have been the last RBMK styled-reactor to be constructed. Its status following the collapse of the central Soviet government is uncertain.

International cooperation. The Kursk station participated in a program sponsored by the World Association of Nuclear Operators (WANO) for East-West exchanges. Personnel visited the Susquehanna nuclear power plant in Pennsylvania to discuss matters such as management programs, safety, training, and operational issues.

In February 1992, Russian officials asked the International Atomic Energy Agency (IAEA) to conduct an ASSET (Assessment of Safety Significant Events Team) review at the Kursk site. The purpose of an ASSET study is to determine the root causes of safety-significant events and to recommend the steps necessary to implement corrections.

Sources

"Around the States." *Nuclear Industry*, Third Quarter 1991.
Source Book: Soviet-Designed Nuclear Power Plants in the Former Soviet Republics and Czechoslovakia, Hungary and Bulgaria. U.S. Council for Energy Awareness, April 1992.

—Karen Bellenir

---

172 ☐

Kursk Nuclear Power Station, Unit 5

307239   Telex: 137185 Alfa
Kurskaya obl.
g. Kurchatov, Russia
Vladimir Gusarov, Plant Mgr.

Owning Utility

Russian Ministry of Atomic Energy   Phone: 095-230-24-20
  (MINATOM)
24/26 B. Ordynka
Moscow, Russia
Victor Mikhailov, Minister

Contact

Vitaliy F. Konovalov, Deputy Minister
Ministry of Atomic Energy (MINATOM)

Novovoronezh

Basic Facts

Unit 5 *Status:* Planned. *Type of Reactor:* Light-water-cooled, graphite-moderated reactor. *Megawatts (electric):* 1000. *Megawatts (thermal):* 3200. *Generator Supplier:* Elektrosila.

Plant Report

See Plant Report for Kursk Nuclear Power Station, Units 1-4.

Novovoronezh Nuclear Power Station, Units 1-2

396072
Voronezhskaya obl
Novovoronezh, Russia
Vyatcheslav A. Vikin, Plant Mgr.

Phone: 073-56-16-08

Owning Utility

Russian Ministry of Atomic Energy (MINATOM)
24/26 B. Ordynka
Moscow, Russia
Victor Mikhailov, Minister

Phone: 095-230-24-20

Contact

Gennady A. Kulakov, Head of Technical Info. Office
Novovoronezh Nuclear Power Plant

Basic Facts

Units 1-2 *Status:* Shut down. *Type of Reactor:* Pressurized light-water reactor. *Megawatts (electric):* Unit 1, 210; Unit 2, 365. *Megawatts (thermal):* Unit 1, 760; Unit 2, 1320. *Generator Supplier:* Elektrosila.

Key Dates

Unit 1 *Ordered:* 1957. *Construction began:* 1957. *Criticality:* December 1963. *First Power:* September 1964. *Commercial Operation:* December 1964. *Shut Down:* February 1988.

Unit 2 *Ordered:* 1964. *Construction began:* 1964. *Criticality:* December 1969. *First Power:* December 1969. *Commercial Operation:* April 1970. *Shut Down:* August 1990.

Plant Report

Three of five units remain in operation. The Novovoronezh nuclear power station, located on the River Don in Voronezh, in the Russian Federation was named in honor of the fiftieth anniversary of the former Soviet Union. The facility has been a major participant in the development of Soviet pressurized water reactor technology. Experimentation and design work conducted at Novovoronezh span the three generations of VVER reactors. The station's five reactors offer a chronological depiction of the emerging nuclear power industry and have played an important role in operator training. For example, the plant's VVER-1000 simulator was used to train operators for the Kozloduy Nuclear Power Plant in Bulgaria.

The first two units at Novovoronezh were first-generation VVER's, a design which has been criticized for its lack of western-styled safety features such as emergency core-cooling systems and containment structures. Unit 1 was a VVER-210 prototype. It began operation in 1964 and was permanently shut down in 1988. The second unit, a VVER-365 prototype operated until 1990.

The remaining three units at Novovoronezh are still in service. Units 3 and 4 are VVER-440 reactors, a design which was first standardized with the V230 model. Unit 5 is a newer third-generation VVER-1000 reactor. Combined, the three operating units give the station a generating capacity of 1,720 megawatts.

Early successes at the station contributed to the Soviet motivation to expand nuclear production. A cost study conducted in 1975 showed that nuclear-produced electricity at Novovoronezh cost less per kilowatt hour than did electricity produced by modern fossil-fueled facilities in the western regions of the Soviet Union.

Public opposition following Chernobyl. Following the 1986 accident at Chernobyl, public opposition to nuclear power grew. Workers at the Novovoronezh facility wrote an open letter to Mikhail Gorbachov, then President of the Soviet Union, blaming biased media reporting for the negative feelings. The letter, printed in the newspaper *Sovetskaya Roissya*, claimed that rumors were being published rather than the truth about the nuclear industry.

International cooperation. The Novovoronezh station has participated in several post-Chernobyl international efforts aimed at improving reactor safety. A special group supervised by the U.S. Department of Energy and the Institute of Nuclear Power Operations (INPO) conducted a study at Units 3 and 4 to determine necessary changes in procedures, training, and management controls that Soviet regulators hoped would ultimately be applicable to all VVER reactors.

Two working groups established under the U.S./Soviet Joint Coordinating Committee on Civilian Nuclear Reactor Safety (JCCNRS) have visited Novovoronezh. Working group 3, assigned to study embrittlement and annealing (a technique that restores the strength of a reactor pressure-vessel), observed the annealing process at Unit 3 during February and March 1991. Working group 11 met three times at Novovoronezh to discuss topics related to improvements necessary at older plants. One goal was to study Novovoronezh as a model for implementing procedures developed for upgrading older Soviet-styled VVER reactors.

The International Atomic Energy Agency (IAEA) has also participated in studies at Novovoronezh. One IAEA assessment criticized the station's training facilities as inadequate, however an Assessment of Safety Significant Events Team (ASSET) review conducted in May 1991 stated that the "safety culture was found generally satisfactory" and that the station's VVER-440 units were operating at a capacity factor above the world average for pressurized water reactors. Another ASSET review is scheduled for 1993.

Improvements. Many improvements have been made at Novovoronezh's older units, including the correction of faulty welds in the shut-off valves and the annealing process. Many additional improvements were scheduled to be made before the

Soviet Union's central government dissolved. Scheduled modifications included the addition of a diagnostics system, replacement of boron-injection pumps and upgrading training and operating procedures. The status of planned upgrades following the Russian Federation's independence is uncertain.

Sources

"Nuclear Power Engineering in the Soviet Union." *The Bulletin*, January 1980.
"Soviet Blackout Looms as Nuclear Disquiet Deepens." *New Scientist*, August 11, 1990.
Mounfield, Peter. *World Nuclear Power*. Routledge, 1991.
Source Book: Soviet-Designed Nuclear Power Plants in the Former Soviet Republics and Czechoslovakia, Hungary and Bulgaria. U.S. Council for Energy Awareness, 1992.

—Karen Bellenir

---

174 •

Novovoronezh Nuclear Power Station, Units 3-5

396072   Phone: 073-56-16-08
Voronezhskaya obl
Novovoronezh, Russia
Vyatcheslav A. Vikin, Plant Mgr.

Owning Utility

Russian Ministry of Atomic Energy   Phone: 095-230-24-20
  (MINATOM)
24/26 B. Ordynka
Moscow, Russia
Victor Mikhailov, Minister

Contact

Gennady A. Kulakov, Head of Technical
  Info. Office
Novovoronezh Nuclear Power Plant

Basic Facts

Units 3-5 *Status:* Operable. *Type of Reactor:* Pressurized light-water reactor. *Megawatts (electric):* Units 3-4, 417; Unit 5, 1000. *Megawatts (thermal):* Units 3-4, 1375; Unit 5, 3000. *Generator Supplier:* Units 3-4, Elektrosila; Unit 5, Elektrotyazhmash.

Key Dates

Unit 3 Ordered: 1965. Construction began: 1967. Criticality: December 1971. First Power: December 1971. Commercial Operation: June 1972.

Unit 4 Ordered: 1965. Construction began: 1967. Criticality: December 1972. First Power: December 1972. Commercial Operation: March 1973.

Unit 5 Ordered: 1969. Construction began: 1974. Criticality: April 1980. First Power: May 1980. Commercial Operation: February 1981.

Plant Report

See Plant Report for Novovoronezh Nuclear Power Station, Units 1-2.

---

175 •

Obninsk Nuclear Power Plant, Units 1-2

Kaluga, Russia

Owning Utility

Russian Ministry of Atomic Energy   Phone: 095-230-24-20
  (MINATOM)
24/26 B. Ordynka
Moscow, Russia
Victor Mikhailov, Minister

Contact

Vitaliy F. Konovalov, Deputy Minister
Ministry of Atomic Energy (MINATOM)

Basic Facts

Units 1-2 *Status:* Operable. *Type of Reactor:* Unit 1 (APS), Light-water-cooled, graphite-moderated reactor; Unit 2 (BR5), Liquid metal fast breeder reactor. *Megawatts (electric):* Unit 1 (APS), 5; Unit 2 (BR5), 15. *Megawatts (thermal):* Unit 1 (APS), 30.

Key Dates

Unit 1 Ordered: 1951. Construction began: 1951. Criticality: 1954. First Power: June 1954. Commercial Operation: December 1954.

Unit 2 Ordered: 1954. Construction began: 1954. Commercial Operation: 1959.

Plant Report

Beginning of nuclear program. The world's first nuclear power plant, Obninsk APS Nuclear Power Plant, generated its first electricity on June 27, 1954 in Obninsk, a town southwest of Moscow. The station had a capacity of 5,000 kilowatts (5 megawatts). Obninsk served as the earliest Soviet nuclear research reactor facility.

In the late 1950's, Soviet researchers began experimental work on small fast breeder reactors at the site. The Obninsk BR5 Nuclear Power Plant became a prototype for further breeder technology.

Sources

"Nuclear Power Industry." *Soviet Life*, February 1986.
Mounfield, Peter. *World Nuclear Power*. Routledge, 1991.

—Karen Bellenir

---

176 ○

Rostov Nuclear Power Plant, Units 1-4

347340   Telex: 178472 Vulkan
Rostovskaya obl.
Volgodonsk-28, Russia
Eduard Mustafinov, Plant Mgr.

177 Smolensk

RUSSIA

Owning Utility

Russian Ministry of Atomic Energy Phone: 095-230-24-20
 (MINATOM)
24/26 B. Ordynka
Moscow, Russia
Victor Mikhailov, Minister

Contact

Valerji Matveev, Deputy Plant Mgr.
Rostov Nuclear Power Plant

Basic Facts

Units 1-4 *Status:* Under construction. Status as of January 1992: Unit 1, 95% complete; Unit 2, 30% complete; Unit 3, 5% complete; Unit 4, 1% complete. *Type of Reactor:* Pressurized light-water reactor. *Megawatts (electric):* 1000. *Megawatts (thermal):* 3000. *Reactor System Supplier:* Atommash. *Generator Supplier:* Elektrosila.

Key Dates

Unit 1 *Ordered:* 1978. *Construction began:* 1978. *Commercial Operation Expected:* Indefinite.

Unit 2 *Ordered:* 1980. *Construction began:* 1980. *Commercial Operation Expected:* Indefinite.

Units 3-4 *Ordered:* 1983. *Construction began:* 1983. *Commercial Operation Expected:* Indefinite.

177 ●

Smolensk Nuclear Power Plant, Units 1-3

216532 Telex: 781432 Atom
Smolenskaya obl.
Roslaviskiy r-n
pos. Desnogorsk, Russia
Evgeny Safrikin, Plant Mgr.

Owning Utility

Russian Ministry of Atomic Energy Phone: 095-230-24-20
 (MINATOM)
24/26 B. Ordynka
Moscow, Russia
Victor Mikhailov, Minister

Contact

Vitaliy F. Konovalov, Deputy Minister
Ministry of Atomic Energy (MINATOM)

Basic Facts

Units 1-3 *Status:* Operable. *Type of Reactor:* Light-water-cooled, graphite-moderated reactor. *Megawatts (electric):* 1000. *Megawatts (thermal):* 3200.

Key Dates

Unit 1 *Ordered:* 1971. *Construction began:* 1975. *Criticality:* September 1982. *First Power:* September 1983. *Commercial Operation:* September 1983.

Unit 2 *Ordered:* 1971. *Construction began:* 1976. *Criticality:* April 1985. *First Power:* May 1985. *Commercial Operation:* July 1985.

Unit 3 *Ordered:* 1981. *Construction began:* 1984. *Criticality:* November 1989. *First Power:* January 1990. *Commercial Operation:* January 1990.

Plant Report

Four unit station. The Smolensk Nuclear Power Plant is a four unit generating station in Desnogorsk, Smolensk in the Russian Federation. Three units are operational. The fourth unit was expected to come online in 1991; however, it isn't known whether the unit officially opened. The four reactors are all 1,000-megawatt RBMK units, the same style as the unit that exploded at Chernobyl.

International agreement. Operators at Smolensk entered into an agreement with Scottish Nuclear to participate in a plant personnel exchange. Staff at the Torness Power Station will share operating experiences with their counterparts at Smolensk over a six-month period

Sources

"Scottish Nuclear." *Nuclear Industry*, First Quarter 1992.
Source Book: Soviet-Designed Nuclear Power Plants in the Former Soviet Republics and Czechoslovakia, Hungary and Bulgaria. U.S. Council for Energy Awareness, April 1992.

—Karen Bellenir

178 ○

Smolensk Nuclear Power Plant, Unit 4

216532 Telex: 781432 Atom
Smolenskaya obl.
Roslaviskiy r-n
pos. Desnogorsk, Russia
Evgeny Safrikin, Plant Mgr.

Owning Utility

Russian Ministry of Atomic Energy Phone: 095-230-24-20
 (MINATOM)
24/26 B. Ordynka
Moscow, Russia
Victor Mikhailov, Minister

Contact

Vitaliy F. Konovalov, Deputy Minister
Ministry of Atomic Energy (MINATOM)

Basic Facts

Unit 4 *Status:* Under construction. *Type of Reactor:* Light-water-cooled, graphite-moderated reactor. *Megawatts (electric):* 1000. *Megawatts (thermal):* 3200.

Key Dates

Unit 4 *Ordered:* 1982. *Construction began:* 1984. *Commercial Operation Expected:* 1991.

Plant Report

See Plant Report for Smolensk Nuclear Power Plant, Units 1-3.

Plant Profiles

179 ●

Sosnovy Bor Nuclear Power Plant, Units 1-4

188537
Leningradskaya obl
g. Sosnoviy Bor, Russia
Anatoli Yeperin, Plant Mgr.

Telex: 121535 CURIE

Owning Utility

Russian Ministry of Atomic Energy (MINATOM)
24/26 B. Ordynka
Moscow, Russia
Victor Mikhailov, Minister

Phone: 095-230-24-20

Contact

Vitaliy F. Konovalov, Deputy Minister
Ministry of Atomic Energy (MINATOM)

Basic Facts

Units 1-4 *Status:* Operable. *Type of Reactor:* Light-water-cooled, graphite-moderated reactor. *Megawatts (electric):* 1000. *Megawatts (thermal):* 3200. *Generator Supplier:* Electrosila. *Architect Engineer:* Atomenergyoprojekt.

Key Dates

Unit 1 Ordered: 1968. Construction began: 1970. Criticality: September 1973. First Power: December 1973. Commercial Operation: November 1974.

Unit 2 Ordered: 1968. Construction began: 1970. Criticality: May 1975. First Power: July 1975. Commercial Operation: February 1976.

Unit 3 Ordered: 1973. Construction began: 1973. Criticality: September 1979. First Power: December 1979. Commercial Operation: June 1980.

Unit 4 Ordered: 1975. Construction began: 1975. Criticality: December 1980. First Power: February 1981. Commercial Operation: August 1981.

Plant Report

Historic generating station. The Sosnovy Bor Nuclear Power Plant (formerly known as the Leningrad Nuclear Power Plant) is a four-unit facility located near St. Petersburg on the Baltic Sea in the Russian Federation. Its two oldest reactors were the first commercial RBMK units to be operated by the Soviets.

Plans for improved operator training. In the fall of 1991, officials with the former Soviet Union, entered into a contract with General Physics International Engineering and Simulation, a company in the United States, to design a simulator for the RBMK-styled reactors at Sosnovy Bor. The $13 million project will produce the first western-styled simulator for an RBMK plant and will help improve operator training. Completion is expected to take three and a half years.

International criticism. The Sosnovy Bor station has been an object of international criticism. A former official with Sweden's Nuclear Power Inspectorate demanded that the two oldest units be shut down. He claimed their quality was inferior to the more modern units. Other European countries have voiced apprehension about the plant's RBMK reactor design.

The Sosnovy Bor plant has also been charged with inadequate provisions for nuclear waste storage. Critics claim that storage buildings have cracks and holes and that radiation surrounding them measures up to 380 times above normal.

Radioactive leak makes world headlines. Sosnovy Bor made international headlines on March 24, 1992, when a serious pipe break in a reactor at the plant suddenly leaked radioactive gas. Although the incident did not cause any harm, it prompted new concern over dangerous Soviet-designed nuclear reactors.

After a sudden loss of pressure in one of the reactor's cooling channels, radioactive steam poured into the machine room, and some of it was vented into the atmosphere. Although personnel were not evacuated, the reactor was shut down by the plant's emergency system. Russian officials insisted the incident posed no health danger to humans. Western nuclear experts claimed that the leak at Sosnovy Bor put more radioactive materials into the atmosphere than were released during the nuclear accident at Three Mile Island in the United States.

The plant has the same kind of reactor as Chernobyl. The March 1992 leakage registered three points on the seven-point scale of the International Atomic Energy Agency. The 1979 Three-Mile Island accident in the United States was a level six; the 1986 Chernobyl disaster was a level seven.

German, Finnish, and U.S. responses. In the former Soviet Union and Eastern Europe, there are a total of 16 Chernobyl-type reactors still in use, including one in Bulgaria, which is regarded as the most dangerous power plant in the world. The March 1992 incident, although not that serious, set off alarm bells all across Europe. The German government issued a command that all Chernobyl-type reactors be closed. In nearby Helsinki, Finnish officials expressed great concern after detecting minute increases of radiation levels at high altitudes.

In the United States, nuclear experts urged the Bush administration to focus not only on nuclear weapons, but to pay more attention to the fate of nuclear power plants in the former East Bloc nations. Experts fear that many Chernobyls are waiting to happen due to the fact that Soviet-built plants are aging, the maintenance is poor, the spare parts are limited, and the staff morale and training is very low.

Healthy developments in Russian nuclear policy. The quick release of information about the radiation leakage in March 1992 was a sign of a more open and responsible attitude among nuclear operators in the former Soviet Union. Unlike the Chernobyl accident, which was first announced by Sweden, this incident was made public to the people of the city and to the world by the government of Russia.

Although officials in the Russian Federation were praised for their prompt release of information during the accident, the nation also received criticism for failing to act promptly in efforts to upgrade or shut down its old reactors.

Sources

"Nuclear Scare in Russia." ABC World News Tonight, March 24, 1992.

"Experts Fault Efforts to Shut Down Old Reactors." *Detroit News*, March 25, 1992.

Source Book: Soviet-Designed Nuclear Power Plants in the Former Soviet Republics and Czechoslovakia, Hungary and Bulgaria. U.S. Council for Energy Awareness, April 1992.

"Former Soviet Union Awakens to a Nuclear Waste Nightmare." *Detroit News*, July 19, 1992.

—Karen Bellenir

Slovenia

COUNTRY REPORT

Political and operational uncertainty

The newly formed political entity of Slovenia acquired a single nuclear power plant when the central Yugoslavian government collapsed. The former Yugoslavia entered the ranks of nations with nuclear power capabilities when the Krsko power plant began commercial production in 1983. The Krsko plant, located in Slovenia on the border of Croatia and Serbia, has a single 660-megawatt pressurized water reactor. Before the splintering of Yugoslavia, Krsko's production accounted for about 20 percent of Slovenia's electrical generation but only about three percent of Yugoslavia's total generating capacity. Due to the unresolved political situation in the former nation of Yugoslavia, available information about Krsko's status and plans for further nuclear efforts may not be current.

In 1985, Yugoslavian officials had planned to construct four additional 1,000 megawattt reactors at other locations throughout the country. Nuclear power proponents favored increased nuclear generation to help Yugoslavia achieve energy independence from the former Soviet Union.

Critics charged that the country could not afford to incur more debt by importing nuclear technology from western nations and that domestic natural gas and hydroelectric sources had not been fully utilized. In response to one of the planned plants, Professor Dragica Ivanovic, President of the Technical University of Belgrade, told the Central Committee of the Federation of Yugoslav Communists: "We cannot allow Yugoslavia's head to be placed in a dangerous noose by the construction of the nuclear reactor at Prevlaka."

The Chernobyl accident in the former Soviet Union also impacted public opinion and increased fear of accidents. Prior to Chernobyl, Yugoslavia's population only mildly opposed nuclear construction. Following the accident, three-quarters of the nation's citizens were against it. In July 1986, Yugoslavian officials set aside plans for further nuclear construction, but Krsko continued to operate.

Operational difficulties

The Krsko power station achieved relatively good ratings in a few areas. For example, statistics concerning the plant's "Equivalent Availability Factor," and "Collective Radiation Exposure" show that INPO goals were surpassed during most years of operation. Operational difficulties, however, have been encountered in other areas. The Krsko station has been plagued by a high number of unplanned automatic scrams, unplanned actuations of the plant's safety system, and forced outages.

In May 1990, officials of the former Yugoslavian government announced a decision to close the plant by 1995. The shut down was welcomed by Slovenia's environmentalists, who had been campaigning for Krsko's closing, and by neighboring Austria, whose government repeatedly expressed concern about the plant's safety.

Sources

"Eastern Europe's Nuclear Power Plans." *World Press Review*, December 1986.

"Yugoslavian Shutdown." *The Wall Street Journal*, May 24, 1990.

Nuclear Power Plant Krsöko 1990 Performance Indicators. Reactor Engineering Department, NEK, n.d.

—*Karen Bellenir*

PLANT PROFILES

180 •
Krsko Nuclear Power Plant

68270 KRSKO
Vrbina 12, Slovenia

Phone: 38-608-21-621
Fax: 38-608-21-528
Telex: 35748 YU NUELKR

Owning Utility

Savske Elektrarne Ljubljana (Slovenia)
   and Elektroprivreda Zagreb (Croatia)
68270 KRSKO
Vrbina 12, Slovenia
Janez Krajnc, Operations Supt.

Phone: 38-608-21-621
Fax: 38-608-21-528
Telex: 35748 YU NUELKR

Contact

Ivan Spiler, Administration Mgr.
Savske Elektrarne Ljubljana (Slovenia)
   and Elektroprivreda Zagreb (Croatia)

Basic Facts

Status: Operable. *Type of Reactor:* Pressurized light-water reactor. *Megawatts (electric):* 664. *Megawatts (thermal):* 1876. *Reactor System Supplier:* Westinghouse Electric Corp. (United States). *Generator Supplier:* Westinghouse Electric Corp. (United States). *Architect Engineer:* Gilbert.

Key Dates

Ordered: 1973. *Construction began:* 1974. *Criticality:* 1981. *First Power:* October 1981. *Commercial Operation:* January 1983. *Shut Down Expected:* 1995.

Spain

COUNTRY REPORT

Background

The nine operating nuclear power plants in Spain produce about 37.5 percent of the country's electricity. In 1991, the plants boasted an overall capacity factor of 86.1 percent.

Nuclear energy is one of Spain's least costly sources of electricity, according to some reports from the Spanish government. In 1990, nuclear power cost 7.4 pesetas per kilowatt-hour (almost 7.6 cents), compared with 20.49 pesetas (20 cents) for oil and natural gas. Only electricity made from imported coal and water cost less than nuclear energy—7.22 and 5.99 pesatas, respectively (7.1 and 5.9 cents).

When in 1989, severe shortages of water cut Spain's hydroelectricity production to less than half the normal level, output from Spain's nuclear plants met Spain's rapidly expanding demand for electricity.

Spain's nuclear protest movement

In the early 1970s, when Spain's first nuclear plants were coming online, no national movement had yet emerged to challenge nuclear power. As a result, most Spanish utilities had large liberties when choosing sites for nuclear development. Critics argue that Spanish electric companies took advantage of the neutral climate by building reactors in poverty stricken and depopulated areas—regions which were unlikely to offer much local resistance.

However, construction of the Lemoniz plant, near Bilbao, did not go forward without problems. Here, Basque separatists vowed to keep the plant from functioning and through terrorist activities in the early 1980s successfully delayed the opening of the plant.

By 1990, anti-nuclear sentiment seemed to be gaining ground. A fire in October 1989 at the Vandellos 1 Nuclear Center in northeastern Spain sparked a storm of public protest and raised many questions about nuclear safety within the government. Vandellos I was permanently closed in May 1990, and the fate of new construction projects was threatened.

New energy plan threatens nuclear development

Late in 1991, the Spanish government adopted a national energy plan which for the first time included targets for cutting pollution. The plan, which will govern policy until the year 2000, was also expected to upset the nuclear industry. By 2000, the government wants the proportion of the country's energy generated by nuclear plants to be reduced. The plant calls for $8.7 billion to be spent over 10 years on non-nuclear power, including gas-fired plants, biomass, and solar generators.

Sources

"Spanish Nuclear Shutdown." *The Wall Street Journal,* May 31, 1990.

Mounfield, Peter. *World Nuclear Power,* Routledge, 1991.

"Spain Gives Itself Elbow Room on Energy Targets." *New Scientist,* September 14, 1991.

"Ole! Nuclear Energy Saves Pesatas." *Nuclear Industry,* First Quarter 1992.

"A Good Year for Nuclear Energy Abroad." *INFO,* U.S. Council for Energy Awareness, April 1992.

PLANT PROFILES

181 •
Almaraz Nuclear Center, Units 1-2

Apartado, 74   Phone: 927-531-250
Navalmoral de la Mat   Telex: 28928
10080 Caceres, Spain
Ignacio Araluce, Plant Mgr.

Owning Utility

Central Nuclear de Almaraz   Phone: 431-42-22
Claudio Coello, 123   Fax: 435-73-10
28006 Madrid, Spain   Telex: 23923 CNAM
Jose Mendez-Villamil, Pub. Info.

Contact

Antonio Calderon, Mgr., Almaraz Info.
   Center
Almaraz Nuclear Power Plant

Basic Facts

Units 1-2 *Status:* Operable. *Type of Reactor:* Pressurized light-water reactor. *Megawatts (electric):* 930. *Megawatts (thermal):* 2686. *Reactor System Supplier:* Westinghouse Electric Corp. (United States). *Generator Supplier:* Westinghouse Electric Corp. (United States). *Architect Engineer:* Empresarios Agrupados (Spain); Gibbs & Hill Espanola SA (Spain).

Key Dates

Unit 1 *Ordered:* 1971. *Construction began:* 1973. *Criticality:* 1981. *First Power:* May 1981. *Commercial Operation:* October 1981.

Unit 2 *Ordered:* 1972. *Construction began:* 1973. *Criticality:* 1983. *First Power:* October 1983. *Commercial Operation:* February 1984.

Plant Report

Background. The Almaraz Nuclear Center is situated within the province of Caceres, just 16.4 kilometers away from Navalmoral de la Mata. The plant is cooled by waters from the Arrocampo River, which flows into the Tajo River.

Almaraz's two pressurized water reactors, which came online between 1981 and 1984, are part of a series of six standardized units ordered from Westinghouse in a coordinated program by Spanish utilities. (The first two in the series—at Lemoniz—have never been put into service because of Basque separatist terrorist activities; the other two are in operation at Asco.)

Steam generator replacement planned at Asco. The six steam generators at the two reactors at Almaraz could be replaced by the mid 1990s. Central Nuclear de Almaraz has approached the Spanish safety authority, Consejo de Seguridad Nuclear (CSN), for the necessary licensing approvals for a program of replacements. Bids were expected from international contractors for the replacement work. The order for the steam generator replacements may be coordinated with the operators of Almaraz's sister plant at Asco.

Like other plants of this kind, the Asco and Almaraz PWRs have required a lot of remedial work to limit tube degradation in the steam generators. Although neither of the plants is close to using up its tube plugging margin, the decision has been made to replace the steam generators in order to reduce the cost in outage time and radiation doses made during inspection and maintenance work.

Sources

Hidroelectrica Espanola: Centrales Nucleares. Informacion Nuclear, May 1985.
"Steam Generator Replacement Planned for Asco, Almaraz." *Nuclear News,* Feburary 1991.

182 •
Asco Nuclear Center, Units 1-2

C.N. Asco   Phone: 77-405000
Asco (Tarragona), Spain   Fax: 77-405181
Luis Coll, Plant Mgr.   Telex: 56751 feas e

Owning Utility

Asociacion Nuclear Asco   Phone: 3-204-04-15
Tres Torres   Fax: 3-204-04-21
08017 Barcelona, Spain   Telex: 51775 fedp e
Juan Alguero, Gen.Mgr.

Contact

Antonio Fernandez Savin, Admin.,
   Pub.Info.
Asco Nuclear Power Plant

Basic Facts

Units 1-2 *Status:* Operable. *Type of Reactor:* Pressurized light-water reactor. *Megawatts (electric):* 930. *Megawatts (thermal):* 2696. *Reactor System Supplier:* Westinghouse Electric Corp. (United States). *Generator Supplier:* Westinghouse Electric Corp. (United States); Empresa Nacional Bazan (Spain). *Architect Engineer:* Bechtel; Empresa Nacional de Ingenieria y Technologia SA (Spain); Informas y Projectas SA (Spain).

Key Dates

Unit 1 *Ordered:* 1972. *Construction began:* 1974. *Criticality:* June 1983. *First Power:* August 1983. *Commercial Operation:* September 1983.

Unit 2 *Ordered:* 1973. *Construction began:* 1975. *Criticality:* September 1985. *First Power:* October 1985. *Commercial Operation:* March 1986.

Plant Report

Background. In July of 1973, the companies of Fuernas Electricas de Cataluna (FECSAO, ENHER), Hydroelectrics of Cataluna (HECSA), and the Hydroelectric Force of Segre (FHSSA) made up the Asociasion Nuclear Asco to own and operate the Asco Nuclear Power Plant. Percentages of participation are as follows: FECSA-FHSSA, 45 percent; ENHER, 40 percent; and HECSA, 15 percent.

Asco's two pressurized water reactors, which came online between 1983 and 1986, are part of a series of six standardized units ordered from Westinghouse in a coordinated program by

Spanish utilities. (The first two in the series—at Lemoniz—have never been put into service because of Basque separatist terrorist activities; the other two are in operation at Almaraz.)

Site selection. Although antinuclear sentiment seemed to be gaining ground in the 1990s, no national movement had emerged in Spain to challenge nuclear power at the time Asco was constructed. Consequently, utilities had a wide open choice of sites when building a nuclear power plant in Spain.

Critics argue that Spanish electric companies took advantage of the neutral climate by building reactors in poverty stricken and depoopulated areas—regions which were unlikely to offer much local resistance. A prime example is the Asco nuclear power plant which was built in the bleak countryside along the lower Ebro River, less than 1.5 miles from the town center and only a half a mile above the point from where much of the town's drinking water is drawn.

Steam generator replacement planned at Asco. The six steam generators at the two reactors at Asco could be replaced by the mid 1990s. Asociacion Nulcear Asco has approached the Spanish safety authority, Consejo de Seguridad Nuclear (CSN), for the necessary licensing approvals for a program of replacements. Bids were expected from international contractors for the replacement work. The order for the steam generator replacements may be coordinated with the operators of Asco's sister plant at Almaraz.

Like other plants of this kind, the Asco and Almaraz PWRs have required a lot of remedial work to limit tube degradation in the steam generators. Although neither of the plants is close to using up its tube plugging margin, the decision has been made to replace the steam generators in order to reduce the cost in outage time and radiation doses made during inspection and maintenance work.

Sources

Hidroelectrica Espanola: Centrales Nucleares. Informacion Nuclear, May 1985.
"Steam Generator Replacement Planned for Asco, Almaraz." Nuclear News, Feburary 1991.
Mounfield, Peter. World Nuclear Power, Routledge, 1991.

183 ●

Cofrentes Nuclear Power Plant

46625 Cofrentes (Valencia), Spain
Mariano Garcia, Plant Mgr.

Phone: 96-219-62-62
Fax: 96-384-50-88
Telex: 64402

Owning Utility

Iberdrola II, S.A.
Hermosilla, 3
28001 Madrid, Spain
Javier Pinedo, Generation Area, Dir.

Phone: 92-577-65-00
Fax: 92-576-67-62
Telex: 23786

Contact

Jesus Cruz Heras, Mgr., Commun. & Info.
Cofrentes Nuclear Power Plant

Basic Facts

Status: Operable. *Type of Reactor:* Boiling water reactor. *Megawatts (electric):* 990. *Megawatts (thermal):* 2952. *Reactor System Supplier:* General Electric Co. (United States). *Generator Supplier:* General Electric Co. (United States). *Architect Engineer:* Empresarios Agrupados (Spain); Sener; Gibbs & Hill, Inc. (United States).

Key Dates

Ordered: 1973. *Construction began:* 1975. *Criticality:* August 1984. *First Power:* October 1984. *Commercial Operation:* March 1985.

Plant Report

Background. Cofrentes Nuclear Power Plant was formerly run by the utility Hidroelectrical Espanola, S.A. In 1992, the company merged with Iberduero, S.A., the utility that operated the Lemoniz Nuclear Power Plant, and the two companies changed their names to Iberdrola II, S.A. and Iberdrola I, S.A., respectively.

The plant is located in the heart of the village of Cofrentes in the Ayora valley of the Spanish province of Valencia. The area is popular with tourists who come to view the large number of Mediterranean works of art found in the region's caves and shelters.

Rainfall causes radioactive leak. On October 19, 1982, a heavy downpour of rain began in the valley of Ayora and continued for 33 hours straight. A total of 11 meters of rain fell, flooding much of the region, including the Cofrentes plant. Flooding inside the plant triggered a leakage of 580 liters of radioactive waste.

Sources

Informacion Nulcear. Central Nuclear de Cofrentes, May 1983.
Valle de Ayora Cofrentes. Hidroelectrtica Espanola, S.A., 1991.

184 ●

Jose Cabrera Nuclear Power Plant

Central Nuclear Jose Cabrera
Almonacid de Zorita
19119 Guadalajara, Spain
Juan Vicente Llinares, Plant Mgr.

Phone: 521-28-74
Fax: 521-28-71
Telex: 23921

Owning Utility

Union Electrica FENOSA, S.A.
Capitan Haya, 53
28020 Madrid, Spain
Pablo Blanc, Subdirector, Nuclear

Phone: 571-37-00
Fax: 570-43-49
Telex: 27412 unel e

Contact

Ramon Barro, Dir., Pub.Info.
Union Electrica FENOSA, S.A.

Basic Facts

Status: Operable. *Type of Reactor:* Pressurized light-water reactor. *Megawatts (electric):* 160. *Megawatts (thermal):* 510. *Reactor System Supplier:* Westinghouse Electric Co. (United States). *Generator Sup-*

plier: Westinghouse Electric Co. (United States). *Architect Engineer:* Gibbs & Hill, Inc. (United States).

Key Dates
Ordered: 1965. *Construction began:* 1965. *Criticality:* 1968. *First Power:* July 1968. *Commercial Operation:* February 1969.

185 ○ Lemoniz Nuclear Power Plant, Units 1-2
Arminza, Spain

Owning Utility
Iberdrola I, S.A.   Fax: 431-8701
Serrano, 26
28001 Madrid, Spain
Jose Antonio Garrido, Gen. Dir.

Basic Facts
Units 1-2 *Status:* Under construction (Unit 1, 97% complete as of January 1992; Unit 2, 57% complete as of January 1992). *Type of Reactor:* Pressurized light-water reactor. *Megawatts (electric):* 930. *Megawatts (thermal):* 2696. *Reactor System Supplier:* Westinghouse Electric Co. (United States). *Generator Supplier:* Westinghouse Electric Co. (United States). *Architect Engineer:* Iberdrola I, S.A; Bechtel Corp. (United States); Sener; Empresa Nacional de Ingeneria y Technologia SA (Spain).

Key Dates
Units 1-2 *Ordered:* 1971. *Construction began:* 1974. *Commercial Operation Expected:* Indefinite.

Plant Report
Ownership. Lemoniz Nuclear Power Plant was formerly run by the utility Iberduero, S.A. In 1992, the company merged with Hidroelectrical Espanola, S.A., the utility that operated the Cofrentes Nuclear Power Plant, and the two companies changed their names to Iberdrola I, S.A. and Iberdrola II, S.A., respectively.

Terrorist activities keep plant closed. The two pressurized water reactors under construction at Lemoniz, near Bilbao, are part of a series of six standardized units ordered from Westinghouse in the early 1970s in a coordinated program by Spanish utilities. Although the other four reactors at Almaraz and Asco came online between 1981 and 1986, the first two in the series—at Lemoniz—have never been put into service because of terrorist activities.

In February 1981, the chief engineer at the Lemoniz plant, Jose Mari Ryan, was assassinated by ETA, the militant Basque separatist organization. ETA, which had vowed to keep the plant from functioning, also killed the plant's director, Angel Pascual, in May of 1982.

Although Lemoniz 1 was 97 percent complete as of January 1992, it is not known when the plant unit will go online.

Sources
"Steam Generator Replacement Planned for Asco, Almaraz." *Nuclear News,* Feburary 1991.
Mounfield, Peter. *World Nuclear Power,* Routledge, 1991.

186 ● Santa Maria de Garona Nuclear Center

Barcina del Barco   Phone: 947-35-70-00
09212 Burgos, Spain   Fax: 947-35-72-00
Francisco Mier, Production Dir.   Telex: 39471-N-SMG-3

Owning Utility
Centrales Nucleares del Norte, S.A.   Phone: 942-22-58-00
 (NUCLENOR)   Fax: 942-31-10-71
Hernan Cortes, 26   Telex: 35640
39003 Santander, Spain
Jose L. Antonanzas, Chairman

Contact
Antonio Cornado Quibus, Pub.Info.   Phone: 947-35-71-94
Santa Maria de Garona Nuclear Power   Fax: 947-35-72-55
 Station

Basic Facts
Status: Operable. *Type of Reactor:* Boiling water reactor. *Megawatts (electric):* 460. *Megawatts (thermal):* 1380. *Reactor System Supplier:* General Electric Co. (United States). *Generator Supplier:* General Electric Co. (United States). *Architect Engineer:* Ebasco.

Key Dates
Ordered: March 1, 1965. *Construction began:* September 1966. *Criticality:* November 15, 1970. *First Power:* March 2, 1971. *Commercial Operation:* May 1971.

Plant Report
Plant origins. In 1955 the demand for electricity in Spain was increasing by approximately 10 percent per year. Although the country began mobilizing to increase the usage of coal and hydroelectric energy sources, two organizations—Iberduero, S.A. and Electra de Viesgo—felt these energy reserves would be insufficient to meet the national demand and began exploring nuclear energy options. The two companies formally associated on March 2, 1957, to create the private stock company Centrales Nucleares del Norte or NUCLENOR with a share capital of five million pesetas.

Being the first Spanish utility to embark on the construction of a nuclear power plant, NUCLENOR began contacting many international organizations and foreign engineering firms and utilities to build a proper foundation of scientific and technical knowledge. Then, the utility searched for a site. Santa Maria de Garona was chosen for the following reasons:

- closeness to large grids and substations
- proximity to the Ebro river to be used in plant cooling
- good meteorological, geological, and seismic conditions
- easy access for construction machinery and large power station components

The plant is comprised of two oblong buildings, one containing the reactor and the other housing the turbogenerator. Over 100,000 cubic meters of concrete, 12,000 tons of steel, and 600,000 meters of cable were used in constructing the power station. The effort required a labor force of over 3,000.

Successful operations. In 1978 and 1981, Santa Maria de Garona obtained excellent capacity factors, placing it among the top five boiling water reactors designed by General Electric Co. in the world. From August 1982 to May 1983, the plant was coupled to the national grid for 277 consecutive days—a world record at that time among reactors of the same type.

In 1992, Santa Maria de Garona was rated second in the world for overall nuclear production next to the Susquhanna Nuclear Power Plant.

For the most part, Santa Maria de Garona has enjoyed a favorable reputation in the eyes of the public due to a long history of normal operations.

Dealing with radiation and the media. Solid wastes generated by the plant are deposited in a transit store after being subjected to a canning process. Liquid wastes pass through a number of treatments to remove radioactive particles before returning to the reactor coolant. If the degree of radioactivity is sufficiently low, it is at times discharged into the Ebro river, under limitations imposed by the Consejo de Sequridad Nuclear.

Radioactive gases, such as Krypton and Xenon which are given off by the reactor core in small quantities, pass through a long pipe where they lose a part of their radioactivity. Later they are passed through several beds of active charcoal, which reduces the radioactivity of the gases even further, before begin taken into the atmosphere through a chimney 100 meters high. Local and national media have protested that these gaseous releases contain dangerous levels of radioactivity. Officials at Santa Maria de Garona have denied the claims.

Radiation levels both inside and outside the power station are monitored continuously. Santa Maria de Garona's radiation monitoring system includes three stations on site, 15 secondary stations located in various villages within a 30 kilometer radius of the plant, and two 100-meter meteorological stations. Each year, an environmental laboratory some 30 kilometers from the Power Station verifies the radiological impact on the surrounding ecosystem by studying vegetable, animal, and river mud samples.

Public is welcome. Group tours of Santa Maria de Garona are available upon request. Over 180 groups visited the nuclear installation in 1990. Thousands of area residents also toured the site.

Sources

Boletin Informativo, Numero 18. NUCLENOR, n.d.
NUCLENOR Corporate Brochure. NUCLENOR, 1987.
Boletin Informativo, Numero 23. NUCLENOR, February 1992.

187 ●

Trillo Nuclear Power Plant

Trillo (Guadalajara), Spain
Victor Sola Gutierrez, Plant Mgr.

Owning Utility

Central de Trillo
Rosario Pino, 14-16
28020 Madrid, Spain

Phone: 572-04-47
Fax: 571-19-09
Telex: 46220 NUTR E

Contact

Jose Luis Sanchez Miro, Mgmt.Dir. & Financial Dir.
Trillo Nuclear Power Plant

Basic Facts

Status: Operable. *Type of Reactor:* Pressurized light-water reactor. *Megawatts (electric):* 1040. *Megawatts (thermal):* 2027. *Reactor System Supplier:* Kraftwerk Union AG (Germany); Equipos Nucleares SA (Spain). *Generator Supplier:* Kraftwerk Union AG (Germany); Empresa Nacional Bazan (Spain). *Architect Engineer:* Empresarios Agrupados (Spain). *Owner:* Union Electrica FENOSA, S.A; Iberdrola I, S.A; Hidroelectrica del Cantabrico. *Operator:* Central de Trillo.

Key Dates

Ordered: 1975. *Construction began:* 1979. *Criticality:* May 1988. *First Power:* May 1988. *Commercial Operation:* August 1988.

188 ○

Valdecaballeros Nuclear Center, Units 1-2

Baldajoz, Spain
Jose M. Arcos, Startup and Operations Mgr.

Phone: 924-643211

Owning Utility

Central Nuclear de Valdecalballeros
Goya 4
28001 Madrid, Spain
Rafael Olavarria, Project Coord.

Phone: 1-4318617
Fax: 1-4318617 x 262
Telex: 45750

Basic Facts

Units 1-2 *Status:* Under construction (Unit 1, 71 percent complete as of January 1992; Unit 58 percent complete as of January 1992). *Type of Reactor:* Boiling water reactor. *Megawatts (electric):* 975. *Megawatts (thermal):* 2894. *Reactor System Supplier:* General Electric Co. (United States). *Generator Supplier:* General Electric Co. (United States). *Architect Engineer:* Empresarios Agrupados (Spain). *Owner:* Compania Sevillana de Electricidad S.A; Iberdrola II. S.A.

Key Dates

Unit 1 *Ordered:* 1974. *Construction began:* 1980. *Commercial Operation Expected:* Indefinite.

Unit 2 *Ordered:* 1975. *Construction began:* 1980. *Commercial Operation Expected:* Indefinite.

Plant Report

Plant in jeopardy. A fire in October 1989 at the Vandellos 1 Nuclear Power Plant in northeastern Spain sparked a storm of protest in the country and questions about nuclear safety. When the government unexpectedly decided to permanently close Vandellos, the fate of the Valdecaballeros Nuclear Power Plant, under construction in central Spain, also came into question.

Jose Porta, an analyst with BSN S.A. in Madrid, said that "permanently closing Vandellos clearly sends a signal that the antinuclear position is gaining ground in the government." Porta also predicted that the Valdercaballeros would never see a startup as a result of the antinuclear sentiment.

Sources

"Spanish Nuclear Shutdown." *The Wall Street Journal*, May 31, 1990.

189 ★
Vandellos I Nuclear Center

Carretera Nacional, 340, km. 1.123
43890 Hospitalet de l'Infant, Tarragona, Spain
Carlos Fernandez Palomero, Plant Mgr.

Phone: 977-82-30-50
Fax: 977-82-00-75
Telex: 56430 Vanen E

Owning Utility

Hispano-Francesa de Energia Nuclear, S.A. (HIFRENSA)
Tuset, 20, 4a planta
08006 Barcelona, Spain
Fernando Roset Cunill, Dir.

Phone: 93-217-92-00
Fax: 93-217-55-24
Telex: 52205 Baren E

Basic Facts

Status: Shut down. *Type of Reactor:* Gas-cooled reactor. *Megawatts (electric):* 496. *Megawatts (thermal):* 1750. *Reactor System Supplier:* Societe des Forges et Ateliers du Creusot (Usines Schneider) (France). *Generator Supplier:* Alsthom; Jeumont-Schneider (France). *Architect Engineer:* Societe pour l'Industrie Atomique (France).

Key Dates

Ordered: 1966. *Construction began:* 1967. *Criticality:* 1972. *First Power:* May 1972. *Commercial Operation:* August 1972. *Shut Down:* May 1990.

Plant Report

Fire prompts government to close plant. Vandellos 1, one of Spain's first nuclear power plants, operated from August 1972 until October 1989, when it was shut down after a fire. When the gas-cooled reactor, located in northeastern Spain, was closed permanently in May 1990, it threw a monkey wrench into the future of Spain's nuclear energy program. Spain's energy minister, Claudio Aranzadi, told the Spanish parliament that repair and modernization of the plant would be far too expensive to rationalize to the consumer. The cost of restarting the power station was estimated at 33 billion pesetas ($315 million).

Even though Spain's Nuclear Safety Commission determined that no radiation had leaked from the plant during the fire, the accident sparked a storm of protest and raised many questions about the plant's safety. The government's decision to permanently close the power station was unexpected.

Jose Porta, an analyst with BSN S.A. in Madrid, said that "permanently closing Vandellos clearly sends a signal that the antinuclear position is gaining ground in the government." Porta also predicted that the Valdercaballeros Nuclear Power Plant, under construction in central Spain, would never see a startup as a result of the antinuclear sentiment.

Sources

"Spanish Nuclear Shutdown." *The Wall Street Journal*, May 31, 1990.

190 ●
Vandellos II Nuclear Center

Aptdo de Correos No. 27
43891 Hospitalet del Infante, Tarragona, Spain
Enrique Cabellos, Plant Mgr.

Phone: 977-810011
Fax: 977-810014
Telex: 56545

Owning Utility

Asociacion Nuclear Vandellos
Travessera de les Corts, 39-43
08028 Barcelona, Spain
Juan Estape, Mgr.

Phone: 93-334-70-00
Fax: 93-240-58-72
Telex: 54202 ANV E

Contact

Fco. Masana, Head of Development & Info.
Vandellos II Nuclear Power Plant

Basic Facts

Status: Operable. *Type of Reactor:* Pressurized light-water reactor. *Megawatts (electric):* 992. *Megawatts (thermal):* 2785. *Reactor System Supplier:* Westinghouse Electric Corp. (United States). *Generator Supplier:* Westinghouse Electric Corp. (United States); Empresa Nacional Bazan (Spain). *Architect Engineer:* Empresa Nacional de Ingenieria y Technologia SA (Spain); Bechtel Corp. (United States).

Key Dates

Ordered: 1975. *Construction began:* 1981. *Criticality:* 1987. *First Power:* 1987. *Commercial Operation:* March 1988.

Plant Report

Valve leaks radioactive steam. Vandellos II nuclear power plant, one of the newest of Spain's nine operating nuclear power plants, came under close scrutiny in 1990 when it was ordered closed due to a valve leak that filled the containment building with radioactive steam. Early investigations by the Spanish Nuclear Security Commission indicated that Asociacion Nuclear Vandellos, the owner of the power plant, delayed in reporting the incident. The Commission allowed Vandellos to start-up four days after the incident after ascertaining that no radiation leaked into the atmosphere.

Production data. Availability statistics are as follows: 1988, 68 percent; 1989, 71 percent; and 1990, 88 percent.

Vandellos

Numbers of scrams: 1988, 11; 1989, 6; and 1990, 0.

Load factor for the years 1988, 1989, and 1990 were, respectively, 67.5 percent, 71.03 percent, and 88.23 percent.

Gross electrical production in gigawatthours: 1988, 4767 GWh; 1989, 6131 GWh; 1990, 7667 GWh.

Sources

"Spanish Nuclear Incident." *The Wall Street Journal*, July 13, 1990.
"Vandellos II: Production Data." *Asociacion Nuclear Vandellos*, 1991.

Sweden

COUNTRY REPORT

Fifth among nations relying on nuclear energy

Nearly 50 percent of Sweden's power is generated by 12 nuclear reactors on four sites throughout the country. Although Sweden ranks fifth among industrialized nations in its reliance on nuclear energy, acceptance of the widespread use of atomic power has not come easily.

Early years of nuclear technology

The relationship between Sweden and nuclear technology began in 1938 in the town of Kungälv where German-Jewish expatriate physicist Lise Meitner and her nephew Otto Frisch, an assistant to famed atomic theorist Niels Bohr, met during the winter holidays. Together they developed a theory of nuclear fission which supported and expanded on the work of Otto Hahn, a friend of Meitner and a leading German scientist. When they reported their findings to Bohr, he shared them with Enrico Fermi who had just won the 1938 Nobel Prize for his work with slow atoms. Fermi soon went on to build the world's first nuclear reactor based on the information Meitner and Frisch had discovered.

Positive response to nuclear energy

Prior to the advent of nuclear energy, Sweden's only domestic power source was water. Environmental groups, however, had long objected to the development of hydroelectric dams, thus restricting the use of water as an energy source and producing Swedish dependence on oil imports. The Swedes were never comfortable relying on foreigners for energy though and continued to search for alternative sources. Nuclear technology seemed the perfect answer because Sweden is rich in uranium—an essential component in the fission process. Although it was not economically prudent to mine Sweden's uranium at that time, the potential for self-sufficiency was appealing.

After World War II, Swedish scientists set to work developing a system for the peaceful use of nuclear energy. In 1954, after nine years of preparation, the Atombolaget Company saw the fruition of these efforts with the completion of a trial nuclear reactor. Successful beyond original expectations, the heat and electricity producing plant prompted Sweden's move into a full scale program to make nuclear energy the source for the future.

In 1955, the major private utility companies in Sweden formed a joint venture devoted to the development of nuclear energy. This group became known as OKG and built the first commercial nuclear reactor in Sweden. Over the next 20 years, plans were laid by OKG and other organizations for the construction of 12 reactors throughout the nation.

Nuclear industry threatened by fear and controversy

Through the 1950s annd 1960s, the move toward nuclear energy enjoyed strong political support. Utility companies created divisions dedicated to the study of nuclear power and formed joint ventures to encourage research and development of the new technology.

This impetus subsided in the 1970s as the opposition to nuclear power on a worldwide scale mounted. The life-and-death struggle for the future of nuclear energy in Sweden began in earnest in 1976. Demonstrations were held with increasing frequency and Parliament began passing tight controls on the nuclear industry.

Swedes vote to phase out nuclear energy

In March 1979, the Three Mile Island accident in the United States sparked international protest over the use of nuclear energy. Sweden experienced a wave of antinuclear sentiment which was reflected in the 1980 referendum on the issue. Three proposals were submitted to the voters; all three inquired not *if*, but *when* nuclear power should be phased out. The major difference among the three was that proposals one and two advocated a long-term phase-out while number three espoused a more abrupt termination of nuclear energy. The moderate views won a majority—proposals one and two garnered more than 70 percent of the total vote. Although the Prime Minister Thorbjörn Fälldin was vehemently antinuclear, he agreed to oversee a slow phase-out plan which called for the decommissioning of all nuclear power facilities by the year 2010.

Time eases public misgivings—phase out abandoned

In effect, the vote for a long-term phase-out simply slowed rather than stopped nuclear development. Government policies enacted in response to the referendum allowed the completion of plants already slated for construction but prohibited the building of any new facilities. By 1982, conservative political parties had replaced the moderates in the majority of the Swedish Parliament and took a position unique within the history of the European nuclear experience: they voted to continue nuclear development aiming at a 50 percent dependence on nuclear energy by 1990, while simultaneously preparing to end nuclear power generation.

Meanwhile, public fear of nuclear power subsided as the results from the Three Mile Island incident proved to be less serious than was originally anticipated. In fact, those who most strongly supported the use of nuclear power were the people who lived nearest to the reactors, and polls in the early 1990s indicated that 54 percent of the population would be willing to accept a nuclear waste site in their area. As a result of this shift in political and public opinion, the Swedish Parliament voted in 1991 to abandon the plan to begin the phase-out in 1995. Instead, a budget of three billion Swedish crowns ($500 million U.S.) was established for research and development of energy-saving technology and other power sources.

Public and private owners of Sweden's nuclear industry

In addition to OKB, which runs three reactors, several other private utilities operate reactors in Sweden. The largest privately owned utility, Sydkraft, supplies power to 20 percent of Sweden's population. The utility operates two of its own reactors and has a controlling interest in OKG. Private facilities, however, are overshadowed by the mammoth state-owned company, Vattenfall. It provides nearly half of the nation's electricity, solely owns and operates four reactors, and jointly operates three others.

Corporate cooperation

In 1972, Vattenfall, Sydkraft, OKG, and Forsmark Kraftgrupp (a fourth utility), formed a company to address the issue of nuclear waste. The company, Svendk Kärnbränslehantering AB (SKB), was commissioned to develop, plan, build, and operate facilities and systems for managing spent nuclear fuel and radioactive waste. This joint venture satisfied the legal requirements placed on the utility companies for the safe disposal of nuclear by-products. Since its inception, SKB has developed such systems as CLAB (a central interim storage facility for spent fuel), a transportation system for spent fuel and core containers, and a final repository for reactor waste.

Rådet För Kärnkraftsäkerhet (RKS), or the Nuclear Safety Board of the Swedish Utilities, was formed in 1980 to facilitate collaboration on safety issues. Among other duties, RKS collects information on nuclear incidents in both Swedish and international plants. The data collected is analyzed and used to improve the safe operations of Swedish nuclear facilities. RKS participates with about 15 other nations in the international clearinghouse for nuclear safety information, NETWORK.

A third cooperative organization formed by the utilities is the training facility, AB Kärnkraftutbildning (AKU). Employees from the 12 nuclear power sites in Sweden and two sites in Finland are trained at this center, which has three full scope simulated nuclear reactors. They are designed to train employees in both normal and emergency operations.

Federal nuclear authorities

The Swedish Nuclear Power Inspectorate, Statens KKärnkraftinspektion (SKI), is responsible for inspecting each nuclear reactor and site to verify its safety. SKI acts as the liaison agency between the nuclear facilities and the Swedish government, reporting regularly to the Ministry of Industry on the state of safety in nuclear power plants.

The Statens Strålskyddsinstitut (SSI) was established to ensure that people and the environment are not subjected to unsafe levels of radiation. By establishing safe radiation limits, approving licenses to work with radiation, publishing safety guidelines, and inspecting work sites, SSI fulfills the duties mandated by the Ministry of Agriculture, the branch of government to which it is accountable.

The National Board for Spent Nuclear Fuel, Statens Kärnbränslenämnd (SKN) monitors the treatment of nuclear waste. Its responsibilities include: developing nuclear waste disposal facilities, overseeing the decommissioning of old reactors, supervising the implementation of waste research and development programs proposed by the utilities, and collecting and administrating the fees paid by utility companies to cover the cost of waste disposal.

Plans for the future

Although political controversy still surfaces on occasion over the use and administration of nuclear power, Swedish confidence in nuclear energy has greatly increased since the early 1980s as is evidenced by the 1991 record output of nearly 52 percent of Sweden's energy from nuclear sources. This growing dependence on nuclear energy (up from 46 percent in 1990) may be due to improvements within the industry itself and to public realization that the results of nuclear accidents may be less devastating than it was originally expected. Thus, while no new reactors are planned and funding has increased for the development of alternative fuel sources, the 2010 nuclear phase-out date seems to be fading from view.

Sources

"The Swedish Referendum: Do Away With It But Not Yet'." *Bulletin of the Atomic Scientists,* June 1980.

Nuclear Sweden VI Swedish Atomic Forum, 1985.

"More Swedes Oppose Nuclear Phase-Out." *Nuclear Industry,* Summer 1990.

Mounfield, Peter. *World Nuclear Power,* Routledge, 1991.

"Swedish Parliament Votes Against Nuclear Shutdown." *Info,*, U.S. Council for Energy Awareness, June 1991.

"USCEA Survey: Fifteen New Nuclear Plants Began Commercial Operation in '90s." *Info,* U.S. Council for Energy Awareness, July 1991.

"Heaviest Users of Nuclear Energy Have Low Per Capita CO2 Emissions." *Info,* U.S. Council for Energy Awareness, August 1991.

"No Nimby' in Sweden." *Nuclear Industry,* Third Quarter 1991.

"Socialist Lawmakers Opposing Swedish Utility's Privatization." *Nucleonics Week,* November 28, 1991.

"Swedes and Nuclear Waste: Yes In My Backyard." *Nuclear Industry,* First Quarter 1992.

"Swedes Keeping Open Mind About Nuclear Waste Repository." *Info,* January 1992.

"Swedish Nuclear Output Record." *Info,* U.S. Council for Energy Awareness, February 1992.

—Linda M. Ross

PLANT PROFILES

191 ●
Barseback Nuclear Power Plant, Units 1-2

Box 524
S-240 21 Loddekopinge, Sweden
Leif Bergstrom, Plant Mgr.

Phone: 046-46-72-40-00
Fax: 046-46-77-57-93
Telex: 322 09 bvt s

Owning Utility

Sydkraft
Karl Bustafs vag 1
S-217 01 Malmo, Sweden
Goran Ahlstrom, Pres.

Phone: 046-40-25-50-00
Fax: 046-40-97-60-69
Telex: 32829 skdmlml s

Contact

Stieg Claesson, Dir., Pub.Info.
Sydkraft

Basic Facts

Units 1-2 *Status:* Operable. *Type of Reactor:* Boiling water reactor. *Megawatts (electric):* 615. *Megawatts (thermal):* 1800. *Reactor System Supplier:* ASEA-Atom (Sweden). *Generator Supplier:* Stal-Laval Turbin AB (Sweden). *Architect Engineer:* ASEA-Atom (Sweden); Stal-Laval Turbin AB (Sweden); Sydkraft AB; AB Vattenbyggnadsbyran (Sweden).

Key Dates

Unit 1 *Ordered:* 1969. *Construction began:* 1971. *Criticality:* 1975. *First Power:* May 1975. *Commercial Operation:* July 1975.
Unit 2 *Ordered:* 1973. *Construction began:* 1973. *Criticality:* 1977. *First Power:* March 1977. *Commercial Operation:* July 1977.

Plant Report

Background. Located in southern Sweden, the Barsebäck nuclear power site supports two Boiling Water Reactors (BWR). Sydkraft, the owning utility, commissioned Barsebäck 1 and 2 for operation in 1975 and 1977 respectively. Although the plants have functioned without serious incident throughout their history, they have been at the crux of some political and environmental controversy.

Proximity to Danish capital causes concern. The location of the two reactors is within a very short distance of the Danish capital of Copenhagen. This as well as the fact that Denmark has been more reluctant to participate in nuclear power made Barsebäck the focal point of resistance to the use of nuclear power. The protest spurred a 1980 public referendum over the issue which resulted in limitations on the development of the nuclear industry in Sweden. Also as a result of the controversy, Sweden realized the need to participate in international cooperative endeavors to assure the safety of nuclear reactors and the development of new safety measures.

Innovative safety filter system developed. The formation of Nordel, a cooperative organization of Nordic countries, helped to encourage the development of nuclear power technology while keeping peace between nations. Another breakthrough in international nuclear power relations was the development of FILTRA by the scientists working at Barsebäck. FILTRA is an advanced filtering system that provides a fifth level of protection on top of the four barriers provided to all western nuclear reactors.

A concrete cylinder 40 meters high and 20 meters in diameter containing 10,000 cubic meters of crushed stone is connected to both Barsebäck reactors. In the event that all other barriers fail, the steam and fuel from the reactor are diverted through pipes to the FILTRA system. The stones cool the steam, causing it to condense, and they catch 99.9 percent of the radioactive particles and iodine on their surface. Gases that cannot be liquified are released through a chimney, but the amount of those gases is insignificant overall.

Since the system is passive, it requires no human involvement for the first 24 hours of its cycle. Thus, even if the normal safety barriers fail, the FILTRA system adds a nearly infallible protection mechanism. The system has gained international attention, especially in light of the 1987 Chernobyl accident.

Sources

Nuclear Power. Sydkraft, Malm, 1988.
Mounfield, Peter. *World Nuclear Power.* Routledge, 1991.
Summary of Operating Experience in Swedish Nuclear Power Plants 1990. Kärnkraftsäkerhet och Utbildning AB, 1991.

—Linda M. Ross

192 ●
Forsmark Nuclear Power Plant, Units 1-3

S-742 03 Osthammar, Sweden
Alf Lindfors, Plant Mgr.

Phone: 46-173-81000
Fax: 46-173-55116
Telex: 76065 SVVKF S

Owning Utility

Statens Vattenfallsverk (SSPB)
S-162 87 Vallingby, Sweden
Carl-Erik Nyquist, Pres.

Phone: 46-8-7395000
Fax: 46-8-370170
Telex: 19653 SVTELVX S

Contact

Helge Jonsson, Dir. of Pub.Info.
Statens Vattenfallsverk (SSPB)

Basic Facts

Units 1-3 *Status:* Operable. *Type of Reactor:* Boiling water reactor. *Reactor System Supplier:* ASEA-Atom (Sweden). *Generator Supplier:* Stal-Laval Turbin AB (Sweden). *Architect Engineer:* ASEA-Atom (Sweden); Statens Vattenfallsverk (SSPB); Stal-Laval Turbin AB (Sweden).

Key Dates

Unit 1 *Ordered:* 1971. *Construction began:* 1973. *Criticality:* 1980. *First Power:* June 1980. *Commercial Operation:* December 1980.
Unit 2 *Ordered:* 1969. *Construction began:* 1975. *Criticality:* 1980. *First Power:* January 1981. *Commercial Operation:* July 1981.
Unit 3 *Ordered:* 1975. *Construction began:* 1979. *Criticality:* 1984. *First Power:* March 1985. *Commercial Operation:* August 1985.

193 Oskarshamn

Plant Report

Forsmark celebrates tenth anniversary. In 1990, the proud owners of Forsmark, the Forsmark Kraftgrupp, celebrated 10 successful years of safe plant operations. The history of the plant began in 1970 when the Swedish Parliament approved Forsmark as a location for three Boiling Water Reactor (BWR) nuclear power plants. Between 1971 and 1977, the government gave hesitant permission for the construction of each of the three reactors, but in 1978, permission to commission the first of the three was postponed over political controversy. After the 1979 Three Mile Island incident in the United States, all nuclear construction was halted in Sweden until a referendum could be held. Finally, in 1980, 10 years after the process began, Forsmark 1 was given permission to begin operations. Forsmark 2 came on board in 1981 and Forsmark 3 followed in 1985.

Throughout 1987 and 1988, Forsmark's capacity continued to increase and their safety standards improved to the point that they received high ratings from nuclear safety experts.

Chernobyl disaster first detected by Forsmark. The three reactors functioned without incident and were the cause of little public attention until one day in 1986 when the censors at Forsmark began to detect something unusual in the atmosphere. Checking their own operations yielded no clue, but upon investigation, Forsmark officials realized that there had been a terrible nuclear accident at the Soviet reactor in Chernobyl. The Soviet government, however, had not released this information, so Swedish authorities informed the world of the disaster.

Final repository to be located at Forsmark. Forsmark has been assigned to be the site for final disposal of Sweden's nuclear byproducts. After the waste has spent time cooling in short term storage at each individual plant, it is transported to a central interim storage facility (CLAB) at the Oskarshamn site. The radioactive material will be contained at CLAB for about 40 years then transported to Forsmark where it will be sealed in special flasks and permanently stored in crystalline underground caverns. This system is considered by experts to be the safest and most efficient method of dealing with nuclear waste that has been proposed by any nation with nuclear power. The Forsmark disposal facility should be in place by 2020 when the first waste materials will be ready for final disposal.

Information on the Swedish system for nuclear waste management can be obtained by writing to SKB, Box 5864, S-102 48 Stockholm, Sweden (telephone: 46 8 665 28 00).

Visitors welcome. In addition to its nuclear facilities, Forsmark is proud of its historical heritage and encourages visits to its estate and English gardens. Information about visits can be obtained by writing the Forsmarks Kraftgrupp Aktiebolag, Kungsträdgårdsgatan 16, S-111 47 Stockholm.

Sources

"Sentry Equipment Helps Utilities Keep Watch on Nuclear Safety." *The Business Journal of Milwaukee*, June 23, 1986.
Nuclear Waste Management. OCED/NEA and Swedish Nuclear Fuel and Waste Management Company, 1987.
Forsmarks Kraftgrupp Aktiebolag Annual Report 1990. Forsmark Kraftgrupp Aktiebolag, 1991.
Forsmark: Facts About Our Energy Supply. Vattenfall, sthammar, Sweden, n.d.

—Linda M. Ross

193 ●

Oskarshamn Nuclear Power Plant, Units 1-3

S-570 93 Figeholm, Sweden
Sven Magnusson, Vice Pres., Unit 1

Phone: 46-491-865-00
Fax: 46-491-860-90
Telex: 43995 elatom s

Owning Utility

OKG Aktiebolag
Simpevarp
S-570 93 Figeholm, Sweden
Georg Bissmarck, V.Pres., Unit 2

Contact

Lars-Goran Wahlberg, V.Pres., Pub.Aff.
OKG Aktiebolag

Basic Facts

Unit 1 *Status:* Operable. *Type of Reactor:* Boiling water reactor. *Megawatts (electric):* 460. *Megawatts (thermal):* 1375. *Reactor System Supplier:* ASEA-Atom (Sweden). *Generator Supplier:* ASEA-Atom; Stal-Laval Turbin AB (Sweden). *Architect Engineer:* ASEA-Atom.

Unit 2 *Status:* Operable. *Type of Reactor:* Boiling water reactor. *Megawatts (electric):* 617. *Megawatts (thermal):* 1800. *Reactor System Supplier:* ASEA-Atom. *Generator Supplier:* ASEA-Atom; Brown Boveri (Switzerland); Stal-Laval Turbin AB (Sweden). *Architect Engineer:* OKG Aktiebolag; ASEA-Atom; Brown Boveri; Stal-Laval AB (Sweden); AB Vattenbyggnadsbryan (Sweden).

Unit 3 *Status:* Operable. *Type of Reactor:* Boiling water reactor. *Megawatts (electric):* 1200. *Megawatts (thermal):* 3300. *Reactor System Supplier:* ASEA-Atom (Sweden). *Generator Supplier:* ASEA-Atom (Sweden); Stal-Laval Turbin AB (Sweden). *Architect Engineer:* ASEA-Atom (Sweden); Stal-Laval Turbin AB (Sweden); OKG Aktiebolag; AB Vattenbyggnadsbryan (Sweden).

Key Dates

Unit 1 *Ordered:* 1965. *Construction began:* 1965. *Criticality:* 1970. *First Power:* July 1971. *Commercial Operation:* February 1972. *Shut Down Expected:* 2010.

Unit 2 *Ordered:* 1969. *Construction began:* 1970. *Criticality:* March 1974. *First Power:* October 1974. *Commercial Operation:* January 1975. *Shut Down Expected:* 2010.

Unit 3 *Ordered:* 1976. *Construction began:* 1980. *Criticality:* 1984. *First Power:* March 1985. *Commercial Operation:* August 1985. *Shut Down Expected:* 2010.

Plant Report

Sweden's first nuclear power plant. In 1965, the cooperative efforts of seven municipal and private power utilities culminated in the formation of OKG Aktiebolag and an order for Sweden's first nuclear power plant: Oskarshamn 1. After nearly 30 years of operation, the Oskarshamn site now supports three reactors and, in 1991, reported a record energy output.

Plant Profiles

Oskarshamn 1 began full power operation in 1972. The third unit, Oskarshamn 3, was ordered in 1976. Construction and commissioning were delayed, however, because of the public debate over the future of nuclear power in Sweden. After the 1980 referendum, approval was given to complete all nuclear facilities that were currently on the drawing board—a group to which Oskarshamn 3 belonged. Thus, in August 1985, the long-awaited plant began full power operations.

The three reactors at the site are the Swedish-designed Boiling Water Reactor (BWR) type, as are six other of the 12 nuclear reactors in Sweden.

Disposal units held in high regard. Out of envrionmental concern, Sweden has taken the issue of nuclear waste disposal quite seriously. Through careful study and innovation, Swedish scientists have constructed a waste disposal system that has received international acclaim. The interim storage component of this system, CLAB, is located at Oskarshamn.

This part of the program consists of a receiving section, an auxiliary section, and a water-filled storage cavern situated 30 meters below ground. It has a maximum storage capacity in its current state of 3,000 tons of uranium with expansion possibilities to contain up to 9,000 tonnes. Nuclear waste is transported to the facility by ship after it has been given sufficient time to cool in the high level radiation containment centers located at each reactor site in the country. The spent fuel is stored at CLAB until the middle range radioactive components of the material have lost their potency, about 40 years.

The nuclear power industry in Sweden is now actively engaged in designing and constructing long-term storage for final disposal of the materials. This facility, to be located at the Foresmark Nuclear Power Plant, will be in place by 2020 when the first waste materials will be ready for final disposal.

Information on the Swedish system for nuclear waste management can be obtained by writing to SKB, Box 5864, S-102 48 Stockholm, Sweden (telephone: 46 8 665 28 00).

Sources

Nuclear Sweden VI, Swedish Atomic Forum, Stockholm, 1985.
Nuclear Waste Management. OCED/NEA and Swedish Nuclear Fuel and Waste Management Company, 1987.
Summary of Operating Experience in Swedish Nuclear Power Plants 1990. Kärnkraftsäkerhet och Utbildning AB, Stockholm, 1991.
OKG Aktiebolag Annual Report 1991. OKG Aktiebolag, Figeholm, Sweden, 1992.

—Linda M. Ross

| 194 | ●

Ringhals Nuclear Power Plant, Units 1-4

S-430 22 Varobacka, Sweden
Hakan Johansson, Plant Mgr.

Phone: 46-340-67000
Fax: 46-340-65184
Telex: 3495 SVVKPR S

Owning Utility

Statens Vattenfallsverk (SSPB)
S-162 87 Vallingby, Sweden
Carl-Erik Nyquist, Pres.

Phone: 46-8-7395000
Fax: 46-8-370170
Telex: 19653 SVTELVX S

Contact

Helge Jonsson, Dir. of Pub.Info.
Statens Vattenfallsverk (SSPB)

Basic Facts

Unit 1 *Status:* Operable. *Type of Reactor:* Boiling water reactor. *Megawatts (electric):* 780. *Megawatts (thermal):* 2270. *Reactor System Supplier:* ASEA-Atom (Sweden). *Generator Supplier:* English Electric Co., Ltd. (United Kingdom). *Architect Engineer:* ASEA-Atom (Sweden); Statens Vattenfallsverk (SSPB) (United States); Stal-Laval Turbin AB (Sweden).

Unit 2 *Status:* Operable. *Type of Reactor:* Boiling water reactor. *Megawatts (electric):* 840. *Megawatts (thermal):* 2432. *Reactor System Supplier:* Westinghouse Electric Corp. (United States). *Generator Supplier:* Stal-Laval Turbin AB (Sweden). *Architect Engineer:* Statens Vattenfallsverk (SSPB); Gibbs & Hill, Inc. (United States).

Units 3-4 *Status:* Operable. *Type of Reactor:* Boiling water reactor. *Megawatts (electric):* 960. *Megawatts (thermal):* 2775. *Reactor System Supplier:* Westinghouse Electric Corp. (United States). *Generator Supplier:* Stal-Laval Turbin AB (Sweden). *Architect Engineer:* AB Vattenbyggnadsbryan (Sweden); Traction-Electricite (Belgium).

Key Dates

Unit 1 *Ordered:* 1968. *Construction began:* 1969. *Criticality:* August 1973. *First Power:* October 1974. *Commercial Operation:* January 1976.

Unit 2 *Ordered:* 1968. *Construction began:* 1970. *Criticality:* 1974. *First Power:* August 1974. *Commercial Operation:* May 1975.

Unit 3 *Ordered:* 1972. *Construction began:* 1973. *Criticality:* 1980. *First Power:* September 1980. *Commercial Operation:* September 1981.

Unit 4 *Ordered:* 1972. *Construction began:* 1974. *Criticality:* 1982. *First Power:* June 1982. *Commercial Operation:* November 1983.

Plant Report

Sweden's largest power plant. Ringhals is the largest nuclear reactor site in Sweden. Ringhals 1 is a Boiling Water Reactor (BWR), but Ringhals 2, 3, and 4 are the only Pressurized Water Reactors (PWR) operating in Sweden. The PWRs were designed in the United States by Westinghouse while the boiling water reactor is of Swedish design.

Construction of Ringhals 1 began in 1969, four years after the ground was broken on the first Swedish nuclear power plant, Oskarshamn 1. Construction on the other three units began between 1970 and 1974. In 1975, Ringhals 2 was the first reactor ready for commercial operations. Ringhals 1 followed one year later, but 3 and 4 were delayed by the controversy over the future of nuclear energy in Sweden. After the 1980 referendum, permission to open Ringhals 3 and 4 was given and the plants opened in 1981 and 1983 respectively. All four plants concurrently operated at full power for the first time in January of 1984, a landmark date for the facility.

Cracks found in Ringhals 2. A routine fueling outage in 1992 uncovered cracks in the reactor vessel of Ringhals 2. Investigations identified five cracks; two that required repair, and three others of no safety significance based on size and location.

Ringhals 2, 3, and 4 are designed with French Inconel-600 components, which tend to be subject to stress corrosion cracking. Ringhals 2 was allowed to resume operations six weeks later, but the plant is now required to conduct regular inspections that specifically check for reactor vessel cracks in the three affected units.

International relationships. One of the outstanding features of Ringhals is its international outlook. Through Nordel, a Nordic nation cooperative organization, Ringhals participates in the exchange of technical information and research projects. In addition, Ringhals was one of the first companies to share safety procedures and upgrade technology with nuclear power plants in Eastern Europe. Especially in light of the Chernobyl incident and the state of nuclear power plants in eastern Germany, Sweden views the safety and development of Eastern European nuclear energy as vital to its own interests.

Visitors Center. Ringhals also takes pride in its community services and industry relations. Information about the safety and operations of the plants as well as community programs offered by Ringhals are available at the Kärnhuset (Visitors' Center) located at the site. Group visits and conferences are welcome with advanced arrangements.

Sources

"Cracks in Vessel Penetrations Found, Fixed; Ringhals 2 Restarted." *Nuclear News*, September 1992.
Ringhals 1990: Annual Report. Vattenfall, 1991.
Summary of Operating Experience in Swedish Nuclear Power Plants 1990. Kärnkraftsäkerhet och Utbildning AB, Stockholm, 1991.
Welcome to Ringhals. Vattenfall, n.d.
Vattenfall Ringhals Nuclear Power Plant—that's us. Vattnefall, n.d.

—Linda M. Ross

Switzerland

COUNTRY REPORT

Comprehensive nuclear program

Switzerland commissioned its first commercial reactor at the Beznau Nuclear Power Plant in 1969. Nuclear construction continued throughout the country during the 1970's and early 1980's. Currently, there are five reactors at four sites. Combined, they provide about 40 percent of Switzerland's electricity. Hydroelectric sources meet most of the remaining demand for electric power.

Antinuclear efforts

Although Switzerland relies heavily on atomic power, anti-nuclear forces have persisted in challenging further development. Swiss voters rejected an anti-nuclear ballot initiative in 1979. Again in 1984, Switzerland's citizens rejected two anti-nuclear proposals. In 1990, two more anti-nuclear initiatives were placed on the ballot, one to place a 10-year moratorium on building new nuclear plants and the other to shut down the country's existing plants. The Swiss Parliament recommended that voters reject both measures; however, only the shut-down order was defeated. The building ban passed, effectively halting further nuclear development until after the turn of the century.

Nuclear waste storage

One of the biggest difficulties facing the Swiss nuclear industry is the management of nuclear waste. Swiss law assigns responsibility for nuclear waste management to its producers and requires that all radioactive waste materials be placed in underground repositories. In 1972, the nation's electric utilities along with medical, industrial, and research users formed the National Cooperative for the Storage of Radioactive Waste (NAGRA) to develop a strategy for handling nuclear waste materials.

Swiss researchers have examined 100 potential sites for short-lived nuclear waste repositories. Of these, four are still under consideration: Bois de La Glaive in the country's southwestern region; Piz Pian Grand in the southeast; and two, Oberbauenstock and Wellenberg, in central Switzerland.

In 1991, the Swiss federal government approved the construction of ZWIBEZ, a temporary interim storage facility, at the Beznau Nuclear Power Plant. ZWIBEZ will accept nuclear waste, including spent fuel and other radioactive materials, from its host plant.

Areas in the northern section of the country are being studied for a permanent high-level waste repository. In 1992, the community of Leuggern, in the Canton of Aargau, volunteered to receive such a facility.

Sources

"Swiss Parliament Says No to Nuclear Ban." *Nuclear Industry*, Summer 1990. "Nuclear Waste Management: Switzerland." OCED/AEN, OECD/NEA and Nagra, March 1991.

"Swiss Limit Nuclear Power." *The New York Times*, September 24, 1990.

"Nuclear Waste Management: Switzerland." OCED/AEN, OECD/NEA and Nagra, March 1991.

Report on the Swiss Nuclear Installations in 1990. Swiss Federal Nuclear Safety Inspectorate, September, 1991.

Beznau Nuclear Power Plant. Nordostschweizerische Kraftwerke AG, 1991.

"Swiss Approve Waste Storage." *Nuclear Industries*, Third Quarter 1991.

"Swiss Community Volunteers as Waste Site." *Nuclear Industry*, First Quarter 1992.

—*Karen Bellenir*

PLANT PROFILES

195 ●
Beznau Nuclear Power Plant, Units 1-2

CH-5312 Doettingen, Switzerland
Hans Wenger, Plant Mgr.

Phone: 056-99-71-11
Fax: 056-99-77-02
Telex: 827 429 kkb ch

Owning Utility

Nordostschweizerische Kraftwerke AG
Parkstrasse 23
CH-5401 Baden, Switzerland
Christoph Tromp, Dir. of Pub. Info.

Phone: 056-20-31-11
Fax: 056-20-37-92
Telex: 52086 nok ch

Contact

Klaus Niederau, Press Office
Beznau Nuclear Power Plant

Basic Facts

Units 1-2 *Status:* Operable. *Type of Reactor:* Pressurized light-water reactor. *Megawatts (electric):* 364. *Megawatts (thermal):* 1130. *Reactor System Supplier:* Westinghouse Electric Corp. (United States). *Generator Supplier:* Brown Boveri et Cie. (Switzerland). *Architect Engineer:* Gibbs & Hill, Inc. (United States); Brown Boveri et Cie. (Switzerland).

Key Dates

Unit 1 *Ordered:* 1965. *Construction began:* 1965. *Criticality:* 1969. *First Power:* July 1969. *Commercial Operation:* December 1969.

Unit 2 *Ordered:* 1967. *Construction began:* 1968. *Criticality:* 1971. *First Power:* October 1971. *Commercial Operation:* March 1972.

Plant Report

Switzerland's first nuclear power plant. When Beznau 1 came online in 1969, it was the first commercial nuclear power plant in Switzerland and a milestone in the history of its owner, Nordostschweizerische Kraftwerke AG (NOK).

NOK was founded in 1914 to be an inter-cantonal generating organization. It is made up of nine participating cantons or the electricity companies owned by them and assures the supply of electricity to the member cantons.

When NOK was founded, it took over two power stations: the low-head river power station at Beznau and the pumped storage station at Lontch in the Canton of Glaris. With these two stations, NOK practiced large-scale interconnected operation for the first time in Switzerland. and World War 2, then the rapid rise following the Second World War, induced NOK to build a number of power stations on their own and to acquire holdings in the share capital of some joint undertakings. The company's decision to build a nuclear power plant had a significant impact in averting the construction of large oil-fired power stations in northeast Switzerland. Today, NOK has nine of its own generating stations and holdings in 22 others.

Heat from Beznau. In addition to supplying electricity, Beznau helps save fuel oil by supplying heat to Refuna, the regional district heating network of the lower Aare valley. Approximately 15,000 people who live in the area obtain heating from this source.

Steam at a temperature of 127 C is drained off between high and low pressure parts of Beznau's turbines, then fed to a heat exchanger. There, the heat in the drained steam is moved to the district heating network, whose waters are subsequently heated to 120 C.

Since both units at Beznau supply the regional heating network, they are never shut down simultaneously.

Waste storage. The Swiss federal government has approved plans for the construction and operation at Beznau of an intermediate storage facility for low-, medium-, and high-level waste. The facility will accept waste from Beznau and from used fuel elements reprocessed in France and the United Kingdom. The project still requires approval by 26 cantons, but only for non-nuclear aspects of the construction.

Information center. An information center and the plant itself are open for guided tours from Monday through Friday from 9:00 a.m. to 12 noon and from 1:00 p.m. to 7:00 pm, Saturdays from 9:00 a.m. to 6:00 p.m., and Sundays from 11:00 a.m. to 6:00 p.m.

For reservations write the NOK Information Center at 5313 Bottstein or call call 056-45-38-15.

Sources

Beznau Nuclear Power Plant. Nordostschweizerische Kraftwerke AG, n.d.
"Swiss Approve Waste Storage." *Nuclear Industry,* Third Quarter 1991.

196 ●
Goesgen Nuclear Power Plant

Postfach 55
CH-4658 Daeniken, Switzerland
Christian Donatsch, Dir.

Phone: 062-65-15-65
Fax: 062-65-22-01
Telex: 981 713 kkg-ch

Owning Utility

Kernkraftwerk Goesgen-Daeniken AG
Postfach 55
CH-4658 Daeniken, Switzerland

Contact

Bruno Elmiger, Pub.Info.
Goesgen Nuclear Power Plant

Basic Facts

Status: Operable. *Type of Reactor:* Pressurized light-water reactor. *Megawatts (electric):* 970. *Megawatts (thermal):* 3002. *Reactor System Supplier:* Kraftwerk Union AG (Germany). *Generator Supplier:* Kraftwerk Union AG (Germany). *Architect Engineer:* Kraftwerk Union AG (Germany).

Key Dates

Ordered: 1973. *Construction began:* 1973. *Criticality:* January 1979. *First Power:* February 1979. *Commercial Operation:* November 1979.

SWITZERLAND

Leibstadt

Plant Report

The site. The Goesgen Nuclear Power Plant was built in the southern foothills of the Swiss Jura, in a bow of the river Aare about half way between the cities of Olten and Aarua, in the municipality of Daeniken, Canton of Solothurn. The site was advantageous for several reasons:

- Proximity to one of the most important junctions of the Swiss high voltage grid.
- Availability of water for use with the cooling tower.
- Good ground conditions—foundation soil of an alluvial gravel layer of 20 to 30 meters, which lies on solid limestone.

Sources

"Goesgen: The First Standard PWR for Switzerland." *Nuclear Engineering International*, February 1980.

Leibstadt Nuclear Power Plant

Kernkraftwerk Leibstadt AG
CH-4353 Leibstadt, Switzerland
Hugo Schumacher, Plant Mgr.
Phone: 056-47-71-11
Fax: 056-47-14-37
Telex: 827 430

Owning Utility

Kernkraftwerk Leibstadt AG
c/o Electrizitats-Gesellschaft Laufenburg AG
CH-4335 Laufenburg, Switzerland
Adolf Gugler, Chairman
Phone: 064-69-63-63
Fax: 064-69-64-50
Telex: 98 22 66

Contact

Leo Erne, Pub.Info.
Leibstadt Nuclear Power Plant

Basic Facts

Status: Operable. *Type of Reactor:* Boiling water reactor. *Megawatts (electric):* 1045. *Megawatts (thermal):* 3138. *Reactor System Supplier:* General Electric Technical Services Co. (United States). *Generator Supplier:* Brown Boveri et Cie. (Switzerland). *Architect Engineer:* Brown Boveri et Cie. (Switzerland); General Electric Technical Services Co. (United States); Electrowatt Ltd. (Switzerland).

Key Dates

Ordered: 1973. *Construction began:* 1975. *Criticality:* March 1984. *First Power:* May 1985. *Commercial Operation:* December 1984.

Plant Report

Switzerland's largest nuclear plant. The Leibstadt nuclear power plant is located on the south bank of the Rhine in northern Switzerland, just 5 kilometers away from the town of Koblenz and where both the Rhine and Aare rivers join together. The plant employs some 400 people, most of whom live in the surrounding towns and countryside.

The plant is owned by a partnership. Each of the 12 partners—all engaged in the generation, distribution, and utilization of electrical energy—receives electricity from Leibstadt based on its shareholdings. The Electrizitats-Gesellschaft Laufenburg AG is contracted by the partners to manage Leibstadt. Leibstadt is connected to the European high-voltage grid via the Laufenburg substation. Approximately 12.5 percent of its electricity is taken by German consumers.

Leibstadt started commercial operation on December 15, 1984, the latest and largest of Switzerland's five nuclear power plants. Twenty years elapsed from the initial planning stages of Leibstadt to its commercial operation—a period marked by multiple obstacles and delays. However, these same delays allowed the original design rating of the reactor to be increased, in step with technical progress in the nuclear industry, from 600 megawatts to 990 megawatts.

With Leibstadt online, Switerland's total amount of electrical energy produced by nuclear energy came to approximately 40 percent. The remaining 60 percent of the country's power is generated by hydro plants. Leibstadt's contribution is about 15 percent, or 7 billion kilowatt-hours of total energy annually.

Sources

Leibstadt Nuclear Power Plant: Technical Description. Kernkraftwerk Leibstadt AG, April 1991.

Muehleberg Nuclear Power Plant

Kernkraftwerk Muehleberg
CH-3203 Muehleberg, Switzerland
G. Markoczy, Plant Mgr.
Phone: 031-754-71-11
Fax: 031-754-71-20
Telex: 911 141 kkm ch

Owning Utility

Bernische Kraftwerke AG
Viktoriaplatz 2
CH-3000 Bern 25, Switzerland
W. Augsburger, Chairman
Phone: 031-40-51-11
Fax: 031-40-56-35
Telex: 912 352 bkwb ch

Contact

M. Pfisterer, Dir. of Pub.Info.
Muehleberg Nuclear Power Plant

Basic Facts

Status: Operable. *Type of Reactor:* Boiling water reactor. *Megawatts (electric):* 336. *Megawatts (thermal):* 967. *Reactor System Supplier:* General Electric Technical Services Co. (United States). *Generator Supplier:* Brown Boveri et Cie. (Switzerland). *Architect Engineer:* Brown Boveri et Cie. (Switzerland); Emch & Berger (Switzerland); General Electric Technical Services Co. (United States).

Key Dates

Ordered: 1966. *Construction began:* 1967. *Criticality:* August 1971. *First Power:* July 1971. *Commercial Operation:* November 1972.

Taiwan

COUNTRY REPORT

Brief history

Taiwan is nearly wholly dependent on foreign imports for its energy supplies, making it vulnerable to the effects of a naval blockade. As early as 1955, the Taiwan Power Company, an agency of the federal government, began investigating the possibility of nuclear power generation as a means of diversifying the nation's energy sources. Because so much less fuel is required to operate nuclear power plants than conventional plants, supplies for two to three years can easily be stockpiled on the small island, forestalling the effects of a blockade. By 1964, a detailed plan was in place calling for three power stations with a total of six reactors to be built on the island between 1972 and 1985.

Since the end of World War II, Taiwan's rapidly expanding economy has been plagued by frequent shortages of energy, forcing the government to ration power, particularly during peak hours in the summer months. By 1985, however, with the completion of the country's third nuclear power station, more than half of Taiwan's power was generated by its nuclear plants and rationing was no longer necessary. The years since 1985 have seen not only a decline in the percentage of energy consumption supplied by nuclear power, but a growing resistance within the nation's population to the use of nuclear power to supply energy. The issues debated are the disposal of nuclear waste, Taipower's ability to manage complex nuclear technology, and the growing demand for energy in Taiwan. Opposition since the early 1980s has focused not so much on the three existing nuclear power plants, but on the proposed fourth nuclear power plant.

Nuclear waste disposal

Taiwan is only in the earliest stages of determining permanent disposition of both its high- and low-level radioactive waste. Low level radioactive waste from the country's three nuclear power stations is shipped to Lan Yu (Orchid Island), where an interim storage facility is located. High-level waste in the form of spent fuel is currently stored in pools located at each of the power stations.

As storage space at the plant sites runs low, Taipower has turned for short-term relief for its high-level waste storage problem to reracking the fuel rods in order to use existing space more advantageously. Reracking at two of the power stations will extend their usefulness until about the year 2000. In 1989, therefore, Taipower embarked on a plan to build additional storage facilities on the plant sites for high-level waste. The first interim storage facility is scheduled to be available in 1997. Taipower is also investigating the feasibility of reprocessing spent fuel in the future.

Taipower is attempting to resolve the problem of declining space for low-level radioactive waste in three ways. First, by constructing a supercompactor and an incinerator at its Kuosheng plant site, the company has reduced by 3500 waste drums its annual accumulation of 12,000 drums of LLW. Second, Taipower is investigating permanent land disposal of LLW to determine site and method.

Third, Taipower plans to expand the capacity of its interim storage facility located on Lan Yu, where current capacity is expected to be reached by the late 1990s. The island's aboriginal inhabitants, however, have requested that Taipower halt construction of the expanded facility, stop shipping nuclear waste to the island, and set up a timetable to remove the dump site from the island. In 1991 the government in mainland China offered to store Taiwan's low-level nuclear waste.

Public confidence in Taipower

Taiwan joined the nuclear age in the 1970s and up until 1985 public confidence in the safety of the state-owned nuclear power stations was reported to be high. But in July 1985 a fire broke out in the turbine generator at the nuclear facility near Maanshan, and in the following years reports of radiation leakage and improper attention to safety factors at the plants—some denied by Taipower officials—came to light, eroding public confidence in Taipower's ability to manage nuclear technology.

Growing international concern over environmental issues, and the recognition that, 20 years after construction began on its first plant, Taipower still does not have a program for permanent disposal of its radioactive waste, has fueled a "green" movement in opposition to nuclear power.

In 1991, the Taiwan Ministry of Economic Affairs hired six international nuclear experts to examine Taipower's management of the three existing power stations; the resulting report indicated that Taipower's management is equivalent to standards in other countries and continues to improve.

Energy demand in Taiwan

Taipower officials predict that the growth in demand for energy during the 1990s will necessitate doubling its capacity by the year 2000, and prefer that the planned fourth nuclear power station meet a significant portion of that demand. A number of physicists and economists question this figure, citing the changeover from an industry-based to a service-based economy in Taiwan, and increased use of co-generation in energy-intensive industries. Some claim that Taipower's threats of power shortages and a stalled economy if plans for the fourth nuclear station do not go forward are mere scare tactics. Edgar Lin, an environmental scientist at Tunghai University in Taiching, noted in 1988, "Taipower has never considered conservation as an alternative to creating energy."

The population of Taiwan has been described as increasingly polarized over the nuclear issue. One anti-nuclear rally in 1991 resulted in the death of a police officer when two protesters ran their mini-bus through a police blockade. The Taiwan government however, remained steadfastly in favor of maintaining nuclear power's percentage of the nation's energy supply. Economic Affairs Minister Vincent Siew summarized the government's position in 1991 when he announced, "We will most likely follow the world trend of using nuclear-generated electricity because we really don't have any alternatives."

Sources

"Nuclear Shutdown." *Far Eastern Economic Review*, June 2, 1988.

"Aborigines blast waste site." *China Post*, February 21, 1991.

"M'land Offer to Store Nuclear Waste Being Considered Locally." *China Post*, February 23, 1991.

"Powerhouse Efficiency Must Be Raised." *China Post*, May 12, 1991.

"Nuclear Qualms." *Far Eastern Economic Review*, July 4, 1991.

"Official Says Taiwan Must Build 4th Nuclear Plant." *China Post*, September 12, 1991.

"Activist Rams Police: 1 Dead." *China Post*, October 4, 1991.

—Mary Gillis

PLANT PROFILES

199 ●
Chinshan Nulcear Power Plant, Units 1-2

12, HsiaoKeng
Chienhua Village
Shihman, Taipei County, Taiwan
S.J. Liao, Supt.
Phone: 02-6381701
Fax: 886-2-6382111
Telex: 33193

Owning Utility

Taiwan Power Co.
242 Roosevelt Rd., Section 3
Taipei 100, Taiwan
E. Lin, V.Pres. Nuclear Operations
Phone: 02-3651234
Fax: 02-3650037
Telex: 11520 TAIPOWER

Contact

L.Y. Chang, Dir., Pub.Aff. Dept.
Taiwan Power Co.
Phone: 02-396-7777

Basic Facts

Units 1-2 *Status:* Operable. *Type of Reactor:* Boiling water reactor. *Megawatts (electric):* 636. *Megawatts (thermal):* 1775. *Reactor System Supplier:* General Electric Co. (United States). *Generator Supplier:* Westinghouse Electric Corp. (United States). *Architect Engineer:* Ebasco.

Key Dates

Unit 1 *Ordered:* 1969. *Construction began:* 1972. *Criticality:* October 1977. *First Power:* November 1977. *Commercial Operation:* December 1978.

Unit 2 *Ordered:* 1970. *Construction began:* 1973. *Criticality:* November 1978. *First Power:* December 1978. *Commercial Operation:* July 1979.

200 ●
Kuosheng Nuclear Power Station, Units 1-2

60, Patou
Kuosheng Village
Wanli, Taipei County, Taiwan
G.C. Lee, Supt.
Phone: 02-4985990
Fax: 886-2-4982624
Telex: 33182

Owning Utility

Taiwan Power Co.
242 Roosevelt Rd., Section 3
Taipei 100, Taiwan
E. Lin, V.Pres. Nuclear Operations
Phone: 02-3651234
Fax: 02-3650037
Telex: 11520 TAIPOWER

Contact

L.Y. Chang, Dir., Pub.Aff. Dept.
Taiwan Power Co.
Phone: 02-396-7777

Basic Facts

Units 1-2 *Status:* Operable. *Type of Reactor:* Boiling water reactor. *Megawatts (electric):* 985. *Megawatts (thermal):* 2894. *Reactor System Supplier:* General Electric Co. (United States). *Generator Supplier:* Westinghouse Electric Co. (United States). *Architect Engineer:* Bechtel.

Key Dates

Unit 1 *Ordered:* 1973. *Construction began:* 1975. *Criticality:* February 1981. *First Power:* May 1981. *Commercial Operation:* December 1981.

Unit 2 *Ordered:* 1973. *Construction began:* 1975. *Criticality:* March 1982. *First Power:* June 1982. *Commercial Operation:* March 1983.

201 ●
Maanshan Nuclear Power Station, Units 1-2

387, Nanwan Rd.
Hengtsun, Pingtung County, Taiwan
S.H. Soong, Supt.
Phone: 08-8893470
Fax: 886-8-8894817
Telex: 71885

Owning Utility

Taiwan Power Co.
242 Roosevelt Rd., Section 3
Taipei 100, Taiwan
E. Lin, V.Pres. Nuclear Operations
Phone: 02-3651234
Fax: 02-3650037
Telex: 11520 TAIPOWER

Contact

L.Y. Chang, Dir., Pub.Aff. Dept.
Taiwan Power Co.
Phone: 02-396-7777

Basic Facts

Units 1-2 *Status:* Operable. *Type of Reactor:* Pressurized light-water reactor. *Megawatts (electric):* 951. *Megawatts (thermal):* 2785. *Reactor System Supplier:* Westinghouse Electric Corp. (United States). *Generator Supplier:* General Electric Co. (United States). *Architect Engineer:* Bechtel.

Key Dates

Unit 1 *Ordered:* 1974. *Construction began:* 1978. *Criticality:* March 1984. *First Power:* May 1984. *Commercial Operation:* July 1984.

Unit 2 *Ordered:* 1977. *Construction began:* 1978. *Criticality:* February 1985. *First Power:* February 1985. *Commercial Operation:* May 1985.

202 ■
Yenliao Nuclear Power Station, Units 1-2

Yenliao, Taiwan

Owning Utility

Taiwan Power Co.
242 Roosevelt Rd., Section 3
Taipei 100, Taiwan
E. Lin, V.Pres. Nuclear Operations
Phone: 02-3651234
Fax: 02-3650037
Telex: 11520 TAIPOWER

Contact

L.Y. Chang, Dir., Pub.Aff. Dept.
Taiwan Power Co.
Phone: 02-396-7777

Basic Facts

Units 1-2 *Status:* On order. *Megawatts (electric):* 950.

Key Dates

Unit 1 *Ordered:* 1992. *Commercial Operation Expected:* 2000.

Unit 2 *Ordered:* 1992. *Commercial Operation Expected:* Indefinite.

Yenliao

Plant Report

National controversy. Taiwan's Fourth Nuclear Power Station has been at the center of a national controversy since 1981 when it was first proposed. The project was stalled for a decade as first worldwide recession and then local environmentalists obstructed Taipower's plan to keep the contribution of nuclear power a significant percentage of the nation's power resources. By 1991, when preliminary safety, feasability and cost reports for the fourth station were submitted and approved, the population was reportedly polarized on the issue, each side becoming more entrenched in its own viewpoint.

The debate over the fourth nuclear power plant was overtly political. Antinuclear activists were backed by the opposition Democratic Progressive Party (DPP), and Taipower, which is governed by the Ministry of Economic Affairs, utilized the government-run media to promulgate the safety of nuclear power, and the need for the additional energy the proposed station would provide. Opposition leaders refuted Taipower's projections for growth in energy demand, citing the changing economy and the need for a national energy conservation program in order to slow the growth in demand. Taipower hopes to have the first of the two reactors at Yenliao operational by the year 2000.

—Mary Gillis

Ukraine

COUNTRY REPORT

Large producer of nuclear power

Ukraine possesses the second largest nuclear generating capacity of all the former Soviet republics (see the former Soviet Union), surpassed only by the Russian Federation. Although the country has a supply of domestic coal and meets 71 percent of its need for electricity through fossil-fueled generation, nuclear power plays a significant role in the country's overall energy program. Ukraine maintains five nuclear generating stations with a total of 15 operating reactors. Nuclear power accounts for 25 percent of Ukraine's energy production.

Changes attributed to Chernobyl

In April 1986, while Ukraine was still part of the former Soviet Union, an explosion at Chernobyl-4 changed the relationship between the Ukrainian population and their nuclear reactors. Before Chernobyl, Ukraine generated 260 billion kilowatt-hours per year, approximately twenty percent of all the power produced in the Soviet Union. At that time, nuclear power accounted for 15 percent of the Ukrainian production. The central Soviet government had plans for substantial increases in Ukraine's nuclear capacity.

Ukraine was also chosen as the site for the Soviet Union's first commercial nuclear plant intended to supply heat as well as electricity. The complex was under construction in the city of Odessa, on the Black Sea. Odessa's population of more than one million was assured that nuclear generation was safe and environmentally benign. Plans included using waste heat for agricultural purposes.

After the explosion at Chernobyl, public protests forced the Soviet government to reconsider nuclear expansion in Ukraine. In 1988, the Ukrainian Academies of Science added its voice to the protest against Soviet projects in the republic. The Soviet government had been proceeding with its intention to build six new 1,000-megawatt reactors at Ukraine's existing nuclear stations. The new construction would have increased to 6,000 megawatts the generating capacity at some of the stations. The Ukrainian Academies of Science claimed that dense concentrations of reactors were being placed too close to major population centers, and that 4,000 megawatts was the maximum acceptable size for a nuclear generating station. The scientists also raised questions about potential earthquakes in the region.

The Ukrainian government demanded a ban on nuclear construction. The prohibition halted work at Khmelnitsky, Zaporozhye, and Rovno. The Soviets were also forced to yield to public pressure and cancel construction efforts at Chernobyl, South Ukraine and on the Crimean peninsula.

International cooperation

Following the Chernobyl disaster, Ukraine began participating in international efforts to make nuclear power generation safer. Plant personnel from all of Ukraine's nuclear stations have participated in exchanges conducted by the World Association of Nuclear Operators (WANO). The country has also hosted activities undertaken by some of the working groups formed by the U.S./Soviet Joint Coordinating Committee on Civilian Nuclear Reactor Safety (JCCNRS). As part of the JCCNRS program, researchers have studied fire prevention efforts at the Zaporozhye station and radiation effects at the Chernobyl site.

Post-Chernobyl power picture

At Chernobyl, where the fourth unit exploded and a fifth was cancelled, three other units continue to produce electricity. Their output accounts for 20 percent of Ukraine's nuclear power. The Soviets originally planned to phase out production and close the station in 1995. Public pressure and technical problems, however, forced officials to move the target date up to 1993.

According to an announcement made in 1991 by the Soviet Ministry of Atomic Power and Industry (MAPI), a power shortage complicated decommissioning plans at Chernobyl. MAPI stated that because Ukraine had no power reserves, a generating plant of some kind would have to be constructed at the site to provide the power necessary to maintain an adequate heat supply for decommissioning. The government's job of increasing Ukraine's electrical output without relying on nuclear power has been hampered by environmental concerns about the pollution caused by fossil-fueled generation.

Effects of the Soviet collapse

Following the August 1991 coup attempt in the Soviet Union, the Ukrainian government began moving toward assuming responsibility for its nuclear power plants. On November 1, 1991, the Ukrainian Parliament declared its ownership of all the nuclear power plants in Ukraine. Feelings of nationalism at some locations prevented Russian experts from entering Ukrainian facilities.

In December 1991, Russia stopped providing funds for security at Ukraine's nuclear plants. Guards went without paychecks for two months while the Ukrainian government began taking steps to establish its own currency.

As Ukrainian sovereignty emerged, officials urged cooperation between Ukraine and Russia in contracting for services necessary to administer autonomous nuclear programs. Before the Soviet break-up, reactors were designed, manufactured and regulated by the central government in Moscow. Training institutes for technicians and engineers were located only in the Russian controlled cities of Moscow and Tomsk, Siberia. Ukraine had no domestic centers to train operators. The republic's officials looked to the Soviet organization MAPI and the Russian agency, Minatom RF for help in establishing its own nuclear programs.

Sources

"Editor's Notes." "Nuclear Power Industry in the Ukraine." *Soviet Life*, February 1986.

"Ukrainian Scientists Protest Nuclear Plans." *Science*, March 4, 1988.

"Many Chernobyls Just Waiting to Happen." *Business Week*, March 16, 1992.

Source Book: Soviet-Designed Nuclear Power Plants in the Former Soviet Republics and Czechoslovakia, Hungary and Bulgaria. U.S. Council for Energy Awareness, 1992.

"Nuclear Black Market Scares Experts." *Detroit News*, July 20, 1992.

—Karen Bellenir

PLANT PROFILES

203 • Chernobyl Nuclear Power Station, Units 1-3

255614
Kievskaya obl.
g. Chernobyl, Ukraine

Telex: 132209 NEON

Owning Utility

Ukratomenergoprom (Ukrainian Nuclear Energy Concern)
8 ul. Lenina
252000 Kiev, Ukraine
Mikhail Uemanets, Chairman

Phone: 044-229-27-57
Fax: 044-229-24-39

Basic Facts

Units 1-3 *Status:* Operable. *Type of Reactor:* Light-water-cooled, graphite-moderated reactor. *Megawatts (electric):* 1000. *Megawatts (thermal):* 3200. *Generator Supplier:* Elektrosila.

Key Dates

Unit 1 *Ordered:* 1971. *Construction began:* 1972. *Criticality:* August 1977. *First Power:* September 1977. *Commercial Operation:* May 1978. *Shut Down Expected:* 1993.

Unit 2 *Ordered:* 1971. *Construction began:* 1973. *Criticality:* November 1978. *First Power:* December 1978. *Commercial Operation:* May 1979. *Shut Down Expected:* 1993.

Unit 3 *Ordered:* 1974. *Construction began:* 1977. *Criticality:* June 1981. *First Power:* November 1981. *Commercial Operation:* June 1982. *Shut Down Expected:* 1993.

Plant Report

The three lesser known reactors. Operation at Chernobyl began in September 1977 when the station's first reactor started generating power. In the next seven years, three more reactors came on line. All four reactors were 1000-megawatt RBMK's. The world's worst nuclear mishap occurred in 1986 at Chernobyl-4 when it exploded.

Following the disastrous accident, the three other reactors were brought back on line. Safety measures adopted included:

- Operating at reduced power with control rods partially inserted
- Installing safety equipment
- Intensive operator training.

The three remaining operating units at Chernobyl generated 2,775 megawatts of electricity (925 net megawatts each), more than 20 percent of the nuclear-generated power in Ukraine. Chernobyl, located in Pripyat, Ukraine, is the nation's only RBMK station.

Units 5 & 6 cancelled. Before the Chernobyl accident, Soviet officials had begun construction work on two additional units at the site. Units 5 and 6 became objects of controversy a month before the Unit 4 explosion when *Literaturna Ukrayina* published charges of poor workmanship at Unit 5. Following the accident, in the face of growing public opposition to reliance on nuclear power, Soviet officials cancelled Chernobyl-5 and -6. Their decision was based on a new Soviet policy not to build anymore RBMK reactors and was apparently unrelated to the prior criticism.

Decision to shut down. Because of the accident in Unit 4, the Ukrainian Parliament demanded that the station be shut down by 1995. In 1991 Chernobyl-2 was closed following a fire in a turbine generator. In response, Ukrainian officials changed the proposed decommissioning of the entire Chernobyl station to 1993.

The action prompted Soviet officials to announce the need to construct a generating plant of some kind at Chernobyl. The Soviets claimed that because the republic had no reserve power, there would be no heat for the plant's buildings during the decommissioning process if additional generating capacity were not built.

Sources

"Soviet Scraps a New Atomic Plant in Face of Protest Over Chernobyl." *The New York Times*, January 28, 1988.
"Soviet Worries About Nuclear Safety After Chernobyl." *Nature*, August 1988.
"Update on Chernobyl — Safety and Safety, and Ever More Safety." *Electric Light and Power*, June 1991.
Source Book: Soviet-Designed Nuclear Power Plants in the Former Soviet Republics and Czechoslovakia, Hungary and Bulgaria. U.S. Council for Energy Awareness, 1992.

—Karen Bellenir

204 ★ Chernobyl Nuclear Power Station, Unit 4

255614
Kievskaya obl.
g. Chernobyl, Ukraine

Telex: 132209 NEON

Owning Utility

Ukratomenergoprom (Ukrainian Nuclear Energy Concern)
8 ul. Lenina
252000 Kiev, Ukraine
Mikhail Uemanets, Chairman

Phone: 044-229-27-57
Fax: 044-229-24-39

Basic Facts

Unit 4 *Status:* Shut down. *Type of Reactor:* Light-water-cooled, graphite-moderated reactor. *Megawatts (electric):* 1000. *Megawatts (thermal):* 3200. *Generator Supplier:* Elektrosila.

Key Dates

Unit 4 *Ordered:* 1974. *Construction began:* 1979. *Criticality:* November 1983. *First Power:* December 1983. *Commercial Operation:* April 1984. *Shut Down:* April 1986.

Chernobyl

UKRAINE

Plant Report

What happened. The world's worst nuclear power plant accident occurred at the Chernobyl Unit 4 reactor in the Soviet Ukraine on April 26, 1986.

Negligent testing procedures by plant technicians and serious flaws in reactor design led to a disaster unprecedented in the annals of nuclear power generation. Two explosions and a fire killed 31 people, spewed radiation around the world to create an environmental holocaust, and forced the evacuation of some 200,000 people.

Chernobyl-4 follows the standard design of Soviet reactors and is very different from the pressurized water and boiling water reactors used almost universally in the West. It is a graphite-moderated, boiling-water-cooled facility (RBMK) that lacks the massive containment structure common to most nuclear energy plants elsewhere in the world.

At 1 a.m. on April 25, slightly more than 24 hours before the accident, technicians set the stage for an experiment to determine if the reactor's steam-driven turbines could supply power to diesel generators in case of an unexpected power cutoff. They began by closing the plant's emergency cooling system, a grave mistake that left the plant in a dangerous condition. The intention was to let the plant supply electricity to the area instead of taking it off line.

Ensuingly, technicians flipped switches that resulted in costly errors. They removed all but a few of the control rods and disconnected the automatic-rod-control system. Even so, the operators had trouble stabilizing the power output, eventually getting it back up to 200 thermal megawatts instead of the 700-megawatt level the test called for. The water and pressure levels continued to fluctuate and operators decided the block the emergency water- and pressure-level signals, bringing the plant to "the maximum point of danger," according to a U.S. nuclear industry expert.

The test commenced at 1:23.04 a.m. on April 26. Operators then made a fatal mistake, shutting off the steam valves to prevent steam from reaching the turbine-generating unit they intended to test. Operators also bypassed a warning signal that would have commanded the plant to shut down.

A runaway reaction ensued after a heat and steam buildup was triggered by less cooling water coming through the reactor core. Soon after, two enormous explosions rocked the plant.

According to nuclear experts, the first explosion (steam) stemmed from the reaction that severely damaged the reactor core and surrounding structures. This damage set the stage for the second explosion (hydrogen), which was much more violent than the first. It blew the 1,000-ton steel lid off the reactor building; exposed the reactor's burning graphite core which reached temperatures of more than 5,000 degrees; and spewed deadly radioactive isotopes into the atmosphere—for 10 days.

Social Impact. The Chernobyl 4 accident stimulated anger, cynicism, doubt and fear about the nuclear industry. Alarming reactions spread while authorities attempted to scope the disaster.

A 1987 Chernobyl study by the antinuclear Worldwatch Institute concluded that governments were uncertain how to cope with the crisis. "Responsible officials in most countries reacted poorly to the accident," the report indicated. "Through weeks of warnings, misstatements, deceptions and corrections, the international community showed itself ill-prepared for a nuclear emergency."

However, a U.S. Nuclear Regulatory Commission (NRC) report noted that many organizations and functional groups participated in the emergency response. The report quoted then Soviet President Mikhail Gorbachev saying, in the future, greater attention will be paid to the reliability of equipment and "questions of discipline, order and organization" at nuclear power plants.

Reaction in the Soviet Union was one of confusion, mistrust and condemnation of Moscow for its shoddy reaction. The accident fueled a growing ecological movement in Russia with "Green World," a Ukranian nuclear opposition party, at the forefront. The group organized a rally of 50,000 people in 1989 in Kiev, which is 80 miles south of Chernobyl, in part to completely close all of Chernobyl's plants.

The accident itself and public mistrust of the nuclear system led to rallies and protests—worldwide. On the accident's first anniversary, hundreds of people protested nuclear energy by marching to the Seabrook nuclear power plant in Seabrook, N.H. The march resulted in approximately 900 arrests. An ABC News/Washington Post survey taken several weeks after the accident showed public opposition toward nuclear power at an all-time high of 80 percent.

Soviet authorities evacuated some 200,000 people from contaminated areas. France withdrew 12 of its nationals from Kiev; the Finnish Embassy sent a special plane to help evacuate more than 100 students and technicians from the area; and Britian evacuated 86 teachers and students from Kiev.

Elsewhere, Eastern Europe showed an antinuclear reaction; large demonstrations occurred in Rome and West Germany; and petition drives in Italy and Switzerland called for national referenda on nuclear programs.

The United States Department of Energy and the NRC published extensive reports. In part, the NRC attempted tto assess the accident's implications as they relate to commercial nuclear safety regulation in the United States. Congress staged hearings to also determine what happened.

A new Soviet public attitude toward nuclear power reflected culturally in "Sarcophagus," a powerful drama inspired by the Chernobyl tragedy.

Environmental Impact. The accident's venting of lethal gases into the atmosphere exacted a heavy toll on the ecology and agriculture of the Soviet Union and Europe. Explosions spewed at least 50 tons of volatile radioactive particles—10 times the fallout at Hiroshima—across Byelorussia, western Russia and the Ukraine.

Humans and animals became exposed to harmful radiation doses, while the agriculture and food industries—in the Soviet Union and neighboring countries—were disrupted. People,

Plant Profiles

Chernobyl 204

UKRAINE

land and food were monitored for radioactive releases of strontium, cesium, and plutonium.

Soviet scientists reported that Chernobyl 4 contained about 190 metric tons of uranium dioxide fuel and fission products. They estimate that about four percent of this material escaped.

Radioactive contamination was scattered irregularly depending on weather conditions. Scientists indicated that about 70 percent of the contamination fell on Byelorussia. But a large area in the Russian Federation was also contaminated, as were parts of the Ukraine. Overall, some 13,100 square miles of agricultural land, dotted with small cities, became contaminated with radioactivity at levels of five or more curies per square kilometer.

The worst fallout settled in the 10-kilometer radius surrounding the plant. Fallout levels in that zone were as high as 130,000 curies per square kilometer.

The primary environmental concern today is contamination of the soil with Cesium-137, which has a half-life of about 31 years. According to Soviet scientists, 28,000 square kilometers are contaminated by Cesium-137 to levels greater than five curies per square kilometer. About 10,500 square kilometers are contaminated to levels more than 15 curies.

The Chernobyl area is heavily agricultural and the radiation fallout damaged the area's food cycle and water supply. After the accident, farm products in the contaminated zone were irradiated and could not be harvested. Food must still be monitored and, in some cases, native produce, milk and meat can't be consumed. Farmland was abandoned, forests razed and the trees buried. Animals were born deformed and cattle grazing on contaminated grass soon produced milk with significant levels of radioactive materials.

The radioactive cloud also created environmental problems outside the USSR. For several weeks, at least 100 million Europeans, confused and frightened, reduced their consumption of fresh vegetables and dairy products. Farmers in some countries had to leave fresh produce to rot in the fields. Pastures for sheep farmers in North Wales and Cumbria in the United Kingdom were littered with radiation; livestock in those areas was destroyed more than two years after the accident. Large numbers of reindeer, an important part of Lapland's economy in Sweden, were measured in excess of permissible radiation levels and declared unfit for human consumption. To protect the Lapps, the Swedish government bought and fed contaminated deer meat for the region's mink and fox farms.

Health Impact. The health consequences from the Chernobyl 4 accident may never be determined. Figures frequently differ, sometimes to the extreme, and some studies may exaggerate the toll if organizations are jockeying for publicity points.

The Soviet's deplorable health situation and Moscow's order that no medical diagnosis may connect an illness with radiation exposure may make it hard to figure if radioactive emissions are to blame for deaths.

Many experts agree that one of the most reliable studies is the International Chernobyl Project (ICP), a United Nations-affiliated multinational report conducted five years after the accident.

The ICP concluded the following: Despite the dangerous radiation emitted to the areas surrounding Chernobyl, no evidence exists of higher rates of leukemia, other cancers, thyroid disorders, birth defects or other health problems. Instead, the report indicated that significant non-radiation related health disorders resulted from high radiation levels.

The Soviet government acknowledged that 31 workers and emergency personnel exposed to extremely high radiation levels on the reactor site died soon after the accident. It was the only recorded time in commercial nuclear history that people were killed. Two people died almost immediately from trauma or burns, and the remaining 29 died over the next two months from radiation effects complicated by severe thermal and beta-radiation burns. Another 200 people, all of whom worked at the reactor or were brought in to deal with the emergency, were hospitalized because of acute radiation sickness.

Of the 600,000 people who participated in cleanup efforts, 4,000 probably died prematurely within five years of the blast, according to Alfred Friendly Jr. and Murray Feshbach, co-authors of "Ecocide in the USSR." The book also cites experts estimating up to 10,000 early deaths from radiation sickness, and 1,000 to 25,000 fatal Chernobyl-induced cancer cases in the next half century. Other estimates range from 170 to 500,000 extra cancer deaths.

Even the Soviets admitted that many people would suffer physically. Thyroid cancer in Russia has been on the rise since the disaster. Health experts expect an increase in mental retardation and genetic defects.

The Worldwatch Institute's 1987 study determined that "the long-term health effects of the fallout cannot be accurately predicted." Of the 200,000 people evacuated by the Soviet government, many received between 20-100 rads (units of measurement of absorbed radiation), a non-lethal dose that nonetheless increases the risk of developing cancer.

"Estimates of resulting cancer deaths by researchers range from fewer than 1,000 to almost 500,000," Worldwatch concluded. "Many cancers will take decades to develop, and since Chernobyl is the first nuclear disaster of its kind, there is little evidence on which to base such estimates. The victims of Hiroshima are still developing cancer 42 years later."

Worldwide scientific, political and economic dimensions. The Worldwatch Institute's 1987 Chernobyl study noted that explosions induced "potentially health-threatening levels of radioactive materials deposited more than 2,000 kilometers from the plant and in at least 20 countries. An estimated 7,000 kilograms of radioactive materials were released—over 1,000 times what was released at Three Mile Island."

Sweden became the first country to suspect a possible nuclear accident. After the explosions, a southeast wind sent abnormally high levels of radiation toward Scandinavia, when Swedish monitoring devices made the detection. Swedish officials found no leaks at their own plants and looked in the direction of the prevailing breezes—toward the Soviet Union. Moscow initially denied the accident, but on April 28, two days after the disaster, the Soviet news agency Tass announced tersely that an accident had occurred at Chernobyl. Experts believe that if Sweden hadn't picked up the radiation on its monitors, the Soviets would not have announced what happened.

The Soviets' silence—43 hours—before Tass's announcement and close control of information afterward infuriated Western Europeans, who were in the eye of the fallout. Adding insult to injury, the Soviets locked up the accident's records for three years, prompting critics to paint Russia immoral with a disdain for human life.

More than 10 European nations produced reports on the radiological effects. Italy, Sweden, and Finland published the most detailed studies, which attempted to assess collective long-term doses in their countries.

After Sweden and Finland reported radiation on April 28, gases persisted across Scandinavian countries and reached central Euroope on April 30, reflecting a northwestern trajectory that subsequently veered west toward the Baltic Sea. From May 1-3, the plume spread to essentially cover Europe.

The accident created domestic and international economic repercussions. Agricultural and other economic losses worldwide topped $4.5 million. The Soviets' local safety measures included sheltering citizens; administering iodine pills; restricting the distribution of milk or dairy products; advising citizens not to drink rainwater; advising people to wash fresh vegetables before eating them; and banning or restricting game or domestic meat. In international trading, eight countries, including the United States, applied economic sanctions on the Soviets to prevent the importation of contaminated food.

Countries other than the Soviet Union were affected. The sale of powdered milk from Europe was banned in Brazil and Thailand. A Brazilian court found that powdered milk imported from western Europe was contaminated with radioactive residue from the accident. A Gambian newspaper accused Europeans of dumping radioactive powdered milk on African countries.

Nuclear power after Chernobyl. Outside assistance is vital in Russia's nuclear modernization efforts. Germany is leading efforts to get Western cooperation on improving nuclear safety in Russia; Sweden is helping Lithuania upgrade its facilities; and the European Community is funding the rehabilitation of two Soviet-built nuclear reactors in Bulgaria.

The crippled Chernobyl 4 reactor was an RBMK with no containment walls. It will remain isolated for several thousand years. (Cleanup efforts included dropping tons of sand on the reactor that remain today.) Two years after the accident, sloppy repair work, drunkenness, and nepotism were still rampant at Chernobyl, according to the Communist Party newspaper Pravda.

In May 1992, the Group of Seven industrial nations considered providing financial help for the 16 RBMKs still operating in Russia, the Ukraine and Lithuania. (Russia relies on nuclear power for 12 percent of its electricity needs.) Estimates range from $6 billion to $50 billion to rehabilitate, decommission or replace the antique RBMKs.

Five planned stations in Russia were cancelled after the accident.

Investigations by the International Atomic Energy Agency and the Nuclear Energy Agency of the Orgganization for Economic Cooperation and Development concluded that the Chernobyl accident was specific to a particular type of Soviet reactor. The reports did not call into question the safety of Western reactors. Western European nations affected by the radioactive debris are continuing to push ahead with their nuclear programs.

Further implications in the Soviet Union. Overall, the accident crippled the Chernobyl region environmentally, economically, psychologically and politically.

Some Western observers viewed Gorbachev's announcement about the accident—however late—the debut of glasnost, or openess. Another theory is that Chernobyl was more a turning point for Russian nationalism. Unlike the Baltic republics or the people of Caucasus, Byelorussia and the Ukraine, non-Russian slavic republics, supported Moscow strongly until the disaster. But the disaster and its aftermath became catalysts for political upheaval along the southwestern edge of the Soviet Union.

In legal ramifications, a Soviet Supreme Court judge sentenced Chernobyl's former director, its chief engineer and his deputy to 10 years each in a labor camp. Three subordinates got lesser terms.

Sources

"The Fear of Nuclear Chaos." *MacLean's*, May 12, 1986.
"After Chernobyl." *Fortune*, May 26, 1986.
"Anatomy of a Catastrophe." *Newsweek*, September 1, 1986.
"Chernobyl Shakes Reindeer Culture of Lapps." *The New York Times*, September 14, 1986.
"Chernobyl and the U.S. Nuclear Industry." *Bulletin of the Atomic Scientists*, November 1986.
Health and Environmental Consequences of the Chernobyl Nuclear Power Plant Accident. U.S. Department of Energy, Washington, D.C., 1987.
Report on the Accident at the Chernobyl Nuclear Power Station. U.S. Nuclear Regulatory Commission, Washington, D.C., 1987.
"Brazil Court Bans Milk From Europe." *The New York Times*, January 12, 1987.
"Senate Panel Hears Russian About Chernobyl." *The New York Times*, January 21, 1987.
Reassessing Nuclear Power: The Fallout From Chernobyl. Worldwatch Institute, March 1987.
"Disaster's Impact on Health Won't Be Known for Years." *The Wall Street Journal*, April 23, 1987.
"Chernobyl's Legacy: Days of Radiation, but Decades of Fears Over Health." *The New York Times*, April 27, 1987.
"Europe After Chernobyl: Cooler Attitudes Toward Nuclear Power." *The New York Times*, April 27, 1987.
"400 Recall Chernobyl in Anti-Seabrook Rally." *Boston Globe*, April 27, 1987.
"Specter of Chernobyl Looms Over Bangladesh." *New York Times*, June 5, 1987.
"Judgement at Chernobyl." *Time*, July 20, 1987.
"Chernobyl: The Lessons the Soviets Learned." *U.S. News & World Report*, July 20, 1987.
"Ten Years in Stir for Chernobyl's Scapegoats." *Newsweek*, August 10, 1987.
"After Chernobyl, Africans Ask if Food Is Hot." *The New York Times*, January 10, 1988.
"Problems Still Plague Chernobyl." *The Washington Post*, April 25, 1988.
"Chernobyl: What Really Happened." *Technology Review*, July 1989.
"The Second Chernobyl Disaster." *The Philadelphia Inquirer Magazine*, January 14, 1990.
"Four Years Later, Soviets Reveal Wider Scope to Chernobyl Horror." *The New York Times*, April 28, 1990.
Mounfield, Peter. *World Nuclear Power*, Routledge, 1991.
"Children of Chernobyl." *Greenpeace Magazine*, January-February 1991.

"As Villagers Return Unfazed, Chernobyl Aims for Tourists." *The New York Times*, February 4, 1991.

Chernobyl Briefing Book. U.S. Council for Energy Awareness, Washington, D.C., March 26, 1991.

"Chernobyl: Five Years Later, the Danger Persists." *The New York Times Magazine*, April 14, 1991.

International Chernobyl Project (Overview). International Advisory Committee, May 1991.

"Stress, Anxiety Primary Health Effects From Chernobyl." *Nuclear Industry*, Third Quarter 1991.

Chernobyl Briefing Book. U.S. Council for Energy Awareness, Washington, D.C., March 26, 1991.

"Chernobyl: Five Years Later, the Danger Persists." *The New York Times Magazine*, April 14, 1991.

—*Michael Richman*

205 ☆

Chernobyl Nuclear Power Station, Units 5-6

255614   Telex: 132209 NEON
Kievskaya obl.
g. Chernobyl, Ukraine

Owning Utility

Ukratomenergoprom (Ukrainian Nuclear   Phone: 044-229-27-57
    Energy Concern)   Fax: 044-229-24-39
8 ul. Lenina
252000 Kiev, Ukraine
Mikhail Uemanets, Chairman

Basic Facts

Units 5-6 *Status:* Cancelled. *Type of Reactor:* Light-water-cooled, graphite-moderated reactor. *Megawatts (electric):* 1000. *Megawatts (thermal):* 3200. *Generator Supplier:* Elektrosila.

Key Dates

Unit 5 *Ordered:* 1981. *Cancelled:* 1986. *Construction began:* 1981.
Unit 6 *Ordered:* 1983. *Cancelled:* 1986. *Construction began:* 1983.

Plant Report

See Plant Report for Chernobyl Nuclear Power Station, Units 1-3.

206 ●

Khmelnitsky Nuclear Station, Unit 1

281093   Telex: 291618 LUCH
Khmel'nitskaya obl. Slavutskiy r-n
g. Neteshin, Ukraine
Nikolai Gabriychuk, Plant Mgr.

Owning Utility

Ukratomenergoprom (Ukrainian Nuclear   Phone: 044-229-27-57
    Energy Concern)   Fax: 044-229-24-39
8 ul. Lenina
252000 Kiev, Ukraine
Mikhail Uemanets, Chairman

Basic Facts

Unit 1 *Status:* Operable. *Type of Reactor:* Pressurized light-water reactor. *Megawatts (electric):* 1000. *Megawatts (thermal):* 3000. *Generator Supplier:* Elektrosila.

Key Dates

Unit 1 *Ordered:* 1976. *Construction began:* 1981. *Criticality:* December 1987. *First Power:* December 1987. *Commercial Operation:* August 1988.

Plant Report

One operating unit. The Khmelnitsky nuclear station, located in Neteshin, Ukraine, generated its first power in December 1987. Unit 1 is a third-generation, Soviet-styled VVER-1000 reactor. Three additional 1,000-megawatt units were under construction. The Ukrainian Parliament halted efforts to complete them in the aftermath of the Chernobyl accident. The Ukrainian government also closed Unit 1 for political reasons.

Soviet officials originally had plans for a fifth unit at Khmelnitsky. Researchers were working on an improved VVER-88 design, which incorporated improvements based on the Chernobyl experience. Planners hoped Unit 5 would begin operation in 1994, but it was judged not economical.

In the fall of 1991, the Soviet Ministry of Atomic Power and Industry (MAPI) announced its plans to complete a second unit at Khmelnitsky in spite of Ukrainian opposition. Unit 2's status following the break-up of the Soviet Union and the inception of Ukrainian independence is unknown.

Sources

Source Book: Soviet-Designed Nuclear Power Plants in the Former Soviet Republics and Czechoslovakia, Hungary and Bulgaria. U.S. Council for Energy Awareness, 1992.

—*Karen Bellenir*

207 ○

Khmelnitsky Nuclear Station, Units 2-4

281093   Telex: 291618 LUCH
Khmel'nitskaya obl. Slavutskiy r-n
g. Neteshin, Ukraine
Nikolai Gabriychuk, Plant Mgr.

Owning Utility

Ukratomenergoprom (Ukrainian Nuclear   Phone: 044-229-27-57
    Energy Concern)   Fax: 044-229-24-39
8 ul. Lenina
252000 Kiev, Ukraine
Mikhail Uemanets, Chairman

Basic Facts

Units 2-4 *Status:* Under construction. *Type of Reactor:* Pressurized light-water reactor. *Megawatts (electric):* 1000. *Megawatts (thermal):* 3000. *Generator Supplier:* Elektrosila.

Rovno

UKRAINE

Key Dates

Unit 2 *Ordered:* 1979. *Construction began:* 1985. *Commercial Operation Expected:* Indefinite.

Unit 3 *Ordered:* 1983. *Construction began:* 1986. *Commercial Operation Expected:* Indefinite.

Unit 4 *Ordered:* 1986. *Construction began:* 1987. *Commercial Operation Expected:* Indefinite.

Plant Report

See Plant Report for Khmelnitsky Nuclear Station, Unit 1.

---

208 ●

Rovno Nuclear Power Plant, Units 1-3

265921   Telex: 167771 ATOM
Rovenskaya obl.
pgt. Kuznetzovask, Ukraine
Vladmir Korovkin, Plant Mgr.

Owning Utility

Ukratomenergoprom (Ukrainian Nuclear   Phone: 044-229-27-57
    Energy Concern)   Fax: 044-229-24-39
8 ul. Lenina
252000 Kiev, Ukraine
Mikhail Uemanets, Chairman

Basic Facts

Units 1-3 *Status:* Operable. *Type of Reactor:* Pressurized light-water reactor. *Megawatts (electric):* Unit 1, 392; Unit 2, 416; Unit 3, 1000. *Megawatts (thermal):* Units 1-2, 1375; Unit 3, 3000. *Generator Supplier:* Elektrosila.

Key Dates

Unit 1 *Ordered:* 1971. *Construction began:* 1976. *Criticality:* December 1980. *First Power:* December 1980. *Commercial Operation:* September 1981.

Unit 2 *Ordered:* 1971. *Construction began:* 1977. *Criticality:* December 1981. *First Power:* December 1981. *Commercial Operation:* July 1982.

Unit 3 *Ordered:* 1979. *Construction began:* 1981. *Criticality:* November 1986. *First Power:* December 1986. *Commercial Operation:* May 1987.

Plant Report

Three operating units. The Rovno nuclear power plant, located in Kuznetsovsk, Ukraine, has three operating units. Rovno-1 and -2 are second-generation Soviet-styled Model V213 VVER-440s. They were designed at the same time the Soviet Union issued its first uniform safety requirements. Unit 3 is a third-generation VVER-1000 reactor.

Additional construction at the Rovno site was halted by the Ukrainian Parliament following the 1986 Chernobyl accident. The Soviet Ministry of Atomic Power and Industry (MAPI) issued an announcement the in fall of 1991 expressing its wishes for Rovno-4's completion. Following the USSR's demise and Ukraine's independence, the construction status at Rovno-4 is uncertain.

Operational studies. The Rovno plant has experienced problems with breaks in its steam-generator tubes. As a result, in 1989 the plant hosted one of the working groups formed under the U.S./Soviet Joint Coordinating Committee on Civilian Nuclear Reactor Safety (JCCNRS). JCCNRS working group 2, which analyzes severe accidents based on experience and computer models, visited Rovno to study hypothetical loss-of-coolant scenarios and situations where operators control a sudden change in temperature without shutting down the reactor.

Another group sponsored by the U.S. Nuclear Regulatory Commission studied the fire prevention techniques in use at the Rovno facility.

International cooperation. In addition to its participation in U.S. working groups, the Rovno plant has cooperated with other international efforts at improving reactor safety. The International Atomic Energy Agency (IAEA) sent an OSART (Operational Safety Review Team) mission to the station in 1988. The OSART objective was to review operational safety practices. The study was the first one of its type conducted in the former Soviet Union. The team's assessment was generally positive. Rovno was commended for staff proficiency, safety awareness, radiation shielding, and ecological considerations. Problems were noted in the quality of equipment arriving at the site and instrument shortages. The OSART panel also recommended that Rovno's station management be given more authority over operations.

In 1992 the Moscow Research Institute of Instrument-Making installed special computers, designed to facilitate plant operations, at one of Rovno's units. The installation marked the first time the system was installed in the former Soviet Union. Plans call for its installation at Rovno's other units.

Sources

"Soviet Nuclear Power Plant Programme Marks Time." *Nature*, January 1989.
Source Book: Soviet-Designed Nuclear Power Plants in the Former Soviet Republics and Czechoslovakia, Hungary and Bulgaria. U.S. Council for Energy Awareness, 1992.

—*Karen Bellenir*

---

209 ○

Rovno Nuclear Power Plant, Unit 4

265921   Telex: 167771 ATOM
Rovenskaya obl.
pgt. Kuznetzovask, Ukraine
Vladmir Korovkin, Plant Mgr.

Owning Utility

Ukratomenergoprom (Ukrainian Nuclear   Phone: 044-229-27-57
    Energy Concern)   Fax: 044-229-24-39
8 ul. Lenina
252000 Kiev, Ukraine
Mikhail Uemanets, Chairman

Plant Profiles

Basic Facts

Unit 4 *Status:* Under construction. *Type of Reactor:* Pressurized light-water reactor. *Megawatts (electric):* 1000. *Megawatts (thermal):* 3000. *Generator Supplier:* Elektrosila.

Key Dates

Unit 4 *Ordered:* 1983. *Construction began:* 1986. *Commercial Operation Expected:* Indefinite.

Plant Report

See Plant Report for Rovno Nuclear Power Plant, Units 1-3.

---

210 ●

South Ukraine Nuclear Power Plant, Units 1-3

329543
Nikolaevskaya obl
Arbuzinskiy r-n
pgt. Konstantinovka, Ukraine
Vladmir Fuks, Plant Mgr.

Telex: 272471 PROTON

Owning Utility

Ukratomenergoprom (Ukrainian Nuclear Energy Concern)
8 ul. Lenina
252000 Kiev, Ukraine
Mikhail Uemanets, Chairman

Phone: 044-229-27-57
Fax: 044-229-24-39

Basic Facts

Units 1-3 *Status:* Operable. *Type of Reactor:* Pressurized light-water reactor. *Megawatts (electric):* 1000. *Megawatts (thermal):* 3000. *Generator Supplier:* Elektrosila.

Key Dates

Unit 1 *Ordered:* 1974. *Construction began:* 1977. *Criticality:* December 1982. *First Power:* December 1982. *Commercial Operation:* October 1983.

Unit 2 *Ordered:* 1974. *Construction began:* 1979. *Criticality:* December 1984. *First Power:* January 1985. *Commercial Operation:* April 1985.

Unit 3 *Ordered:* 1981. *Construction began:* 1985. *Criticality:* September 1989. *First Power:* September 1989. *Commercial Operation:* December 1989.

Plant Report

Background. The South Ukraine Nuclear Power Plant is located in Konstantinovka, Ukraine. The station consists of three third-generation VVER-1000 reactors. The design incorporated some western practices pertaining to plant safety. The status of a fourth reactor under construction at the site is unknown.

International cooperation. The South Ukraine station participated in a plant exchange program sponsored by the World Association of Nuclear Operators (WANO). In 1991 plant personnel visited the Waterford nuclear plant in Louisiana.

A French firm, Framatome, manufactured specialized equipment to remove bolts from steam-generator manhole covers. The South Ukraine station was one of three VVER-1000s in the former USSR to receive the machinery.

Sources

Source Book: Soviet-Designed Nuclear Power Plants in the Former Soviet Republics and Czechoslovakia, Hungary and Bulgaria. U.S. Council for Energy Awareness, 1992.

—Karen Bellenir

---

211 ○

South Ukraine Nuclear Power Plant, Unit 4

329543
Nikolaevskaya obl
Arbuzinskiy r-n
pgt. Konstantinovka, Ukraine
Vladmir Fuks, Plant Mgr.

Telex: 272471 PROTON

Owning Utility

Ukratomenergoprom (Ukrainian Nuclear Energy Concern)
8 ul. Lenina
252000 Kiev, Ukraine
Mikhail Uemanets, Chairman

Phone: 044-229-27-57
Fax: 044-229-24-39

Basic Facts

Unit 4 *Status:* Under construction. *Type of Reactor:* Pressurized light-water reactor. *Megawatts (electric):* 1000. *Megawatts (thermal):* 3000. *Generator Supplier:* Elektrosila.

Key Dates

Unit 4 *Ordered:* 1983. *Construction began:* 1987. *Commercial Operation Expected:* Indefinite.

Plant Report

See Plant Report for South Ukraine Nuclear Power Plant, Units 1-3.

---

212 ●

Zaporozhye Nuclear Power Station, Units 1-5

332608
Zaporozhskaya obl.
pos. Energodar, Ukraine
Vladmir Bronnikov, Plant Mgr.

Telex: 127445 ATOM

Owning Utility

Ukratomenergoprom (Ukrainian Nuclear Energy Concern)
8 ul. Lenina
252000 Kiev, Ukraine
Mikhail Uemanets, Chairman

Phone: 044-229-27-57
Fax: 044-229-24-39

Zaporozhye

Basic Facts
Units 1-5 *Status:* Operable. *Type of Reactor:* Pressurized light-water reactor. *Megawatts (electric):* 1000. *Megawatts (thermal):* 3200. *Generator Supplier:* Elektrosila.

Key Dates
Unit 1 *Ordered:* 1978. *Construction began:* 1980. *Criticality:* September 1984. *First Power:* December 1984. *Commercial Operation:* April 1985.

Unit 2 *Ordered:* 1978. *Construction began:* 1981. *Criticality:* June 1985. *First Power:* September 1985. *Commercial Operation:* October 1985.

Unit 3 *Ordered:* 1980. *Construction began:* 1982. *Criticality:* April 1986. *First Power:* December 1986. *Commercial Operation:* January 1987.

Unit 4 *Ordered:* 1980. *Construction began:* 1984. *Criticality:* December 1987. *First Power:* December 1987. *Commercial Operation:* January 1988.

Unit 5 *Ordered:* 1983. *Construction began:* 1985. *Criticality:* 1989. *First Power:* August 1989. *Commercial Operation:* October 1989.

Plant Report

A five-unit nuclear station. The Zaporozhye Nuclear Power Station, located in Energodar, Ukraine has five reactors. All of them are Soviet-styled third-generation VVER-1000's.

As a consequence of the 1986 Chernobyl accident, the Ukraine Parliament imposed a ban on further construction at the site. In 1991 Soviet officials hoped to resume construction and complete an additional unit by 1995. The status of another potential reactor at Zaporozhye is uncertain following the dissolution of the Soviet Union.

Steam generator difficulties. Zaporozhye plant officials reported corrosion problems in the plant's steam generators. The steam generators in all five units may have to be replaced.

International cooperation. Prior to the Soviet break-up, the United States/Soviet Joint Coordinating Committee on Civilian Nuclear Reactor Safety (JCCNRS) established twelve separate working groups. Two of the groups have performed studies at Zaporozhye. Group 1, which examines safety regulation issues, visited the plant in 1989. Group 4, in its efforts to study methods necessary to safely shut down a plant after a fire, visited Zaporozhye's local fire brigade training and research center in 1990.

The Zaporozhye plant also participated in a personnel exchange program sponsored by the World Association of Nuclear Operators (WANO). A team from Zaporozhye visited the Catawba plant in South Carolina.

A French firm, Framatome, manufactured specialized equipment to remove bolts from steam-generator manhole covers. The Zaporozhye station was one of three VVER-1000's in former USSR to receive the machinery.

Sources
Source Book: *Soviet-Designed Nuclear Power Plants in the Former Soviet Republics and Czechoslovakia, Hungary and Bulgaria.* U.S. Council for Energy Awareness, 1992.

—*Karen Bellenir*

Zaporozhye Nuclear Power Station, Unit 6

332608
Zaporozhskaya obl.
pos. Energodar, Ukraine
Vladmir Bronnikov, Plant Mgr.

Telex: 127445 ATOM

Owning Utility
Ukratomenergoprom (Ukrainian Nuclear Energy Concern)
8 ul. Lenina
252000 Kiev, Ukraine
Mikhail Uemanets, Chairman

Phone: 044-229-27-57
Fax: 044-229-24-39

Basic Facts
Unit 6 *Status:* Under construction. *Type of Reactor:* Pressurized light-water reactor. *Megawatts (electric):* 1000. *Megawatts (thermal):* 3200. *Generator Supplier:* Elektrosila.

Key Dates
Unit 6 *Ordered:* 1983. *Construction began:* 1986. *Commercial Operation Expected:* Indefinite.

Plant Report
See Plant Report for Zaporozhye Nuclear Power Station, Units 1-5.

Union of Soviet Socialist Republics

COUNTRY REPORT

The Soviet Union in history

The Soviet Union (USSR), as a political entity, lasted 74 years. Its 15 republics dissolved in 1991 at which time 11 of them entered into an alliance called the Commonwealth of Independent States (CIS). During its tenure as a world power, the USSR played a major role in the development of nuclear power. Its legacy is carried on by the newly-formed, individual governments.

Development of nuclear power

The Soviet Union began its involvement with nuclear power development in the early 1940's. Soviet scientists achieved the nation's first controlled chain reaction on December 24, 1946. Most of the initial work on thermal reactor design in the USSR was done at the I.V. Kurchatov Institute of Atomic Energy. The Radium Institute in St. Petersburg developed the nuclear fuel cycle for reactors. The first Soviet nuclear power plant began operation on July 27, 1954 in Obnisk, near Moscow. It had a generating capacity of 5,000 kilowatts. The Soviets believed that long-term nuclear capacity would be achieved through the use of breeder reactors designed to make more fuel than they consumed. In the late 1950's the Soviet researchers experimented with sodium-cooled, fast-breeder reactors at the Obnisk site.

Following their initial success with generating electrical power and establishing fast-breeder technology, the Soviets built a military nuclear station at Novotroitsk to produce weapons-grade plutonium. The station consisted of six 100 megawatt reactors constructed between 1958 and 1963.

The first commercial nuclear plants, using graphite-moderated, boiling-water-reactor technology, were built during the 1960s in Beloyarsk and Novovoronezh. Early industrial scale fast-breeder reactors were constructed during the 1970s at Shevenchenko and Beloyarsk. The Soviets originally planned to have large capacity breeder reactors developed by the early 1980s, however, difficulties forced the expected introduction of commercially economical fast-breeder reactors to be delayed until the end of the century.

The Soviet government created the Ministry of Atomic Power and Industry (MAPI) to oversee the construction, operation and safety of nuclear plants. Following the accident at Chernobyl in 1986 (see Chernobyl, Ukraine), responsibility for nuclear safety was transferred to a separate agency, Gospromatomnadzor (GPAN) with offices in Moscow and Kiev.

Types of Soviet reactors - RBMK

The Soviets developed several styles of reactors. The two primary types used to generate electricity are the RBMK reactor, a water-cooled, graphite-moderated unit; and the VVER reactor, a water-cooled, water-moderated reactor.

RBMK's produced 40 percent of the nuclear-generated electricity in the Soviet Union. The key advantage to the RBMK design is that it can be refueled without shutting down the reactor. The main equipment for early RBMKs was built at ordinary engineering plants which permitted more immediate production. In conceiving the design, Kremlin officials focused on the ability to generate large amounts of power and the capability to produce plutonium for weapons. Safety was not a primary consideration and the design has many disadvantages: The RBMK has no containment building and no redundant safety systems. The graphite used as a moderator can burn. Loss of coolant results in a faster, unstable nuclear chain reaction. There are no RBMK reactors operating outside the former Soviet Union.

The USSR, however, built 17 RBMKs, most of them with a capacity of 1,000 megawatts. The nuclear accident at Chernobyl occurred in an RBMK reactor. The RBMKs constructed in Lithuania were designed as 1,500 megawatt reactors but power levels have been reduced for safety reasons.

Types of Soviet reactors - VVER

The second style of Soviet reactor, the VVER, evolved over thrree decades. First-generation VVERs were produced between 1956 and 1970. The Soviets constructed prototypes at Novovorenezh in Russia. Novovoronezh-1 (VVER-210) operated until 1984. Novovoronezh-2 (VVER-365) operated until 1990. The design was first standardized with the VVER-440 unit which was built at Novovoronezh-3 and -4, and also constructed in Kola, Armenia, Bulgaria, Czechosolvakia, and in the former East Germany.

The VVER reactors were built at two specialized atomic engineering facilities, Atommash in Volgodonsk and Izhorski Works in Leningrad (St. Petersburg). The design is similar to western-styled pressurized water reactors but with significant differences.

Benefits of first-generation VVERs include:

— Radiation levels are reported to be lower than those at many Western plants.

— The design allows for larger amounts of power production.

— Six coolant loops enable more coolant to circulate

— Isolation valves enable operators to shut down individual loops for maintenance without shutting down the reactor.

In spite of these advantages, first-generation VVER reactors had significant deficiencies. They used an "Accident Localization System (ALS)" in place of western-styled containment buildings. The ALS was capable of controlling only one 4-inch pipe rupture. If a larger problem developed, the ALS was designed to vent radiation directly into the environment. Also of concern was the fact that the reactor top was not enclosed in the ALS. The reactors did not have emergency core-cooling systems or auxiliary feedwater systems. Their reactor pressure vessels were plagued with "embrittlement," a gradual weakening due to radiation. VVER instruments, safety systems, material quality, construction, and training all fell below western standards.

Second generation VVERs were designed between 1970 and 1980. Units were constructed in Russia, Ukraine, Hungary, Czechoslovakia, East Germany, and Finland. These 440 megawatt units ameliorated some of the problems experienced in the first-generation VVERs. Improvements included:

— Upgrading the Accident Localization System.

— Installing a vapor-suppression containment structure.

— Adding emergency core-cooling and auxiliary feedwater systems.

— Lining the reactor pressure vessel with stainless steel to reduce embrittlement.

— Improving the coolant pump.

— Standardizing components.

Unresolved problems included substandard instruments, controls and safety systems. A lack of documentation led to concerns about the quality of equipment used. Operator training and emergency preparedness varied significantly.

Third generation VVERs were designed between 1975 and 1985. These 1,000 megawatt units were built in Russia, Ukraine and Bulgaria. They used western-styled containment structures, employed standardization, and included safety improvements. Third-generation VVERs reduced the number of coolant loops to four and eliminated the loop-isolation valves. Redesigned fuel assemblies permitted better coolant flow, and control rods were refined.

Third-generation VVERs still fell below western standards in several areas. Instruments and controls were inferior. Because emergency systems were not isolated from operating systems, a malfunction in one system could prevent the operation of a back-up safety system. VVER plant fire protection systems were inadequate, and operating procedures and emergency preparedness were deficient.

Before the Soviet Union disintegrated, design improvements were being appraised, but researchers did not consider safety improvements economical. In 1989, the USSR and Finland started joint work on the VVER-91, a 1,000 megawatt unit that would have been one of the most advanced light-water reactors in the world. Finnish experts continue developing a VVER-92 design.

Combined heat and power units (CHPs)

In addition to building reactors for electrical generation and plutonium, the Soviets developed smaller nuclear stations to provide heat in urban areas. A prototype Combined Heat and Power plant (CHP) was constructed in Gorky, and another was built in Voronenezh. Before the Chernobyl accident, these heat-supplying reactors were considered innocuous enough to build near densely populated areas. However, following Chernobyl, public opposition forced officials to reexamine safety issues. One CHP project planned for Minsk was changed to a fossil-fueled heat generating station.

Location problems

Mismatched distribution between the Soviet population and the country's natural resources led to a clustering of nuclear power plants near major urban areas in the western regions of the USSR. Western Soviet regions, with their relatively high population density, demanded the most electrical power. Western Soviet regions also contained most of the Soviet's agricultural land and industrial base. However, an estimated 80 percent of all energy resources lay east of the Volga-Baltic Canal. Because of the great distances between these resources and the population centers, Kremlin officials considered nuclear power more economical than transporting fossil fuel from the far reaching parts of the USSR.

As a result of the increasing number of nuclear plants being built in the area, environmentally concerned researchers raised questions about the "ecological capacity" of the region. They claimed that heat discharges, the loss of water through evaporation (estimated at more than two cubic kilometers per year), and the accumulation of nuclear waste would ruin valuable land.

Radiation problems

Although it was not disclosed at the time, problems with radioactive waste emerged early in the Soviet nuclear program. During the 1940's and 1950's, radioactive waste materials were dumped in both urban and rural areas, in rivers and lakes. The handling of radioactive waste continues to raise questions. Critics claim that modern waste facilities are poorly built and that they are quickly reaching their full capacity. Investigators at one building, used to store nuclear waste products, measured radiation leakage up to 380 times normal amounts.

Nuclear picture in USSR before Chernobyl

From 1954 until 1970, the former Soviet Union had placed 1.5 million kilowatts of nuclear generating capacity in service. By 1980, two million kilowatts of generating capacity were being added annually. Plans called for the expansion of nuclear power through the 1980's, and officials anticipated that by 1990 five to eight million kilowatts of new nuclear generating capacity would be added per year. This would be achieved by building increasingly large plants, putting more generating units at a site, and upgrading existing plants to a capacity of four to seven million kilowatts. Nuclear power was considered cost-effective, reliable and safe for the environment.

The Draft Guidelines for the Economic and Social Development of the USSR for the five-year period of 1986 through 1990 called for a five to seven-fold increase in the amount of electricity generated by nuclear power plants. Construction projects underway in more than 20 regions of the USSR were expected to bring nuclear-generated electricity up to a total of 390 billion kilowatt-hours. By the turn of the century, Soviet officials expected to generate up to 30 percent of the country's electricity with nuclear power.

Post-Chernobyl issues

Before the Chernobyl nuclear accident in 1986, there was virtually no discussion about any hazards related to radiation. Safety issues were not raised except by underground sources. The undeniable events at the Chernobyl-4 reactor helped open discussion about public health and protection.

In 1987 the International Atomic Energy Agency (IAEA) began receiving an increasing number of requests from countries with Soviet-designed nuclear reactors to inspect their plants. In 1988 Soviet officials asked the IAEA to review a Soviet plant. In 1990 the IAEA began an investigation of VVER 440 reactors and reported that nearly 60 percent of the safety issues raised required immediate attention.

A Soviet report, examining reactor performance during the first six months of 1990, called safety standards "unsatisfactory." The report cited two leaks of radioactive matter, 70 emergency shutdowns, and a 15 percent increase in occupational accidents and deaths over the previous year. Vadim Malyshev, chairman of the commission responsible for the report, attributed the problems to "the slackening of discipline, the drop in staff (occupational) standards, irresponsible management, antiquated equipment, out-dated technology, and reduction in the number of departmental safety standards." Most of the problems enumerated concerned the VVER reactors, which are considered safer than RBMKs.

The Soviet agency, MAPI, adopted a plan to upgrade safety at all VVERs. In addition, an increasing number of western organizations began to express concern about safety improvements at Soviet-designed plants. Some efforts at renovating Soviet plants included:

— Entering into a contract with Singer Link-Miles for training simulators

— Signing a contract with Electricité de France for risk-assessment software

— Retaining Bechtel Power Corp to upgrade seismic safeguards

— Receiving proposals from Kraftwerk Union (West Germany) to help upgrade nuclear plants

The Commission of the European Communities (CEC) assumed responsibility for coordinating international efforts. Potential funding sources included the European Bank for Reconstruction and Development (EBRD) and the European Investment Bank (EIB). Although the World Bank had refused to provide money to build new plants, many were hopeful it would grant capital necessary to make safety improvements.

Increased U.S./Soviet cooperation

The post-Chernobyl era also marked the beginning of increased international cooperation between the United States and the USSR in issues surrounding nuclear power generation. The United States and the former Soviet Union entered into an agreement in 1988 to exchange information about plant operation and operator proficiency. Other issues such as accident analysis, plant-life extension, and the health effects of radiation were also addressed. As a result of these agreements, the U.S./Soviet Joint Coordinating Committee on Civilian Nuclear Reactor Safety (JCCNRS) established 12 separate working groups:

— Group 1 examines safety regulation issues.

— Group 2 analyzes severe accidents based on experience and computer models to determine how a plant will behave during an accident.

— Group 3 studies embrittlement and annealing, a technique that restores the sstrength of a reactor pressure-vessel.

— Group 4 researches methods to safely shut down a plant after a fire.

— Group 5 examines ways to improve and modernize old plants.

— Group 6 investigates ways to prevent and mitigate severe accidents.

— Group 7 studies the health and environmental effects of the Chernobyl accident.

— Group 8 discusses how Eastern and Western operators can learn from each other's experiences.

— Group 9 appraises the ability of operators to operate a plant.

— Group 10 explores elements that could corrode pipes and other plant components.

— Group 11 shares operating experiences, personnel training programs and management systems.

— Group 12 examines issues related to plant aging and extending plant life.

Although these groups were functioning before the break-up of the Soviet Union, the U.S. government expressed hope that political changes would not disrupt the ongoing work of the groups.

In addition, the U.S. government approved the sale of six advanced computers and previously restricted nuclear safety software to the Soviets in 1990. The Soviet central government in Moscow agreed to keep the computers from being used for military purposes. The computers represented a significant increase in the level of technology sold to the USSR. The Cyber 962 computers, built by Control Data Corp, would help Soviet scientists upgrade safety standards at existing power plants, design safer new reactors and better cope with accidents. The Soviets planned to develop commercial "inherently safe breeder reactors" similar in style to some of their existing military reactors. These units use lead or lead-bismuth, which does not evaporate quickly or explode, as a coolant. The Cyber 962s would also permit Soviet scientists to share data with other western computers.

World Association of Nuclear Operators (WANO)

In response to the Chernobyl accident, delegates from around the world representing 144 electric utilities with nuclear power plants met in Moscow. They formed the World Association of Nuclear Operators (WANO) patterned after American Institute of Nuclear Power Operations (INPO). WANO believes that the plant operator is ultimately responsible for reliable operation and safety. To improve plant operator capabilities, an early priority was to arrange personnel exchanges between power plants. By 1991, teams from every nuclear plant in Eastern Europe and the former Soviet Union had visited Western plants.

Power crisis preceded USSR break-up

Some analysts feel that the power crisis in the Soviet Union, exacerbated by post-Chernobyl events, was a contributing factor in the break up of the USSR. Prior to the accident, the Soviet Union had 56 operating nuclear units, 26 additional units were being constructed, and dozens more were on the drawing board. Following Chernobyl, public opposition rose and forced the cancellation of nuclear projects in the Soviet Republics of Azerbaijan, Georgia, Armenia, Lithuania, Byelorussia, Ukraine and Russia as well as in Hungary and the former East Germany.

After analyzing the situation, Soviet officials announced that they would not build any more RBMKs. Existing RBMKs were directed to operate at reduced power levels with their control rods partially inserted. Modifications, including reducing the amount of graphite in the fuel channels and developing procedures to reduce the time required to shut down a plant in the event of an emergency were ordered.

In the aftermath of Chernobyl an estimated 100,000 megawatts of generating capacity were lost through the combined effects of plant shutdowns, the cancellation of nuclear construction projects, and power reductions at RBMKs. The crisis left the USSR with only 45 operating nuclear units and virtually halted construction efforts. Soviet and Western researchers debated whether to fix the RBMKs or to permanently close them. Many Western experts believed the upgrades necessary to meet Western standards for safety would be too expensive, but Soviet officials stressed their dependence on nuclear power generation and lack of alternatives.

Every regional power system (except Siberia) in the former Soviet Union was anticipating power shortages by the mid 1990's. Many states had formally requested the construction of more nuclear power units. At the beginning of 1991, power reserves in the USSR were estimated at only two to four percent. Only four of the 15 republics were self-sufficient in power. Of these, only three were able to export electricity to other republics.

Plant aging contributed to the anticipated power shortages. Nuclear plants built during the late 1950's and 1960's were reaching the end of their productive lives. Other types of generating plants, including fossil fueled plants and hydroelectric plants, were equally old. Officials gave SpetsAtom the responsibility of investigating issues related to repairing, updating, or decommissioning old plants. The organization also undertook research into the development of robots and remote-controlled equipment to work in areas of excessive radiation.

For information about the nuclear industry in related countries, refer to Country Reports and Plant Profiles for Armenia, Kazakhstan, Lithuania, Russia, and Ukraine.

Sources

"Nuclear Power Engineering in the Soviet Union." *The Bulletin*, January 1980.

"Five-Year Plan." *Scientific American*, October 1983.

"Nuclear Power Engineering and International Security." *Soviet Life*, February 1986.

"Editor's Notes." *Soviet Life*, February 1986.

"Kraftwerk Gave Soviets Nuclear-Plant Proposals." *The Wall Street Journal*, January 15, 1987.

"Soviet Scraps a New Atomic Plant in Face of Protest Over Chernobyl." *The New York Times*, January 28, 1988.

"Lithuanians Protest at Soviet Nuclear Plants." *New Scientist*, October 8, 1988.

"Soviet Nuclear Power Plant Programme Marks Time." *Nature*, January 26, 1989.

"Soviet Citizens Not Impressed by IAEA." *Nature*, March 1989.

"Soviet Blackout Looms as Nuclear Disquiet Deepens." *New Scientist*, August 11, 1990.

"Soviets Buy Advanced Computers." *The Washington Post*, August 22, 1990.

"A Russian Tragedy in the Making." *Nuclear News*, February 1991.

"Fallout From Chernobyl Accident Still Clouds Soviet Nuclear Plans." *Journal of Commerce and Commercial New York*, April 19, 1991.

"Artificial Intelligence." *Omni*, June 1991.

"Nuclear Perils in Eastern Europe." *The Economist*, July 27, 1991.

"Soviet Nuclear Energy Plants Raise Concern as Republics Gain Control." *The Wall Street Journal*, August 30, 1991.

Mounfield, Peter. *World Nuclear Power*. Routledge, 1991.

"The End of an Empire." *Maclean's*, January 6, 1992.

"Many Chernobyls Just Waiting to Happen." *Business Week*, March 16, 1992.

"The Safety of Soviet-Designed Nuclear Plants: A U.S. Industry Perspective." U.S. Council for Energy Awareness, April 22, 1992.

Source Book: Soviet-Designed Nuclear Power Plants in the Former Soviet Republics and Czechoslovakia, Hungary and Bulgaria. U.S. Council for Energy Awareness, 1992.

"Former Soviet Union Awakens to a Nuclear Waste Nightmare." *The Detroit News*, July 19, 1992.

"Nuclear Black Market Scares Experts." *The Detroit News*, July 20, 1992.

"Russia Tries to Destroy Nuclear Arms." *The Detroit News*, July 21, 1992.

—Karen Bellenir

United Kingdom

COUNTRY REPORT

Early lead in power plant construction

In the early fifties, Britain lead the way in nuclear power plant technology. After opening the world's first atomic generating station at Calder Hall, she embarked on an ambitious program of new construction. The future looked bright for the British gas-graphite technology. Along with a dozen new plants planned for England, the Magnox reactor appeared at Latina in Italy and Tokai Mura in Japan. At the same time, Britain's Central Electric Generating Board (CEGB) began closing 175 outdated coal-fired plants.

By 1992, nuclear power provided about 20 percent of all electricity generated in England and Wales. In Scotland, it produced over 50 percent. At that time, Britain's 39 actively generating reactors, spread among 19 stations in 15 separate locations, had a capacity of over 12,800 MWe.

Protests and problems

However, the gas-graphite technology of the Magnox reactor and Advanced Gas Cooled Reactor (AGR) proved expensive and unreliable. Spiralling costs left them open to criticism.

Nuclear protestors attacked the industry on two fronts: cost -effectiveness and safety. Labor leaders began to perceive the nuclear industry as a threat. Atomic power cleaned up much of the fossil fuel pollution that had blanketed England's industrial heartland. It also improved the efficiency of electrical generation. However, it cost an estimated 8,000 jobs in the coal fields.

As designers expanded the size of reactors, they discovered increasingly sophisticated problems. The simple natural uranium of the early Magnox gave way to exotic blends of uranium oxide and plutonium. The cores grew hotter, melting normal metals and creating buffeting high pressure gales that tore gas ducting apart.

The first Magnox reactor took under four years to complete; the first AGR took over 18 years. Each successive AGR plant became a new prototype as researchers experimented with a technology not yet fully understood. By the late 1970s, the British government was ready to adopt American technology. It turned to the Pressurized Water Reactor (PWR), at least until researchers could perfect the next phase of gas-graphite reactor. It hoped by using the proven international design in a succession of similar plants, it could bring the cost of nuclear power back in line.

Early nuclear accident early

The Windscale plutonium pile accident, in 1957, focused public attention on Sellafield and its poor safety record. Medical reports of excessive childhood leukemia around the reprocessing center led journalists to dub the surrounding farming community a "green and poisoned land." The image quickly spread to the industry as a whole, even though Britain's nuclear power plants have never experienced a serious accident.

In 1982, the CEGB decided to launch its first PWR, Sizewell B. When it convened the public planning hearing, it expected disorganized, ineffective opposition from anti-nuclear groups. Instead, it found itself increasingly forced to defend questionable cost calculations, and the safety of the imported technology. Eventually, it won approval for the construction after Britain's longest public commission. However, planners could no longer hope to justify new plants by claiming nuclear power to be cheaper than coal.

Privatization announcement

In 1987, Prime Minister Margaret Thatcher decided it was time for the nuclear industry to pay its own way. She announced all electric generation would be privatized, sold to private investors, by 1992. Two companies, National Power (UK) and PowerGen (UK), would operate the generating stations, including reactors, while a third company would operate the National Power Grid.

Over the next three years, the industry and the government wrestled with the real cost of nuclear power generation. With the Magnox plants approaching 30 years of age, the prospect of decommissioning worn-out plants loomed large. Investors expressed little interest in the scheme without open-ended support from the government for decommissioning costs.

In November 1989, Thatcher withdrew all nuclear plants from the sell-off. Instead of privatizing the industry, she created two new companies, Nuclear Electric in England and Wales and Scottish Nuclear in Scotland. They would operate all existing atomic power plants, but the government would not approve any new reactors until a policy review in 1994. At the same time, it cut off money for reactor research.

In 1990, the second PWR, Hinkley Point C, did receive planning approval after a year-long inquiry, but Energy Secretary John Wakeham withheld funding.

Security and diversity of energy supply

During the inquiry, the government introduced a new concept. Instead of justifying nuclear construction by cost savings, it declared atomic power necessary for diversity and security of supply. The "Non-Fossil Fuel Obligation (NFFO)" compelled private electric distributors to purchase a quota of nuclear generated electricity. It imposed a "nuclear levy" on power from coal, oil and gas to offset the difference in production cost.

The new government policy creates an industry responsive to market pressures, trimming waste while promoting competition with other power sources. However, the lack of government subsidies may kill future development of gas-graphite reactors and the even more exotic Prototype Fast Reactor (PFR). Only those designs that prove economically viable will receive planning approval.

Nuclear agencies restructure

Meanwhile, the old players in the nuclear game jockey for new positions.

The original reactor development agency, United Kingdom Atomic Energy Authority (UKAEA) finds itself without an operating reactor. Government funding cutbacks closed

Winfrith and threaten Dounreay. It will team up with Rolls Royce to develop a new PWR, and it will explore international decommissioning under its new name, AEA Technology.

In the early 1950s, the UKAEA spun off British Nuclear Fuels Ltd (BNFL) to produce fuel rods for the Magnox reactor. In the 1990s, BNFL sees that business about to disappear. Consequently, it has developed new methods of dealing with radioactive waste and reprocessing fuels, while developing fuels for other reactor designs. The utilities complain that since the 1970s BNFL has raised the cost of nuclear fuel and reprocessing exorbitantly in order to improve its own profitability at their expense.

BNFL will also apply to build PWR's at Calder Hall and Chapelcross after 1994. It argues it can build the plants cheaper than Nuclear Electric. Another card in the companies favor might be the mothballing of its eight Magnox reactors at those two sites. Currently, the British military draws all its plutonium from BNFL. Unlike the utility reactors, BNFL's new design might produce plutonium and tritium.

The original utilities, Central Electric Generating Board and South of Scotland Generating Board, received new names and lost most of their non-nuclear generating capacity. Nuclear Electric retained a hydro-electric plant near Trawsfynydd. Each company will now compete with other electricity generators for consumers' dollars. In reaction, both companies have dropped research programs and closed aging plants. Hinkley C may never receive funding while the price of Sizewell B may continue to explode. That one plant may have to recover costs originally factored over four. Critics now claim it would be cheaper not to finish the reactor.

Without a renewal of government support in 1994, Britain's era of nuclear power generation may be nearing its end.

Sources

Patterson, Walter C. *Nuclear Power*. New York: Penguin Books Ltd., 1977.

Burn, Duncan. *Nuclear Power and the Energy Crisis*. New York: New York University Press, 1978.

"Power Complex at the CEGB." *Management Today*, May 1984.

"A Report on Sizewell." *Bulletin of the Atomic Scientists*, June 1984.

Bupp, Irvin C. and Jean-Claude Derian. *Light Water*. New York: Basic Books, 1987.

Collier, John, and Geoffrey F. Hewitt. *Introduction to Nuclear Power*. Washington: Hemisphere Publishing Corporation, 1987.

Lohr, Steve. "British Inquiry Gives Go-Ahead on Nuclear Power." *The New York Times*, January 27, 1987.

"Missing the Hinkley Point." *New Scientist*, September 30, 1989.

"Nobody's Fuel." *New Scientist*, November 18, 1989.

"Marshall explains why nuclear will not be privatized." *Nuclear News*, January 1990.

"Green and Poisoned Land?" *New Statesman & Society*, April 6, 1990.

"Thatcher's Failed Romance with Nuclear Power." *Bulletin of the Atomic Scientists*, April 1990.

"What to Pay for Nuclear Power?" *Nature*, July 5, 1990.

"Pressurized Accounting." *Nature*, July 5, 1990.

"Nuclear Power Reappears on the Agenda." *New Scientist*, February 2, 1991.

"Nuclear Industry Aims to Improve its Image." *New Scientist*, April 20, 1991.

Nuclear Electric Report and Accounts 1990-91, Nuclear Electric plc. Gloucester, July 1991.

"All Pals in Pall Mall." *Power in Europe*, September 12, 1991.

Mounfield, Peter. *World Nuclear Power*. Routledge, 1991.

—*Al Cook*

PLANT PROFILES

214 ▲
Berkeley Power Station, Units 1-2

Berkeley, Glos. GL13 9PA, United Kingdom
A. Bland, Site Mgr.
Phone: 0453-810431
Fax: 0453-810047
Telex: 43112 (BERKPS G)

Owning Utility

Nuclear Electric plc
Barnett Way
Barnwood, Glos. GL4 7RS, United Kingdom
John G. Collier, Chm. & CEO
Phone: 0452-652222
Fax: 0452-652776
Telex: 43501

Contact

S. Moore, Dir., PR
Nuclear Electric plc

Basic Facts

Units 1-2 *Status:* Decommissioned. *Type of Reactor:* Gas-cooled reactor. *Megawatts (electric):* 167. *Megawatts (thermal):* 585. *Reactor System Supplier:* The Nuclear Power Group (United Kingdom). *Generator Supplier:* Associated Electric Industries. *Architect Engineer:* United Kingdom Atomic Energy Authority (UKAEA).

Key Dates

Unit 1 *Ordered:* 1956. *Construction began:* 1957. *Criticality:* August 1961. *First Power:* June 1962. *Commercial Operation:* July 1962. *Shut Down:* March 1989. *Decommissioned:* 1989.

Unit 2 *Ordered:* 1956. *Construction began:* 1957. *Criticality:* March 1962. *First Power:* June 1962. *Commercial Operation:* October 1962. *Shut Down:* October 1988. *Decommissioned:* 1989.

Plant Report

Decommissioning. Berkeley Power Station, on the Severn River near Gloucester, began sending electricity to the British national power grid on June 12, 1962. It provided reliable service until October 1988, when it became Britain's first commercial nuclear power plant to reach the end of its useful life.

The facility dates from the beginning of nuclear power development in the UK, when the British-designed Magnox reactor began to evolve.

The term, Magnox, derived from the magnesium alloy canister used to contain the natural-uranium fuel elements. The system used a graphite-block core within a 3-inch thick, 67-foot-diameter steel pressure vessel. For the safety of the operators, a concrete biological shield enclosed the entire structure. The system's unique design allowed a special machine to refuel the core every two or three days without the need of a shutdown. This gave the Magnox a distinct advantage in reliability over other reactors, most of which need annual shutdowns of three or four weeks. In operation, it resembled the American High Temperature Gas Cooled Reactor (HTGR), but used carbon dioxide rather than helium as the coolant.

The Central Electric Generating Board (CEGB) chose the location because it met the three main criteria set up in 1955 by the government. Its stable geology could support the massive building and facilities. A high demand in the surrounding area provided a ready market and need. Its proximity to the sea guaranteed a ready source of coolant and a high volume of water to dilute small quantities of radioactive effluent. Its two Magnox reactors provided a high of 302 megawatts electric for Southern England and Wales through the national grid system.

The British government first proposed Berkeley in 1955 as one of a series of twelve stations to produce a total of 2000 MWe. They were to answer an increasing post-war demand for power and a need to replace dirty and inefficient coal-fired generating plants. After much re-evaluation, the program delivered only nine Magnox installations. Construction began on Jan. 7, 1957, and the CEGB officially opened the plant on April 5, 1963.

Design changes cost contractor. The United Kingdom Atomic Energy Authority (UKAEA) did the basic design work for the Magnox reactor but contractors completed much of the developmental. In Berkeley's case, that was Associated Electrical Industries. Design refinement resulted in a considerable loss for that contractor.

Nuclear Electric plc, inherited the plant from the CEGB on March 31, 1990, when the British government privatized all power generation except nuclear. It also inherited the liability for decommissioning the plant.

All Magnox stations subject to Long Term Safety Reviews. The designers, expecting a useful life of 20 to 25 years, originally licensed the plant till 1990. However, the reliability of the Magnox and the escalating cost of new facilities prompted a reassessment. Nuclear Electric petitioned the Nuclear Installations Inspectorate (NII) to extend the operating licenses of all its Magnox stations to 30 or 40 years. Without the Magnox stations, southern England and Wales could face shortfalls of 11,000 to 12,000 megawatts electric of generating capacity by the turn of the century.

High repair costs. The Long Term Safety Review, issued by the Inspectorate, cited several safety deficiencies. Welds in the steel pressure vessel and the high pressure gas ducting were dangerously brittle from more than 20 years of exposure to high radiation levels. The utility considers repairs to the welds, located inside the reactor containment vessel, impractical. In addition, to meet international safety standards, the plant would need new electronic control circuits. Unlike other Magnox stations with similar problems, the cost of repair for Berkeley was too high for the CEGB.

Decommissioning to take 100 years. Shutdown of the plant began in October 1988, in preparation for the three phase decommissioning process.

Once the second reactor ceased generation in March 1989, defuelling could begin. In that process, the utility will remove all fuel rods from the reactor, the cooling ponds and the store houses. Packed into 43-metric ton forged-steel flasks, they travel by rail to the British Nuclear Fuels reprocessing plant at Sellafield in Cumbria. The utility expects to complete that phase by 1995.

Nuclear Electric notes it has been transporting radioactive material by rail for more than 30 years without serious incident. In that time, it has moved 12,000 loads of spent fuel rods over more than three million miles of track. In 1984, it demonstrated the safety of the steel flask transportation vessel by crashing a train, traveling at 100 miles per hour, into one. The impact destroyed the train, but the flask remained intact.

With all radioactive material cleared from the site, demolition and dismantling of the structures and equipment outside the biological shield of the reactor begins. That will take another five years to complete.

The utility will allow a one-hundred-year waiting period for radiation in the core to decay before dismantling the reactor. Eventually, the pieces, classified as "intermediate-level waste," will arrive at BFNL's new underground depository at Sellafield.

In its 1990-91 Financial Report, Nuclear Electric termed its estimate of the final cost of closing the plant as "subject to considerable uncertainty." It could only estimate expenses that will span the next century.

Sources

"Five Reactors Must Prove Safety or Shut." *The Guardian*. September 10, 1991.
"Conditional Reprieve for Magnox Stations." *Financial Times*, September 12, 1991.
Mounfield, Peter. *World Nuclear Power*. Routledge, 1991.
Nuclear Electric Report and Accounts 1990-91, Nuclear Electric plc, Gloucester, n.d.

—Al Cook

215 •

Bradwell Power Station, Units 1-2

Bradwell-on-Sea
Southminster, Essex CM0 7HP, United Kingdom
P. Maycock, Stat.Mgr.

Phone: 0621-76331
Fax: 0621-76331 x 252

Owning Utility

Nuclear Electric plc
Barnett Way
Barnwood, Glos. GL4 7RS, United Kingdom
J.G. Collier, Chm. & CEO

Phone: 0452-652222
Fax: 0452-652776
Telex: 43501

Contact

S. Moore, Dir., PR
Nuclear Electric plc

Basic Facts

Units 1-2 *Status:* Operable. *Type of Reactor:* Gas-cooled reactor. *Megawatts (electric):* 173. *Megawatts (thermal):* 500. *Reactor System Supplier:* The Nuclear Power Group (United Kingdom). *Generator Supplier:* C.A. Parsons & Co., Ltd. (United Kingdom). *Architect Engineer:* The Nuclear Power Group (United Kingdom).

Key Dates

Unit 1 *Ordered:* 1956. *Construction began:* 1957. *Criticality:* August 1961. *First Power:* July 1962. *Commercial Operation:* July 1962.
Unit 2 *Ordered:* 1956. *Construction began:* 1957. *Criticality:* April 1962. *First Power:* July 1962. *Commercial Operation:* November 1962.

Plant Report

Nuclear Electric's smallest nuclear power plant resists old age. Bradwell Power Station remains one of Britain's oldest and smallest operating nuclear power plant, despite fears of developing cracks in its pressurized containment vessel. Only the prototype reactors operated by British Nuclear Fuels Ltd (BNFL) at Calder Hall and Chapelcross can claim longer service records. Its sister plant, Hunterston A in Scotland, began decommissioning in 1988.

Located on the Blackwater estuary, 1.5 miles from Bradwell-on-Sea, it began contributing electricity to the British national power grid on July 1, 1962.

The facility dates from the beginning of nuclear power development in the UK, when the British-designed Magnox reactor began to evolve.

Hot reactor refueled every two or three days without shutdown. The term, Magnox, derives from the magnesium alloy canister used to contain the natural-uranium fuel elements. The system uses a graphite-block core within a 3-inch thick, 67-foot-diameter steel pressure vessel. For the safety of the operators, a concrete biological shield encloses the entire structure. The system's unique design allows a special machine to refuel the core every two or three days without the need of a shutdown. This gives the Magnox a distinct advantage in reliability over other reactors, most of which need annual shutdowns of three or four weeks. In operation, it resembles the American High Temperature Gas Cooled Reactor (HTGR), but uses carbon dioxide rather than helium as the coolant.

The government-owned utility, Central Electric Generating Board (CEGB), chose Bradwell's location because it met the three main criteria set up in 1955 by the government. Its stable geology could support the massive building and facilities. A high demand in the surrounding area provided a ready market and need. Its proximity to the sea guaranteed a plentiful source of coolant and a high volume of water to dilute small quantities of radioactive effluent. The use of sea water avoided the need for unsightly cooling towers. Its two Magnox reactors provide 245 megawatts electric for Southern England and Wales through the national grid system.

Twelve nuclear power plants planned in aftermath of war. The Magnox construction program began in 1955 with a proposal for twelve stations to produce a total of 2000 MWe. They answered an increasing post-war demand for power and a need to eliminate dirty and inefficient coal-fired generating plants. After much re-evaluation, the program delivered only nine Magnox installations.

The Nuclear Electric plc, inherited the plant from its predecessor, the CEGB, on March 31, 1990, when the British government privatized all power generation except nuclear.

The designers, expecting a useful life of 20 to 25 years, originally licensed the plant till 1992, but the high level of reliability achieved by the Magnox has prompted Nuclear Electric to petition the Nuclear Installations Inspectorate (the NII) to extend the operating license of the plant to 30 or 40 years. The station won a temporary clearance to operate until 1995, subject to another safety review at the end of 1992 and a program of inspections and specific upgrades to be in place by then.

The NII issued the license extension after conducting a Long Term Safety Review (LTSR) which identified nine key areas of concern. In particular, welds in the steel pressure vessel and the high pressure gas ducting were dangerously brittle from more than 20 years of exposure to high radiation levels. The utility considers repairs to the welds, located inside the reactor containment vessel, impractical. Nuclear Electric hopes to be able to prove the welds safety with a new series of tests on the actual critical areas. In addition, to meet international safety standards, the plant will need new electronic control circuits. Unlike other Magnox stations with similar, but more expensive problems, Bradwell will be renovated at a cost of £9 million to £15 million.

Corrosion problems in Magnox systems first surfaced in 1968. Inspectors at Bradwell discovered deterioration of bolts and other fasteners near the gas ducts. At that time, the NII ordered all Magnox stations except Calder Hall, Chapelcross and Berkeley to reduce operating temperature and pressure by 10 to 20 percent.

Transporting nuclear waste raises fears despite safety record and dramatic safety demonstration. Other safety concerns center on the transportation of spent fuel rods. British Nuclear Fuels Ltd (BNFL) reprocesses 96 percent of that material for the CEGB as new fuel. That makes it a valuable resource for the power utility, but the high radiation level brings unique handling problems.

Cooling ponds on site store used fuel assemblies for at least 90 days, allowing short-lived radioactivity to decay. Then, packed into 43-metric ton forged-steel flasks, they travel by rail to the BNFL reprocessing plant at Sellafield in Cumbria.

Local community officials have expressed concern that some shipments stop on sidings throughout the West Midlands area, particularly in the Bescot shunting yards. They also complain that keeping shipment times secret hinders the effectiveness of their emergency disaster planning. Nuclear Electric insists all shipments go directly to Sellafield and agreed to provide complete time tables in September 1991.

The utility notes it has been transporting radioactive material by rail for 30 years without serious incident. In that time, it has moved 12,000 loads of spent fuel rods over three million miles of track. In 1984, they demonstrated the safety of their £1 million steel-flask transportation vessel by crashing a train, traveling at 100 miles per hour, into one. The impact destroyed the train, but the flask remained intact.

Missing fuel rods tracked to London. Although the spent fuel has been transported without incident, Bradwell has had problems with its supply of new rods. In November, 1966, thieves diverted a shipment of 20 natural-uranium elements from the station, but London police recovered the canisters intact. Thefts like that remain rare since the unenriched material used at Magnox plants like Bradwell has little monetary or strategic value. Refining it to weapons grade requires extensive and expensive technology.

Educational programs. Nuclear Electric operates an educational program for primary and secondary schools at Bradwell in conjunction with the Essex Education Authority. Study groups examine the science of a nuclear facility and the latest management techniques used in British industry.

Up to 5000 visitors view the plant yearly on pre-arranged two or three hour tours. Visitors can arrange excursions by calling 0621 76331.

Sources

Patterson, Walter C. *Nuclear Power*. New York: Penguin Books Ltd., 1977.
Bradwell Power Station. Nuclear Electric plc. Gloucester, n.d.
Mounfield, Peter. *World Nuclear Power*. Routledge, 1991.
"Five Reactors Must Prove Safety or Shut." *The Guardian*, September 10, 1991.
"Conditional Reprieve for Magnox Stations." *Financial Times*, September 12, 1991.
"Meeting to Draw up Nuclear-Trains Timetable." *Birmingham Post*, September 17, 1991.

—Al Cook

Calder Hall Nuclear Power Station, Units 1-4

BNFL Sellafield
Seascale, Cumbria CA20 1PG, United Kingdom
George Ayres, Asst. Dir.

Phone: 09467-28333
Fax: 09467-28987
Telex: 64237 BNFLWWG

Owning Utility

British Nulcear Fuels plc (BNFL)
Risley
Warrington, Cheshire WA3 6AS, United Kingdom
Joffroy Proocо, Dir., Info. Svcs.

Phone: 0925-832000
Fax: 0925-822711
Telex: 627581

Contact

Jake Kelly, Media Relations Mgr.
Calder Hall Nuclear Power Station

Basic Facts

Units 1-4 *Status:* Operable. *Type of Reactor:* Gas-cooled reactor. *Megawatts (electric):* 61. *Megawatts (thermal):* 270. *Reactor System Supplier:* United Kingdom Atomic Energy Authority. *Generator Supplier:* C.A. Parsons & Co., Ltd. (United Kingdom). *Architect Engineer:* United Kingdom Atomic Energy Authority.

Calder

UNITED KINGDOM

Key Dates

Unit 1 *Ordered: 1953. Construction began: 1953. Criticality: 1956. First Power: 1956. Commercial Operation: October 1956. Shut Down Expected: 1996.*

Unit 2 *Ordered: 1953. Construction began: 1953. Criticality: 1957. First Power: 1957. Commercial Operation: March 1957. Shut Down Expected: 1996.*

Unit 3 *Ordered: 1953. Construction began: 1953. Criticality: 1958. First Power: 1958. Commercial Operation: April 1959. Shut Down Expected: 1996.*

Unit 4 *Ordered: 1953. Construction began: 1953. Criticality: 1958. First Power: 1959. Commercial Operation: May 1959. Shut Down Expected: 1996.*

Plant Report

World's first nuclear power plant. The world knew the destructive potential of nuclear energy, but on Oct. 17, 1956, Calder Hall demonstrated its creative side. With a flick of a switch, Queen Elizabeth II began the era of nuclear power generation, when the world's first commercial reactor began to feed electricity into Britain's national grid.

But that was not what the designers, The United Kingdom Atomic Energy Authority (UKAEA), envisioned as the purpose of the station. They set out to produce plutonium for nuclear weapons. The production of electricity evolved as a use for the waste heat given off by the fission process. Over time, power generation and the development of reactor design supplanted the military purpose of the facility.

The primary concerns in choosing the site for Calder Hall and its sister, Chapelcross, centered on military security and secrecy. Formal licensing procedures did not begin to evolve until the late 1950s when the disastrous accidental release of fission products from the Windscale pile in 1957 underscored the threat of radioactive isotopes of iodine to the environment. Those guidelines only applied to reactors built for the government utilities like the Central Electric Generating Board (CEGB) and the South of Scotland Electricity Board (SSEB). British Nuclear Fuels Ltd. (BNFL), operator of Calder Hall, began as a department of the Ministry of Supply after World War II. Its main function was to produce a British atomic bomb, but when the concept of "atoms for peace" became popular in the 1950s, it became the production arm of the UKAEA, developing new reactors and reprocessing methods for spent nuclear fuel.

The beginning of the Magnox line built in three years. The first reactor at Calder Hall, and the threee that followed, formed the base point for the evolution of the Magnox and Advanced Gas Cooled Reactor (AGR) that dominated the nuclear industry in Britain. Built in just over three years by Taylor Woodrow plc, under the presidency of Lord Frank Taylor, the original reactor worked better than planned. Instead of the expected 168 MWe, the four reactor set produces over 200 megawatts electric at a full capacity efficiency rate of about 85 percent. It regularly operates at over 98 percent efficiency during the winter months when demand is at a steady high.

In addition to producing enough electricity to power a town the size of Brighton, Calder Hall also supplies process steam for the fuel handling complex at neighboring Sellafield. That makes it the world's first industrial complex to fully utilize nuclear energy. Meanwhile, the development of the Magnox reactor and its successors continues.

The term, Magnox, derives from the magnesium alloy canister used to contain the natural uranium fuel elements. The system uses a graphite-block core within a 3-inch thick, 67-foot-diameter steel pressure vessel. A concrete biological shield encloses the entire structure. In operation, it resembles the American High Temperature Gas Cooled Reactor (HTGR), but uses carbon dioxide rather than helium as the coolant.

Yearly check-ups, a lifelong habit for aging plant. One major advantage of the Magnox over other designs, like the Pressurized Water Reactor (PWR), explains its historically higher efficiency and reliability ratings. A special machine refuels the core every two or three days without the need of a shutdown. Other systems need weeks or months of down-time to refuel on a regular cycle. All commercial Magnox and AGR reactors in the British utility system enjoy this advantage. However, Calder Hall remains an experimental station, where scientists observe the reactor-physics characteristics of the Magnox. Consequently, each reactor shuts down yearly for inspection and refueling.

BNFL reprocesses 96 percent of the spent fuel from the four Magnox reactors at Sellafield, leaving three percent as highly active nuclear waste and one percent as plutonium.

Originally, the Magnox designers expected the reactor to last 20 to 25 years, but the high reliability of the design and the increasing cost of new facilities prompted BNFL to petition the Nuclear Installations Inspectorate (NII) to extend the operating licenses to 30 or more years.

Cracks in pressure vessel warrant close scrutiny. The Long Term Safety Review, issued by the Inspectorate, cited a number of deficiencies. In particular, welds in the steel alloy pressure vessel and the high pressure gas ducting had become dangerously brittle from more than 30 years of exposure to high levels of radiation. The Inspectorate considers repairs to the welds, located inside the reactor containment vessel, impractical. BNFL initiated a regular program of ultrasonic inspection of those welds in 1991, and installed new electronic safety circuits to replace aging systems. The NII extended the operating license for the facility to 1996.

Noting the extreme age of the plant, BNFL opened a 170 megawatts electric gas turbine generating station on the site in 1993 to ensure continued supply of electricity and process steam to Sellafield. In the meantime, it sought and received approval-in-principal for replacement of the Calder Hall reactors with a 1500 megawatts electric PWR. BNFL hopes to build the new power plant in cooperation with the government-owned utility, Nuclear Electric plc. (formerly CEGB).

New BNFL chairman prepares company for 21st Century. Calder Hall shares a one square mile site on the Cumberland shore in Lakeland with the Sellafield reprocessing complex and a decommissioning AGR reactor called Windscale. Over the years, the site has attracted nuclear protest starting with the

Windscale plutonium pile fire in October 1957 and continuing until 1986 when Christopher Harding became chairman of BNFL. He worked to change the image of BNFL, earned in the shoddy days when nuclear waste could be locked in an abandoned building and forgotten, to one of publicly accessible, technologically advanced competence in the handling of nuclear material. As a result, nuclear protest has evaporated. In 1991, Greenpeace, which had been trying to have the facility closed, received a court order to pay $150,000 in legal fees to BNFL.

The company ran three major advertising campaigns designed to inform the public of what Sellafield did and to invite tourists to the Visitors Centre opened in 1988 by His Royal Highness the Duke of Edinburgh. Over 137,000 people responded, many arriving at Sellafield's own tiny rail station on the historic "Flying Scotsman" steam locomotive which once regularly plied the route from London to Edinburgh.

Sources

"How Far Are You From A Reactor?" *New Statesman*, May 16, 1986.
"Britain's Nervous Nuclear Reaction." *Management Today*, April 1988.
"Briton Lord Taylor Strengthens Local Operations of Realty Empire." *San Francisco Business Times*, May 23, 1988.
Calder Hall, British Nuclear Fuels, September 1988.
Sellafield, British Nuclear Fuels plc, September 1988.
United Kingdom Atomic Energy Authority Annual Report and Accounts 1989-1990, AEA Technology, London, December 1990.
Mounfield, Peter. *World Nuclear Power*. Routledge, 1991.
British Nuclear Fuels Plc Annual Report and Accounts 1990-1991, British Nuclear Fuels, August 1991.
"The Greening of Sellafield." *Nuclear Industry*, First Quarter 1992.
United Kingdom Atomic Energy Authority Annual Report and Accounts 1989-1990, AEA Technology, London, December 1990.
Mounfield, Peter. *World Nuclear Power*. Routledge, 1991.
British Nuclear Fuels Plc Annual Report and Accounts 1990-1991, British Nuclear Fuels, August 1991.

—Al Cook

Chapelcross Nuclear Power Station, Units 1-4

Chapelcross Works, Annan
Dumfries & Galloway
Scotland DG12 6RF, United Kingdom
Peter Jenkinson, Supt.

Phone: 0461-202835
Fax: 0461-208497
Telex: 77249 BNFLCXG

Owning Utility

British Nuclear Fuels plc (BNFL)
Risley
Warrington, Cheshire WA3 6AS, United Kingdom
Harold Butler, Dir. of Corp. Aff.

Phone: 0925-832000
Fax: 0925-822711
Telex: 627581

Contact

Jeffrey Preece, Dir., Info. Svcs.
British Nuclear Fuels plc (BNFL)

Basic Facts

Units 1-4 *Status:* Operable. *Type of Reactor:* Gas-cooled reactor. *Megawatts (electric):* 60. *Megawatts (thermal):* 260. *Reactor System Supplier:* United Kingdom Atomic Energy Authority. *Generator Supplier:* C.A. Parsons & Co., Ltd. (United Kingdom). *Architect Engineer:* United Kingdom Atomic Energy Authority.

Key Dates

Unit 1 *Ordered:* 1953. *Construction began:* 1956. *Criticality:* 1958. *First Power:* 1959. *Commercial Operation:* March 1959. *Shut Down Expected:* 1999.
Unit 2 *Ordered:* 1953. *Construction began:* 1956. *Criticality:* 1958. *First Power:* 1959. *Commercial Operation:* August 1959. *Shut Down Expected:* 1999.
Unit 3 *Ordered:* 1953. *Construction began:* 1956. *Criticality:* 1959. *First Power:* 1959. *Commercial Operation:* December 1959. *Shut Down Expected:* 1999.
Unit 4 *Ordered:* 1953. *Construction began:* 1956. *Criticality:* 1959. *First Power:* 1960. *Commercial Operation:* March 1960. *Shut Down Expected:* 1999.

Plant Report

Fishing village turned nuclear center. During the nineteenth century, Annan, near Dumfriesshire enjoyed the reputation as one of the most important ports in Southwest Scotland. However, silting of the harbor destroyed the local shipbuilding trade, reducing the town once more to a fishing village dependant on salmon and shrimp. By the early 1950s, its now remote, thinly populated location made it ideal as a site for Chapelcross, Scotland's first commercial nuclear power station.

The plant's designers, The United Kingdom Atomic Energy Authority (UKAEA), created it to produce plutonium and tritium for nuclear weapons. The production of electricity evolved as a use for the waste heat given off by the fission process. Over time, power generation and the development of reactor design supplanted the military purpose of the facility.

The primary concerns in choosing the site for Chapelcross and its sister, Calder Hall, centered on military security and secrecy. Formal licensing procedures did not begin to evolve until the late 1950s when the disastrous accidental release of fission products from the Windscale plutonium piles in 1957 underscored the threat of radioactive isotopes of iodine to the environment. Those guidelines only applied to reactors built for the government utilities like the Central Electric Generating Board (CEGB) and the South of Scotland Electricity Board (SSEB). British Nuclear Fuels Ltd. (BNFL), operator of Chapelcross, began as a department of the Ministry of Supply after World War II. Its main function was to produce a British atomic bomb, but when the concept of "Atoms for Peace" became popular in the mid-1950s, it became the production arm of the UKAEA, developing new reactors and reprocessing methods for spent nuclear fuel.

The beginning of the Magnox line completed in five years. Calder Hall, in Cumbria, began the evolution of the Magnox reactor. The UKAEA copied the design when it started construction of Chapelcross in October 1955, one year before completion of the Calder Hall station.

The reactors at Chapelcross, like the earlier ones at Calder Hall, formed the base point for the evolution of the Magnox and Advanced Gas Cooled Reactor (AGR) that dominated the nuclear industry in Britain. Taylor Woodrow plc, under the presidency of Lord Frank Taylor built each station in under five years. In each case, the first reactor came on line just over three years after construction began. The design worked better than planned. Instead of the expected 168 MWe, the four reactor set produces over 200 megawatts electric at a full capacity efficiency rate of about 85 percent. It regularly operates at over 98 percent efficiency during the winter months when demand is at a steady high.

Originally, the electricity from the plant went to Southern Scotland. However by 1988, with the demand for electricity shifting from the north to the south of England, Chapelcross diverted its output into the NORWEB grid serving North West England. In addition, the station supplies the power needs of the BNFL facilities at Springfields and Capenhurst. Meanwhile, the development of the Magnox reactor and its successors continues.

The term, Magnox, derives from the magnesium alloy canister used to contain the natural uranium fuel elements. The system uses a graphite-block core within a 3-inch thick, 67-foot diameter steel pressure vessel. A concrete biological shield encloses the entire structure. In operation, it resembles the American High Temperature Gas Cooled Reactor (HTGR), but uses carbon dioxide rather than helium as the coolant.

Yearly check-ups, a lifelong habit for aging plant. One major advantage of the Magnox over other designs, like the Pressurized Water Reactor (PWR), explains its historically higher efficiency and reliability ratings. A special machine to refuel the core every two or three days without the need of a shutdown. Other systems need weeks or months of down-time to refuel on a regular cycle. All commercial Magnox and AGR reactors in the British utility system enjoy this advantage. However, Chapelcross remains an experimental station, where scientists observe the reactor-physics characteristics of the Magnox. Consequently, each unit shuts down annually for inspection and refueling.

BNFL reprocesses 96 percent of the spent fuel from the four Magnox reactors, leaving three percent as highly active nuclear waste and one percent as plutonium and tritium.

Transporting nuclear waste raises fears despite safety record and dramatic safety demonstration. That reprocessed fuel represents a valuable resource, but the high level of radioactivity brings unique handling problems. Cooling ponds on site store used fuel assemblies for at least 90 days, allowing short-lived radioactivity to decay. Then, packed into 43-metric ton forged-steel flasks, they travel by rail to the BNFL reprocessing plant at Sellafield in Cumbria.

BNFL notes it has been transporting radioactive material by rail since the early 1950s without an incident that resulted in the escape of radioactive material. The specially built containment flasks cost the company £1 million each. To prove the safety of the devices, the CEGB intentionally crashed a train, traveling at 100 miles per hour, into one. The impact destroyed the locomotive, but the flask remained intact.

The original reactor designers conceived the Magnox with an expected life of 20 to 25 years, bbut the high reliability of the design and the increasing cost of new facilities prompted BNFL to petition the Nuclear Installations Inspectorate (NII) to extend the operating licenses to 30 or more years.

Cracks in pressure vessel warrant close scrutiny. The Long Term Safety Review, issued by the Inspectorate, cited a number of deficiencies. In particular, welds in the steel alloy pressure vessel and the high pressure gas ducting had become dangerously brittle from more than 30 years of exposure to high levels of radiation. The Inspectorate considers repairs to the welds, located inside the reactor containment vessel, impractical. BNFL initiated a regular program of ultrasonic inspection of those welds in 1991, and installed new electronic safety circuits to replace aging systems. The NII extended the operating license for the facility to 1999.

Recognizing the advancing age of the station and the expected shortfall in generating capacity by the turn of the century, BNFL has sought and received approval in principal for its replacement with a 1500 megawatts electric PWR. BNFL hopes to build the new power plant in cooperation with the government-owned utility, Nuclear Electric plc. (formerly CEGB).

Sources

"How Far Are You From A Reactor?" *New Statesman*, May 16, 1986.
"Britain's Nervous Nuclear Reaction." *Management Today*, April 1988.
"Briton Lord Taylor Strengthens Local Operations of Realty Empire." *San Francisco Business Times*, May 23, 1988.
Chapelcross. British Nuclear Fuels, September 1988.
Sellafield. British Nuclear Fuels plc, September 1988.
Transport of Spent Nuclear Fuel. British Nuclear Fuels, November 1989.
United Kingdom Atomic Energy Authority Annual Report and Accounts 1989-1990. AEA Technology, London, December 1990.
Mounfield, Peter. *World Nuclear Power*. Routledge, 1991.
British Nuclear Fuels Plc Annual Report and Accounts 1990-1991. British Nuclear Fuels, August 1991.
Transport of Spent Nuclear Fuel. British Nuclear Fuels, November 1989.
United Kingdom Atomic Energy Authority Annual Report and Accounts 1989-1990. AEA Technology, London, December 1990.
Mounfield, Peter. *World Nuclear Power*. Routledge, 1991.

—Al Cook

Dounreay Prototype Fast Breeder Reactor (PFR)

Thurso
Caithness KW14 7TZ, United Kingdom
R. N. James, Site Mgr.

Phone: 0847-802121
Fax: 0847-802697
Telex: 75297

Owning Utility

AEA Technology
329 Harwell Laboratory
Harwell, Didcot, Oxon. OX11 ORA,
United Kingdom
Andrew Munn, Pub. Info.

Phone: 0235-821111
Fax: 0235-832591
Telex: 83135

Plant Profiles

Dounreay

Contact
A. McCree, Pub.Info.
Dounreay Prototype Fast Breeder
   Reactor (PFR)

Basic Facts
Status: Operable. *Type of Reactor:* Liquid metal fast breeder reactor. *Megawatts (electric):* 270. *Megawatts (thermal):* 600. *Reactor System Supplier:* United Kingdom Atomic Energy Authority; The Nuclear Power Group (United Kingdom). *Generator Supplier:* English Electric Co., Ltd. (United Kingdom). *Architect Engineer:* AEA Technology.

Key Dates
Ordered: 1966. *Construction began:* 1967. *Criticality:* March 1974. *First Power:* February 1975. *Commercial Operation:* August 1976.

Plant Report

Original nuclear research station becomes industries last hope. Far to north, near the very tip of Scotland, sits the last bastion of the United Kingdom Atomic Energy Authority (UKAEA). At one time, Dounreay housed Britain's hopes for nuclear power in the 21st century. However, with the push for privatization of the electric generation industry and the failure of British researches to perfect nuclear technology, the next century may see the total abandonment of atomic power research in Britain.

Originally, Britain built gas-graphite reactors but switched momentarily to a heavy water design before finally opting for American pressurized light water units. She started building these families of thermal reactors in the mid 1950s supply electricity to her expanding industrial base. They allowed her to mothball outdated coal-fired plants.

However, nuclear researchers saw them as a stop-gap. Thermal plants provided needed electricity, but also reactor experience and the fuel required for the evolving Fast Breeder Reactor (FBR).

World's first fast breeder reactor built at Dounreay. On March 1, 1954, Whitehall announced plans to build an experimental FBR at Dounreay. On November 14, 1959, the new reactor went critical, becoming the first operating fast breeder reactor in the world.

Over the next 18 years, it operated successfully, generating 15 megawatts electric for the national grid. By then, researchers felt confident enough in the technology to launch a 250 megawatts electric replacement.

The Prototype Fast Breeder Reactor (PFR) went on line on February 2, 1975, achieving full power in February 1977.

However, this remained well behind schedule. Originally, the UKAEA proposed to build a working prototype by 1965. Later that slipped to 1970, then 1971. The first full-scale 1000 megawatts electric commercial plant, originally scheduled for 1974, receded to a distant maybe for 1980.

France gains lead in fast breeder technology. Actually, in 1985, France built the first commercial fast breeder plant (FBR), the 1,200 megawatts electric Superphenix.

In 1988, the British government decided to cut its losses, announcing the end of funding for Dounreay's reactor program as of 1994. Money for reprocessing research would vanish in 1997.

Between 1954 and 1988, the FBR program consumed £4,000 million. Now, researchers estimate the development of the reactor has struggled half-way through a 60-year evolution.

Ideally, an FBR solves many of the problems of nuclear energy. Thermal reactors burn very little of their uranium fuel. The highlyradioactive, leftover material must be stored or reprocessed and eventually disposed of in depositories that may stay dangerous for thousands of years. However, the FBR can turn that deadly problem into useful energy.

Along with plutonium, also produced in thermal reactors, the FBR burns the depleted uranium many times. Eventually, most the original material burns away leaving a small amount of waste. In concert with thermal reactors, an FBR system can recover 50-60 times as much energy from nuclear fuel as current systems do.

However, no shortage of uranium exists today. That fact makes cheaper, if less efficient, thermal systems more cost effective. Baring changes in the world supply of uranium, the British government estimates the FBR will remain uncompetitive until at least 2020.

Atomic bombs could generate future electricity. One change that could dramatically affect that equation, and nuclear energy's image, may be the scrapping of atomic weapons by major world powers. The sudden influx of a plentiful supply of weapons-grade plutonium and the need to find a peaceful use for the material might, once again, spur FBR research in Britain.

In the meantime, the UKAEA struggles to keep its reactor and reprocessing plant afloat despite funding cutbacks.

If approved, the Authority may use fuel from the mothballed German FBR, SNR 300. Even so, it will need about £5 million a year of government money to operate.

The plant earns £14 million per year from the sale of electricity. That does not come close to covering the current 60 million operating budget.

International pressure mounts to regain British commitment. Pressure from European countries and from Japan may help to change the government's mind. In those countries FBR research continues to forge ahead. They see the loss of the British effort as a serious blow to international cooperation.

It could jeopardize the development of the European Fast Reactor (EFR). Currently, Britain contributes £10 million yearly to the project, but analysts expect her part of the costs to skyrocket to £350 million during peak construction years.

The reprocessing plant may fare better. In August 1989, it won a £2 million contract to reprocess German fuel that may keep the facility running beyond 1997.

Isolated station could attract terrorists. However, Dounreay's security causes many concern. Its remote location on the exposed northern coast and its own tiny air field could make it a prime target for hightech terrorists.

The UKAEA evacuated Dounreay on Dec. 14, 1972, after an anonymous caller warned of bombs planted on the site. Security staff found two suspicious packages, which the army bomb squad disposed of and later pronounced harmless.

Another concern focused on the shipment of plutonium between Windscale and Dounreay. One shipment of 100 kilograms travels the 600 kilometers of public roads between the plants each month. Officials express confidence in the security for the shipments, but decline to detail the type of arrangements.

However, in December 1991, the plant did manage to misplace 10 kg of enriched uranium. The lose of the material surfaced during a visit by inspectors from the European Atomic Energy Community (Euratom). With the plant closed, UKAEA will conduct an investigation and issue a report in 1992.

Does Dounreay cause cancer?. For some, the escape of radioactive material from the plant comes as no surprise. In at least six cases, plant workers died of cancer, years after exposure to such material at Dounreay. After decades of fruitless legal action by the widows of those workers, UKAEA announced a compensation program in 1987. Similar to an earlier scheme operated by British Nuclear Fuels Ltd. (BNFL), the plan will pay pensions to families of workers who develop cancer.

A 1986 Medical Research Council study found a "significant excess" in the death rate from leukemia and thyroid cancer. However, medical researchers hesitate to draw a direct connection between the operation of the plant and cancer in children. Searching for a reason for elevated childhood rates, the Scottish Health Services conducted a case-control study on 69 Caithness children. In its report, published in JAMA in 1991, it found no connection between the children's cancer and their father's work at Dounreay. It did cite the use of beaches in the area as a possible contributing factor.

Sources

Patterson, Walter C. *Nuclear Power.* New York: Penguin Books Ltd., 1977.
Burn, Duncan. *Nuclear Power and the Energy Crisis.* New York: New York University Press, 1978.
"Breeder reactor politics in Europe." *Bulletin of the Atomic Scientists,* May 1986.
Bupp, Irvin C., and Jean-Claude Derian. *Light Water,* New York: asic Books, 1987.
Collier, John, and Geoffrey F. Hewitt. *Introduction to Nuclear Power.* Washington: Hemisphere Publishing Corporation, 1987.
"Dounreay's widows." *New Statesman,* November 13, 1987.
"Europe could rescue Britain's fast breeder." *New Scientist,* August 12, 1989.
"Reactor judged not so fast." *Nature,* August 2, 1990.
United Kingdom Atomic Energy Authority Report and Accounts 1989-90. AEA Technology, London: December 1990.
"AEA cutbacks jeopardize neutron, isotope outputs." *Nuclear News,* February 1991.
"Case-Control Study of Leukaemia and Non-Hodgkin's Lymphoma in Children in Caithness Near the Dounreay Nuclear Installation." *Journal of the American Medical Association,* August 7, 1991.
"Dounreay loses uranium." *Nature,* December 12, 1991.
"Nuclear chiefs urge stay of execution for Dounreay." *New Scientist,* December 21, 1991.
"AEA cutbacks jeopardize neutron, isotope outputs." *Nuclear News,* February 1991.
"Case-Control Study of Leukaemia and Non-Hodgkin's Lymphoma in Children in Caithness Near the Dounreay Nuclear Installation." *Journal of the American Medical Association,* August 7, 1991.
"Dounreay loses uranium." *Nature,* December 12, 1991.

—Al Cook

Dungeness A Power Station, Units 1-2

Romney Marsh, Kent TN29 9PP, United Kingdom
N. Callaghan, Stat.Mgr.

Phone: 0679-20461
Fax: 0679-20461 x 253

Owning Utility

Nuclear Electric plc
Barnett Way
Barnwood, Glos. GL4 7RS, United Kingdom
J.G. Collier, Chm. & CEO

Phone: 0452-652222
Fax: 0452-652776
Telex: 43501

Contact

S. Moore, Dir., PR
Nuclear Electric plc

Basic Facts

Units 1-2 *Status:* Operable. *Type of Reactor:* Gas-cooled reactor. *Megawatts (electric):* 285. *Megawatts (thermal):* 720. *Reactor System Supplier:* The Nuclear Power Group (United Kingdom). *Generator Supplier:* C.A. Parsons & Co., Ltd. (United Kingdom); Associated Electric Industries Ltd. (United Kingdom). *Architect Engineer:* The Nuclear Power Group (United Kingdom).

Key Dates

Unit 1 *Ordered:* 1959. *Construction began:* 1960. *Criticality:* June 1965. *First Power:* September 1965. *Commercial Operation:* December 1965.
Unit 2 *Ordered:* 1959. *Construction began:* 1960. *Criticality:* September 1965. *First Power:* November 1965. *Commercial Operation:* December 1965.

Plant Report

End of the line draws near for aging Magnox Station. Dungeness A heralded the beginning of the atomic age as one of the original five nuclear power stations in the UK. The 26-year-old facility dates from Britain's first commercial nuclear development program, which chose the British-designed Magnox reactor in 1955. Now, as the plant approaches the end of its useful life, the technology faces an uncertain future in the 21st Century.

The term, Magnox, derives from the magnesium alloy canister used to contain the natural-uranium fuel elements. The system uses a graphite-block core within a 3-inch thick, 67-foot-diameter steel pressure vessel. For the safety of the operators, a concrete biological shield encloses the entire structure. The system's unique design allows a special machine to refuel the core every two or three days without the need of a shutdown. This gives the Magnox a distinct advantage in reliability over other reactors, most of which need annual shutdowns of three or four weeks. In operation, it resembles the American High Temperature Gas Cooled Reactor (HTGR), but uses carbon dioxide rather than helium as the coolant.

Site choice critical to success. The 225-acre site on the south coast of Kent accommodates both the Magnox A station and a more modern Advanced Gas Cooled Reactor (AGR) B station. The United Kingdom Atomic Energy Authority (UKAEA) chose the location because it met the three main criteria set up in 1955 by the government. Its stable geology could support the massive building and facilities. A high demand in the surrounding area provided a ready market and need. Its proximity to the sea guaranteed a ready source of coolant and a high volume of water to dilute small quantities of radioactive effluent. Both stations draw their coolant water from the English Channel.

The British government first proposed Dungeness A in 1955 as one of a series of twelve stations to produce a total of 2000 MWe. They were to answer an increasing post-war demand for power and a need to replace dirty and inefficient coal-fired generating plants. After much re-evaluation, the program delivered only nine Magnox installations. Dungeness A-1 first fed electricity to the national grid on Sept. 21, 1965. Once fully operational, its two Magnox reactors provided 424 megawatts electric for south east England through the grid system.

The government-owned utility, Nuclear Electric plc, inherited the facility from the Central Electric Generating Board (CEGB) on March 31, 1990, when the British government privatized all power generation except nuclear.

All Magnox stations subject to Long Term Safety Reviews. The designers, expecting a useful life of 20 to 25 years, originally licensed the plant till 1990. However, the reliability of the Magnox and the escalating cost of new facilities prompted a reassessment. Nuclear Electric petitioned the Nuclear Installations Inspectorate (NII) to extend the operating licenses of all its Magnox stations to 30 or 40 years. Without Dungeness, and other Magnox stations like it, southern England and Wales could face a shortfall of 11,000 to 12,000 megawatts electric of generating capacity by the turn of the century.

The NII will conduct a Long Term Safety Review (LTSR) of Dungeness A before issuing a license extension. Earlier reviews gave Calder Hall and Chapelcross an extra 10 years each, but shut down Berkeley and Hunterston A because upgrading would cost too much.

The LTSRs identified specific concerns with the Magnox design. In particular, welds in the steel alloy pressure vessel and the high pressure gas ducting may have become dangerously brittle from more than 20 years of exposure to high levels of radiation. The utility considers repairs to the welds, located inside the reactor containment vessel, impractical. Nuclear Electric hopes to prove the welds safety with a new series of tests on the actual critical areas. The Inspectorate will allow the station to operate at reduced pressure until it does.

Transporting nuclear waste raises fears despite safety record and dramatic safety demonstration. The handling of nuclear waste, especially spent fuel rods, commands special attention. British Nuclear Fuels Ltd (BNFL) reprocesses 96 percent of that material as new fuel. That makes it a valuable resource for the power utility, but the high radiation level brings unique handling problems.

Cooling ponds on site store used fuel assemblies for at least 90 days, allowing short-lived radioactivity to decay. Then, packed into 43-metric ton forged-steel flasks, they travel by rail to the BNFL reprocessing plant at Sellafield in Cumbria.

Local community officials have expressed concern that some shipments stop on sidings throughout the West Midlands area, particularly in the Bescot shunting yards. They also complain that keeping shipment times secret hinders the effectiveness of their emergency disaster planning. Nuclear Electric insists all shipments go directly to Sellafield and agreed to provide complete time tables in September 1991.

The utility notes it has been transporting radioactive material by rail for 30 years without serious incident. In that time, it has moved 12,000 loads of spent fuel rods over three million miles of track. In 1984, they demonstrated the safety of their £1 million steel-flask transportation vessel by crashing a train, traveling at 100 miles per hour, into one. The impact destroyed the train, but the flask remained intact.

Site of Special Scientific Interest has visitors looking up. The Nature Conservancy Council designated Dungeness a Site of Special Scientific Interest (SSSI). Many rare plants and insect colonies flourish in the preserve, including the white spot moth. Nuclear Electric established a public observatory for viewing the many species of migratory birds that pass over the site.

Information Center. Interactive videos and models tell the story of electricity in the high technology information center. It operates seven days a week from April to September and Sunday through Friday from October to May. The visitor can arrange a tour of the facility for any day of the week by calling 0679-20461.

Sources

Dungeness A. Nuclear Electric plc. Gloucester, n.d.
"Five reactors must prove safety or shut." *The Guardian*, September 10, 1991.
"Hinkley cleared to 1995." *Power in Europe*, September 12, 1991.
"Conditional reprieve for Magnox stations." *Financial Times*, September 12, 1991.
"Can Hinkley keep its place in the Generation Game?" *Western Morning News*, September 16, 1991.
"Meeting to draw up nuclear-trains timetable." *Birmingham Post*, September 17, 1991.
Mounfield, Peter. *World Nuclear Power.* Routledge, 1991.
Nuclear Electric Report and Accounts 1990-91, Nuclear Electric plc, Gloucester, n.d.

—*Al Cook*

Dungeness B Power Station, Units 1-2

Romney Marsh, Kent TN29 9PX, United Kingdom
G.M. Sutherland, Stat.Mgr.

Phone: 0679-20551
Fax: 0679-21331
Telex: 966128 (CEGBDNB G)

Owning Utility

Nuclear Electric plc
Barnett Way
Barnwood, Glos. GL4 7RS, United Kingdom
J.G. Collier, Chm. & CEO

Phone: 0452-652222
Fax: 0452-652776
Telex: 43501

Contact

S. Moore, Dir., PR
Nuclear Electric plc

Basic Facts

Units 1-2 *Status:* Operable. *Type of Reactor:* Advanced gas-cooled reactor. *Megawatts (electric):* 660. *Megawatts (thermal):* 1480. *Reactor System Supplier:* Atomic Power Construction Ltd. (United Kingdom). *Generator Supplier:* C.A. Parsons & Co., Ltd. (United Kingdom). *Architect Engineer:* Atomic Power Construction Ltd. (United Kingdom).

Key Dates

Unit 1 *Ordered:* 1965. *Construction began:* 1966. *Criticality:* December 1982. *First Power:* April 1983. *Commercial Operation:* April 1985.

Unit 2 *Ordered:* 1965. *Construction began:* 1966. *Criticality:* December 1985. *First Power:* December 1985. *Commercial Operation:* January 1986.

Plant Report

Facility powers all of southeast England. The Dungeness facility, located in the Romney Marsh on the south coast of Kent, has supplied electricity for southern England's needs since 1965. In the early 1980s, the addition of a second, larger station, satisfied the energy thirst of the whole of southeast England. On a typical day, the two stations provide 30 million kilowatt hours of electricity.

The new plant's Advanced Gas Cooled Reactor (AGR) evolved from the earlier Magnox design used in the original station. Instead of the Magnox's steel-alloy pressure vessel, the AGR uses prestressed concrete lined with steel. The pressure vessel protectively contains the boilers. The graphite core remains, but uranium dioxide pellets replace the natural-uranium fuel rods. As with the Magnox, the AGR's design allows a special machine to refuel the core every two or three days without a shutdown. This gives the AGR a distinct advantage in reliability over other reactors, most of which need annual shutdowns of three or four weeks. In operation, it resembles the American High Temperature Gas Cooled Reactor (HTGR), but uses carbon dioxide rather than helium as the coolant. The system's designers believed the use of a concrete containment vessel instead of the Magnox's steel vessel made the AGR safer.

New reactor proves disappointing. Confident in the new technology, the British government began the AGR program in April 1964, before completing the last of the Magnox plants. It started with Dungeness B. This second program projected a target of 5,000 megawatts electric by 1970-75. Design changes and construction difficulties created massive cost overruns, delaying the first plant opening, Hinkley B, until 1976. Dungeness B, first ordered in 1965, took 20 years to complete. It began sending electricity to the national grid on April 3, 1983. Once fully operational on Dec. 29, 1985, its two AGR reactors provided 720 megawatts electric for south east England through the grid system.

Many experts now argue the AGR represented too large a technological step from the Magnox. Designers continued to push the size and operating pressure and temperature of successive reactors, hoping to increase thermal efficiency. New materials and fuel compounds evolved, but each improvement brought new problems. In 1966, scientist's noted that the carbon dioxide coolant gas corroded the pressure vessel's mild-chrome-steel liner. They stopped that with small amounts of methane. However, the methane caused carbon to form on fuel assemblies reducing efficiency.

Even though the contractor, Atomic Power Constructions (APC), also built Wylfa, the largest and most advanced Magnox station, design errors and poor workmanship continually delayed Dungeness B.

APC attempted to install a steel liner for the containment vessel that was too small. Attempts to attach ducting deformed the liner visually, prompting replacement.

The high temperature and pressure of the gas within the boilers caused more unforseen problems. The original design did not balance the need for rigid tubes to withstand the buffeting of the high pressure gas and the need for flexibility to compensate for thermal expansion. Construction halted while APC redesigned the boiler system.

Another three-year delay happened in the early 1970s. Seals in the gas circulators could not withstand the operating conditions.

Meanwhile, more corrosion problems threatened the pressure vessel liner. As a precaution, the company covered nine percent of the metal with stainless steel. The process involved disassembling a portion of the graphite block core.

None of the AGRs, including Dungeness B, have approached the operational reliability of the Magnox reactor. However, in its 1990-91 annual report, Nuclear Electric noted a general increase in AGR efficiency.

The design of the AGR avoids the problem of cracks in the steel containment vessel that the Magnox is subject to. However, recent tests by the United Kingdom Atomic Energy Authority (UKAEA) on the decommissioning AGRs, Windscale and Winfrith, uncovered potential problems. Irradiation of the graphite-core bricks has changed the thermal expansion characteristics of the material. Unexpected internal stress may cause early failure of the core.

In 1979, the British government announced any new reactors built under a third program would use the more internationally common Pressurized Water Reactor (PWR) design.

The government-owned utility, Nuclear Electric plc, inherited the facility from the Central Electric Generating Board (CEGB) on March 31, 1990, when the British government privatized all power generation except nuclear.

China Syndrome possible if everything goes wrong. In 1988, the CEGB publicly conceded that an AGR core could catch fire and melt 30-to-35 seconds after a serious accident. The admission came during the Hinkley C public inquiry in answer to American physicist, Richard Webb. He claimed the resulting explosion could release more radiation than the 1986 Chernobyl accident. While acknowledging the potential, Brian George, CEGB senior engineer, noted, "Sixteen separate contactors and relays would have to fail." He pointed out the safety systems designed to minimize the danger. Only nine out of the available 81 boron control rods would be enough to contain the reactor. The rods of the AGR fall by their own weight into the graphite core if anything interrupts the system control power.

Transporting nuclear waste raises fears despite safety record and dramatic safety demonstration. The handling of nuclear waste, especially spent fuel rods, commands special attention. British Nuclear Fuels Ltd (BNFL) reprocesses 96 percent of used Magnox material as new fuel. BNFL is spending £4 million to build a high technology processing plant to handle AGR and pressurized light-water reactor fuel. The Thermal Oxide Reprocessing Plant (THORP), at Sellafield, should begin operation in 1992. That makes nuclear waste a valuable resource for the power utility, but the high radiation level brings unique handling problems.

Cooling ponds on site store used fuel assemblies for at least 90 days, allowing short-lived radiation to decay. Then, packed into 43-tonne forged-steel flasks, they travel by rail to the BNFL reprocessing complex at Sellafield in Cumbria.

The utility notes it has been transporting radioactive material by rail for 30 years without serious incident. In that time, it has moved 12,000 loads of spent fuel rods over more than three million miles of track. In 1984, they demonstrated the safety of their £1 million steel-flask transportation vessel by crashing a train, traveling at 100 miles per hour, into one. The impact destroyed the train, but the flask remained intact.

Site of Special Scientific Interest has visitors looking up. The Nature Conservancy Council designated Dungeness a Site of Special Scientific Interest (SSSI). Many rare plants and insect colonies flourish in the preserve, including the white spot moth, unique to the area. Nuclear Electric established a public observatory for viewing the many species of migratory birds that pass over the site.

Information Center. Interactive videos and models tell the story of electricity in the high technology information center. The visitor can arrange a tour of the facility for any day of the week by calling 0679-20551.

Sources

Patterson, Walter C. *Nuclear Power.* New York: Penguin Books Ltd., 1977.

Burn, Duncan. *Nuclear Power and the Energy Crisis.* New York: New York University Press, 1978.
"How far are you from a reactor?" *New Statesman,* May 16, 1986.
Bupp, Irvin C., and Jean-Claude Derian. *Light Water.* New York: Basic Books, 1987.
Collier, John, and Geoffrey F. Hewitt. *Introduction to Nuclear Power,* Washington: Hemisphere Publishing Corporation, 1987.
"CEGB admits to possible meltdown." *New Scientist,* January 1989.
"Nuclear Safety: Some like it hot." *New Scientist,* November 4, 1989.
Dungeness B. Nuclear Electric plc. Gloucester, n.d.
Nuclear Electric Report and Accounts 1990-91. Nuclear Electric plc, Gloucester, n.d.
United Kingdom Atomic Energy Authority Annual Report and Accounts 1989-1990. AEA Technology, London, December 1990.
Mounfield, Peter. *World Nuclear Power.* Routledge, 1991.
Dungeness B. Nuclear Electric plc. Gloucester, n.d.

—Al Cook

Hartlepool Power Station, Units 1-2

Tees Rd.
Hartlepool, Cleveland TS25 2BZ, United Kingdom
A.C. Capp, Stat.Mgr.

Phone: 0429-265841
Fax: 0429-265085

Owning Utility

Nuclear Electric plc
Barnett Way
Barnwood, Glos. GL4 7RS, United Kingdom
J.G. Collier, Chm. & CEO

Phone: 0452-652222
Fax: 0452-652776
Telex: 43501

Contact

S. Moore, Dir., PR
Nuclear Electric plc

Basic Facts

Units 1-2 *Status:* Operable. *Type of Reactor:* Advanced gas-cooled reactor. *Megawatts (electric):* 666. *Megawatts (thermal):* 1500. *Reactor System Supplier:* National Nuclear Corporation (United Kingdom). *Generator Supplier:* General Electric Co. (United Kingdom). *Architect Engineer:* National Nuclear Corporation (United Kingdom).

Key Dates

Unit 1 *Ordered:* 1967. *Construction began:* 1968. *Criticality:* 1981. *First Power:* August 1983. *Commercial Operation:* May 1984.

Unit 2 *Ordered:* 1967. *Construction began:* 1968. *Criticality:* 1982. *First Power:* October 1984. *Commercial Operation:* January 1985.

Plant Report

England's most northerly nuclear power plant dominates landscape. The assembling of Hartlepool station marked the mid-point of Britain's second nuclear power plant program. The government-owned utility, Central Electric Generating Board (CEGB), placed it near the mudflats and sand dunes of the Tess estuary, in Cleveland County. Now the massive blockhouse containing the Advanced Gas Cooled Reactor (AGR) towers over tiny fishing boats scuttling down the estuary to the deeper waters of the North Sea.

Hartlepool

Nuclear Electric plc, inherited the plant from the CEGB on March 31, 1990, when the British government privatized all power generation except nuclear.

The AGR, Britain's second generation of reactor, evolved from the earlier Magnox design. It replaced the Magnox's steel-alloy pressure vessel with one of prestressed concrete lined with steel and moved the boilers to within the vessel. The graphite core remained, but uranium dioxide pellets replaced the natural-uranium fuel rods. As with the Magnox, the AGR's design allows a special machine to refuel the core every two or three days without a shutdown. This gives the Magnox a distinct advantage in reliability over other reactors, most of which need annual shutdowns of three or four weeks. In operation, it resembles the American High Temperature Gas Cooled Reactor (HTGR), but uses carbon dioxide rather than helium as the coolant. The system's designers believed the use of a concrete containment vessel instead of the Magnox's steel vessel made the AGR safer.

New reactor proves disappointing. Confident in the new technology, the British government began the AGR program in April 1964, before completing the last of the Magnox plants. This second program projected a target of 5,000 megawatts electric by 1970-75, but design changes and construction difficulties created massive cost overruns, delaying the first plant opening until 1976. Hartlepool suffered a severe setback when the government watchdog, Nuclear Installations Inspectorate (NII), rejected its pressure vessel design as unsafe. It finally began sending electricity to the national grid on August 1, 1983.

None of the AGRs, including Hartlepool, have approached the operational reliability of the Magnox reactor. However, in its 1990-91 annual report, Nuclear Electric commended the station for its "greatly improved operational reliability."

The design of the AGR avoids the problem of cracks in the steel containment vessel that the Magnox is subject to. However, recent tests by the United Kingdom Atomic Energy Authority (UKAEA) on the decommissioning AGRs, Windscale and Winfrith, uncovered potential problems. Irradiation of the graphite-core bricks has changed the thermal expansion characteristics of the material. Unexpected internal stress may cause early failure of the core.

In 1979, the British government announced any new reactors built under a third program, would use the more internationally common Pressurized Water Reactor (PWR) design.

Early safety standards relaxed. Mud flats, salt marshlands and sand dunes surround the facility, and it lies four miles south of a major city, Hartlepool. Even so, the site meets the NII requirements for the locating of a nuclear plant. Its stable geology could support the massive building and facilities. A reasonable demand in the surrounding area provided a ready market and need. Its proximity to the sea guaranteed a ready source of coolant and a high volume of water to dilute small quantities of radioactive effluent.

However, the NII had to adjust its earlier regulations defining maximum population densities to permit a reactor so close to a major urban center. With the addition of the concrete pressure vessel, the NII believed the design was safe enough and approved the site as "relaxed."

China Syndrome possible if everything goes wrong. In 1988, the CEGB publicly conceded that an AGR core could catch fire and melt 30-to-35 seconds after a serious accident. The admission came during the Hinkley C public inquiry in answer to American physicist, Richard Webb. He claimed the resulting explosion could release more radiation than the 1986 Chernobyl accident. While acknowledging the potential, Brian George, CEGB senior engineer, noted, "sixteen separate contactors and relays would have to fail." He pointed out the safety systems designed to minimize the danger. A minimum of nine, out of the available 81, boron control rods would be enough to contain the reactor. The rods of the AGR fall by their own weight into the graphite core if anything interrupts the system control power.

Of all the British nuclear power plants, Hartlepool has the largest population living within a 19-mile radius, 987,000. A large chemical complex also operates near by. During the Chernobyl incident, 49,000 people evacuated a 19-mile radius of that plant.

Hartlepool's two reactors produce a total of 1250 megawatts electric for the national grid, twice as much energy as Cleveland County can use. Energy-hungry Southern England and Wales eagerly soak up the rest. Power plants in the south produce 40 percent of Britain's electricity, but the region uses 55 percent of the power supplied by the grid. Nuclear Electric imports the shortfall from more northerly stations and from France.

Recycling big business for nuclear industry. The handling of nuclear waste, especially spent fuel rods, commands special attention. British Nuclear Fuels Ltd (BNFL) reprocesses ninety-six percent of used Magnox material as new fuel. BNFL is spending £4 million to build a high technology processing plant to handle AGR and pressurized light-water reactor fuel. The Thermal Oxide Reprocessing Plant (THORP), at Sellafield, should begin operation in 1992. That makes nuclear waste a valuable resource for the power utility, but the high radiation level brings unique handling problems.

Cooling ponds on site store used fuel assemblies for at least 90 days, allowing short-lived radiation to decay. Then, packed into 43-tonne forged-steel flasks, they travel by rail to the BNFL reprocessing complex at Sellafield in Cumbria.

Local community officials have expressed concern that some shipments stop on sidings throughout the West Midlands area, particularly in the Bescot shunting yards. They also complain that keeping shipment times secret hinders the effectiveness of their emergency disaster planning. Nuclear Electric insists all shipments go directly to Sellafield and agreed to provide complete time tables in September 1991.

The utility notes it has been transporting radioactive material by rail for 30 years without serious incident. In that time, it has moved 12,000 loads of spent fuel rods over more than three million miles of track. In 1984, they demonstrated the safety of their £1 million steel-flask transportation vessel by crashing a train, traveling at 100 miles per hour, into one. The impact destroyed the train, but the flask remained intact.

Natural sanctuary provides educational opportunities. The salt marshes, mudflats and dunes surrounding Hartlepool form three Sites of Special Scientific Interest (SSSI). Many species of birds and animals shelter there. To help preserve the sanctuaries and make them accessible to the public, Hartlepool constructed the Teesmouth Field Centre in 1970. The center has attracted more than 26,000 visitors per year. It operates as an educational facility in conjunction with the Cleveland County Education Authority and the Nature Conservancy Council. The visitor can view the inner workings of the station by remote video or take a half-day hike on the sand dune nature trail. The plant offers special open weekends or the visitor can arrange a tour of the facility for any day of the week by calling 045 652618.

Sources

Burn, Duncan. *Nuclear Power and the Energy Crisis.* New York: New York University Press, 1978.
"How far are you from a reactor?" *New Statesman*, May 16, 1986.
"CEGB admits to possible meltdown." *New Scientist*, January 1989.
United Kingdom Atomic Energy Authority Annual Report and Accounts 1989-1990. AEA Technology, London, December 1990.
Mounfield, Peter, *World Nuclear Power.* Routledge, 1991.
Hartlepool. Nuclear Electric plc. Gloucester, July 1991.
Nuclear Electric Report and Accounts 1990-91. Nuclear Electric plc, Gloucester, July 1991.
"Meeting to draw up nuclear-trains timetable." *Birmingham Post*, September 17, 1991.

—Al Cook

Heysham A Power Station, Units 1-2

Heysham, PO Box 4
Morecambe, Lancs. LA3 2XQ, United Kingdom
M.T. Hardy, Stat.Mgr.
Phone: 0524-53131
Fax: 0524-55104

Owning Utility

Nuclear Electric plc
Barnett Way
Barnwood, Glos. GL4 7RS, United Kingdom
J.G. Collier, Chm. & CEO
Phone: 0452-652222
Fax: 0452-652776
Telex: 43501

Contact

S. Moore, Dir., PR
Nuclear Electric plc

Basic Facts

Units 1-2 *Status:* Operable. *Type of Reactor:* Advanced gas-cooled reactor. *Megawatts (electric):* 666. *Megawatts (thermal):* 1510. *Reactor System Supplier:* National Nuclear Corporation (United Kingdom). *Generator Supplier:* General Electric Co. (United Kingdom). *Architect Engineer:* National Nuclear Corporation (United Kingdom).

Key Dates

Unit 1 *Ordered:* 1967. *Construction began:* 1970. *Criticality:* 1983. *First Power:* July 1983. *Commercial Operation:* April 1984.

Unit 2 *Ordered:* 1967. *Construction began:* 1970. *Criticality:* 1983. *First Power:* October 1984. *Commercial Operation:* November 1984.

Plant Report

Second-generation reactors provide plentiful power for British northwest. Heysham A took 13 years to build, but now its Advanced Gas Cooled Reactors (AGRs) produce enough electricity to power three cities the size of Liverpool.

The AGR, Britain's second generation of reactor, evolved from the earlier Magnox design. It replaced the Magnox's steel-alloy pressure vessel with one of prestressed concrete lined with steel and moved the boilers to within the vessel. The graphite core remained, but uranium dioxide pellets replaced the natural-uranium fuel rods. As with the Magnox, the AGR's design allows a special machine to refuel the core every two or three days without a shutdown. This gives the Magnox a distinct advantage in reliability over other reactors, most of which need annual shutdowns of three or four weeks. In operation, it resembles the American High Temperature Gas Cooled Reactor (HTGR), but uses carbon dioxide rather than helium as the coolant. The system's designers believed the use of a concrete containment vessel instead of the Magnox's steel vessel made the AGR safer.

New reactor proves disappointing. Confident in the new technology, the British government began the AGR program in April 1964, before completing the last of the Magnox plants. This second program projected a target of 5,000 megawatts electric by 1970-75, but design changes and construction difficulties created massive cost overruns, delaying the first plant opening until 1976. Construction of Heysham A began in October 1970. The first reactor began sending electricity to the national grid on July 9, 1983. Once fully operational on Oct. 11, 1984, its two AGR reactors provided 1320 megawatts electric for north west England through the grid system.

None of the AGRs, including Heysham A, have approached the operational reliability of the Magnox reactor. However, in its 1990-91 annual report, the government-owned utility, Nuclear Electric plc, noted a general increase in AGR efficiency.

The design of the AGR avoids the problem of cracks in the steel containment vessel that the Magnox is subject to. However, recent tests by the United Kingdom Atomic Energy Authority (UKAEA) on the decommissioning AGRs, Windscale and Winfrith, uncovered potential problems. Irradiation of the graphite-core bricks has changed the thermal expansion characteristics of the material. Unexpected internal stress may cause early failure of the core.

Britain abandons gas-cooled technology for American light-water design. In 1979, the British government announced any new reactors built under a third program would use the more internationally common Pressurized Water Reactor (PWR) design. The first pressurized light-water reactor is due to open in mid-1994. A second received planning approval in 1990, but actual construction must await the government's review of nuclear power generation due in 1994.

Heysham

Nuclear Electric inherited Heysham from the Central Electric Generating Board (CEGB) on March 31, 1990, when the British government privatized all power generation except nuclear.

Second plant built on same site. The facility sits on Morecambe Bay in northwest Lancashire, less than fifty miles from Liverpool and Manchester. The Nuclear Installations Inspectorate (NII) approved the 223-acre site because it met the government's three main criteria set up in 1955. Its stable geology could support the massive 1.05 million tonne building and facilities. A high demand in the surrounding area provided a ready market and need. Its proximity to the Irish Sea guaranteed a ready source of coolant and a high volume of water to dilute small quantities of radioactive effluent. The location proved so ideal, the CEGB built a second AGR station on the same site. The two stations draw five million liters of coolant water from the harbor per minute.

However, the NII had to adjust its earlier guidelines for maximum population densities to permit a reactor so close to a major urban center. With the concrete pressure vessel, the NII believed the design was safe enough and approved the site as "relaxed."

China Syndrome possible if everything goes wrong. In 1988, the CEGB publicly conceded that an AGR core could catch fire and melt 30-to-35 seconds after a serious accident. The admission came during the Hinkley C public inquiry in answer to American physicist, Richard Webb. He claimed the resulting explosion could release more radiation than the 1986 Chernobyl accident. While acknowledging the potential, Brian George, CEGB senior engineer, noted, "sixteen separate contactors and relays would have to fail." He pointed out the safety systems designed to counter the danger. A minimum of nine, out of the available 81, boron control rods would be enough to contain the reactor. The rods of the AGR fall by their own weight into the graphite core if anything interrupts the system control power.

Heysham has a population 540,000 living within a 19-mile radius of its four reactors. During the Chernobyl incident, 49,000 people evacuated a 19-mile radius of that plant.

Transporting nuclear waste raises fears despite safety record and dramatic safety demonstration. The handling of nuclear waste, especially spent fuel rods, commands special attention. British Nuclear Fuels Ltd (BNFL) reprocesses 96 percent of used Magnox material as new fuel, but stores spent AGR fuel. With an eye to the future, the company will invest £4 million to build a high technology processing plant to handle AGR and pressurized light-water reactor fuel. The Thermal Oxide Reprocessing Plant (THORP), at Sellafield, should begin operation in 1992. That makes nuclear waste a valuable resource for the power utility, but the high radiation level brings unique handling problems.

Cooling ponds allow radiation to decay for safer handling. Cooling ponds on site store used fuel assemblies for at least 90 days, allowing short-lived radiation to decay. Then, packed into 43-tonne forged-steel flasks, they travel by rail to the BNFL reprocessing complex at Sellafield in Cumbria. Local community officials have expressed concern that some shipments stop on sidings throughout the West Midlands area, particularly in the Bescot shunting yards. They also complain that keeping shipment times secret hinders the effectiveness of their emergency disaster planning. Nuclear Electric insists all shipments go directly to Sellafield and agreed to provide complete time tables in September 1991.

The utility notes it has been transporting radioactive material by rail for 30 years without serious incident. In that time, it has moved 12,000 loads of spent fuel rods over more than three million miles of track. In 1984, they demonstrated the safety of their 1 million steel-flask transportation vessel by crashing a train, traveling at 100 miles per hour, into one. The impact destroyed the train, but the flask remained intact.

New construction project returns land to nature. After decades of massive construction on the site, Nuclear Electric is now restoring the land. A Nature Conservancy Management Committee is working to create a publicly-accessible natural environment with a field study center, nature trails, and bird hides.

Information Center. Interactive videos and models tell the story of electricity in Heysham's information center. An estimated 15,000 people view the two plants each year. The visitor can arrange a tour of the facility for any day of the week by calling 0452-652618.

Sources

"How far are you from a reactor?" *New Statesman*, May 16, 1986.
"CEGB admits to possible meltdown." *New Scientist*, January 1989.
United Kingdom Atomic Energy Authority Annual Report and Accounts 1989-1990. AEA Technology, London, December 1990.
Heysham 1. Nuclear Electric plc. Gloucester, July 1991.
Nuclear Electric Report and Accounts 1990-91. Nuclear Electric plc, Gloucester, July 1991.
"Meeting to draw up nuclear-trains timetable." *Birmingham Post*, September 17, 1991.
Mounfield, Peter. *World Nuclear Power*. Routledge, 1991.

—*Al Cook*

Heysham B Power Station, Units 1-2

Heysham
Morecambe, Lancs. LA3 2XN, United Kingdom
K.S. White, Stat.Mgr.

Phone: 0524-859101
Fax: 0524-55712

Owning Utility

Nuclear Electric plc
Barnett Way
Barnwood, Glos. GL4 7RS, United Kingdom
J.G. Collier, Chm. & CEO

Phone: 0452-652222
Fax: 0452-652776
Telex: 43501

Contact

S. Moore, Dir., PR
Nuclear Electric plc

Basic Facts

Units 1-2 *Status:* Operable. *Type of Reactor:* Advanced gas-cooled reactor. *Megawatts (electric):* 660. *Megawatts (thermal):* 1552. *Reactor System Supplier:* National Nuclear Corporation (United Kingdom). *Generator Supplier:* Northern Engineering Industries (United Kingdom). *Architect Engineer:* Nuclear Electric plc.

Key Dates

Unit 1 *Ordered:* 1980. *Construction began:* 1980. *Criticality:* June 1988. *First Power:* July 1988. *Commercial Operation:* September 1988.

Unit 2 *Ordered:* 1980. *Construction began:* 1980. *Criticality:* November 1988. *First Power:* November 1988. *Commercial Operation:* December 1988.

Plant Report

England's last AGR wins world record for speed of delivery. The designers of Heysham B learned from their mistakes with the earlier Advanced Gas Cooled Reactors (AGRs) like Heysham A. Tighter design control and contract awarding eliminated the many costly delays of the earlier plants. As a result, Heysham B, England's last AGR, earned a world record by bringing its second reactor on-line just 44 days after the first.

The AGR, Britain's second generation reactor, evolved from the earlier Magnox design. It replaced the Magnox's steel-alloy pressure vessel with one of prestressed concrete lined with steel and moved the boilers to within the vessel. The graphite core remained, but uranium dioxide pellets replaced the natural-uranium fuel rods. As with the Magnox, the AGR's design allows a special machine to refuel the core every two or three days without a shutdown. This gives the AGR a distinct advantage in reliability over other reactors, most of which need annual shutdowns of three or four weeks. In operation, it resembles the American High Temperature Gas Cooled Reactor (HTGR), but uses carbon dioxide rather than helium as the coolant. The system's designers believed the use of a concrete containment vessel instead of the Magnox's steel vessel made the AGR safer.

New reactor proves disappointing. Confident in the new technology, the British government began the AGR program in April 1964, before completing the last of the Magnox plants. This second program projected a target of 5,000 megawatts electric by 1970-75, but design changes and construction difficulties created massive cost overruns, delaying the first plant opening until 1976.

Heysham B's first reactor began sending electricity to the national grid on July 12, 1988. The second followed on November 11. The two new reactors doubled the output of the Heysham facility providing an additional 1320 megawatts electric for northwest England through the grid system. The two plants combined generate enough electricity to supply all of Lancashire and most of Greater Manchester.

None of the AGRs, including Heysham B, have approached the reliability of the Magnox reactor. However, in its 1990-91 annual report, the government-owned utility, Nuclear Electric plc, noted a general increase in AGR efficiency.

The design of the AGR avoids the problem of cracks in the steel containment vessel that the Magnox is subject to. However, recent tests by the United Kingdom Atomic Energy Authority (UKAEA) on the decommissioning AGRs, Windscale and Winfrith, uncovered potential problems. Irradiation of the graphite-core bricks has changed the thermal expansion characteristics of the material. Unexpected internal stress may cause early failure of the core.

In 1979, the British government announced any new reactors built under a third program would use the more internationally common Pressurized Water Reactor (PWR) design. The first pressurized light-water reactor is due to open in mid-1994. A second received planning approval in 1990, but actual construction must await the government's review of nuclear power generation due in 1994.

Nuclear Electric inherited Heysham from the Central Electric Generating Board (CEGB) on March 31, 1990, when the British government privatized all power generation except nuclear.

Site choice fits the bill. The facility sits on Morecambe Bay in northwest Lancashire, less than fifty miles from Liverpool and Manchester. The Nuclear Installations Inspectorate (NII) approved the 223-acre site because it met the government's three main criteria set up in 1955. Its stable geology could support the massive 1.05 million-tonne building and facilities. A high demand in the surrounding area provided a ready market and need. Its proximity to the Irish Sea guaranteed a ready source of coolant and a high volume of water to dilute small quantities of radioactive effluent. Heysham B shares the site with the earlier AGR, Heysham A. The two stations draw five million liters of coolant water from the harbor per minute.

However, the NII had to adjust its earlier guidelines for maximum population densities to permit a reactor so close to a major urban center. With the concrete pressure vessel, the NII believed the design was safe enough and approved the site as "relaxed."

China Syndrome possible if everything goes wrong. In 1988, the CEGB publicly conceded that an AGR core could catch fire and melt 30-to-35 seconds after a serious accident. The admission came during the Hinkley-C public inquiry, in answer to American physicist, Richard Webb. He claimed the resulting explosion could release more radiation than the 1986 Chernobyl accident. While acknowledging the potential, Brian George, CEGB senior engineer, noted, "sixteen separate contactors and relays would have to fail." He pointed out the safety systems designed to counter the danger. A minimum of nine, out of the available 81, boron control rods would be enough to contain the reactor. The rods of the AGR fall by their own weight into the graphite core if anything interrupts the system control power.

Heysham has a population 540,000 living within a 19-mile radius of its four reactors. During the Chernobyl incident, 49,000 people evacuated a 19-mile radius of that plant.

New technology prepares for AGR and pressurized light-water reactor recycling. The handling of nuclear waste, especially spent fuel rods, commands special attention. British Nuclear Fuels Ltd (BNFL) reprocesses 96 percent of used Magnox material as new fuel, but stores AGR rods in cooling ponds. With an eye to the future, the company will invest £4 million to build a high technology processing plant to handle AGR and pressurized light-water reactor fuel. The Thermal Oxide Reprocessing Plant (THORP), at Sellafield, should begin operation in 1992. That makes nuclear waste a valuable resource for the power utility, but the high radiation level brings unique handling problems.

Cooling ponds on site store used fuel assemblies for at least 90 days, allowing short-lived radiation to decay. Then, packed into 43-tonne forged-steel flasks, they travel by rail to the BNFL reprocessing complex at Sellafield in Cumbria.

Local community officials have expressed concern that some shipments stop on sidings throughout the West Midlands area, particularly in the Bescot shunting yards. They also complain that keeping shipment times secret hinders the effectiveness of their emergency disaster planning. Nuclear Electric insists all shipments go directly to Sellafield and agreed to provide complete time tables in September 1991.

The utility notes it has been transporting radioactive material by rail for 30 years without serious incident. In that time, it has moved 12,000 loads of spent fuel rods over more than three million miles of track. In 1984, they demonstrated the safety of their £1 million steel-flask transportation vessel by crashing a train, traveling at 100 miles per hour, into one. The impact destroyed the train, but the flask remained intact.

Environmental restoration becomes the next big project. After decades of massive construction on the site, Nuclear Electric is now restoring the land. A Nature Conservancy Management Committee is working to create a publicly-accessible natural environment with a field study center, nature trails, and bird hides.

Information Center. Interactive videos and models tell the story of electricity in Heysham's information center. An estimated 15,000 people view the two plants each year. The visitor can arrange a tour of the facility for any day of the week by calling 0452-652618.

Sources

"Power complex at the CEGB." *Management Today*, May 1984.
"How far are you from a reactor?" *New Statesman*, May 16, 1986.
"CEGB admits to possible meltdown." *New Scientist*, January 1989.
United Kingdom Atomic Energy Authority Annual Report and Accounts 1989-1990. AEA Technology, London, December 1990.
Heysham 1. Nuclear Electric plc. Gloucester, July 1991.
Nuclear Electric Report and Accounts 1990-91. Nuclear Electric plc, Gloucester, July 1991.
"Nuclear output almost covers NW energy needs." *Business North West*, September 1991.
Gregan, Paul. "Meeting to draw up nuclear-trains timetable." *Birmingham Post*, September 17, 1991.
Mounfield, Peter. *World Nuclear Power.* Routledge, 1991.
Nuclear Electric Report and Accounts 1990-91. Nuclear Electric plc, Gloucester, July 1991.
"Nuclear output almost covers NW energy needs." *Business North West*, September 1991.
Gregan, Paul. "Meeting to draw up nuclear-trains timetable." *Birmingham Post*, September 17, 1991.

—Al Cook

Hinkley Point A Power Station, Units 1-2

Nr. Bridgwater TA5 1UD, United Kingdom
P. Welsh, Stat.Mgr.

Phone: 0278-652461
Fax: 0278-653304
Telex: 46261 (HINPPS G)

Owning Utility

Nuclear Electric plc
Barnett Way
Barnwood, Glos. GL4 7RS, United Kingdom
J.G. Collier, Chm. & CEO

Phone: 0452-652222
Fax: 0452-652776
Telex: 43501

Contact

S. Moore, Dir., PR
Nuclear Electric plc

Basic Facts

Units 1-2 *Status:* Operable. *Type of Reactor:* Gas-cooled reactor. *Megawatts (electric):* 282. *Megawatts (thermal):* 925. *Reactor System Supplier:* English Electric Co., Ltd. (United Kingdom); Babcock & Wilcox Co. (United States); Taylor & Woodrow Construction Ltd. (United Kingdom). *Generator Supplier:* English ELectric Co., Ltd. (United Kingdom). *Architect Engineer:* English Electric Co., Ltd. (United Kingdom); Babcock Power Ltd. (United Kingdom); Taylor & Woodrow Construction Ltd. (United Kingdom).

Key Dates

Unit 1 *Ordered:* 1957. *Construction began:* 1957. *Criticality:* 1964. *First Power:* February 1965. *Commercial Operation:* May 1965.
Unit 2 *Ordered:* 1957. *Construction began:* 1957. *Criticality:* 1964. *First Power:* March 1965. *Commercial Operation:* May 1965.

Plant Report

Distinctive design wins world record for continuous operation. Until Friday, May 10, 1991, Hinkley Point A power station held the world record for continuous operation with 700 days and seven hours logged. The 26-year-old facility dates from the first commercial nuclear development program in the UK, which chose the British-designed Magnox reactor.

The term, Magnox, derives from the magnesium alloy canister used to contain the natural-uranium fuel elements. The system uses a graphite-block core within a 3-inch thick, 67-foot-diameter steel pressure vessel. For the safety of the operators, a concrete biological shield encloses the entire structure. The system's unique design allows a special machine to refuel the core every two or three days without the need of a shutdown. This gives the Magnox a distinct advantage in reliability over other reactors, most of which need annual shutdowns of three or four weeks. In operation, it resembles the American High Temperature Gas Cooled Reactor (HTGR), but uses carbon dioxide rather than helium as the coolant.

Site choice critical to success. The Hinkley facility occupies a 65-acre site near Bridgwater in Somerset. It sits on the Bristol Channel, five miles west of the River Parret estuary. The government-owned utility Central Electricity Generating Board (CEGB), chose the location because it met the three main criteria set up in 1955 by the government. Its stable geology could support the massive building and facilities. A high demand in the surrounding area provided a ready market and need. Its proximity to the sea guaranteed a plentiful source of coolant and a high volume of water to dilute small quantities of radioactive effluent. The use of sea water avoided the need for unsightly cooling towers. The two Magnox reactors of the A station provide 470 megawatts electric for Southern England and Wales through the national grid system.

The Magnox construction program began in 1955 with a proposal for twelve stations to produce a total of 2000 MWe. They answered an increasing post-war demand for power and a need to eliminate dirty and inefficient coal-fired generating plants. After much reevaluation, the program delivered only nine Magnox installations.

Nuclear Electric plc, inherited the plant from the CEGB on March 31, 1990, when the British government privatized all power generation except nuclear.

Bid to extend life renews opposition to nuclear power. The designers, expecting a useful life of 20 to 25 years, originally licensed the plant till 1992, but the high level of reliability achieved by the Magnox has prompted Nuclear Electric to petition the Nuclear Installations Inspectorate (the NII) to extend the operating license of the plant to 30 or 40 years. The station won a temporary clearance on Sept. 11, 1991, to operate until 1995, subject to another safety review at the end of 1992 and a program of inspections and specific upgrades to be in place by then.

The NII issued the license extension after conducting a Long Term Safety Review (LTSR) which identified nine key areas of concern. In particular, welds in the steel pressure vessel and the high pressure gas ducting were dangerously brittle from more than 20 years of exposure to high radiation levels. The utility considers repairs to the welds, located inside the reactor containment vessel, impractical. Nuclear Electric hopes to be able to prove the welds safety with a new series of tests on the actual critical areas. In addition, to meet international safety standards, the plant would need new electronic control circuits.

Nuclear Electric estimates the cost of upgrading the plant at 15 million. The directive marks the ffifth in a series of LTSR's examining each of the UK's 22 Magnox reactors.

The Inspectorate will allow the station to operate at reduced pressure until it does, but the environmental group, Greenpeace, would like to see the plant closed immediately. It cites 24 incidents at Hinkley Point A since 1977 as proof of what it calls "irreparable technical defects" and "worn-out and damaged components and safety systems." It points to a 1987 report that names Hinkley the worst polluting nuclear power station in the UK and to a 1988 health authority report that shows a higher than average incidence of childhood leukemia in an eight-mile radius of the facility.

Corrosion problems in Magnox systems first surfaced in 1968. Inspectors at Bradwell discovered deterioration of bolts and other fasteners near the gas ducts. At that time, the NII ordered all Magnox stations except Calder Hall, Chapelcross, and Berkeley to reduce operating temperature and pressure by 10-to-20 percent.

Transporting nuclear waste raises fears despite safety record and dramatic safety demonstration. A major area of concern centers on the transportation of spent fuel rods. British Nuclear Fuels Ltd (BNFL) reprocesses 96 percent of that material for the CEGB as new fuel. That makes it a valuable resource for the power utility, but the high radiation level brings unique handling problems.

Cooling ponds on site store used fuel assemblies for at least 90 days, allowing short-lived radioactivity to decay. Then, packed into 43-metric ton forged-steel flasks, they travel by rail to the BNFL reprocessing plant at Sellafield in Cumbria.

Local community officials have expressed concern that some shipments stop on sidings throughout the West Midlands area, particularly in the Bescot shunting yards. They also complain that keeping shipment times secret hinders the effectiveness of their emergency disaster planning. Nuclear Electric insists all shipments go directly to Sellafield and agreed to provide complete time tables in September 1991.

The utility notes it has been transporting radioactive material by rail for 30 years without serious incident. In that time, it has moved 12,000 loads of spent fuel rods over three million miles of track. In 1984, they demonstrated the safety of their 1-million steel-flask transportation vessel by crashing a train, traveling at 100 miles per hour, into one. The impact destroyed the train, but the flask remained intact.

Site of Special Scientific Interest attracts variety. Many species of birds and animals find a home in the natural woods and rocky shore of Hinkley Point. In recognition of this, the British government designated the area as a Site of Special Scientific Interest (SSSI). Today, nature trails provide access for naturalists year round. Nearby Taunton Museum displays details of a 1906 excavation, on the Hinkley grounds, of a Bronze Age burial mound. The grave site dates back to 1500 BC. Locals refer to it as the Pixies' Mound.

Visitors Center. Interactive videos and models the story of electricity in the Hinkley Point visitor center. Nuclear Electric operates it in conjunction with Somerset Education Authority as an educational resource.

The visitors center operates seven days a week from April to September and Sunday through Friday from October to May. The visitor can arrange a tours of the facility for any day of the week by calling 0278-652-461.

Sources

Patterson, Walter C. *Nuclear Power*. New York: Penguin Books Ltd., 1977.
Hinkley Point. Nuclear Electric plc. Gloucester, n.d.

Mounfield, Peter. *World Nuclear Power.* Routledge, 1991.
"U.K. Nuclear Plant Breaks World Record." *INFO,* U.S. Council for Energy Awareness, May 1991.
"Five reactors must prove safety or shut." *The Guardian,* September 10, 1991.
"Conditional reprieve for Magnox stations." *Financial Times,* September 12, 1991.
"Hinkley cleared to 1995." *Power in Europe,* September 12, 1991.
"Can Hinkley keep its place in the Generation Game?" *Western Morning News,* September 16, 1991.
"Meeting to draw up nuclear-trains timetable." *Birmingham Post,* September 17, 1991.
"Conditional reprieve for Magnox stations." *Financial Times,* September 12, 1991.
"Hinkley cleared to 1995." *Power in Europe,* September 12, 1991.
"Can Hinkley keep its place in the Generation Game?" *Western Morning News,* September 16, 1991.

—Al Cook

Hinkley Point B Power Station, Units 1-2

Nr. Bridgwater TA5 1UD, United Kingdom
P. Welsh, Stat.Mgr.

Phone: 0278-652461
Fax: 0278-653304
Telex: 46261 (HINPPS G)

Owning Utility

Nuclear Electric plc
Barnett Way
Barnwood, Glos. GL4 7RS, United Kingdom
J.G. Collier, Chm. & CEO

Phone: 0452-652222
Fax: 0452-652776
Telex: 43501

Contact

S. Moore, Dir., PR
Nuclear Electric plc

Basic Facts

Units 1-2 *Status:* Operable. *Type of Reactor:* Advanced gas-cooled reactor. *Megawatts (electric):* 660. *Megawatts (thermal):* 1493. *Reactor System Supplier:* The Nuclear Power Group (United Kingdom). *Generator Supplier:* Associated Electric Industries Ltd. (United Kingdom); General Electric Co. (United Kingdom). *Architect Engineer:* The Nuclear Power Group (United Kingdom).

Key Dates

Unit 1 *Ordered:* 1967. *Construction began:* 1967. *Criticality:* 1974. *First Power:* October 1976. *Commercial Operation:* October 1978.
Unit 2 *Ordered:* 1967. *Construction began:* 1967. *Criticality:* 1976. *First Power:* February 1976. *Commercial Operation:* September 1976.

Plant Report

Britain's first AGR. On February 5, 1976, Hinkley Point B became the first Advanced Gas Cooled Reactor (AGR) power station to generate electricity and the tenth operational nuclear generating plant in Britain.

The AGR, Britain's second generation of reactor, evolved from the earlier Magnox design. It replaced the Magnox's steel alloy pressure vessel with one of prestressed concrete lined with steel, and it moved the boilers into the pressure vessel. The graphite core remained, but uranium dioxide pellets replaced the natural-uranium fuel rods. Like the Magnox, the AGR's design allows a special machine to refuel the core every two or three days without the need of a shutdown. This gives the AGR a distinct advantage in reliability over other reactors, most of which need annual shutdowns of three or four weeks. In operation, it resembles the American High Temperature Gas Cooled Reactor (HTGR), but uses carbon dioxide rather than helium as the coolant. The use of a concrete containment vessel instead of the Magnox's steel vessel is believed to make the AGR a safer design.

New reactor proves disappointing. With full confidence in the new technology, the British government began the AGR program in April 1964, before completing the last of the Magnox plants. This second program projected a target of 5,000 megawatts electric by 1970-75, but design changes and construction difficulties created massive cost overruns and delayed the opening of the first plant until 1976. None of the AGRs, including Hinkley Point B, have approached the operational reliability of the Magnox reactors.

The design of the AGR avoids the problem of cracks in the steel containment vessel that the Magnox is subject to. However, recent tests by the United Kingdom Atomic Energy Authority (UKAEA) on the decommissioning AGRs, Windscale and Winfrith, uncovered potential problems. Irradiation of the graphite-core bricks has changed the thermal expansion characteristics of the material. Unexpected internal stress may cause early failure of the core.

In 1966, scientist's noted that the carbon dioxide coolant gas corroded the pressure vessel's mild-chrome-steel liner. They stopped that with small amounts of methane. However, the methane caused carbon to form on fuel assemblies reducing efficiency.

To forestall corrosion problems, the contractor, The Nuclear Power Group, installed stainless steel plates over portions of the mild steel presure vessel liner in 1972. To do so, it dismantled part of the graphite core.

Inneffective insulation forced a rerouting of some ducting around the core. The resulting bends created noise and vibration greater than a jet's tailpipe.

The refueling system needed perfecting. When inserting new rods with the reactor running, the pressure changes in the core caused control problems in the boilers.

On Nov. 19, 1978, a fuel assembly being withdrawn from reactor R4, snagged. After disengaging the assembly, workers discovered damage to the graphite sleeve and damage to the fuel pins caused by lack of coolant. Investigators blamed the failure on small cracks in the fuel element sleeves which caused the casing to fail from the buffeting of high pressure gas. The utility halted on-load refueling pending new quality control inspections. Refueling now takes place at low-power levels only.

In 1979, the British government announced any new reactors built under a third program, would use the more internationally common Pressurized Water Reactor (PWR) design.

Millions of gallons of water needed to quench Southern England's energy thirst. The Hinkley facility occupies a 65-

acre site near Bridgwater in Somerset. It sits on the Bristol Channel, five miles west of the River Parret estuary. The government owned utility, Central Electric Generating Board (CEGB) chose the location because it met the three main criteria set up in 1955 by the government. Its stable geology could support the massive building and facilities. A high demand in the surrounding area provided a ready market and need. Its proximity to the sea guaranteed a plentiful source of coolant and a high volume of water to dilute small quantities of radioactive effluent. The use of sea water avoided the need for unsightly cooling towers.

Its two AGR reactors produce a total of 1230 megawatts electric for the national grid. Energy-hungry Southern England and Wales use up all they can produce. Power plants in the area produce 40 percent of Britain's electricity, but the region uses 55 percent of the power supplied by the grid. The utiltiy imports the shortfall from France.

Hinkley Point B uses up to 24 million gallons of filtered water per hour from the Bristol Channel to cool its reactors. That water is returned to the Channel with its temperature raised by 25 degrees Fahrenheit.

Nuclear Electric plc, inherited the plant from the CEGB on March 31, 1990, when the British government privatized all power generation except nuclear.

Hinkley high on antinuclear groups' hit list. Recent attempts by Nuclear Electric to extend the operating life of the two Magnox reactors at Hinkley Point A, and to secure permission to build a Pressurized Water Reactor (PWR) as Hinkley Point C, have attracted the criticism of environmental groups like Greenpeace. They point to a 1987 report that names Hinkley the most polluting nuclear power station in the UK and to a 1988 health authority report that shows a higher than average incidence of childhood leukemia in an eight-mile radius of the facility.

A major area of concern centers on the transportation of spent fuel rods. British Nuclear Fuels Ltd (BNFL) reprocesses ninety-six percent of used Magnox material as new fuel. BNFL is spending £4 million to build a high technology processing plant to handle AGR and pressurized light-water reactor fuel. The Thermal Oxide Reprocessing Plant (THORP), at Sellafield, should begin operation in 1992. That makes nuclear waste a valuable resource for the power utility, but the high radiation level brings unique handling problems.

Cooling ponds on site store used fuel assemblies for at least 90 days, allowing short-lived radiation to decay. Then, packed into 43-tonne forged-steel flasks, they travel by rail to the BNFL reprocessing complex at Sellafield in Cumbria.

Local community officials have expressed concern that some shipments stop on sidings throughout the West Midlands area, particularly in the Bescot shunting yards. They also complain that keeping shipment times secret hinders the effectiveness of their emergency disaster planning. Nuclear Electric insists all shipments go directly to Sellafield and agreed to provide complete time tables in September 1991.

The utility notes it has been transporting radioactive material by rail for 30 years without serious incident. In that time, it has moved 12,000 loads of spent fuel rods over more than three million miles of track. In 1984, they demonstrated the safety of their £1 million steel-flask transportation vessel by crashing a train, traveling at 100 miles per hour, into one. The impact destroyed the train, but the flask remained intact.

Natural sanctuary provides educational opportunities. Many species of birds and animals find a home in the natural woods and rocky shore of Hinkley Point. In recognition of this, the British government designated the area as a Site of Special Scientific Interest (SSSI). Today, nature trails provide access for naturalists year round. Nearby Taunton Museum displays details of a 1906 excavation, on the Hinkley grounds, of a Bronze Age burial mound. The grave site dates back to 1500 BC. Locals refer to it as the Pixies' Mound.

Visitors Center. Interactive videos and models the story of electricity in the Hinkley Point visitor center. Nuclear Electric operates it in conjunction with Somerset Education Authority as an educational resource.

The visitors center operates seven days a week from April to September and Sunday through Friday from October to May. The visitor can arrange a tour of the facility for any day of the week by calling 0278-652-461.

Sources

Hinkley Point. Nuclear Electric plc. Gloucester, n.d.
Patterson, Walter C. *Nuclear Power*, New York: Penguin Books Ltd., 1977.
Burn, Duncan. *Nuclear Power and the Energy Crisis.* New York: New York University Press, 1978.
"How far are you from a reactor?" *New Statesman*, May 16, 1986.
Bupp, Irvin C., and Jean-Claude Derian. *Light Water.* New York:Basic Books, 1987.
Collier, John, and Geoffrey F. Hewitt. *Introduction to Nuclear Power.* Washington: Hemisphere Publishing Corporation, 1987.
"Nuclear Safety: Some Like it Hot." *New Scientist*, November 4, 1989.
Mounfield, Peter. *World Nuclear Power.* Routledge, 1991.
"Can Hinkley keep its place in the Generation Game?" *Western Morning News*, September 16, 1991.
"Meeting to draw up nuclear-trains timetable." *Birmingham Post*, September 17, 1991.
"Nuclear Safety: Some Like it Hot." *New Scientist*, November 4, 1989.
Mounfield, Peter. *World Nuclear Power.* Routledge, 1991.

—Al Cook

Hinkley Point C Power Station

Nr. Bridgwater TA5 1UD, United Kingdom
P. Welsh, Stat.Mgr.

Phone: 0278-652461
Fax: 0278-653304
Telex: 46261 (HINPPS G)

Owning Utility

Nuclear Electric plc
Barnett Way
Barnwood, Glos. GL4 7RS, United Kingdom
J.G. Collier, Chm. & CEO

Phone: 0452-652222
Fax: 0452-652776
Telex: 43501

227 Hunterston

UNITED KINGDOM

Contact
S. Moore, Dir., PR
Nuclear Electric plc

Basic Facts
Status: Planned. *Type of Reactor:* Pressurized light-water reactor. *Megawatts (electric):* 1250. *Megawatts (thermal):* 3411.

Plant Report

Battle over nuclear energy flares over Hinkley Point C. Even though construction has not yet begun, Hinkley Point C has become the battleground for pro- and anti-nuclear forces in a war which will probably be decided in 1994. That is when the British government will review its nuclear option and either continue construction of Pressurized Water Reactors (PWR) or begin closing down its existing plants.

The Central Electric Generating Board (CEGB) proposed Hinkley C in April 1987, shortly after construction of Britain's first pressurized light-water reactor began at Sizewell in Anglia. At that time, it had the backing of Margaret Thatcher's Conservative government despite the growing trend towards privatization of the power industry. That was because the project fulfilled a government policy of reducing dependency on fossil fuels. However, that policy, the Non-Fossil Fuel Obligation (N-FFO), fell from grace during the year-long Hinkley Inquiry. Opponents of nuclear power pointed to safety concerns raised by Chernobyl, cost overruns at Sizewell and evidence that the cost of electricity generated by nuclear stations was not competitive with fossil fuel plants.

Inquiry sparks chain reaction. The £10 million Hinkley Inquiry issued its 3000-page report on Sept. 6, 1990, and set off a chain of events that stunned the nuclear industry. The government decided to continue its plan of privatizing the power generation industry, but withdrew its nuclear reactors from the sell-off of assets. It disbanded the CEGB and created Nuclear Electric plc to look after those plants. It envisioned the utility as a "swing producer" designed to stabilize the newly-created private power industry. Then, it approved the construction of a pressurized light-water reactor at Hinkley Point, but withheld funding until after the 1994 review. Two other PWRs were cancelled. Shortly after, Lord Marshall of Goring, former chairman of the CEGB, resigned his position as head of Nuclear Power noting that he had agreed to run a utility that would "construct, own and operate nuclear power stations in England and Wales.and that is not now to be the case." He had spent the previous quarter century lobbying for the pressurized light-water reactor design.

The case for the new plant. If construction were to begin shortly after 1994, the 1200 megawatts electric plant could begin operating by 2002, at a cost of £1.7 billion. Nuclear Electric claims the plant would create 10,000 jobs over seven years and pump £300 million into the Southwest England economy. It would provide power to over one million people.

In his report, Mr. Michael Barnes QC predicted a shortfall of 11,000 to 12,000 megawatts electric of generating capacity, about one fifth of the current total, by the end of the century. However, Energy Secretary John Wakeham hopes to stretch available power with conservation and extensions of the operating lives of existing Magnox stations like Hinkley Point A.

Barnes argues the use of nuclear power plants like Hinkley Point C in the future will reduce the burning of fossil fuels and the production of greenhouse gases.

Opposition grows. During the inquiry, over 20,000 people officially registered as objectors. They raised questions of relative cost between nuclear and fossil fuel plants, emergency planning, apparently missing plutonium, and the effect of a fifth nuclear reactor on the Hinkley Point site to the local ecology and residents. Inspector Barnes dismissed the last concern noting the site of Hinkley Point was classified as "remote" from major population areas and that the risk of death from radiation for a person living five kilometers from the plant, under normal operating conditions, was less than four in 10 million per year. He did call for a systematic investigation of childhood leukemia clusters around the site as a follow up to the 1988 health authority report. That study showed a higher-than-average incidence of the disease in an eight mile radius of Hinkley Point and in the Sellafield area.

The environmental group, Friends of the Earth, labeled the decision "bizarre, dated and irrelevant," noting the plant cannot proceed without funding. They pledge to join other environmental groups, local authorities and residents to continue to lobby the government to refuse that funding. The Labour party has already declared its opposition to the government's decision.

Sources
"Marshall explains why nuclear will not be privatized." *Nuclear News*, January 1990.
"Decision gives cold comfort to nuclear lobby." *Guardian*, September 7, 1990.
"Hinkley ruling stirs up fresh nuclear protest." *The Daily Telegraph*, September 7, 1990.
"Labour fury over N-plant." *Daily Post*, September 7, 1990.
"N-plant decision disquiets industry." *Resources*, September 7, 1990.
"Wakeham delays Hinkley C go-ahead." *Guardian*, September 7, 1990.

—Al Cook

227 ▲

Hunterston A Nuclear Power Station, Units 1-2

West Kilbride
Ayrshire
Scotland KA23 9QJ, United Kingdom
Gordon Hodgson, Decommissioning Mgr.

Phone: 0294-822311
Fax: 0294-822311 x 2307
Telex: 778483

Owning Utility
Scottish Nuclear Ltd.
3 Redwood Crescent, Peel Park
East Kilbride
Scotland G74 5PR, United Kingdom
Robin Jeffrey, CEO

Phone: 03552-62000
Fax: 03552-62626
Telex: 777700

Hunterston

Contact
Richard Marshall, PR Mgr.
Scottish Nuclear Ltd.

Basic Facts
Units 1-2 *Status:* Decommissioned. *Type of Reactor:* Gas-cooled reactor. *Megawatts (electric):* 169. *Megawatts (thermal):* 344.5. *Reactor System Supplier:* General Electric Co. (United Kingdom). *Generator Supplier:* General Electric Co. (United Kingdom).

Key Dates
Unit 1 *Ordered:* 1956. *Construction began:* 1957. *Criticality:* 1963. *First Power:* 1964. *Commercial Operation:* May 1964. *Shut Down:* March 1990. *Decommissioned:* 1990.

Unit 2 *Ordered:* 1956. *Construction began:* 1957. *Criticality:* 1964. *First Power:* 1964. *Commercial Operation:* September 1964. *Shut Down:* December 1989. *Decommissioned:* 1990.

Plant Report

Scotland's first nuclear power station shuts down in 1990. HRH The Queen Mother opened Hunterston A power station in 1964. For the next 26 years, the plant served the realm faithfully. From 1965 till 1980, it held the record as the best performing nuclear power station in the world. From then until it started its shutdown program in 1988, it never dropped below a top-ten ranking.

The facility dates from the beginning of nuclear power development in the UK, when the British-designed Magnox reactor began to evolve. The South of Scotland Electricity Board (SSEB) copied the design from Scotland's first experimental station, Chapelcross.

The term, Magnox, derived from the magnesium alloy canister used to contain the natural-uranium fuel elements. The system used a graphite-block core within a 3-inch thick, 67-foot-diameter steel pressure vessel. For the safety of the operators, a concrete biological shield enclosed the entire structure. The system's unique design allowed a special machine to refuel the core every two or three days without the need of a shutdown. This gave the Magnox a distinct advantage in reliability over other reactors, most of which need annual shutdowns of three or four weeks. In operation, it resembled the American High Temperature Gas Cooled Reactor (HTGR), but used carbon dioxide rather than helium as the coolant.

The facility sits on the shore of the Firth of Clyde, about 35 miles south west of Glasgow. Twin round towers house the reactors within 60-meter-high glass walls. At night, the illuminated spires form a distinctive landmark and act as a beacon for sailors. As another Scottish first, the SSEB added an Advanced Gas Cooled Reactor (AGR) station to the site in 1976.

The United Kingdom Atomic Energy Authority (UKAEA) chose the location because it met the three main criteria set up in 1955 by the government. Its stable geology could support the massive building and facilities. A high demand in the surrounding area provided a ready market and need. Its proximity to the sea guaranteed a plentiful source of coolant and a high volume of water to dilute small quantities of radioactive effluent. The use of sea water avoided the need for unsightly cooling towers.

Its two Magnox reactors provided 300 megawatts electric primarily for Glasgow and the manufacturing areas of west Scotland.

The British government first proposed Hunterston A in 1955 as one of a series of twelve stations to produce a total of 2000 MWe. They were to answer an increasing post-war demand for power and a need to replace dirty and inefficient coal-fired generating plants. After much reevaluation, the program delivered only nine Magnox installations. Construction of Hunterston A began in 1957.

Scottish Nuclear Limited inherited the plant from the SSEB on March 31, 1990, when the British government privatized all power generation except nuclear. It also inherited the liability for decommissioning the plant.

All Magnox stations subject to Long Term Safety Reviews. The designers, expecting a useful life of 20 to 25 years, originally licensed the plant till 1989. However, the reliability of the Magnox and the escalating cost of new facilities prompted a reassessment. Scottish Nuclear petitioned the Nuclear Installations Inspectorate (NII) to extend the operating license to 30 or 40 years.

High repair costs doom plant. The Long Term Safety Review, issued by the Inspectorate, cited several safety deficiencies. Welds in the steel pressure vessel and the high pressure gas ducting were dangerously brittle from more than 20 years of exposure to high radiation levels. The utility considered repairs to the welds, located inside the reactor containment vessel, impractical. In addition, to meet international safety standards, the plant would need new electronic control circuits. After careful assessment of the costs of repair, the SSEB decided to proceed with decommissioning.

Decommissioning will take 100 years. Shutdown of the plant began in 1988, in preparation for the three phase decommissioning process.

Once the second reactor ceased generation in March 1990, defuelling began. In that process, the utility removes all fuel rods from the reactor, the cooling ponds and the store houses. Packed into 43-metric ton forged-steel flasks, the rods travel by rail to the British Nuclear Fuels (BNFL) reprocessing plant at Sellafield in Cumbria. Scottish Nuclear expects to complete this phase by 1995.

BNFL notes it has been transporting radioactive material by rail for more than 30 years without serious incident. In that time, it has moved 12,000 loads of spent fuel rods over more than three million miles of track. In 1984, the English utility, Nuclear Electric, demonstrated the safety of the steel-flask transportation vessel. They intentionally crashed a train, traveling at 100 miles per hour, into one. The impact destroyed the train, but the flask remained intact.

With all radioactive material cleared from the site, demolition and dismantling of the structures outside the biological shield of the reactor can begin. That will take another five years to complete.

The utility will allow a one-hundred-year waiting period for radiation in the core to decay before dismantling the reactor. Eventually, the pieces, classified as "intermediate-level waste," will arrive at BFNL's new underground depository at Sellafield.

In its 1990-91 Financial Report, Scottish Nuclear termed its estimate of the final cost of closing the plant as subject to considerable "uncertainty." It could only estimate expenses that will span the next century.

Sources

Mounfield, Peter. *World Nuclear Power*. Routledge, 1991.
Hunterston Power Station. Scottish Nuclear Limited, Glasgow, n.d.
Scottish Nuclear Annual Report and Accounts 1990-91, Scottish Nuclear Limited, Glasgow, June 27, 1991.
"Five Reactors Must Prove Safety or Shut." *The Guardian*. September 10, 1991.
"Conditional Reprieve for Magnox Stations." *Financial Times*, September 12, 1991.

—Al Cook

228 •

Hunterston B Nuclear Power Station, Units 1-2

West Kilbride
Ayrshire
Scotland KA23 9QJ, United Kingdom
Peter Robson, Stat.Mgr.

Phone: 0294-822311
Fax: 0294-822311 x 2307
Telex: 778483

Owning Utility

Scottish Nuclear Ltd.
3 Redwood Crescent, Peel Park
East Kilbride
Scotland G74 5PR, United Kingdom
Robin Jeffrey, CEO

Phone: 03552-62000
Fax: 03552-62626
Telex: 777700

Contact

Richard Marshall, PR Mgr.
Scottish Nuclear Ltd.

Basic Facts

Units 1-2 *Status:* Operable. *Type of Reactor:* Advanced gas-cooled reactor. *Megawatts (electric):* 660. *Megawatts (thermal):* 1496. *Reactor System Supplier:* The Nuclear Power Group (United Kingdom). *Generator Supplier:* C.A. Parsons & Co., Ltd. (United Kingdom). *Architect Engineer:* The Nuclear Power Group (United Kingdom).

Key Dates

Unit 1 *Ordered:* 1967. *Construction began:* 1967. *Criticality:* 1976. *First Power:* February 1976. *Commercial Operation:* June 1976.
Unit 2 *Ordered:* 1967. *Construction began:* 1967. *Criticality:* 1976. *First Power:* February 1977. *Commercial Operation:* March 1977.

Plant Report

Second plant introduces advanced technology to Scottish nuclear industry. The South of Scotland Electricity Board (SSEB) already had a success story with Hunterston A. That Magnox station held the world record for reliable performance. By building a bigger and better plant on the same site, using Advanced Gas Cooled Reactors (AGR), the SSEB hoped to continue that achievement. In 1976, Hunterston B became Scotland's first AGR station and the second AGR commissioned in the United Kingdom.

The AGR, Britain's second generation of reactor, evolved from the earlier Magnox design. It replaced the Magnox's steel-alloy pressure vessel with one of prestressed concrete lined with steel and moved the boilers to within the vessel. The graphite core remained, but uranium dioxide pellets replaced the natural-uranium fuel rods. As with the Magnox, the AGR's design allows a special machine to refuel the core every two or three days without a shut-down. This gives the AGR a distinct advantage in reliability over other reactors, most of which need annual shut-downs of three or four weeks. In operation, it resembles the American High Temperature Gas Cooled Reactor (HTGR), but uses carbon dioxide rather than helium as the coolant.

Hot gas makes for safer design. The system's designers believed the use of a concrete containment vessel instead of the Magnox's steel vessel made the AGR safer. Many scientists now rate the HTGR safer than the more-common Pressurized Water Reactor (PWR). Unlike the HTGR, the pressurized light-water reactor must maintain coolant in the reactor core to prevent meltdown.

Confident in the new technology, the British government began the AGR program in April 1964, before completing the last of the Magnox plants. This second program projected a target of 5,000 megawatts electric by 1970-75, but design changes and construction difficulties created massive cost overruns, delaying the first plant opening, Hinkley B, until 1976.

However, the AGR technology never approached the over-90-percent efficiency of the Magnox. In 1991, Hunterston B ranked as the most productive AGR station in the UK with an efficiency of 79.1 percent. It produces 1200 megawatts electric primarily for Glasgow and the manufacturing areas of west Scotland, but recurring technical problems continue to take the plant off-line. The replacement of the liners for the refueling standpipes guarantee scheduled maintenance shut-downs until at least 1993. Earlier problems included corrosion of the mild-steel pressure vessel by the carbon dioxide coolant gas, carboning of fuel rods from methane gas used to control corrosion, excessive noise and vibration from rerouted gas ducts, and ineffective core insulation.

Hunterston's most serious accident happened on Oct. 11, 1977. Following a shutdown on October 22, 8000 liters of seawater entered the reactor vessel. The water entered through a temporary waste duct installed during construction. It was meant to exhaust contaminated demineralized water until repairs to defective circulator seals could be made. When the reactor shut down, pressure in the system allowed the water flow to reverse, channelling seawater into the reactor.

In 1979, the British government announced any new reactors built under a third program would use the more internationally common pressurized light-water reactor design. The first pressurized light-water reactor is due to open in mid-1994. A second received planning approval in 1990, but actual con-

Plant Profiles

struction must await the government's review of nuclear power generation due in 1994.

Nuclear lighthouses beckon sailors. Hunterston sits on the shore of the Firth of Clyde, about 35 miles south west of Glasgow. Two glass wings atop the rectangular blockhouse of Hunterston-B mimic the earlier station's twin curved-glass towers. At night, the illuminated spires form a distinctive landmark and act as a beacon for sailors.

The reactors themselves, sit within 5.8-meter-thick concrete walls. Edinburgh Castle's walls, at their thickest only reach 4.5 meters. More than 1000 kilometers of steel wire, embedded in Hunterston B's concrete, meet in a stressing gallery. Workers at the plant constantly adjust the tension on the wires to optimize the integrity of the containment system.

The SSEB chose the location because it met the three main criteria set up in 1955 by the government. Its stable geology could support the massive building and facilities. A high demand in the surrounding area provided a ready market and need. Its proximity to the sea guaranteed a plentiful source of coolant and a high volume of water to dilute small quantities of radioactive effluent. The use of sea water avoided the need for unsightly cooling towers.

Scottish Nuclear inherited Hunterston from the SSEB on March 31, 1990, when the British government privatized all power generation except nuclear.

Utility may abandon recycling in favor of burial. Cooling ponds, on site, store used fuel assemblies for at least 90 days. This procedure allows short-lived radiation to decay. The next step is to pack the cooled rods into 43-tonne forged-steel flasks and send them by rail to the British Nuclear Fuels Ltd (BNFL) reprocessing complex at Sellafield in Cumbria. However, the escalating cost of recycling at BNFL has Scottish Nuclear concerned. That cost played an important part in their decision to close Hunterston A.

Consequently, Scottish Nuclear would rather not send its spent fuel to Sellafield. Instead it has sought permission from the Nuclear Installations Inspectorate (NII) to build a dry storage facility on site. The utility would encase spent fuel rods in copper tubes and store them in air-cooled concrete vaults for up to 50 years. It could then either reprocess the material or bury it for long-term storage. Such storage would reduce the need to transport radioactive material and allow the utility some flexibility and bargaining power in negotiating future reprocessing contracts. Scottish Nuclear hopes to complete one dry store at Hunterston B and another at Torness by 1994 when its contract with BNFL expires.

Sources

Patterson, Walter C. *Nuclear Power.* New York, Penguin Books Ltd., 1977.
Burn, Duncan. *Nuclear Power and the Energy Crisis.* New York, New York University Press, 1978.
Bupp, Irvin C., and Jean-Claude Derian. *Light Water,* New York, Basic Books, 1987.
Collier, John, and Geoffrey F. Hewitt. *Introduction to Nuclear Power,* Washington, Hemisphere Publishing Corporation, 1987.
"Cloud over Sellafield's Reprocessing Unit." *New Scientist,* April 22, 1989.

"Measures to Make Nuclear Power Safer than ever Before." *New Scientist,* November 4, 1989.
"Nuclear Safety: Some Like it Hot." *New Scientist,* November 4, 1989.
Mounfield, Peter. *World Nuclear Power.* Routledge, 1991.
Scottish Nuclear Annual Report and Accounts 1990-1991. Scottish Nuclear, Glasgow, June 27, 1991.
British Nuclear Fuels Plc Annual Report and Accounts 1990-1991. British Nuclear Fuels, August 1991.
Hunterston. Scottish Nuclear, Glasgow, n.d.
Mounfield, Peter. *World Nuclear Power.* Routledge, 1991.

—Al Cook

229 ●

Oldbury Power Station, Units 1-2

Oldbury Naite
Thornbury, Avon BS12 1RQ, United Kingdom
J. Holland, Stat.Mgr.

Phone: 0454-416631
Fax: 0454-414125

Owning Utility

Nuclear Electric plc
Barnett Way
Barnwood, Glos. GL4 7RS, United Kingdom
J.G. Collier, Chm. & CEO

Phone: 0452-652222
Fax: 0452-652776
Telex: 43501

Contact

S. Moore, Dir., PR
Nuclear Electric plc

Basic Facts

Units 1-2 *Status:* Operable. *Type of Reactor:* Gas-cooled reactor. *Megawatts (electric):* 313. *Megawatts (thermal):* 750. *Reactor System Supplier:* The Nuclear Power Group (United Kingdom). *Generator Supplier:* Associated Electric Industries Ltd. (United Kingdom); C.A. Parsons & Co., Ltd. (United Kingdom). *Architect Engineer:* The Nuclear Power Group (United Kingdom).

Key Dates

Unit 1 *Ordered:* 1961. *Construction began:* 1962. *Criticality:* 1967. *First Power:* November 1967. *Commercial Operation:* January 1968.
Unit 2 *Ordered:* 1961. *Construction began:* 1962. *Criticality:* 1967. *First Power:* April 1968. *Commercial Operation:* May 1968.

Plant Report

Distinctive design wins world record for continuous operation. On Friday, May 10, 1991, Oldbury-on-Severn power station earned the world record for continuous operation with 700 days and seven hours logged. The 23-year-old facility dates from the first commercial nuclear development program in the UK, which chose the British-designed Magnox reactor.

The term, Magnox, derives from the magnesium alloy canister used to contain the natural uranium fuel elements. The system's unique design allows a special machine to refuel the core every two or three days without the need of a shutdown. This gives the Magnox a distinct advantage in reliability over other reactors, most of which need annual shutdowns of three or four weeks. In operation, it resembles the American High Temperature Gas

Cooled Reactor (HTGR), but uses carbon dioxide rather than helium as the coolant.

Design changes improve safety. The United Kingdom Atomic Energy Authority (UKAEA) created the basic design for the Magnox reactor, but each new power plant project developed the concept further. Most Magnox reactors use a graphite-block core within a 3-inch thick, 67-foot-diameter steel pressure vessel surrounded by a concrete biological shield. Oldbury was the first Magnox to replace the steel containment vessel with pre-stressed concrete lined with steel and to move the boilers into the containment vessel. The pressure vessel has its own cooling system of constantly circulating demineralized water to maintain an even temperature.

Site choice critical to success. The 175-acre facility sits on the River Severn in Somerset, 15 miles north of Bristol. The Severn flows into the Bristol Channel. The UKAEA chose the location because it met the three main criteria set up in 1955 by the government. Its stable geology could support the massive building and facilities. A high demand in the surrounding area provided a ready market and need. Its proximity to the sea guaranteed a plentiful source of coolant and a high volume of water to dilute small quantities of radioactive effluent. The use of sea water avoided the need for unsightly cooling towers.

A rock shelf formation in the river creates a natural 380-acre reservoir holding an estimated 416 million gallons of cooling water. Oldbury sucks in up to 11 million gallons of filtered water per hour, but a special recovery plant built into the intake system diverts large numbers of salmon smolt making their first journey to the ocean. The water is returned to the estuary 1.5 miles upstream with its temperature raised by 17 degrees Fahrenheit.

The Magnox construction program began in 1955 with a proposal for twelve stations to produce a total of 2000 MWe. They answered an increasing post-war demand for power and a need to eliminate dirty and inefficient coal-fired generating plants. Only nine Magnox plants were actually built for that program.

The Minister of Technology, the Rt Hon. Anthony Wedgewood Benn MP, officially opened Oldbury on June 10, 1969, but the plant first contributed electricity to the grid in January 1968. Its two Magnox reactors provide 435 megawatts electric for Southern England and Wales.

The government owned utility, Nuclear Electric plc, inherited the plant from the Centraal Electric Generating Board (CEGB) on March 31, 1990, when the British government privatized all power generation except nuclear.

All Magnox stations subject to Long Term Safety Reviews. The designers, expecting a useful life of 20 to 25 years, originally licensed the plant till 1995, but the high level of reliability achieved by the Magnox and the escalating cost of new facilities prompted Nuclear Electric to petition the Nuclear Installations Inspectorate (the NII) to extend the operating licenses of all its Magnox stations to 30 or 40 years. Without Oldbury and other Magnox stations like it, southern England and Wales could face a shortfall of 11,000 to 12,000 megawatts electric of generating capacity by the turn of the century.

The NII will conduct a Long Term Safety Review (LTSR) of Oldbury before issuing a license extension. Some older plants have already been reviewed with mixed results. Calder Hall and Chapelcross won extensions, but Berkeley and Hunterston A shut down because the cost of upgrading was too high.

Oldbury's advanced design exempts it from the Inspectorate's main concern. In earlier Magnoxes, welds in the steel alloy pressure vessel and the high pressure gas ducting may have become dangerously brittle from more than 20 years of exposure to high levels of radiation. However, the Inspectorate may still require new electronic safety circuits to be installed.

United Nations safety inspection approves plant operations. The plant has already passed inspections by the United Nations International Atomic Energy Agency in 1989 and 1990, making it the first British nuclear power station to be reviewed by the Operational Safety Review Team (OSART).

Nuclear waste. As with all nuclear facilities, the handling of waste, especially spent fuel rods, commands special attention. 96 percent of that material can be reprocessed as new fuel, making it a valuable resource for the power utility, but the high level of radioactivity brings unique handling problems.

Cooling ponds on site store used fuel assemblies for at least 90 days, allowing short-lived radiation to decay. Then, packed into 43- metric ton forged-steel flasks, they travel by rail to the British Nuclear Fuels (BNFL) reprocessing plant at Sellafield in Cumbria.

Local officials from surrounding communities have expressed concern that some shipments stop on sidings throughout the West Midlands area, particularly in the Bescot shunting yards, and that keeping shipment times secret hinders the effectiveness of their emergency disaster planning. Nuclear Electric insists all shipments go directly to Sellafield and agreed to provide complete time tables in September 1991.

The utility notes it has been transporting radioactive material by rail for longer than 30 years without serious incident. In that time, it haas moved 12,000 loads of spent fuel rods over more than three million miles of track. In 1984, they demonstrated the safety of their 1 million steel-flask transportation vessel by crashing a train, traveling at 100 miles per hour, into one. The impact destroyed the locomotive, but the flask remained intact.

Visitors center opens to fanfare of world record. Oldbury opened its new visitors center on September 16, 1991, in time to commemorate its new world record. Interactive videos and models tell the story of electricity every day from 10 till 4. From October to March the center closes on Saturdays. The visitor can arrange a tour for any day of the week by calling 0454-416631.

Sources

Mounfield, Peter. *World Nuclear Power*. Routledge, 1991.
Nuclear Electric Report and Accounts 1990-91. Nuclear Electric plc, Gloucester, n.d.
Oldbury. Nuclear Electric plc. Gloucester, n.d.
"U.K. Nuclear Plant Breaks World Record." *INFO*, U.S. Council for Energy Awareness, May 1991.

"Five reactors must prove safety or shut." *The Guardian*, September 10, 1991.
"Conditional reprieve for Magnox stations." *Financial Times*, September 12, 1991.
"Meeting to draw up nuclear-trains timetable." *Birmingham Post*, September 17, 1991.

—Al Cook

Sizewell A Power Station, Units 1-2

Leiston, Suffolk IP16 4UE, United Kingdom
S.J. Price, Station A Mgr.

Phone: 0728-830444
Fax: 0728-832195

Owning Utility

Nuclear Electric plc
Barnett Way
Barnwood, Glos. GL4 7RS, United Kingdom
J.G. Collier, Chm. & CEO

Phone: 0452-652222
Fax: 0452-652776
Telex: 43501

Contact

S. Moore, Dir., PR
Nuclear Electric plc

Basic Facts

Units 1-2 *Status:* Operable. *Type of Reactor:* Gas-cooled reactor. *Megawatts (electric):* 325. *Megawatts (thermal):* 800. *Reactor System Supplier:* English Electric Co., Ltd. (United Kingdom); The Babcock & Wilcox Co. (United States); Taylor Woodrow Construction Ltd. (United Kingdom). *Generator Supplier:* English ELectric Co., Ltd. (United Kingdom). *Architect Engineer:* English Electric Co., Ltd. (United Kingdom); Babcock Power Ltd. (United Kingdom); Taylor Woodrow Construction Ltd. (United Kingdom).

Key Dates

Unit 1 *Ordered:* 1960. *Construction began:* 1961. *Criticality:* 1965. *First Power:* January 1966. *Commercial Operation:* March 1966.
Unit 2 *Ordered:* 1960. *Construction began:* 1961. *Criticality:* 1965. *First Power:* April 1966. *Commercial Operation:* May 1966.

Plant Report

Older Magnox station makes room for foreign technology. Sizewell nuclear facility, on the North Sea's Suffolk coast, showcases both the old and the new of Britain's nuclear power-generation industry. The contrast also marks a break with British reactor technology in favor of American. By 1994, Britain's first Pressurized Water Reactor (PWR) should begin operation. Earlier systems, like Sizewell's A-station, used a British version of High Temperature Gas Cooled Reactors (HTGR).

Sizewell-A dates from the first commercial nuclear development program in the UK, which chose the British-designed Magnox. The Magnox construction program began in 1955 with a proposal for 12 stations to produce a total of 2000 megawatts electric. They answered an increasing post-war demand for power and a need to eliminate dirty and inefficient coal-fired generating plants. After much re-evaluation, the program delivered only nine Magnox installations.

The term Magnox derives from the magnesium alloy canister used to contain the natural-uranium fuel elements. The system uses a graphite-block core within a 4-inch thick, 64-foot-diameter steel pressure vessel. For the safety of the operators, a concrete biological shield encloses the entire structure. The system's unique design allows a special machine to refuel the core every two or three days without the need of a shut-down. This gives the Magnox a distinct advantage in reliability over other reactors, most of which need annual shut-downs of three or four weeks. In operation, it resembles the American High Temperature Gas Cooled Reactor (HTGR), but uses carbon dioxide rather than helium as the coolant.

East Anglia provides perfect site. The facility occupies a 245-acre site near Ipswich in Anglia. It sits on the North Sea, about 75 miles northeast of London. The Central Electric Generating Board (CEGB) chose the location because it met the three main criteria set up in 1955 by the government. Its stable geology, 160 feet of Norwich Crag on top of firm London clay, could support the massive building and facilities. A high demand in the surrounding area provided a ready market and need. Its proximity to the sea guaranteed a plentiful source of coolant and a high volume of water to dilute small quantities of radioactive effluent. Using 122 million liters of filtered sea water each hour avoided the need for unsightly cooling towers. The two Magnox reactors of the A-station provide 420 megawatts electric primarily for the East Anglia area.

The government owned utility, Nuclear Electric plc, inherited the plant from the CEGB on March 31, 1990, when the British government privatized all power generation except nuclear.

Continued operation needed into 21st century. The designers, expecting a useful life of 25 to 30 years, originally licensed the plant until 1996. However, the reliability of the Magnox and the escalating cost of new facilities prompted a reassessment. Nuclear Electric petitioned the Nuclear Installations Inspectorate (NII) to extend the operating licenses of all its Magnox stations to 40 years. Without stations like Sizewell-A, southern England and Wales could face 11,000-to-12,000 megawatts electric shortfalls in generating capacity by the turn of the century.

The Inspectorate will examine the physical condition of the plant in a Long Term Safety Review (LTSR) before extending the plant's license. LTSRs at other Magnox plants revealed several safety deficiencies. Welds in the steel pressure vessel and the high pressure gas ducting were dangerously brittle from more than 20 years of exposure to high radiation levels. The utility considers repairs to the welds, located inside the reactor containment vessel, impractical. In 1970, the inspectorate ordered most Magnox plants to operate at reduced power when it discovered corrosion of pressure vessels from the carbon dioxide cooling gas. In addition, international safety standards demand new electronic control circuits.

A major area of concern centers on the transportation of spent fuel rods. British Nuclear Fuels Ltd (BNFL) reprocesses 96 percent of that material for the Nuclear Electric as new fuel. That makes it a valuable resource for the power utility, but the high radiation level brings unique handling problems.

Special containers move used uranium by rail. Cooling ponds on site store used fuel assemblies for at least 90 days, allowing short-lived radiation to decay. Then, packed into 43 metric ton forged-steel flasks, the rods travel by rail to the BNFL reprocessing plant at Sellafield in Cumbria.

Local community officials have expressed concern that some shipments stop on sidings throughout the West Midlands area, particularly in the Bescot shunting yards. They also complain that keeping shipment times secret hinders the effectiveness of their emergency disaster planning. Nuclear Electric insists all shipments go directly to Sellafield and agreed to provide complete time tables in September 1991.

The utility notes it has been transporting radioactive material by rail for 30 years without serious incident. In that time, it has moved 12,000 loads of spent fuel rods over more than three million miles of track. In 1984, they demonstrated the safety of their £1,000,000 steel-flask transportation vessel by crashing a train, traveling at 100 miles per hour, into one. The impact destroyed the train, but the flask remained intact.

Natural surroundings receive boost from utility. Sizewell sits within an ecologically important region. The shore line, a Heritage Coast, is also an Area of Outstanding Natural Beauty (AONB). Further inland, the Nature Conservancy Council (NCC) designated two separate Sites of Special Scientific Interest (SSSI). Sand and shingle dune systems stretch on both north and south sides of the plants. To the north and west, fresh water grazing marshes provide homes to rare plants, insects and birds.

Nuclear Electric planted conifer woodlands on former heathland to promote populations of deer and birds. The company also purchased Kenton and Goose Hill conifer woods and planted broadleaf trees within the old stands. The company hopes the new woods and the reconstruction of the dunes damaged during the construction of Sizewell-B will help to integrate the power plant visually with its natural surroundings.

Visitors Center. More than 12,000 people tour Sizewell-A and its visitors center each year. Interactive videos and models tell the story of electricity, but a high-tech diorama provides the most popular focal point. That exhibit shows cutaway models of Magnox and pressurized light-water reactor cores, while a light and sound show plays in the background. The visitor can arrange a tour of the facility for any day of the week by calling 0452 652618.

Sources

Mounfield, Peter. *World Nuclear Power.* Routledge, 1991.
"Weakened welds shut down reactor." *New Scientist,* Feb. 9, 1991.
Sizewell A. Nuclear Electric plc. Gloucester, July 1991.
"Five reactors must prove safety or shut." *The Guardian,* September 10, 1991.
"Hinkley cleared to 1995." *Power in Europe,* September 12, 1991.
"Conditional reprieve for Magnox stations." *Financial Times,* September 12, 1991.
"Can Hinkley keep its place in the Generation Game?" *Western Morning News,* September 16, 1991.
"Meeting to draw up nuclear-trains timetable." *Birmingham Post,* September 17, 1991.

—Al Cook

231 ○
Sizewell B Power Station

Leiston, Suffolk IP16 4UE, United Kingdom
D. Joynson, Station B Mgr.

Phone: 0728-830444
Fax: 0728-832195

Owning Utility

Nuclear Electric plc
Barnett Way
Barnwood, Glos. GL4 7RS, United Kingdom
John G. Collier, Chm. & CEO

Phone: 0452-652222
Fax: 0452-652776
Telex: 43501

Contact

S. Moore, Dir., PR
Nuclear Electric plc

Basic Facts

Status: Under construction (50% complete as of January 1992). *Type of Reactor:* Pressurized light-water reactor. *Megawatts (electric):* 1258. *Megawatts (thermal):* 3411. *Reactor System Supplier:* PWR Power Projects (United Kingdom). *Generator Supplier:* General Electric Co. (United Kingdom). *Architect Engineer:* National Nuclear Corporation (United Kingdom).

Key Dates

Ordered: 1987. *Construction began:* 1988. *Criticality expected* December 1993. *Commercial Operation Expected:* May 1994.

Plant Report

American technology under pressure. The world's first commercial nuclear reactor, Calder Hall, commissioned in 1956, took less than four years to complete. By 1987, construction of Britain's first Pressurized Water Reactor (PWR), Sizewell B, had taken four years to get started.

The pressurized light-water reactor eventually became the technology of choice for Britain's third program of nuclear power plant construction, but not without decades of debate and government indecision.

In December 1973, the government utility, Central Electricity Generating Board (CEGB), proposed building 32 Westinghouse 1300-megawatts electric pressurized light-water reactor generating stations starting in 1982.

The ambitious program grew out of fears of future energy shortages. The OPEC oil embargo had shaken the western world, prompting a search for alternative power sources. However, electricity demand in Britain fell throughout the later 70s, invalidating the CEGB's estimates.

On Dec. 18, 1979, David Howell, secretary of state for energy, issued a statement in the House of Commons. He suggested a construction campaign of one new pressurized light-water reactor reactor every year for 10 years, starting in 1982.

With excess generating capacity exceeding 30 percent even with the shutdown of 171 older plants, mostly coal-fired, the argument of capacity for future demand became harder to justify. Economics became the issue, and Sizewell B, the first test of the nuclear industry's assertion that it had a monopoly on clean, safe, efficient electricity production.

Sizewell Inquiry shows growing opposition to nuclear power. At the Sizewell inquiry, which started on Jan. 11, 1983, the CEGB argued new pressurized light-water reactor plants would produce electricity cheaper than existing fossil fuel systems. That made nuclear the logical choice for future generating stations.

After a four-year inquiry, which sifted 55 metric tons of evidence to produce a 3,000-page report, the project received a hesitant go-ahead on Jan 26, 1987. Sir John Layfield, inquiry chairman, admitted he was leery of the CEGB's estimates of cost savings from nuclear power compared with fossil fuel generation, but accepted PWRs as safe, efficient and necessary for future demand. Three years later, the Hinkley Inquiry would reject even those claims.

At an initially-projected cost of $1700 million, construction should finish in 1993. However, the CEGB based that projection on a program of four similar 1,170 megawatts electric plants built by the same contractors. When the government cancelled two of those projects and placed the third on hold, in 1990, the contractors promptly raised their fees by $350 million.

Moratorium on nuclear power dims hopes for future plants but Sizewell carries on. In June, 1990, with $990 million needed to finish Sizewell, Energy Secretary John Wakeham announced construction would continue. He claimed switching to a gas-fired generator would not reduce the cost.

Nuclear Electric plc inherited the plant from the CEGB on March 31, 1990, when the British government privatized all power generation except nuclear.

Sizewell's contractor, National Nuclear Corporation (NCC) based its proposed design on the American Trojan plant. However, the Nuclear Installations Inspectorate (NII) insisted on modifications to meet British safety standards. Those changes would have made the plant 30 percent more expensive than its American cousin. Despite charges of compromised safety, the CEGB managed to purge the extra costs and meet most of the NII requirements.

Safety of micro-processor technology to be proven at Sizewell. One concern that remains may delay the 1994 opening. The plant's computerized reactor protection system, designed by Westinghouse, remains unproven. It uses new microprocessor technology not found anywhere else. Consequently, the NII insists Nuclear Electric demonstrate the probability of a system failure, and that an independent contractor verify the results.

Nuclear Electric has agreed to operate the system for 12 months before loading nuclear fuel into the reactor. Meanwhile, it has scheduled that loading for November 1993.

French computer dropped as too complex and expensive. Even so, more computer problems plague the project. The main system, built by Cegelec PIC of France, was to control the day-to-day activities of the plant. The Controlbloc P-20 proved too complex, prompting Electricité de France to abandon it, delaying its own Chooz-B pressurized light-water reactor by at least four years. The CEGB selected the P-20 because the system was to have been in operation for two years before Sizewell started its reactor. With construction scheduled to finish in just 25 months, Nuclear Electric decided to switch to a Westinghouse system in June 1991.

Westinghouse will now supply both the reactor protection system and the operating system. The change will necessitate rearranging the control room layout. The $100-million package involves additional hardware and needs more ventilation and cooling than the French system.

Despite the setback, Brian George, Sizewell B project manager, insists the contractor will finish on time and within budget.

Utility confident of completion date - looks to brighter future. In its 1990-91 Annual Report, published in July 1991, Nuclear Electric claimed construction was progressing well. The venture remained eight months ahead of schedule and under the cost ceiling negotiated with the government in 1990. At that time, Sizewell's price tag grew to $2030 million.

By the end of 1990, NCC had poured 80 percent of the structural concrete needed, installed giant concrete tunnels for water intake and exhaust, and completed the cylindrical portion of the reactor containment building. In August 1991, it used Britain's largest polar crane to lift the 435-tonne reactor pressure vessel into place within the containment building. Framatome built the vessel in Chalon, France.

NCC will top the concrete structure with a three-piece dome larger than the roof of St. Paul's Cathedral.

John Collier, Nuclear Elecctric chairman, sees the successful completion of Sizewell B, on time and to budget, as essential to the future of nuclear power generation in Britain. He wants the plant operating by the time the government holds its review of the nuclear industry in 1994. Plans to duplicate the pressurized light-water reactor at Hinkley Point C await the outcome of that review.

Delicate ecology protects public image. The facility occupies a 245-acre site, overlooking the North Sea, near Ipswich in Anglia. About 75 miles northeast of London, the location's delicate ecology demands careful handling. The shore line, a Heritage Coast, is also an Area of Outstanding Natural Beauty (AONB). Further inland, the Nature Conservancy Council (NCC) designated two separate Sites of Special Scientific Interest (SSSI). Sand and shingle dune systems stretch on both north and south sides of the plant. To the north and west, fresh water grazing marshes provide homes to rare plants, insects and birds.

UNITED KINGDOM

Nuclear Electric planted conifer woodlands on former heathland to promote populations of deer and birds. The company also purchased Kenton and Goose Hill conifer woods and planted broadleaf trees within the old stands. The company hopes the new woods and the reconstruction of the dunes damaged during the construction of Sizewell-B will help to integrate the power plant visually with its natural surroundings.

Visitors Center. Since construction began in August 1988, more than 30,000 people have toured Sizewell B and its visitors center each year. Interactive videos and models tell the story of electricity, but a high-tech diorama provides the most popular focal point. That exhibit shows cutaway models of Magnox and pressurized light-water reactor cores, while a light and sound show plays in the background. The visitor can arrange a tour of the facility for any day of the week by calling 0728 642178.

Sources

"A report on Sizewell." *Bulletin of the Atomic Scientists*, June 1984.
"British Inquiry Gives Go-Ahead on Nuclear Power." *The New York Times*, January 27, 1987.
"Software row dogs nuclear power plans." *New Scientist*, April 1, 1989.
"Missing the Hinkley Point." *New Scientist*, September 30, 1989.
"A shabby tale." *Economist*, June 30, 1990.
"What to pay for nuclear power?" *Nature*, July 5, 1990.
"Pressurized accounting." *Nature*, July 5, 1990.
"Safety concerns could delay Sizewell start-up." *New Scientist*, September 15, 1990.
Mounfield, Peter. *World Nuclear Power*. Routledge, 1991.
"Edf drops programmable controllers at Chooz b-1." *Nuclear News*, February 1991.
"Company Given Contract By U.K.'s Nuclear Electric." *The Wall Street Journal*, June 11, 1991.
"Sizewell faces delay as French computers are scrapped." *New Scientist*, June 15, 1991.
Nuclear Electric Report and Accounts 1990-91. Nuclear Electric plc. Gloucester, July 1991.
Sizewell B. Nuclear Electric plc. Gloucester, July 1991.
"Britain's first PWR means heavy lifting." *Nuclear Industry*, Fourth Quarter, 1991.
"Sizewell faces delay as French computers are scrapped." *New Scientist*, June 15, 1991.

—Al Cook

232 ●

Torness Nuclear Power Station, Units 1-2

Dunbar
East Lothian
Scotland EH42 1QZ, United Kingdom
Bill Doig, Stat.Mgr.

Phone: 0368-63500
Fax: 0368-62562
Telex: 72386

Owning Utility

Scottish Nuclear Ltd.
3 Redwood Crescent, Peel Park
East Kilbride
Scotland G74 5PR, United Kingdom
Robin Jeffrey, CEO

Phone: 03552-62000
Fax: 03552-62626
Telex: 777700

Contact

Richard Marshall, PR Mgr.
Scottish Nuclear Ltd.

Basic Facts

Units 1-2 *Status:* Operable. *Type of Reactor:* Advanced gas-cooled reactor. *Megawatts (electric):* 682. *Megawatts (thermal):* 1555. *Reactor System Supplier:* National Nuclear Corporation (United Kingdom). *Generator Supplier:* General Electric Co. (United Kingdom). *Architect Engineer:* National Nuclear Corporation (United Kingdom).

Key Dates

Unit 1 *Ordered:* 1980. *Construction began:* 1980. *Criticality:* September 1987. *First Power:* May 1988. *Commercial Operation:* May 1988.

Unit 2 *Ordered:* 1980. *Construction began:* 1980. *Criticality:* December 1988. *First Power:* January 1989. *Commercial Operation:* April 1989.

Plant Report

Scotland's newest plant starts slowly. In 1988, the South of Scotland Electricity Board (SSEB) commissioned its newest nuclear power plant, Torness. By the end of 1991, technical problems and design changes to the station's two Advanced Gas Cooled (AGR) reactors still delayed full-power operation.

The AGR, Britain's second generation of reactor, evolved from the earlier Magnox design. It replaced the Magnox's steel-alloy pressure vessel with one of prestressed concrete lined with steel and moved the boilers to within the vessel. The graphite core remained, but uranium dioxide pellets replaced the natural-uranium fuel rods. As with the Magnox, the AGR's design allows a special machine to refuel the core every two or three days without a shut-down. This gives the AGR a distinct advantage in reliability over other reactors, most of which need annual shut-downs of three or four weeks. In operation, it resembles the American High Temperature Gas Cooled Reactor (HTGR), but uses carbon dioxide rather than helium as the coolant.

The system's designers believed the use of a concrete containment vessel instead of the Magnox's steel vessel made the AGR safer. Many scientists now rate the HTGR safer than the more-common Pressurized Water Reactor (PWR). Unlike the HTGR, the pressurized light-water reactor must maintain coolant in the reactor core to prevent meltdown.

Government rushes to try new reactors. Confident in the new technology, the British government began the AGR program in April 1964, before completing the last of the Magnox plants. This second program projected a target of 5,000 megawatts electric by 1970-75, but design changes and construction difficulties created massive cost overruns, delaying the first plant opening, Hinkley B, until 1976.

However, the AGR technology never approached the over-90-percent efficiency of the Magnox. In 1991, Torness could only achieve an operating efficiency of 38.6 percent. New government requirements in March 1990 forced a redesign of the plant's spent-fuel handling process and equipment. That work will guarantee low operating efficiency for years to come. Torness can supply 1350 megawatts electric to Edinburgh and the central valley Scotland. When available, the plant can sell excess capacity to England through the national power grid.

Plant Profiles

In 1979, the British government announced any new reactors built under a third program would use the more internationally common pressurized light-water reactor design. The first pressurized light-water reactor is due to open in mid-1994. A second received planning approval in 1990, but actual construction must await the government's review of nuclear power generation due in 1994.

Construction project boon to Scottish industry and labor. Torness perches on the Lothian coast of the North Sea, about 35 miles east of Edinburgh. Construction cost $1.8 billion and employed more than 6,500 workers at its peak.

The reactors themselves, sit within 5.8-meter-thick walls. Edinburgh Castle's walls, at their thickest, only reach 4.5 meters. To further strengthen the reactor vessel, the designers wove more than 1000 kilometers of steel wire through the concrete. The strands come together in a stressing gallery where workers constantly adjust the tension to optimize the integrity of the containment system.

The SSEB chose the location because it met the three main criteria set up in 1955 by the government. Its stable geology could support the massive building and facilities. A high demand in the surrounding area provided a ready market and need. Nearby connections in the Norweb power grid provided ready access to the northern England market. Its proximity to the sea guaranteed a plentiful source of coolant and a high volume of water to dilute small quantities of radioactive effluent. The use of sea water avoided the need for unsightly cooling towers.

Scottish Nuclear inherited Torness from the SSEB on March 31, 1990, when the British government privatized all power generation except nuclear.

Fuel handling problems yet to be overcome. Throughout 1991, Torness struggled to perfect its fuel cycle. With the reactors fully loaded with 228 metric tons of uranium dioxide pellets, workers next turned their attention to the process of disposal of the nuclear waste. Starting in 1992, cooling ponds on site will store used fuel assemblies for at least 90 days. This procedure allows short-lived radiation to decay.

For all other nuclear reactors in Britain, the next step is to pack the cooled rods into 43-tonne forged-steel flasks. The flasks travel by rail to the British Nuclear Fuels Ltd (BNFL) reprocessing complex at Sellafield in Cumbria. However, the escalating cost of recycling at BNFL has Scottish Nuclear concerned. That cost played an important part in their decision to close the Magnox station, Hunterston A.

BNFL reprocesses 96 percent of used Magnox material as new fuel, but stores AGR rods in cooling ponds. The company invested $1.85 billion to build a high technology processing plant to handle AGR and pressurized light-water reactor fuel. Originally scheduled to open in 1987, the Thermal Oxide Reprocessing Plant (THORP), at Sellafield, should begin operation in 1993. Meanwhile, at $250 million over budget, BNFL's investment in the technology continues to grow.

Scottish Electric will break with English recycler. Consequently, Scottish Nuclear would prefer not to send its spent fuel to Sellafield. Instead, it has sought permission from the Nuclear Installations Inspectorate (NII) to build a dry storage facility on site. The utility would encase spent fuel rods in copper tubes and store them in air-cooled concrete vaults for up to 50 years. Then, it could either reprocess the material or bury it for long-term storage. Such storage would reduce the need to transport radioactive material and allow the utility some flexibility and bargaining power in negotiating future reprocessing contracts. Scottish Nuclear hopes to complete one dry store at Torness and another at Hunterston B by 1994 when its contract with BNFL expires.

Torness Visitor Centre. In May 1991, Scottish Nuclear opened the Torness Visitor Centre as part of its "Come and See" program designed to make the nuclear industry in Scotland open to public scrutiny. A coastal sea wall protects the plant from the North Sea. Visitors can stroll along its walkway for a closer look at many of the species of sea birds native to the area.

Sources

"Cloud Over Sellafield's Reprocessing Unit." *New Scientist*, April 22, 1989.
"Measures to Make Nuclear Power Safer than ever Before." *New Scientist*, November 4, 1989.
Mounfield, Peter. *World Nuclear Power*. Routledge, 1991.
Scottish Nuclear Annual Report and Accounts 1990-1991. Scottish Nuclear, Glasgow, June 27, 1991.
British Nuclear Fuels Plc Annual Report and Accounts 1990-1991. British Nuclear Fuels, August 1991.
Torness. Scottish Nuclear, Glasgow, n.d.

—Al Cook

Trawsfynydd Power Station, Units 1-2

Blaenau Ffestiniog　　　　　　　　　　Phone: 0766-87331
Gwynedd LL41 4DT, United Kingdom　　Fax: 0766-87267
J. Moares, Stat.Mgr.

Owning Utility

Nuclear Electric plc　　　　　　　　Phone: 0452-652222
Barnett Way　　　　　　　　　　　　Fax: 0452-652776
Barnwood, Glos. GL4 7RS, United　　Telex: 43501
　Kingdom
John G. Collier, Chm. & CEO

Contact

S. Moore, Dir., PR
Nuclear Electric plc

Basic Facts

Units 1-2 *Status:* Operable. *Type of Reactor:* Gas-cooled reactor. *Megawatts (electric):* 290. *Megawatts (thermal):* 860. *Reactor System Supplier:* Atomic Power Construction Ltd. (United Kingdom). *Generator Supplier:* Richardsons Westgarth Ltd. (United Kingdom). *Architect Engineer:* Atomic Power Construction Ltd. (United Kingdom).

Key Dates

Unit 1 *Ordered:* 1958. *Construction began:* 1959. *Criticality:* 1964. *First Power:* January 1965. *Commercial Operation:* February 1965.

Trawsfynydd

Unit 2 *Ordered: 1958. Construction began: 1959. Criticality: 1964. First Power: February 1965. Commercial Operation: April 1965.*

Plant Report

Distinctive location makes Snowdon Nuke visible reminder of nuclear age. The builders of Trawsfynydd nuclear power station chose more than just an unusual name. They also selected a distinctive site, putting the plant on the shore of a man-made lake within Snowdonia National Park.

During the heyday of nuclear plant construction, a government official remarked that Trawsfynydd showed the future of clean, safe energy production. He suggested every national park should have one. Not everyone agreed; environmentalists quickly dubbed the plant, the Snowdon Nuke.

The Central Electric Generating Board (CEGB) chose the location because it met the three main criteria set up in 1955 by the British government. Its stable geology could support the massive building and facilities. Access to the national power grid nearby provided a ready market and demand. Trawsfynydd Lake guaranteed an ample source of coolant and a high volume of water to dilute small quantities of radioactive effluent. The utility built the lake in 1928 to supply water for the Maentrog hydro-electric generating station.

British-designed reactor proves successful. Trawsfynydd uses two Magnox reactors to produce 390 megawatts electric for Southern England and Wales. The facility dates from the beginning of nuclear power development in the UK, when the British-designed Magnox reactor began to evolve.

The term Magnox derives from the magnesium alloy canister used to contain the natural-uranium fuel elements. The system uses a graphite-block core within a 3-inch thick, 67-foot-diameter steel pressure vessel. For the safety of the operators, a concrete biological shield encloses the entire structure. The system's unique design allows a special machine to refuel the core every two or three days without the need of a shut-down. This gives the Magnox a distinct advantage in reliability over other reactors, most of which need annual shut-downs of three or four weeks. In operation, it resembles the American High Temperature Gas Cooled Reactor (HTGR), but uses carbon dioxide rather than helium as the coolant.

The British government first proposed Trawsfynydd in 1955 as one of a series of 12 stations to produce a total of 2000 megawatts electric. They were to answer an increasing post-war demand for power and a need to replace dirty and inefficient coal-fired generating plants. After much re-evaluation, the program delivered only nine Magnox installations.

Design changed with every station built. The United Kingdom Atomic Energy Authority (UKAEA) did the basic design work for the Magnox reactor, but contractors completed the development. As a result, each plant evolved unique characteristics. Even the specifications of the fuel rod assemblies vary with each reactor.

Nuclear Electric plc, inherited the plant from the CEGB on March 31, 1990, when the British government privatized all power generation except nuclear.

All Magnox stations subject to Long Term Safety Reviews. The designers, expecting a useful life of 25 to 30 years, originally licensed the plant until 1995. However, the reliability of the Magnox and the escalating cost of new facilities prompted a reassessment. Nuclear Electric petitioned the Nuclear Installations Inspectorate (NII) to extend the operating licenses of all its Magnox stations to 40 years. Without the Magnox stations, southern England and Wales could face shortfalls of 11,000 to 12,000 megawatts electric of generating capacity by the turn of the century.

The Inspectorate will examine the physical condition of the plant in a Long Term Safety Review (LTSR) before extending the plant's license. LTSRs at other Magnox plants revealed several safety deficiencies. Welds in the steel pressure vessel and the high pressure gas ducting were dangerously brittle from more than 20 years of exposure to high radiation levels. The utility considers repairs to the welds, located inside the reactor containment vessel, impractical. In 1970, the inspectorate ordered most Magnox plants to operate at reduced power when it discovered corrosion of pressure vessels from the carbon dioxide cooling gas. In addition, international safety standards demand new electronic control circuits.

Safety modeling predicts Chernobyl-type accidents. Nuclear Electric closed Trawsfynydd in February 1991 pending the LTSR. Originally, the shutdown was to be for three weeks. However, the NII remains concerned with the integrity of the pressure vessel. In particular, two types of catastrophic accidents appear in its safety modeling.

If the pressure vessel collapsed from a failed weld in the main body, a graphite fire and core meltdown could result. Nuclear Electric has proposed procedures and tests that would give early warning of that type of failure. However, similar tests for welds between the vessel and the high-pressure ducting do not give adequate warning for the NII. If welds failed there, the rapid change in pressure from the loss of cooling gas could distort the graphite core. Such distortion could prevent the lowering of boron rods needed to control the nuclear reaction. The result would be a meltdown and graphite core fire.

Nuclear Electric has petitioned the NII to allow the re-opening of Trawsfynydd for a six-month trial period. The plant would operate at even further reduced pressure than before. If the Inspectorate refuses, or if the reduced operating efficiency proves uneconomical, the plant will close permanently.

Problem of plant-closing will belong to future generations. Closing a nuclear power plant, especially one prominently placed in a national park, will create challenges and spark debate. Nuclear Electric is developing a long-term strategy to deal with the problem, but both the cost and the risks appear enormous.

Once the reactors cease generation, defuelling begins. In that process, the utility will remove all fuel rods from the reactor, the cooling ponds and the store houses. Then it will pack the mate-

Plant Profiles

rial into 43-tonne forged-steel flasks and send it by rail to the British Nuclear Fuels (BNFL) reprocessing plant at Sellafield in Cumbria. Defuelling will take five years.

With all radioactive material cleared from the site, demolition and dismantling of the structures and equipment outside the biological shield can begin. The utility estimates another five years for that task.

The final stage creates the real problems. Nuclear Electric plans to allow a 100-year waiting period for radiation in the core to decay. Only then would it dismantle the reactor and send the pieces, classified as "intermediate-level waste," to BNFL's new underground depository at Sellafield.

Man-made mountains may mark final resting place. Environmentalists argue it would be safer and cheaper to decommission the Snowdon Nuke *in situ*, without opening the containment structure. That would reduce the chance of radiation leakage either at the site or during transport to Sellafield.

In the spring of 1991, Nuclear Electric began reassessing its strategy. One option proposed burying nuclear reactors after the hundred-year waiting period instead of taking them apart. Trawsfynydd would disappear under a mountain of slate more than 100 feet high and 600 feet across at its base. It would be the same height as the man-made mound of Silbury Hill in Wiltshire, but much wider. The utility proposed filling the reactor building with concrete and preparing footings around the site to support the massive weight of the mountain. The slate would come from the discard heaps of local quarries providing at least temporary jobs and income to the region. That could go along way to allay local concerns in an area suffering from the worst unemployment rate in Wales. The power plant employs more than 600 people on site and its visitors center attracts about 7000 tourists each year.

Burying a containment building should cut the cost of decommissioning by 80 percent, according to Fred Passant, Nuclear Energy's Waste and Decommissioning Manager. He estimates the price tag for Trawsfynydd at $70 million in 1991 money.

The CEGB may decide to decommission other plants using a similar method. Britain built all its nuclear power plants except Trawsfynydd on coastal sites. The utility could suck up large quantities of sand and gravel from the sea bed to create similar mountains. However, the circumstances of each location will determine the viability of the burial method, according to Passant. The utility is only considering Magnox stations for the method. It might not work for AGRs, according to Passant.

Sources

Nuclear Electric Report and Accounts 1990-91. Nuclear Electric plc, Gloucester, n.d.
Mounfield, Peter. *World Nuclear Power*. Routledge, 1991.
"Dead and Buried." *National Parks Today*, Spring 1991.
"When the wind blows." *National Parks Today*, Spring 1991.
"Weakened welds shut down reactor." *New Scientist*, February 9, 1991.
Trawsfynydd. Nuclear Electric plc. Gloucester, July 1991.
"Five Reactors must prove safety or shut." *The Guardian*. September 10, 1991.

Windscale | 234

"Conditional reprieve for Magnox stations." *Financial Times*, September 12, 1991.

—Al Cook

234 ★

Windscale Advanced Gas Cooled Reactor (WAGR)

Cumbria, United Kingdom

Owning Utility

AEA Technology
329 Harwell Laboratory
Harwell, Didcot, Oxon. OX11 ORA,
 United Kingdom
Andrew Munn, Pub. Info.

Phone: 0235-821111
Fax: 0235-832591
Telex: 83135

Basic Facts

Status: Shut down. *Type of Reactor:* Advanced gas-cooled reactor. *Megawatts (electric):* 36. *Megawatts (thermal):* 110. *Reactor System Supplier:* United Kingdom Atomic Energy Authority (UKAEA). *Generator Supplier:* English Electric Co., Ltd. (United Kingdom). *Architect Engineer:* AEA Technology.

Key Dates

Ordered: 1958. *Construction began:* November 1958. *Criticality:* August 1962. *First Power:* February 1963. *Commercial Operation:* March 1963. *Shut Down:* April 1981.

Plant Report

Miscalculation causes world's worst nuclear accident until Chernobyl. On Oct. 8, 1957, a physicist in charge of a routine operation at the Windscale plutonium piles made a fatal error. He was performing a Wigner release, raising and lowering the temperature of the reactor core to coax distorted graphite blocks back into proper shape. His reading of the instruments said the effect was not complete; he needed more heat. When he withdrew additional boron control rods, at least one fuel rod ignited, beginning the worst nuclear accident in British history. Until Chernobyl, it held the title of worst in the world.

The fire raged out of control for five days, until Tom Touhy, Windscale General Manager, personally doused the core with 5 million liters of water. Despite the risk, the core did not explode from the force of super-heated steam or the hydrogen bubble created. Filters installed in the exhaust stacks as an after-thought caught 50 percent of the released radiation. However, scientists identified one isotope that escaped, Iodine-131, as deadly to humans. The government promptly bought all the milk produced by cows on the surrounding farmland. In a dramatic gesture, it dumped the milk into rivers and streams leading to the sea, creating a lingering, pervasive stench.

Public relations damage control campaign initiated in wake of fire. Some experts call the action a waste. Iodine-131 has a short half-life. Drying the milk and storing it for two or three weeks would have eliminated all danger. However, public image concerned the government and the United Kingdom

● Operable   ○ Under Construction   △ Indefinitely Deferred   ▲ Decommissioning   □ Planned   ■ On Order   ☆ Cancelled   ★ Shut Down

Winfrith

Atomic Energy Authority (UKAEA) most. They quickly closed and sealed the two plutonium producing piles. The facility would continue to produce plutonium by other means, and it would develop new fuels and reprocessing methods, but not as Windscale. Shortly after the fire, the UKAEA adopted the wartime name of the munitions factory that operated on the site, Sellafield.

However, Windscale did not disappear. In the same year as the fire, the British government approved the construction of a prototype Advanced Gas Cooled Reactor (AGR). The 30 megawatts electric reactor would launch the second program of nuclear power-plant construction in Britain. Built next to the newly-opened Magnox station, Calder Hall, and within sight of Windscale's sealed stacks, the new plant took an old name. Officially designated WAGR, its silvery globe-shaped containment building became one of the most photographed industrial structures in Britain. Everyone called it Windscale.

New reactor meant to test and prove advanced design. The design, completed in March 1958, was to advance the earlier evolution of the Magnox stations. Particularly, researchers wanted to increase the thermal efficiency of the reactor by operating it at higher temperatures and pressures. That meant new fuel design and new metals for use inside the core.

Eventually, the UKAEA used Windscale to justify its choice of AGRs for Britain's second program of reactors. However, critics claim the Authority did not make its scientific case. The prototype provided less than two years of operating experience before the Authority's deadline for final AGR-designs.

Windscale did not use a concrete containment vessel as Oldbury and Wylfa did. Like early Magnox plants, it sealed the reactor within a steel containment vessel, with the boilers outside. All AGRs would use concrete.

Wylfa, the largest Magnox station, did not begin operation until after construction began on Windscale. High operating temperatures and pressures caused chronic problems at that plant.

However, the UKAEA's performance projections drew the most fire. It based its predictions of the new 600-megawatts electric AGRs operating at 650 psi on the relatively tiny, 30-megawatts electric Windscale, operating at 150 psi.

First AGRs became real prototypes. The critics concern would prove warranted as construction problems and design changes mired the AGR program in decades-long delays.

For Windscale, construction proceeded smoothly. Crews broke ground in November 1958. In August 1962, the reactor went critical. The plant started sending electricity to the grid in December 1962, reaching full power by February 1963. Designed to produce only 30 megawatts electric, Windscale exceeded expectations, reaching a steady output of 41 megawatts electric. It operated without serious mishap until 1981, when the UKAEA shut it down.

The prototype reactor will now teach researchers the next phase of reactor design, decommissioning. With more than 200 reactors, worldwide, awaiting dismantling, the UKAEA and its partner, British Nuclear Fuels Ltd (BNFL), will use Windscale to develop safe methods of permanent disposal of radioactive materials. In 1991, BNFL won a contract with the U.S. Department of Energy for decommissioning studies for the Shoreham reactor on Long Island, New York.

Reactor top cracked like hard-boiled egg. To begin, engineers sliced the top off Windscale's 45-tonne reactor dome. They removed the 60-tonne concrete biological shield in one piece in preparation for the dismantling of the steel pressure vessel. A special robot arm will cut up the core and store the pieces in concrete boxes. Eventually, the boxes, classified as intermediate waste, will find a permanent home at Sellafield. The new underground depository should open in 1995.

Meanwhile, the original Windscale will suffer the same fate. On Oct. 6, 1987, the UKAEA declared it would begin dismantling the plutonium piles. After a 30-year wait, radioactivity in the burnt-out core has dropped to one-hundredth of what it was. The clean-up will take 10 years and cost tens of millions of dollars, according to UKAEA chairman John Collier.

The announced goal of the UKAEA is to return the site to a green field, leaving no trace of either Windscale.

Sources

Patterson, Walter C. *Nuclear Power.* New York: Penguin Books Ltd., 1977.
Burn, Duncan. *Nuclear Power and the Energy Crisis.* New York: New York University Press, 1978.
"Windscale accident: 24-year perspective." *Science News,* September 5, 1981.
Bupp, Irvin C., and Jean-Claude Derian. *Light Water.* New York:Basic Books, 1987.
Collier, John, and Geoffrey F. Hewitt. *Introduction to Nuclear Power.* Washington: Hemisphere Publishing Corporation, 1987.
"Britain to Clean Atom Plant, Site of Disastrous Fire in 1957." *The New York Times,* October 6, 1987.
"Britain's Nervous Nuclear Reaction." *Management Today,* April 1988.
"Nuclear Safety: Some like it Hot." *New Scientist,* November 4, 1989.
United Kingdom Atomic Energy Authority Report and Accounts 1989-90. AEA Technology, London: December 1990.
"Breakers decapitate disused reactor vessel." *New Scientist,* October 19, 1991.
Mounfield, Peter. *World Nuclear Power.* Routledge, 1991.
"Nuclear Safety: Some like it Hot." *New Scientist,* November 4, 1989.

—*Al Cook*

Winfrith Steam Generating Heavy Water Reactor (SGHWR)

Dorchester, Dorset DT2 8DH, United Kingdom
R.S. Peckover, Dir.

Phone: 0305-251888
Fax: 0305-25188 x 3479
Telex: 41231

Owning Utility

AEA Technology
329 Harwell Laboratory
Harwell, Didcot, Oxon. OX11 0RA, United Kingdom
Andrew Munn, Pub. Info.

Phone: 0235-821111
Fax: 0235-832591
Telex: 83135

Winfrith

Contact
John Bentley, Pub.Info.
Winfrith Steam Generating Heavy Water Reactor (SGHWR)

Basic Facts
Status: Shut down. *Type of Reactor:* Heavy-water/light-water reactor. *Megawatts (electric):* 100. *Megawatts (thermal):* 330. *Reactor System Supplier:* United Kingdom Atomic Energy Authority (UKAEA). *Generator Supplier:* Associated Electric Industries Ltd. (United Kingdom);. *Architect Engineer:* AEA Technology.

Key Dates
Ordered: 1963. *Construction began:* 1963. *Criticality:* December 1967. *First Power:* December 1967. *Commercial Operation:* February 1968. *Shut Down:* September 1990.

Plant Report

British Government chooses future of nuclear industry. By 1974, the British government realized it had made a disastrous choice when it selected the Advanced Gas Cooled Reactor (AGR) for its second program of nuclear power stations. The technological jump from the Magnox to the AGR proved too great. The Windscale prototype did not prepare the industry for the challenge of large reactors operating at high temperatures and pressures. For the third program, the government resolved not to repeat its error.

Three designs came to the forefront: the Light Water Reactor (LWR), the High Temperature Reactor (HTR) and the Steam Generating Heavy Water Reactor (SGHWR).

The American LWR dominated the industry with 157 plants operating in 19 countries. The term includes both pressurized water reactors (PWR) and boiling water reactors (BWR). For Britain to adopt an LWR design, it meant abandoning its own ambitions of exporting the gas-graphite technology it pioneered since the forties. In 1974, it was not ready to do that.

Third generation gas reactor proves too sophisticated. The HTR represented the next step in gas-graphite evolution, after the Magnox and the AGR. The reactor operates at even higher temperatures and pressures than the AGR and uses helium rather than carbon-dioxide as the coolant. With the chronically-delayed opening of the first AGR plant still 10 years off, the HTR appeared unlikely to produce a workable design in the near future.

The SGHWR resembles the Canadian Deuterium Uranium reactor (CANDU). Like the Candu, it uses a calandria filled with heavy water moderator covering zircaloy pressure tubes. Each tube contains one fuel rod. Light water circulates through the pressure tubes, generating steam to drive a turbo-generator. Unlike the Candu, the SGHWR uses uranium oxide fuel enriched to about 2 percent uranium-235. That makes the system compact. In addition, designers envisioned a modular plant. That expansibility made the concept cheaper to use over a wider range of applications.

Winfrith ready to go, whichever British design chosen. Winfrith Atomic Energy Establishment, in Dorset, already had a prototype of each of the two British designs operating on its grounds.

In 1957, the high temperature problems of gas-graphite reactors prompted a search for more appropriate core materials and fuel configurations. The Organization for Economic Cooperation and Development (OECD) conceived the Dragon HTR in 1964. Its super-hot core used no metals, only ceramics. For fuel, researchers fed the Dragon uranium oxide enriched with 93 percent uranium-235. Rather than pierce the core with rods, they sprinkled the graphite with tiny spheres of uranium coated with carbon and silicon carbide. Deceptively, the Dragon's core formed a hexagon only 1.6 meters tall. Being an experimental station, designers never harnessed the high-tech beast to an electric generator.

HTR too hot to handle. Early successes vanish in puff of smoke. American and German firms advanced the technology into full-scale commercial operations. Peach Bottom-1, near Philadelphia became the world's first operating HTR power station in 1965. However, orders for 10 more General Atomic plants, similar to Ft. Vrain in Colorado, vanished as operating problems surfaced. By 1975, Britain abandoned the project in favor of the SGHWR.

The United Kingdom Atomic Energy Authority (UKAEA) built its prototype SGHWR at Winfrith in 1967 to explore other options than the gas-graphite reactor. Its experiment appeared vindicated in 1974, when the Secretary of State for Energy, Eric Varley, announced the government's directive to the British and Scottish electricity boards. The next program of power plant construction must use the SGHWR.

Prototype SGHWR renews hopes of holding American technology. A major reason for the decision was the existence of the 100-megawatts electric prototype at Winfrith. Some experts argued the modular design meant the experience gained from the small unit applied directly to large commercial plants. That avoided the AGR mistake of Windscale.

However, that expectation quickly proved unfounded. Design difficulties steadily pushed up construction costs. Instead of producing electricity competitively with LWR stations, an SGHWR would cost as much as 35 percent more to operate.

By 1979, with no commercial SGHWR planned, the government reversed its decision, opting instead for pressurized light-water reactor reactors for future nuclear power plants.

Winfrith continued to operate successfully, generating £10.9 million in revenue from the sale of its electricity to Southern Electricity in 1989. Despite a shutdown in November and December, the plant achieved a 53.5 percent load factor, just short of its lifetime load factor of 58 percent. After 22 years of operation, the plant looked set to continue productively for many more years.

However, the Authority closed the reactor in September 1990, after the government reduced funding for reactor research. The closing preserved resources for the development of its last experimental reactor, the Fast Breeder Reactor (FPR) at Dounreay.

Wylfa

Rolls-Royce may build next British reactor and sell it to United States. However, the UKAEA has plans for Winfrith. The site may house another new radical design experiment. In the wake of government privatization of electrical generation, UKAEA has teamed up with Rolls-Royce Associates, Combustion Engineering, and Stone and Webster to create a new generation of miniature pressurized light-water reactor power plants.

Its 300-megawatts electric design calls for a pressure vessel buried in an underground silo. Rather than being pushed by expensive and complicated pumps, water will circulate by convection. Designers call the arrangement a "passive safe system." The reactor draws its name from that feature, Safe Integral Reactor (SIR).

Although the British government remains unlikely to fund the project, the U.S. Department of Energy will probably contribute $50 million. The consortium hopes to sell similar plants in the United States.

Sources

Bupp, Irvin C., and Jean-Claude Derian. *Light Water.* New York: Basic Books, 1987.
Patterson, Walter C. *Nuclear Power.* New York: Penquin Books Ltd., 1977.
Burn, Duncan. *Nuclear Power and the Energy Crisis.* New York: New York University Press, 1978.
Collier, John, and Geoffrey F. Hewitt. *Introduction to Nuclear Power.* Washington: Hemisphere Publishing Corporation, 1987.
"UKAEA tries to bury nuclear power." *New Scientist,* January 21, 1989.
United Kingdom Atomic Energy Authority Report and Accounts 1989-90, AEA Technology, London: December 1990.
"AEA cutbacks jeopardize neutron, isotope outputs." *Nuclear News,* February 1991.
Mounfield, Peter. *World Nuclear Power.* Routledge, 1991.

—Al Cook

236 ●
Wylfa Power Station, Units 1-2

Camaes Bay
Anglesey
Gwynedd LL67 0DH, United Kingdom
W.M. Williams, Stat.Mgr.

Phone: 0407-710471
Fax: 0407-710406

Owning Utility

Nuclear Electric plc
Barnett Way
Barnwood, Glos. GL4 7RS, United Kingdom
John G. Collier, Chm. & CEO

Phone: 0452-652222
Fax: 0452-652776
Telex: 43501

Contact

S. Moore, Dir., PR
Nuclear Electric plc

Basic Facts

Units 1-2 *Status:* Operable. *Type of Reactor:* Gas-cooled reactor. *Megawatts (electric):* 655. *Megawatts (thermal):* 1600. *Reactor System Supplier:* English Electric Co., Ltd. (United Kingdom); Babock & Wilcox Co. (United States); Taylor Woodrow Construction Ltd. (United Kingdom). *Generator Supplier:* English Electric Co., Ltd. (United Kingdom). *Architect Engineer:* English Electric Co., Ltd. (United Kingdom); Babcock Power Ltd. (United Kingdom); Taylor Woodrow Construction Ltd. (United Kingdom).

Key Dates

Unit 1 *Ordered:* 1963. *Construction began:* 1963. *Criticality:* 1969. *First Power:* March 1971. *Commercial Operation:* November 1971.
Unit 2 *Ordered:* 1963. *Construction began:* 1963. *Criticality:* 1970. *First Power:* March 1971. *Commercial Operation:* January 1972.

Plant Report

Britain's last Magnox station goes on-line in 1971. By the time construction finished on Wylfa nuclear power station, the British government had already begun building the next generation of reactor. However, in 1991, Wylfa's outdated technology earned it recognition for record-high output and reliable performance. Meanwhile, the new Advanced Gas Cooled Reactors (AGR) proved disappointingly unproductive. Some experts argue the AGR was too large a technological jump from the original British-designed Magnox reactor used for Wylfa.

The term, Magnox, derives from the magnesium alloy canister used to contain the natural-uranium fuel elements. The system's unique design allows a special machine to refuel the core every two or three days without the need of a shut-down. This gives the Magnox a distinct advantage in reliability over other reactors, most of which need annual shut-downs of three or four weeks. In operation, it resembles the American High Temperature Gas Cooled Reactor (HTGR), but uses carbon dioxide rather than helium as the coolant.

The Magnox construction program began in 1955 with a proposal for 12 stations to produce a total of 2000 megawatts electric. They answered an increasing post-war demand for power and a need to eliminate dirty and inefficient coal-fired generating plants. After much re-evaluation, the program delivered only nine Magnox installations.

Design changes improve safety. The United Kingdom Atomic Energy Authority (UKAEA) designed the original Magnox, but each new project developed the concept further. Most use a graphite-block core within a 3-inch-thick, 67-foot-diameter steel pressure vessel surrounded by a concrete biological shield. For Wylfa, the contractor replaced the steel containment vessel with prestressed concrete lined with mild-steel. In addition, the new design moved the boilers into the containment vessel. Wylfa's pressure vessel has its own cooling system of constantly circulating demineralized water to maintain even temperature and stress.

The government utility, the Central Electric Generating Board (CEGB) placed the 120-acre facility on the north coast of Angsley Island. It chose the location because it met the three main criteria set up in 1955 by the government. The island's stable geology could support the massive building and facilities. The proximity of the national power grid provided a ready market and demand. The surrounding Irish Sea guaranteed a plentiful source of coolant and a high volume of water to dilute small quantities of radioactive effluent. Using 241 million liters each hour of filtered sea water avoided the need for unsightly cooling towers.

Wylfa generates 960 megawatts electric for north west England and Wales, more than doubling the output of any other Magnox station.

Nuclear Electric plc inherited the plant from the CEGB on March 31, 1990, when the British government privatized all power generation except nuclear.

All Magnox stations subject to Long Term Safety Reviews. The designers, expecting a useful life of 25 to 30 years, originally licensed the plant till 2001. However, the high level of reliability achieved by the Magnox and the escalating cost of new facilities prompted Nuclear Electric to petition the Nuclear Installations Inspectorate (NII) to extend the operating licenses of all its Magnox stations to 40 years. Without stations like Wylfa, southern England and Wales could face 11,000-to-12,000 megawatts electric shortfalls in generating capacity by the turn of the century.

The NII will conduct a Long Term Safety Review (LTSR) of Wylfa before issuing a license extension. Earlier LTSRs on older plants revealed serious safety concerns. Calder Hall and Chapelcross won extensions, but Berkeley and Hunterston A shut down because of the $15 million cost of upgrading.

Wylfa's advanced design exempts it from the Inspectorate's main concern. In earlier Magnoxes, welds in the steel-alloy pressure vessel and the high-pressure gas ducting become dangerously brittle from more than 20 years of exposure to high levels of radiation. That cannot happen with Wylfa's concrete pressure vessel. However, the Inspectorate may still require new electronic safety circuits.

Special handling recycles burnt fuel rods for future use. Other safety concerns center on the transportation of spent fuel rods. British Nuclear Fuels Ltd (BNFL) reprocesses 96 percent of that material for the CEGB as new fuel. That makes it a valuable resource for the power utility, but the high radiation level brings unique handling problems.

Cooling ponds on site store used fuel assemblies for at least 90 days, allowing short-lived radiation to decay. Then, packed into 43 metric ton forged-steel flasks, the rods travel by rail to the BNFL reprocessing plant at Sellafield in Cumbria.

Local community officials have expressed concern that some shipments stop on sidings throughout the West Midlands area, particularly in the Bescot shunting yards. They also complain that keeping shipment times secret hinders the effectiveness of their emergency disaster planning. Nuclear Electric insists all shipments go directly to Sellafield and agreed to provide complete time tables in September 1991.

The utility notes it has been transporting radioactive material by rail for 30 years without serious incident. In that time, it moved 12,000 loads of spent fuel rods over more than three million miles of track. In 1984, they demonstrated the safety of their £1,000,000 steel-flask transportation vessel by crashing a train, traveling at 100 miles per hour, into one. The impact destroyed the train, but the flask remained intact.

Uranium theft still unsolved. Although spent fuel moves without incident, Wylfa has had problems with its supply of new fuel assemblies. Five unused rods disappeared mysteriously from the plant. Authorities cannot explain the incident and have not recovered the material. The only other reported case of nuclear-fuel theft in Britain happened at Bradwell station in 1966. Thefts like that remain rare since the unenriched material used at Magnox plants like Wylfa and Bradwell has little monetary or strategic value. Refining it to weapons grade requires extensive and expensive technology.

Information Center. More than 15,000 people visit Wylfa's information center each year. The facility tour includes interactive videos and models, telling the story of electricity at Wylfa. The visitor can arrange a tour for any day of the week by calling 0407-710471.

Sources

"Nuclear Safety: Some like it Hot." *New Scientist*, November 4, 1989.
Mounfield, Peter. *World Nuclear Power*. Routledge, 1991.
Nuclear Electric Report and Accounts 1990-91. Nuclear Electric plc, Gloucester, n.d.
Wylfa. Nuclear Electric plc. Gloucester, February, 1991.
"Five reactors must prove safety or shut." *The Guardian*, September 10, 1991.
"Conditional reprieve for Magnox stations." *Financial Times*, September 12, 1991.
"Meeting to draw up nuclear-trains timetable." *Birmingham Post*, September 17, 1991.

—Al Cook

United States

COUNTRY REPORT

History of nuclear power development

The industry that introduced nuclear-generated electricity to the world is just over 50 years old. Its genesis occurred when Enrico Fermi, an Italian-born physicist, created the first self-sustaining nuclear chain reaction at the University of Chicago in 1942. Nine years later, on December 20, 1951, the first useable electricity from the atom was obtained at the Argonne National Laboratory in Idaho using the Experimental Breeder Reactor (EBR-1), a sodium-cooled breeder with a capacity of 150 kilowatts.

In the United States, the first nuclear plant to supply electricity to a utility's power grid was the Shippingport Nuclear Station in Pennsylvania. Shippingport, a 68-megawatt pressurized water reactor (PWR) designed and built by Westinghouse Electric Corporation, began operation on December 2, 1957 after a four-year construction period. The reactor was modeled after the light water reactor design used in the U.S. submarine program.

Development of regulatory agencies

As the nuclear power industry in the United States began to grow, the government assumed regulatory responsibilities. On August 1, 1946, President Harry Truman signed the Atomic Energy Act that created the Atomic Energy Commission (AEC). The AEC was given the task of developing the nation's nuclear energy capabilities and exploring peaceful uses of atomic energy. In 1954, the Atomic Energy Act was revised to permit private ownership of nuclear power. In 1974, President Gerald Ford signed the Energy Reorganization Act, abolishing the AEC and creating two separate agencies: the Energy Research and Development Administration and the Nuclear Regulatory Commission (NRC). In 1977, the Energy Research and Development Administration and the Federal Energy Administration were combined to create the Department of Energy (DOE).

Worldwide regulation and direction were also necessary. In 1953, President Dwight Eisenhower delivered a speech known as "Atoms for Peace" to the United Nations in New York. Eisenhower urged all countries to work together to promote peaceful uses of atomic energy and to share information and materials. In response, on October 1, 1957, the UN established the International Atomic Energy Agency (IAEA). The IAEA's charter was signed by 62 countries.

Past trends in the development of nuclear power plants

Nuclear power expanded rapidly in the United States in the 1960s. The first nuclear power plant financed entirely by a utility, the Dresden Nuclear Power Station operated by Chicago's Commonwealth Edison, began commercial operation in 1960. By 1966 28 other utilities had built a total of 38 new nuclear units.

During the early 1970s utilities continued announcing plans for nuclear construction. By 1975, 48 units had entered service. That same year the NRC issued its General Design Criteria that codified safety requirements and established standardized plant safety review procedures.

The rapid growth of nuclear construction decreased by the mid-1970s. The oil embargo of 1973 and subsequent economic downturn resulted in smaller increases in demand for electricity. High inflation made large-scale construction impractical. Political opposition and regulatory changes increased costs dramatically: in the early 1970s nuclear plants cost approximately $200 per kilowatt of capacity to construct; by the end of the decade this had risen to $750 per kilowatt and continued to escalate. Plants still under construction ten years later came with price tags as high as $3,500 per kilowatt of capacity.

In 1979, the United States suffered its worst nuclear accident at Three Mile Island (TMI), increasing public wariness of nuclear power and further weakening industry credibility. There have been no new domestic orders for nuclear power plants since 1978. Following TMI, U.S. reactor construction decreased significantly, with more reactors cancelled (120) than completed (112). Losses attributed to abandoned nuclear projects are estimated at $10 billion.

Development of antinuclear groups

Operational problems with nuclear plants fueled, but did not create, antinuclear sentiments, for opposition to nuclear power programs had surfaced almost as soon as their construction started. In the 1950s, the site selected for a fast-breeder reactor (Fermi), located 25 miles from Detroit, was opposed by local trade unions. The first public demonstration against a civilian nuclear plant occurred in 1957 when the first civilian nuclear plant, Shippingport, began operation. In California, projects were challenged on the basis of seismic dangers. In New York, Consolidated Edison was forced to back away from its plans to construct two pressurized light-water reactors 1.5 miles from Central Park.

Until the middle 1970s opposition to nuclear power focused on local sites. In 1974, Ralph Nader founded the Critical Mass Energy Project to supply opposition groups with information and expert sources necessary to challenge nuclear construction in court. Nader's group also helped propel the debate onto the national stage. Initially, antinuclear groups relied on a strategy of legal challenges that forced construction delays and resulted in significant cost overruns. In 1976, the Clamshell Alliance, protesting the Seabrook plant in New Hampshire, introduced civil disobedience to the nuclear protest movement. In one demonstration 1,414 people were arrested during a site occupation attempt. Other antinuclear groups with similar philosophies were formed, such as the Abalone Alliance opposing Diablo Canyon. Many of the civil disobedience groups, however, failed to generate sustained support and most antinuclear activity remains focused on legal opposition.

Future trends

Despite a lack of unqualified public acceptance, many energy experts think that a new emphasis on nuclear power will emerge during the mid 1990s, in part because the U.S. Energy Information Administration expects electricity sales to continue rising at approximately two percent per year through 2010. Some researchers predict that the mid-Atlantic states, the Southeast, parts of California, and sections of the Midwest may experience blackouts if no new generating capacity is constructed. Economic analysts also look to nuclear power as a means to reduce reliance on oil from the Middle East.

Sources

"An Industry Still in Disarray." Time, April 11, 1983.

"Revising the Nuclear Dream." USA Today, July 1987.

Atoms to Electricity. U.S. Department of Energy, November 1987.

Completing the Task: Decommissioning Nuclear Power Plants. U.S. Council for Energy Awareness, October 1988.

Rudig, Wolfgang. Anti-Nuclear Movements: A World Survey of Opposition to Nuclear Energy. Longman Group UK, distributed in U.S. and Canada by Gale Research Inc., 1990.

Kruschke, Earl R. and Byron M. Jackson. Contemporary World Issues: Nuclear Energy Policy. ABC-CLIO, Inc., 1990.

"A Time to Choose." Time, April 29, 1991.

USCEA 1990 International Nuclear Plant Survey. News Release, U.S. Council for Energy Awareness, June 26, 1991.

Electricity from Nuclear Energy, 1991-92 Edition, U.S. Council for Energy Awareness, September 1991.

Nuclear Power Plant Components: Wearing Out? No . Wearing Well. U.S. Council for Energy Awareness, October 1991.

Strategic Plan for Building New Nuclear Power Plants: First Annual Update. Nuclear Power Oversight Committee, November 1991.

"40 Years After Generating First Electricity Nuclear Energy Light More Than a Few Bulbs." INFO, U.S. Council for Energy Awareness, January 1992.

"Voluntary Approach Paying Dividends For U.S. Nuclear Waste Negotiator." INFO, U.S. Council for Energy Awareness, April 1992.

USCEA 1991 International Nuclear Wrap-up. News Release, U.S. Council for Energy Awareness, May 26, 1992.

"License Renewal: Continuing the Benefits of America's Operating Nuclear Energy Plants." Energy Update, U.S. Council for Energy Awareness.

Plant Profiles

PLANT PROFILES

237 ☆
Allen's Creek Nuclear Power Plant, Units 1-2
Wallis, TX

Owning Utility

Houston Lighting and Power Company
PO Box 1700
Houston, TX 77251
D.P. Hall, Group V.Pres., Nuclear

Phone: (713)228-9211
Fax: (713)220-5016

Contact

R.L. Waldrop, V.Pres., Pub.Aff.
Houston Lighting and Power Company

Basic Facts

Units 1-2 *Status:* Cancelled. *Type of Reactor:* Boiling water reactor. *Megawatts (electric):* 1150. *Reactor System Supplier:* General Electric Co. (United States). *Owner:* Houston Lighting and Power. *Operator:* Houston Lighting and Power.

Key Dates

Unit 1 Ordered: 1973. Cancelled: 1982.
Unit 2 Ordered: 1973. Cancelled: 1976.

238 ●
Arkansas Nuclear One Steam Electric Station, Units 1-2 (ANO)

Rte. 3, Box 137G
Russellville, AR 72801
R.A. Sessoms, Plant Mgr.

Phone: (501)964-5000
Fax: (501)964-8800

Owning Utility

Entergy Operations, Inc.
PO Box 31995
Jackson, MS 39286-1995
William Cavanaugh III, Chairman & CEO

Phone: (601)984-9000
Fax: (601)984-9817

Contact

Phil R. Miracle, Dir., Corp. Commun.
Entergy Operations

Basic Facts

Unit 1 *Status:* Operable. *Type of Reactor:* Pressurized light water reactor. *Megawatts (electric):* 833. *Megawatts (thermal):* 2568. *Reactor System Supplier:* The Babcock & Wilcox Co. (United States). *Generator Supplier:* Westinghouse Electric Corp. (United States). *Architect Engineer:* Bechtel. *Owner:* Arkansas Power and Light Company. *Operator:* Entergy Operations, Inc.

Unit 2 *Status:* Operable. *Type of Reactor:* PWR. *Megawatts (electric):* 897. *Megawatts (thermal):* 2515. *Reactor System Supplier:* Combustion Engineering, Inc. (United States). *Generator Supplier:* General Electric Co. (United States). *Architect Engineer:* Bechtel. *Owner:* Arkansas Power and Light. *Operator:* Entergy Operations, Inc.

Key Dates

Unit 1 Ordered: 1967. Construction began: 1968. Criticality: August 1974. First Power: August 1974. Commercial Operation: December 1974. Shut Down Expected: December 2008.

Unit 2 Ordered: 1970. Construction began: 1972. Criticality: December 1978. First Power: December 1978. Commercial Operation: March 1980. Shut Down Expected: December 2012.

Operating Costs

1975	$4,109,000	1983	$66,175,000
1976	$6,015,000	1984	$75,818,000
1977	$8,379,000	1985	$77,771,000
1978	$12,125,000	1986	$94,484,000
1979	$18,925,000	1987	$123,784,000
1980	$31,161,000	1988	$142,497,000
1981	$54,423,000	1989	$139,701,000
1982	$54,496,000		

Plant Report

ANO's history. The Arkansas Nuclear One Steam Electric Station (ANO) is a two unit nuclear generating station located on an 1,100 acre site in Pope County, Arkansas, on the northern bank of the Dardanelle Reservoir. The station is owned by Arkansas Power & Light (AP&L) and operated by Entergy Operations, Inc. Both AP&L and Entergy Operations are wholly owned subsidiaries of Entergy Corporation, a public utility holding company. Entergy Operations was formed in 1990 for the purpose of operating the three nuclear stations owned by Entergy Corporation: ANO, Waterford-3 in Louisiana, and Grand Gulf in Mississippi. AP&L serves 600,000 customers in Arkansas and southeastern Missouri. Power generated at ANO accounted for two-thirds of the electricity supplied by AP&L during 1991.

Construction costs at both ANO units totaled $907 million. The units were designed and built before the Three Mile Island accident and modifications were necessary to meet changing regulations.

Comparison of two units. Although both units are pressurized water reactors and both containment buildings are 200 feet high and 120 feet in diameter, there are significant differences between them. ANO-1 is an 836-megawatt unit with a Westinghouse supplied turbine-generator. ANO-2 is an 858-megawatt unit with a General Electric supplied turbine-generator. Both reactor vessels contain 177 fuel assemblies but they use different types of uranium fuel. ANO-1 uses U-235 enriched to 3.2 percent. U-235 enrichment at ANO-2 is 3.4 percent. The reactor vessel in ANO-1 also weighs 30 tons more than the reactor vessel in ANO-2.

Perhaps the most notable distinction between the two units is in their cooling systems. Unit one cools its condenser with a flow-through system using water from the Dardanelle Reservoir. Unit two employs a 447-foot natural draft cooling tower. The cooling tower for Unit 2 was constructed to avoid the impact of thermal pollution on Lake Dardanelle.

The Dardanelle Reservoir is a 34,300-acre lake used by residents and tourists for water sports. ANO-1's cooling syystem circulates 760,000 gallons per minute through the reservoir resulting in temperature increases of about 15 ° F. The lake is monitored by ANO personnel, researchers from two universities, and scientists from state environmental agencies. Entergy claims that during years of studying ANO's effect on the lake and surrounding environment, no negative impact has been identified.

NRC Fine. In 1991 the NRC levied a $50,000 fine against Entergy Operations for not testing a back up system that would have been necessary in the event of an emergency. The agency, however, commended the company for addressing problems, taking prompt corrective action, and improving its operating philosophy.

Fueling outages in 1992 affect both units. ANO-1 shut down for its tenth refueling on February 29, 1992. During the outage workers replaced 60 fuel assemblies and performed numerous maintenance tasks including sleeving 302 tubes in the steam generators, replacing water cooling coils and pipe, and installing 25 modifications to plant systems. The plant's heat exchangers and motor-operated valves were also tested. Refueling at ANO-2 began in August 1992.

On-going training. Entergy operates the Reeves E. Ritchie Nuclear Training Center at the ANO site to provide on-going training for plant employees. The training center has two control room simulators, one for each of ANO's units. Reactor operators are required to spend a week in training every six weeks.

Sources

"NRC Okays Nuclear Merger at Entergy Corp." *Electrical World*, February 16, 1990.
"Entergy 1991 System Profile." Entergy Corporation, n.d.
"News Release: ANO has best year ever in 1991." Entergy Operations, January 31, 1992.
"News Release: Unit 1 refueling outage starts Saturday at ANO." Entergy Operations, February 27, 1992.
"Arkansas Nuclear One." Entergy Operations, n.d.
"ANO Technical Fact Sheet." Entergy Operations, n.d.

—*Karen Bellenir*

239 ☆

Atlantic Nuclear Power Plant, Units 1-2

Atlantic City, NJ

Owning Utility

Public Service Electric and Gas Company
80 Park Plaza
Newark, NJ 07101
Steven E. Miltenberger, V.Pres. & Chief Nuclear Officer

Phone: (201)430-7000

Contact

Michaele L. Camp, Mgr., Nuclear Commun.
Public Sevice Electric and Gas Company

Basic Facts

Units 1-2 *Status:* Cancelled. *Type of Reactor:* Pressurized light-water reactor. *Megawatts (electric):* 1212. *Megawatts (thermal):* 3425. *Reactor System Supplier:* Unit 1, Offshore Power Systems; Unit 2, Westinghouse Electric Corp. (United States). *Owner:* Public Service Electric and Gas Co; Jersey Central Power and Light; Atlantic City Electric. *Operator:* Public Service Electric and Gas Co.

Key Dates

Units 1-2 *Ordered:* 1972. *Cancelled:* 1978.

240 ☆

Bailly Nuclear Power Plant

Bailleytown, IN

Basic Facts

Status: Cancelled (construction less than 1% complete). *Type of Reactor:* Boiling water reactor. *Megawatts (electric):* 684. *Megawatts (thermal):* 1931. *Reactor System Supplier:* General Electric (United States). *Generator Supplier:* General Electric Co. (United States). *Architect Engineer:* Sargent and Lundy Engineers (United States). *Owner:* Northern Indiana Public Service Co. *Operator:* Northern Indiana Public Service Co.

Key Dates

Ordered: 1971. *Cancelled:* 1981. *Construction began:* 1974.

Plant Report

Project cancelled, millions lost, state-wide repercussions. Northern Indiana Public Service Co. began planning the construction of Bailly Nuclear Power Plant in the late 1960s, and placed an order for the plant in 1971. They were issued a construction permit in 1974, but decided to cancel the project in 1981 with construction less than 1 percent complete.

Northern Indiana Public Service Co. obtained permission to increase rates to make up for the $190.7 million lost on the canceled Bailly project. However, early in 1985 an Indiana appeals court ruling overturned the state commission's approval saying that utilities may not charge their customers for canceled nuclear power plants.

The court ruling had repercussions for other state utilities, including the Public Service Company of Indiana who was attempting to amortize its own $2.3 billion loss on the canceled Marble Hill Nuclear Power Plant. Some business analysts felt the court decision would shift the utility debate away from the question of who must pay for canceled nuclear plants to a new focus on how the utility can remain a solvent, quality provider of electric service.

Sources

"PSI Stability Threatened." *Indianapolis Business Journal*, January 14, 1985.

Plant Profiles

241 ☆
Alan R. Barton Nuclear Power Plant, Units 1-4
Clanton, AL

Basic Facts

Units 1-4 *Status:* Cancelled. *Type of Reactor:* Boiling water reactor. *Megawatts (electric):* 1250. *Megawatts (thermal):* 3583. *Reactor System Supplier:* General Electric Co. (United States). *Owner:* Alabama Power Company. *Operator:* Alabama Power Company.

Key Dates

Units 1-2 *Ordered:* 1972. *Cancelled:* 1977.
Units 3-4 *Ordered:* 1974. *Cancelled:* 1975.

242 ●
Beaver Valley Power Station, Units 1-2

PO Box 4 Phone: (412)393-6000
Shippingport, PA 15122-0004 Fax: (412)643-4671
K.L. Ostrowski, Operations Supt., Unit 1

Owning Utility

Duquesne Light Company Phone: (412)393-6000
301 Grant St. Fax: (412)393-6448
One Oxford Centre
Pittsburgh, PA 15279
John D. Sieber, V.Pres., Nuclear

Contact

J. M. Sasala, Dir., Nuclear Commun.
Duquesne Light Company

Basic Facts

Units 1-2 *Status:* Operable. *Type of Reactor:* Pressurized light-water reactor. *Megawatts (electric):* 360. *Megawatts (thermal):* 2652. *Reactor System Supplier:* Westinghouse Electric Corp. (United States). *Generator Supplier:* Westinghouse Electric Corporation. *Architect Engineer:* Stone & Webster Engineering Corp. (United States).

Key Dates

Unit 1 *Ordered:* 1967. *Construction began:* 1969. *Criticality:* 1976. *First Power:* July 1976. *Commercial Operation:* October 1976.
Unit 2 *Ordered:* 1971. *Construction began:* 1974. *Criticality:* August 1987. *First Power:* August 1987. *Commercial Operation:* November 1987.

Operating Costs

Units 1-2

Year	Cost	Year	Cost
1976	$1,777,000	1983	$68,156,000
1977	$14,692,000	1984	$69,462,000
1978	$22,681,000	1985	$52,844,000
1979	$22,907,000	1986	$65,136,000
1980	$34,771,000	1987	$66,506,000
1981	$35,837,000	1988	$97,048,000
1982	$64,444,000	1989	$135,774,000

Plant Report

Anticipated need for nuclear power. The Beaver Valley Power Station is a two unit nuclear power plant located in western Pennsylvania near the Ohio boarder. The units were built under the auspices of the Central Area Power Coordination (CAPCO) group, an association of utilities that shares ownership in 10 power plants. Five utilities make up CAPCO: Duquesne Light Co (of Pittsburgh), Pennsylvania Power Co., Cleveland Electric Illuminating (CEI), Ohio Edison of Akron, and Toledo Edison.

The plant was built in response to an anticipated increase in the amount of electricity required. A blackout in the northeastern part of the United States in 1965 was seen as evidence for the need to expand capacity. In 1967, federal officials estimated that the need for electricity would increase 200 percent by 1980. In 1967, CAPCO began a plant construction program that originally estimated that nuclear plants would cost $300 million each.

Rising expense, rate battles, and cost containment. Early cost estimates were woefully short of reality. For example, Beaver Valley-2, completed in 1987, came with a $4.5 billion price tag. The CAPCO utilities scrambled to arrange creative financing and make their cases before Public Utility Commissions for rate increases.

To help cut costs and trim a rate increase request, Duquesne Light sold part of their interest in Beaver Valley-2. Investors paid $538 million for an ownership interest under an agreement whereby the generating capacity is leased back to Duquesne. The utility used the cash it raised by the sale to retire some of its debt and reduce its rate increase request from $383 million to $232 million. The controversial increase was granted and Duquesne, with one of the highest electric rates in the United States, agreed not to ask for another increase until 1993.

In a similar move, Centerior Power Corporation, the parent company of Toledo Edison and CEI originally owned 44 percent of Beaver Valley-2. In 1987, they sold part of their interest to investors and also leased back the power capacity.

In 1989, as part of an ongoing effort to reduce expenses, Centerior Power formed a fuel leasing company, Centerior Fuel Corporation. Under the plan, Centerior Fuel buys the raw material for nuclear fuel (mainly uranium oxide) and leases it to CEI and Toledo Edison. The two utilities then pay Centerior Fuel as the fuel is consumed at Beaver Valley and other the nuclear plants in which Centerior has an interest. By consolidating the fuel financing, the two utilities expect to pass on a $10 million per year savings to their customers.

Safety precautions. In addition to working toward a solution to financial problems, the Beaver Valley Power Plant complex is involved in settling concerns about safety. Duquesne Light conducts annual practice drills in emergency procedures using computer generated "worse-case scenarios." Every two years the drills include the surrounding community.

In May 1985, plant officials declared an "unusual event" when tornadoes approached the area. The plant typically experiences two "unusual events" per year, which is in line with industry averages.

Fined by NRC. In 1991, Duquesne Light Company was fined $25,000 by the NRC for a violation of inspection codes at Beaver Valley-1. An unlisted pipe weld (which meant that it would not have been inspected) was discovered. The NRC fined Duquesne Light for failing to search for other unlisted pipe welds until the following month. When the search was conducted, an additional 147 unlisted welds were discovered.

Law suit against Westinghouse. Further concerns about safety were raised regarding the six steam generators at the Beaver Valley complex. Each generator is 67 feet high and weighs 326 tons. They make electricity by heating water to produce steam which turns a turbine. The steam comes from non-radioactive water which is heated when it passes over special tubes containing radioactive water. In all, the six generators contain more than 20,300 water tubes. These tubes were found to be subject to corrosion and cracking. Reactor owners claimed the defective tubes did not pose a safety hazard because they were inspected regularly and replaced when necessary. Critics claimed that if cracks escaped detection and a tube ruptured, radioactive materials could leak into the atmosphere.

In 1991, the five utilities who share ownership at Beaver Valley filed a suit against Westinghouse Electric Corporation, the supplier of the tubes, claiming that the company knew about the problem. They asked the court to order Westinghouse to fix the generators and reimburse them for associated losses. Westinghouse denied the charges.

Beaver Valley-2 receives recognition. In 1991, Beaver Valley-2 ranked among the top U.S. nuclear plants, as determined by capacity factor. Capacity factor is a measurement of a plant's electrical output stated as a percentage of the theoretical maximum. The average capacity factor of U.S. reactors in 1991 was 69.3 percent. Beaver Valley 2, one of only 15 U.S. nuclear plants with capacity factors greater than 90 percent, achieved 91.8 percent.

Sources

"Managing Crisis: How 7 Local Companies Cope." *Pittsburgh Business Times & Journal*, March 31, 1986.

"Troubles in Nuclear Power Industry Hurt Centerior." *Crains Cleveland Business*, May 12, 1986.

"Duquesne Light Moving Ahead with Power Plant Sale." *Pittsburgh Business Times & Journal*, August 17, 1987.

"Lights Out: Unproductive CAPCO Energy Consortium Fading." *Pittsburgh Business Times & Journal*, August 31, 1987.

"A Regulator's Nuclear-Age Balancing Act." *The Wall Street Journal*, August 31, 1987.

"Talks Set on Splitting Cost of Reactors Between Centerior Holders, Ratepayers." *The Wall Street Journal*, February 17, 1988.

"Out of the Darkness? Duquesne Light's Von Schack Looks to Future." *Pittsburgh Business Times & Journal*, August 7, 1989.

"Utilities Dealt a Blow on Nuclear Costs." *The Washington Post*, January 12, 1989.

"Duquesne Light Will Pay $25,000 Fine by the NRC." *The Wall Street Journal*, October 10, 1991.

"U.S. Nuclear Plants Shatter Records for Output, Performance." INFO, U.S. Council for Energy Awareness, February 1992.

"Out of the Darkness? Duquesne Light's Von Schack Looks to Future." *Pittsburgh Business Times & Journal*, August 7, 1989.

"Utilities Dealt a Blow on Nuclear Costs." *The Washington Post*, January 12, 1989.

—Karen Bellenir

Bellefonte Nuclear Power Plant, Units 1-2

P.O Box 2000　　　　　　　　　Phone: (205)574-8000
Hollywood, AL 35752　　　　　Fax: (205)574-8704
John Garrity, Site V.Pres.

Owning Utility

Tennessee Valley Authority　　Phone: (615)751-0011
6N 38A Lookout Pl.　　　　　　Fax: (615)751-4904
1101 Market St.　　　　　　　　Telex: 361951
Chattanooga, TN 37402-2801
J. R. Bynum, V.Pres., Nuclear Operations

Contact

Dick Salisbury, Pub.Info.
Bellefonte Nuclear Power Plant

Basic Facts

Units 1-2 *Status:* Indefinitely deferred. As of August 1991: Unit 1, construction 89% complete; Unit 2, construction 58% complete. *Type of Reactor:* Pressurized light-water reactor. *Megawatts (electric):* 1263. *Megawatts (thermal):* 3620. *Reactor System Supplier:* The Babcock & Wilcox Co. (United States). *Generator Supplier:* Brown Boveri et Cie (Switzerland). *Architect Engineer:* Tennessee Valley Authority. *Owner:* Tennessee Valley Authority. *Operator:* Tennessee Valley Authority.

Key Dates

Units 1-2 *Ordered:* 1970. *Construction began:* 1974. *Commercial Operation Expected:* Indefinite.

Plant Report

Potential power for Alabama. The Bellefonte Nuclear Plant is located on Guntersville Lake in northeast Alabama, about seven miles east of Scottsboro.

When construction is completed by the operating utility, the Tennessee Valley Authority (TVA), the plant will be able to produce enough electricity to supply approximately 534,000 homes a day, or one city the size of Nashville.

Facts and features. Bellefonte houses two pressurized water reactors, manufactured by Babcock & Wilcox. As of August 1991, Bellefonte 1 was 89 percent completed, and Bellefonte 2 was 58 percent completed.

Each reactor will contain 205 fuel assemblies within a steel pressure vessel, about 16 feet in diameter and 47 feet high, with walls about seven inches thick.

The condenser will use cooling water from the Tennessee River and will be stored in two cooling towers. Each tower will be 535 feet high and 400 feet in diameter at the base.

A history of delays. Construction on Bellefonte began in late 1974, and projected online dates were 1989 for Bellefonte 1 and 1992 for Bellefonte 2. TVA began to cut back its power construction program in 1979, as it grew uncertain about the need for additional power capacity.

In 1985, TVA decided to slow down their construction schedule at Bellefonte. This decision was made because TVA determined that Bellefonte would not be needed until the mid-1990s, and that early operation of the plant would not provide any operational cost savings.

The online dates were changed to 1994 for Bellefonte 1 and 1996 for Bellefonte 2. TVA decided that delaying the construction would minimize rate increases to its ratepayers, decrease expenditures at the plant, and increase TVA's flexibility to match the supply of capacity with requirements for capacity.

By 1987, TVA had spent $4 billion on Bellefonte. In 1988, TVA suspended construction because of lower-than-expected load forecasts for the near future, cost-cutting efforts to improve TVA's financial position, and a desire to hold electric rates constant for a specific period of time.

TVA had three options for Bellefonte: they could finish the units as nuclear facilities, convert them to gas-fired units, or convert them to pulverized coal units.

Construction permits expire on July 1, 1994 for Bellfonte 1 and July 1, 1996 for Bellefonte 2, and the NRC performed periodic inspections of the "lay-up" program while construction was suspended.

During fiscal year 1992, TVA plans to review systems and components at Bellefonte, conduct pilot studies, and inform the NRC of plans for a possible restart. By the end of the year, TVA expects to increase the labor force at the plant from 400 to nearly 1,000.

Sources

Bellefonte Nuclear Plant. Tennesse Valley Authority, n.d.

"Lost in the Dark: TVA Stumbling Blindly over its Nuclear Program." *Public Citizen,* January 1987.

"Generating Woe: if TVA were Private, it'd be Bankrupt." *Barron's,* May 2, 1988.

"TVA Projects." *U.S. NRC Annual Report 1990,* U.S. Nuclear Regulatory Commission, n.d.

"Six Units Still in Pipeline; Extension of Perry Construction Permit Sought." *Inside NRC,* December 2, 1991.

—Charles Coplocha

---

244 •
Big Rock Point Nuclear Power Plant

10269 U.S. 31 N
Charlevoix, MI 49720
William L. Beckman, Plant Mgr.

Phone: (616)547-6537

Owning Utility

Consumers Power Company
212 W. Michigan Ave.
Jackson, MI 49201
David P. Hoffman, V.Pres., Nuclear Operations

Phone: (517)788-2552
Fax: (517)788-0045
Telex: 223454

Contact

Charles E. MacInnis, Dir., News & Info.
Consumers Power Company

Basic Facts

Status: Operable. *Type of Reactor:* Boiling water reactor. *Megawatts (electric):* 75. *Megawatts (thermal):* 240. *Reactor System Supplier:* General Electric Co. (United States). *Generator Supplier:* General Electric. *Architect Engineer:* Bechtel.

Key Dates

Ordered: 1959. *Construction began:* 1960. *Criticality:* September 1962. *First Power:* December 1962. *Commercial Operation:* March 1963. *Shut Down Expected:* 2002.

Plant Report

Electricity for northern lower Michigan. Big Rock Point Nuclear Power Plant is located on a 600-acre site in Big Rock Point, Michigan. The site is about four miles northeast of Charlevoix on the shore of Lake Michigan.

Big Rock Point houses a boiling water reactor and produces enough electricity to power a city of almost 50,000 people. The site services the northwest corner of Michigan's Lower Penninsula.

A first for Michigan. In operation since 1962, Big Rock Point was the first plant in Michigan, and the fifth in the United States. The plant was among three proposed in 1957 in the third round of the U.S. Power Demonstration Reactor Programme (PDRP).

Realizing that utilties could not justify to their shareholders the investment costs required to build nuclear power plants, the Atomic Energy Commission and Congress devised the Power Demonstration Reactor Programme in 1955 to help launch the commercial nuclear industry in the United States. The PDRP offered cooperating utilities with financial incentives such as research and development assistance, financing of the reactor system, and a waiver of normal fuel charges during the first five years of plant operation.

The plant took two and a half years to build at a cost of $27 million.

Features of Big Rock Point. Big Rock Point uses 10 tons of uranium oxide as fuel and generates 67,000 net kilowatts. The voltage of electricity produced is 13,800, increased to 138,000 volts in the outdoor substation. The reactor containment vessel is a steel sphere, 11 stories high and 130 feet in diameter. The plant also has its own control room simulator, designed to train operators.

Record accomplishments. In 1967, the plant had a record annual electric output of 531,103,000 kilowatt-hours.

The plant's longest generating run was 343 days, from August 13, 1976, through July 22, 1977, a world record for a boiling water reactor.

In 1992, Big Rock Point was honored by GE Nuclear Energy as being one of 23 U.S.-supplied boiling water reactors that operated at 75 percent efficiency or above during 1991.

Economic impact. Big Rock Point benefits the Charlevoix-Petoskey-Boyne City-East Jordan area with a buying power of approximately $7 million. At least $5.8 million is paid annually in taxes and payroll to approximately 185 plant employees and security personnel.

The plant has supported various community activities, financially and/or with employee participation. These activities include: the Little Traverse Conservancy, four area fire departments (training), several local United Ways, a youth counseling center, and an annual marksmanship competition for area law enforcement personnel.

Plant safety. Big Rock Point's plant employees have received the National Safety Council Award of Merit and the Consumers Power President's Award several times. However, in 1987, the Nuclear Regulatory Commission (NRC) proposed a $25,000 fine for the owner, Consumers Power, for security violations. The NRC found an unsecured opening that they felt may have allowed unauthorized individuals undetected access into a vital area. The fine was half of the traditional penalty because of the plant's good safety record.

License expires. Big Rock Point is among the majority of plants in the United States whose 40-year operating licenses will expire over the next 25 years. The plant's license expires in 2002, and Congress is considering the allowance of 20-year renewals.

Open to the public. Tours of Big Rock Point are available for groups of no more than 50 people and can be arranged by calling Timothy D. Petrosky, plant public affairs director, at (616) 547-6547.

Sources

Big Rock Point Nuclear Plant, Consumers Power, n.d.
A Good Friend and Neighbor since 1962, Big Rock Nuclear Power Plant, Consumers Power, n.d.
"NRC Proposes $25,000 Fine for Nuclear Site Violation." *The Wall Street Journal*, Dec. 14, 1987.
Mounfield, Peter. *World Nuclear Power*. Routledge, 1991.
"And the Winner is." *INFO*, U.S. Council for Energy Awareness, May 1992.

—Charles Ceplecha

245 ☆
Black Fox Nuclear Power Plant, Units 1-2
Inola, OK

Basic Facts

Units 1-2 *Status:* Cancelled (Unit 1, construction less than 1% complete). *Type of Reactor:* Boiling water reactor. *Megawatts (electric):* 1225. *Megawatts (thermal):* 3579. *Reactor System Supplier:* General Electric Co. (United States). *Owner:* Public Service Company of Oklahoma. *Operator:* Public Service Company of Oklahoma.

Key Dates

Unit 1 *Ordered:* 1973. *Cancelled:* 1982. *Construction began:* 1978.
Unit 2 *Ordered:* 1973. *Cancelled:* 1982.

246 ☆
Blue Hills Nuclear Power Plant, Units 1-2
Mayflower, TX

Owning Utility

Gulf States Utilities Company Phone: (409)838-6631
350 Pine St. Fax: (409)839-3077
PO Box 2951 Telex: 779312
Beaumont, TX 77704
J.E. Booker, Mgr., Nuclear Industry Relations

Contact

K.R. McMurray, Mgr., Pub.Aff.
Gulf States Utilities Company

Basic Facts

Units 1-2 *Status:* Cancelled. *Type of Reactor:* Pressurized light-water reactor. *Megawatts (electric):* 950. *Megawatts (thermal):* 2825. *Reactor System Supplier:* Combustion Engineering, Inc. (United States). *Owner:* Gulf State Utilities Co. *Operator:* Gulf State Utilities Co.

Key Dates

Unit 1 *Ordered:* 1973. *Cancelled:* 1978.
Unit 2 *Ordered:* 1974. *Cancelled:* 1978.

247 ☆
Bodega Bay Nuclear Power Plant
Bodega Bay, CA

Owning Utility

Pacific Gas and Electric Company Phone: (415)973-7000
77 Beale St.
San Francisco, CA 94106
Gregory M. Rueger, Sr.V.Pres., Nuclear Power Generation

Basic Facts

Status: Cancelled. *Type of Reactor:* Boiling water reactor. *Megawatts (electric):* 315.

Key Dates

Ordered: October 1961. *Cancelled:* October 1964.

Plant Profiles **Braidwood** 248

Plant Report

Why build at Bodega Bay?. On October 4, 1961, Pacific Gas and Electric (PG&E) filed an application with the California Public Utilities Commission requesting a certificate of public convenience and necessity to construct a nuclear power station at Bodega Bay, Sonoma County, approximately 50 miles north of San Francisco. PG&E chose the site for the following reasons:

- proximity to load centers
- excellent dispersion of the warm water discharge
- a solid granite type of rock providing good foundations
- isolation from population centers
- a harbor making possible water transportation of the heavy components during construction and for shipping of the irradiated fuel elements for processing.

At 315 MGe, PG&E's proposed plant would be much larger than any previously attempted. Also, PG&E was intending to build without federal assistance—an uncommon occurence in the country's young nuclear industry.

Protested to the state supreme court. Although many state and county organizations claimed the plant would benefit the local community through increased employment, tax revenues, and business, the public vigorously protested the granting of the application. The Sierra Club spearheaded objection to Bodega Bay on the basis of recreation and conservation. Others were concerned about safety as the reactor would operate nearly a quarter of a mile from the San Andreas fault zone.

Despite the opposition, the California Public Utilities Commission on November 2, 1962, granted PG&E an interim certificate of public convenience and necessity, a decision repeatedly protested, but ultimately upheld by the California Supreme Court.

Problems at the federal level lead to cancellation of project. After receiving the go-ahead at the state level, PG&E filed a formal application for a construction permit with the U.S. Atomic Energy Commission (USAEC) in December of 1962. A two-year dialog between PG&E and the USAEC ensued, resulting in the submission of nine amendments to the application.

During this process the U.S. Geological Survey expressed concern that PG&E's initial plant design was insufficient in the face of a maximum credible earthquake or possible tsunami. Although PG&E and its consultants held a differing view, they agreed to adjust their proposed design to meet higher seismic criteria.

On October 20, 1964, the Advisory Committee on Reactor Safeguards told the USAEC that the power station as proposed could be constructed at the Bodega Bay site without undue hazard to the health and safety of the public. However, in a separate report, the Regulatory Staff of the Division of Reactor Licensing said the site was not suitable and that the construction effort would be based on unverified engineering principles.

PG&E withdrew its application to build a nuclear power plant at Bodega Bay on October 30, 1964.

Sources

Mounfield, Peter. *World Nuclear Power*. Routledge, 1991.

248 ●
Braidwood Station, Units 1-2

R.R. No. 1, Box 84 Phone: (815)458-2801
Braceville, IL 60407
Kurt Kofron, Stat.Mgr.

Owning Utility

Commonwealth Edison Company Phone: (312)294-4321
PO Box 767 Fax: (312)294-2995
Chicago, IL 60690
Cordell Reed, Sr.V.Pres., Nuclear Operations

Contact

John F. Hogan, Dir., Commun. Svcs.
Commonwealth Edison Company

Basic Facts

Units 1-2 *Status:* Operable. *Type of Reactor:* Pressurized light-water reactor. *Megawatts (electric):* 1175. *Megawatts (thermal):* 3411. *Reactor System Supplier:* Westinghouse Electric Corp. (United States). *Generator Supplier:* Westinghouse Electric Corp. (United States). *Architect Engineer:* Sargent & Lundy Engineers (United States).

Key Dates

Unit 1 *Ordered:* 1973. *Construction began:* 1975. *Criticality:* May 1987. *First Power:* May 1987. *Commercial Operation:* July 1988.

Unit 2 *Ordered:* 1973. *Construction began:* 1975. *Criticality:* January 1988. *First Power:* February 1988. *Commercial Operation:* October 1988.

Operating Costs

Units 1-2
1988 $90,531,000
1989 $127,044,000

Plant Report

An Illinois power facility. Braidwood Station occupies a 2,600-acre site near Braidwood, Illinois, in Will County. The site is about 20 southwest of Joliet.

Braidwood Station is owned and operated by Commonwealth Edison. The plant is identical to the utility's Byron Station plant near Byron, Illinois.

The initial project cost for Braidwood Station was $5.10 billion for both units.

Features. The pressure vessel at Braidwood Station is 44 feet high and 14 1/2 feet in diameter. The walls are made of manganese-molybdenum steel and are 8 1/2 inches thick with a stainless steel lining. The vessel's total weight, including controls, is 452 tons. The design pressure is 2500 pounds per square inch.

Braidwood Station utilizes 193 fuel bundles containing 264 rods apiece. Each rod is 12 feet long and .360 inches in

● Operable ○ Under Construction △ Indefinitely Deferred ▲ Decommissioning ☐ Planned ■ On Order ☆ Cancelled ★ Shut Down

diameter. The total core contains 88.2 metric tons of fuel, and between one-third and one-quarter of it is replaced approximately every 18 months during scheduled outages.

The turbine generator is driven by steam at 1800 revolutions per minute. It requires 15 million pounds of steam at 1000 pounds per square inch pressure and 550 degrees Fahrenheit. Generators produce electricity at 25,000 volts, stepped up to 345,000 volts for transmission.

The condenser cooling system at Braidwood Station is lake. It has a total pond surface area of 2540 acres, with an average depth of 7.4 feet. The maximum flow to the condenser is 1.46 million gallons of water per minute, with a minimum water residence time of almost three days.

Costly problems. When construction began on Braidwood Station in 1975, Commonwealth Edison estimated a five-year period for completion at a cost of $934 million. Construction continued until 1979, when technical and economic problems forced the utility to stop work at the plant. The Illinois Commerce Commission (ICC) had granted the utility a 1.9 percent rate hike, which was only one-fifth the amount they had requested. The ICC granted a more generous rate hike six months later, but it took the utility a full 17 months to rehire a skilled workforce, and to get the project back up to speed.

By November 1982, initial cost estimates had roughly tripled, and the program had run behind schedule. Some factors that contributed to this were costly design changes that the NRC required after the accident at the Three Mile Island plant and high interest rates that drove up costs.

Between 1984 and 1986, delays caused estimated costs at Braidwood Station to rise about 40 percent to roughly $5.05 billion. In April 1986, the ICC weighed whether to withdraw Commonwealth Edison's Certificate of Public Convenience and Necessity for the plant. The utility needed this certificate to put Units 1 and 2 into the rate base, and losing it could have stopped construction.

In July, the utility was asked to submit its economic justification for completing Braidwood's Unit 2 to the ICC in response to a petition for cancellation filed by opponents of the plant. The petition was filed by Business and Professional People in the Public Interest, which felt that the Chicago area didn't need the additional generating capacity, and that ratepayers shouldn't have had to pay for power that they didn't need.

In August 1986, a study released by the Chicago Department of Planning predicted massive job losses if Commonwealth Edison was allowed the rate hikes needed to pay for finishing Unit 2, which was 75 percent completed, as well as Unit 1 and Byron Station's Unit 2. The utility had already spent $2.5 billion on Unit 1. The study, which was prepared by Chase Econometrics Inc., predicted that if the utility raised rates 4.8 percent per year for 11 years to phase in the costs of bringing Units 1 and 2 and Byron Station's Unit 2 on-line, the city would lose more than 112,000 jobs as companies fled the area to escape the escalating power tab.

Rate debates. In June 1987, Commonwealth Edison proposed a package to the ICC, which included a rate increase of $660 million, or 9.6 percent, to cover the costs of Units 1 and 2, and Byron Station's Unit 2. The increase was to be followed by a five-year rate freeze.

The package also included a proposal by the utility to establish a $450 million fund to finance the retirement of the three plants, setting aside $150 million for each. The utility had charged customers more than $175 million to pay for nuclear decommissioning, but, with state approval, had already spent the money.

The utility had also proposed transferring ownership of the plants to a wholly owned subsidiary, then buying back the power.

In July, the ICC turned down the entire package, leaving Commonwealth Edison with an investment of $7.1 billion in the three plants and no immediate way to pay for them. The utility countered by proposing a rate increase of 27 percent, or $1.4 billion.

In August, the utility asked the ICC to reconsider its original proposal. It had expected to begin full operations of Unit 1 in September. Unit 2 was 95 percent completed at the time.

Possible sale of Braidwood. In November 1987, negotiations took place between Commonwealth Edison and the city of Chicago concerning a city-wide electrical bill that totalled $2 billion annually, or one-third of the utility's sales. To reduce the costs to its customers, the city was considering either buying one-third of the utility's infrastructure, including parts of Units 1 and 2 and Byron Station's Unit 2, or unplugging itself from the utility's system. The 1990 expiration of the then current electric supply franchise agreement between the city and the utility forced the action.

Audit affects the rate debate. In February 1988, Commonwealth Edison conceded that the previously proposed $660-million rate figure was negotiable, and negotiations began. It could not collect revenues for the plants because of Illinois's Public Utilities Act, which stipulated that new power had to pass a prudency audit to determine how much of the cost was justifiable and how much was wasted. The audit for Unit 1 was done by O'Brien-Kreitzburg & Associates, and stated that $872 million, or 26 percent, of the construction cost was unjustifiable. The audit for Unit 2 was not yet completed.

Pump inspection. In the fiscal year 1990, the NRC assigned augmented inspection teams to determine facts regarding several operating reactor events. The team investigated a residual heat removal pump suction relief valve that was stuck open at Unit 1 in December 1990.

An end to the rate debate?. In December 1991, two hearing examiners of the ICC proposed that Commonwealth Edison be granted a seven percent rate increase, worth $361 million, to cover the commissioning of Units 1 and 2, and Byron Station's Unit 2. The examiners considered all the previously done audits, which charged that $1.4 billion of the $7.1 billion spent on all three reactors was spent imprudently, and based their recommendations on a $700-million disallowance.

Plant Profiles

Accomplishments. For 1990, Braidwood Station was ranked sixth in the United States for lowest-cost nuclear electricity by the Utility Data Institute (UDI). The UDI, an independent research and data base publishing company in Washington, D.C., found that the plant had produced 14,821,165 megawatts at $14.10 per megawatt during the year.

Sources

"The High Cost of Letting Edison do it." *Chicago*, November 1982.
"Hard Times in Chicago." *Forbes*, April 21, 1986.
"Why ComEd Faces Trouble at Braidwood." *Crains Chicago Business*, August 11, 1986.
"Commonwealth Edison Co. Proposes Fund for 3 Plants." *The Wall Street Journal*, June 18, 1987.
"Commonwealth Edison's Plan to Form Nuclear-Plant Unit Rejected by Agency." *The Wall Street Journal*, July 3, 1987.
"Why City-ComEd Fight will end in Compromise." *Crains Chicago Business*, November 9, 1987.
"How Pressures Changed ComEd Rate Tactics." *Crains Chicago Business*, February 22, 1988.
"Despite Export Boom, Trade Deficit Takes Toll on Market." *Crains Chicago Business*, April 18, 1988.
Braidwood Nuclear Station Fact Sheet, Commonwealth Edison Co., May 1988.
"Operational Safety Assessment." *U.S. NRC 1990 Annual Report*, U.S. Nuclear Regulatory Commission, n.d.
Commonwealth Edison 1990 Annual Report, Commonwealth Edison Co, n.d.
"ICC Staff Proposes Small Boost for CECo." *Nuclear News*, February 1991.
"Wisconsin Electric's Point Beach Lowest Cost Nuclear Plant in '90." *Info*, U.S. Council for Energy Awareness, July 1991.
"Operational Safety Assessment." *U.S. NRC 1990 Annual Report*, U.S. Nuclear Regulatory Commission, n.d.
Commonwealth Edison 1990 Annual Report, Commonwealth Edison Co, n.d.
"ICC Staff Proposes Small Boost for CECo." *Nuclear News*, February 1991.

—Charles Ceplecha

---

249 •

Browns Ferry Nuclear Plant, Units 1-3

PO Box 2000　　　　　　　　　Phone: (205)729-2000
Decatur, AL 34502　　　　　　 Fax: (205)729-2170
O. J. Zeringue, V.Pres., Browns Ferry
　Operations

Owning Utility

Tennessee Valley Authority　　　Phone: (615)751-0011
6N 38A Lookout Pl.　　　　　　Fax: (615)751-4904
1101 Market St.　　　　　　　 Telex: 361951
Chattanooga, TN 37402-2801
J. R. Bynum, V.Pres., Nuclear
　Operations

Contact

Craig Beasley, Pub.Info.
Browns Ferry Nuclear Plant

Basic Facts

Units 1-3 *Status:* Operable. *Type of Reactor:* Boiling water reactor. *Megawatts (electric):* 1098. *Megawatts (thermal):* 3293. *Reactor System Supplier:* General Electric Co. (United States). *Generator Supplier:* General Electric Co. (United States). *Architect Engineer:* Tennessee Valley Authority.

Key Dates

Unit 1 *Ordered:* 1966. *Construction began:* 1967. *Criticality:* 1973. *First Power:* October 1973. *Commercial Operation:* August 1974.

Unit 2 *Ordered:* 1966. *Construction began:* 1967. *Criticality:* 1974. *First Power:* August 1974. *Commercial Operation:* March 1975.

Unit 3 *Ordered:* 1967. *Construction began:* 1968. *Criticality:* 1976. *First Power:* September 1976. *Commercial Operation:* March 1977.

Operating Costs
Units 1-3

Year	Cost	Year	Cost
1975	$6,626,000	1983	$108,946,000
1976	$16,104,000	1984	$129,996,000
1977	$19,305,000	1985	$99,913,000
1978	$45,921,000	1986	$143,268,000
1979	$55,589,000	1987	$178,661,000
1980	$66,968,000	1988	$149,512,000
1981	$85,470,000	1989	$90,378,000
1982	$92,273,000		

Plant Report

TVA's first plant. Browns Ferry Nuclear Plant is located on Wheeler Lake, 10 miles southwest of Athens, Alabama.

Browns Ferry houses three 1,098-megawatt boiling water reactors, each capable of producing enough electricity to supply approximately 200,000 customers, or the number of homes in the North Alabama areas of Huntsville, Madison County, Decatur, and Athens.

Major construction began on Browns Ferry in 1967, making it the first plant built by the operating utility, the Tennesse Valley Authority (TVA). At its initial time of operaton, Browns Ferry was the world's largest commercial nuclear plant and was built at a cost of $1 billion. The plant's workforce totals 2,400 employees.

Plant features. Each reactor contains 764 fuel assemblies within a steel pressure vessel, about 73 feet high and 21 feet in diameter, with walls more than six inches thick. These reactors can produce more than 13 million pounds of steam per hour each, at 560 degrees Fahrenheit and 1020 pounds per square inch.

Three low-pressure and one high-pressure turbines on a single shaft spin a generator at 1800 revolutions per minute. After water is condensed, the condensers are cooled by river water in a separate system. The water can be cooled further in the plant's five cooling towers before being returned to Wheeler Lake.

A near catastrophe. In 1975, while Browns Ferry was at peak power, a workman, searching for air leaks in a cable room, accidently started a fire. The workman had been using a candle to check for air flows between cables and walls, and ignited polyurethane insulation. The fire burned for seven and one half hours, partly because of a misconception by plant personnel that water could not be used to extinguish it.

The fire occurred below the control room for Units 1 and 2. Unit 2 was shut down immediately, but the burning cables prevented a quick shutdown of Unit 1, which continued to operate, unchecked.

● Operable　　○ Under Construction　　△ Indefinitely Deferred　　▲ Decommissioning　　□ Planned　　■ On Order　　☆ Cancelled　　★ Shut Down

The fire damaged 1600 cables and circuits, which controlled the emergency core-cooling systems of the reactor. The plant was out of service for 18 months, and more than $100 million in damages occurred, costing TVA customers approximately $213 million.

Fire affects the industry. The fire at Browns Ferry revealed weaknesses in fire protection at nuclear plants nationwide. The NRC's investigation of the plant concluded that where cables controlling redundant safety systems were too close together, one fire could render both systems useless. The NRC also found that parts of the plant were not designed according to safety rules, and that some of the cables were not the kind that regulations required.

As a direct result of the fire, the NRC, in 1981, issued stronger fire-protection regulations. These included: separating cables, not using candles to check for air leaks, and the banning of polyurethane insulation.

Fines. Between 1980 and 1985, the NRC fined TVA $1 million for various violations at Browns Ferry. In 1984, a $100,000 fine was assessed after 13 workers had to get medical treatment for radiation exposure from leaking reactor coolant.

Problems amount to a shutdown. In September 1984, TVA shut down Unit 2 for repairs. Just five months later, in February 1985, discrepancies arose while workers were starting up Unit 3. Instruments measuring levels of cooling water in the reactor's core were producing conflicting readings. Unit 3 was started anyway, and it was not discovered until three weeks later that monitors were not working correctly. In November, more problems arose when 12 out of 15 licensed reactor operators flunked their competency tests. TVA decided to shut down Unit 3 for repairs, and five months later, Unit 1 was added to the list, closing the entire plant.

A partial restart. TVA planned to restart Unit 2 before the others. In January 1990, the reactor was defueled to facilitate inventory of special nuclear material and ongoing plant modifications. Later, in April, the NRC issued a Safety Evaluation Report and instituted a major inspection program to deal with the restart plans.

In May 1991, the NRC approved the restart of Unit 2, after a six-year shutdown, and the reactor was restarted. Up to the time of the approval, TVA had spent $1.3 billion on repairing the reactor. They retrained operators, replaced computer equipment, repaired fuel bundles, and modernized cooling pipes and other vital components.

Critics protest restart. In approving the restart of Unit 2, the NRC granted TVA exemptions from their 1981 fire-protection rules, outraging nuclear power critics. Most plants were required to have sophisticated automatic fire-suppression equipment in backup control rooms, but Browns Ferry was allowed to have only fire extinguishers. Also, most plant were required to have fire-fighting equipment for key pumps, but Browns Ferry was not.

The road to recovery. In August 1991, Browns Ferry successfully completed 75 days of power ascension testing, following extensive modification work, without an unplanned automatic shutdown. As of 1991, Units 1 and 3 were still inoperable, pending further modificatons.

Sources

Browns Ferry Nuclear Plant, Tennesse Valley Authority, n.d.
"The Cult of the Atom." *The New Yorker*, November 1, 1982.
"Candle in a Nuclear Plant: Ten Years Later." *Technology Review*, February/March 1985.
"Can a Chastened TVA Reform its Nuclear Ways?" *Business Week*, July 15, 1985.
"Troubled Times for TVA." *Newsweek*, January 27, 1986.
"Lost in the Dark: TVA Stumbling Blindly Over its Nuclear Program." *Public Citizen*, January 1987.
"Generating Woe: If TVA Were Private, it'd be Bankrupt." *Barron's*, May 2, 1988.
"TVA Projects." *U.S. NRC Annual Report 1990*, U.S. Nuclear Regulatory Commission, n.d.
"U.S. Gives Backing to Start Reactor Shut for Six Years." *The New York Times*, May 3, 1991.
"Lost in the Dark: TVA Stumbling Blindly Over its Nuclear Program." *Public Citizen*, January 1987.
"Generating Woe: If TVA Were Private, it'd be Bankrupt." *Barron's*, May 2, 1988.
"TVA Projects." *U.S. NRC Annual Report 1990*, U.S. Nuclear Regulatory Commission, n.d.

—Charles Ceplecha

Brunswick Nuclear Plant, Units 1-2

PO Box 10429
Southport, NC 28461
R. B. Richey, V.Pres., Brunswick Nuclear Project

Phone: (919)457-9521
Fax: (919)457-2150

Owning Utility

Carolina Power & Light Company
PO Box 1551
Raleigh, NC 27602
R. A. Watson, Sr.V.Pres., Nuclear Generation

Phone: (919)546-6111
Fax: (919)546-7678

Contact

R. J. White, V.Pres., Corp. Commun.
Carolina Power & Light Company

Basic Facts

Units 1-2 *Status:* Operable. *Type of Reactor:* Boiling water reactor. *Megawatts (electric):* 849. *Megawatts (thermal):* 2436. *Reactor System Supplier:* General Electric Co. (United States). *Generator Supplier:* General Electric Co. (United States). *Architect Engineer:* United Engineers & Constructors (United States). *Owner:* Carolina Power & Light Company; North Carolina Eastern Municipal Power Agency. *Operator:* Carolina Power & Light Company.

Key Dates

Unit 1 *Ordered:* 1968. *Construction began:* 1970. *Criticality:* October 1976. *First Power:* December 1976. *Commercial Operation:* March 1977. *Shut Down Expected:* February 2010.

Plant Profiles   Brunswick 250

Unit 2 *Ordered:* 1968. *Construction began:* 1970. *Criticality:* March 1975. *First Power:* April 1975. *Commercial Operation:* November 1975. *Shut Down Expected:* February 2010.

Operating Costs
Units 1-2

Year	Cost	Year	Cost
1975	$4,473,000	1983	$109,815,000
1976	$10,518,000	1984	$126,560,000
1977	$25,378,000	1985	$143,045,000
1978	$26,633,000	1986	$123,825,000
1979	$34,206,000	1987	$119,058,000
1980	$57,516,000	1988	$126,969,000
1981	$73,150,000	1989	$131,240,000
1982	$112,236,000		

Plant Report

The facility. The Brunswick Nuclear Plant occupies a 1200-acre site along the Cape Fear River 2.5 miles upstream from Southport, North Carolina. Its two boiling water reactors have a combined capacity of 1580 megawatts. Together, they play a major role in the generating ability of Carolina Power & Light Company (CP&L).

CP&L supplies electricity to customers in a 30,000 square mile service area in North and South Carolina. Sixteen power stations give the utility a combined generating capacity of 9,613 megawatts. CP&L operates Brunswick and is its major owner, holding an 81.67 interest in the plant. The North Carolina Eastern Municipal Power Agency (Power Agency) owns the remaining interest in the facility. The Power Agency includes a majority of CP&L's municipal wholesale customers. CP&L and the Power Agency are each entitled to shares of generating capacity and output equal to their respective ownership interests. Each company is also responsible to pay additional construction costs, fuel purchases, and operating expenses based on the percentage of its ownership.

Construction information and plant statistics. Construction at Brunswick lasted seven years. Workers used 336,000 cubic yards of concrete, enough to build a two-lane highway 103 miles long. The containment buildings' walls were designed to withstand winds up to 360 miles per hour.

Brunswick's reactors use uranium fuel in the form of uranium dioxide pellets. Each pellet, about the size of a half-inch long piece of chalk, produces an amount of power equivalent to 3/4 ton of coal. The pellets are put together in 12-foot tubes. Sixty-four tubes are bundled together to form a fuel assembly. Each of Brunswick's reactors contains 560 assemblies. A coal-fired plant would require 12,400 tons of coal (156 average-sized railroad cars) per day to produce an equivalent amount of power. The reactors are on an 18 month operating cycle. This means that about every 18 months approximately one-third of the fuel must be replaced.

Water inside the Brunswick reactors boils to produce steam to turn the turbines and generators. Used steam, which is radioactive, then goes through a condenser where it is cooled, transformed back into to water, and recycled through the reactor. Water for the condensers is kept in a completely separate system. It is drawn from the Cape Fear River and discharged via a pumping station through pipes extending 2,000 feet into the ocean where tidal action dissipates the waste heat.

Training facility. The Brunswick training center opened in 1983. It is housed in a 39,000 square-foot building and is used to prepare plant employees, including reactor operators, for duty. The facility has 11 classrooms, a library, and a control room simulator which is a replica of Unit 2's control room. The training program at Brunswick is accredited by the National Academy for Nuclear Training.

Performance. Although the Brunswick plant has never declared a General Emergency or Site Area Emergency, the plant received the second largest penalty assessed by the NRC in 1983. The regulatory agency levied a $600,000 fine against CP&L after discovering that the utility had operated Brunswick 1 & 2 for seven years without testing certain safety systems and components.

To improve operating efficiency, CP&L switched to an "off-season" refueling schedule. The program, which also includes maintenance tasks and the repair of non-safety systems, permits the utility to plan outages during the fall and spring months when demand for electricity is traditionally at its lowest levels. In 1991 the off-season refueling schedule enabled CP&L to operate all four of its nuclear generating plants through the month of July when demand peaked. The utility's aggregate capacity factor, however, is only 66 percent.

Decommissioning. In 1990 the NRC issued regulations requiring utilities operating nuclear power units to set aside funds for eventual decommissioning. CP&L established an external decommissioning trust to hold funds which will be collected through electric rates. The utility plans to decommission the plant using the prompt dismantlement method. The project is expected to cost CP&L $176.7 million (1989 dollars) at Brunswick 1 and $162.4 million (in 1989 dollars) at Brunswick 2. The decommissioning estimates do not include the Power Agency's portion.

Visitors Center. CP&L maintains a visitors center at Brunswick. The center includes displays about energy and an auditorium for film and slide presentations. It is open 9:00 a.m. to 4:00 p.m. Monday through Friday (except holidays). In the summer months the center is also open Sundays 1:00 p.m. to 4:00 p.m. and on the 4th of July from 1:00 p.m. - 4:00 p.m. For more information contact: Carolina Power & Light Company, Brunswick Visitors Center, P.O. Box 10408, Southport, NC 28461. (919) 457-6041.

Sources

Brunswick Nuclear Plant. Carolina Power and Light Company, n.d.
"NRC Strikes Out." *Critical Mass Bulletin*, September 1984.
Carolina Power & Light Company 1990 Annual Report. Carolina Power and Light Co., n.d.
Carolina Power & Light Company 1991 Annual Report. Carolina Power and Light Co., n.d.
Nuclear News World Directory. N.p., 1992.

—*Karen Bellenir*

251 Byron

251 •
Byron Station, Units 1-2

44450 N. German Church Rd. Phone: (815)234-5441
Byron, IL 61010
Richard Pleniewicz, Stat.Mgr.

Owning Utility

Commonwealth Edison Company Phone: (312)294-4321
One First National Plaza Fax: (312)294-2995
PO Box 767
Chicago, IL 60690
Cordell Reed, Sr.V.Pres., Nuclear
  Operations

Contact

John F. Hogan, Dir., Commun. Svcs.
Commonwealth Edison Company

Basic Facts

Units 1-2 *Status:* Operable. *Type of Reactor:* Pressurized light-water reactor. *Megawatts (electric):* 1175. *Megawatts (thermal):* 3411. *Reactor System Supplier:* Westinghouse Electirc Corp. (United States). *Generator Supplier:* Westinghouse Electric Corp. (United States). *Architect Engineer:* Sargent & Lundy Engineers (United States).

Key Dates

Unit 1 *Ordered:* 1971. *Construction began:* 1975. *Criticality:* February 1985. *First Power:* March 1985. *Commercial Operation:* September 1985.

Unit 2 *Ordered:* 1973. *Construction began:* 1975. *Criticality:* January 1987. *First Power:* Feburary 1987. *Commercial Operation:* August 1987.

Operating Costs

Units 1-2
1985 $34,431,000
1986 $55,513,000
1987 $88,135,000
1988 $115,838,000
1989 $117,328,000

Plant Report

Electricity for Illinois. Byron Station occupies a site at Byron, Illinois, approximately 17 miles southwest of Rockford. The station is owned and operated by Commonwealth Edison and is identical to the utility's Braidwood Station plant; both plants were patterned after another Commonwealth Edison plant, Zion Station.

The initial project cost for Byron Station was $4.543 billion, and construction was completed in 1985.

Features. The pressure vessel at Byron Station is 44 feet high and 14 1/2 feet in diameter. The walls are made of manganese-molybdenuim steel and are 8 1/2 inches thick, with a stainless steel liner. The pressure vessel weighs 452 tons, including controls, and the design pressure is 2250 pounds per square inch.

The plant contains 193 fuel assemblies of 264 rods each. The rods are 12 feet long and .360 inches in diameter. There are about 88 metric tons of fuel in the total core, and between one-third and one-quarter of it is replaced approximately every 18 months during scheduled outages.

The turbine-generator is driven by steam at 1800 revolutions per minute. It requires 15 million pounds of steam at 1000 pounds per square inch pressure and 500 degrees Fahrenheit.

Generators produce electricity at 25,000 volts, stepped up by transformer to 345,000 volts for transmission.

In the condenser cooling system, the cooling towers are 495 feet high, 272 feet in diameter at the top, and 605 feet in diameter at the bottom. Circulation is 1.26 million gallons per minute.

Licensing problems for Unit 1. In January 1984, in an unprecedented move, the NRC unconditionally denied Commonwealth Edison an operating license for Byron Station's Unit 1. Grounds for the denial were that the utility failed to assure the quality of construction and safe operations at the plant. It was predicted that the delay from the denial would add $100 million to Byron Station's cost and cause Commonwealth Edison to raise rates.

The Atomic Safety and Licensing Board (ASLB) of the NRC was unanimous in its decision, citing numerous construction problems at the plant. The ASLB said Commonwealth Edison had failed in its responsibility to assure that its contractors had carried out their delegated quality assurance tasks. The ASLB's particular citations include:

- The fraudulent performance of the electrical contractor, The Hatfield Electrical Co.
- The fraudulent and ineffective quality assurance programs of the supplier of safety-related and control equipment, The Systems Control Corp.
- The failure to maintain a reliable method of identifying non-conforming conditions by the piping contractor, The Hunter Corp.

The citations arose from NRC inspector's testimony given to the ASLB in August 1983, which detailed more than 60 allegations of defects in construction and falsification of records by quality-control inspectors at Byron Station. The ASLB's ruling was the first time that questions about construction safety caused the unconditional denial of an operating license.

In November 1984, the ASLB reversed its decision and decided to permit the fueling and low-power startup of Unit 1. The board said it was satisfied that Commonwealth Edison had taken the necessary steps to insure the quality of the plant.

More problems for Unit 1. In June 1987, the Illinois Supreme Court ruled that state regulators had to reconsider a $495 million rate increase that the Illinois Commerce Commission (ICC) had granted Commonwealth Edison in 1984 to cover the cost of Unit 1. The court ordered the commission to re-audit the plant and refund any unnecessary costs to the utility's customers. The re-audit was completed in April 1988, by Arthur Young & Co., and concluded that between $134 million and $169 million of the plant's cost was unreasonable.

In August 1989, an ICC order disallowed $200 million additional construction costs for Unit 1, excluding these costs from Commonwealth Edison's rate base. The ICC ordered the utility to refund approximately $190 million to its customers.

In December 1989, the ICC ordered the utility to reduce its rates by $43.2 million for illegal rates collected for cost overruns on Unit 1.

Rate debates begin for Unit 2. Byron Station's Unit 2 was completed in 1987. In June, Commonwealth Edison proposed a package to the ICC, including a rate increase of $660 million, or 9.6 percent, to cover the costs of Unit 2 and the utility's Braidwood Station's Units 1 and 2. The increase was to be followed by a five-year rate freeze.

The package included a proposal by the utility to establish a $450 million fund to finance the retirement of the three plants, setting aside $150 million for each. The utility had charged customers more than $175 million to pay for nuclear decommissioning, but, with state approval, had already spent the money.

The utility had also proposed transferring ownership of the plants to a wholly owned subsidiary, then buying back the power.

In July, the ICC turned down the entire package, leaving the utility with an investment of $7.1 billion in the three plants and no immediate way to pay for them. Commonwealth Edison countered by proposing a rate increase of 27 percent, or $1.4 billion, and in August, asked the ICC to reconsider its original proposal.

In February 1988, the utility conceded that the $660-million figure was negotiable, and negotiations began. It could not collect revenues for the plants because of Illinois's Public Utilities Act, which stipulated that new power had to pass a prudency audit to determine how much of the cost was justifiable and how much was wasted. The audit for Unit 2 was slated for completion in the summer.

Possible sale of Unit 2. In November 1987, negotiations took place between Commonwealth Edison and the city of Chicago. The negotiations concerned a citywide electrical bill that totalled $2 billion annually, or one-third of the utility's sales. To reduce that cost to its customers, the city was considering either buying one-third of the utility's infrastructure, including parts of Byron Station's Unit 2 and Braidwood's Units 1 and 2, or unplugging itself from the utility's system. The 1990 expiration of the then current electric supply franchise agreement between the city and the utility forced the negotiations.

Workers fired. In February 1988, Commonwealth Edison announced that it had terminated 12 fire surveillance workers employed by the Wackenhut Corp. of Coral Gables, Florida, and suspended 35 others at Byron Station. The utility took the action after a continuing investigation showed some discrepancies between logs kept by the workers, and computer printouts that register the time they enter and exit electronically monitored doors. Fire surveillance workers patrol the plant to ensure that fire safety specifications are maintained.

Audit proves costly. In April 1988, the ICC had an audit performed on Byron Station's Unit 2. The auditors concluded that Commonwealth Edison had spent $400 million too much on the plant. The audit was done by Arthur Young & Co. and found that poor management had slowed completion of Unit 2 by 23 months, adding $400 million to the plant's final cost. The auditors said that the utility slowed construction by diverting workers to another unit at the plant. They also said that the utility failed to draw up an adequate long-term schedule and was unable to reasonably predict when the project would be completed.

An end to the rate debate?. In December 1991, two hearing examiners of the ICC proposed that Commonwealth Edison be granted a seven percent rate increase, worth $361 million, to cover the commissioning of Byron Station's Unit 2 and both Braidwood Station units. The examiners considered a report by outside auditors that charged that $1.4 billion of the $7.1 billion spent in the three reactors was spent imprudently, and based their recommendations on a $700-million disallowance.

Accomplishments of Byron Station

- During 1989, Byron Station's Unit 1 ranked among the world's top 25 producers of electricity.

- In 1991, Byron Station ran non-stop for 258 days, setting a new endurance record for U.S. dual-reactor plants.

- During 1991, Byron's Unit 2 had a capacity factor of 89.6 percent, the 16th highest factor in the United States.

Sources

"Bryon Blasted." *Critical Mass Bulletin*, January 1984.
"News Updates." *Science News*, November 10, 1984.
"Man Bites Dog." *Forbes*, December 3, 1984.
"Commonwealth Edison Proposes Fund for 3 Plants." *The Wall Street Journal*, June 18, 1987.
"Commonwealth Edison's Plan to Form Nuclear-Plant Unit Rejected by Agency." *The Wall Street Journal*, July 3, 1987.
"Why City-ComEd Fight Will End in Compromise." *Crains Chicago Business*, November 9, 1987.
"Chicago Utility Terminates Workers at Plant." *The Wall Street Journal*, February 11, 1988.
"How Pressures Changed ComEd Rate Talk Tactics." *Crains Chicago Business*, February 22, 1988.
"Commonwealth Edison is Criticized by Auditors Again." *The Wall Street Journal*, April 19, 1988.
Byron Station Fact Sheet, Commonwealth Edison Co., May 1988.
"Electric Company Ordered to Reduce Rates by Jan. 1." *The Wall Street Journal*, December 26, 1989.
Commonwealth Edison 1990 Annual Report, Commonwealth Edison, n.d.
"Nine U.S. Nuclear Plants Made the World's Top 25 Last Year, and That's not all the Good News." *Nuclear Industry*, First Quarter 1990.
"ICC Staff Proposes Small Boost for CECo." *Nuclear News*, February 1991.
"World List of Nuclear Power Plants." *Nuclear News*, August 1991.
"Non-Stop Operating Record Falls Three Times in Two Months." *Nuclear Industry*, Fourth Quarter 1991.
"U.S. Nuclear Plants Shatter Records for Output, Performance." *INFO*, U.S. Council for Energy Awareness, February 1992.
"ICC Staff Proposes Small Boost for CECo." *Nuclear News*, February 1991.
"World List of Nuclear Power Plants." *Nuclear News*, August 1991.
"Non-Stop Operating Record Falls Three Times in Two Months." *Nuclear Industry*, Fourth Quarter 1991.

—*Charles Ceplecha*

Callaway Nuclear Plant, Unit 1

PO Box 620
Fulton, MO 65251
Garry L. Randolph, V.Pres., Nuclear Operations

Phone: (314)676-8000
Fax: (314)676-5384

Owning Utility

Union Electric Company
PO Box 149
St. Louis, MO 63166
Donald F. Schnell, Sr.V.Pres., Nuclear

Phone: (314)621-3222
Fax: (314)554-3558
Telex: 231-1890

Contact

Michael B. Cleary, Supv., Nuclear Info.
Callaway Nuclear Plant

Basic Facts

Unit 1 *Status:* Operable. *Type of Reactor:* Pressurized light-water reactor. *Megawatts (electric):* 1219. *Megawatts (thermal):* 3565. *Reactor System Supplier:* Westinghouse Electric Corp. (United States). *Generator Supplier:* General Electric Company (United States). *Architect Engineer:* Bechtel.

Key Dates

Unit 1 *Ordered:* 1973. *Construction began:* 1975. *Criticality:* October 1984. *First Power:* October 1984. *Commercial Operation:* December 1984.

Operating Costs

Unit 1

1985	$57,674,000
1986	$85,233,000
1987	$86,011,000
1988	$70,498,000
1989	$80,807,000

Plant Report

Background. Union Electric supplies electricity to more than one million customers in Missouri, Illinois, and Iowa. The 24,000 square-mile service area includes metropolitan St. Louis. Nuclear power plays a major role in Union Electric's generating capacity. About 25 percent of the electricity supplied by UE comes from the Callaway Nuclear Plant. The rest is generated from coal-powered plants with a small percentage coming from hydro-electric sources.

About 1,000 people work at the Callaway Nuclear Plant, a 1,150 megawatts pressurized water reactor located in Callaway County, Missouri, 10 miles southeast of Fulton. It is one of only two plants in the United States built using a standardized design (the other is at Wolf Creek, Kansas). The plant was one of the last nuclear plants to go on-line in the United States.

Construction history. UE announced the site in 1973 after studying 70 sites in four states. A limited work authorization was granted by the NRC in August 1975. The plant was completed nine years later.

Although Callaway's construction went better than many other nuclear plants, it was not without its problems and cost overruns. Callaway was originally intended to have two units. Early cost estimates were for $1.3 billion and the plant was scheduled for completion in early 1983. Construction of Unit No. 2 at Callaway was cancelled in 1981 forcing UE to write off a loss of $50 million. The completed single-unit reactor went on-line in December 1984 at a total cost of $3 billion.

Political opposition threatens plant. Political resistance to the plant began in 1976 when Missouri voters approved a measure preventing any cost of building the plant from being passed on in electric rates until the plant was completed. This forced UE to borrow funds at high interest rates for construction. The total indebtedness was not retired until 1988.

Callaway became embroiled in politics again in 1980, but this time the vote went with the plant. Missouri's citizens turned down an initiative that would have prohibited the operation of any nuclear power stations until federally approved permanent storage sites for nuclear wastes were available.

Antinuclear groups circulated petitions again in 1984. They collected 83,000 signatures in an effort to get a proposition on the ballot that would prevent the Callaway plant from operating and ban all future nuclear plants.

Rate increases follow plant construction. When Callaway became fully operational, the Public Service Commission approved a 45 percent rate increase for costs related to its construction. The increase was to have been phased in over a six-year period. However, the last three years of hikes were eliminated when the PSC concluded that Union Electric's return on investment had become excessive.

The cost of decommissioning the Callaway plant is estimated at $336 million. This cost is included in electric rates and held in a trust fund. Electric rates also include a charge of one mill per kilowatt-hour, as required under the Nuclear Waste Policy Act of 1982, to pay for the disposal of spent nuclear fuel.

Fuel and waste. Callaway uses a fuel made of ceramic pellets of uranium dioxide. A single pellet is about the size of a cigarette filter. Each fuel pellet provides as much energy as:

- 1,780 pounds of coal;
- 149 gallons of oil;
- 17,000 cubic feet of natural gas.

The pellets are made into 12-foot fuel rods which are arranged in bundles called "fuel assemblies." The reactor contains 193 fuel assemblies; each one weighs about 1,140 pounds. The plant holds 110 tons of fuel. About half the fuel must be replaced every 18 months.

Currently, used fuel is stored underwater in a stainless steel-lined pool. Although Callaway's administrators plan to ship this waste fuel to a reprocessing plant or underground storage site, there are currently no such facilities available in the United States. In 1989 Callaway had 121 million metric tons of high-level radioactive waste stored in its pool.

A good safety record. The Callaway plant was built to withstand earthquakes, tornadoes, and other natural hazards. Its containment building consists of a steel-reinforced concrete

Plant Profiles

building. The walls are four feet thick and the dome is three feet thick.

Operators at the plant must complete a year-long training program before they can be licensed. After initial licensing, each operator is required to take six additional weeks of training annually and be relicensed every six years.

NRC evaluators have consistently given the Callaway plant "superior performance" and "good performance" ratings. It has repeatedly been recognized for top performance and safety. In 1991, for the fourth time, the plant received recognition from the NRC for "sustained outstanding safety performance."

Callaway, however, has had a high number of unplanned shutdowns, called "trips," for things such as equipment failures and human errors. The industry average is 2.3 reactor trips per year. Callaway experienced six trips in 1987. The NRC report called for an improvement in this area.

Even with the trips, the plant's productivity remains high. In 1985 it set a record for a first-year U.S. nuclear unit, operating at 81 percent of its capacity. Callaway frequently places among the top 10 producing plants in the country.

Callaway Plant Visitors Center. The actual power plant facilities occupy only a small fraction of the 7,200 acre site. The Missouri Department of Conservation oversees 6,800 acres as the Reform Wildlife Management Area. Part of this land is farmed and the resulting income is used to fund wildlife management and public recreation activities.

For more information, write or call: The Callaway Plant Visitors Center, Union Electric Company, P.O. Box 620, Fulton, Missouri, 65251. (314) 676-8155.

Sources

"Referendums: Rising Impatience." *Time*, November 17, 1980.
"Grassroots." *Critical Mass Bulletin*, October 1984.
"Union Electric Expects 15 Percent Earnings Drop." *St. Louis Business Journal*, March 24, 1986.
"Union Electric Stock a Buy for Long Term, Analysts Say." *St. Louis Business Journal*, March 28, 1988.
"UE Retires All of Callaway Plant's Debt." *St. Louis Business Journal*, July 11, 1988.
"Callaway Gets High Ratings from Nuclear Commission." *St. Louis Business Journal*, November 21, 1988.
"Union Electric Company/Callaway Plant Fact Sheet." *Union Electric*, August 1990.
"Notes to Financial Statements." *Union Electric 1990 Annual Report*, Union Electric, n.d.
Union Electric's Callaway Nuclear Plant., Union Electric, 1991.
"Kathleen Lally: Corporate Presentation of Union Electric." *The New York Society of Security Analysts, Inc.*, June 26, 1991.
"Consumer Group's Report Flatters Region's Utilities." *Kansas City Star*, July 18, 1991.
"Palo Verde Units 1 & 2 Top American Producers in May." *Business Wire*, July 12, 1991. Also see February 15, 1991; March 13, 1991; and September 19, 1990.

—Karen Bellenir

| 253 | ☆ |

Callaway Nuclear Plant, Unit 2

PO Box 620 Phone: (314)676-8000
Fulton, MO 65251 Fax: (314)676-5384
Garry L. Randolph, V.Pres., Nuclear Operations

Owning Utility

Union Electric Company Phone: (314)621-3222
PO Box 149 Fax: (314)554-3558
St. Louis, MO 63166 Telex: 231-1890
Donald F. Schnell, Sr.V.Pres., Nuclear

Contact

Michael B. Cleary, Supv., Nuclear Info.
Callaway Nuclear Plant

Basic Facts

Unit 2 *Status:* Cancelled (construction less than 1% complete). *Type of Reactor:* Pressurized light-water reactor. *Megawatts (electric):* 1174. *Megawatts (thermal):* 3411. *Reactor System Supplier:* Westinghouse Electric Corp. (United States). *Generator Supplier:* General Electric Co. (United States). *Architect Engineer:* Bechtel.

Key Dates

Unit 2 *Ordered:* 1973. *Cancelled:* 1981. *Construction began:* 1976.

Plant Report

See Plant Report for Callaway Nuclear Plant, Unit 1.

| 254 | ● |

Calvert Cliffs Nuclear Power Plant, Units 1-2

1650 Calvert Cliffs Pkwy. Phone: (410)260-4600
Lusby, MD 20657 Fax: (410)260-4787
R.E. Denton, Plant Gen.Mgr.

Owning Utility

Baltimore Gas & Electric Company Phone: (410)234-5000
PO Box 1475
Baltimore, MD 21203
G.C. Creel, V.Pres., Nuclear Energy

Contact

R.L. Wenderlich, Supt., Nuclear Operations
Calvert Cliffs Nuclear Power Plant

Basic Facts

Units 1-2 *Status:* Operable. *Type of Reactor:* Pressurized light-water reactor. *Megawatts (electric):* 900. *Megawatts (thermal):* 2700. *Reactor System Supplier:* Combustion Engineering, Inc. (United States). *Generator Supplier:* Unit 1, General Electric Co. (United States); Unit 2, Westinghouse Electric Corp. (United States). *Architect Engineer:* Bechtel.

Key Dates

Unit 1 *Ordered:* 1967. *Construction began:* 1968. *Criticality:* October 1974. *First Power:* December 1974. *Commercial Operation:* May 1975. *Shut Down Expected:* January 2015.

Unit 2 *Ordered:* 1967. *Construction began:* 1968. *Criticality:* 1976. *First Power:* December 1976. *Commercial Operation:* April 1977.

Operating Costs

Units 1-2

1975	$4,241,000	1983	$52,772,000
1976	$8,984,000	1984	$62,343,000
1977	$20,158,000	1985	$72,293,000
1978	$25,997,000	1986	$72,908,000
1979	$36,398,000	1987	$77,290,000
1980	$41,627,000	1988	$84,873,000
1981	$50,409,000	1989	$118,409,000
1982	$61,971,000		

Plant Report

Background. The Calvert Cliffs nuclear power plant is located in Lusby, Maryland, about 40 miles southeast of Washington on the western shore of the Chesapeake Bay, which supplies the plant with 2.5 million gallons of cooling water a minute. Calvert Cliffs' two 825-megawatt reactors supply 50 percent of Baltimore Electric's generating capacity, enough to meet the needs of about 400,000 Maryland houses.

Major accomplishments. Calvert Cliffs claims to have saved its customers more than $2.9 billion in its first nine years of operation compared with traditional energy sources.

The plant cost $178 million to build and quickly became a model of success as it was cited by industry analysts as one of the premier nuclear plants in the world. In August 1987 Calvert Cliffs was the site for the first U.S. visit by the International Atomic Energy Agency which gave the plant a favorable review.

In 1988 the company earned $303 million, making it one of the most profitable utilities in the nation.

Accidents and technical problems. Despite these achievements, Calvert Cliffs has had a history of accidents and technical problems. Some of these are listed below in chronological order.

During an NRC audit on February 1, 1988, a fire started in an equipment closet which caused a plant alert and drew criticism from the NRC.

On September 15, 1988, Ashton King, a Calvert Cliffs worker assigned to fix a sensor inside a 36-foot deep tank of fresh water used in the plant's steam system, climbed eight feet down into the partially filled tank. No air sample had been taken of the tank, and King entered it without a respirator. When he encountered a cloud of nitrogen gas put in the tank to prevent corrosion, King passed out. Gary Proffit, a relief diver attempting to rescue him, dove into the tank without a safety line or a breathing apparatus. Proffit also passed out and disappeared. King was pulled from the tank by four people and revived. Proffit, however, drowned. His body was recovered at 1:18 p.m., three hours after his rescue attempt.

On March 8, 1989, a steam leak developed in Unit 2 in a line below the reactor's containment area in the 37-foot-high, seven-foot-wide pressurizer. NRC staff began to pressure the utility to shut the reactor down and fix the problem. However, plant managers tried to keep the plant on-line until a scheduled refueling shutdown. By that time, the pinhole-sized leak had grown substantially and Baltimore Electric was forced to shut the plant down on March 17, just before the refueling period. Two years would pass before the unit would operate again.

Unit 1 shut down May 1989 for inspection to ensure that similar steam leaks did not exist. Baltimore Electric agreed to correct all problems and seek permission from the NRC before restarting. The outage was extended several times as workers discovered problems with a system that pumps cooling water through the plant's turbines and were unable to complete hundreds of maintenance chores on schedule. Finally on April 13, 1990, after almost a year hiatus, Unit 1 was restarted. (Outages like this forced Baltimore Electric to purchase more expensive power from other sources. Based on 1985 estimates, replacement power costs the utility $600,000 a day if both reactors are down.)

In September 1989, Michael Mariotte, executive director of the Nuclear Information and Resource Service, an antinuclear interest group, told state lawmakers that the older Calvert Cliffs reactor vessel was one of 12 in the country that had become too brittle and did not meet NRC standards. During an accident the vessel could crack, potentially leading to the inability to cool down a melting core. This problem prompted the NRC to raise its brittleness threshold nearly 30 percent to allow the 12 plants, including Calvert Cliffs, to continue operating.

In March 1991, 1,900 gallons of Unit 2 reactor cooling water was sprayed into the containment building because a valve had been connected incorrectly. Start up permission for Unit 2 came from the NRC on April 5, 1991, after being shut down for two years; the company estimates that replacement power cost it $415 million.

During an NRC inspection between March 25 and May 31, 1990, an unplanned one-hour long release of gases from one of the three waste decay tanks that store reactor coolants occurred. On a five-point scale with one the most severe, the violation was a level four. The severity of the incident was minimized because the reactors were shut down for a planned eight-week maintenance check-up. The release was blamed on communication errors within the operations department. Calvert Cliffs said it probably wouldn't have made much difference if the plant was running because it has low radiation activity compared to other nuclear plants. The average worker is annually exposed to .224 REM (roentgen-equivalent man) compared to the industry average of .390 or 11 chest x-rays to 20 chest x-rays.

To correct deficiencies in operations, maintenance, and other processes, Baltimore Electric developed a long-term Performance Improvement Plan. In June 1989 the NRC set up a Calvert Cliffs Assessment Panel to evaluate the plant, and inspections by the NRC in 1989 found improved performance levels in most areas.

Personnel and management issues. In addition to its accidents and technical problems, Calvert Cliffs has also had a history of personnel and management problems. In 1986 Baltimore Electric transferred its key nuclear plant operations from its headquarters in the Baltimore area to the plant site. The authority at Calvert Cliffs was spread among four senior managers, none of whom had overall decision-making responsibility, and operations suffered as a result. NRC inspections in the following three years revealed such problems as:

- A lack of cooperation between departments.
- A lack of attention to detail, insufficient control and supervision, and a lack of confidence among some workers with decisions made by maintenance managers.
- A tendency for employees to go through the motions when checking equipment; they aimed for technical competence rather than actual reliability.
- Significant discipline problems with plant staff failing to adhere to procedures.
- Declining pride in workmanship which resulted in poor quality of repairs and maintenance; primary emphasis was on power production rather than safety.
- A failure by managers to stay up to date with advances in the nuclear industry.

Responding to pressure from the NRC to bolster their management and staff, Baltimore Electric increased spending in 1989 by $27.2 million or nearly 37 percent. A large increase went to maintenance and engineering staffs. The budget for its vice president and managers was increased by 100 percent to $4 million, and $1.3 million was added for new computer systems.

Calvert Cliffs' troubles with management quality and personnel performance earned them the rating of the worst managed nuclear plant in the country, according to a comparison of scores by Public Citizen, a citizen advocacy group.

Other problems involving plant employees at Calvert Cliffs include the following.

In November 1990 when Audrey W. Erickson of West Virginia filed a $2.2 million sexual discrimination suit against Baltimore Electric claiming the company denied her the right to work as a subcontractor at Calvert Cliffs because of her sex, although she had worked at other nuclear power plants.

On August 8, 1991, William J. Flanagin, 37, a resident of Lexington Park and a mid-level security supervisor at Calvert Cliffs, was found tending 898 marijuana plants, valued at $170,000, in a remote 50-by-25-foot field outside the security fence at the 2,200-acre facility. Flanagin, a 14-year employee, was charged with several felony and misdemeanor counts.

Safety, operational, and procedural violations. In April 1985, after an NRC staff inspection in October 1984, the Union of Concerned Scientists (UCS), a respected watchdog group, expressed concern about the ability of Calvert Cliffs' safety equipment to function during an accident. The group maintained that the NRC could have no basis for claiming the safety equipment was environmentally qualified due to lacking documentation from Calvert Cliffs.

In July 1985 the NRC responded saying "inadequate documentation does not necessarily mean that the equipment is not qualified." However, in the spring of 1988 the NRC imposed a $300,000 fine for the environmental qualification (EQ) deficiencies.

Other fines imposed on Calvert Cliffs by the NRC include the following:

In September 1988 the NRC fined Baltimore Electric $150,000 for improper operation of two emergency systems at Calvert Cliffs: The emergency diesel generator and the reactor protection system.

In the first nine months of 1988, the NRC assessed Baltimore Electric $450,000 in fines, mostly for safety violations.

In December 1988 the plant was placed on the Nuclear Regulatory Commission's (NRC) watch list of problem plants. In the Summer of 1990 the plant was upgraded to Category 2 on the watch list.

On March 7, 1990, the NRC fined Calvert Cliffs $100,000 after concluding, from a special inspection between October 1, 1989, and November 30, 1989, that Baltimore Electric officials failed for years to implement procedures needed to ensure the plant's nuclear reactor vessels did not crack under low temperatures. The usual $50,000 fine was doubled because the utility repeatedly had been reminded since 1977 about the need for safeguards but had not corrected them.

In September 1990, Baltimore Electric was fined $12,500 when it was discovered that a security supervisor had allowed about 100 employees enter the plant without being searched after three metal detectors stopped working. The fine was mitigated 75 percent because of good, past security performance.

Waste disposal. As of December 1987, Calvert Cliffs had generated 436 metric tons of high level waste, ranking it 13th in the nation in radioactive content of waste. In 1987 the plant disposed of 9,023 cubic feet of nuclear waste, and in 1988 they dumped 5,300 cubic feet. Annually, Baltimore Electric spends approximately $1 million to dispose low-level waste. A 1989 decision by the Calvert County commissioners that allowed Calvert Cliffs to store radioactive waste above ground, drew much public criticism—especially from the Maryland Safe Energy Coalition.

Federal policies allow Baltimore Electric to legally dump low-level radioactive waste generated from Calvert Cliffs into landfills, sewers, incinerators or recycle it into commercial products such as baby toys. Under the NRC policy, about one-third of the waste disposed of at the plant could be deregulated, as long as it doesn't expose any person to a radiation dose of more than one millirem per year. Although the policy could mean a savings of about $333,000 for Baltimore Electric, the company has decided to follow the industry's decision and not deregulate low-level radiation waste.

Rate increases and finances. In May 1989 Baltimore Electric requested a $120 million rate hike to cover rising operating costs at Calvert Cliffs. The request was made after the state People's Counsel asked the Public Service Commission to cut the utility's rates by $83.6 million due to over-earning by the utility.

The People's Counsel stated Baltimore Electric's shareholders should pay the mounting costs for management mistakes at Calvert Cliffs, not the customers. Baltimore Electric countered that the increasing costs are necessary to meet higher regulatory standards and should be paid by ratepayers.

The argument is not new and has surrounded most of the rate increases requested by Baltimore Electric, including the following:

From September 1989 to May 1990 Baltimore Electric requested five rate increases in an attempt to pass on to consumers $347 million spent to buy power during outages at Calvert Cliffs.

From April 1989 to April 1990 Baltimore Electric was granted a rate increase of $273 million, of which about $200 million is attributed to the problems at Calvert Cliffs.

In December 1989 the Public Service Commission allowed Baltimore Electric to pass on $18.3 million in new charges to customers because the utility was forced to spend $26.5 million to remedy problems cited by the NRC at the plant.

In April 1990 Baltimore Electric requested a $74.5 million rate increase to offset the cost of purchased power. If approved, rates would increase by nearly $6.87 on an average monthly bill. As of April 6, 1991, the decision concerning the utility's claim, that its ratepayers should help finance the long outages, was still pending. However, the utility set aside $35 million as the amount of money the regulator is likely to say it will have to pay.

In May 1990 the utility filed an action seeking a 12.1 percent base rate hike to cover higher operating and maintenance expenses plus capital costs.

Also in May 1990, the Public Service Commission issued an interim ruling that ordered Baltimore Electric not to pass along to consumers a portion of added fuel purchase costs.

In December 1990, the Public Service Commission approved a two-step rate increase for Baltimore Electric, a $77 million increase that took effect January 1, 1991, and an additional $72 million in June 1991—increases of about 4.7 percent and 4.4 percent, and about $121 million less than the utility requested.

Also in December 1990, the commission refused to let Baltimore Electric recover about $21 million worth of interest costs on money it borrowed to buy power while the plant was down, and they denied the utility a $24 million rate increase to cover the plant's costs.

As for the Baltimore Electric's financial status, in 1990 the utility earned $213.2 million or $2.10 a share, down 23 percent from the 1989 total of $276.3 million or $2.81 a share. Revenue, however, rose the same year to $1.68 billion from $1.52 billion.

Sources

"Calvert County Finds Marijuana Farm at Nuclear Power Plant." *The Washington Post*, August 10, 1991.

"BG&E's Calvert Cliffs Back on Line, Overcomes Setback." *Baltimore Business Journal*, May 13, 1991.

"Warm Winter Chilled BG&E Profits, Investors are Told." *Baltimore Business Journal*, April 29, 1991.

"Calvert Cliffs Permitted to Resume Full Power." *The Washington Post*, April 6, 1991.

"Four Fines Proposed, One Imposed, One Fought." *Nuclear News*, February 1991.

"Baltimore Gas & Electric Co." *The Wall Street Journal*, Janunuary 21, 1991.

"BG&E Rate Hike Under Fire." *Baltimore Business Journal*, December 24, 1990.

"BG&E's Nuclear Plant Policy Discriminates, Lawsuit Claims." *Baltimore Business Journal*, December 17, 1990.

"Nuclear Waste in Baby Toys? U.S. Says Yes, BG&E Says No." *Baltimore Business Journal*, July 9, 1990.

"Calvert Cliffs Improves, But NRC Notes Another Violation." *Baltimore Business Journal*, June 18, 1990.

"'Watch List' Pared By NRC." *Nuclear Industry*, Summer 1990.

"Consumers vs. Stockholders Over Calvert Cliffs." *The Washington Post*, May 30, 1990.

"Size of BG&E Rate Request Caught Observers Off Guard." *Baltimore Business Journal*, May 28, 1990.

"With the Calvert Cliffs Plant, It's the Customers Who Bear the Burden." *The Washington Post*, April 19, 1990.

"No-Nuke Group to Turn Up Hear at BG&E Annual Meeting." *Baltimore Business Journal*, April 16, 1990.

"BG&E Restarts Calvert Cliffs Plant." *Baltimore Business Journal*, April 16, 1990.

"Calvert Cliffs Allowed to Start One Reactor." *The Washington Post*, April 12, 1990.

"Don's Corporate Cronies." *Regardies The Business of Washington*, April 1990.

"Calvert Cliffs is Fined For Safety Violations." *The Washington Post*, March 8, 1990.

"NRC May Fine Calvert Cliffs Plant." *The Washington Post*, February 16, 1990.

"The Calvert Syndrome." *Regardies The Business of Washington*, February 1990.

"Special Cases." *NRC Annual Report*, U.S. Nuclear Regulatory Commission, 1990.

"Idled BG&E Reactor Costing Consumers $1 Million a Day." *Baltimore Business Journal*, December 11, 1989.

"Experts Warn Aging Calvert Cliffs' Costs May Jump 15 Percent." *Baltimore Business Journal*, September 25, 1989.

"Leaks to Idle Calvert Cliffs Well Into 1990, BG&E Says." *Baltimore Business Journal*, August 21, 1989.

"BG&E to Pour Millions Into Calvert Cliffs Reactor Woes." *Baltimore Business Journal*, June 26, 1989.

"Calvert Cliffs Puts Production Over Safety, NRC Says." *The Washington Post*, May 25, 1989.

"BG&E Report Details Problems at Calvert Cliffs." *Baltimore Business Journal*, April 24, 1989.

"Calvert Cliffs Fails in Latest Safety Review." *Baltimore Business Journal*, February 20, 1989.

"NRC Staff Proposes Baltimore Gas Pay Penalty of $150,000." *The Wall Street Journal*, September 22, 1988.

"Calvert Cliffs Puts Production Over Safety, NRC Says." *The Washington Post*, May 25, 1989.

"BG&E Report Details Problems at Calvert Cliffs." *Baltimore Business Journal*, April 24, 1989.

—Sharon Miscavage

Carolinas-Virginia Tube Reactor (CVTR)

Parr, SC

Basic Facts

Status: Decommissioned. *Type of Reactor:* Pressurized heavy-water reactor. *Megawatts (electric):* 17. *Megawatts (thermal):* 65. *Reactor System Supplier:* Westinghouse Electric Corp. (United States). *Architect Engineer:* Stone & Webster Engineering Corp. (United States). *Owner:* Carolinas Virginia Nuclear Power Associates. *Operator:* Carolinas Virginia Nuclear Power Associates.

Key Dates

Ordered: 1955. *Construction began:* 1960. *Criticality:* March 1963. *First Power:* March 1963. *Commercial Operation:* December 1963. *Shut Down:* January 1967.

Plant Report

A project of the Power Demonstration Reactor Programme (PDRP). Realizing that utilties could not justify to their shareholders the investment costs required to build nuclear power plants, the Atomic Energy Commission and Congress devised the Power Demonstration Reactor Programme in 1955 to help launch the commercial nuclear industry in the United States. The PDRP offered cooperating utilities with financial incentives such as research and development assistance, financing of the reactor system, and a waiver of normal fuel charges during the first five years of plant operation.

The Carolinas-Virginia Tube Reactor (CVTR), a heavy-water-moderated pressure-tube plant in Parr, South Carolina, was one of the last nuclear plants built under the PDRP. The small plant was shut down in 1967 and subsequently mothballed.

Sources

Mounfield, Peter. *World Nuclear Power*. Routledge, 1991.

---

256 ☆
Carroll County Nuclear Power Plant, Units 1-2

Savanna, IL

Owning Utility

Commonwealth Edison Company Phone: (312)294-4321
One First National Plaza Fax: (312)294-2995
PO Box 767
Chicago, IL 60690
Cordell Reed, Sr.V.Pres., Nuclear
  Operations

Contact

John F. Hogan, Dir., Commun. Svcs.
Commonwealth Edison Company

Basic Facts

Units 1-2 *Status:* Cancelled (plants were ordered in 1978). *Type of Reactor:* Pressurized light-water reactor. *Megawatts (electric):* 1175. *Reactor System Supplier:* Westinghouse Electric Corp. (United States). *Owner:* Commonwealth Edison Company. *Operator:* Commonwealth Edison Company.

Key Dates

Unit 1 *Ordered:* 1970. *Construction began:* 1974. *Criticality:* 1985. *First Power:* January 1985. *Commercial Operation:* June 1985.
Unit 2 *Ordered:* 1970. *Construction began:* 1974. *Criticality:* 1986. *First Power:* 1986. *Commercial Operation:* September 1986.

---

257 ●
Catawba Nuclear Station, Units 1-2

PO Box 256 Phone: (803)831-3000
Clover, SC 29710
M.S. Tuckman, V.Pres., Catawba Site

Owning Utility

Duke Power Company Phone: (702)373-4011
PO Box 33189
Charlotte, NC 28242
H.B. Tucker, Sr.V.Pres., Nuclear
  Generation

Contact

T.J. Pettit, Mgr., Community Relations
Catawba Nuclear Station

Basic Facts

Units 1-2 *Status:* Operable. *Type of Reactor:* Pressurized light-water reactor. *Megawatts (electric):* 1205. *Megawatts (thermal):* 3411. *Reactor System Supplier:* Westinghouse Electric Corp. (United States). *Generator Supplier:* General Electric Co. (United States). *Architect Engineer:* Duke Power. *Owner:* Duke Power; Piedmont Municipal Power Agency; Saluda River Electric Cooperative; North Carolina Electric Membership Corporation; North Carolina Municipal Power Agency. *Operator:* Duke Power.

Key Dates

Unit 1 *Ordered:* 1970. *Construction began:* 1974. *Criticality:* 1985. *First Power:* January 1985. *Commercial Operation:* June 1985.
Unit 2 *Ordered:* 1970. *Construction began:* 1974. *Criticality:* 1986. *First Power:* 1986. *Commercial Operation:* September 1986.

Operating Costs

Units 1-2
1985 $35,968,000
1986 $85,372,000
1987 $118,799,000
1988 $136,370,000
1989 $132,983,000

Plant Report

Many owners. The Catawba Nuclear Station is the newest of three nuclear stations operated by Duke Power Company. It has two pressurized water reactors and is located approximately 19 miles southwest of Charlotte, North Carolina. Although the station was built and is operated by Duke Power Company, Duke's nuclear program encountered financial difficulties during the 1970s and the utility sold interests in both units to regional electric corporations. As of December 31, 1990, Duke retained only a 12.5 percent ownership interest. Other owners were: the North Carolina Munuicipal Power Agency (37.5 percent), the North Carolina Electric Membership Corporation (28.125 per cent), the Piedmont Municipal Power Agency (12.5 percent) and the Saluda River Electric Cooperative (9.375 percent).

Construction challenged. Duke had estimated that demand for electric power would grow 10 percent a year when the company originally made plans for Catawba in 1973. The Carolina Environmental Study Group (CESG) claimed that the additional capacity was not necessary and attempted to force cancellation of the plant. The North Carolina Utilities Commission (NCUC), however, ruled that the construction was "prudent and reasonable."

Sources

"The Best." *Forbes*, February 11, 1983.
"'Prudent' Duke Gets 100 percent of Catawba-1 Cost in Rate Base." *Electrical World*, October 1985.
Duke Power Company Annual Report 1990. Duke Power Co., n.d.
Duke Power Company. "Corporate Communications." August 1992.

—Karen Bellenir

258 ☆
Cherokee Nuclear Power Plant, Units 1-3
Blacksburg, SC

Owning Utility

Duke Power Company Phone: (702)373-4011
PO Box 33189
Charlotte, NC 28242
H.B. Tucker, Sr.V.Pres., Nuclear Generation

Contact

Andy Thompson, Dir. of Issues Mgmt.
Duke Power Company

Basic Facts

Units 1-3 *Status:* Cancelled (Unit 1, construction 17% complete). *Type of Reactor:* Pressurized light-water reactor. *Megawatts (electric):* 1343. *Megawatts (thermal):* 3800. *Reactor System Supplier:* Combustion Engineering, Inc. (United States).

Key Dates

Units 1-3 *Ordered:* 1974. *Cancelled:* 1982.

Plant Report

Reduced demand forced cancellation. In response to anticipated increases in demand for electricity, Duke Power Company announced its plans to construct the Cherokee Nuclear Station near Gaffney, South Carolina, in February 1974. The utility proposed constructing three 1,280-megawatt units; however, revised projections and regulatory uncertainties led to the plant's cancellation in 1982.

Duke Power sold the plant site in 1985 for $3 million to Earl Owensby. Mr. Owensby anticipated turning the half-finished plant into a movie studio.

Sources

"Has North Carolina's Silver-Screen Image Tarnished?" *Business-North Carolina*, April 1988.
Duke Power Company. "Corporate Communications." August 1992.

—Karen Bellenir

259 ☆
Clinch River Breeder Reactor Project (CRBRP)
Oak Ridge, TN

Owning Utility

Tennessee Valley Authority Phone: (615)751-0011
6N 38A Lookout Pl. Fax: (615)751-4904
1101 Market St. Telex: 361951
Chattanooga, TN 37402-2801
J.R. Bynum, V.Pres., Nuclear Operations

Basic Facts

Status: Cancelled. *Type of Reactor:* Liquid metal fast breeder reactor. *Megawatts (electric):* 375. *Megawatts (thermal):* 1175. *Reactor System Supplier:* Westinghouse Electric Corp. (United States). *Architect Engineer:* Burns & Roe, Inc. (United States). *Owner:* Tennessee Valley Authority; U.S. Department of Energy. *Operator:* Tennessee Valley Authority.

Key Dates

Ordered: 1972. *Cancelled:* 1983.

Plant Report

A technological turkey. In what has been called a "technological turkey," the 375-megawatt breeder reactor experiment at Clinch River, Tennessee, may well be one of the biggest flops ever undertaken by the United States government. Although the project was designed to show that a breeder could be feasible in the commercial nuclear make-up of the U.S. electrical industry, the development process was fraught with controversy, and the project eventually folded.

What is a breeder?. A breeder is a nuclear reactor that "breeds" more fuel than it uses. It utilizes Uranium-235 in the core and during the process of the chain reaction, Plutonium-239 is "bred" or produced as a by-product. The plutonium can then either replace the U-235 in the core, or be used in the production of nuclear weapons. Due to the efficiency of reusing the fuel, the breeder can produce 60 times as much energy as a normal reactor.

However, plutonium is not exactly a safe substance. A particle the size of a speck of dust can cause lung cancer when inhaled. It is so highly reactive that small chips and shavings can spontaneously ignite. The instability of plutonium is one of the highly debatable controversies over the use of breeders.

Applying laboratory technology to the real world. The idea of operating a commercial breeder reactor was conceived in 1969 when it was believed that the U.S. was going to consume huge amounts of uranium due to increased electricity usage. Since there was a shrinking supply of uranium that could only last for a couple more decades, a breeder reactor became very appealing. But by 1971, new deposits of natural uranium were discovered, which eased fears that the industry would be suffering from a lack of nuclear fuel by the late 1980s. With the discovery of the new deposits, estimates of the occurrence of a shortage were pushed back to the year 2050 or later.

Major players. The idea of building a breeder reactor was a cooperative effort of the federal government and private industry. Major government agencies involved included the U.S. Atomic Energy Commission (AEC), now the Nuclear Regulatory Commission (NRC), the U.S. Department of Energy (DOE), and the government-owned Tennessee Valley Authority, which would be the operator of the plant and the recipient of its power.

Private interests groups were comprised of 753 utilities and 149 companies, including Westinghouse Electric Corporation, the reactor manufacturer; Stone & Webster Engineering Corporation, the primary contractor; and Burns & Roe, the architects. In 1980, over 4,000 people in 24 states were involved in one way or another on the project, and by 1983, 3,500 people in 32 states were involved.

Although, at first, the utility officials thought they would have a say in the reactor's operation, in actuality the project was a cooperative effort to which the government held all the cards. The government would maintain sole-ownership and control, while the utilities were to gain enormous knowledge from the learning experience.

Costs kept escalating. Because the technology was so new, many companies were skeptical and afraid of how big of a commitment they were making. Therefore, in 1972, in order to keep the project on track, Congress was persuaded to require that the federal government would incur all cost overruns associated with Clinch River. This pledge is what has cost U.S. taxpayers millions of dollars on Clinch River.

Originally estimated to cost $700 million by 1983, the DOE estimated that initial operating costs would be in the neighborhood of $3.85 billion. Of this, due to the agreement, private industry had only shelled out $257 million. As of July 1983, when a new plan was introduced for continued financing of Clinch River, the project had already cost the DOE $1.5 billion, which represented 90 percent of the costs incurred.

Why the problems began. Overall, problems in completing the first commercial breeder began with the government's commitment of providing unlimited financing. Back in 1977, when former President Jimmy Carter called for the cancellation of the project, Congress continued appropriating funds to keep it moving forward. Carter also made it impossible for Clinch river to be issued a construction permit by the NRC. To get around this, manufacturers of major components and design work housed the competed work in warehouses in Memphis and Oak Ridge, Tennessee.

There were also problems with how the plutonium from Clinch River was going to be reprocessed or separated from the uranium in the spent fuel. When the contract was signed in 1971, there appeared to be an industry for reprocessing spent fuel. Construction on a new reprocessing plant at Morris, Illinois, was completed in 1974, and construction on another plant at Barnwell, South Carolina, was under way when the contract was signed for Clinch River. But by 1983, the industry was nonexistent because of the threat it posed for weapons proliferation.

Because of the debate between Carter and Congress on the worthiness of the project, the scheduled completion date of 1985 was pushed back to 1988. During the 3-1/2 year stalemate, between 1977 and 1980, the program incurred an additional $600 million in costs. But the project continued for one reason: proponents knew that the changing political climate was in favor of the breeder.

Work proceeds without construction permit. By the end of 1982, work completed for the project, located on 1,364 acres in Oak Ridge, Tennessee, included design work that was 75 percent complete. In addition, the land had been cleared for construction, and the reactor vessel itself was being stored in a shed in Mt. Vernon, Indiana. By 1983, the engineering designs were 90 percent complete, research and development was 98 percent complete, and the value of equipment in storage was $750 million. But construction had not yet begun at the site even though it had been authorized in October 1982.

Opposition to completing the breeder. Opponents of the breeder reactor cited three major obstacles to constructing a useful piece of technology. First and foremost was the issue of non-proliferation. Since the "bred" substance would be highly reactive plutonium, by abandoning the project, the U.S. would show its commitment to a plutonium-free economy. By doing this, they felt that other countries would follow suit, thus reducing the spread of nuclear weapons. The second argument involved project costs. Although the DOE had spent $1.5 billion as of July 1983, and it expected that by the time work was complete, it would cost a total of $3.85 billion, other estimates were not so conservative. A report issued by the General Accounting Office (GAO) anticipated the end cost would reach $8.5 billion. Also, even if the plant had been shut down at that time, another $500 million would have been required to scrap the plant.

The last criticism of the Clinch River project was that its basic design was outdated (hence the term "technological turkey"). Opponents argued that the "loop" design is inferior to a "pool-shaped design" and that contractors had failed to make changes.

Proponents refute objections. In response to criticisms of the project, proponents of the breeder reactor argued that the circumstances surrounding the reasons for stopping construction had changed. Proliferation of nuclear power has always been a threat, but, say proponents, the industry and the government are always striving to improve the safety of its facilities from any illegal weapons production.

They also pointed out that with any conceptual project, the cost was bound to escalate. Revised government safety standards were also cited as carrying some of the responsibility of the increased cost. By the time a new project is completed, proponents argued that some of the technology is bound to be outdated. This is why contractors were continually updating the breeder's design.

One other factor noted by proponents was the industry-wide need of the continual development process. Experts have predicted that by the year 2000, electricity usage will have increased by more than 70 percent over 1980 levels. At this rate, breeders would need to be commercially available in the 21st century and could only be possible with continuing research.

Last ditch effort to save Clinch River. In June 1983, the House of Representatives voted to cut all funding for Clinch River as of October 1, 1983; the Senate, however, did not concur. To keep the project moving forward, a coalition of electric utilities and Wall Street investment firms proposed an alternative financing plan to raise money for the project. Backed by President Ronald Reagan, the plan proposed that private industry contribute another $1 billion to the project whereas taxpayers would have to bear another $1.4 billion.

The plan itself consisted of two parts. The first part would have required that companies designing and building the breeder would have to contribute $325 million, but in the process, the companies would become part owners. These companies were expected to recoup approximately $150 million of the $325 million investment through tax write-offs.

The second part of the plan consisted of an investment by private banks lending $700 million to participating nuclear power companies. To pay off the loans, the companies would be required to sell special bonds to private investors who would then be paid off by raising more money.

However, there were problems associated with the plan. First, the Treasury would have to bear all responsibility and cover most of the costs. This means that if the breeder failed, U.S. taxpayers would have to pay 100 percent of the breeder's debts. But if it achieved its goal, all of the income would go to the private sector to pay off bond debts, and the government, which shelled out most of the money, wouldn't get any of the profits.

The breeder demise after 13 years of debate. The controversy over the breeder reactor demonstration project at Clinch River ended on October 26, 1983, when the Senate voted 56-40 to defeat the plan and stop funding altogether. As of November 7, 1983, $1.7 billion had been spent on the project and only a hole in the ground existed to show for all of the effort. Those against the project said that it never made economic sense.

Breeder reactors in other countries. The United States was not the first country to experiment with commercial breeders, even though it was the country that developed the technology. The United Kingdom, Germany, Japan, France, and the Commonwealth of Independent States all have advancing breeder programs. In the midst of the troubles with Clinch River, France was in the process of upgrading its 250-megawatt Phenix plant to the 1200-megawatt Superphenix. Although the French breeders have been uneconomical, they now have the knowledge and expertise to construct other breeders.

Sources

"Stalemate Along Clinch River." *Forbes*, October 13, 1980.
"Half-Life at Clinch River." *Newsweek*, January 12, 1981.
"Kettles of Fission," *The New Republic*, November 11, 1981.
"Break the Clinch." *The New Republic*, October 11, 1982.
"Wrong in the Clinch." *National Review*, November 26, 1982.
"The Breeder's Birth Pains." *Nation's Business*, December 1982.
"Dream Machine." *The Atlantic Monthly*, April 1983.
"Clinch River Begins It's Final Meltdown." *Newsweek*, November 7, 1983.
"DOE Budget: Material Gains and Energy Losses." *Science News*, July 23, 1983.
"Hanging on at Clinch River." *Industry Week*, July 25, 1983.
"And Now the Clincher." *The Progressive*, October 1983.
"Clinch Crunch." *Critical Mass Bulletin*, November 1983.
"1.7 Billion-Dollar Hole in the Ground." *U.S. News & World Report*, November 7, 1983.
"Hanging on at Clinch River." *Industry Week*, July 25, 1983.
"And Now the Clincher." *The Progressive*, October 1983.
"Clinch Crunch." *Critical Mass Bulletin*, November 1983.

—Yvette M. Costa

260 ●
Clinton Nuclear Power Station, Unit 1

Box 678
Clinton, IL 61727
John G. Cook, Mgr.

Phone: (217)935-8881
Fax: (217)935-8294

Owning Utility

Illinois Power Company
500 S. 27th St.
Decatur, IL 62525
J. Stephen Perry, V.Pres.

Phone: (217)424-6600
Fax: (217)424-6978

Contact

Rodney A. Smith, V.Pres., Pub.Aff.
Illinois Power Company

Basic Facts

Unit 1 *Status:* Operable. *Type of Reactor:* Boiling water reactor. *Megawatts (electric):* 985. *Megawatts (thermal):* 2894. *Reactor System Supplier:* General Electric Co. (United States). *Generator Supplier:* General Electric Co. (United States). *Architect Engineer:* Sargent & Lundy Engineers (United States). *Owner:* Illinois Power Company; Soyland Power Cooperative. *Operator:* Illinois Power Company.

Key Dates

Unit 1 *Ordered:* 1973. *Construction began:* 1975. *Criticality:* February 1987. *First Power:* April 1987. *Commercial Operation:* April 1987.

Operating Costs

Unit 1
1987 $65,156,000
1988 $72,016,000
1989 $91,227,000

Plant Report

Construction problems and cost overruns. Construction of the Clinton Nuclear Power Station, a project undertaken by Illinois Power Company (IP), began in 1970 and was completed in the spring of 1987. Clinton's initial construction budget was estimated at $500 million, but the final cost totaled more than $4 billion. Expenses incurred at Clinton caused IP to become one of the most expensive suppliers of electricity in the Midwest.

Construction on a second unit at Clinton began in 1975, but it was cancelled in 1983 because of declining demand for power in IP's service area. Some materials from the scratched unit were sold and some were retained for use at Clinton 1.

Accident hurts three employees. Operation at Clinton has not been without difficulty. In June 1988, three workers were injured when they removed a glass shield from one of the plant's evaporators. The error resulted in a spray of hot, contaminated waste water. One worker suffered serious burns and the other two incurred minor burns. The accident did not disrupt the plant's production and no radioactivity escaped to the air inside the plant or to the environment.

Financial impact. More than operational mishaps, however, the Clinton plant has suffered financial stresses. Construction efforts were plagued with difficulties and cost overruns. Critics charged that the nuclear plant was too large for the modest demand in IP's service area. They also asserted that the plant represented too large a percentage of IP's assets and budget. While Clinton accounts for only 19 percent of IP's generating capacity, it represents 67 percent of the utility's assets and consumes 40 percent of IP's operating, maintenance, and depreciation expenses.

To help pay for Clinton, IP requested a 30 percent rate increase, representing $256 million, to be phased in over an 11-year period. In March 1989, the Illinois Commerce Commission (ICC) granted an increase of only 6.9 percent ($61 million). The ICC claimed that only 27 percent of Clinton's capacity was needed, and that the plant was creating surplus power supplies of 58 percent. The ICC also disallowed the recovery of $615 million in "unreasonable" expenses.

The reduction in IP's rate request forced the utility to suspend dividend payments on common stock. Utility analysts called 1989 the worst year in IP's 66-year history.

IP requested a second rate hike. The utility asked for a 23 percent, one-time increase representing $215 million. In 1990, the ICC again granted an increase significantly lower than the request. Although the ICC found Clinton 61 percent "used and useful," it permitted an increase of only 7.7 percent ($75 million).

Future prospects. Critics charge that Clinton was a mistake. IP officials continue to justify the utility's investment in the plant, insisting its generating capacity is vital for the future power needs of its service area. IP claims nuclear generation will serve its customers best interests because of negative backlash anticipated from the Clean Air Act in the mid-1990s regarding its five coal-fired plants.

IP's plan for financial recovery calls for court challenges of the low rate increases, new rate hike requests, a reduction of operating costs, and the expansion of economic development within the company's 15,000 square-mile service area.

Sources

"Grassroots." *Critical Mass Bulletin*, December 1983.
"ICC Report Triggers Stock Slide, Earnings Worries at IP." *Crains Chicago Business*, May 2, 1988.
"Three Hurt in Accident at Illinois Nuclear Plant." *The Wall Street Journal*, June 24, 1988.
"Illinois Power is Denied Boost It Asked in Rates." *The Wall Street Journal*, March 31, 1989.
"Illinois Power Launches Turnaround Plan." *St. Louis Business Journal*, April 16, 1990.
"Rate Request Key for Illinois Power." *Crains Chicago Business*, May 14, 1990.
"Illinois Power Gets $75 Million Rate Increase." *The Wall Street Journal*, June 7, 1990.
"Illinois Power Users Paying for the Future." *St. Louis Business Journal*, October 15, 1990.

—Karen Bellenir

261 ☆
Clinton Nuclear Power Station, Unit 2

Box 678
Clinton, IL 61727
John G. Cook, Mgr.

Phone: (217)935-8881
Fax: (217)935-8294

Owning Utility

Illinois Power Company
500 S. 27th St.
Decatur, IL 62525
J. Stephen Perry, V.Pres.

Phone: (217)424-6600
Fax: (217)424-6978

Contact

Rodney A. Smith, V.Pres., Pub.Aff.
Illinois Power Company

Basic Facts

Unit 2 *Status:* Cancelled. *Type of Reactor:* Boiling water reactor. *Megawatts (electric):* 985. *Megawatts (thermal):* 2894. *Reactor System Supplier:* General Electric Co. (United States). *Generator Supplier:* General Electric Co. (United States). *Architect Engineer:* Sargent & Lundy Engineers (United States). *Owner:* Illinois Power Company; Soyland Power Cooperative. *Operator:* Illinois Power Company.

Key Dates

Unit 2 *Ordered:* 1973. *Cancelled:* 1983. *Construction began:* 1975.

Plant Report

See Plant Report for Clinton Nuclear Power Station, Unit 1.

262 ●
Comanche Peak Nuclear Power Plant, Unit 1

PO Box 1002
Glen Rose, TX 76043
J.J. Kelly Jr., Plant Mgr.

Phone: (817)897-4856

Owning Utility

TU Electric
Skyway Tower
400 N. Olive, LB81
Dallas, TX 75201
A.B. Scott, V.Pres., Nuclear Operations

Phone: (214)812-8200

Contact

R.L. Ramsey, Dir., Public Commun.
TU Electric

Basic Facts

Unit 1 *Status:* Operable. *Type of Reactor:* Pressurized light-water reactor. *Megawatts (electric):* 1150. *Megawatts (thermal):* 3411. *Reac-*

tor System Supplier: Westinghouse Electric Corp. (United States). Generator Supplier: Allis-Chambers (United States); Power Contractors Incorporated. Architect Engineer: Gibbs & Hill, Inc. (United States).

Key Dates
Unit 1 Ordered: 1972. Construction began: 1974. Criticality: 1990. First Power: 1990. Commercial Operation: August 1990.

Plant Report

Long-delayed project finally nears completion. On February 8, 1990, Comanche Peak Unit 1, owned by Texas Utilities (TU), became the 112th commercial nuclear reactor licensed in the United States. After 17 years of construction delays and design changes, the Nuclear Regulatory Commission granted a low-power testing license.

Despite a serious operator error in March, which prematurely actuated a safety system, the NRC issued a full power license on April 17. The 1,150 megawatts electric Pressurized Water Reactor began full commercial operation on August 13. Comanche Peak Unit 2 should begin loading fuel in December 1992.

Even so, TU's problems did not end with the opening of the plant, despite favorable reports from the NRC.

One week after Comanche Peak received its full-power licence, the Texas Public Utility Commission reported that TU had "imprudently" overspent at least $713 million in construction costs. By August 9, that estimate had grown to $1.38 billion. The ruling by the commission meant TU could not pass those costs on to Texas consumers in the form of higher electricity rates. The company must absorb the charge as an operating loss.

Shareholders sue Texas Utilities' management team. Two shareholders took exception to the loss. Nancy King of Massachusetts and Rodney Shields of Arizona filed suit against TU's top officers and directors. They charged the executives displayed "gross and intentional negligence" in allowing construction costs to balloon from the original estimate of $712.6 million in 1974 to over $9.4 billion and still rising in 1990. The stockholders seek to recover damages from the executives for the company.

Originally, TU hoped to have Comanche Peak up and running by 1980. It missed that target by three years, applying in 1983 for the initial NRC low-power license. The group, Citizens' Association for Sound Energy (CASE), headed by Juanita Ellis, opposed the application. She had developed a critique of the power station design and construction history with the help of several plant workers. In response, the NRC condemned what it called a sloppy design and poor workmanship, ordering reinspection of the facility and major reconstruction. In January 1985, it suspended licensing procedures for the plant until TU could show it had addressed the quality and safety concerns.

With licensing on hold and huge construction costs looming, TU's minority partners decided to abandon the project. Brazos Electric Cooperative bailed out in May 1985, followed by Tex-La Electric in May 1986.

Critics suggested TU should do the same, but the utility refused. It noted its growing customer base. With over two million customers spread over the northern third of Texas, not having Comanche Peak on-line forced the utility to import 800 to 1,000 megawatts of electricity each summer. When operational, the plant, located 75 miles southwest of Dallas, will produce 2,300 MWe.

Problems continue to stall construction. Concerns about the reactor's safety continued throughout the 1980s. As the next licensing hearing approached, the NRC acknowledged a backlog of 20,000 problems, while TU lost critical documents. CASE continued to publicize problems in its newsletter, but a decade of activism had worn Ellis down.

Leaked details of $10 million payment raise questions of plant safety throughout country. On June 30, 1988, Ellis announced CASE would withdraw its objections to the licensing of the plant. In return, the organization received a seat on TU's Oversight Review Committee until 1993, giving it a say in future safety concerns. Later, an informant revealed a $10 million payment from TU to CASE. CASE kept $4.5 million to offset expenses, both past and future, while the other $5.5 million went to 12 whistle blowers at the plant. Payment of the money depended on a favorable ruling from the NRC.

In 1989, a report for a Senate Subcommittee on Nuclear Regulation indicated the practice of "restrictive settlements" had become common in the nuclear industry. It raised the concern that plant management and safety may be being misrepresented to the NRC.

On October 16, 1989, the NRC convened a special inspection team in response to allegations from an anonymous group of agency inspectors. The inspectors suggested documents and reports had been altered to remove information that might slow the licensing process.

Even so, licensing did continue. Last-minute attempts to block the process by the Fort Worth-based Citizens for Fair Utility Regulation had no effect. The NRC rejected its action as too late.

After the hearing which granted the full-power license on April 17, 1990, Ellis raised concerns about the plant's concrete foundations. Along with the license to operate, TU received approval for a 10 percent increase in electricity rates to cover the cost of Comanche Peak Unit 1. It expects to receive another 10 percent increase when Unit 2 comes on line in 1993. TU will spend another $800 million to complete the second reactor, on top of the $10.7 billion already consumed by the project.

Over the year following the granting of the license, Comanche Peak Unit 1 shut down 11 times to correct malfunctions. NRC inspections rated the plant as improving and moved to reduce future inspections.

Comanche Peak has its own Karen Silkwood. However, the plants problems are not over yet. On March 8, 1992, CBS News—60 Minutes revealed allegations of unsafe working conditions during the construction of Comanche Peak. It likened the plight of painting supervisor Linda Porter to that of Karen Silkwood, the nuclear plant worker who died mysteriously after questioning plant safety in Oklahoma.

Plant Profiles

Porter claimed she and her crew worked with inadequate equipment in clouds of asbestos paint dust. Complaints to her supervisors brought threats and intimidation. Exposure to the dust made her physically and chronically ill. Eventually, tumors developed in her throat. After four years, the construction contractor, Brown and Root, laid Porter off as part of a large staff cutback.

Dr. Saul Wilen treated 150 Comanche Peak workers with similar symptoms. He blames the work environment.

Porter and 2,000 other construction workers are suing TU. Their appeals for help from the Occupational Safety and Health Administration went unanswered.

Sources

"Downgraded Credit Ratings Can Be Costly." *Dallas Business Courier*, November 11, 1985.
"The Troublesom Trends Worrying Texas Utilities." *Dallas Business Courier*, July 28, 1986.
"Utility Again Pushes Back Start-Up of Comanche Peak." *The Wall Street Journal*, December 21, 1987.
"Texas Plant Comes Under Scrutiny As Cover-Up of Problems Charged." *The New York Times*, October 17, 1989.
"The Co-opting of CASE." *Nation*, December 4, 1989.
U.S. NRC 1990 Annual Report. U.S. Nuclear Regulatory Commission, n.d.
"Comanche Peak Unit 1 Licensed for Low-Power." *Nuclear Industry*, First Quarter 1990.
"Texas Utilities Gets Low-Power License at Comanche Peak." *The Wall Street Journal*, February 9, 1990.
"Comanche Can Operate at Its Peak." *Dallas Times Herald*, April 17, 1990.
"Report Criticizes Texas Utilities Unit on Nuclear Plant." *The Wall Street Journal*, April 23, 1990.
"Shareholders Sue TU Managers." *Dallas Times Herald*, May 3, 1990.
"Ever-rising Rates Tax Homeowner Patience." *Dallas Times Herald*, September 16, 1990.
"NRC Report Gives Plant High Marks." *Dallas Times Herald*, April 18, 1991.
"Texas Utilities Taking a Charge of $1 Billion." *The Wall Street Journal*, August 9, 1991.
1991 Nuclear Power Review. U.S. Council for Energy Awareness, December 19, 1991.
"Another Karen Silkwood?" *CBS News—60 Minutes*, March 8, 1992.

—Al Cook

263 ○
Comanche Peak Nuclear Power Plant, Unit 2

PO Box 1002 Phone: (817)897-4856
Glen Rose, TX 76043
J.J. Kelly Jr., Plant Mgr.

Owning Utility

TU Electric Phone: (214)812-8200
Skyway Tower
400 N Olive, LB81
Dallas, TX 75201
A.B. Scott, V.Pres., Nuclear Operations

Contact

R.L. Ramsey, Dir., Public Commun.
TU Electric

Basic Facts

Unit 2 *Status:* Under construction. *Type of Reactor:* Pressurized light-water reactor. *Megawatts (electric):* 1150. *Megawatts (thermal):* 3411. *Reactor System Supplier:* Westinghouse Electric Corp. (United States). *Generator Supplier:* Allis-Chambers (United States); Power Contractors Incorporated. *Architect Engineer:* Gibbs & Hill, Inc. (United States).

Key Dates

Unit 2 *Ordered:* 1972. *Construction began:* 1974. *Commercial Operation Expected:* 1993.

Plant Report

See Plant Report for Comanche Peak Nuclear Power Plant, Unit 1.

264 ●
Donald C. Cook Nuclear Plant, Units 1-2

One Cook Pl. Phone: (616)465-5901
Bridgman, MI 49106
L.S. Gibson, Asst. Plant Mgr.

Owning Utility

Indiana Michigan Power Co. Phone: (614)223-1000
c/o American Electric Power Service Corp.
One Riverside Plaza
Columbus, OH 43215
E.L. Draper, Pres.

Contact

L.M. Feck, Sr.V.Pres., Pub.Aff.
Indiana Michigan Power Co.

Basic Facts

Unit 1 *Status:* Operable. *Type of Reactor:* Pressurized light-water reactor. *Megawatts (electric):* 1056. *Megawatts (thermal):* 3250. *Reactor System Supplier:* Westinghouse Electric Corp. (United States). *Generator Supplier:* General Electric Co. (United States). *Architect Engineer:* Indiana Michigan Power Co.

Unit 2 *Status:* Operable. *Type of Reactor:* Pressurized light-water reactor. *Megawatts (electric):* 1100. *Megawatts (thermal):* 3411. *Reactor System Supplier:* Westinghouse Electric Corp. (United States). *Generator Supplier:* Brown Boveri et Cie. (Switzerland). *Architect Engineer:* Indiana Michigan Power Co.

Key Dates

Unit 1 *Ordered:* 1967. *Construction began:* 1969. *Criticality:* 1975. *First Power:* February 1975. *Commercial Operation:* August 1975.

Unit 2 *Ordered:* 1967. *Construction began:* 1969. *Criticality:* 1978. *First Power:* March 1978. *Commercial Operation:* July 1978.

Operating Costs

Units 1-2

1975	$1,662,000	1983	$59,516,000
1976	$7,047,000	1984	$80,435,000
1977	$10,012,000	1985	$102,604,000
1978	$15,707,000	1986	$101,294,000
1979	$26,751,000	1987	$117,545,000
1980	$32,409,000	1988	$116,628,000
1981	$37,967,000	1989	$126,644,000
1982	$50,858,000		

Plant Report

Background. The Cook Nuclear Plant, located in southwestern Michigan at Bridgman, 11 miles south of Benton Harbor and St. Joseph, was named for Donald C. Cook, a Michigan native who served as chief executive officer of Indiana Michigan Power Co. (I&M) and American Electrical Power Service Corporation (AEP) from 1961 to 1976.

The 2.1-million-kilowatt plant is one of America's largest nuclear power stations. It is owned by Indiana Michigan Power Co., one of eight operating subsidiaries of American Electrical Power Service Corporation.

Based in Columbus, Ohio, AEP is an investor-owned electric utility holding company, which owns the common stock of its operating subsidiariees. The major electric facilities of the operating companies—their 22 major power plants and principal transmission lines—are interconnected and coordinated to function as a single, integrated electric-utility system. The AEP system provide electric service to more than 2,560,000 customers in 3,200 communities in parts of seven east-central states.

I&M is responsible for supplying power to 231 communities in parts of 25 counties in northern and east-central Indiana as well as six counties in southwestern Michigan. The company has more than 470,600 customers. I&M operaties four major steam-electric generating plants, three of which are coal-fired. The fourth, Cook Plant, is the only nuclear-power plant on the AEP system.

The two-reactor Cook plant supplies 35 percent of I&M's generating capacity. Both units are currently licensed to operate until 2009.

Cook 1 reaches 100 billion kilowatt mark. On July 28, 1991, Cook 1 and Oconee 1 in South Carolina became the first nuclear energy plants in the U.S. to produce 100 billion kilowatt-hours of electricity. That's enough electricity to supply the yearly needs of 10 million American homes, which on average use 10,000 Kwh a year.

Radioactive waste storage. AEP spent some $4.5 million to build a concrete storage facility at the Cook Station capable of handling up to 25,000 curies of radioactive wastes. Scheduled for completetion in April 1992, the facility will have 12 cells divided by one-foot-thick concrete walls, a truck bay, a dry active waste storage area, and a service area. The exterior walls are 30 inches thick. The facility is capable of being doubled in size. The Cook Station will store all of its low-level radioactive waste in the new facility.

Shutdown at Unit 2. In September 1992, the emergency generator for Unit 2 automatically shut down when the lube oil pressure became too low. An oil leak had been discovered the previous May, but workers failed to correct the problem. The NRC has proposed a fine of $37,500 be paid by the Michigan Power Co. because of the incident.

Information center. The Cook Energy Information Cener is open mid-January through mid-December (except holidays). Summer hours (June through September) are Wednesday-Saturday 10 a.m. to 5 p.m., Sunday 11 a.m. to 5 p.m. Visiting hours from mid-January through May and October through mid-December are 10 a.m. to 5 p.m., Monday-Friday.

For more information or to arrange special tours, call 616-465-6101 or 616-983-2028.

Sources

"Oconee 1, Cook 1, Reach 100 Billion Kwh Mark." *INFO*, U.S. Council for Energy Awareness, August 1991.

"Regulators, LLW Generators Trying to Sort Out On-Site Storage Duties." *Inside N.R.C.*, December 2, 1991.

Cook Nuclear Plant. Cook Energy Information Center, AEP Indiana Michigan Power, n.d.

---

265 ●

Cooper Nuclear Power Plant

PO Box 98　　　　　　　　　　　Phone: (402)825-3811
Brownville, NE 68321　　　　　　Fax: (401)825-5211
John M. Meacham, Nuclear Operations
　Div. Mgr.

Owning Utility

Nebraska Public Power District　　Phone: (402)564-8561
PO Box 499　　　　　　　　　　Fax: (401)563-5551
Columbus, NE 68602-0499
Guy R. Horn, Nuclear Power Group
　Mgr.

Contact

Ron C. Bogus, Pub.Info.Mgr.
Nebraska Public Power District

Basic Facts

Status: Operable. *Type of Reactor:* Boiling water reactor. *Megawatts (electric):* 801. *Megawatts (thermal):* 2381. *Reactor System Supplier:* General Electric Co. (United States). *Generator Supplier:* Westinghouse Electric Corp. (United States). *Architect Engineer:* Burns & Roe, Inc. (United States).

Key Dates

Ordered: 1966. *Construction began:* 1968. *Criticality:* February 1974. *First Power:* May 1974. *Commercial Operation:* July 1974.

Plant Profiles | Cooper

Operating Costs

1974	$2,691,000	1982	$23,482,000
1975	$7,386,000	1983	$30,893,000
1976	$10,211,000	1984	$25,699,000
1977	$10,218,000	1985	$40,304,000
1978	$8,606,000	1986	$48,092,000
1979	$10,232,000	1987	$50,242,000
1980	$19,004,000	1988	$54,875,000
1981	$20,455,000	1989	$60,818,000

Plant Report

A power center in Nebraska. Cooper Nuclear Station occupies a 1,351-acre site, about three miles south of Brownville, Nebraska, on the west bank of the Missouri River. The Station employs 550 people for plant operation.

The cost of original plant construction was $313 million, and excavation involved the moving of over 760,000 cubic yards of earth. Over 90,000 cubic yards of concrete and 10,000 tons of steel were used in the construction.

Power customers. Power from Cooper Station flows into a vast system of extra high voltage (EHV) transmission lines. One 345,000 volt line links the plant with the NPPD's Sheldon Station plant near Hallam, Nebraska. The line continues west, where it connects with another 345,000 volt line, which extends from Fort Thompson, South Dakota. The line is owned by the U.S. Bureau of Reclamation.

Iowa Power of Des Moines, Iowa, which contracted to purchase up to one half of Cooper Station's production, has its own line which extends from the plant to eastern Iowa.

Contractual agreements provide for the Lincoln Electric System to receive up to 12 percent of Cooper Station's output. There are also transmission connections with systems of other regional utilities, including the Mid-Continent Area Power Pool.

Facts and features of Cooper Station. The reactor building is approximately seven stories tall, with a rectangular outer wall made of steel and concrete.

The reactor vessel weighs 684.5 tons and is made of low alloy steel, clad with a stainless steel wall, which ranges up to 7 1/2 inches thick. The vessel is 69 feet high and 18 feet across, with an inside diameter of 18.2 feet. The design pressure is 1250 pounds per square inch at 575 degrees Fahrenheit. The core thermal power is 2,381 megawatts.

The containment is a GE Mark I primary containment, consisting of a steel drywell which is encircled by a wetwell (torus). The drywell, which houses the reactor vessel, is a 3/4 inch thick steel pressure vessel with a spherical lower portion that is 60 feet in diameter and a cylindrical upper portion that is 35 feet 7 inches in diameter. The drywell is about 100 feet high, enclosed in reinforced concrete more than six feet thick.

The fuel assemblies are made up of 12,330,000, or approximately 100 metric tons, of uranium dioxide. Enrichment ranges from 2.5 percent to 3 percent U-235 by weight. The reactor contains 548 fuel assemblies, consisting of 32,880 rods apiece. Each rod is made of Zircalloy 2 and measures 12.5 feet long.

The reactor coolant system is a direct cycle, water-cooled system with a core coolant flow rate of 73,500,000 pounds per hour. The feedwater flows from the Missouri River at a rate of 9,520,000 pounds per hour at 367 degrees Fahrenheit. The steam capacity is 9,560,000 pounds per hour with a steam outlet temperature of 541 degrees Fahrenheit.

The generator runs at 1800 revolutions per minute, creating 22,000 volts of electricity. The tandem compound four flow turbines drive the generator to produce a net capacity of about 778 megawatts of electrical power.

Recirculation is accomplished by two variable speed pumps, each driven by a motor that is capable of producing 550 horsepower, while turning at 1680 revolutions per minute. Each pumps up to 42,200,000 gallons per minute.

The plant contains more than 50 piping systems with pipe up to 13 feet in diameter, and an electrical system made up of about 1,100 miles of electric cable.

Refueling and waste disposal. The first partial refueling of Cooper Station was completed in November 1976. Since then, the reactor has been partially refueled on a periodic basis, with 20 percent of the fuel being removed and replaced.

The spent fuel that is removed from the reactor is placed in the plant's fuel storage pool. FFrom there, it is loaded into shipping casks for shipment to a spent fuel storage facility.

Low-level radioactive waste (LLRW) is monitored and packaged for shipment to a LLRW licensed disposal facility.

Settlement. In 1990, the NPPD and General Electric reached a settlement concerning containment and spent fuel from Cooper Station. As a part of the settlement, the utility was to receive discounts on future purchases of certain equipment and services for the plant, and credits and discounts as an amendment to an existing fuel fabrication contract. The utility was amortizing the entire amount of benefits allocated to operations over a two-year period, which began in 1990. These benefits were to be recognized over the next 18 years. This difference resulted in an increase in revenues during the amortization period and increased costs thereafter.

Decommissioning fund. In July 1990, the NPPD established an external trust fund, segregated from their assets, to pay for the future decommissioning of Cooper Station. The NRC's minimum amount to be accumulated in the fund was, in 1986 dollars, $125.4 million, but this amount did not include the cost of removal and disposal of spent fuel and nonradioactive structures, or materials necessary to terminate the plant's operating license. The utility had estimated the total decommissioning cost at, in 1988 dollars, approximately $316 million.

The NPPD planned for the costs of decommissioning to be funded from revenues, reserve funds, and surplus funds derived from the ownership and operations of Cooper Station. It also planned to continue to review costs and methods of funding as a result of the changing conditions and requirements for decommissioning.

Accomplishments of 1990. Cooper Station's gross generation was 5.3 billion kilowatt-hours during 1990. The annual refueling and maintenance outage was the shortest in recent history, and there was only one forced outage during the year.

Cooper Station earned the highest rating possible from the Institute of Nuclear Power Operations, making it the first nuclear facility, operated by a public power district, to earn the top rating.

Cooper Station's plant personnel earned the Board of Directors' Gold Safety Award for working more than one million hours, between July 1987 and August 1990, without a disabling injury.

Environmental studies. Environmental studies are constantly being conducted in the vicinity of Cooper Station. A radiation monitoring program measures the natural background radiation levels in samples of air, soil, vegetation, milk, river water, well water, and wildlife. Also, a meteorological program is conducted to determine wind speed and direction, temperature, and precipitation at the site.

The NPPD had previously conducted an extensive environmental program on the Missouri River, which measured water temperature, chemistry and biology. The data was used in assessing the condition of the environment, and it was determined that no adverse impact would be produced by Cooper Station.

Sources

Cooper Station Facts. The Nebraska Public Power District, October 10, 1989.
A Tour of Cooper Nuclear Station. The Nebraska Public Power District, 1990.
NPPD 1990 Annual Report: A Powerful Responsibility. The Nebraska Public Power District, n.d.

—Charles Ceplecha

Crystal River Energy Complex, Unit 3

PO Box 219   Phone: (904)795-6486
Crystal River, FL 32623-0219
Paul F. McKee, Dir., Nuclear Plant
  Operations

Owning Utility

Florida Power Corporation   Phone: (813)866-5151
PO Box 14042
St. Petersburg, FL 33733
Percy M. Beard, Sr.V.Pres., Nuclear
  Operations

Contact

Raymond P. Blush, Dir., Corp. Commun.
Florida Power Corporation

Basic Facts

Unit 3 *Status:* Operable. *Type of Reactor:* Pressurized light-water reactor. *Megawatts (electric):* 860. *Megawatts (thermal):* 2544. *Reactor System Supplier:* The Babcock & Wilcox Co. (United States). *Generator Supplier:* Westinghouse Electric Corp. (United States). *Architect Engineer:* Gilbert Associates.

Key Dates

Unit 3 *Ordered:* 1967. *Construction began:* 1968. *Criticality:* January 1977. *First Power:* January 1977. *Commercial Operation:* March 1977. *Shut Down Expected:* 2016.

Operating Costs

Unit 3

1977	$8,444,000	1984	$84,681,000
1978	$15,613,000	1985	$68,228,000
1979	$23,992,000	1986	$72,645,000
1980	$39,842,000	1987	$79,381,000
1981	$47,014,000	1988	$80,303,000
1982	$51,996,000	1989	$97,750,000
1983	$75,060,000		

Plant Report

Overview. The Crystal River Energy Complex is one of nine owned by the Florida Power Corporation. Generating nearly half the energy supplied by Florida Power, the Crystal River facility contains five separate power stations. Units 1 and 2 are the original coal-fired plants, together generating almost one million kilowatts of energy. Unit 3, a pressurized water nuclear reactor, has the capacity to generate 825,000 kilowatts of electricity. Units 4 and 5 are Florida Power's newest plants and are considered among the nation's top 10 coal plants for efficiency, generating a total of almost 1,500,000 kilowatts.

Safety features and the effect of Three Mile Island. Designed by the same firm that designed the reactor at Three Mile Island, Crystal River experienced a similar, though less serious, accident less than a year after the disastrous accident at Three Mile Island. The nuclear reactor at Crystal River became operational in 1977. Just three years later the plant shut down due to a class-B emergency that involved a malfunction in the operation of the safety mechanism, resulting in the spilling of 43,000 gallons of contaminated water onto the containment floor.

Due to a temporary blackout, the control system in Unit 3 mistakenly reported that the level of coolant in the reactor core was dangerously low. In accordance with automatic safety features installed in all nuclear plants in the United States, control rods were automatically inserted into the core to halt the chain reaction, and the core was flooded with coolant. Since the level of coolant was not low, the radioactive water filling the reactor core overflowed into the quench tank underneath the containment structure. Ultimately, 43,000 gallons of contaminated water were spilled onto the containment floor.

Plant personnel were evacuated and the NRC was notified when radiation monitors in the reactor building registered 50 rems per hour due to the spill. No radiation was leaked to the environment however and in the weeks after the accident pumps were employed to move the contaminated water to an auxiliary building where it would be filtered and recyled back into the reactor.

Melinda Beck of *Newsweek* summarized the attitude of the NRC to this near disaster when she stated, "The fact remains that at Crystal River the fail-safe system worked." NRC officials praised the reactions of operators within the Crystal River facility to the

Plant Profiles

emergency, and credited their efficiency with the increased training mandated by the recent Three Mile Island disaster.

However, critics of nuclear power cited this incident as another warning of the dangers of nuclear power, and some called for an investigation of Babcock and Wilcox, which had designed two reactors that within a year had suffered accidents due to the design of their safety features.

The 1980s: Small problems and business as usual. The nuclear reactor at Crystal River was shut down in early January 1987 while operating at 40 percent capacity, due to a weakened seal in one of the coolant pipes. The reactor had resumed operation on December 25 at limited capacity after having been shut down in November for undisclosed reasons.

The 1988 annual report of the Florida Progress Corporation, whose principal subsidiary, the Florida Power Corporation, owns and operates the Crystal River facility, hinted that economic problems within the parent corporation might affect the operation of the nuclear plant at Crystal River. In this report, the board of directors informed its shareholders that it had sold off a number of its holdings in 1988 due to their failure to maintain profit standards.

Two proposals were offered for proxy voting at the time the annual report was issued. The first was the election of the 13-member board of directors. The second was a request to devote additional resources to develop renewable energy programs with commercial potential and to phase out the nuclear power plant at Crystal River. The board of directors recommended shareholders vote against this proposal.

In August 1991, the nuclear plant along with one of Crystal River's four coal-run generators was shut down due to a mass of seaweed, part of an eighty-mile long mass extending into the Gulf, clogging one of the plant's main water intake pipes. Crystal River, which is situated along the coast of the Gulf of Mexico, draws in water from the Gulf at a rate of 693,000 gallons per minute for use as coolant in its power plants. The water is returned to the Gulf uncontaminated and slightly warmed.

Environmental concerns and information center. Crystal River officials are proud of their efforts to minimize the effect of their power plant on the environment and to educate the public about energy. Situated on 4700 acres near the Gulf of Mexico, the Crystal River Energy Complex—one of the largest electric generating sites in the United States—provides a home for a variety of local wildlife species, including alligators, pigs, and a variety of tropical birds. The slightly warmed water of the discharge canal offers refuge to the manatee, an endangered species, during the winter months. Further, Crystal River employees yearly recycle more than 350 tons of paper trash, as well as smaller amounts of coal fly ash, copper wire, aluminum conductors, motor oil, streetlights, and laser cartridges as part of a company-wide program to reduce and recycle waste.

The facility performs a number of environmental and educational services, including the operation of a fish hatchery on its grounds. Thousands of fingerling-size fish are produced at the hatchery in eight one-acre fish ponds, spawning tanks, and a public display area, and released into Florida's waters to replenish the Florida ecosystem. Small mesh screens positioned at the mouth of the intake pipes minimize the impact of the plant's intake canal on small marine life in the Gulf. The Crystal River site also sponsors an energy information center called The Power Place, where people of all ages learn more about nuclear power, fossil fuels, electricity, promising new technologies, and other energy-related topics.

Sources

"At Crystal River, The System Worked." *Newsweek,* March 10, 1980.
"Atom Plant in Florida is Closed Down Again." *The New York Times,* January 4, 1987.
"Seaweed Clog in Water Pipe Halts Florida Nuclear Plant." *The New York Times,* August 13, 1991.
Crystal River. Corporate Communications, Florida Power Corporation, 1991.

—Mary Gillis

267 ●

Davis-Besse Nuclear Power Plant, Unit 1

5501 N. Star Rte. 2 Phone: (419)249-5000
Oak Harbor, OH 43449 Fax: (419)249-2342
Louis F. Storz, Plant Mgr.

Owning Utility

Toledo Edison Company Phone: (419)249-5000
300 Madison Ave. Fax: (419)249-5165
Toledo, OH 43652
Donald C. Shelton, V.Pres., Nuclear, Davis-Besse

Contact

Richard A. Kelly, Mgr., Toledo Pub.Aff.
Toledo Edison Company

Basic Facts

Unit 1 *Status:* Operable. *Type of Reactor:* Pressurized light-water reactor. *Megawatts (electric):* 925. *Megawatts (thermal):* 2772. *Reactor System Supplier:* The Babcock and Wilcox Co. (United States). *Generator Supplier:* General Electric Co. (United States). *Architect Engineer:* Bechtel.

Key Dates

Unit 1 *Ordered:* 1968. *Construction began:* 1969. *Criticality:* August 1977. *First Power:* August 1977. *Commercial Operation:* November 1977.

Operating Costs
Unit 1

1977	$607,000	1984	$57,507,000
1978	$14,096,000	1985	$82,346,000
1979	$21,725,000	1986	$112,478,000
1980	$44,631,000	1987	$125,423,000
1981	$41,415,000	1988	$161,604,000
1982	$59,956,000	1989	$124,281,000
1983	$49,327,000		

● Operable ○ Under Construction △ Indefinitely Deferred ▲ Decommissioning □ Planned ■ On Order ☆ Cancelled ★ Shut Down

Davis-Besse

Plant Report

Early safety reports substandard. The Davis-Besse Plant is a single unit, 860-megawatt reactor located east of Toledo in Port Clinton, Ohio. The plant was constructed and originally operated by Toledo Edison. Its early years were plagued with safety-related problems, culminating in 1985 when NRC fines totaled $900,000.

These safety violations gave Davis-Besse notoriety as one of America's worst nuclear plants. The NRC had noted design flaws at the plant as early as 1975 but had allowed Toledo Edison to postpone improvements. Critics charged that Toledo Edison treated NRC recommendations in a cursory manner and left many problems uncorrected. In 1979 the NRC called for the plant to install a $1.2 million backup emergency pump. This task was not completed until after a serious accident forced the plant to close in 1985.

An accident occurs. On June 9, 1985, the main feedwater pump at Davis-Besse shut down, and an operator error turned off the water to the steam generators. The system began to overheat. Technicians took 16 minutes to get things under control and were only moments away from needing to take emergency measures. If operators had not discovered the problem in time, the radioactive core would have been in danger of a meltdown. No radiation was released but the plant was shut down immediately.

The day after the Davis-Besse mishap, the NRC instituted a new program for investigating accidents. NRC chief of staff at the time, William Dicks, assembled experts with no involvement at Davis-Besse. He called the group the "Incident Investigation Team" (IIT). Their job was to find out what caused the problem and recommend corrective action.

At Davis-Besse, the IIT report cited a failure to properly care for plant equipment and a tendency to treat past problems in a superficial manner. This led to equipment failures and the loss of redundant safety systems. The accident has been called "the most serious nuclear mishap since Three Mile Island."

After the accident. Davis-Besse remained shut down for 18 months. To accomplish the necessary renovations, Toledo Edison hired retired Admiral Joe Williams, Jr. He fired several managers, imposed new disciplinary measures, and started random drug testing. He also pledged to open company records to the NRC. Williams is credited with bringing Toledo Edison out of a "fossil mentality" and into modern nuclear technology.

The plant re-opened in December 1986 under the auspices of a newly formed holding company, Centerior Power Corporation. Because the NRC had cited Toledo Edison for not having enough funds to operate Davis-Besse safely, Toledo Edison and Cleveland Electric Illuminating (CEI) merged to form Centerior. Improvements at the plant led to greater efficiency. In 1989 Davis-Besse, operating at 95.14 percent, placed among the top 25 plants worldwide as measured by capacity factor.

A leader in cost containment. In an attempt to keep costs down and rate increases reasonable, Davis-Besse pioneered a new cost tracing system to reduce expenses incurred in hiring outside contractors. This type of specialized accounting is necessary because many jobs at nuclear power plants, such as those related to re-fueling, special maintenance, and keeping up with changing regulations, are done by outside contractors.

The ongoing effort to reduce expenses led to the formation of Centerior Fuel Corporation in 1989. The fuel leasing company buys the raw material for nuclear fuel (mainly uranium oxide) and leases it to CEI and Toledo Edison. The two utilities then pay Centerior Fuel as the fuel is consumed at Davis-Besse and the other nuclear plants in which Centerior has an interest. By consolidating the fuel financing, the two utilities expect to pass on a $10 million per year savings to their customers.

Sources

"Holy Toledo!" *The Progressive*, August 1985.
"A SWAT Team for Nuclear Accidents." *Science*, February 18, 1986.
"Analysts Turn Bearish on Centerior Due to Heavy Nuclear Commitment." *Crains Cleveland Business*, September 8, 1986.
"Nukes of the Ohio." *The Nation*, November 15, 1986.
"America's Big Risk." *Newsweek*, April 27, 1987.
"Lights Out: Unproductive CAPCO Energy Consortium Fading." *Pittsburgh Business Times & Journal*, August 31, 1987.
"The Key to Davis Besse: Cost Control." *Electrical World*, January 1988.
"Northern Ohio Electricity Users to Save $10 Million a year." *Business Wire*, October 30, 1989.
"The Eureka File." *Nuclear Industry*, First Quarter 1990.
"Lights Out: Unproductive CAPCO Energy Consortium Fading." *Pittsburgh Business Times & Journal*, August 31, 1987.
"The Key to Davis Besse: Cost Control." *Electrical World*, January 1988.
"Northern Ohio Electricity Users to Save $10 Million a year." *Business Wire*, October 30, 1989.

—Karen Bellenir

Davis-Besse Nuclear Power Plant, Units 2-3

5501 N. Star Rte. 2
Oak Harbor, OH 43449
Louis F. Storz, Plant Mgr.

Phone: (419)249-5000
Fax: (419)249-2342

Owning Utility

Toledo Edison Company
300 Madison Ave.
Toledo, OH 43652
Donald C. Shelton, V.Pres., Nuclear, Davis-Besse

Phone: (419)249-5000
Fax: (419)249-5165

Contact

Richard A. Kelly, Mgr., Toledo Pub.Aff.
Toledo Edison Company

Basic Facts

Units 2-3 *Status:* Cancelled. *Type of Reactor:* Pressurized light-water reactor. *Megawatts (electric):* 925. *Megawatts (thermal):* 2772. *Reactor System Supplier:* The Babcock and Wilcox Co. (United States). *Generator Supplier:* General Electric Co. (United States). *Architect Engineer:* Bechtel.

Key Dates

Units 2-3 *Ordered:* 1973. *Cancelled:* 1980.

Plant Profiles

Plant Report
See Plant Report for Davis-Besse Nuclear Power Plant, Unit 1.

---

269 •
Diablo Canyon Nuclear Power Plant, Units 1-2

PO Box 56
Avila Beach, CA 93424
Phone: (805)545-3100
Fax: (805)545-4985
John D. Townsend, V.Pres. and Plant Mgr.

Owning Utility

Pacific Gas and Electric Company
77 Beale St.
San Francisco, CA 94106
Phone: (415)973-7000
Gregory M. Rueger, Sr.V.Pres., Nuclear Power Generation

Contact

C. Brad Thomas, News Rep.
Diablo Canyon

Basic Facts

Units 1-2 *Status:* Operable. *Type of Reactor:* Pressurized light-water reactor. *Megawatts (electric):* 1125 (Unit 1); 1130 (Unit 2). *Megawatts (thermal)* 3338 (Unit 1); 3411 (Unit 2). *Reactor System Supplier:* Westinghouse Electric Corp. (United States). *Generator Supplier:* Westinghouse Electric Corp. (United States). *Architect Engineer:* Pacific Gas and Electric Company; Bechtel.

Key Dates

Unit 1 *Ordered:* 1967. *Construction began:* 1968. *Criticality:* April 1984. *First Power:* November 1984. *Commercial Operation:* May 1985. *Shut Down Expected:* 2025.

Unit 2 *Ordered:* 1968. *Construction began:* 1970. *Criticality:* August 1985. *First Power:* October 1985. *Commercial Operation:* March 1986. *Shut Down Expected:* 2026.

Operating Costs

Units 1-2
1985 $41,217,000
1986 $135,795,000
1987 $146,563,000
1988 $200,072,000
1989 $200,023,000

Plant Report

Background. The site of the Diablo Canyon nuclear power plant covers 735 acres of land 12 miles southwest of San Luis Obispo, a small city located midway between San Francisco and Los Angeles, California. The ambitious 2190 megawatt plant comprised of two units generated little more than controversy and enormous bills for its owners, Pacific Gas and Electric (PG&E), from the start of construction in the late 1960s until it went into commercial operation in the mid-1980s. The combination of an unusual rate settlement, however, and a consistently high level of productivity has since made the Diablo Canyon facility one of the top 10 in its class in the United States.

The construction history of Diablo Canyon is one of repeated delays and escalating costs due as much to evolving NRC regulations as to miscalculation and inexperience of the engineers and managers at PG&E. Construction began on the site in 1968, and PG&E originally projected a start-up date early in the 1970s, estimating the total cost of construction at $350 million. By 1981, after a series of delays and unexpected costs due to such occurences as having to replace miles of copper tubing when it was discovered that its runoff was killing marine life near the plant, the final cost of construction was estimated at $2.3 billion. In 1988, when the method for paying off the debt PG&E accumulated during the plant's construction was negotiated, the final estimate was more than $5 billion.

The discovery of Hosgri Fault and the beginning of trouble. PG&E began having difficulty meeting the ambitious schedule and budget set for its first nuclear power plant as soon as construction began. It became clear that building a nuclear facility was far more complicated than building a conventional power-generating plant. The plant's construction was subject to continual modification, due in part to the inexperience of PG&E's engineers who were trained for conventional generating plant design rather than for nuclear power. The problems with design eventually resulted in 30 volumes of Quick Fixes, modifications of the official design that are neither incorporated into official plant drawings nor analyzed for their cumulative effect on the reactor's safety. The problem was viewed by some as industry wide. NRC Commissioner Victor Gilinsky remarked: "Many utilities underestimated the inherent dangers and technical problems, and some did not have the managerial competence to handle large projects as sophisticated as nuclear plants."

One of the most far-reaching problems PG&E faced, however, was one that had little to do with plant management. For many, the location of the plant near the Hosgri Fault was the most significant defect in the utility's plan. Located approximately three miles from the Diablo Canyon site, the Hosgri Fault placed the stability of the plant in question in the event of an earthquake. In 1977, the NRC ordered PG&E to upgrade the plant's ability to withstand earthquakes to 7.5 on the Richter scale. The resulting cost was approximately $100 million and added three years to the construction schedule.

Although PG&E subsequently determined that it had met NRC requirements regarding the safety of the plant in the event of an earthquake, it remained a question in the minds of many for years afterwards. In 1981, thousands of demonstrators descended on the plant site, surrounding it by land and sea to protest the planned start-up of the reactor at Diablo Canyon, many naming the presence of the earthquake fault as the primary reason for their continued protests. Involving more than 3,000 activists, the 1981 demonstrations were the culmination of 15 years of opposition to the power plant by a coalition of protest groups called the Abalone Alliance. The involvement of such prominent entertainment figures as singer/musician Jackson Browne and actor Robert Blake ensured the attention of the national media. Most media sources described the demonstrations as spectacular but nonetheless a failure—as the expected low-level operating license was issued by the NRC despite the efforts of the Abalone Alliance. One of the protesters concluded: "The NRC has decided that people in this area are expendable."

However, in an equally spectacular debacle that occurred just as the fuel was being loaded into one of the reactors, it was discovered that the plant's engineers had read some blueprints backwards, placing pipes intended for Unit One in Unit Two. This embarrassing incident sparked an NRC investigation that uncovered a number of additional errors in design and calculation at the plant. In November 1981, the NRC voted unanimously to suspend the operating license for Diablo Canyon and ordered an independent audit of the plant's earthquake protection system.

Mistakes, reverses, and the Reagan effect. When Ronald Reagan became president of the United States in 1980, the nuclear power industry anxiously awaited results from campaign promises regarding deregulation and the importance of lessening the country's dependence on foreign oil—both issues that would potentially affect the troubled industry. During the 1970s, a number of plans for nuclear power plants had been cancelled, and the NRC had placed a ban on approving new plants until a plan for permanent disposal of high-level radioactive waste could be devised. And in 1979, in reaction to the accident at Three Mile Island, the NRC temporarily suspended licensing of all nuclear facilities while it upgraded and revised its standards for equipment and operating procedures. Upgrading Diablo Canyon to meet these new standards cost $70 million and set back the start-up of the plant another two years. Clearly, some help was needed from the federal government if nuclear power was going to obtain a substantial place among power sources in the United States.

The 1981 discovery that fixtures in vital areas of Diablo Canyon had literally been misplaced did not mark the end of Diablo Canyon's problems. It may be said, however, that it did mark the last time a serious deficiency in plant management was severely reprimanded by the NRC. As late as 1984, Diablo Canyon employees, along with inspectors from some of its supply companies, testified to, as one put it, "unreliable and illegal engineering practices" at the plant. In February 1984, 516 specific allegations concerning quality assurance breakdown at the facility were conveyed to the NRC on behalf of Diablo Canyon employees by an independent group dedicated to protecting whistleblowers called the Government Accountability Project (GAP). In May of the same year, Isa Yin, one of the NRC's own inspectors, concluded: "Unit 1 of the Diablo Canyon facility should not be permitted to go critical."

Nevertheless, Diablo Canyon was permitted to begin low-power testing that May. Within a matter of hours a stuck valve began channeling coolant water into a holding tank, but testing resumed after a short delay. Amid reports of on-site drug use by twenty-one Diablo Canyon employees, and sabotage of a coolant pipe leading to the reactor in one of the units attributed to an "insider," the plant began commercial operation in 1985, thirteen years behind schedule and at seventeen times the original estimated cost. President Reagan's commitment to nuclear power as the most viable alternative to foreign oil was finally paying off.

Who pays for Diablo Canyon?. The distribution of the debt accumulated by PG&E during construction of the Diablo Canyon facility has been viewed as a further indirect consequence of Reagan's support for nuclear energy. Typically, the state public utilities commission determines what percentage of the cost of building a power generating plant may be passed along to the utility's customers through raising its rates, and what percentage must be absorbed by the company's owners, its stockholders. In February, 1988 however, the Sacramento, California *Business Journal* reported that the California Public Utilities Commission, overwhelmingly appointed by Republican governor George Deukmejian, planned to reevaluate their regulatory procedures. The implication was that any change in utility regulation would be toward loosening regulatory strictures in an attempt to increase competitiveness and efficiency—a move in keeping with Reagan's policies. The decision regarding who would pay for the cost of building Diablo Canyon, its shareholders or its ratepayers, was predicted to re-set the industry norm across the country.

As soon as Diablo Canyon went into commercial operation, the debate over who should pay for the cost PG&E had incurred began. In 1987, the Division of Ratepayer Advocates, part of the California Public Utilities Commission, recommended that PG&E be allowed to recover only $1 billion of its costs by raising its rates to residential customers. The recommendation was clearly a criticism of PG&E's management of the plant's construction, and a company spokesperson called the commission's report "misleading, ill founded, and irresponsible." PG&E demanded public hearings to decide the issue, and meanwhile instituted an accounting change which reduced its 1987 earnings by $470 million, hoping to offset the damage a large negative settlement could have on its financial stability.

Instead of going into extended public hearings to determine where the fault lay for the enormous cost overruns incurred by Diablo Canyon, a deal was struck between the California Public Utilities Commission Division of Ratepayer Advocacy, the California Attorney General, and PG&E that determined, in effect, that ratepayers would pay only for the electricity generated at Diablo Canyon, rather than for the cost of building the plant itself. While original estimates of the amount PG&E would be forced to absorb ran to $2 billion, it was widely held that the decision favored the utility, which had successfully avoided public hearings debating the "reasonableness and prudence" of its management of the plant's construction. The $2 billion estimated disallowance represented a new record for amounts a utility was barred from earning back through rate increases, but it was nonetheless far lower than the amount recommended by ratepayer advocates.

The estimated amount the company would not be able to earn back from ratepayers in higher rates was based on the average efficiency of nuclear power plants, which at that time was 58 percent of capacity. However, since the plant had gone into operation in 1985, it had consistently operated at higher than average efficiency—65-70 percent of capacity—and the utility hoped that the amount of unrecovered debt would be much lower than $2 billion. The spokesperson for Mothers for Peace, based in San Luis Obispo, commented that the settlement created "a profit incentive for PG&E, not one of safety." However, the decision also created an independent safety committee to ensure that the utility did not cut back on safety measures or proper maintenance in order to improve its profitability.

John Curti, a California-based utilities analyst, noted that the decision was a consequence of the trend toward deregulation of the utility industry. By removing Diablo Canyon from the complex rate structures imposed on other power companies, he continued, "they're almost turning Diablo Canyon into an independent power producer."

The response from PG&E was clearly one of relief. "It gets this horrible uncertainty off our backs," remarked Richard A. Clarke, PG&E chief operating officer, in a public response to the announcement. Summarizing the effects of the agreement early in 1989, Clarke continued: "While the Diablo Canyon settlement resulted in adverse consequences for PG&E and its shareholders last year, it resolved at long last a complex and controversial issue which has clouded the company's financial situation. The settlement offers the prospect of substantial future earnings if the plant continues the outstanding performance it has achieved since it began operating, and if cost levels remain as expected."

PG&E looks toward the future. By 1988, when the rate settlement was reached, the plant had been ranked by the Institute of Nuclear Power Operations as one of the six best nuclear plants in the United States. The discovery later in 1988 that Diablo Canyon, along with numerous other power plants across the United States, had been duped into buying substandard equipment such as broken or old circuit breakers refurbished as new, was the cause of some consternation in the industry but did not otherwise affect the operation if the plant. Indeed, the plant's long history of embarrassing and potentially dangerous mishaps has apparently failed to impact negatively on its ability to produce energy for its customers or profits for the parent company.

Due to changes in the economy and increased awareness of environmental concerns on the part of its customers, PG&E announced in 1990 its intention to shift some of its resources into research on wind and solar generating facilities, and expanding the market for natural gas as a fuel for automobiles. A significant indicator of PG&E's changed status in the eyes of environmental activists as a result of this new commitment came in 1991 with unexpected praise from the Sierra Club, widely recognized as the most powerful environmental group in the United States. Michael Fischer, executive director of Sierra Club announced: "After Diablo Canyon, PG&E and Sierra Club became archenemies of the first order. But now we're looking at a PG&E that says 'Nuclear power plants, never again.' That's a PG&E we haven't seen before."

Sources

"When a Nuclear Plant Strangles in Red Tape." *U.S. News & World Report*, April 6, 1981.
"Showdown at Diablo Canyon." *Newsweek*, August 10, 1981.
"Defying Diablo." *Maclean's*, September 28, 1981.
"Diablo Canyon: The Assault that Failed." *Newsweek*, September 28, 1981.
"Diablo Canyon Only the Opening Gun." *U.S. News & World Report*, September 28, 1981.
"On D-Day at Diablo Canyon, It Was Jackson Browne If By Land and Robert Blake If By Sea." *People Weekly*, October 5, 1981.
"Radiation Sickness: America's Atomic Program Is Ailing, But Reagan Wants to Try to Cure It." *Time*, October 28, 1981.
"Diablo Canyon Loses Its License." *Newsweek*, November 30, 1981.
"Spotlight on Nuclear Safety." *Dun's Business Month*, January, 1982.
"Grassroots." *Critical Mass Bulletin*, January, 1984.
"Comment." *Critical Mass Bulletin*, May, 1984.
"Washington Watch." *Critical Mass Bulletin*, May, 1984.
"Sorry, Wrong Number." *The Progressive*, May, 1984.
"Testing and Protesting." *Time*, May, 1984.
"What Next?" *Critical Mass Bulletin*, September, 1984.
"Diablo Canyon Cost Report." *The New York Times*, May 15, 1987.
"PUC Likely to Ease Back and Re-Evaluate Its Role." *The Business Journal-Sacramento*, February 1, 1988.
"Diablo Canyon Pact Calls for PG&E to Pay Full Cost." *Los Angeles Times*, June 28, 1988.
"Diablo Pack Could Leave Utility Wary." *The New York Times*, June, 28, 1988.
"Rate Accord Saves Utility from Inquiry." *San Francisco Business Times*, July 4, 1988.
"Sham Circuit Breakers: They're Showing Up in Nuclear Plants." *Barron's*, August 8, 1988.
"Rate Plan Approved at Diablo Canyon Unit." *The New York Times*, December 20, 1988.
"PG&E Announces Financial Results." *Business Wire*, January 18, 1989.
"PG&E Is Ready for the 90s, Chairman Tells Shareholders." *Business Wire*, April 18, 1990.
"Michael Fischer Spurs Coalitions Between Sierra Club, Companies." *San Francisco Business Times*, April 19, 1991.
"Rate Plan Approved at Diablo Canyon Unit." *The New York Times*, December 20, 1988.
"PG&E Announces Financial Results." *Business Wire*, January 18, 1989.
"PG&E Is Ready for the 90s, Chairman Tells Shareholders." *Business Wire*, April 18, 1990.

—Mary Gillis

Douglas Point Nuclear Power Plant, Units 1-2

Nanjemoy, MD

Basic Facts

Units 1-2 *Status:* Cancelled. *Type of Reactor:* Boiling water reactor. *Megawatts (electric):* 1206. *Megawatts (thermal):* 3579. *Reactor System Supplier:* General Electric Co. (United States). *Owner:* Potomac Electric Power. *Operator:* Potomac Electric Power.

Key Dates

Units 1-2 *Ordered:* 1972. *Cancelled:* 1977.

Dresden

Dresden Nuclear Station, Unit 1

R.R. No. 1
Morris, IL 60450
Charles W. Schroeder, Stat.Mgr.
Phone: (815)942-2920

Owning Utility

Commonwealth Edison Company
One First National Plaza
PO Box 767
Chicago, IL 60690
Cordell Reed, Sr.V.Pres., Nuclear Operations
Phone: (312)294-4321
Fax: (312)294-2995

Contact

John F. Hogan, Dir., Commun. Svcs.
Commonwealth Edison Company

Basic Facts

Unit 1 *Status:* Shut down. *Type of Reactor:* Boiling water reactor. *Megawatts (electric):* 210. *Megawatts (thermal):* 700. *Reactor System Supplier:* General Electric Co. (United States). *Generator Supplier:* General Electric Co. (United States). *Architect Engineer:* Bechtel.

Key Dates

Unit 1 *Ordered:* 1955. *Construction began:* 1956. *Criticality:* October 1959. *First Power:* April 1960. *Commercial Operation:* July 1960. *Shut Down:* October 1978.

Operating Costs

Units 1-3

Year	Cost	Year	Cost
1968	$1,673,000	1979	$44,579,000
1969	$1,788,000	1980	$38,130,000
1970	$2,294,000	1981	$40,359,000
1971	$3,639,000	1982	$43,740,000
1972	$9,142,000	1983	$47,133,000
1973	$9,050,000	1984	$65,921,000
1974	$16,731,000	1985	$67,523,000
1975	$32,895,000	1986	$82,639,000
1976	$30,092,000	1987	$92,743,000
1977	$26,999,000	1988	$117,413,000
1978	$33,932,000	1989	$116,765,000

Plant Report

Power for Illinois. Dresden Nuclear Station occupies a 2000-acre site near Morris, Illinois. The site is located at the confluence of the Kankakee and Des Plaines rivers, about 50 miles southwest of Chicago.

Dresden Nuclear Station houses three boiling water reactors, manufactured by General Electric. Unit 1 generated 207 megawatts of electricity until it was retired in August 1984. Units 2 and 3 generate 794 megawatts each.

Unit 1: First in the United States. Unit 1 was the first privately-financed nuclear generating station built in the United States. The plant was among five projects undertaken in 1957, in response to the first round of the Power Demonstration Reactor Programme. The PDRP was devised by Congress and the United States Atomic Energy Commission to encourage the construction of several nuclear power stations. Along with another plant, Indian Point, Unit 1 was undertaken by its utility without direct government financial involvement.

Facts and features of Unit 1. The initial project cost of Unit 1 was $36.1 million, fixed price. The plant's first generation of electricity to Illinois was in August 1960. Along with two other plants, Shippingport and Yankee, Unit 1 took part in accounting for virtually all of the nuclear power station capacity in operation in the United States at the end of 1961.

The pressure vessel for Unit 1 was 41 feet high and 12 feet 2 inches in diameter, with walls about 5 1/2 inches thick. The cylinder heads were 9 inches thick and the design pressure was 1,250 pounds per square inch gauge.

A call for the decommissioning of Unit 1. In 1978, Unit 1 was shut down when radioactive sediment was found in its pipes. In December 1983, the Public Citizen's Critical Mass Energy Project and a coalition of energy, labor, and consumer groups representing about 500,000 ratepayers called for Commonwealth Edison to decommission Unit 1. At the time, the utility was asking for a $1 billion rate increase to build five new plants. The coalition felt that the decommissioning of Unit 1 should have taken precedence.

Up to 1983, there had not been any major plants decommissioned in the United States, so it would have been difficult for Commonwealth Edison to estimate the cost of decommissioning Unit 1. The groups felt that Commonwealth Edison's estimates were far too low and could have left the utility, which planned on having 12 plants on-line by 1986, billions of dollars short of the amount needed to decommission these plants.

As of 1990, no plant, including Unit 1, had been decommissioned. Only small experimental plants had undergone the process, and true costs were not available. The earliest date Commonwealth Edison could decommission would be 2004, starting with the three Dresden plants.

Construction costs of Units 2 and 3. The initial cost total for Dresden 2 and 3 was $464 million, fixed price. Unit 3's operating license is up for renewal in 2009. As of June 1991, the NRC was considering a 20-year renewal.

Features of Units 2 and 3. The pressure vessels for Units 2 and 3 are 68 feet 7 5/8 inches high and 20 feet 11 inches in diameter. The walls have a thickness of 6 7/16 inches, with a 1/8 inch stainless steel overlay. The cylinder heads are 8 7/16 inches thick, with a 1/8 inch stainless steel overlay. The design pressure is 1,250 pounds per square inch.

Units 2 and 3 contain 724 fuel assemblies, 114 inches long each. These assemblies represent 140 tons of uranium at two percent enrichment in Unit 2, and 138 tons at two percent enrichment in Unit 3. Between one-third and one-quarter of the total core is replaced approximately every 18 months during scheduled outages.

The turbine generator is driven by steam at 1,800 revolutions per minute. Generators produce electricity at 1,800 volts, stepped up by transformer to 345,000 volts for transmission.

Plant Profiles   Dresden | 272

The condenser cooling system is lake. The total pond surface area is 1,280 acres, with an average depth of 10 feet, varying between three and 20 feet. In the spray canals, the supply canal is 11,292 feet long, 57 feet wide, and 10 feet deep. The return canal is 13,000 feet long, 57 feet wide, and 10 feet deep. Diffusers blow down 50,000 gallons of hot water per minute, and at the same time, 68,000 gallons of cool water is taken from the lake. The 18,000 gallon difference is due to seepage and evaporation during the cooling process.

Problems for Units 2 and 3. In January 1990, the NRC assigned augmented inspection teams to determine facts regarding the loss of a reserve auxiliary transformer at Unit 2. The incident resulted in a loss of off-site power. The teams also investigated an incident of inoperable high-pressure coolant injection systems at Unit 3, which had occurred in October 1989.

Also in 1990, the AEOD initiated a program to enhance the NRC staff's understanding of factors affecting human performance during reactor events. Under the program, teams of staff, contractors, and specialists performed a study of an event which occurred at Unit 2 in August. While the plant was at 87 percent power, operators manually scrammed it, after trying unsuccessfully to shut a Target Rock relief that failed open. During the next hour, the plant cooldown rate exceeded the technical specification normal cooldown rate limit. The findings of the team concerned performance, procedures, communications, and training.

In February 1991, the NRC assessed a $27,000 fine to Commonwealth Edison regarding a containment air sampling cooling system used at Units 2 and 3 on a daily basis from 1987 to 1990. The NRC said that the system could have created a radioactive release pathway to the environment because it was not geared to shut off automatically in the event of an accident. A plant employee discovered the condition in June 1990 while reviewing a change in in the sampling procedure. New, acceptable systems were installed at both units, and the utility paid the fine.

Report on Unit 2. In December 1991, the Advisory Committee on Reactor Safeguards (ACRS) issued a report on Commonwealth Edison's request for a full-term operating license for Unit 2. The unit had been under a provisional operating license for the last 21 years. The ACRS believed that there was reasonable assurance of continued safe operations at the plant.

Sources

"Dresden: Pulling the Plug." *Critical Mass Bulletin*, December 1983.
Dresden Nuclear Station Fact Sheet. Commonwealth Edison Co., May 1988.
U.S. NRC Annual Report 1990. U.S. Nuclear Regulatory Commission, n.d.
"The Cost of Closing Nukes." *Crains Chicago Business*, May 7, 1990.
Mounfield, Peter. *World Nuclear Power*. Routledge, 1991.
"ACRS Favors Full-term OLs for Palisades, Dresden 2." *Nuclear News*, February 1991.
"Four Fines Proposed, One Imposed, One Fought." *Nuclear News*, February 1991.

"Due Up for License Renewal: The Future of Nuclear Power." *The New York Times*, June 24, 1991.

—*Charles Ceplecha*

272 ●

Dresden Nuclear Station, Units 2-3

R.R. No. 1
Morris, IL 60450
Charles W. Schroeder, Stat.Mgr.
Phone: (815)942-2920

Owning Utility

Commonwealth Edison Company
One First National Plaza
PO Box 767
Chicago, IL 60690
Cordell Reed, Sr.V.Pres., Nuclear Operations
Phone: (312)294-4321
Fax: (312)294-2995

Contact

John F. Hogan, Dir., Commun. Svcs.
Commonwealth Edison Company

Basic Facts

Units 2-3 *Status:* Operable. *Type of Reactor:* Boiling water reactor. *Megawatts (electric):* 828. *Megawatts (thermal):* 2527. *Reactor System Supplier:* General Electric Co. (United States). *Generator Supplier:* General Electric Co. (United States). *Architect Engineer:* Sargent & Lundy Engineers (United States).

Key Dates

Unit 2 *Ordered:* 1965. *Construction began:* 1966. *Criticality:* January 1970. *First Power:* April 1970. *Commercial Operation:* June 1970.

Unit 3 *Ordered:* 1965. *Construction began:* 1966. *Criticality:* January 1971. *First Power:* July 1971. *Commercial Operation:* November 1971.

Operating Costs
Units 1-3

Year	Cost	Year	Cost
1968	$1,673,000	1979	$44,579,000
1969	$1,788,000	1980	$38,130,000
1970	$2,294,000	1981	$40,359,000
1971	$3,639,000	1982	$43,740,000
1972	$9,142,000	1983	$47,133,000
1973	$9,050,000	1984	$65,921,000
1974	$16,731,000	1985	$67,523,000
1975	$32,895,000	1986	$82,639,000
1976	$30,092,000	1987	$92,743,000
1977	$26,999,000	1988	$117,413,000
1978	$33,932,000	1989	$116,765,000

Plant Report

See Plant Report for Dresden Nuclear Station, Unit 1.

● Operable   ○ Under Construction   △ Indefinitely Deferred   ▲ Decommissioning   □ Planned   ■ On Order   ☆ Cancelled   ★ Shut Down

Duane Arnold Energy Center (DAEC)

3277 DAEC Rd.
Palo, IA 52324
David L. Wilson, Plant Supt., Nuclear
Phone: (319)851-7611
Fax: (319)851-7323

Owning Utility

Iowa Electric Light & Power Company
PO Box 351
Cedar Rapids, IA 52406
John F. Franz Jr., V.Pres., Nuclear
Phone: (319)398-4411
Fax: (319)398-8192

Contact

Colleen R. Dykes, Mgr., Corp. Commun.
Iowa Electric Light & Power Company

Basic Facts

Status: Operable. *Type of Reactor:* Boiling water reactor. *Megawatts (electric):* 565. *Megawatts (thermal):* 1658. *Reactor System Supplier:* General Electric Co. (United States). *Generator Supplier:* General Electric Co. (United States). *Architect Engineer:* Bechtel. *Owner:* Iowa Electric Light & Power; Central Iowa Power Cooperative; Corn Belt Power Cooperative. *Operator:* Iowa Electric Light & Power.

Key Dates

Ordered: 1968. *Construction began:* 1970. *Criticality:* March 1974. *First Power:* May 1974. *Commercial Operation:* February 1975. *Shut Down Expected:* June 2010.

Operating Costs

1974	$1,485,000	1982	$29,238,000
1975	$3,839,000	1983	$45,949,000
1976	$7,050,000	1984	$34,587,000
1977	$10,727,000	1985	$56,023,000
1978	$11,916,000	1986	$35,463,000
1979	$9,530,000	1987	$51,792,000
1980	$18,397,000	1988	$54,279,000
1981	$21,957,000	1989	$38,854,000

Plant Report

Site and ownership. The Duane Arnold Energy Center (DAEC), a 530-megawatt boiling water reactor, is Iowa's only nuclear-fueled electric power plant. It is located on the Cedar River about 10 miles northwest of downtown Cedar Rapids. Iowa Electric Light and Power Company (IE), based in Cedar Rapids, serves as operator and owns 70 percent of the plant. Two rural cooperatives, Central Iowa Power Cooperative of Marion (20 percent) and Corn Belt Power Cooperative of Humboldt (10 percent) own the remainder of the plant.

Operation. In capacity performance, DAEC is fourth in the world among General Electric boiling water reactors. Capacity factor is the ratio of a plant's actual output in relation to its potential full power output during a year. A refueling outage had a negative impact on DAEC's production during 1990, resulting in a capacity factor of 67 percent. The decrease in nuclear generation led to increased fuel costs for IE's fossil-fueled operations and increases in costs associated with purchasing replacement power.

In 1991, however, DAEC did not experience a refueling outage and its capacity factor jumped to a record 91.8 percent. The 1991 annual generator record also surpassed previous annual electric generator records. By the end of 1991, four million megawatt hours had been generated, a half million more than the previous record.

In 1991, the Institution of Nuclear Power Operation's (INPO) chemical performance index indicated that DAEC's index rating was the best among 114 plants examined. INPO found that DAEC's primary system water chemistry was the best in the United States. In addition, DAEC has consistently performed in the lowest quartile of radwaste volume among world nuclear operators. Since 1979, the plant has not released any radioactive water.

Safety and training. DAEC performance in safety ranks within the nation's top quartile of plants in operation. In June 1992, DAEC surpassed two million man hours without a time lost accident.

DAEC created a control room simulator and a training facility to provide instruction to its personnel. The plant maintains 11 programs accredited by the National Academy of Nuclear Training, which offer its employees on-going instruction. On- and off-site programs offer workers the opportunity to pursue degrees and training in their respective fields.

In addition to educational efforts, several plant renovations led to changes which improved operations. The modifications included the elimination of some single point main turbine trips, which eliminated the impact of switch failures. Also DAEC recently upgraded internal and external electrical power supplies and added on- and off-site electric distribution sites.

Pleasant Creek Reservoir. The Cedar River provides cooling water for DAEC. About 11,000 gallons of river water are required per minute, much of which is evaporated during the cooling process. About 4,000 gallons per minute flow back into the river. A reserve water facility was created in the event of inadequate river flow. If water flow in the Cedar River falls below a designated rate, water drawn by DAEC would be replaced from the Pleasant Creek Reservoir.

The Pleasant Creek Reservoir forms a 410-acre lake and is part of a 1,900-acre recreation facility. The Iowa Department of Natural Resources operates the reservoir and adjoining Pleasant Creek Park where summer-time visitors can swim, sail, wind surf, fish, and camp. In the winter the facility is open to cross-country skiing and snow mobiling.

The Pleasant Creek facility also contains 30 acres of newly-seeded prairie. The DNR and employees of Linn County initiated the program in 1990 in an attempt to replicate an old Iowa prairie. Over 60 species of plants, including various flowers, have been planted in order to promote a unique visual and environmentally conscious attraction.

Decommissioning. DAEC has an estimated operating life span of 36 years. IE plans to restore the property to its original state when DAEC is decommissioned. IE's share of the cost is estimated at $194 million (in 1991 dollars). In order to provide for

Plant Profiles

eventual decommissioning, funds are collected through rates and held in external trust funds and in an internal decommissioning reserve.

Sources

"Balance of Strenghts: IES Industries, Inc." *IES Industries, Inc.*, IES Industries, Inc., n.d.

Corporate Communications Office. Iowa Electric Light and Power Company. Cedar Rapids, Iowa, n.d.

—Karen Bellenir

274 ▲
Elk River Nuclear Power Plant
Elk River, MN

Basic Facts
Status: Decommissioned. *Type of Reactor:* Boiling water reactor. *Megawatts (electric):* 23. *Megawatts (thermal):* 58.2. *Reactor System Supplier:* Allis-Chalmers (United States); Sargent & Lundy Engineers (United States). *Generator Supplier:* Elliot. *Architect Engineer:* Sargent & Lundy Engineers. *Owner:* U.S. Department of Energy. *Operator:* Rural Coooperative Power Association.

Key Dates
Ordered: 1958. Construction began: 1959. Criticality: November 1962. First Power: August 1963. Commercial Operation: July 1964. Shut Down: February 1968.

Plant Report

A project of the Power Demonstration Reactor Programme (PDRP). Realizing that utilties could not justify to their shareholders the investment costs required to build nuclear power plants, the Atomic Energy Commission and Congress devised the Power Demonstration Reactor Programme in 1955 to help launch the commercial nuclear industry in the United States. The PDRP offered cooperating utilities with financial incentives such as research and development assistance, financing of the reactor system, and a waiver of normal fuel charges during the first five years of plant operation.

The small plant at Elk River, Minnesota, was initiated under the PDRP, shut down in 1968, and subsequently dismantled.

Sources
Mounfield, Peter. *World Nuclear Power.* Routledge, 1991.

275 ●
Joseph M. Farley Nuclear Power Station, Units 1-2

P.O. Drawer 470
Ashford, AL 36312
Dave Morey, Plant Mgr.

Phone: (205)899-5156
Fax: (205)899-6049

Owning Utility
Southern Nuclear Operating Company
PO Box 1295
Birmingham, AL 35201
R. Patrick McDonald, Exec.V.Pres., Nuclear

Phone: (205)868-5000
Fax: (205)870-6194

Contact
Ed Crosby, Mgr., Pub.Aff.
Southern Nuclear Operating Company

Basic Facts
Units 1-2 *Status:* Operable. *Type of Reactor:* Pressurized light-water reactor. *Megawatts (electric):* 860. *Megawatts (thermal):* 2652. *Reactor System Supplier:* Westinghouse Electric Corp. (United States). *Generator Supplier:* Westinghouse Electric Corp. *Architect Engineer:* Alabama Power Company; Bechtel. *Owner:* Alabama Power Company. *Operator:* Southern Nuclear Operating Company.

Key Dates
Unit 1 Ordered: 1969. Construction began: 1970. Criticality: August 1977. First Power: August 1977. Commercial Operation: December 1977.

Unit 2 Ordered: 1970. Construction began: 1970. Criticality: May 1981. First Power: May 1981. Commercial Operation: July 1981.

Operating Costs
Units 1-2

1977	$462,000	1984	$76,822,000
1978	$12,207,000	1985	$97,054,000
1979	$22,542,000	1986	$106,148,000
1980	$25,735,000	1987	$122,146,000
1981	$41,426,000	1988	$118,717,000
1982	$52,489,000	1989	$131,233,000
1983	$60,275,000		

Plant Report

Background. The Joseph M. Farley twin nuclear power plant station located near Dothan, Alabama, is owned and operated by Alabama Power, a subsidiary of Southern Company. The Nuclear Regulatory Commission (NRC) has approved Southern Company to transfer daily operations and control of Farley to its new subsidiary, Southern Nuclear Operating Company (SONOPCO), in order to bring the nuclear operations of Southern Company under a single unit.

Named for the former president of Alabama Power, who is the current chair of the American Nuclear Energy Council and the executive vice president for nuclear at Southern Company, the plants at Farley have turned in consistently high efficiency ratings. In 1989, Farley 2 operated for 453 consecutive days whereby 365 days is considered outstanding, and in 1991, it reached a capacity utilization factor of 92.8 percent. In addition, Farley 1 has also reported efficient ratings, operating at 95 percent of capacity in 1980.

Sources
"Southern Company President Addresses the Company's 44th Annual Meeting of Stockholders." *Business Wire*, May 23, 1990.

"Southern Announces Approval From Securities and Exchange Commission to Form a New Subsidiary." *Business Wire*, December 18, 1990.

"Power Briefs." *Nuclear News*, February 1991.
"Another Rave Review for 1990." *Nuclear Industry*, Fourth Quarter 1991.
"St. Lucie 2 Completes Record Run, Leads World Efficiency." *INFO*, April 1992.

—Yvette M. Costa

Enrico Fermi 1 Fast Breeder Reactor Project

6400 N. Dixie Highway
Newport, MI 48166
Phone: (313)586-5300
Telex: 313-586-4530

Owning Utility

Detroit Edison
2000 Second Ave.
Detroit, MI 48226
Phone: (313)237-8000
Fax: (313)237-8055

Contact

Robert M. Vergiels, Media Rep.
Detroit Edison Public Affairs Office
Phone: (313)237-8850

Basic Facts

Status: Decommissioned. *Type of Reactor:* Liquid metal fast breeder reactor. *Megawatts (electric):* 65. *Megawatts (thermal):* 200. *Reactor System Supplier:* Combustion Engineering, Inc. (United States). *Operator:* Power Reactor Development Company.

Key Dates

Ordered: 1955. *Construction began:* 1956. *Criticality:* 1963. *First Power:* 1966. *Commercial Operation:* 1966. *Shut Down:* 1972. *Decommissioned:* December 31, 1975.

Plant Report

Reprint. Much of the following essay was reprinted from *Enrico Fermi 1 Fast Breeder Reactor Project: History, Contributions, Significance* with permission from the Detroit Edison Public Affairs Office. Source information given below.

It started with Enrico Fermi. In 1945 Enrico Fermi, the nuclear physicist who first achieved a controlled nuclear reaction, predicted that the country which first developed a breeder reactor would have a great competitive advantage in atomic energy. His premonition was shared by others in the U.S. government.

In the early 1950's the United States Atomic Energy Commission (AEC) invited industry groups to look into the possibility of using nuclear fuels for the economic generation of electric power. (The AEC invitation began to fracture a national tradition which saw the nuclear field as strictly a governmental matter.) One such industry advisory group which responded to the AEC request consisted of Detroit Edison and Dow Chemical Company. James W. Parker, president of Detroit Edison, served as chairman; Walker Cisler, an AEC official and ardent exponent of industry participation in the nuclear field, served as Secretary.

After intensive study, the Dow-Edison Industry Advisory Group proposed the development of a fast breeder reactor—a long-range goal but one nevertheless that should be started by industry as soon as possible. Walker Cisler began bringing together a group of private utilities, manufacturers, and research organizations to support the construction of the reactor. He also began working with legislators and other leaders in the federal government for a revision of the Atomic Energy Act of 1948, which prohibited private companies to own nuclear facilities or nuclear fuels.

Official announcement in Geneva. In 1954 the Atomic Energy Act was revised to permit private ownership of nuclear facilities, but not of nuclear fuels, which for another 10 years were solely owned by the AEC. In January 1955, the AEC invited companies to submit proposals for nuclear power plants; four proposals were submitted by April of that year. In March 1955, a nonprofit research organization, Atomic Power Development Associates (APDA), was formally organized for the development of a fast breeder reactor. On August 8, 1955, at the First International Conference on the Peaceful Uses of Atomic Energy, in Geneva, official announcement was made of the Fermi 1 Project. Later that month, another nonprofit organization, Power Reactor Development Company (PRDC), was formed to design, build, and operate Fermi 1. Laura Fermi, wife of the late Dr. Enrico Fermi, personally gave Walker Cisler permission to use her husband's name for the plant. Cisler was made President of both APDA and PRDC.

A year later ground was broken for the world's first commercial size demonstration breeder reactor to operate as a power plant in an electric utility system. The plant, to be cooled by sodium and operated essentially at atmospheric pressure, was designed for a maximum capability of 430 thermal megawatts (MWt)—although the initial fuel loading was designed for and limited by the operating license to a maximum of 200 MWt.

Protest. On August 4, 1956, the AEC issued PRDC a construction permit on a provisional basis. Work began immediately. However, on August 31, several labor unions filed a Petition to Intervene, contending that the requisite showing of safety and financial qualifications had not been made. The case went through the Commission and the courts as construction work slowly proceeded, and finally reached the Supreme Court of the United States. On June 12, 1961, the Supreme Court issued a seven-to-two majority opinion sustaining the AEC and PRDC on every point presented, ruling that the safety and site findings which the Commission had made were adequate in all particulars and complied fully with the requirement of the statute and the regulations.

The construction of Fermi 1 seemed to enjoy general support from the public. At no time during the long construction permit hearings did Fermi 1 experience any organized or individual opposition other than the formal intervenors.

Crucial non-nuclear tests. On December 1, 1960, Fermi 1, the world's largest liquid-metal-cooled reactor system at the time, was filled to its operating level with 345,000 pounds of sodium. Early in 1961, all major components, monitoring devices, and instrumentation were successfully operated in sodium at the refueling temperature of 500° F; no evidence of corrosion of leaks was found. Here, a major technological step forward was taken: Fermi 1 confirmed that operation of equipment in a sodium environment presented no major difficulties.

In May of 1961, one of the most severe tests on the reactor and primary coolant system was performed: the system temperature was raised to 1000° F for a period of approximately one week. This temperature was well in excess of maximum anticipated operation condition. During this important period, the primary sodium pumps, control rods, safety rods, safety rod extensions, and all other reactor mechanisms and instrumentation operated satisfactorily. However, the high-temperature condition caused extensive deterioration of the graphite shielding material external to the reactor vessel. The graphite was replaced with improved quality material. Also, an inert, nitrogen atmosphere and monitoring system were provided as additional protection for the graphite shield.

The non-nuclear testing program, which continued for a couple years, was surrounded with both successes and problems. Troubles encountered during this period affected the hold down mechanism, hold down fingers, bent dummy core subassemblies, and lower core support plate (erosion damage)—all of which were costly in terms of schedule. However, solutions that were developed have contributed significantly to the advancement of Liquid Metal Fast Breeder Reactor (LMFBR) technology.

Criticality. During July and August 1963, loading to criticality progressed smmoothly with the exception of some mechanical difficulties with the cask car. On August 23, 1963, the reactor was made critical and the low-power nuclear tests were initiated and performed at increasing power levels up to 1 MWt until December 1965. The low power tests established the reactor characteristics and the soundness of the Fermi 1 design. No instabilities were expected at higher power.

Reactor operators carried out high-power tests during the first nine months of 1966 and completed the program through 100 MWt operation, including a sustained 60-hour test run that generated more than a million kilowatt hours of electricity for the Detroit Edison system. With the exception of the steam generators, which required many repairs, all systems and components performed as expected.

Fuel melting incident—a set back. On October 5, 1966, during a slow increase in power to a planned level of 74 MWt, a fuel melting incident occurred in portions of two fuel assemblies. The reactor was shut down from a power level of 31 MWt. From monitoring the on-site radiation levels, it was determined that the incident had not caused any public safety hazard. Consequences of the incident were well within the safety envelope provided in the plant design.

A year of investigation was required to determine that the melting was caused by a zirconium segment that had become detached from the conical flow guide beneath the core and was forced against the lower support plate by the coolant flow resulting in partial flow blockage of four subassemblies. Almost four years later the reactor was again taken to criticality following the removal of all six zirconium segments, addition of flow blockage devices, installation of computerized plant monitoring capabilities to detect abnormal conditions that precede fuel melting, and a careful reassessment by the AEC. Fermi 1 was relicensed on February 10, 1970.

Licensed power of 200 MWt was first achieved on October 16, 1970 and was followed by a series of varying power runs through 1971, all with good results.

Lack of funds lead to decommissioning. The reactor fuel was now just short of its allowable burn-up. To continue significant power operation of the Fermi 1 reactor would require another core loading. PRDC developed a 6-year R&D program centered around an advanced type of fuel loading composed of uranium oxide. The cost of this program was about $50 million. Some $30 million had been promised or pledged by private utility systems in the U.S. and by groups in Japan and Western Europe, but further efforts to obtain the remainder of the financial support were not fruitful.

On November 16, 1972, the PRDC executive committee decided to decommission the reactor. Three years later the decommissioning was complete. Nuclear components of Fermi 1 are in safe-store condition: fuel, primary coolant, and many components have been removed. The unit's turbine-generator was converted and served a 175-MW oil-fired peaker until closing in 1978.

Major contributions of Fermi 1. In the history of technology, Fermi 1 is one of the first major efforts to commercially harness the vast potential energy resources in uranium. From 1963 to 1966 and from 1970 to 1972, it was the largest operating LMFBR in the world.

Fermi 1 set precedence in many areas of reactor technology. Before Fermi 1 there was no prior experience in the design of large LMFBR power plants; Fermi established many fundamentals of reactor design that were foundational in the development of the next generation of nuclear power plants around the world. Fermi 1 also pioneered in areas of nuclear and plant test experience, operating experience, component and system experience, maintenance of large sodium systems, the prediction of nuclear characteristics, and guidelines for decommissioning.

One of the principal purposes behind the Fermi 1 Project was the training of personnel along with a strong spirit of international cooperation. More than 250 U.S. and foreign personnel obtained practical experience in nuclear power plants through the Fermi 1 project. The technological background invested in these people were applied to LMFBR projects around the world, including the Rapsodie and Phenix projects of France; the PEC project in Italy; SNR 300 project in West Germany, Belgium and Holland; the Joyo and Monju projects in Japan; as well as the Fast Flux Test Facility (FFTF) and other demonstration plant projects in the U.S.

An industry-wide cooperative of the scale of the Fermi 1 project was the first of its kind for the utility industry. Some 23 utility and industrial companies initially committed $23 million to the project, paving the way for the unified action by the utility industry of pooling resources and coordinating efforts for the accomplishment of major projects.

The series of court cases stemming from the labor union protest of the construction of Fermi 1 established many legal precedents that are still recognized today. The U.S. Supreme Court's historic decision in 1961, sustaining AEC and PRDC on every point of contention, paved the way for future nuclear power development by private industry.

Sources

Meyers, James E., and Earl M. Page. *Enrico Fermi 1 Fast Breeder Reactor Project: History, Contributions, Significance.* Power Reactor Development Company, November 1975.

"1991 Electric Utility of the Year: Detroit Edison Delivers Competitive Culture Changes." *Electric Light & Power.* PenWell Publishing Company, November 1991.

277 •
Enrico Fermi 2 Nuclear Power Plant

6400 N. Dixie Highway Phone: (313)586-5300
Newport, MI 48166 Telex: 313-586-4530
R. McKeon, Plant Mgr.

Owning Utility

Detroit Edison Phone: (313)237-8000
2000 Second Ave. Fax: (313)237-8055
Detroit, MI 48226
William S. Orser, Sr.V.Pres., Nuclear Generation

Contact

Robert M. Vergiels, Media Rep. Phone: (313)237-8850
Detroit Edison Public Affairs Office

Basic Facts

Status: Operable. *Type of Reactor:* Boiling water reactor. *Megawatts (electric):* 1154. *Megawatts (thermal):* 3292. *Reactor System Supplier:* General Electric Co. (United States). *Generator Supplier:* General Electric Co. (United Kingdom). *Architect Engineer:* Detroit Edison.

Key Dates

Ordered: July 1968. *Construction began:* July 1969. *Criticality:* June 1985. *First Power:* September 1986. *Commercial Operation:* January 1988.

Operating Costs

1988 $138,804,000
1989 $153,727,000

Plant Report

A major electricity generator for Southeastern Michigan. Fermi 2 is located on a 1,120-acre site along the western shore of Lake Erie in Frenchtown Township, Monroe County, Michigan. The site is 30 miles southwest of Detroit and eight miles northeast of the city of Monroe.

The 93rd nuclear power plant to be licensed in the U.S., Fermi 2 produces enough electricity to serve a city of about one million people. It plays a significant generating role among Detroit Edison's plants that supply some 1,927,466 customers in southeastern Michigan. As of November 1991, Detroit Edison's generation was 88 percent coal-fired, 11 percent nuclear, and 1 percent gas- and oil-fired.

Fermi 2 shares the site with Fermi 1, an historic developmental fast breeder reactor (see separate entry). The nuclear portion of Fermi 1 was decommissioned in 1975; the unit's turbine generator was converted and served a 175 megawatts oil-fired peaker until closing in 1978.

Construction: Why did it take so long and cost so much?. The road to commercial operation was long and costly for Fermi 2. Detroit Edison announced plans for construction in July 1968, anticipating the plant to be in service by July 1974. However, Fermi 2's first power was not generated until 1986, and the plant finally began commercial operation in 1988.

The construction budget for Fermi 2 was announced in 1969 as $228.8 million. But by the late 1970s, the project's cost was estimated at $1.3 billion. On Dec. 31, 1987, Fermi 2's total expense was figured at $4.565 billion.

Construction delays and cost increases were due in part to major changes in federal regulations for nuclear plants, particularly following the 1979 incident at Three Mile Island in Pennsylvania. For example, from 1972 to 1983 nearly 2,200 regulations affecting nuclear plants under construction were issued by the U.S. Nuclear Regulatory Commission (NRC). In many cases, completed work had to be redone under the new regulations. Other factors included inflation, which increased the cost of materials, labor, and borrowing money, and the economic recession of the mid-1970s, which prompted Detroit Edison to stop construction of all power plants for three years.

A chronological list of the major events causing Fermi 2's construction delays follows:

- February 1969. Construction budget announced as $228.8 million.

- June 1969. Site preparation started. Expected construction time: five years. Plant to be in service by February 1974.

- November 1970. Announced commitment to build cooling towers at added cost of $24 million to avoid heating up Lake Erie (four years before the U.S. Environmental Protection Agency made closed cooling systems mandatory).

- July 1971. Fermi 2 environmental status adversely affected with U.S. Court of Appeals ruling that Atomic Energy Commission (AEC) (now the Nuclear Regulatory Commission) failed to satisfy environmental requirements for Calvert Cliffs nuclear power plant, owned by Baltimore Gas and Electric Co. Fermi had to report on environmental consequences of all alternatives in plans; site preparation had to be partially suspended until AEC could determine what additional environmental review would be required.

- April 1972. Nuclear commitment by Detroit Edison enhanced with announcement of three more nuclear units: Fermi 3 and Greenwood 2 and 3, all with financial commitments limited to engineering and licensing work.

- September 26, 1972. Construction permit for Fermi 2 issued by AEC, 41 months after it was filed. Target date for commercial operation reset for April 1977.

Plant Profiles — Fermi

- November 1974. Construction halted at Fermi 2, except for cooling towers, for financial reasons. Halt would last for 28 months.
- June 1975. Announced cancellation of Fermi 3.
- February 1977. Northern Michigan Electric Cooperative, Inc. and Wolverine Electric Cooperative, Inc. together purchased a 20 percent ownership interest in Fermi 2.
- August 1977. Nuclear Regulatory Commission amended Fermi 2 construction permit, delaying completion date to January 1, 1982.
- December 1977. Construction pace increased rapidly as work force reached 1,700. Plant 57 percent complete.
- March 1979. Three Mile Island accident.
- April 1979. Detroit Edison formed its own 25-member Fermi 2 Safety Review Task Force as a result of Three Mile Island accident.
- July 1979. Management received NRC Task Force's short-term recommendations for changes as a result of Three Mile Island accident. NRC schedule required implementation of those changes before operating license is issued.
- August 1979. Fermi 2 Safety Review Task Force issued its recommendations, as a result of Three Mile Island accident, for immediate implementation.
- August 1979. NRC Task Force issued its long-term recommendations, as a result of Three Mile Island accident, requiring additional work that would delay completion of construction into mid-1980.
- December 1979. Commercial operation delayed to March 1982 because of regulatory uncertainty and at least 15 months of additional work brought on by the Three Mile Island event. "A blizzard of new regulations and (safety) requirements followed (TMI)" from the NRC. The Fermi 2 Safety Review Task Force issued more than 100 recommendations for procedural changes, engineering reviews, and design modifications—all costly adjustments.
- June 1981. Commercial operation delayed to November 1983 because of new safety requirements and cost of additional work.
- May 1983. Delay in commercial operation extended another seven months, delaying Fermi 2 to June 1984.

Features of Fermi 2. The construction of Fermi 2 was an enormous undertaking. A look at some of the plant components provides further insight:

- Fermi 2 contains 300,000 cubic yards of concrete, 20,000 tons of steel, 1,220 miles of electric wire, and 70 miles of conduit.
- Fermi 2's water reservoir can store about 30 million gallons.
- Fuel costs for Fermi 2 are about half those of the most efficient coal-fired plants.
- Each of Fermi 2's 400-foot cooling towers has the capacity to pump 450,000 gallons of water per minute.
- The reactor vessel is 73 feet high and 22 feet wide, weighs 769 tons, and has a 6-inch steel shell.
- The reactor's steam output is 14 million pounds per hour at 545 degrees fahrenheit and 1,000 pounds per square inch.
- The reactor contains 185 control rods and 764 zirconium alloy assemblies holding 17 million uranium pellets.
- The reactor's primary containment is a one-inch thick, 115-foot high, steel drywell with an outside backing of between 18 inches and eight feet of concrete for structural support and radiation shielding.
- The 68-foot drywell base is surrounded by a large doughnut-shaped torus, or wetwell; should the reactor rupture, any escaping steam would vent to the torus.
- With walls 4-6 feet thick, the reactor building is constructed to withstand severe earthquakes and tornado winds up to 300 miles per hour; it is operated at a negative air pressure to help confine radioactive particles within the building.
- The electricity produced by one fuel pellet, roughly the size of a pencil eraser, is equal to the amount produced by 4-1/2 barrels of oil or one ton of coal.
- Fermi 2 operates on uranium oxide-enriched U-235 fuel and consumes about one ton of uranium annually.

A rough start in operations. As was its construction history, Fermi 2's operational beginnings were delayed due to problems. In March 1985, Fermi 2 received its NRC license for low-power, start-up testing (5 percent of plant capacity). On July 1, 1985, a control room operator caused a premature nuclear chain reaction by withdrawing control rods too quickly and in the wrong sequence. The Nuclear Regulatory Commission did not become aware of the incident until July 15th—after they had issued a full-power license to Fermi 2. On July 16th, the NRC limited the plant to 5 percent power again. A regional investigator for the NRC ordered investigators to determine whether Detroit Edison intentionally withheld details of the July 1st incident until it received its license—a charge Edison officials denied.

Later that month, the plant was shut down after a pump that provides water to cool the reactor stopped working during a test. Then in September of the same year, a valve and pipe designed to check leaks of radioactive material were left open and unplugged for more than two months. The error could have permitted radioactive gases or steam to escape out of containment into the plant and eventually into the outside atmosphere.

These incidents in 1985 were the first in a string of equipment breakdowns and personnel errors to occur at Fermi 2 over the next two years. In 1988, Detroit Edison was fined $200,000 for two violations of safety standards at Fermi 2, becoming one of the most-penalized utilities in the nuclear industry. (Edison previously had been fined $625,000 for various penalties since it obtained an operating license.)

A 1988 report by Public Citizen, founded by consumer advocate Ralph Nader, ranked Fermi 2 fifth worst among nuclear power plants in the natiion. The Public Citizen report looked at federal and industry data since 1985 in 10 categories. Fermi 2 was ranked among the worst plants in four categories: first in amount of fines and emergency shutdowns, sixth in management performance, and seventh in mishap reports.

A comeback, Fermi 2 off NRC's watch list. From the time Fermi 2 first sustained a nuclear reaction in 1985, it has been on the NRC's watch list of plants needing extra oversight. This status lasted until June 1989, when the NRC, citing improved operation, removed Fermi 2 from its list.

One of the main reasons for the change in status was Detroit Edison's Nuclear Operations Improvement Plan (NOIP), established in 1986 and culminating in the plant's commercial start-up and major improvements in its operations.

For instance, Fermi 2's first refueling took place in September-December 1989 and lasted 103 days—considered at the time to be a good first-outage time for a boiling water reactor. The second refueling outage, March-June 1991, lasted a record 73 days.

Fermi 2's performance has been affected by organizational enhancements as well as operational improvements. In 1987, the plant adopted Detroit Edison's Fermi 2 Business Plan, which builds incentive compensation around the mission, goals, and strategies for nuclear operations. Also, in 1988 a new Corrective Action System was implemented and has dramatically reduced the number of problems being addressed in the organization.

Both the Nuclear Regulatory Commission and the Institute for Nuclear Power Operations—in their regular reviews of Fermi 2's operational and organizational performance—have noted continued significant improvement.

Operations overview. A chronological list of Fermi 2's major mishaps as well as accomplishments follows:

- August 6, 1986. Fermi 2 shutdown due to smoke coming from electric circuitry that controlled a motor-operated valve. Alert declared.
- August 12, 1986. NRC concern raised about potential wiring problem with Motor Operated Valves (MOV). Company decides to delay testing and keep reactor pressure below 150 psi until MOV concern is resolved.
- August 20, 1986. MOV issue resolved; reactor pressure increased.
- June 26, 1987. Inadvertent mode change event (temperature increase); trainee operator failed to notice rise in temperature. Detroit Edison eventually pays $75,000 fine to the NRC.
- July 31, 1987. Reactor shutdown to make repairs to feedwater check valves; Unusual Event declared.
- August 2, 1987. Reactor water level lowered accidentally from 20 to 15 feet during outage; accident occured during two-week NRC probe of equipment and personnel problems at Fermi 2.
- October 9, 1987. Feedwater check valves repaired, outage completed, and reactor restarted.
- March 18, 1988. Unusual Event declared when anonymous bomb threat call is received. Call is hoax.
- June 22, 1988. NRC announced $200,000 fine for two violations—one on the primary containment radiation monitoring system and a second involved interpretation of operability requirements for a redundant air compressor for plant safety systems.
- July 23, 1988. Unusual Event declared when leak of radioactive steam in drywell increased past allowable limits. Plant shutdown required. Reactor restarted August 6.
- August 20, 1988. Unusual Event declared when a valve failed to close on a reactor recirculation pump.
- August 28, 1988. Unusual Event declared when, for the second time, a valve failed to close on a reactor recirculation pump. Plant enters month-long outage to correct the problem.
- December 21, 1988. Detroit Edison and intervenors reach agreement on a $29.5 million rate increase to cover final Fermi 2 costs and operational and maintenance expenses.
- December 22, 1988. NRC announced that Fermi 2 will remain on its list of plants that require close monitoring in 1989.
- December 29, 1988. Plant personnel completed two major commitments to the NRC—upgrade of 4,656 procedures and review of 934 surveillance article reviews of tech spec improvement program.
- December 29, 1988. NRC announced it was decreasing by $25,000 a $200,000 fine announced last summer.
- February 9, 1989. Unusual Event declared; problems with control center heating, ventilating, and air-conditioning system.
- March 1985, 1989. NRC released Systematic Assessment of Licensee Performance (SALP) report—the plant's best since 1985.
- April 17, 1989. Unusual Event declared; fire discovered in a switchgear room of the Availability Improvement Building. Contractor building cleaner later terminated; fire started because he was smoking in the room.
- June 1, 1989. Citing improved operation, NRC removed Fermi 2 from its close-monitoring list.
- September 4, 1989. Plant shut down due to hydrogen inleakage to the stator water cooling system of main turbine generator. Plant completed longest run of 169.55 days—third longest in industry prior to first refueling. First refueling outage begins.
- September 22, 1989. Monroe County Sheriff arrested two men in labor dispute on Point aux Peaux Road. More than 20 cars "stall" on road, blocking traffic and protesting the contractor brought in for refueling work.
- November 30, 1989. Members of the Professional Security Officers Association turned down company's one-year contract offer and authorized a strike if no agreement is reached.
- December 6, 1989. Reactor placed in start-up mode. Security officers voted to accept company's one-year contract offer.
- February 16, 1990. NRC issued Detroit Edison a Notice of Violation for false statements made during an NRC investigation from 1984 to 1986; no fine was levied.

Plant Profiles — Fermi 278

- February 21, 1989. Detroit Edison became sole owner of plant, purchasing the remaining 11 percent from Wolverine Power Cooperative, Inc.
- March 22, 1990. Small amount of lubricating oil ignites on emergency diesel generator 14. Although it was put out with a hand-held fire extinguisher and caused no damage, a Canadian reporter caused a stir by taking pictures of trash burning at nearby state game area and claimed they were pictures of generator fire.
- March 1990. Plant generated a record 782,916 megawatthours of electricity.
- April 10, 1990. Plant automatically shut down when four inboard main steam isolation valves shut due to a partial loss of power to reactor protection system. Number 5 north feedwater heater leaked nearly 60,000 gallons of slightly radioactive water onto turbine building floor.
- May 30, 1990. Plant worker injured; plant's safety streak came to an end at 5.2 million hours, a company record, retroactive to November 4, 1989.
- October 18, 1990. Engineer Judith Cumbow is killed when she fell into an air-supply fan she was testing inside the turbine building.
- November 10, 1990. South Carolina, Nevada, and Washington refused to accept low-level waste from Michigan generators, alleging that the state was not making progress toward siting a low-level repository in Michigan.
- January 16, 1991. Members of the Professional Security Officers Association voted to disband the union. Because of a tie vote, a labor mediator eventually would uphold decertification of the union.
- May 1, 1991. NRC SALP report noted overall improvement in most areas reviewed; report was the best plant had ever received.
- June 19, 1991. A U.S. District Judge ruled that three states which had been refusing to accept Michigan's low-level waste must again start accepting it.
- August 1991. Plant set new electrical generation record with 784,169 megawatthours of electricity and a net capacity factor of 99.4 percent.

Where low- and high-level radioactive waste goes. Late in 1990 Fermi 2 had to begin storing its own low-level radioactive waste as three states that had traditionally accepted waste from Michigan generators—Nevada, South Carolina, and Washington—halted access from Michigan. The states felt Michigan was not making progress in siting and constructing a low-level repository. (By 1993, all low-level waste generated in Michigan and six other states in the Midwest Compact was supposed to be disposed of at a Michigan site yet to be chosen). The cut-off is expected to have little immediate effect on Fermi 2, since the plant's on-site storage space is predicted to be sufficient until 1998.

Enough space remains in Fermi's spent-fuel pond for storing high-level radioactive waste until 2000. Detroit Edison is considering options for expanding its onsite storage capacity because of delays in establishing the federal government's permanent repository for high-level radioactive waste.

Radioactivity, plant personnel, and pheasants. In the 1987 Business Plan, Fermi had as the number one safety goal to minimize personnel radiation exposure. As a result, the plant's first refueling outage in 1989, radiation exposure was 199 person-rem; in the plant's second refueling outage in 1991, 175 person-rem was achieved—a top ranked figure for low-radiation exposures at U.S.

Sources

Fermi 2 Construction Chronology, Detroit Edison Public Affairs Office, n.d.
Fermi 2 Operating Chronology, Detroit Edison Public Affairs Office, n.d.
Tour Fermi 2, Fermi 2 Visitors Center, n.d.
Your Electric Connection, Detroit Edison Public Affairs Office, n.d.
How Fermi 2 Works, Detroit Edison Public Affairs Office, n.d.
"Fermi Shut Down After Pump Fails." *Detroit Free Press*, July 26, 1985.
"Fermi II Test Valve Left Open 2 Months." *Detroit Free Press*, September 7, 1985.
"Another Fermi 2 Incident is Probed." *Detroit Free Press*, August 14, 1987.
"Edison Pays $75,000 Fine for Latest Fermi Incident." *Detroit Free Press*, October 29, 1987.
"Fermi 2 Called 5th-Worst Nuclear Plant." *Detroit Free Press*, April 27, 1988.
"Fermi II Hit with $200,000 Safety Fine." *Detroit News*, June 23, 1988.
Detroit Edison 1990 Annual Report. Detroit Edison, n.d.
"1991 Electric Utility of the Year: Detroit Edison Delivers Competitive Culture Changes." *Electric Light & Power*, PenWell Publishing Company, November 1991.
"Fermi 2 Called 5th Worst Nuclear Plant." *Detroit Free Press*, April 27, 1988.
"Fermi II Hit with $200,000 Safety Fine." *Detroit News*, June 23, 1988.
Detroit Edison 1990 Annual Report. Detroit Edison, n.d.

278 ☆

Enrico Fermi 3 Nuclear Power Plant

Newport, MI

Owning Utility

Detroit Edison　　　　　　　　　　　Phone: (313)237-8000
2000 Second Ave.　　　　　　　　　Fax: (313)237-8055
Detroit, MI 48226
William S. Orser, Sr.V.Pres., Nuclear Generation

Contact

Robert M. Vergiels, Media Rep.　　Phone: (313)237-8850
Detroit Edison Public Affairs Office

Basic Facts

Status: Cancelled. *Type of Reactor:* Boiling water reactor. *Megawatts (electric):* 1220. *Megawatts (thermal):* 3579. *Reactor System Supplier:* General Electric Co. (United States). *Architect Engineer:* Ebasco.

Plant Report

Milestones. Following are some important events in the history of Fermi 3:

- May 1971: Detroit Edison Board of Directors approved building of Fermi 3, an 1220-megawatt Boiling Water Reactor intended for commercial operation in 1979; unit to be identical to Fermi 2.

UNITED STATES

- June 1971: Detroit Edison Board authorized new containment and reactor blueprint for Fermi-3; plant no longer duplicate of Fermi 2 and in-service date set back to 1981.
- May 1974: Due to financial constraints, Detroit Edison Board ordered delays in Company's entire construction program, including indefinite deferral of Fermi-3 project.
- June 1975: Acting on recommendations of management, Detroit Edison Board of Directors canceled project.
- November 8, 1975: Michigan Public Service Commission hosted public hearing on Detroit Edison request to recover $6,653,549 in costs of Fermi 3 activities.
- 1976: Rate relief granted.

Sources

Fermi-3 Chronology, Detroit Edison Public Affairs Office, n.d.

James A. Fitzpatrick Nuclear Power Plant

PO Box 41
Lycoming, NY 13093
Harry P. Salmon Jr., Resident Mgr.
Phone: (315)342-3840

Owning Utility

New York Power Authority
123 Main St.
White Plains, NY 10601
Ralph E. Beedle, Exec.V.Pres., Nuclear Generation
Phone: (914)681-6200

Contact

Elwood Berzins, Mgr. of Commun.
James A. FitzPatrick Nuclear Power Plant

Basic Facts

Status: Operable. *Type of Reactor:* Boiling water reactor. *Megawatts (electric):* 849. *Megawatts (thermal):* 2436. *Reactor System Supplier:* General Electric Co. (United States). *Generator Supplier:* General Electric Co. *Architect Engineer:* Stone & Webster Engineering Corp. (United States).

Key Dates

Ordered: 1966. *Construction began:* 1968. *Criticality:* November 1974. *First Power:* February 1974. *Commercial Operation:* July 1975.

Operating Costs

1975	$6,902,000	1983	$43,171,000
1976	$10,700,000	1984	$53,797,000
1977	$17,383,000	1985	$56,859,000
1978	$19,045,000	1986	$54,596,000
1979	$25,132,000	1987	$73,875,000
1980	$33,303,000	1988	$84,183,000
1981	$36,679,000	1989	$74,759,000
1982	$31,505,000		

Plant Report

Part of a complex power picture. The James A. Fitzpatrick plant, a 800 megawatt reactor, was originally built by Niagara Mohawk. The plant was purchased by the New York Power Authority, but Niagara Mohawk continued to operate it until the Authority finished training a staff and was able to take over in 1977.

The New York Power Authority is the nation's largest non-federal public power organization. It supplies more than one third of the electricity used in New York. Seventy-five percent of the power supplied by the Authority is generated from hydroelectric sources. Most of the remaining power is generated by two nuclear plants, FitzPatrick and Indian Point 3.

The Authority sells power from the FitzPatrick plant to municipal electric systems, rural cooperatives, designated industries, and the state's major private electric utilities for resale to their customers without profit. Power is also sold to public utility agencies for economic development purposes.

FitzPatrick was the first nuclear plant owned by the New York Power Authority. It is located on the shore of Lake Ontario in Scriba, New York and shares a site with Niagara Mohawk's Nine Mile Point nuclear stations. The three plants are linked to a common switchyard.

The fuel and disposal concerns. The FitzPatrick plant uses a fuel made of enriched uranium in the form of pellets about the size of a pencil eraser. The fuel pellets are assembled into 12-foot-long fuel rods. 62 or 63 fuel rods are put together to form a bundle. The FitzPatrick reactor core contains 560 fuel rod bundles.

Most of the radioactive materials produced at the plant are confined in the fuel pellets. Used fuel rods are currently stored temporarily on site in underwater storage tanks. In 1983 the Power Authority entered into a contract with the U. S. Department of Energy to begin permanent storage of spent fuel no later than January 31, 1998. The DOE, however, will be unable to accept this material until 2005 or later because construction of a permanent storage facility has not yet begun.

Plans for decommissioning. In 1988, NRC regulations required that utilities set aside funds for the future decommissioning of nuclear plants. The amount was determined by estimating the total required to demolish the plant and restore the site. In 1990 the Authority established a decommissioning trust fund and deposited $125,136,000.

Environmental precautions. The New York Power Authority and Niagara Mohawk work together to monitor the environmental impact of the Fitzpatrick and Nine Mile Point nuclear stations. They study the plants' effect on aquatic and terrestrial ecosystems. Samples of air, water, milk and vegetation are monitored and the results of studies are reported to the NRC.

To reduce the effects of heat contamination in Lake Ontario, the FitzPatrick plant uses a system whereby the warm water is diffused and released through six pairs of underwater nozzles. This system ensures that the surrounding surface temperature is not increased more than 3 ° F.

Visitors welcome. Visitors are welcome at "The Energy Center," an information station operated jointly by Niagara Mohawk and the New York Power Authority. It serves visitors to all three of the nuclear plants at the site. The center features hands-on exhibits about electricity and nuclear power. Guided tours are available in July and August.

Sources

James A. Fitzpatrick Nuclear Power Plant. New York Power Authority, n.d.
"James A. Fitzpatrick Nuclear Power Plant: Fact Sheet". New York Power Authority, n.d.
"News for Further Information: James A. Fitzpatrick Nuclear Power Plant." New York Power Authority, n.d.
The Power of Nine Mile Point. Niagara Mohawk, n.d.
Put Energy Into Your Future. New York Power Authority, n.d.
"U.S. Grants Upstate A-Plant A License to Run at Full Power." *The New York Times,* July 2, 1987.
Annual Report for 1990. New York Power Authority, n.d.

—Karen Bellenir

---

280 ☆
Forked River Nuclear Power Plant
Forked River, NJ

Basic Facts
Status: Cancelled (construction 5% complete). *Type of Reactor:* Pressurized light-water reactor. *Megawatts (electric):* 1123. *Megawatts (thermal):* 3410. *Reactor System Supplier:* Combustion Engineering, Inc. (United States). *Owner:* Jersey Central Power and Light. *Operator:* General Public Utilities Nuclear Corporation.

Key Dates
Ordered: 1969. Cancelled: 1980. Construction began: 1973.

Plant Report

Cancelled plant causes problems. The Forked River nuclear power plant was abandoned before its completion. Owned by Jersey Central Power & Light Co., a unit of General Public Utilities Corp. of Parsippany, New Jersey, the plant has created numerous court battles, hearings, and financial woes over the costs it incurred on the abandoned project.

Financial difficulties for Jersey Central. Under Federal Energy Regulatory Commission (FERC) policy, that part of the investment deemed prudent may be amortized, but any unamortized portion may not. This blanket policy allocates incurred losses to investors and makes it possible to recoup any lost investment.

Under the Policy, the FERC refused a Jersey Central request to earn a return on its $400 million invvestment at Forked River. Instead the costs were to be placed on the shoulders of its shareholders. In 1984, a District of Columbia appeals court ruling upheld the decision by the FERC. However, in August 1985, after reviewing the case, the court reversed itself in a 2 to 1 decision that sent the case back to the FERC. The court declared that the FERC's ruling without a hearing was unjust.

The appellate court stated that its earlier decision was based on a misinterpretation of a 1944 Supreme Court case ruling in FPC vs. Hope Natural Gas Company in which stated that in a fair rate case, the decision should be based on calculating a return on invested assets as well as calculating a proper rate base. In the end, Jersey Central was allowed to amortize $397 million of the investment in Forked River over a period of 15 years, but the request for placing the unamortized portion into the rate base to cover debt and preferred stock was denied.

Since most state utility commissions have adopted the policy outlined by the FEERC, utility companies are worried about how the mixed message will be interpreted. On the one hand, utilities may be reluctant to get involved in a major project. And on the other hand, utilities may finish a questionable project in order to avoid a confrontation with its regulator over recovering its investment.

In a related case, a federal appeals court ruled that the FERC must hear Jersey Central out before making a decision as to whether or not to reduce rates. One of the judges said the FERC failed to find an equitable solution to benefit both investors and consumers, and that Jersey Central's claims of financial hardships warranted further investigation.

Sources

"The Canceled-Plant Question." *Electrical World,* September 1985.
"GPU Unit is Victor in Appeal of Decision Involving Rate Base." *The Wall Street Journal*, February 4, 1987.

—Yvette M. Costa

---

281 ●
Fort Calhoun Generating Station

PO Box 399   Phone: (402)426-4011
Fort Calhoun, NE 68023
T.L. Patterson, Mgr.

Owning Utility

Omaha Public Power District   Phone: (402)636-2000
Energy Plaza
444 S. 16th St. Mall
Omaha, NE 68102-2247
W.C. Jones, Sr.V.Pres., Nuclear

Contact

H.F. Sterba, Div.Mgr., Corp. Commun.
Omaha Public Power District

Basic Facts

Status: Operable. *Type of Reactor:* Pressurized light-water reactor. *Megawatts (electric):* 510. *Megawatts (thermal):* 1500. *Reactor System Supplier:* Combustion Engineering, Inc. (United States). *Generator Supplier:* General Electric Co. (United States). *Architect Engineer:* Gibbs and Hill, Inc. (United States).

Key Dates

Ordered: 1966. Construction began: 1968. Criticality: August 1973. First Power: August 1973. Commercial Operation: September 1973. Shut Down Expected: June 2008.

Operating Costs

Year	Cost	Year	Cost
1973	$592,000	1982	$18,935,000
1974	$3,413,000	1983	$23,859,000
1975	$5,962,000	1984	$25,239,000
1976	$7,449,000	1985	$30,459,000
1977	$8,493,000	1986	$35,767,000
1978	$8,116,000	1987	$42,965,000
1979	$8,503,000	1988	$71,205,000
1980	$14,332,000	1989	$105,760,000
1981	$11,472,000		

Plant Report

Omaha Public Power District goes nuclear. Omaha Public Power District (OPPD), a publicly owned regional power supplier, provides electricity to customers in a 5,000-square-mile service area in eastern Nebraska. In the mid-1960's, the utility decided to increase its generating capacity by building a nuclear station. Officials came to this conclusion after considering studies showing that nuclear power was safe, reliable and environmentally clean. Approval to purchase the site for the proposed plant was granted on October 14, 1965 and construction of the Fort Calhoun Generating Station began in 1968.

Fort Calhoun is a 510 (gross) megawatt pressurized water reactor located on the west bank of the Missouri River about 19 miles north of Omaha. During its operating life, Fort Calhoun has supplied 41.8 percent of the electricity generated by OPPD. The balance of the electricity produced by the utility is generated with oil and gas at four other power plants.

Fuel use and storage. The fuel used at Fort Calhoun is uranium dioxide in the form of ceramic pellets. Each pellet is 0.43 inches long and 0.37 inches in diameter. A single pellet can produce as much electricity as 2,400 pounds of coal. The pellets are assembled in 11 1/3-foot long tubes called "fuel pins." 176 fuel pins are bundled together to form a fuel assembly. Fort Calhoun's reactor consists of 133 fuel assemblies. The strength of the nuclear reaction is moderated by control rods which are positioned over the reactor. If power is lost for any reason at Fort Calhoun, the rods will drop under the force of gravity and shut down the reactor in 2.5 seconds.

Every 12 to 18 months the reactor is shut down for refueling. Workers remove the oldest batch fuel, about 44 assemblies, and install new fuel. Used fuel assemblies are stored on vertical racks in a pool about 43 feet deep lined with stainless steel. The pool has the capacity to store 729 fuel assemblies.

The Nuclear Waste Policy Act of 1982 as amended in 1987 requires that permanent storage of used nuclear fuel be made at a federal repository. The first shipment of spent fuel to the yet-to-be-established storage site is scheduled for 2010, two years after Fort Calhoun's operating license will expire. Consequently, OPPD is reracking its spent fuel storage pool to increase the capacity. This will enable the pool to hold all the fuel the plant will use during its licensed lifetime.

Containment building. One of the most important structures at the Fort Calhoun site is the containment building, which houses the nuclear reactor. The building's purpose is to prevent any radiation from entering the outside environment. The building is designed to endure emergencies within and without. It can withstand an internal pressure greater than could be created if the largest primary pump in the reactor cooling system ruptured.

During an incident in 1990, a leak developed in the primary cooling system; all radioactivity was successfully kept within the containment building.

The outside of the structure is designed to withstand earthquakes and tornadoes. For example, the steel-reinforced walls can endure the impact of a 4,000 pound car hurled by a 500 mph wind. The structure is also designed to sustain a direct hit by a large aircraft.

Emergency procedures. In accordance with NRC regulations, OPPD established an emergency procedure for residents within a 10-mile radius of the plant. A siren system, which can be heard throughout the entire area, would sound for 3 to 5 minutes. The alarm would alert residents to tune to a designated radio station for official information.

The NRC describes four types of emergencies in ascending levels of threat to the public safety. An unusual event is the least serious. It means that there is a chance that the plant's safety may be reduced. The second, an "Alert" means that there is a possibility of substantial reduction in plant safety or that plant safety has actually been reduced. The third level, a "Site Area Emergency" means that there has been or is likely to be a major failure in operations necessary for public protection. The fourth and highest alert level, a "General Emergency" is one in which actual or imminent serious damage to the reactor occurs. The siren system is not activated unless the third or fourth warning levels are declared.

"Alert" declared. Although neither of the two most severe types of accidents have ever been reported at the Fort Calhoun facility, OPPD declared an "Alert" on July 3, 1992. A malfunction in an electrical component caused the reactor to automatically shutdown. A pressure relief valve failed to close and 20,000 gallons of water flowed into the sump of the containment building. All radiation was successfully controlled within the building. The outage lasted twenty days.

Environmental safeguards. In addition to procedures to protect the human population surrounding the plant, measures are taken to safeguard the environment. Workers regularly take air samples to see if any radiation has been released. Water (from wells and from the Missouri River), local vegetation, wildlife, farm animals, crops and soil are also routinely studied.

Record set for longest run. The Fort Calhoun Station set a world record for the longest continuous run achieved by a light water reactor. The plant operated 477 days between June 8, 1987 and September 27, 1988. The record stood for three years.

NRC evaluations. In spite of its record-setting performance the NRC put Fort Calhoun on its "problem plant" list in 1988. The agency felt that OPPD was not spending enough money to maintain the plant's high level of performance. OPPD responded by making numerous improvements, including the opening of a new training center with a computer-driven simulator identical to the actual station control room. The plant had originally been constructed at a cost of $178.3 million; by the end of 1990 the cost of modifications raised the total cost to $354.4 million.

Increased operating and maintenance costs at the Fort Calhoun Station, however, caused the plant to become one of the mosst expensive in the country on the basis of dollars-per-megawatt-hour. OPPD instituted a "Cost Effectiveness Review Program" encouraging employees to submit suggestions for enhancing efficiency. As a result of improvements made, the plant operated below budget in 1991. In 1991, the NRC gave the plant "superior" and "good" ratings in all seven of the operating areas reviewed by the agency.

Decommissioning. The operating license at Fort Calhoun will expire in 2008. Decommissioning costs are estimated at $114,900,000 in 1991 dollars. The NRC requires that decommissioning costs be funded and periodically reviewed and adjusted if necessary. OPPD holds decommissioning funds in an external trust. At the end of 1991, the balance held for eventual decommissioning was $42,396,000.

Sources

"O&M Costs Worry OPPD, Cloud Plant's Future." *Nuclear News*, February 1991.
"Fact Sheet: OPPD Fort Calhoun Nuclear Station." Omaha Public Power District, October 31, 1991.
Fort Calhoun Nuclear Power Station. Omaha Public Power District, December 1991.
"Around the States." *Nuclear Industry*, Fourth Quarter 1991.
1991 OPPD Annual Report. Omaha Public Power District, n.d.
Fort Calhoun Nuclear Station Emergency Planning Information. Omaha Public Power District. July 1992.
"1992 OPPED Second Quarter Financial Report." Omaha Public Power District, n.d.

—Karen Bellenir

282 ☆
Fulton Nuclear Power Plant, Units 1-2
Lancaster, PA

Owning Utility
Philadelphia Electric Company (PECO) Phone: (215)841-4000
2301 Market St. Fax: (215)841-4188
PO Box 8699
Philadelphia, PA 19101
Dickinson M. Smith, Sr.V.Pres., Nuclear

Contact
Neil J. McDermott, Dir., Media & PR
Philadelphia Electric Company

Basic Facts
Units 1-2 *Status:* Cancelled. *Type of Reactor:* Gas-cooled reactor. *Megawatts (electric):* 1200. *Megawatts (thermal):* 3000. *Reactor System Supplier:* General Atomic. *Owner:* Philadelphia Electric. *Operator:* Philadelphia Electric.

Key Dates
Units 1-2 *Ordered:* 1971. *Cancelled:* 1975.

283 ●
Robert Emmet Ginna Nuclear Power Plant
1503 Lake Rd. Phone: (315)524-4446
Ontario, NY 14519
Joseph A. Widay, Plant Mgr.

Owning Utility
Rochester Gas and Electric Corporation Phone: (716)546-2700
89 E Ave. Fax: (716)546-1511
Rochester, NY 14649
Robert C. Mecredy, V.Pres., Ginna Nuclear Production

Contact
John W. Edmunds, Dept. Mgr., Pub.Aff.
Rochester Gas and Electric Corporation

Basic Facts
Status: Operable. *Type of Reactor:* Pressurized light-water reactor. *Megawatts (electric):* 498. *Megawatts (thermal):* 1520. *Reactor System Supplier:* Westinghouse Electric Corp. (United States). *Generator Supplier:* Westinghouse Electric Corp. (United States). *Architect Engineer:* Gilbert.

Key Dates
Ordered: 1965. *Construction began:* 1966. *Criticality:* November 1969. *First Power:* December 1969. *Commercial Operation:* March 1970. *Shut Down Expected:* April 2006.

Operating Costs
1970	$3,199,000	1980	$18,924,000
1971	$4,391,000	1981	$22,481,000
1972	$4,082,000	1982	$29,570,000
1973	$3,536,000	1983	$26,955,000
1974	$5,391,000	1984	$32,679,000
1975	$6,597,000	1985	$31,609,000
1976	$7,356,000	1986	$37,389,000
1977	$7,942,000	1987	$37,763,000
1978	$9,819,000	1988	$44,427,000
1979	$12,817,000	1989	$63,572,000

Plant Report

Serious steam generator tube rupture. In January 1982, the Robert Emmet Ginna Nuclear Power Plant, located in western New York state, suffered what some term the "most important event in the history of nuclear power since the Three Mile Island accident." A steam generator tube ruptured causing 8,000 gallons of radioactive coolant to leak onto the plant floor. Although measurements of radioactivity both on-site and off-site were low, unnecessary personnel had to be evacuated from the plant.

Officials at the facility declared a "Site Area Emergency." Under federal guidelines a "Site Area Emergency" is the third highest in a series of four ascending alert levels. First is an

"Unusual Event" which means that there is a possibility that the plant's safety will be reduced. The second, an "Alert" means that a potential exists for substantial reduction in plant safety or that plant safety has actually been reduced. The third, a "Site Area Emergency" means there has been or is likely to be a major failure in the plant functions that are necessary for public protection. The highest alert level, a "General Emergency," is defined as one that threatens "actual or imminent serious damage to the reactor."

Rochester Gas and Electric, Ginna's owner, was commended for its openness in dealing with the public during the crisis. The company even set up a rumor-control room to help ensure the accuracy of information being circulated in the community.

The accident, however, worried many about the safety of pressurized water reactors. An NRC analysis of the tubes predicted that tube ruptures could happen once every forty years. The rupture at Ginna was one of four to occur in a seven year period. It was blamed on a loose object apparently left behind by repair workers in 1976.

In addition to the tube rupture, other equipment failed to respond in a helpful manner as operators tried to cope with the problem. A pilot-operated relief valve, like the valve that failed at Three Mile Island, initially would not open. Then it stuck open. A steam bubble formed in the reactor vessel, and as operators tried to compress it, they brought the pressure too high. As a result radioactive steam escaped into the atmosphere. NRC investigators concluded that the operators had responded appropriately. If the steam bubble had not been compressed, it might have grown and uncovered the core. This chain of events could have resulted in a meltdown.

Workers exposed to radiation. Plant employees had to work in highly radioactive areas to repair the generator tubes. Workers could not stay inside the steam generator for more than a couple minutes to accomplish some of the most "hot" tasks. Workers were again exposed to high levels of radioactivity in 1989 when crews had to work inside the generator to change the "snubbers."

Snubbers replaced. A "snubber" is designed to support pipe systems and equipment and to compensate for changes in size during extreme fluctuations in temperature. For example, when the reactor is in a "cold shutdown" stage, the water is 200 ° F. At "hot standby," the water is 547 ° F. At Ginna, this temperature change causes the generator to move away from the reactor by a distance up to 1 3/4 inches.

Some of the snubbers at Ginna were located 40 feet above the floor in hard to reach areas. Three out of four sets per generator were replaced with struts to reduce maintenance costs and to reduce the risk to plant personnel who had to expose themselves to radiation during repairs. Strut installation was completed on schedule with no safety or radiation consequences.

Labor unrest. In addition to safety and design concerns, Ginna has been involved in labor disputes. Unions protested Rochester Gas and Electric's decision to hire a non-union company for maintenance at the plant. RG&E refuted charges that it was anti-union and pointed to other contracts with union companies.

Unions expanded their efforts by launching a public relations campaign featuring newspaper and television advertisements. In May 1980, 800 people attended a rally sponsored by the Allied Building Trades Council to protest non-union contracts and urge union supporters to attend RG&E's annual meeting.

Recognized for performance. In 1991 The Nuclear Industry Performance Compendium recognized Ginna for "consistently superior performance" during a five year period (1985-1990). Ginna earned the honor by achieving a capacity factor greater than 70 percent and for operating at below-average costs. Ginna continued to improve and reached an operating capacity factor of 84.4 percent in 1991, well above the average U.S. nuclear capacity factor of 69.3 percent for that year.

License questions. Ginna, like many nuclear plants, originally received a 40-year license to operate. Its license will expire in 2009. Details of the renewal process are uncertain. Some groups claim that licenses should not be renewed because older plants can never be as safe as they were when new. In addition, the NRC has not yet decided whether old plants seeking license renewals should have to meet requirements established after they were first built.

Sources

"Fighting the Epidemic of Nuclear Plant Leaks." *USA Today*, March 1983.
"RG&E Officials Nixed Bausch Office Venture." *Rochester Business Journal*, December 18, 1989.
"Struts Replace Snubbers at Ginna, Improve Safety." *Electrical World*, December 1989.
"PR Campaign Replacing Strike as Key Weapon in Labor Arsenal." *Rochester Business Journal*, September 3, 1990.
"RG&E Assesses Damage From Storm PR Errors." *Rochester Business Journal*, March 18, 1991.
"Due Up for License Renewal: The Future of Nuclear Power." *The New York Times*, June 24, 1991.
"New York A-Plant Shut Down by Staff as Instruments Fail." *The New York Times*, August 14, 1991.
"Around the States." *Nuclear Industry*, Fourth Quarter 1991.
"If You Hate People Looking Over Your Shoulder, Don't Even Think About Working in a Nuclear Power Plant." *Nuclear Industry*, Fourth Quarter 1991.
"U.S. Nuclear Plants Shatter Records for Output, Performance." *INFO*, U.S. Council for Energy Awareness, February 1992.
"New York A-Plant Shut Down by Staff as Instruments Fail." *The New York Times*, August 14, 1991.
"Around the States." *Nuclear Industry*, Fourth Quarter 1991.

—Karen Bellenir

Grand Gulf Nuclear Station, Unit 1

PO Box 756
Port Gibson, MS 39150
William T. Cottle, V.Pres., Operations—
  Grand Gulf

Phone: (601)437-2800
Fax: (601)437-2322

Plant Profiles   Grand 284

Owning Utility

Entergy Operations, Inc. Phone: (601)984-9000
PO Box 31995 Fax: (601)984-9817
Jackson, MS 39286-1995
William Cavanaugh III, Chairman & CEO

Contact

Phil R. Miracle, Corp. Commun.
Entergy Operations, Inc.

Basic Facts

Unit 1 *Status:* Operable. *Type of Reactor:* Boiling water reactor. *Megawatts (electric):* 1306. *Megawatts (thermal):* 3833. *Reactor System Supplier:* General Electric Co. (United States). *Generator Supplier:* Allis-Chalmers (United States). *Architect Engineer:* Bechtel. *Owner:* System Energy Resources, Inc; Mississippi Electric Power Association. *Operator:* Entergy Operations, Inc.

Key Dates

Unit 1 *Ordered:* 1972. *Construction began:* 1974. *Criticality:* August 1982. *First Power:* October 1984. *Commercial Operation:* July 1985. *Shut Down Expected:* September 2014.

Operating Costs

Unit 1
1985 $68,350,000
1986 $108,391,000
1987 $98,327,000
1988 $81,388,000
1989 $105,984,000

Plant Report

Grand Gulf overview. The Grand Gulf Nuclear Station occupies a 2,300-acre site, 50 miles southwest of Jackson on the Mississippi River's east bank. The single 1,250 megawatt boiling water reactor boasts a 520-foot natural draft cooling tower, the tallest concrete structure in Mississippi. The largest ownership interest in Grand Gulf, 90 percent, is held by System Energy Resources, Inc. (SERI), a subsidiary of Entergy Corporation. Mississippi Electric Power Association owns the remaining 10 percent.

The nuclear plant is one of 50 power stations contributing to Entergy Corporation's 15,200 megawatts of generating capacity. Through its various operating companies, Entergy supplies electricity to 1.7 million customers in a 91,000 square-mile service area covering portions of Arkansas, Louisiana, Mississippi and Missouri.

Plans for Grand Gulf were first made public in January 1972 when Mississippi Power and Light (MP&L) announced its intention to build a two-unit nuclear station. MP&L, the plant's original operator, was one of four utilities connected with Middle South Energy (MSE), a subsidiary of Middle South Utilities (MSU). The other three sister utilities were: Arkansas Power and Light Company (AP&L), Louisiana Power & Light Company (LP&L), and New Orleans Public Service Inc. (NOPSI).

Through a series of corporate reorganizations, MSE became System Energy Resources, Inc. (SERI) and MSU became Entergy Corporation. The organizational restructuring also resulted in transfers of Grand Gulf's operating license, first to MSU and then in 1990 to the newly formed Entergy Operations. Entergy Operations was established for the purpose of operating three nuclear plants owned by the four utilities in the Entergy system: Grand Gulf, Arkansas Nuclear One (ANO), and Waterford 3. Together, these nuclear plants account for 45.8 percent of Entergy's generating capacity.

Construction. Construction at Grand Gulf began in 1974. Delays attributed to design modifications, inflation, high interest, and regulatory changes following the Three Mile Island accident, led to cost overruns and caused the project to fall years behind its construction schedule. Initial fuel loading did not begin until the summer of 1982, and commercial operation began at Grand Gulf 1 on July 2, 1985. Grand Gulf's owners defended the costs incurred claiming that the plant's price of $2,800 per kilowatt of capacity was less than the national average of $3,000 per kilowatt of capacity for nuclear plants of similar size constructed during the same time.

Construction on Grand Gulf 2 was halted in 1985. The second unit was only one-third complete and cost estimates had risen to $3.4 billion with commercial operation not expected until April 1990. In September 1989, MSE officially cancelled the unit and wrote off $910 million in expenses attributed to Grand Gulf 2.

Cost allocations and rate issues. When Grand Gulf 1 came on line in 1985, the four utilities supplied by System Energy Resources, Inc. (AP&L, LP&L, MP&L and NOPSI) requested rate hikes totaling $1.5 billion, an increase of 50 percent. Regulatory agencies in each of the states served by the sister utilities fought the cost allocation formula, attempting to pass on Grand Gulf's expenses to others. The Federal Energy Regulatory Commission (FERC) stepped in and distributed Grand Gulf's $3.8 billion price tag among the utilities as follows: 36 percent to AP&L; 14 percent to LP&P; 33 percent to MP&P and 17 percent to NOPSI. Although the Mississippi Supreme Court ruled against the FERC formula claiming that MP&L had 85 percent more power than it needed without the nuclear capacity generated at Grand Gulf, the U.S. Supreme Court upheld the allocation formula in December 1987.

Rate issues were further complicated when the New Orleans city council considered buying NOPSI in an attempt to protect its residents and businesses from rate increases expected to be as high as 150 percent. A franchise agreement made in 1923 granted the city council the right to purchase NOPSI's assets at their base rate value (purchase price less depreciation) by a majority vote. When a legal analysis suggested that NOPSI's contract with Grand Gulf would be broken if the utility were bought out by the city, a task force was established to study possible municipalization. The city council finally decided against the buy-out in May 1990, but legal proceedings continued concerning whether NOPSI's interest in Grand Gulf had been prudently incurred. A state appeals court ruled against NOPSI in April 1991, disallowing $476.9 million rate increase charging that the utility had imprudently accepted too large a share in Grand Gulf. NOPSI appealed to the U.S. Supreme Court.

● Operable   ○ Under Construction   △ Indefinitely Deferred   ▲ Decommissioning   □ Planned   ■ On Order   ☆ Cancelled   ★ Shut Down

Charges about safety. In addition to controversies surrounding cost allocations and rate increases, Grand Gulf's safety was challenged. Critics claimed that the plant was built prematurely and costs were cut at the expense of public protection. Charges made against the plant included allegations of design flaws, inadequate cooling systems, and potentially high radiation exposure to workers. The plant's location near the New Madrid fault line also led to questions about its ability to withstand earthquakes.

Robert Pollard, a nuclear safety expert with the Union of Concerned Scientists, also charged that the plant's safety was compromised because of problems related to its GE Mark III reactor. Grand Gulf was the first nuclear station to receive a 1,250 megawatt GE Mark III reactor. GE's previous Mark II reactor had been only an 800 megawatt reactor. Pollard claimed that MP&L pressured the NRC into letting construction at Grand Gulf begin before design problems with the new-sized reactor were fully understood. According to Pollard, in the event of a nuclear accident, the Mark III would require a much larger quantity of water for steam suppression than Grand Gulf's original design could accommodate. Although additional water reservoirs were built into the upper levels of the containment building, questions remained concerning building's ability to withstand the resulting internal pressures.

Accomplishments. In spite of the controversies surrounding Grand Gulf, the plant has been recognized for its performance records. On February 18, 1987, the station set a record for the most power generated by a boiling water reactor in a 24 hour period — 30,930 megawatts. Grand Gulf topped this record on January 27, 1988 by generating 31,130 megawatts of power. The station also set a world record for the longest run achieved by a boiling water reactor in its second fuel cycle when it completed 171 days of continuous operation. In January 1991, Grand Gulf generated 931,527 gross megawatt-hours of electricity and earned a place on the list of America's top 10 producers.

Visitors Center. Entergy maintains a visitors center, Energy Central, at the Grand Gulf site. Energy Central features exhibits about radiation detection and a control room simulator. It is open from 8:00 a.m. to 5:00 p.m. Monday through Friday and admission is free. For more information contact t: Grand Gulf Nuclear Station at (601) 437-6317.

Sources

"Middle South's Stockholders: What's Ahead for Them?" *New Orleans Business*, February 7, 1985.
"Middle South CEO Effectively Defends Grand Gulf." *Jacksonville Journal of Business*, May 1985.
"Mississippi Court Rules Middle South Can't Raise Rates to Cover Nuclear Unit." *The Wall Street Journal*, February 26, 1987.
"GE Questioned Safety of Its Own Reactors." *New Orleans CityBusiness*, June 22, 1987.
"Middle South Unit Asks FERC to Act on Gulf 1 Plan." *The Wall Street Journal*, July 20, 1987.
"Big Bucks or Small on Tale in City Buy Out?" *New Orleans CityBusiness*, August 3, 1987.
"High Court Rules Against Illinois Law on Notification of Abortions by Minors." *The Wall Street Journal*, December 15, 1987.
"Entergy Corp." *The Wall Street Journal*, November 8, 1989.
"NRC Okays Nuclear Merger at Entergy Corp." *Electrical World*, February 16, 1990.
"Power Company is Rated Top in Southeast." *Mississippi Business Journal*, September 10, 1990.
"Utility Exec Puts Energy into N.O." *New Orleans CityBusiness*, December 31, 1990.
"Palo Verde Units 3 & 3 Top Producers in United States for January." *Business Wire*, March 13, 1991.
"Entergy's Nopsi Unit is Set Back on Plans for Rate Increases." *The Wall Street Journal*, April 5, 1991.
"Around the States." *Nuclear Industry*, Third Quarter 1991.
"Entergy: 1991 System Profile." Entergy Corporation, n.d.
"Grand Gulf, Entergy Operations and Energy." Entergy Corporation, n.d.
"Facts about the Grand Gulf Nuclear Station." Entergy Corporation, n.d.
"Grand Gulf Nuclear Station: Chronology of Significant Events." Entergy Corporation, n.d.
"Entergy: 1991 System Profile." Entergy Corporation, n.d.
"Grand Gulf, Entergy Operations and Energy." Entergy Corporation, n.d.

—Karen Bellenir

---

285 ☆
Grand Gulf Nuclear Station, Unit 2

PO Box 756 Phone: (601)437-2800
Port Gibson, MS 39150 Fax: (601)437-2322
William T. Cottle, V.Pres., Operations—
 Grand Gulf

Owning Utility

Entergy Operations, Inc. Phone: (601)984-9000
PO Box 31995 Fax: (601)984-9817
Jackson, MS 39286-1995
William Cavanaugh III, Chairman &
 CEO

Contact

Phil R. Miracle, Corp. Commun.
Entergy Operations, Inc.

Basic Facts

Unit 2 *Status:* Cancelled (construction 33% complete). *Type of Reactor:* Boiling water reactor. *Megawatts (electric):* 1306. *Megawatts (thermal):* 3833. *Reactor System Supplier:* General Electric Co. (United States). *Generator Supplier:* United Engineers & Constructors (United States). *Architect Engineer:* Bechtel.

Key Dates

Unit 2 *Ordered:* 1973. *Cancelled:* 1989. *Construction began:* 1974.

Plant Report

See Plant Report for Grand Gulf Nuclear Station, Unit 1.

---

286 ☆
Greene County Nuclear Power Plant

Cementon, NJ

Plant Profiles   Haddam 288

Owning Utility

New York Power Authority   Phone: (914)681-6200
123 Main St.
White Plains, NY 10601
Ralph E. Beedle, Exec.V.Pres., Nuclear
   Generation

Basic Facts

Status: Cancelled. *Type of Reactor:* Pressurized light-water reactor. *Megawatts (electric):* 1289. *Megawatts (thermal):* 1300. *Reactor System Supplier:* The Babcock & Wilcox Co. (United States). *Owner:* New York Power Authority. *Operator:* New York Power Authority.

Key Dates

Ordered: 1974. *Cancelled:* 1979.

---

287 ☆

Greenwood Nuclear Power Plant, Units 2-3

Port Huron, MI

Owning Utility

Detroit Edison   Phone: (313)237-8000
2000 Second Ave.   Fax: (313)237-8055
Detroit, MI 48226
William S. Orser, Sr. V.Pres., Nuclear
   Generation

Contact

Robert M. Vergiels, Media Rep.   Phone: (313)237-8850
Detroit Edison Public Affairs Office

Basic Facts

Units 2-3 Status: Cancelled. *Type of Reactor:* Pressurized light-water reactor. *Megawatts (electric):* 1344. *Megawatts (thermal):* 3800. *Reactor System Supplier:* The Babcock and Wilcox Co. (United States). *Generator Supplier:* General Electric Co. (United States). *Architect Engineer:* Bechtel.

Key Dates

Units 2-3 Ordered: April 1972. *Cancelled:* 1980.

Plant Report

Detroit Edison cancels nuclear plans. Detroit Edison first announced plans to build Greenwood Units 2 and 3 (along with Fermi 3) in April of 1972. Greenwood 2 was to be built in Port Huron, Michigan; and Greenwood 3 was to be built in St. Claire, Michigan. Greenwood 1, originally a residual oil plant, operates as a dual fuel oil and gas plant.

Financial committments were invested in engineering and licensing efforts for Greenwood 2 & 3, but Detroit Edison never began construction work. The units were cancelled in 1980.

Sources

Fermi-2 Construction Chronology. Detroit Edison, n.d.
"1991 Electric Utility of the Year: Detroit Edison Delivers Competitive Culture Changes." *Electric Light & Power,* PennWell Publishing Company, November 1991.

---

288 ●

Haddam Neck Nuclear Power Plant

362 Injun Hollow Rd.   Phone: (203)267-2556
Haddam Neck, CT 06424   Fax: (203)267-3501
John P. Stetz, Station Dir.

Owning Utility

Connecticut Yankee Atomic Power Co.   Phone: (203)665-5000
PO Box 270   Fax: (203)665-5884
Hartford, CT 06141-0270
John F. Opeka, Exec.V.Pres., Nuclear

Contact

Anthony J. Castagno, Mgr., Nuclear
   Info.
Connecticut Yankee Atomic Power Co.

Basic Facts

Status: Operable. *Type of Reactor:* Pressurized light-water reactor. *Megawatts (electric):* 609. *Megawatts (thermal):* 1825. *Reactor System Supplier:* Westinghouse Electric Corp. (United States). *Generator Supplier:* Westinghouse Electric Corporation. *Architect Engineer:* Stone & Webster Engineering Corp. (United States).

Key Dates

Ordered: 1963. *Construction began:* 1964. *Criticality:* August 1967. *First Power:* August 1967. *Commercial Operation:* January 1968. *Shut Down Expected:* June 2007.

Plant Report

Background. The Haddam Neck Nuclear Power Plant, also known as Connecticut Yankee Nuclear Power Plant is situated southwest of Hartford on the Connecticut River, from which it draws cooling water. It is located in the town of Haddam Neck, Connecticut, the largest town in the state that is totally dependent on private wells for water.

The 582-megawatt pressurized-water reactor is the oldest and smallest of the four nuclear plants Northeast Utilities operates in Connecticut. It began commercial operation on January 1, 1968.

Northeast Utilities. Northeast Utilities, the largest utility in New England, through a 44 percent interest in the Connecticut Yankee Atomic Power Co., operates and owns the primary share of the Haddam Neck plant. The utility also owns and operates the Millstone plant; and owns a percentage of four other New England nuclear units: Yankee Rowe, Vermont Yankee, Maine Yankee and Seabrook.

Northeast Utilities was formed as a holding company in 1966 through the union of several utilities, principally Connecticut Light & Power Company, Hartford Electric Light Company, and Western Massachusetts Electric Company.

The company uses nuclear energy for about two-thirds of its electricity. In 1989, even though Haddam Neck and the three Millstone units were out of service for refueling and maintenance at various times, their composite capacity factor was 69.6 percent, a cut above the industry average of 65.3 percent.

● Operable   ○ Under Construction   △ Indefinitely Deferred   ▲ Decommissioning   □ Planned   ■ On Order   ☆ Cancelled   ★ Shut Down   p. 371

As of 1990, Northeast had approximately $2.2 billion in sales, just over 1.25 million customers, a 5,890-square-mile service area, and approximately 7,000 megawatts of generating capacity.

Haddam Neck makes the list for longest non-stop operating runs. Two times within a four year period Haddam Neck was one of the U.S. nuclear plants with the longest non-stop operating runs. In 1985, the Haddam Neck plant operated for 417 consecutive days and in 1989 they operated for 463 consecutive days.

Any run of longer than 365 consecutive days in considered exceptional for a power plant.

Fuel assembly dislodges at Haddam Neck. In February 1986, a fuel assembly became dislodged during refueling and fell back into the core, settling at a 45-degree angle. Subsequent inspections showed there was no significant damage to the fuel.

Radioactive gas released from Haddam Neck. On February 18, 1989, an undetermined amount of radioactive gas was inadvertently released from the Haddam Neck plant at 10:15 a.m. The release occurred when plant technicians were removing a gauge from a connection to a plant-system cooler and lasted approximately four minutes, with the gas coming out of a pipe 10 inches long and one-half-inch in diameter.

According to plant spokesperson, Anthony Nericcio, it was a "small, unplanned release of radioactive gas" and the amount was probably well below hazardous levels.

Haddam Neck discovers damage to fuel rods. The damage was apparently caused by debris that had been left inside the vessel from earlier work on the plant's thermal shield. Since the thermal shield is not essential and had been damaged, Northeast decided to take it out.

Haddam Neck expected to pay proportionally less in taxes. The town of Haddam has long relied on the Haddam Neck plant to provide the funds for many town services. In the fiscal year of 1991, it was estimated that the plant would pay about $5 million of the town's $10.5 million budget, but starting in 1990 the plant's share of the municipal finances would begin to decline dramatically.

In April 1991 Haddam, Neck was in the middle of a real estate revaluation and the relative value of the plant was expected to drop from almost one-half the town's grand list to one-third or even one-quarter.

It was expected at that time that the revaluation would result in increases of 200 percent or more in the official values of residents' houses and other real property, while the power plant's value would change only slightly, causing townspeople and businesses to have to pay more just to maintain services.

According to Haddam's assessor, Robert Coates, the bulk of the assessment of a nuclear plant is in personal property, such as equipment, rather than real property, such as land. Personal property is revalued every year, so it keeps pace with inflation while real property is revalued every 10 years and it falls behind.

As a result, when a town undergoes revaluation, the official value of real property rises steeply in that year, and the value of the personal property remains about the same, so its importance to the tax base is diminished.

According to Coates, the more graced a town is with an industrial base, which keeps taxes down, the bigger the shift is in that 10th year. Fortunately for the townspeople, Northeast was willing at least to discuss some arrangement that might east the impact on the town.

Haddam Neck takes steps to ensure the community's safety. Every year, the towns within a 10-mile radius of the Haddam Neck and Millstone plants participate in emergency preparedness drills. These procedures have been helpful in other events including floods, transportation events involving hazardous materials and in 1985 they were used for Hurricane Gloria.

Northeast looks to the future. To prepare for the 1990s Northeast Utilities has searched for ways to cut costs without compromising performance and has come up with downsizing. The Institute of Nuclear Power Operations (INPO) backs up the company's decision because those U.S. plants with high NRC ratings tend to have smaller staffs and use less contract labor.

As a result, Northeast trimmed its nuclear engineering and operations staff from 2,427 in 1987 to 2,124 in 1990. The reduction was achieved through attrition.

Northeast's operating license to expire in 2007. In the year 2007, the 40-year operating license for Haddam Neck, which was expended to reflect time the plant was under construction, is due to expire.

Sources

"Businessperson of the Year: William B. Ellis: An Expectation of Excellence." *New England Business Journal*, December 7, 1987.
"Radioactive Gas Is Released At Connecticut Power Plant." *The New York Times*, February 19, 1989.
"New England's Nuclear Powerhouse." *Nuclear Industry*, Summer 1990.
"Nuclear Mishap Monitored." *Newsday*, October 6, 1990.
"As Power Plant's Tax Share Drops, Town Seeks Ways to Compensate." *The New York Times*, April 7, 1991.
"St. Lucie 2 Completed Record Run, Leads World in Efficiency." *INFO*, U.S, Council for Energy Awareness, April 1992.

—Sharon Miscavage

Hallam Nuclear Power Plant

Hallam, NE

Basic Facts

Status: Shut down. *Type of Reactor:* Sodium-graphite. *Megawatts (electric):* 80. *Megawatts (thermal):* 256. *Reactor System Supplier:* Baldwin Lima Hamilton (United States). *Generator Supplier:* Westinghouse Electric Corp. (United States). *Architect Engineer:* Bechtel. *Owner:* U.S. Department of Energy; Nebraska Public Power District. *Operator:* Nebraska Public Power District.

Key Dates

Ordered: 1957. Construction began: 1960. Criticality: 1962. First Power: May 1963. Commercial Operation: November 1963. Shut Down: September 1964.

Plant Report

A project of the Power Demonstration Reactor Programme (PDRP). Realizing that utilities could not justify to their shareholders the investment costs required to build nuclear power plants, the Atomic Energy Commission and Congress devised the Power Demonstration Reactor Programme in 1955 to help launch the commercial nuclear industry in the United States. The PDRP offered cooperating utilities with financial incentives such as research and development assistance, financing of the reactor system, and a waiver of normal fuel charges during the first five years of plant operation.

Hallam was among the initial projects undertaken in response to the PDRP. It was shut down in 1964 and subsequently entombed.

Sources

Mounfield, Peter. *World Nuclear Power.* Routledge, 1991.

290

Shearon Harris Nuclear Power Plant, Unit 1

PO Box 165
New Hill, NC 27562
G.E. Vaughn, V.Pres., Harris Nuclear Project
Phone: (919)362-8891
Fax: (919)362-6950

Owning Utility

Carolina Power and Light Company
PO Box 1551
Raleigh, NC 27602
R.A. Watson, Sr.V.Pres., Nuclear Generation
Phone: (919)546-6111
Fax: (919)546-7678

Contact

R.J. White, V.Pres., Corp. Commun.
Carolina Power and Light
PO Box 1551
Raleigh, NC 27601
Phone: (919)546-6111
Fax: (919)546-7678

Basic Facts

Unit 1 *Status:* Operable. *Type of Reactor:* Pressurized light-water reactor. *Megawatts (electric):* 950. *Megawatts (thermal):* 2775. *Reactor System Supplier:* Westinghouse Electrical Corp.(United States). *Generator Supplier:* Westinghouse Electrical Corp. (United States). *Architect Engineer:* Ebasco. *Owner:* Carolina Power and Light Company; North Carolina Eastern Municipal Power Agency. *Operator:* Carolina Power and Light Company.

Key Dates

Unit 1 *Ordered:* 1971. *Construction began:* 1978. *Criticality:* January 1987. *First Power:* January 1987. *Commercial Operation:* May 1987. *Shut Down Expected:* January 2018.

Operating Costs

Unit 1
1987 $69,421,000
1988 $67,116,000
1989 $62,408,000

Plant Report

Four units originally planned. The Shearon Harris Plant is the newest of four nuclear units operated by Carolina Power and Light Company (CP&L). The 900 megawatt pressurized water reactor is jointly owned by CP&L and the North Carolina Eastern Municipal Power Agency (Power Agency). CP&L, the major partner, owns 83.83 percent of the plant. The utility serves customers in a 30,000 square-mile area covering parts of North and South Carolina. Harris is one of 16 power plants, including fossil-fueled and hydroelectric units, in CP&L's generating mix. Forty-one percent of the electricity supplied by CP&L comes from nuclear sources.

CP&L announced plans to build Harris in 1971. The original project included four new units. Construction at the 10,723 acre site in Wake County, North Carolina, located 20 miles southwest of Raleigh, began in 1978. CP&L abandoned Harris 3 and Harris 4 in December 1981. In December 1983 Harris 2 was only four percent complete and estimates to finish the unit stood at $2 billion. When it appeared that cost estimates would rise even further, CP&L also cancelled Harris 2.

Unit One finished. Construction at the remaining unit continued, but not without controversy. Commissioners in neighboring Chatham County withdrew their support for the proposed emergency evacuation plan following the Chernobyl accident in 1986. Their action threatened to delay the licensing process until further concerns were addressed. Another group protesting the plant, Coalition for Alternatives to Shearon Harris (CASH), proposed converting it to a coal-fired generating station as had been done at the Zimmer facility in Ohio.

Protesters failed to influence policy-makers. The NRC granted Harris 1 an operating license and commercial operation began May 2, 1987.

Rate controversies. In 1988 the North Carolina Utilities Commission and the South Carolina Public Service Commission ordered CP&L to exclude some of the costs related to Harris 1 from their rate base. The agencies judged that the controversial costs were for equipment originally designed to serve all four Harris units. Consequently, CP&L was ordered to treat the expenses as abandoned plant costs. Although the decisions were challenged in both states, courts upheld the regulators' decisions.

Legal challenges. The Power Agency filed a lawsuit against CP&L in 1989 alleging that the utility had failed to disclose design and management problems prior to their purchase of an interest in the plant. The suit sought damages and a recision of the sale. The Superior Court of Wake County issued two separate rulings in 1991 which struck all references to recision from the complaint and dismissed four of the remaining seven counts against CP&L. The Power Agency immediately appealed.

In other legal action, CP&L filed suit against Westinghouse Electric, the supplier of Harris's steam generators. The suit alleges that the generators were faulty. CP&L claims that although the defects pose no safety hazard, they will result in additional maintenance expenses and will shorten the generators' expected life span.

In spite of its legal difficulties, the Harris plant was recognized in 1990 as one of the 10 lowest-cost producers of nuclear electricity in the U.S. The plant also achieved a capacity factor of 84 percent in 1990.

Decommissioning. In 1990 the NRC issued regulations requiring utilities operating nuclear power units to set aside funds for eventual decommissioning. CP&L established an external decommissioning trust to set aside funds which will be collected through electric rates. The utility plans to decommission the plant using the prompt dismantlement method. At Harris decommissioning costs are estimated at $173.2 million (in 1987 dollars) not including the Power Agency's share of the expenses.

Water cycles and Harris Lake. To operate the plant safely, Harris 1 uses three separate water cycles. Hot water inside the reactor vessel is kept under pressure so that it will not boil. The plant's design isolates this radioactive water from the other water cycles. Its heat is transferred, by means of the steam generators, to water in the second cycle, the steam cycle. As water in the second cycle boils, it produces the steam needed to turn the plant's turbines and thereby generate electricity. The steam is cooled when it passes through a condenser and its heat is transferred to water in the third cycle, the cooling cycle. Excess heat in the cooling cycle is discharged into the atmosphere via Harris's 526-foot cooling tower. Water lost to evaporation during the cooling cycle is replaced by water drawn from Harris Lake.

Harris Lake is a man made body of water created by a 1500-foot dam across Buckhorn Creek two miles above Cape Fear River. It covers 4100 acres, has 40 miles of shore line and an average depth of 18 feet. CP&L constructed Harris Lake to provide water for the nuclear plant but it is not a cooling lake; its function is only to replace water lost through evaporation. The lake and surrounding areas are made available to the public for boating, fishing, hiking, and hunting.

Environmental concerns. CP&L monitors Harris Lake and the adjoining area to make sure the plant does not have an adverse affect on the environment. Portions of the lake and approximately 5,000 acres of land are in the North Carolina Wildlife Resource Commission's Game Lands Program. Of the acres around the lake, 2600 have been certified as an Urban Wildlife Sanctuary including 90 acres set aside to help protect the red-cockaded woodpecker, a federally-listed endangered species.

Harris Energy and Environmental Center. The Harris Energy and Environmental Center houses testing laboratories, training facilities, and the Harris Visitors Center. CP&L also maintains the White Oak Nature Trail which is adjacent to the Visitors Center.

For more information contact: Harris Visitors Center, Rt. 1, Box 327, New Hill, North Carolina, 27562. (919) 362-3261 or (800) 443-8395.

Sources

"Grassroots." *Critical Mass Bulletin*, January 1984.
"Who Wants a Nuke After Chernobyl?" *Sierra*, November/December 1986.
The Harris Plant. Carolina Power and Light Company, May 1989.
Carolina Power and Light Company 1990 Annual Report. Carolina Power and Light Co., n.d.
Carolina Power and Light Company 1991 Annual Report. Carolina Power and Light Co., n.d.
"Wisconsin Electric's Point Beach Lowest Cost Nuclear Plant in '90." *Info*, July 1991.
"The Harris Lake." Carolina Power and Light Company, December 1991.

—Karen Bellenir

Shearon Harris Nuclear Power Plant, Units 2-4

PO Box 165
New Hill, NC 27562
G.E. Vaughn, V.Pres., Harris Nuclear Project

Phone: (919)362-8891
Fax: (919)362-6950

Owning Utility

Carolina Power and Light Company
PO Box 1551
Raleigh, NC 27602
R.A. Watson, Sr.V.Pres., Nuclear Generation

Phone: (919)546-6111
Fax: (919)546-7678

Contact

R.J. White, V.Pres., Corp. Commun.
Carolina Power and Light
PO Box 1551
Raleigh, NC 27601

Phone: (919)546-6111
Fax: (919)546-7678

Basic Facts

Units 2-4 *Status:* Cancelled (Unit 2, construction 4% complete; Units 3-4, construction 1% complete). *Type of Reactor:* Pressurized light-water reactor. *Megawatts (electric):* 950. *Megawatts (thermal):* 2775. *Reactor System Supplier:* Westinghouse Electrical Corp.(United States). *Generator Supplier:* Westinghouse Electrical Corp. (United States). *Architect Engineer:* Ebasco. *Owner:* Carolina Power and Light Company; North Carolina Eastern Municipal Power Agency. *Operator:* Carolina Power and Light Company.

Key Dates

Unit 2 *Ordered:* 1971. *Cancelled:* 1983. *Construction began:* 1978.
Units 3-4 *Ordered:* 1971. *Cancelled:* December 1981. *Construction began:* 1978.

Plant Report

See Plant Report for Shearon Harris Nuclear Power Plant, Unit 1.

Plant Profiles

292 ☆
Hartsville Nuclear Power Plant, Units A1, A2, B1, & B2
Hartsville, TN

Owning Utility

Tennessee Valley Authority
6N 38A Lookout Pl.
1101 Market St.
Chattanooga, TN 37402-2801
J. R. Bynum, V.Pres., Nuclear Operations

Phone: (615)751-0011
Fax: (615)751-4904
Telex: 361951

Basic Facts

Units A1-A2 *Status:* Cancelled (Unit A1, construction 44% complete; Unit A2, construction 34% complete). *Type of Reactor:* Boiling water reactor. *Megawatts (electric):* 1269. *Megawatts (thermal):* 3579. *Reactor System Supplier:* General Electric Co. (United States).

Units B1-B2 *Status:* Cancelled (Unit B1, construction 17% complete; Unit B2, construction 7% complete). *Type of Reactor:* Boiling water reactor. *Megawatts (electric):* 1269. *Megawatts (thermal):* 3579. *Reactor System Supplier:* General Electric Co. (United States).

Key Dates

Units A1-A2 *Ordered:* 1972. *Cancelled:* 1984.
Units B1-B2 *Ordered:* 1972. *Cancelled:* 1982.

Plant Report

Cost projections close down construction of new plant. The Hartsville Nuclear Power Plant occupied a 2,200-acre site in Hartsville, Tennessee, near Nashville. It would have been the world's largest nuclear power plant, but the operating utility, the Tennessee Valley Authority (TVA), cancelled the entire plant in 1984, partially due to high cost projections for operation and maintenance.

TVA had spent $2.4 billion on the plant before it was cancelled. At the time of the deferral, unemployment in the area soared as high as 23 percent.

Cancelled, but still useful. Although the plant was cancelled, TVA still used the site's over 130 buildings as a warehouse facility and distribution center.

In 1989, a regional development board, the Four Lakes Regional Industrial Development Authority, launched a business incubation center at the site. The board hoped to hatch a variety of industries in the economically underdeveloped five-county area around the site.

Sources

"Lost in the Dark: TVA Stumbling Blindly over its Nuclear Program." *Public Citizen,* January 1987.
"Area Board has Sights on Development." *Nashville Business Journal,* May 8, 1989.
"Incubation Center Sends First Pupil on Solo Flight." *Nashville Business Journal,* February 4, 1991.

—*Charles Ceplecha*

293 ●
Edwin I. Hatch Nuclear Power Plant, Units 1-2

PO Box 439
Baxley, GA 31513
H.L. Sumner, Gen.Mgr.

Phone: (912)367-7861
Fax: (912)367-7851

Owning Utility

Georgia Power Company
PO Box 4545
Atlanta, GA 30302
Joseph M. Farley, Chairman, Southern Nuclear

Phone: (404)526-6526

Contact

David Altman, Dir., Corp. Commun.
Georgia Power Co.

Basic Facts

Units 1-2 *Status:* Operable. *Type of Reactor:* Boiling water reactor. *Megawatts (electric):* 810. *Megawatts (thermal):* 2436. *Reactor System Supplier:* General Electric Co. (United States). *Generator Supplier:* General Electric Co. *Architect Engineer:* Georiga Power; Bechtel.

Key Dates

Unit 1 *Ordered:* 1967. *Construction began:* 1968. *Criticality:* September 1974 *First Power:* November 1974. *Commercial Operation:* December 1975.

Unit 2 *Ordered:* 1967. *Construction began:* 1972. *Criticality:* July 1978. *First Power:* September 1978. *Commercial Operation:* September 1979.

Operating Costs

Units 1-2

1976	$5,867,000	1983	$107,802,000
1977	$13,691,000	1984	$139,787,000
1978	$13,936,000	1985	$143,946,000
1979	$27,092,000	1986	$195,755,000
1980	$38,489,000	1987	$134,735,000
1981	$62,011,000	1988	$106,688,000
1982	$67,685,000	1989	$136,127,000

Plant Report

Efficient operations despite 1984 problem. Georgia Power, a subsidiary of Southern Company, is part owner of the Edwin I. Hatch twin nuclear power plant that operates near Baxley, Georgia. Other owners include Oglethorpe Power Corp., the Municipal Electric Authority of Georgia, and the city of Dalton. At the end of 1992, Southern Company transferred daily operations of Hatch over to its new subsidiary, Southern Nuclear Operating Company (SONOPCO) where the needs of the nuclear power industry will be its sole focus. (The address for Southern Nuclear is P.O. Box 1295, Birmingham, AL 35201; Telephone (205) 868-5000.)

Despite the discovery of a four and a half inch crack in a steam pipe of a critical safety system in 1984, Hatch has proved that it operates very efficiently. In 1989, Hatch 1 operated at an impressive 95.5 percent of capacity, and in 1990 it operated for

| 294 | Haven

423 consecutive days—any amount over 365 days is considered outstanding.

Hatch 2 reached 95 percent of capacity in 1990, and together with Hatch 1 the station has set some records for nuclear power plants. The Hatch plants set an endurance record for dual-reactor plants for operating 251 consecutive days. The station also achieved 16 million manhours without a lost-time accident which is the best personnel safety record for nuclear plants operating in the United States.

Sources

"Grassroots." *Critical Mass Bulletin*, February 1984.
"The Eureka File." *Nuclear Industry*, First Quarter 1990.
"Southern Company President Addresses the Company's 44th Annual Meeting of Stockholders." *Business Wire*, May 23, 1990.
"Southern Announces Approval From Securities and Exchange Commission to Form a New Subsidiary." *Business Wire*, December 18, 1990.
"Power Briefs." *Nuclear News*, February 1991.
"Another Rave Review for 1990." *Nuclear Industry*, Fourth Quarter 1991.
"Non-Stop Operating Record Falls Three Times in Two Months." *Nuclear Industry*, Fourth Quarter 1991.
"St. Lucie 2 Completes Record Run, Leads World in Efficiency." *INFO*, U.S. Council for Energy Awareness, April 1992.

—Yvette M. Costa

| 294 | ☆

Haven Nuclear Power Plant, Units 1-2

Haven, WI

Owning Utility

Wisconsin Electric Power Company Phone: (414)221-2345
231 W. Michigan St. Fax: (414)221-2010
Milwaukee, WI 53201
Robert E. Link, V.Pres., Nuclear Power

Contact

Jeff Rauh, Nuclear Info. Coord.
Wisconsin Electric Power Company

Basic Facts

Units 1-2 *Status:* Cancelled. *Type of Reactor:* Pressurized light-water reactor. *Megawatts (electric):* 960. *Megawatts (thermal):* 2785. *Reactor System Supplier:* Westinghouse Electric Corp. (United States). *Owner:* Wisconsin Electric Power Co; Wisconsin Public Service Corp; Wisconsin Power and Light. *Operator:* Wisconsin Electric Power Co.

Key Dates

Unit 1 *Ordered:* 1974. *Cancelled:* 1980.

Unit 2 *Ordered:* 1973. *Cancelled:* 1978.

| 295 | ●

Hope Creek Nuclear Generating Station, Unit 1

PO Box L Phone: (609)935-7400
Hancocks Bridge, NJ 08038 Fax: (609)339-3640
Joseph J. Hagan, Gen.Mgr., Hope Creek Operations

Owning Utility

Public Service Electric and Gas Phone: (201)430-7000
Company
80 Park Plaza
Newawk, NJ 07101
Steven E. Miltenberger, V.Pres. & Chief Nuclear Officer

Contact

Michaele L. Camp, Commun. Mgr., Nuclear
Public Service Electric and Gas Company

Basic Facts

Unit 1 *Status:* Operable. *Type of Reactor:* Boiling water reactor. *Megawatts (electric):* 1170. *Megawatts (thermal):* 3338. *Reactor System Supplier:* General Electric Co. (United States). *Generator Supplier:* General Electric Co. (United States). *Architect Engineer:* Bechtel. *Owner:* Public Service Electric and Gas Company; Atlantic City Electric. *Operator:* Public Service Electric and Gas Company.

Key Dates

Unit 1 *Ordered:* 1969. *Construction began:* 1974. *Criticality:* June 1986. *First Power:* August 1986. *Commercial Operation:* December 1986. *Shut Down Expected:* April 2026.

Operating Costs

Unit 1
1987 $69,421,000
1988 $85,027,000
1989 $88,493,000

Plant Report

The facility. The Hope Creek Nuclear Generating Station sits next to the twin Salem units on Artificial Island in Lower Alloways Creek, New Jersey. The island, 30 miles southwest of Philadelphia in the Delaware River, is three miles long and one mile wide. Hope Creek's cooling tower, the tallest concrete structure in New Jersey, overlooks the 713 acre shared site at the south end of the island.

Construction at the site began in 1976. The original project called for the completion of two units, however Hope Creek 2 was cancelled in December 1981. Unit 1 began commercial operation on December 20, 1986. Construction costs totaled $4.5 billion.

Hope Creek is owned by two New Jersey utilities. Public Service Electric and Gas (PSE&G), the plant's operator, owns 95 percent; Atlantic Electric owns the remaining five percent. PSE&G is the largest utility in New Jersey and provides electricity and gas to more than two million customers.

Construction issues. During the construction period at Hope Creek, Bechtel Power Corporation, instituted a system for assigning computer codes to all the equipment at the nuclear station. The data base provided PSE&G with all the necessary information to meet NRC requirements to maintain a "Master Equipment List." The MEL enabled Hope Creek's start up team to efficiently transfer Bechtel's data into their own computer system and to maintain accurate records through the start up period. The level of detail for each item on the list focuses on what information is required to maintain and operate each piece of equipment in the plant. Programmers expect the MEL to serve for the life of the plant.

Not every event during the construction phase went as smoothly. Bechtel hired a California company to produce a special sealant for Hope Creek's reactor. One of the supplier's employees falsified test records and sold defective sealant to the contractor. If an assistant hadn't reported the fraud to the FBI, the faulty sealant could have clogged pumps vital to the cooling system during a nuclear accident. The incident raised concerns about industry-wide testing procedures.

Early problems. During its first year in operation Hope Creek was plagued by shutdowns. Its longest continuous run, 150 days, was achieved during the first six months of operation, then the plant shut down on July 30, 1987 because of electrical problems. It shut down for two consecutive days in August because of a feed-pump problem. On August 29, 1987 a valve malfunction forced the plant out of service for a day. An outage begun in September for plant inspections lasted almost a month. In October a failure in the main generator shut the plant for nearly two weeks. On December 9th, as its one year anniversary approached, Hope Creek shut down again.

Recognized as an efficient producer. The operating efficiency at Hope Creek has improved in recent years. Since 1987 it has helped PSE&G exceed national capacity factor averages. In 1991 Hope Creek was one of only 12 BWR's operated by American electric utilities (and one of only 23 GE-supplied BWR's in the world) to operate at 75 percent or above.

Emergency preparedness. In accordance with NRC requirements, PSE&G maintains emergency plans for the safety of 25,000 residents who live within 10 miles of the nuclear plants on Artificial Island. The utility sends out annual calendars to provide updated emergency information. It also maintains and tests a siren system which, in the event of a nuclear accident, would alert residents to tune to an Emergency Broadcast System radio station for news and instructions. The establishment of an emergency news center is also tested with working journalists and journalism students participating. Emergency drills are held several times a year, one of which is observed and graded by the NRC.

NRC regulations describe four types of emergencies in ascending levels of threat to the public safety. An unusual event is the least serious. It means that there is a chance that the plant's safety may be reduced. The second, an "Alert" means that there is a possibility of substantial reduction in plant safety or that plant safety has actually been reduced. There have been two "Alerts" at the Hope Creek station. The siren system is not activated unless the third or fourth warning levels are declared. The third level, a "Site Area Emergency" means that there has been or is likely to be a major failure in operations necessary for public protection. The fourth and highest alert level, a "General Emergency" is one in which actual or imminent serious damage to the reactor occurs. There has never been a "Site Area Emergency" or "General Emergency" declared at Hope Creek.

Environmental impact. The NRC and the New Jersey Department of Environmental Protection are among the agencies charged with the responsibility of enforcing regulations to keep harmful effects to the environment at a minimum. PSE&G monitors air and water quality paying close attention to a 20-mile stretch of the Delaware River where the plant draws and discharges a continuous supply of water.

Water is also discharged to the atmosphere as vapor from Hope Creek's cooling tower. Neighboring farmers expressed concern about possible salt deposition in the soil and vegetation from particles carried by the vapor. Consequently, PSE&G monitors salt deposition in many locations up to 8 miles away. A four-year study found the heaviest salt deposits in a parking lot directly below the tower. The closest farm, 3.5 miles away, received less than one-third of a pound per acre. Ocean front land, by way of comparison, receives 32 pounds per acre.

PSE&G is proud of its environmental record at Hope Creek. One official, contrasting nuclear power generation with fossil-generated power, called nuclear production, "an environmentally benign baseload power source."

Sources

"Facts About the Hope Creek Nuclear Generating Station." PSE&G, n.d.
"Nuclear Illusion of Safety." *Mother Jones*, August 1983.
"Hope Creek's MEL Controls Spare Parts." *Electrical World*, September 1985.
"Nuclear Plant Closes for 6th Time in 6 Months." *The New York Times*, December 10, 1987.
PSE&G News. PSE&G, April 1, 1989.
Public Service Enterprise Group Incorporated Annual Report for 1991. Public Service Enterprise Group, Inc., n.d.
"And the Winner Is." *INFO*, U.S. Council for Energy Awareness, May 1992.

—Karen Bellenir

Hope Creek Nuclear Generating Station, Unit 2

PO Box L
Hancocks Bridge, NJ 08038
Joseph J. Hagan, Gen.Mgr., Hope Creek Operations

Phone: (609)935-7400
Fax: (609)339-3640

297 Humboldt

Nuclear Power Plants Worldwide, First Edition

Owning Utility

Public Service Electric and Gas Company
80 Park Plaza
Newawk, NJ 07101
Phone: (201)430-7000
Steven E. Miltenberger, V.Pres. & Chief Nuclear Officer

Contact

Michaele L. Camp, Commun. Mgr., Nuclear
Public Service Electric and Gas Company

Basic Facts

Unit 2 *Status:* Cancelled (construction 19% complete). *Type of Reactor:* Boiling water reactor. *Megawatts (electric):* 1118. *Megawatts (thermal):* 3293. *Reactor System Supplier:* General Electric Co. (United States). *Owner:* Public Service Electric and Gas Company; Atlantic City Electric. *Operator:* Public Service Electric and Gas Company.

Key Dates

Unit 2 *Ordered:* 1969. *Cancelled:* December 1981. *Construction began:* 1976.

Plant Report

See Plant Report for Hope Creek Nuclear Generating Station, Unit 1.

---

297 ★ Humboldt Bay Nuclear Power Plant, Unit 3

Eureka, CA

Owning Utility

Pacific Gas and Electric Company
77 Beale St.
San Francisco, CA 94106
Phone: (415)973-7000
Gregory M. Rueger, Sr.V.Pres., Nuclear Power Generation

Basic Facts

Unit 3 *Status:* Shut down. *Type of Reactor:* Boiling water reactor. *Megawatts (electric):* 65. *Megawatts (thermal):* 210. *Reactor System Supplier:* General Electric Co. (United States). *Generator Supplier:* General Electric Co. *Architect Engineer:* Bechtel.

Key Dates

Unit 3 *Ordered:* 1958. *Construction began:* 1960. *Criticality:* February 1963. *First Power:* April 1963. *Commercial Operation:* August 1963. *Shut Down:* July 1976.

---

298 ★ Indian Point Nuclear Generating Station, Unit 1

Broadway & Bleakley Aves.
Buchanan, NY 10511
Phone: (914)526-5527
Stephen Quinn, Gen.Mgr., Nuclear Power Generation

Owning Utility

Consolidated Edison Co. of New York, Inc.
4 Irving Pl.
New York, NY 10003
Phone: (212)460-4600
Daniel J. Walden, Dir., Pub. Info.

Contact

Michael Spall, Mgr., Nuclear Pub.Info.
Indian Point 2 Nuclear Power Plant

Basic Facts

Unit 1 *Status:* Shut down. *Type of Reactor:* Pressurized light-water reactor. *Megawatts (electric):* 277. *Megawatts (thermal):* 615. *Reactor System Supplier:* The Babcock & Wilcox Co. (United States). *Generator Supplier:* Westinghouse Electric Corp. (United States). *Architect Engineer:* Consolidated Edison Co. of New York, Inc.

Key Dates

Unit 1 *Ordered:* 1955. *Construction began:* 1956. *Criticality:* August 1962. *First Power:* September 1962. *Commercial Operation:* January 1963. *Shut Down:* October 1974.

Operating Costs

Units 1-2

Year	Cost	Year	Cost
1968	$2,831,000	1979	$32,643,000
1969	$2,713,000	1980	$32,965,000
1970	$3,498,000	1981	$54,506,000
1971	$3,962,000	1982	$68,664,000
1972	$6,950,000	1983	$49,911,000
1973	$14,854,000	1984	$96,839,000
1974	$12,737,000	1985	$58,851,000
1975	$13,195,000	1986	$89,015,000
1976	$18,285,000	1987	$109,832,000
1977	$16,525,000	1988	$75,175,000
1978	$28,167,000	1989	$130,579,000

Plant Report

See Plant Report for Indian Point Nuclear Generating Station, Unit 2.

---

299 ● Indian Point Nuclear Generating Station, Unit 2

Broadway & Bleakley Aves.
Buchanan, NY 10511
Phone: (914)526-5527
Stephen Quinn, Gen.Mgr., Nuclear Power Generation

Plant Profiles

Owning Utility
Consolidated Edison Co. of New York, Inc.
4 Irving Pl.
New York, NY 10003
Daniel J. Walden, Dir., Pub. Info.

Phone: (212)460-4600

Contact
Michael Spall, Mgr., Nuclear Pub.Info.
Indian Point 2 Nuclear Power Plant

Basic Facts
Unit 2 *Status:* Operable. *Type of Reactor:* Pressurized light-water reactor. *Megawatts (electric):* 900. *Megawatts (thermal):* 2758. *Reactor System Supplier:* Westinghouse Electric Corp. (United States). *Generator Supplier:* General Electric Co. (United States). *Architect Engineer:* United Engineers & Constructors (United States).

Key Dates
Unit 2 *Ordered:* 1965. *Construction began:* 1966. *Criticality:* May 1973. *First Power:* June 1973. *Commercial Operation:* August 1974.

Operating Costs
Units 1-2

Year	Cost	Year	Cost
1968	$2,831,000	1979	$32,643,000
1969	$2,713,000	1980	$32,965,000
1970	$3,498,000	1981	$54,506,000
1971	$3,962,000	1982	$68,664,000
1972	$6,950,000	1983	$49,911,000
1973	$14,854,000	1984	$96,839,000
1974	$12,737,000	1985	$58,851,000
1975	$13,195,000	1986	$89,015,000
1976	$18,285,000	1987	$109,832,000
1977	$16,525,000	1988	$75,175,000
1978	$28,167,000	1989	$130,579,000

Plant Report

The three units at Indian Point. Indian Point 1, owned by Consolidated Edison, began operating in 1963. It was closed in 1974 after the utility decided it was not practical to add an emergency core cooling system required by new federal regulations. Indian Point 2, which opened in 1973, is owned and operated by Consolidated Edison.

On December 31, 1975, the New York Power Authority purchased the partially completed Indian Point 3 from Consolidated Edison for $385 million. The final cost of the plant, including the purchase price, was $854 million. Indian Point 3 began producing electricity in 1976. In 1978, the NRC transferred the operating license to the New York Power Authority and in 1991 allowed an increase in the operating level to 980 megawatts.

Indian Point 3 supplies electricity to more than 80 governmental customers in New York City and Westchester County. It also provides electricity to Consolidated Edison.

Electricity from Indian Point 2 is distributed to Con Edison's 2.9 million customers in New York City and Westchester County. At full power, the plant generates enough electricity to supply power to approximately one-half million homes. In 1989, nuclear fuel usage by Con Edison customers accounted for 26 percent of the total electricity produced by the utility and cost about one-fifth as much as the oil or gas requirements for an equal amount of electricity.

Radioactive waste. Indian Point 2 is shut down approximately every 20 months for refueling. At that time, one third of the reactor's 193 fuel assemblies are replaced. The replaced fuel assemblies are stored underwater at the site until the U.S. Department of Energy develops a national, permanent high-level radioactive waste repository.

Low-level radioactive waste (LLRW) produced at Indian Point 2 is shipped and subsequently buried as solid waste at licensed disposal sites in Barnwell, South Carolina, Nevada, and Washington. Workers at the plant are trained to reduce, to the greatest extent possible, the amount of LLRW produced at the plant.

Design flaws corrected. In April 1987, Consolidated Edison discovered a fault in the auxiliary cooling system at Indian Point 2. The system's operation relied on a single electrical switch, violating a principle that every piece of safety equipment have redundant features. The switch had not originally been listed in an equipment category requiring redundant features. Its status had been changed by regulations made in the wake of the Three Mile Island accident in 1979. Consolidated Edison avoided a shut down at Indian Point 2 by meeting NRC requirements to fix the problem within 72 hours.

Consolidated Edison reported the problem to the New York Power Authority so that a similar problem could be fixed on the Indian Point 3 plant. At the time, Unit 3 was shut down for routine re-fueling. The repairs were accomplished during the shut down.

Point of controversy. In the 1980s, the three reactors at Indian Point were the subject of the longest series of hearings in the history of the Nuclear Regulatory Commission (NRC). The topic of debate was whether or not proper emergency plans were in place for the surrounding area. Leading the protest was Joan Holt of the New York Public Interest Research Group (NYPIRG).

In 1989, this same group chose not to pursue any opposition against Indian Point 2 when Con Edison was trying to obtain permission to increase the power level by 12.9 percent which would increase the output at the plant from 873 megawatts to 986 megawatts. NYPIRG felt opposition would be futile as the changes being request by Con Edison were supported by Governor Mario Cuomo.

Cracks in steam generators. When workers began the update to increase the power level at Indian Point 2, they discovered cracks in the metal vessels of two steam generators. In February 1990, Con Edison closed down Indian Point 2 to investigate. The inspection was supposed to last four weeks, but continued into June as the condition discovered was worse than expected. Some cracks were three-quarters of an inch long in a wall that is only three and a half inches thick. The possible reasons for the cracks were: metal stress from repeated heating and cooling, variation in metal composition, and/or water chemistry.

Indian

UNITED STATES

With the increase in power output at Indian Point 2, the plant runs the risk of moving closer to damage limits to the core and, in case of an accident, a reduction in the safety margin.

Sources

All About Con Edison. Consolidated Edison, n.d.
Indian Point 2 Nuclear Generating Station. Consolidated Edison, n.d.
Fuel. Consolidated Edison, n.d.
Low Level Radioactive Waste. Consolidated Edison, n.d.
"Design Flaw Reported at 2 Westchester Reactors." *The New York Times,* August 16, 1987.
"Plan to Increase Power at Indian Point is Delayed." *The New York Times,* June 13, 1990.

—Karen Bellenir, Yvette M. Costa

Indian Point Nuclear Power Plant, Unit 3

Box 215 Phone: (914)739-9048
Buchanan, NY 10511
J. Russell, Resident Mgr.

Owning Utility

New York Power Authority Phone: (914)681-6200
123 Main St.
White Plains, NY 10601
Ralph E. Beedle, Exec.V.Pres., Nuclear Generation

Contact

James Steets, Mgr. of Commun. Phone: (914)736-8080
Indian Point 3

Basic Facts

Unit 3 *Status:* Operable. *Type of Reactor:* Pressurized light-water reactor. *Megawatts (electric):* 1000. *Megawatts (thermal):* 3025. *Reactor System Supplier:* Westinghouse Electrical Corp. (United States). *Generator Supplier:* Westinghouse Electrical Corp. *Architect Engineer:* United Engineers & Constructors (United States).

Key Dates

Unit 3 *Ordered:* 1967. *Construction began:* 1969. *Criticality:* 1976. *First Power:* April 1976. *Commercial Operation:* August 1976. *Shut Down Expected:* August 2009.

Operating Costs

Unit 3

Year	Cost	Year	Cost
1976	$2,460,000	1983	$48,683,000
1977	$12,654,000	1984	$55,982,000
1978	$23,318,000	1985	$78,283,000
1979	$28,886,000	1986	$56,857,000
1980	$50,357,000	1987	$58,378,000
1981	$58,175,000	1988	$48,090,000
1982	$82,543,000	1989	$91,608,000

Plant Report

The three units at Indian Point. Indian Point 1, owned by Consolidated Edison, began operating in 1963. It was closed in 1974 after the utility decided it was not practical to add an emergency core cooling system required by new federal regulations. Indian Point 2, which opened in 1973, is also owned and operated by Consolidated Edison.

On December 31, 1975, the New York Power Authority purchased the partially completed Indian Point 3 from Consolidated Edison for $385 million. The final cost of the plant, including the purchase price, was $854 million. Indian Point 3 began producing electricity in 1976. In 1978, the NRC transferred the operating license to the New York Power Authority and in 1991 allowed an increase in the operating level to 980 megawatts.

Indian Point 3 supplies electricity to more than 80 governmental customers in New York City and Westchester County. It also provides electricity to Consolidated Edison.

Safety concerns higher than most. The Indian Point plants, located in Buchanan NY, are in a densely populated area only 35 miles from mid-Manhattan. Because of this location, the plants receive greater scrutiny than many U.S. nuclear power plants and they have more safety features. For example, welds in the containment liner at Unit 3 have special systems installed in them to detect and prevent leakage of radioactivity.

Emergency plans for the plants are administered by officials of four surrounding counties: Westchester, Rockland, Orange and Putnam. Although the emergency plans have not been used for an accident at Indian Point, parts of the plans have proven useful in coping with other incidents, including chemical fires and severe weather problems.

Design flaws corrected. In April 1987, Consolidated Edison discovered a fault in the auxiliary cooling system at Indian Point 2. The system's operation relied on a single electrical switch, violating a principle that every piece of safety equipment have redundant features. The switch had not originally been listed in an equipment category requiring redundant features. Its status had been changed by regulations made in the wake of the Three Mile Island accident in 1979. Consolidated Edison avoided a shut down at Indian Point 2 by meeting NRC requirements to fix the problem within 72 hours.

Consolidated Edison reported the problem to the New York Power Authority so that a similar problem could be fixed on the Indian Point 3 plant. At the time, Unit 3 was shut down for routine re-fueling. The repairs were accomplished during the shut down.

Minor problems reported. In January 1987, the NRC fined the New York Power Authority $50,000 for mispositioned switches at Indian Point 3. The next month, the plant experienced an automatic shut down when one of two pumps providing water to the generator malfunctioned. The utility stressed that the event posed no danger to the public and that no radiation had been released.

Another $50,000 fine was levied in 1990 when NRC investigators claimed that on September 14 two operators weren't fully attentive. The NRC inspector charged that one operator had his feet on his desk with eyes closed and that another operator had his head tilted back and his eyes closed. The Power Authority denied the charges claiming that both operators had performed documented activities only moments before the supposed in-

Plant Profiles

fraction and that one of the operators had a minor back problem and was merely leaning back in his chair to stretch his back.

Fuel accidentally pulled from reactor. On September 15, 1990 Indian Point 3 shut down for a scheduled refueling. When workers lifted the top off the reactor using an overhead crane, two fuel bundles were accidentally pulled up. Although the fuel was not damaged, the fuel assemblies could have triggered a release of radioactive gases inside the plant if they had fallen back into the core and damaged the protective sleeve around the fuel. Although the situation was considered serious, the Power Authority did not declare an "Unusual Event" and no radiation was released.

Sources

"News for Further Information: New York Power Authority." New York Power Authority, n.d.

"News for Further Information: Indian Point 3 Nuclear Power Plant." New York Power Authority, n.d.

"News for Further Information: Indian Point 3 Nuclear Power Plant Background." New York Power Authority, n.d.

"NRC to Fine Agency in New York $50,000 for Errors at a Plant." *The Wall Street Journal*, January 22, 1987.

"Pump Failure Causes Indian Pt. Shutdown." *The New York Times*, February 2, 1987.

"Design Flaw Reported at 2 Westchester Reactors." *The New York Times*, August 16, 1987.

"Nuclear Mishap Monitored." *Newsday*, October 6, 1990.

"NRC Seeks $50,000 Fine Against New York Utility." *The Wall Street Journal*, December 11, 1990.

"Due Up for License Renewal: The Future of Nuclear Power." *The New York Times*, June 24, 1991.

Indian Point 3: 15 Years of Service. New York Power Authority, August 1991.

"Nuclear Mishap Monitored." *Newsday*, October 6, 1990.

"NRC Seeks $50,000 Fine Against New York Utility." *The Wall Street Journal*, December 11, 1990.

—Karen Bellenir

301 ☆
Isolte Nuclear Power Plant
Arecibo, PR

Basic Facts
Status: Cancelled. *Type of Reactor:* Pressurized light-water reactor. *Megawatts (electric):* 625. *Owner:* Puerto Rico Water Resources Authority.

Key Dates
Units 1-2 *Ordered:* 1973. *Cancelled:* 1980.

302 ☆
Jamesport Nuclear Power Plant, Units 1-2
Jamesport, NJ

Owning Utility
Long Island Lighting Co. (LILCO) Phone: (516)933-4590
175 E. Old Country Rd
Hicksville, NY 11801

Contact
J.W. McDonnell, V.Pres. of Pub.Aff.
Long Island Lighting Co. (LILCO)

Basic Facts
Units 1-2 *Status:* Cancelled. *Type of Reactor:* Pressurized light-water reactor. *Megawatts (electric):* Unit 1, 1229; Unit 2, 1182. *Megawatts (thermal):* 3425. *Reactor System Supplier:* Westinghouse Electric Corp. (United States). *Owner:* Long Island Lighting Co. *Operator:* Long Island Lughting Co.

Key Dates
Units 1-2 *Ordered:* 1973. *Cancelled:* 1980.

303 ●
Kewaunee Nuclear Plant
North 490, Highway 42 Phone: (414)388-2560
Kewaunee, WI 54216-9510
Mark L. Marchi, Plant Mgr.

Owning Utility
Wisconsin Public Service Corporation Phone: (414)433-1598
600 N. Adams St.
PO Box 19002
Green Bay, WI 54307-9002
Clark R. Steinhardt, Sr.V.Pres., Nuclear Power

Contact
Doug E. Day Nuclea, Wisconsin Public Phone: (414)433-5528
Service Corporation Fax: (414)433-5741

Basic Facts
Status: Operable. *Type of Reactor:* Pressurized light-water reactor. *Megawatts (electric):* 563. *Megawatts (thermal):* 1650. *Reactor System Supplier:* Westinghouse Electric Corp. (United States). *Generator Supplier:* Westinghouse Electric Corporation. *Architect Engineer:* Pioneer Service and Engineering Company. *Owner:* Wisconsin Public Service Corporation; Wisconsin Power and Light; Madison Gas and Electric. *Operator:* Wisconsin Public Service Corporation.

Key Dates
Ordered: 1967. *Construction began:* November 1967. *Criticality:* 1974. *First Power:* April 1974. *Commercial Operation:* June 1974

Operating Costs
1974	$7,222,000	1982	$21,979,000
1975	$8,945,000	1983	$23,926,000
1976	$10,727,000	1984	$27,829,000
1977	$10,924,000	1985	$31,605,000
1978	$10,430,000	1986	$32,603,000
1979	$11,323,000	1987	$38,989,000
1980	$14,844,000	1988	$49,798,000
1981	$19,334,000	1989	$50,669,000

● Operable ○ Under Construction △ Indefinitely Deferred ▲ Decommissioning □ Planned ■ On Order ☆ Cancelled ★ Shut Down

Kewaunee

Plant Report

Electricity for Wisconsin. The Kewaunee Nuclear Plant occupies a 900-acre site along Lake Michigan in Carlton, Wisconsin. The site is about nine miles south of Kewaunee and 35 miles east of Green Bay, Wisconsin.

Kewaunee houses one, two-loop pressurized water reactor, capable of producing 535,000 kilowatts of electricity. It was the fourth nuclear plant built in Wisconsin, and the 44th in the United States.

The plant was built at an initial cost of $212 million, by a construction workforce with an average of 450 workers. About 400 people now work at the plant's Nuclear Department, or in Wisconsin Public Service's offices in Green Bay.

Wisconsin Public Service (WPS) owns 41.2 percent of Kewaunee, which supplies 20 percent of the utility's electricity to customers in northeast Wisconsin. WPS operates the plant for the two other owners, Wisconsin Power and Light, which owns 41 percent and serves central and southern Wisconsin, and Madison Gas and Electric, which owns 17.8 percent and serves the Madison area.

Features of Kewaunee. The concrete shield building has an inside diameter of 105 feet, and an inside height of 212 feet. The building houses the reactor vessel and core, steam generator, pressurizer and coolant pump.

The reactor vessel is 39 feet high, and weighs 229 tons. The walls have a thickness of 6.5 inches. The core contains 121 fuel assemblies, made up of 5 million uranium pellets. The assemblies are 12 feet long, and one foot in diameter. Each year, one-third of the core, which is 37 assemblies, is replaced.

Westinghouse supplied the steam supply equipment and the turbine/generator for the plant. The steam generator is 65 feet high, and contains 3,388 U-tubes. The pressurizer is designed to keep the primary coolant system under 2,200 pounds per square inch pressure.

The reactor coolant pump is a 6,000-horsepower electric pump. It pumps water from Lake Michigan at 92,500 gallons per minute. When the plant is at full power, approximately 413,000 gallons are circulated through the condenser each minute.

Waste management. Low-level waste from Kewaunee is sent to South Carolina, where it is isolated at a regulated disposal site. High-level waste is being temporarily stored at the plant's spent fuel pools until the U.S. Deparment of Energy builds a national, permanent high-level radioactive waste repository.

Accomplishments of 1990. During the year, Kewaunee operated for more than 200-plus consecutive days. This marked the seventh time in the plant's 16-year history that it achieved the feat.

Kewaunee received good marks in two major assessments during the year. The NRC praised the plant for "sustained good performance," putting it in the top five percent of U.S. plants, and an evaluation by the Institute of Nuclear Power Operations resulted in a Category 1 rating for the plant.

Decommissioning. External trust funds are maintained to cover the estimated future decommissioning costs of Kewaunee. These costs are recovered from WPS customers as rates, and deposited in the trust as depreciation expenses. As of 1991, WPS's share of future decommissioning costs was estimated at $275 million.

Environmental programs. A wooded portion of the Kewaunee site has been set aside as an outdoor education laboratory. Formerly known as Kafta's Grove, the six-acre area is located on the shore of Lake Michigan, just south of the plant. It is used by Kewaunee area school students for outdoor environmental and nature studies, and features a shelter, an outdoor classroom and nature trails.

A wide-range monitoring program is ongoing at Kewaunee. As a part of this program, plant workers collect milk samples from area dairy farmers to analyze and insure the milk's purity.

Milestones. Following are important events in the history of Kewaunee:

- November 28, 1967: Ground breaking at Kewaunee.
- October 25, 1973: Objections to the plant's construction are withdrawn.
- December 21, 1973: AEC release, full power, and full-term license.
- July 16, 1974: Dedication ceremony.
- May 25, 1975: Because of Kewaunee, WPS customers bill down $2 million.
- December 1, 1976: Plant record established for 85-day run.
- December 22, 1977: Another record ends at 127 days as plant is shut down.
- July 24, 1977: NRC issues $10,000 fine (later reduced to $7,000).
- January 8, 1980: Small electrical fire interrupts plant operatons.
- August 20, 1980: Lightening outage at Kewaunee.
- April 9, 1982: Refueling and a 306-day record run.
- July 28, 1982: NRC rates Kewaunee among top plants.
- October 6, 1982: Sensing line capped off, but Kewaunee is eventually fined $30,000.
- July 14, 1983: NRC rates Kewaunee "very good."
- February 7, 1984: Kewaunee has lowest radiation exposure of nuclear plant workers.
- February 20, 1984: Plant simulator is unveiled.
- March 31, 1986: NRC reduces inspections at Kewaunee.
- December 11, 1986: Kewaunee gets national reliability record.
- November 23, 1987: Kewaunee receives second INPO award.
- February 28, 1989: Kewaunee receives perfect score on SALP report.
- June 1, 1989: Kewaunee gets five-year license extension.

- February 27, 19990: NRC rates Kewaunee "consistently good."
- March 11, 1991: Annual outage and record 325-day run.

Sources

Historical Sketch of Kewaunee Nuclear Plant. Wisconsin Public Service Corporation, n.d.
Kewaunee Nuclear Plant. Wisconsin Public Service Corporation, n.d.
Kewaunee Nuclear Plant Fact Sheet. Wisconsin Public Serivce Corporation, n.d.
Kewaunee Nuclear Power Plant. Wisconsin Public Service Corporation, n.d.
Generating Electricity.at Wisconsin Public Service. Wisconsin Public Service Corporation, November 1987.
Wisconsin Public Service Corporation 1990 Annual Report. Wisconsin Public Service Corporation, n.d.

—Charles Ceplecha

304 •
Lasalle County Nuclear Station, Units 1-2

R.R. No. 1 Phone: (815)357-6761
Box 220
Marseilles, IL 61341
Gerald Diederich, Stat.Mgr.

Owning Utility

Commonwealth Edison Company Phone: (312)294-4321
One First National Plaza Fax: (312)294-2995
PO Box 767
Chicago, IL 60690
Cordell Reed, Sr.V.Pres., Nuclear Operations

Contact

John F. Hogan, Dir., Commun. Svcs.
Commonwealth Edison Company

Basic Facts

Units 1-2 *Status:* Operable. *Type of Reactor:* Boiling water reactor. *Megawatts (electric):* 1078. *Megawatts (thermal):* 3323. *Reactor System Supplier:* General Electric Co. (United States). *Generator Supplier:* General Electric Company. *Architect Engineer:* Sargent & Lundy Engineers (United States).

Key Dates

Unit 1 *Ordered:* 1970. *Construction began:* 1973. *Criticality:* June 1982. *First Power:* August 1982. *Commercial Operation:* December 1982.

Unit 2 *Ordered:* 1970. *Construction began:* 1972. *Criticality:* March 1984. *First Power:* April 1984. *Commercial Operation:* October 1984.

Operating Costs
Units 1-2

1982	$4,820,000	1986	$80,417,000
1983	$35,381,000	1987	$101,455,000
1984	$59,982,000	1988	$137,789,000
1985	$74,801,000	1989	$123,010,000

Plant Report

A power station for central Illinois. LaSalle County Nuclear Station is located on a 3,055-acre site in LaSalle County, about six miles southwest of Marseilles, Illinois. The site is about three miles south of the Illinois River and 70 miles southwest of Chicago.

The initial project cost for both units of LaSalle County Station was $2.3 billion. In July 1984, Commonwealth Edison was allowed to pass along a $282 million rate increase to cover costs of Unit 2.

Features. LaSalle County Station's pressure vessels measure 72 feet high with a diameter of 20 feet 11 inches. The walls have a thickness of 6 3/4 inches with a 3/16 inch stainless steel overlay. The design pressure is 1250 pounds per square inch absolute.

There are 764 fuel assemblies, containing 62 fuel rods each. Each rod is 150 inches long and .483 inches in diameter. The total core represents 154 tons of uranium at an average enrichment of 1.9 percent U235. Between one-third and one-fourth of the total core is replaced approximately every 18 months during scheduled outages.

The turbine generators are driven by steam at 1800 revolutions per minute. The generators produce electricity at 25,000 volts, stepped up by transformer to 345,000 for transmission.

The condenser cooling system is a lake, with a pond surface area of 2058 acres. The average depth is 13 feet, and the lake holds approximately 13 billion gallons of water. To maintain cooling water level and quality, 60,000 gallons of water per minute are circulated from the Illinois River.

Incident causes a shutdown. In March 1988, a huge oscillation in speed of a nuclear reaction caused an automatic shutdown of LaSalle County Station. A worker accidently caused the shutdown of two pumps that circulated water in the core. This triggered a condition in which the core alternated between being extremely productive to the nuclear reaction and hostile to it, causing the oscillation. No damage was done to the plant, and no radiation was released.

Prior to LaSalle County Station's opening in 1984, analyses predicted that such an oscillation could not occur, so little guidance and training were provided for operator detection and response. Operators responded unsuccessfully by trying to restart the pumps, which could have made the problem worse.

The NRC feared that such a runaway nuclear reaction could cause a plant to overheat, melting fuel and releasing radiation. The incident also raised serious doubts on whether LaSalle County Station ever met its requirements for stability.

In June 1988, the NRC gave Commonwealth Edison, and the owners of 35 other plants manufactured by General Electric, 60 days to analyze the vulnerability of their plants to the kind of instability that occurred at LaSalle County Station.

Low-cost producer. In 1984, the plant's costs were estimated at $1,153 a kilowatt, a generally lower cost than many other U.S. plants. In the same year, the plant was predicted to be producing power for 5.9 cents a kilowatt, showing a generation of some of the lowest-cost nuclear power in the United States.

According to the Utility Data Institute (UDI), an inspection, research and data base publishing company in Washington, D.C., LaSalle County Station was the seventh lowest-cost producer of nuclear electricity in the United States in 1990. The plant had a production cost of $14.65 per megawatt-hour of electricity at 14,821,165 megawatt-hours.

Award winner. In 1992, LaSalle's Unit 2 was honored by GE Nuclear Energy for being one of 23 U.S.-supplied boiling water reactors that operated at 75 percent efficiency or above during 1991. The plant had a capacity factor of 91.6 percent.

Sources

"Man Bites Dog." *Forbes*, December 3, 1984.
LaSalle County Nuclear Station Fact Sheet, Commonwealth Edison Co., May 1988.
"Incident at Illinois Reactor Prompts Inquiry on Dangers." *The New York Times*, July 10, 1988.
"Wisconsin Electric's Point Beach Lowest Cost Nuclear Plant in '90." *Info*, U.S. Council for Energy Awareness, July 1991.
"U.S. Nuclear Plants Shatter Records for Output, Performance." *Info*, U.S. Council for Energy Awareness, February 1992.
"And the Winner is." *Info*, U.S. Council for Energy Awareness, May 1992.

—Charles Ceplecha

---

305 ●
Limerick Generating Station, Units 1-2

Evergreen & Sanatoga Rds. Phone: (215)327-1200
PO Box A Fax: (215)327-1200
Sanatoga, PA 19464
Jay Doering, Plant Mgr.

Owning Utility

Philadelphia Electric Company Phone: (215)841-4000
2301 Market St. Fax: (215)841-4188
PO Box 8699
Philadelphia, PA 19101
Dickinson M. Smith, Sr.V.Pres., Nuclear

Contact

Neil J. McDermott, Dir., Media & PR
Philadelphia Electric Company

Basic Facts

Units 1-2 *Status:* Operable. *Type of Reactor:* Boiling water reactor. *Megawatts (electric):* 1098. *Megawatts (thermal):* 3293. *Reactor System Supplier:* General Electric Co. (United States). *Generator Supplier:* General Electric Co. *Architect Engineer:* Bechtel.

Key Dates

Unit 1 *Ordered:* 1969. *Construction began:* 1974. *Criticality:* December 1984. *First Power:* April 1985. *Commercial Operation:* February 1986. *Shut Down Expected:* October 2024.
Unit 2 *Ordered:* 1969. *Construction began:* 1974. *Criticality:* August 1989. *First Power:* 1989. *Commercial Operation:* January 1990.

Operating Costs
Unit 1
1986 $48,687,000
1987 $135,986,000
1988 $124,916,000
1989 $192,083,000

Plant Report

Construction of two units separated. The Limerick Generating Station, located 21 miles northwest of Philadelphia in Limerick Township, has two 1,100 megawatt boiling water reactors. Philadelphia Electric Company (PECO) owns and operates both units. Nuclear power generates approximately 65 percent of the electricity supplied to PECO's customers.

The NRC issued construction licenses for both units in June 1974. The project experienced numerous cost overruns and in 1982 Pennsylvania's Public Utility Commission (PUC) ordered PECO not to borrow any more funds for Unit 2 until Unit 1 was finished. The borrowing moratorium stopped virtually all construction on Unit 2. PECO had already spent nearly $1 billion on the second unit and it was about 30 percent complete.

Unit One construction and operating history. Construction continued, however, at Limerick 1 and in February 1986 commercial operation commenced. The final cost figures totaled $3.8 billion. Of this amount, the PUC ruled that $369 million had been imprudently incurred. The regulatory agency refused to allow PECO to recapture the imprudent expenses through electric rates.

During its first year in operation, Limerick 1 set a world record for a boiling water reactor in its first cycle — 198 consecutive days of operation. The NRC's Systematic Assessment of Licensee Performance (SALP) report covering the plant's first year, February 1, 1986 through January 31, 1987, gave Limerick 1 the highest possible marks in eight of 10 categories.

Unit Two construction. Although construction on Limerick 2 was halted for two years, PECO had continued developing engineering plans for the plant. On December 5, 1985 after controversial rate hike hearings and charges that PECO did not need the additional generating capacity, the PUC issued an order giving PECO permission to resume construction. The regulatory agency, however, imposed a cost cap of $3.2 billion on the project.

When construction was able to proceed, PECO targeted the fourth quarter of 1990 for starting commercial operation. Having thorough plans in hand enabled work to progress at a faster rate than was customarily realized during nuclear plant construction. It took 40 months to bring the unit from 30 percent completion to 100 percent.

PECO officials estimate that the pre-planned system saved 24 months of construction time which reduced interest expenses. The savings made up for the $240 million interest expense on outstanding debt incurred during the work hiatus. Construction was finished nine months ahead of time and $382 million under

Plant Profiles • Limerick

budget. Limerick 2 is the only U.S. nuclear plant to be finished after an extended work stoppage.

Limerick 2 start up. In preparation for starting Limerick 2, PECO needed to connect several systems the unit would share in common with Limerick 1. The utility planned to accomplish the work during a refueling outage at Unit 1 scheduled to begin January 13, 1989. During the refueling workers discovered that twice as much fuel had to be replaced as originally planned. Refueling at Limerick 1 was performed by reconstituting the fuel bundles and by using some of the fuel originally slated for Limerick 2. In spite of the difficulties, Limerick 1's refueling was finished only five days over schedule. Work on Limerick 2 was completed and on May 4, 1989 the NRC Advisory Committee on Reactor Safeguards recommended authorization of low-power testing.

Fuel loading at Limerick 2 was planned for June. Limerick Ecology Action (LEA) an intervenor against construction of the plant, filed a lawsuit to stop the licensing procedure. LEA claimed the utility had inadequate provisions for evacuating a state prison and had failed to install safety systems necessary to mitigate severe accidents. In response, the NRC authorized fuel loading but barred low-power testing until after the court ruling.

Low-power testing began on August 11, 1989. PECO applied for full-power license and LEA threatened further court action. PECO and LEA reached a settlement whereby LEA agreed to give up the right to appeal the NRC's decision and PECO agreed to:

- make a $2 million modification to plant safety systems
- allow an LEA representative access to the plant for inspections
- provide LEA copies of all NRC documents related to licensing
- contribute to educational programs
- pay LEA's legal expenses.

Controversial water diversion. Legal disputes over the Limerick Generation Station also came from Del-AWARE, an environmentalist group, protesting a pumping station on the Delaware River. The Point Pleasant Pumping Station, located north of the plant in Bucks County, was vital to Limerick's supplementary cooling system. Court challenges forced delays in construction and the $21 million station was completed only four weeks before the NRC voted on Limerick 2's full-power license. Less than a month after the license was granted, the primary water supply fell below acceptable levels during a dry spell. This event activated the controversial supplemental water system.

Commercial operation. Limerick 2 began commercial production in January 1990. Its start-up was completed in 200 days, setting a record for a U.S. boiling water reactor. The unit operated at 80 percent capacity during its first year. The completion of the project marked the end of a twenty-year construction program through which PECO increased its generating capacity by 6,800 megawatts at a cost of $9 billion.

In July 1989 PECO, asked for a $549 million rate increase to cover the cost of Limerick 2. The PUC allowed only $242 million (44 percent of the request) stating that 399 megawatts represented excess capacity.

Limerick Station's honors. The Limerick Station's commitment to an "ALARA" (as low as reasonably achievable) policy has helped the plant set records for the lowest radiation exposure received by workers. Radiation is measured in "rems." Rem stands for Roentgen-Equivalent-Man; a rem is a standard measurement of radiation absorption. In 1990, radiation exposure at Limerick was 175 man-rem. The national average for plants with two boiling water reactors was 880 man-rem. The station also received recognition for having the lowest three-year collective exposure, 177 man-rem per year.

In 1990, Limerick 1 was praised for achieving a long operating run. It completed 378 days of continuous operation, earning the unit a place on the top twenty list for long runs in the U.S. In 1991, Limerick 2 received recognition as one of only 12 boiling water reactors operated by American electric companies to operate at 75 percent capacity or higher.

Radon Gas identified. The Limerick Generating Station is also known for its role in helping to increase understanding and awareness regarding radon gas. In 1984, a worker at the Limerick complex unexpectedly tripped radioactivity detectors. Because the plant was inactive, health specialists tested the worker's house. There they found radon levels equivalent to almost 500,000 chest X-rays. Radon gas is a byproduct of radioactive substances found naturally in soil and rock. Normally it escapes from the earth and dissipates into the atmosphere without causing any ill-effects. When the gas becomes trapped in an enclosed space, such as a home, it can cling to dust particles and enter the lungs. Radon is the second highest contributor to lung cancer in America. Informational radon seminars are held at Limerick's visitor's center.

Visitor's Center. Limerick's twin 507-foot high natural draft cooling towers create a dramatic backdrop to the Limerick Energy Information Center. Visitors to the center can view exhibits about the construction and operation of the generating station, see a solar energy display, and view demonstrations in a 100-seat auditorium. Plant tours are available which include the Limerick Training Center where a mock control room is on display. PECO also maintains the Brooke Evans Creek Nature Trail which boasts more than a hundred labeled trees and plants as well as snakes and turtles.

The Limerick Energy Information Center is open Tuesdays through Saturdays (except holidays) from 10:00 a.m. until 4:00 p.m. For more information write the Limerick Energy Information Center at 298 Longview Road, Linfield, PA 19468; or phone (215) 495-6767 or (215) 495-7018.

Sources

"The Electric Company in Shock." *Philadelphia*, November 1984.
"PECO Shakes Up Its Management." *Focus*, February 10, 1988.
"PECO's Problem and Pride." *Focus*, February 17, 1988.
"The Peach Bottom Syndrome." *The New York Times*, March 27, 1988.
"Austin Firm Hopes to Cash in on Concerns About Radon Gas." *Austin Business Journal*, May 23, 1988.

"Building Reactors the New Way." *The New York Times*, July 17, 1989.
"Paradise Lost . Paradise Regained." *Nuclear Industry*, First Quarter 1990.
Philadelphia Electric Company Annual Report 1990. Philadelphia Electric Co., n.d.
"PE Reports Continued Low Worker Exposure to Radiation at Limerick." Press Release: Philadelphia Electric Company, February 5, 1991.
"Explore Energy: Visit the Limerick Energy Information Center." Philadelphia Electric Company, n.d.
"Limerick Generating Station: Units 1 & 2 Fact Sheet." Philadelphia Electric Company, n.d.
"St. Lucie 2 Completes Record Run, Leads World in Efficiency." *Info*, U.S. Council for Energy Awareness, April 1992.
"And the Winner Is." *INFO*, U.S, Council for Energy Awareness, May 1992.
"Explore Energy: Visit the Limerick Energy Information Center." Philadelphia Electric Company, n.d.
"Limerick Generating Station: Units 1 & 2 Fact Sheet." Philadelphia Electric Company, n.d.
"St. Lucie 2 Completes Record Run, Leads World in Efficiency." *Info*, U.S. Council for Energy Awareness, April 1992.

—Karen Bellenir

306 • Maine Yankee Nuclear Power Plant

PO Box 408 Phone: (207)882-6321
Wicasset, ME 04578
R.W. Blackmore, Plant Mgr.

Owning Utility

Maine Yankee Atomic Power Company Phone: (207)622-4868
Edison Dr.
Augusta, ME 04336
C.E. Monty, Chairman of the Board

Contact

M.D. Murphy, Dir., Pub.Aff.
Maine Yankee Nuclear Power Plant

Basic Facts

Status: Operable. *Type of Reactor:* Pressurized light-water reactor. *Megawatts (electric):* 890. *Megawatts (thermal):* 2700. *Reactor System Supplier:* Combustion Engineering, Inc. (United States). *Generator Supplier:* Westinghouse Electric Corporation (United States). *Architect Engineer:* Stone & Webster Engineering Corp. (United States).

Key Dates

Ordered: 1967. *Construction began:* 1968. *Criticality:* 1972. *First Power:* October 1972. *Commercial Operation:* December 1972. *Shut Down Expected:* 2008.

Operating Costs

Year	Cost	Year	Cost
1973	$4,034,000	1982	$28,556,000
1974	$5,232,000	1983	$21,556,000
1975	$6,301,000	1984	$32,495000
1976	$5,261,000	1985	$35,760,000
1977	$8,418,000	1986	$21,439,000
1978	$10,817,000	1987	$48,323,000
1979	$9,973,000	1988	$46,471,000
1980	$14,028,000	1989	$35,655,000
1981	$20,575,000		

Plant Report

Background. Located on Montsweag Bay in Wiscasset, Maine, Yankee is the state's only nuclear power plant. Maine Yankee has been in operation since 1972 and currently has an approximate capacity of 850,000 kilowatts, producing about 5.2 billion kilowatt-hours (Kwh) of electricity each year. Utility companies in Maine own 50 percent of the pressurized water reactor's output, which provides roughly one quarter of the state's total electricity supply. Power produced at Maine Yankee costs an average of less than three cents per kilowatt-hour and saves the burning of eight million barrels of fuel oil each year.

The Maine Yankee Atomic Power Company was formed in January of 1966, and heavy construction began on the plant in 1968. The Atomic Energy Commission issued an operating license authorizing the plant to operate at 75 percent capacity in September of 1972, and the plant went critical that October. Through the '70s and '80s, the majority of the plant's history has been one of award-winning production and safety levels, steady improvement of plant equipment in order to increase output, and surviving three state-wide referendums proposing the closure of the plant. As of December 31, 1991, Maine Yankee's net lifetime electricity production exceeded 89 million Kwh. The operating license for Maine Yankee expires in 2008.

Waste Disposal: Low Level. Maine Yankee, like nuclear plants all over the U.S., has become increasingly concerned about disposal sites for both low-and high-level radioactive waste (LLRW and HLRW). As of 1990, three years before federally licensed low-level disposal sites in South Carolina, Nevada, and Washington state will no longer be required to accept LLRW from other states, Maine had yet to make other arrangements that would allow it to continue shipping its LLRW out of state. Pointing to the small amount of LLRW produced in Maine as a percentage of that produced by the nation as a whole, Maine Yankee officials endorse the state's desire to continue shipping LLRW out of state. In 1991, as the deadline for finding an alternative LLRW dump site drew near however, Maine Yankee formally volunteered its site for consideration.

Waste Disposal: High Level. In the absence of a permanent national repository, Maine Yankee disposes its high-level radioactive waste in a spent fuel pool on the plant site. In the early 1980s, the federal government, which retains sole legal responsibility for locating a permanent disposal site for HLRW, issued the Nuclear Waste Policy Act, which sets forth a complete program to build the nation's first permanent high-level radioactive waste repository, scheduled to be operational by the year 2010. In 1987, Congress designated Yucca Mountain in Nevada as the only site to be investigated further for this purpose. In the fall of 1990, a federal appellate court ruled that the Department of Energy (DOE) could proceed with its on-site investigations despite the objections of the Nevada governor, a ruling upheld by the U.S. Supreme Court. As Maine Yankee entered the 1990s, it reported having 1009 spent fuel assemblies in storage, about two-thirds of its total capacity of 1,476, in its spent fuel pool.

Referendums 1 and 2: In the wake of Three Mile Island. In the fall of 1980, Maine citizens voted on the first antinuclear referendum in the U.S. since the accident at Three Mile Island; five other states had nuclear power issues on their ballots that same fall. A sculptor and farmer, Raymond Shadis, began the movement to put the fate of Maine Yankee before the state's voters after the Three Mile Island disaster lead him to investigate the safety of the nuclear power plant in his own backyard. With concerns ranging from waste disposal to unreported radioactive emissions, Shadis and the antinuclear coalition, the Maine Nuclear Referendum Committee, contended that closing the plant would have little negative impact on the surrounding community. The power lost would quickly be compensated for by conservation efforts and alternative sources of power, particularly hydropower.

Opponents of the plant closing focused on economic issues, contending that the cost of replacing the lost power generated by Maine Yankee would amount to $140 million a year. Also, Wiscasset would lose millions in tax revenues from both the plant and unemployed workers. Utility companies across the country were reportedly watching this battle closely in the belief that the closing of one nuclear plant could toll the death knell for others.

An explosion in a Titan missile silo in Arkansas in September ought to have aided the antinuclear cause in Maine in November 1980. However, the Save Maine Yankee Committee, a coalition of utility companies and nuclear-related industries across the country, outspent the campaign to shut down the plant by a large margin, employing a media campaign that extolled the safety and cost-effectiveness of nuclear power. In the largest turnout to date for a single-issue referendum in the state's history, the citizens of Maine voted 230,000 to 160,000 to keep Maine Yankee open. Acknowledging that a three-to-two margin did not represent overwhelming support for nuclear energy, Carl Walske, president of the Atomic Industrial Forum, remarked: "We have not dispelled the notion that nucler power presents an unusual risk for the public." A similar referendum held in 1982 yielded similar results: voters supported the continued operation of Maine Yankee by a vote of 56 to 44 percent.

Had the antinuclear initiative passed, it would have been the first to close a plant already in operation. This unprecedented action would have set in motion what all predicted would be a protracted legal battle over the question of whether the power to close a nuclear plant resides in state or federal hands.

Referendum 3: In the wake of Chernobyl. Against the background of the recent nuclear disaster at Chernobyl in the former Soviet Union, and with the DOE scouting locations for a permanent high-level radioactive waste disposal site in Maine and elsewhere, feelings were running high against nuclear energy in 1987. Arguing that if the state were not producing high-level waste it would be less likely to be chosen as the site for a permanent high-level radioactive waste repository, antinuclear activists collected more than enough signatures to place the issue on the ballot within a single day of circulating the petition. The question itself, which asked: "Do you want to let any power plant like Maine Yankee operate after July 4th 1988 if it makes high level nuclear waste?" was controversial, requiring a "no" vote in order to show support for the plant.

Media analyses of the effects of a successful campaign by the antinuclear forces focused again on the legal and economic ramifications of a plant closure. Estimates of the cost to compensate the plant owners for closing their plant ran between $2 and 4 billion. Robert Deis, spokesperson for People for Maine Yankee's Electricity, a political action group that spent a reported $4.7 million in advertising on behalf of the plant's continuing operation, claimed that 4500 people, including 400 directly employed by Maine Yankee, would lose their jobs due to the plant closing. An additional $1-2 billion would be spent over the next 20 years for replacement power. Alva Morrison, spokesperson for the Maine Nuclear Referendum Committee, countered that the cost of Maine Yankee electricity was climbing as the plant aged, and that $1 billion would be saved over the next 20 years in nuclear waste disposal costs if the plant were shut down. On the legal front, out-of-state owners of Maine Yankee threatened to sue the state on behalf of their customers, and the company vowed to invoke the Fifth Amendment to the U.S. Constitution in order to get reparation for the loss of their business.

In 1987, as in 1980 and 1982, the pro-nuclear forces won over the antinuclear forcees with nearly two-thirds of the vote. By some estimates, the pro-nuclear coalition outspent the antinuclear coalition nearly 10-to-1. In 1988, Congress banned any further efforts by the DOE to search for a site in the eastern part of the country for a national high-level radioactive waste repository, putting to rest the fears of some that the continued operation of Maine Yankee would attract the disposal of the nation's HLRW.

1991: Fire!. In the week following the closing of two nuclear power plants in New England due to accidents in which no radiation was leaked to the atmosphere, an explosion and fire at Maine Yankee closed down the Wiscasset plant as well. The accident occurred at 6:30 p.m. on April 30, 1992, and is believed to have been caused by a short circuit in the transformer outside the plant, causing an immense internal explosion that split the pipes carrying hydrogen to the generator. The source of the gas was cut off minutes after the explosion, but the fire continued to burn for four more hours for safety reasons. The main electrical transformer buckled under the force of the explosion, rupturing seals and spilling thousands of gallons of lubricating oil, at least 200 gallons of which leaked into the Back River.

No people were injured in the explosion or the fire, and the emergency system automatically shut down the reactor, located approximately 100-200 feet away, as soon as it broke out. The reactor was undamaged by the accident. In a news conference the day after the fire, the president of Maine Yankee, Charles Frizzle, emphasized the non-nuclear nature of the accident: "I don't want to downgrade the seriousness of the event, but it's not a nuclear event. From my perspective, this should not reflect negatively on the use of nuclear power."

The Maine Nuclear Referendum Committee remarked that the accident might have been avoided if the warnings implicit in recent problems with the generator had been heeded. During the nine months prior to the explosion, Maine Yankee had been unexpectedly shut down nine times due in large part to the advancing age of the plant, according to the Maine Nuclear Referendum Committee. Nevertheless, the plant was operating at near capacity during the two months before the accident. A week after the explosion, plant officials reported that they planned to reopen the plant by the end of the month, after installing a spare transformer.

Information Center. The Maine Yankee Energy Information Center is open in fall, winter, and spring from 12 noon to 4 p.m. seven days a week. During June, July, and August, the Center is open from 10 a.m. to 5 p.m. Monday through Saturday and 12 noon to 5 p.m. Sunday. For more information call P.G. Whitten, Information Center Supervisor, at 1-800-458-0066.

Sources

"The Battle of Maine Yankee." *Newsweek,*, September 22, 1980.
"Yankee, Yes: Maine Keeps Nuclear Power." *Time,* October 6, 1980.
"Nuclear Power Clears a Hurdle." *U.S. News & World Report,* October 6, 1980.
"Tough Fight Predicted for Referendum That Could Close Maine Yankee." *New England Business Journal,* August 3, 1987.
"Big Vote Down East." *Time,* November 2, 1987.
"Maine Voters to Decide Fate of Nuclear Plant." *The New York Times,* November 2, 1987.
"Shut Down of Nuclear Plant is Rejected by Maine Voter." *The New York Times,* November 4, 1987.
"The Maine Yankee Message." *The Wall Street Journal,* November 5, 1987.
"Fire Damages a Nuclear Plant in Maine." *The New York Times,* May 1, 1991.
"'No Bottom Line Impact Expected from Plant Fire.'" *The Wall Street Journal,* May 2, 1991.
"Nuclear Plant in the Back Yard? Fine." *The New York Times,* May 2,1991.
"Sticken Nuclear Plant Was at Capacity." *The New York Times,* May 2, 1991.
"Maine Yankee Plant to Be In Service by End of Month." *The New York Times,* May 7, 1991.
Maine Yankee Facts 1-3. Maine Yankee Energy Information Center, n.d.
Maine Yankee: Reliable Electricity for Maine Since 1972. Maine Yankee Energy Information Center, n.d.
"Sticken Nuclear Plant Was at Capacity." *The New York Times,* May 2, 1991.

—Mary Gillis

307 ☆

Malibu Nuclear Power Plant

Corral Canyon
Malibu, CA

Basic Facts
Status: Cancelled. *Type of Reactor:* Pressurized light-water reactor. *Megawatts (electric):* 490. *Owner:* Los Angeles Department of Water and Power.

Key Dates
Ordered: November 1963. *Cancelled:* 1966.

Plant Report

A project of the Power Demonstration Reactor Programme (PDRP). Realizing that utilities could not justify to their shareholders the investment costs required to build nuclear power plants, the United States Atomic Energy Commission (USAEC) and Congress devised the Power Demonstration Reactor Programme (PDRP) in 1955 to help launch the commercial nuclear industry in the United States. The PDRP offered cooperating utilities with financial incentives such as research and development assistance, financing of the reactor system, partial support of design costs, and a waiver of normal fuel charges during the first five years of plant operation. Malibu was put forward in 1963 after three rounds of the PDRP, two in 1955 and one in 1957.

Construction application opposed. In November 1963, the Department of Water and Power of the City of Los Angeles (LADW&P) filed an application to construct and operate a nuclear reactor at Corral Canyon, Malibu, California. Corral Canyon is approximately 29 miles west of Los Angeles. In 1963, there were some 9,500 residents within a 6 mile radius of the site, and 3.1 million people within a 30 mile radius.

A group of individuals forming the Malibu Citizens Group resisted the project, arguing that the plant would change the residential character of Malibu, depress land and property values, and be hazardous to the health and safety of the people in the surrounding area. On November 4, 1963, they took their case to a public hearing of the Los Angeles Regional Planning Commission, but the Commission approved the zoning exemption for the site.

The Group continued their protest campaign before the Board of Water and Power Commissioners. The Board, however, also approved the nuclear power plant. Next, the Malibu Citizens Groups appealed to the County Board of Supervisors, which temporarily reversed the Regional Planning Commission's decision.

The case before the Atomic Energy Commission. After LADW&P formally applied for a construction license, the United States Atomic Energy Commission held a series of public hearings throughout 1965. The major players in the proceedings who opposed the granting of a construction license were the Marblehead Land Company, the Malibu Citizens for Conservation Inc., and a private individual. The State of California also participated by presenting evidence.

Much of the Malibu Hearings centered around the geology and seismicity of the reactor site. There was a wide range of disagreement between the various interested parties with regard to the nature and significance of faulting at the site. In the end, the USAEC felt the probability of faulting was so low as to be negligible. On July 14, 1966, the USAEC granted a construction permit with the proviso that the design criteria be adapted to the possibility of ground displacement at the reactor site from earthquake activity.

Despite the positive ruling from the USAEC, the proceedings influenced LADW&P to decide not to commence construction of the power plant, and they cancelled the project shortly thereafter.

Plant Profiles

Sources

Mounfield, Peter. *World Nuclear Power*. Routledge, 1991.

---

308 ☆

Marble Hill Nuclear Power Plant, Units 1-2

Paynesville, IN

Basic Facts

Units 1-2 *Status:* Cancelled (Unit 1, construction 60% complete; Unit 2, construction 37% complete). *Type of Reactor:* Pressurized light-water reactor. *Megawatts (electric):* 1190. *Megawatts (thermal):* 3411. *Reactor System Supplier:* Westinghouse Electric Corp. (United States). *Generator Supplier:* Westinghouse Electric Corp. *Architect Engineer:* Sargent & Lundy Engineers (United States). *Owner:* Public Service of Indiana; Wabash Valley Power Association. *Operator:* Public Service of Indiana.

Key Dates

Units 1-2 *Ordered:* 1973. *Cancelled:* January 1984. *Construction began:* 1976.

Plant Report

Unfulfilled expectations. Intended to function as a 1,130 megawatt nuclear power plant, Marble Hill wastes away along the banks of the Ohio River in southwest Indiana, just outside the city of Madison in Jefferson County. Power lines hang in vain. Massive amounts of concrete and steel are cared for only by maintenance workers fighting to keep the weeds at bay.

Work on Marble Hill was halted on January 16, 1984 after years of construction delays and cost overruns. The planned nuclear plant earned the honor of being the most expensive cancellation in terms of total investment and percentage toward completion; however, it lost this dubious distinction a mere two days later when the Zimmer plant, under construction in Ohio, was removed from the nuclear roster and converted to coal. Critics claimed the name "Marble Hill" would become synonymous for "super corporate boondoggle."

The key players. Public Service of Indiana (now known as PSI Energy, Inc., whose parent company is PSI Resources, Inc.) was Marble Hill's principal builder. Wabash Valley Power Association (WVPA) purchased a 17 percent interest in the nuclear power plant.

PSI is the largest supplier of electricity in Indiana, providing power to 554,000 customers in 69 of Indiana's 92 counties. It serves industrial, agricultural, commercial and residential users. Ninety-nine percent of the electricity generated by PSI is from coal mined in Indiana, Kentucky, and Illinois. PSI is the seventh largest publicly held corporation in the state.

Wabash Valley Power Association (WVPA), formed in 1963, is a non-profit utility company comprised of 24 Rural Electric Membership Corporations. The utility's board of directors consists of volunteers who receive $100 per board meeting plus travel expenses. It serves approximately 400,000 people in northern Indiana and southern Michigan. In addition to its share of the defunct Marble Hill plant, WVPA owns 25 percent of a coal-fired electrical plant and approximately 250 miles of transmission line.

Rationale for going nuclear. The saga of Marble Hill began in 1970 when Congress passed the Clean Air Act. PSI's coal burning plants were a source of controversy because they burned a high-sulphur coal which was thought to contribute to the acid rain problem. PSI considered the comparative costs of switching to low-sulphur coal, installing scrubbers and building a nuclear power plant. The company opted to go nuclear and began purchasing land in 1973.

WVPA estimated it could save $128 million over 13 years by investing in the plant. Wabash borrowed funds to pay for its portion of Marble Hill from the Rural Electrification Administration (REA), a federal lending agency created in 1935 for the purpose of providing jobs and loaning money to rural electric associations. Instead of saving money, the project eventually forced the utility into bankruptcy.

Construction halted after the Three Mile Island accident. The Nuclear Regulatory Commission (NRC) issued a full construction permit in April 1978. Less than a year later, on March 28, 1979, the accident at Three Mile Island sent shock waves through the entire nuclear industry. Repercussions were felt almost immediately at Marble Hill. New regulations made cost overruns inevitable. The NRC halted construction until all the new safety regulations were met. As a result of TMI, the number of permits necessary to build a nuclear plant increased to 1,180.

Marty Irwin, former director of Hoosiers for Economic Development Committee, explained that "Essentially what happened was that Marble Hill was built and rebuilt three times."

More problems. Other problems have plagued Marble Hill almost from its inception. A construction delay in May 1976 caused the first $60 million cost overrun. Thirty-one demonstrators were arrested at the site in October 1978. Builders repeatedly failed to repair air pockets in concrete poured to house the reactor. PSI suspended work again in 1983 because of changes in the structural steel requirements. These delays caused projected commercial operation of the two units at Marble Hill to be postponed until 1988 and 1990.

Marble Hill was also afflicted by the country's economic woes. Interest rates on construction loans rose to 19 percent. The recession of the late 1970's and early 1980's severely impacted Indiana's big users of electricity, namely automobile and steel manufactures. This, along with new efforts aimed at conserving energy, caused predictions concerning future power needs to plunge.

In addition to construction problems and concerns about environmental and safety issues, questions were raised about the costs of decommissioning the plant. A nuclear plant has an approximate life of 30 years. After that time, it must be taken out of service. These costs were never considered.

Rising costs and governmental investigation. Marble Hill came with an original price tag of $1.4 billion. PSI revised this

● Operable   ○ Under Construction   △ Indefinitely Deferred   ▲ Decommissioning   □ Planned   ■ On Order   ☆ Cancelled   ★ Shut Down

figure to $3.4 billion in July 1980. In October 1981 the figures were corrected to $4.3 billion. By July 1982 cost estimates had risen to $5.13 billion.

In July 1983, to cope with ever rising costs, PSI submitted a request for an 8 percent rate increase to help pay for on-going construction. Indiana's Governor Orr appointed a task force to investigate. The task force estimated it would take $7.7 billion to complete the project and recommended that it be abandoned.

Marble Hill plant abandoned. PSI officially stopped building Marble Hill on January 16, 1984 making it the 100th nuclear power plant to have been abandoned since 1974. The company had already spent $2.7 billion. The actual loss to the company, however, was $1.3 billion after subtracting gains from the sale of equipment and considering the tax savings resulting from the loss.

Economic fallout: Jefferson County, PSI, and Wabash Valley. Almost half the workers at Marble hill lived in the city of Madison or Jefferson County. Officials were left scrambling to rebuild their communities when work stopped. Repercussions affected the local employment picture, property values, and area schools.

When the project was officially abandoned 9,000 workers were idled. Unemployment in Jefferson County rose from 8.3 percent in 1983 to 24.3 percent in January 1984.

The housing market saw dramatic effects. Lenders braced for a 300 percent increase in foreclosures and turnbacks. Residential property values decreased as much as 22 percent. Demand fell 30 to 40 percent. Apartment complexes that had maintained waiting lists found themselves only 60 percent occupied.

Public schools experienced a sharp decrease in enrollment. Reassessment of property values caused a drop in the school tax base forcing the Southwestern Consolidated Schools district to consider raising its tax rate from $2.66 per $100 assessed value to $18.

The Marble Hill disaster left PSI with the lowest equity base of any electrical utility in the United States. To cope with severe cash flow problems, PSI was granted an emergency rate hike. The increase went into effect even though it was challenged by various groups in court. The government tried to assure that PSI's stockholders and not Indiana's electrical consumers would pay for the money spent on Marble Hill. To achieve this end, the company was prohibited from paying dividends on its common stock for three years. In addition, top executives received pay reductions and other employees' wages were frozen.

Although PSI and Jefferson County are making economic comebacks, the fate of Wabash Valley Power Authority is less promising. WVPA had invested a total of $478.8 million in Marble Hill. Most of this amount was borrowed from the Rural Electrification Administration (REA). After defaulting twice on payments, WVPA filed for bankruptcy. WVPA claims it was forced into bankruptcy when the Justice Department, citing a 1799 law, threatened to hold the board of directors personally liable for the utility's debt.

WVPA's debt, with accumulated interest, stood at approximately $1 billion in 1990. Efforts at hammering out a repayment plan had all failed. Cases were still pending with the court and expected to go before the Indiana Supreme Court.

Hugh "Al" Barker, the former CEO of PSI, calls Marble Hill "a great waste of resources." He contends however, that building the plant was not a mistake but that it did represent a great "economic tragedy" not applicable only to Marble Hill but to the nuclear industry in the United States as a whole. Without generating a watt of electricity, Marble Hill succeeded in generating controversy and economic fallout that has not yet completely settled.

Sources

"Nuclear Fissures." *Time*, January 30, 1984.
"The Jolts of January." *Electrical World*, March 1984.
"Marble Hill Pull-Out: Are the Shock Waves Settling?" *Indiana Business Magazine*, February 1985.
"Who's holding the Bag?" *Forbes*, February 11, 1985.
"Wabash Valley: Control of Its Future Slips Away." *Indianapolis Business Journal*, June 10, 1985.
"PSI: Closing the Book on Marble Hill." *Indianapolis Business Journal*, March 17, 1986.
"The Long Haul: PSI Rebuilds Finances After Marble Hill." *Indianapolis Business Journal*, September 29, 1986.
"PSI's Hugh Barker Responds to Critics' Questions." *Indianapolis Business Journal*, December 29, 1986.
"PSI: The Electricity is Back." *Indiana Business Magazine*, September 1988.
"Wabash Valley Bankruptcy: At Square One After Appellate Decision." *Indianapolis Business Journal*, February 5, 1990.
"Polishing the Corporate Image." *Indiana Business Magazine*, June 1990.
"PSI's Hugh Barker Responds to Critics' Questions." *Indianapolis Business Journal*, December 29, 1986.

—*Al Cook*

309 ☆

Mayport Nuclear Power Plant, Units 1-2

Mayport, FL

Basic Facts

Units 1-2 *Status:* Cancelled (plants were ordered in 1973). *Type of Reactor:* Pressurized light-water reactor. *Megawatts (electric):* 1150. *Reactor System Supplier:* Offshore Power Systems. *Owner:* Jacksonville Electric Authority. *Operator:* Jacksonville Electric Authority.

Key Dates

Unit 1 *Ordered:* 1969. *Construction began:* 1971. *Criticality:* 1981. *First Power:* September 1981. *Commercial Operation:* December 1981.

Unit 2 *Ordered:* 1969. *Construction began:* 1971. *Criticality:* 1983. *First Power:* May 1983. *Commercial Operation:* March 1984.

310 ●

McGuire Nuclear Station, Units 1-2

12700 Hagers Ferry Rd.   Phone: (704)875-4000
Huntersville, NC 28078-8985
T.C. McMeekin, V.Pres., McGuire Site

Plant Profiles

Owning Utility

Duke Power Company Phone: (704)373-4011
PO Box 33189
Charlotte, NC 28242
H.B. Tucker, Sr.V.Pres., Nuclear
  Generation

Contact

M.A. Mullen, Mgr., Community Relations
McGuire Nuclear Station

Basic Facts

Units 1-2 *Status:* Operable. *Type of Reactor:* Pressurized light-water reactor. *Megawatts (electric):* 1220. *Megawatts (thermal):* 3411. *Reactor System Supplier:* Westinghouse Electric Corp. (United States). *Generator Supplier:* Westinghouse Electric Corp. *Architect Engineer:* Duke Power Company.

Key Dates

Unit 1 *Ordered:* 1969. *Construction began:* 1971. *Criticality:* 1981. *First Power:* September 1981. *Commercial Operation:* December 1981.
Unit 2 *Ordered:* 1969. *Construction began:* 1971. *Criticality:* 1983. *First Power:* May 1983. *Commercial Operation:* March 1984.

Operating Costs

Units 1-2

Year	Cost	Year	Cost
1981	$2,718,000	1986	$136,992,000
1982	$37,258,000	1987	$145,665,000
1983	$56,030,000	1988	$156,205,000
1984	$77,439,000	1989	$137,896,000
1985	$102,464,000		

Plant Report

Owned and operated by Duke Power Company. The McGuire Nuclear Station consists of two 1,150-megawatt pressurized water reactors. It is located about 17 miles northwest of Charlotte, North Carolina on Lake Norman. Lake Norman, the largest body of fresh water in North Carolina, covers 32,500 acres and has 520 miles of shoreline. It was created in 1963 by Duke Power Company for the Cowans Ford Hydroelectric Station. The McGuire station uses its waters for cooling purposes. Duke estimates that it saved almost $2 million by using the lake for cooling rather than constructing and operating cooling towers.

McGuire, built at a cost of $2 billion, is owned and operated by Duke, the nation's seventh largest investor-owned utility. Duke supplies electricity to more than 1.6 million customers in a 20,000 squire-mile service area. Nuclear power accounts for more than 40 percent of the utility's generating capacity. The remaining power is supplied from fossil-fueled plants and hydroelectric sources.

In 1991 increased maintenance costs at McGuire had a negative impact on Duke's earnings. However, also in 1991, the station's second unit was one of only 15 plants in the United States to achieve a capacity factor of more than 90 percent. Capacity factor is a percentage of a plant's output compared to its maximum ability. McGuire-2 achieved a capacity factor of 92.3 percent.

Training at McGuire. The McGuire Nuclear Station's operator training program received accreditation from the Institute of Nuclear Power Operations (INPO) in May 1985. All Duke employees who will be nuclear operators or technicians receive their initial training at the Technical Training Center located at the McGuire site.

Sources

Duke Power Company: Annual Report 1990. Duke Power Co., n.d.
Caring for the Environment. A Duke Power Way of Life for Most of a Century. Duke Power Company, n.d.
"U.S. Nuclear Plants Shatter Records for Output, Performance." *INFO,* U.S. Council for Energy Awareness, February 1992.
Duke Power Company. Corporate Communications. August 1992.
"Nuclear Training: A Duke Power Commitment." Duke Power Company, n.d.

—Karen Bellenir

Midland Nuclear Power Plant, Units 1-2

Midland, MI

Owning Utility

Consumers Power Company Phone: (517)788-2552
212 W. Michigan Ave. Fax: (517)788-0045
Jackson, MI 49201 Telex: 223454
David P. Hoffman, V.Pres., Nuclear
  Operations

Contact

Charles E. MacInnis, Dir., News & Info.
Consumers Power Company

Basic Facts

Units 1-2 *Status:* Cancelled (construction 85% complete). *Type of Reactor:* Pressurized light-water reactor. *Megawatts (electric):* 526 (Unit 1); 852 (Unit 2). *Megawatts (thermal):* 2468. *Reactor System Supplier:* The Babcock & Wilcox Co. (United States). *Generator Supplier:* The Babcock & Wilcox Co. *Architect Engineer:* Bechtel.

Key Dates

Units 1-2 *Ordered:* 1968. *Cancelled:* 1984. *Construction began:* 1974.

Plant Report

A failed attempt. In 1967 Consumers Power Company and Dow Chemical Company announced plans to build a nuclear reactor within half a mile of downtown Midland, a mid-Michigan town of 37,000, on a 1,000-acre site that was once swamp and farmland. Originally, Midland nuclear power plant was slated to be on line by 1975 at a cost of $276 million. However, this never happened due to many unforeseen problems including:

- A lack of construction funds. Shoddy workmanship and construction delays sent costs soaring.

- The 1979 Three Mile Island accident in Pennsylvania changed federal rules governing nuclear plants. The Nuclear Regulatory Commission, the federal agency that oversees nuclear plants, drafted new standards for reactors both planned and under construction.

- The 1983 recession resulted in a drop in electricity demand: the state didn't need the electricity Midland would provide.
- Political opposition to nuclear power grew with double-digit inflation.
- Part of the nuclear power plant was built on shifting soil and was sinking. It was discovered that the builders did not compact the soil properly before they erected five safety-related buildings on it.
- Strong public resistance threatened the plant. One leader of this opposition was environmental activist Mary Sinclair, a technical writer and teacher of Midland, Michigan.

Due to these reasons, construction was halted in 1984. The Midland nuclear power plant had cost $4.2 billion and was 85 percent complete at this time.

The aftermath—results from the abandoned Midland plant. When the plant closed in 1984, things were looking down for everyone involved. Consumers Power was on the verge of bankruptcy. During the 1984 fiscal year, Consumers' net income declined to $221.1 million from $347.8 million in 1983. Earnings per share of common stock declined from $3.12 in 1983 to $1.14 in 1984. At this point the stock was down to around $4 a share and Consumers had stopped paying common stock dividends. In 1985 they had a loss of $270 million on revenues of nearly $3.3 billion due to the Midland plant closing.

Although Consumers Power employees agreed to defer $16.9 million in salaries and benefits during the 1984 fiscal year, the financial consequences of having more than 40 percent of the company assets tied up in the failed Midland plant forced even more drastic change. In 1985, Consumers Power announced a company-wide reorganization that eliminated 350 positions and cut its management team by 20 percent, dropping the Consumers work force to 9,700 employees. Most of the reductions were made through attrition and a retirement incentive program.

In addition to this, things were not going well for Dow Chemical, Consumer's partner in the Midland nuclear project. In 1986, Dow Chemical, Midland's largest employer, was in the process of attempting to regain the public's confidence after the dioxin scare. Dioxin is an unwanted by-product of the manufacturing of some pesticides. To top this off, lawsuits between Dow Chemical and Consumers Power resulted from the failed nuclear plant with each company asking $520 million in damages from the other.

Tension was high, not only between the partners, but also for the residents of the central Michigan town. Following the plant closure, unemployment, which had been nearly nonexistent, skyrocketed near 18 percent. Some of the town's businesses reported a 10 percent decrease in sales. Housing values plummeted, and approximately 2,000 families, most with ties to the plant, left Midland. The town experienced another downfall in 1986 when a flood wiped out many homes and businesses.

The shape of things to come for the idle Midland plant. In 1985, the Michigan Public Service Commission saved Consumers from bankruptcy by granting a 5.6 percent rate increase for six years, amounting to approximately half a billion dollars, on condition that not one cent go to the Midland plant. This surcharge, initially $99 million annually was cut to $79 million a year in 1989 when the company sold its nuclear fuel and salvaged equipment from the Midland project. From 1985 to 1989 the company collected approximately $434 million and cut its debt by $1.9 million.

In November 1985, William T. McCormick Jr., who has a doctorate in nuclear engineering from the Massachusetts Institute of Technology, left his position as chairman of the Detroit-based American Natural Resources Company to replace the retired John D. Selby as chairman, president and chief executive officer of Consumers Power.

In his first year he assembled a youthful management team at what had been an aging company, conducted a high-energy public relations campaign to improve Consumers' tarnished image, launched a refinancing and restructuring program, and in April 1986 announced plans to convert Midland to a natural gas-fueled, combined-cycle cogeneration facility—one which uses both natural gas and the combustion by-product steam, to produce electricity.

Consumers Power estimated it could salvage $993 million of equipment, nearly 25 cents on every dollar spent on Midland. However, converting the plant would cost $434 million and Consumers wanted a rate increase of 25 percent over five years to pay for it.

The solution: The Midland Cogeneration Venture. The $675 million plan McCormick announced, named the Midland Cogeneration Venture (MCV), involved adding gas turbines to the steam turbine already in place for nuclear power generation, to produce electricity and steam.

The Midland plant is the world's largest cogenerator, which means it is a plant that produces both steam and electricity and is owned by businesses that use a portion of the power themselves and sell the rest to local utilities. There is an advantage to being a cogenerator. According to a 1978 federal law, if an independently owned cogenerator can generate electricity for less, it is allowed to keep the extra profit rather than returning it to the ratepayers as a regulated utility would be required to do.

On January 27, 1987 Consumers Power Company and the Dow Chemical Company signed a definitive agreement to form the Midland Cogeneration Venture and convert part of the never-completed nuclear power plant to gas fuel. It was also at this time that the two companies dropped previous lawsuits against one another filed during the closing of the plant.

Consumers Power contributed $1.5 billion of its Midland assets and has a 49 percent stake in the joint venture, the most it could legally hold and avoid state regulation of profits. It will receive $1.27 billion in interest-bearing notes and will have the option of receiving a $103 million payment from the partnership and will also get $16 million a year for the first nine years of commercial operation of the plant. Dow has a 10 percent position in the project.

In April 1987 Fluor Corporation was chosen to design and build the new plant. The contract between MCV and Fluor has a ceiling of $468.9 million and includes the purchase of all equipment for the project. The venture plans to re-design a portion of Midland nuclear power plant to operate as a gas-fueled co generation facility. Fluor will also invest $25 million in the project giving it a three percent stake in the venture.

Other partners in the venture include Asea Brown Boveri Inc., which invested $50 million, provided the Swiss-built turbines, and received a six percent stake in the project. The remaining 32 percent is divided among Combustion Engineering Inc., and two gas pipeline companies that bring in the gas, Costal Corporation and Panhandle Eastern Corporation.

Is the Midland Cogeneration Venture the best solution?. The question arises concerning whether the plant is necessary, and whether Consumers should be allowed to convert the plant using public money specifically intended to keep the company from going under. In response to this question, Consumers Power conducted a comprehensive study of four available options to determine the best use for the Midland facility. The study was conducted by Consumers engineers with assistance of experts from 10 other companies. The four options were:

- Total abandonment, which would have value to the company in the form of tax deductions and would include the sale of any salvageable parts.
- Conversion to a coal-fired plant, which has the drawback of utilizing only a modest portion of the existing Midland plant and would also require a substantial capital investment for mining expenses and pollution control equipment.
- Completion of the plant as an all-electric nuclear facility which would utilize a major portion of the existing facility but has the liabilities of regulatory and technological uncertainties and Midland's past problems.
- Conversion to an electric/gas combined-cycle plant which, according to Consumers, requires the least capital investment, utilizes proven technology and would have the least negative environmental impact. However, with this option Consumers would have to obtain an exemption from the Department of Energy from the Federal Power Plant and Industrial Fuel Use Act of 1978 which prohibits the use of natural gas in new electric power plants.

Consumers Power maintained that converting the Midland plant to a combined-style natural gas facility was the best solution and McCormick argued that Michigan's power needs were expected to increase by 1.9 percent a year and that converting Midland to gas was the cheapest and most environmentally acceptable way to meet demand.

However, not everyone agreed with Consumers Power. Critics of the plan, including Michigan Citizens Lobby (MCL) said the cost of conversion is two to three times what a new plant would cost to build from scratch. There is evidence to support this claim. The cogeneration plant's total cost is $2.1 billion, enough to build three new plants.

In March 1990, Detroit's Big Three Auto Makers argued that the Midland conversion would cause a big jump in their electricity costs, another supported claim. As a result of the high cost of the plant, the electricity produced by the cogeneration plant will cost 27 percent more than the power that CMS Energy (in May 1987 Consumers Power was reorganized to form a holding company called CMS Energy Corporation) produces at its other plants, which are mostly fueled by coal. McCormick did not dispute the fact that building a new plant would have been cheaper than converting Midland, but he argued CMS's overall rates for industrial customers, including power from the new plant will still be 20 percent lower than those charged by Detroit Edison Company.

A further complaint comes from some small power producers who contend that Midland will help Consumers corner the market on energy in Michigan and put them out of business. But Consumers and state regulators dispute that claim and allege that around two dozen small power producers are expected to supply power to Consumers in addition to Midland.

In January 1990 critics claimed Consumers would earn profits while commercial and residential customers would pay too much for the $2.3 billion Midland project. McCormick disputed even with the rate increase, Consumers would have reasonable rates. According to recent studies by the Edison Electric Institute and the National Association of Regulatory Utility Commissioners, Consumers has lower rates than 80 percent of the nation's utilities.

The finished product. After four years since its inception, in March 1990, the Midland Cogeneration Venture came on line and became the nation's first failed nuclear plant to be converted to another fuel source. It uses 12 natural-gas and two steam turbines to generate 1.35 million pounds of steam per hour and 1,370 megawatts of power, enough to light a city of one million. Most of the electricity is sold to Consumers Power. The primary cycle uses 12 gas turbines to power the generators; the secondary cycle uses the heat of their exhaust to make steam, some of which powers a giant turbine left from the original nuclear power plant. The rest is piped over the Tittabawassee River to supply process heat to Dow's chemical factory.

In 1990 the 25 percent rate increase Consumers Power requested to help pay for the Midland Cogeneration Venture was denied. That increase would have allowed Consumers to recover $2.14 billion. An administrative law judge for Michigan's Public Service Commission announced a proposal that would allow CMS to recover $111.5 million with an annual electric rate increase over a 10-year period. As of February 1991, CMS was still in litigation appealing the decision.

Milestones. Following are important events in the history of Midland Nuclear Power Plant:

- December 14, 1967: Consumer Power and Dow Chemical announce plans to build Midland Nuclear Power Plant, a twin-reactor cogeneration project. Project cost: $260 million.
- January 30, 1980: Dow, citing cost overruns and lengthy delays, says it may back out of contract to buy electricity and steam from Midland.

- August 26, 1981: State sues to block the sale of $363 million in securities by Consumers Power to finance Midland plant.
- April 1983: Consumers says its steam-producing unit at Midland won't be finished until August 1985.
- July 1983: Dow, citing costs and long delays, sues Consumers to break agreement to buy steam from plant.
- November 1983: Burdened financially by Midland, Consumers announces plan to cut operating and construction budget by 10 percent through layoffs, wage freezes, and cancellation of some construction and operating programs.
- November 1983: Consumers announces a 15-month delay in the completion of Midland, adding another $300 million to $4.4 billion project.
- July 16, 1984: Consumers halts construction on Midland because of political opposition and soaring costs.
- 1985: Public Service Commission orders Consumers to stop spending money on Midland plant without its approval.
- April 1986: Consumers suggests the $4.2-billion Midland nuclear plant be completed as a natural gas plant; asks commission to lift spending ban.
- August 1986: Consumers' biggest customers say it has failed to prove need to complete nuclear plant. Association of Businesses Advocating Tariff Equity (ABATE) asks state to continue ban on construction spending.
- January 27, 1987: Consumers and Dow sign definitive agreement to convert Midland to a natural gas plant, forming Midland Cogeneration Venture (MCV).
- October 12, 1987: Ground broken for Midland conversion.
- September 12, 1988: The first gas turbine generator for MCV arrives from Switzerland.
- January 31, 1989: The commission limits amount of power Consumers can buy from Midland.
- May 9, 1989: Construction begins on 26-mile pipeline that will carry natural gas to Midland.
- November 10, 1989: Last of Midland's 12 stacks in place; construction more than 80 percent finished.
- December 22, 1989: Commission increases amount of electricity Consumers can buy from Midland to 870 megawatts from 638 megawatts.
- December 28, 1989: Construction 85 percent complete, start-up testing 82 percent finished and six of 12 gas turbines successfully test-fired.
- February 1990: First eight gas turbines and unit 1 steam turbine from former nuclear plant set for commercial operation.
- May 1990: Remaining four gas turbines expected to go into commercial operation.

Sources

"One More Renewable Energy Source: Nuke Plants." *Business Week*, September 16, 1991.
"CMS Energy Takes Big Write-Downs On Nuclear Project." *The Wall Street Journal*, February 28, 1991.
"CMS Charge Of $450 Million Set for Quarter." *The Wall Street Journal*, December 21, 1990.
"Rescue of a Failed Nuclear Plant: Mothballed for Years, Michigan Facility Operates on Natural Gas." *The Washington Post Journal*, April 10, 1990.
"Consumers Power Proposes Ending Surcharge." *Detroit News*, March 13, 1990.
"Sick Nuclear Plant Gets New Energy." *The Wall Street Journal*, March 9, 1990.
"Powering Up: 23 Years Later, Midland's Would-Be Nuclear Plant Has Electricity to Sell." *Detroit News*, January 29, 1990.
"CMS Energy Corp." *The Wall Street Journal*, July 7, 1987.
"Fluor Unit Selected To Outfit Midland For Cogeneration." *The Wall Street Journal*, April 29, 1987.
"Dow Chemical, Utility in Pact." *The New York Times*, January 28, 1987.
"In Whose Hands?" *Michigan Business Journal*, October 1986.
"Finding New Life For a Mothballed Nuke." *Fortune*, July 21, 1986.
"The $4 Billion White Elephant on Bill McCormick's Back." *Business Week*, June 9, 1986.
"The Whiz Kid Who Has to Rescue Consumers." *Crains Detroit Business Journal*, April 28, 1986.
"Consumers Power Reorganizes, Cutting More Than 350 Jobs." *Grand Rapids Business Journal*, July 15, 1985.
"Under the Dow Volcano." *Audubon*, July 1983.

—Sharon Miscavage

Millstone Nuclear Power Plant, Units 1-3

PO Box 128
Waterford, CT 06385
Stephen E. Scace, Station Dir.

Phone: (203)447-1791

Owning Utility

Northeast Utilities
PO Box 270
Hartford, CT 06141-0270
John F. Opeka, Exec.V.Pres., Nuclear

Phone: (203)665-5000
Fax: (203)665-5884

Contact

Steven L. Jackson, Nuclear Info. Supv.
Millstone Nuclear Power Plant

Basic Facts

Unit 1 *Status:* Operable. *Type of Reactor:* Boiling water reactor. *Megawatts (electric):* 684. *Megawatts (thermal):* 2011. *Reactor System Supplier:* General Electric Co. (United States). *Generator Supplier:* General Electric Co. *Architect Engineer:* Ebasco.

Unit 2 *Status:* Operable. *Type of Reactor:* Pressurized light-water reactor. *Megawatts (electric):* 888. *Megawatts (thermal):* 2700. *Reactor System Supplier:* Combustion Engineering, Inc. (United States). *Generator Supplier:* General Electric Co. *Architect Engineer:* Bechtel.

Unit 3 *Status:* Operable. *Type of Reactor:* Pressurized light-water reactor. *Megawatts (electric):* 1209. *Megawatts (thermal):* 3579. *Reactor System Supplier:* Westinghouse Electric Corp. (United States). *Generator Supplier:* General Electric Co. *Architect Engineer:* Stone & Webster Engineering Corp. (United States).

Key Dates

Unit 1 *Ordered:* 1965. *Construction began:* 1966. *Criticality:* October 1970. *First Power:* November 1970. *Commercial Operation:* March 1971.

Unit 2 *Ordered:* 1968. *Construction began:* 1969. *Criticality:* October 1975. *First Power:* November 1975. *Commercial Operation:* December 1975.

Plant Profiles   **Millstone** | 312 |

Unit 3 *Ordered:* 1974. *Construction began:* 1975. *Criticality:* January 1986. *First Power:* February 1986. *Commercial Operation:* April 1986.

Operating Costs
Unit 1

1971	$3,256,000	1979	$23,060,000
1972	$7,677,000	1980	$24,783,000
1973	$7,635,000	1981	$33,272,000
1974	$9,808,000	1982	$33,463,000
1975	$12,065,000	1985	$48,878,000
1976	$14,040,000	1986	$45,640,000
1977	$12,637,000	1987	$71,370,000
1978	$16,448,000		

Plant Report

The structure of the Millstone nuclear power plant. In 1986, 58 percent of the electricity used by Connecticut was generated by nuclear power. At times, in off peak hours, up to 93 percent comes from nuclear sources such as the Millstone nuclear power plant in Waterford, Connecticut.

The location of the Millstone plant on the Long Island Sound is the ideal site for a nuclear reactor. From 1830 until 1963 this area was a granite quarry over 100 feet deep and 1,000 feet long. The quarry is now used as cooling pond, holding the heated water after it is discharged from the plant's closed cooling loop. The warmed water, which is 10 to 15 degrees warmer than the Sound, leaves the old quarry and is fed back into the Sound, making Millstone Point one of the state's best fishing spots.

The security system at the Millstone plant includes a metal detector, a bomb detection device, video cameras, ultra sonic motion detectors, electronic devices and patrols to protect the perimeter.

The Millstone plant also contains a $55 million simulator and training facility. The building, located near the three Millstone units, houses four exact replicas of the control rooms of the three units and the Connecticut Yankee plant. Each year every operator must under go a minimum of eight weeks training for requalification.

Northeast Utilities. The Millstone nuclear power plant is owned by Northeast Utilities, the most successful and largest operator in New England, who holds a 65 percent interest in the plant. The utility owns and operates the Millstone plant, operates and is principal owner of Haddam Neck Nuclear Plant, and owns a percentage of four other New England nuclear units: Yankee Rowe, Vermont Yankee, Maine Yankee, and Seabrook.

Northeast Utilities was formed as a holding company in 1966 through the union of several utilities, principally Connecticut Light & Power Company, Hartford Electric Light Company, and Western Massachusetts Electric Company.

The company uses nuclear energy for about two-thirds of its electricity. Millstone 3 is equivalent to about one-third of the utility's nuclear capacity.

As of 1990, Northeast had approximately $2.2 billion in sales, just over 1.25 million customers, a 5,890-square-mile service area, and approximately 7,000 megawatts of generating capacity.

The three units of the Millstone nuclear power plant. The Millstone nuclear power plant, which consists of three units, is located on a peninsula just west of New London in Waterford, Connecticut. In 1966 construction began on Millstone 1. It took about five years to build at a cost of $100 million and came on line in 1970. Millstone 2 has been in service since 1975. It took about five and one-half years to complete and cost about $430 million.

Millstone 3, also a pressurized water reactor, is a 1,153-megawatt unit and is the region's largest and most expensive construction project in the state's history. It cost about the same to build as the Panama Canal. The original projected cost was $400 million but the final cost after 12 years was $3.8 billion. In August 1974 the construction permit for Millstone 3 was issued with the completion set for April 1978. However, there were many delays to the construction of Millstone 3 including:

- The utility was denied the full rate increases it had requested in the 1970s. Northeast credit ratings plunged and the company was forced to slow construction work in 1975, 1977, and 1979.

- In 1976 Millstone received a major rate decrease. As a result, the company did not have the money to continue building, so construction stopped right after it was started and all the equipment had been ordered.

- Inflation pushed interest rates to double digits, and the company was forced to pay rates in excess of 20 percent for their money.

- Costs went up and safety requirements increased in 1979 after the Three Mile Island accident. As a result the company had to do a lot of re-designing.

- In May 1983 the Connecticut legislature, reacting to escalating costs, caused partly by the serious delay in the construction schedule, imposed a cost cap of $3.54 billion.

As costs for the project continued to increase, the utility's earnings dropped and investor confidence decreased. By 1976, Northeast had borrowed as much money allowed under federal regulations.

Construction was finally restarted in 1982 and on April 23, 1986 Millstone 3 came on line after the utility reached an agreement with the state attorney general and state Consumer Counsel on a cost that Northeast could use to get the plant into its rate base $3.4 billion. Although the utility had to take a $119 million after-tax write-off for Millstone 3, as of the summer of 1990 it had 80 percent of the plant's construction costs in its rate base with the remainder to be phased in over the three years following.

Millstone plant accused of making New England's power system less reliable. In June 1986, approximately two months after Millstone 3 came on line, the New England Power Pool (NEPOOL), a private organization of 93 New England utilities that distributes power in New England and sets the reserve requirements, claimed that the Millstone and Seabrook plants, in their early years, would make New England's electric power system less reliable because of frequent shutdowns and the large size of the plants. In turn, they claimed reserves must be higher in order to cope.

As a result, in 1985 a NEPOOL committee voted to increase the reserve margin, from 20 to 23 percent for the 1985-1986 year, and to add another one percent for each new nuclear plant that came on line during the year. Since Millstone 3 came on line during that time, the system requirement in 1986 was 24 percent, with the one percent increase amounting to about 178 megawatts.

In addition, the NEPOOL study concluded that after five years of operation, the performance of new plants would improve to the point where reserve requirements would steady at about 20 percent of plant capacity, or 230 megawatts for Millstone.

Northeast goes to court and wins. In 1986 Northeast asked state regulators to approve a $155.5 million rate increase with $133 million to be set aside to pay for the utility's share of the Millstone 3 plant. The increase was denied and Northeast was ordered to place $46.5 million in a fund designed to offset rate increases in 1987.

Northeast contested in court the creation of the excess profits fund and won. The settlement was approved July 1, 1986. The agreement called for the following provisions:

- Connecticut Light & Power Company agreed to reduce by $200 million the amount it could charge customers for its $2 billion share of the cost of Millstone 3.

- Northeast agreed to accept a cap of 16 percent as the maximum return for shareholders for any 12 month period.

- Rate increases to pay for Millstone 3 would be phased in over five years, beginning in 1988, a longer period than Northeast originally proposed.

Central Maine power company writes off investment in Millstone 3. On January 29, 1987 the Central Maine Power Company said it would write off $17.2 million of its investment in the Millstone 3 nuclear power plant in Connecticut. This amount translates to $11 million after taxes, or 48 cents a share on 1986 earnings. In 1986 Central Maine and state regulators agreed to a rate order that did not permit recovery of 15 percent of the company's total Millstone 3 investment of $114.9 million.

Rate increase requests bring adverse publicity to Millstone. In response to a request from Northeast for a $56.3 million rate increase in 1976, the regulators instead awarded a $21.6 million rate cut due to pressure from ratepayers protesting higher prices.

However, in 1977 Northeast was granted $35 million of a requested $90 million and in 1979, $86 million was granted of $131 million requested. The utility continued to request annual rate increases, bringing more adverse publicity. Despite these requests, throughout the early 1980s Northeast's rates remained below the average for similar utilities in the northeastern United States.

Northeast Utilities fined by the NRC. On April 13, 1988 the Nuclear Regulatory Commission (NRC) fined Northeast Utilities $50,000 because the company failed to monitor reactor pressure properly at Millstone 3 when all three automatic systems were partly or completely unavailable during a refueling shutdown between January 16 and 19. No radioactivity was released.

According to a draft report, after finding 49 violations of federal rules at the three-plant Millstone complex, the NRC downgraded its security rating. During the 19 months ending December 31, 1987, 20 violations were found at Millstone 1 and 2.

Also detailed in the report were 29 breaches of security that led to a $25,000 fine in 1988. The security problems included three breaches of "vital areas" and one time when a weapon was reportedly taken into the complex.

Northeast utilities helps to finance research of earthquakes. Northeast Utilities was part of an all-out effort to determine the cause of more than 175 small earthquakes that occurred near the town of Moodus, Connecticut from September 17 to October 22, 1988.

The research effort was financed by power companies who wanted to assess whether a severe earthquake would shake the nuclear plants. However, despite their effort, no generally accepted explanation emerged for the quakes in Moodus.

Northeast utilities fined again by the NRC. On September 5, 1989 the NRC proposed a fine of $25,000 against Northeast Utilities for an alleged violation of NRC rules relating to the handling of radiation-contaminated parts from its Millstone nuclear station. Northeast said the fine would be paid by shareholders, not customers.

Northeast's nuclear plants a cut above the rest. Over the years, Northeast's plants have continually demonstrated their capabilities. The following is a list of some of their accomplishments:

- In 1985 Millstone 1 was one of the U.S. nuclear plants with the longest non-stop operating runs, operating for 374 consecutive days.

- In 1988 Millstone 1 ranked eighth internationally for its load factor and was the number one boiling water reactor in the world.

- In 1988 Northeast's four nuclear units recorded an average capacity factor of nearly 80 percent, about 15 percentage points above the industry average. In 1989 their composite capacity factor was 69.6 percent, compared to the industry average of 65.3 percent.

- In 1989, despite downtime for refueling, Millstone Units 1 and 3 recorded capacity factors of 80.4 percent and 70.6 percent respectively.

- In the NRC's October 1989 Systematic Assessment of Licensee Performance (SALP) report on Millstone Units 1 andd 2, the plant's "continued good performance and safe operations" were noted. At Millstone 3, the NRC described plant performance as "careful and safe"' noting "diligent attention to performance at all levels."

- According to *Nucleonics Week*, which compiles free-world generating statistics monthly, Millstone 3 was the fifth in the top 10 nuclear units in the United States during July 1990 with 872,472 MWH generated. In May 1990, Millstone 3 was eighth in the top 10 American nuclear units with 881,732 MWH of electricity generated.
- According to the 1991 Nuclear Industry Performance Compendium, published by Temple, Barker & Sloane, an international management consulting firm, the Millstone plant was one of the 10 nuclear plants that showed "consistently superior performance" during the five years between 1986 to 1991 by achieving a capacity factor greater than 70 percent and by turning in below-average costs each year.

Millstone takes steps to ensure the community's safety. Every year, the towns within a 10-mile radius of the Connecticut Yankee and Millstone plants participate in emergency preparedness drills. These procedures have been helpful in other events including floods, transportation events involving hazardous materials, and in 1985 they were used for Hurricane Gloria.

Northeast looks to the future. To prepare for the 1990s, Northeast has searched for ways to cut costs without compromising performance and has come up with downsizing. The Institute of Nuclear Power Operations (INPO) backs up the company's decision because those U.S. plants with high NRC ratings tend to have smaller staffs and use less contract labor.

As a result, Northeast trimmed its nuclear engineering and operations staff from 2,427 in 1987 to 2,124 in 1990. The reduction was achieved through attrition.

Northeast's operating license to expire in 2010. In the year 2010, Northeast's 40-year operating license for Millstone 1, which was extended to reflect time the plant was under construction, is due to expire.

Northeast oversees nature. Northeast Utilities has built osprey nesting platforms at its Millstone nuclear power plant. As a result, in 1990, 10 percent of all the osprey born in Connecticut were hatched at Millstone.

The utility also studies lobsters. They tag selected crustaceans and then monitor them to estimate the age distribution of the lobster population as a whole. In addition, the utility monitors the flounder population in Niantic Bay.

Millstone keeps the public informed. Each year between 14,000 to 15,000 people tour the Millstone Energy Center, located in the center of Niantic. The information available at the center includes:

- A variety of models and lighted displays that show how and why Millstone operates.
- A Slide presentation that explains the mechanics of electrical power generation.
- An environmental exhibit with tanks displaying the various types of fish and marine life that thrive in the Long Island Sound near Millstone, which shows the careful monitoring

done by scientists and marine biologists at the Millstone Environmental Lab on the plant site.

Sources

"Nuke Plants Make Grid Less Reliable, Hike Demand." *Boston Business Journal*, June 16, 1986.
"Nutmeg Nukes." *The Business Times*, October 1986.
"Maine Utility." *The New York Times*, January 30, 1987.
"Nuclear Plant Shuts Down." *The New York Times*, June 15, 1987.
"Businessperson of the Year: William B. Ellis: An Expectation of Excellence." *New England Business*, December 7, 1987.
"Nuclear Plant Shuts Down." *The New York Times*, March 13, 1988.
"Utility Fined $50,000 for Reactor Problem." *The New York Times*, April 14, 1988.
"Connecticut Tremblors Defy Research on Cause." *The New York Times*, May 22, 1988.
"Northeast Utilities Is Fined." *The Wall Street Journal*, September 5, 1989.
"New England's Nuclear Powerhouse." *Nuclear Industry*, Summer 1990.
"Palo Verde Unit 3 is No. 1 in U.S. Electric Generation." *Business Wire*, September 19, 1990.
"Due Up for License Renewal: The Future of Nuclear Power." *The New York Times*, June 24, 1991.
"Palo Verde Units 1 & 2 Top American Producers in May; Unite 2 Continues Record Run." *Business Wire*, July 12, 1991.
"Another Rave Review for 1990." *Nuclear Industry*, Fourth Quarter 1991.
"St. Lucie 2 Completes Record Run, Leads World in Efficiency." *Info*, U.S. Council for Energy Awareness, April 1992.
"Due Up for License Renewal: The Future of Nuclear Power." *The New York Times*, June 24, 1991.

—Sharon Miscavage

Montague Nuclear Power Plant, Units 1-2

Montague, MA

Owning Utility

Northeast Utilities
PO Box 270
Hartford, CT 06141-0270
John F. Opeka, Exec.V.Pres., Nuclear

Phone: (203)665-5000
Fax. (203)665-5884

Contact

Anthony J. Castagno, Mnager, Nuclear Info.
Northeast Utilities

Basic Facts

Units 1-2 *Status:* Cancelled. *Type of Reactor:* Boiling water reactor. *Megawatts (electric):* 1150. *Reactor System Supplier:* General Electric Co. (United States). *Generator Supplier:* General Electric Co. *Architect Engineer:* Stone & Webster Engineering Corp. (United States). *Owner:* Northeast Utilities. *Operator:* Northeast Utilities.

Key Dates

Units 1-2 *Ordered:* 1974. *Cancelled:* 1980.

Plant Report

New England utilities cancel planned plants. Anticipating a high level of growth in the levels of demand for electricity in the region, several New England utilities planned in the early 1970s to build more nuclear power plants. Many of these plants were opposed by nuclear industry regulators and utility financial advisors, forcing the utilities to eventually cancel the plants. Among the plants that were cancelled were Pilgrim II and III, Seabrook II, Sears Island, and Montague I and II.

According to Paul F. Levy, 1987 chairman of the Massachusetts Department of Public Utilities, had all these plants been built it would have given the region an extra 8000 or 9000 megawatts of capacity and caused an immense strain on the regional economy.

Sources

"Utility Chief Not a Man Hungry for Power: Paul F. Levy." *Boston Business Journal*, January 26, 1987.

Monticello Nuclear Generating Plant

PO Box 600
Monticello, MN 55362
Doug Antony, Site Gen.Mgr.

Phone: (612)295-5151
Fax: (612)295-1017

Owning Utility

Northern States Power Company
414 Nicollet Mall
Minneapolis, MN 55401
Leon Eliason, V.Pres., Nuclear Generation

Phone: (612)330-5500
Fax: (612)330-2900

Contact

Margaret Papin, Pub.Info.
Monticello Nuclear Generating Plant

Basic Facts

Status: Operable. *Type of Reactor:* Boiling water reactor. *Megawatts (electric):* 576. *Megawatts (thermal):* 1670. *Reactor System Supplier:* General Electric Co. (United States). *Generator Supplier:* General Electric Co. *Architect Engineer:* Bechtel.

Key Dates

Ordered: 1966. *Construction began:* 1967. *Criticality:* December 1970. *First Power:* March 1971. *Commercial Operation:* June 1971. *Shut Down Expected:* September 2010.

Operating Costs

Year	Cost	Year	Cost
1971	$1,429,000	1981	$18,261,000
1972	$2,567,000	1982	$30,799,000
1973	$5,006,000	1983	$22,628,000
1974	$5,179,000	1984	$41,844,000
1975	$8,729,000	1985	$30,266,000
1976	$6,609,000	1986	$36,337,000
1977	$11,109,000	1987	$42,406,000
1978	$9,136,000	1988	$41,356,000
1979	$10,585,000	1989	$50,248,000
1980	$21,415,000		

Plant Report

A Minnesota power center. The Monticello Nuclear Generating Plant occupies a 1,400-acre site about three miles northwest of Monticello, Minnesota, and 40 miles northwest of Minneapolis-St. Paul. It produces enough electricity to serve 500,000 homes. Along with its sister plant, Prairie Island, in Red Wing, Minnesota, Monticello accounts for 36 percent of NSP's electric generation. The plant cost about $119 million to design and construct.

World's best producer. Monticello first produced electricity on June 30, 1971. As of January 31, 1991, the plant generated more than 69 million megawatt-hours of electricity, more than any other nuclear plant, under 600 megawatts, in the world.

License renewal. Monticello's plant license expires in 2010. The plant was the pilot extension candidate under a U.S. Department of Energy and Electric Power Research Institute program, and the first to file a formal renewal application. In 1990, the NRC considered Monticello a "lead plant" in the licensee renewal program, and required NSP to submit its application by December 1991. The commission expected to complete its review of the application within two years after it was received.

Community impact. Monticello benefits its nearby communties, including the roughly 5,000 residents of Monticello, Minnesota. In 1990, NSP's property tax payment of about $10.4 million benefited the city, Wright County and the schools, with $2.1 million going to the city, $2.8 million to the county, $5.2 million to the schools, and the remainder to the hospital district.

Monticello specifications. The reactor building stands about 40 feet tall, and the reactor holds 484 fuel assemblies, or 190,000 pounds of uranium. Each assembly contains 62 fuel rods, which measure 13 feet long, and can produce up to 47 million kilowatt-hours of electricity. About 121 assemblies are refueled every 18 months, and spent fuel is stored at the plant in a water-filled, 40-foot-deep pool.

Absorption of free neutrons in the core is controlled by 121 movable control rods, made of boron carbide or hafnium and encased in stainless steel.

The net station heat rate is 10,600 British thermal units per kilowatt-hour, and the maximum fuel-pellet temperature measures 2,750 degrees Fahrenheit.

The plant utilizes water from the Mississippi River for its circulating water system, with two pumps sending 280,000 gallons per minute to the condenser. Water and steam have an average temperature of 545 degrees Fahrenheit and a pressure of 1,000 pounds per square inch in the reactor core. The temperature of the water after it cools the condenser ranges from 86 degrees Fahrenheit in summer, to 37 degrees Fahrenheit in winter, measured 1,000 feet from the discharge point.

The tandem-compound (single shaft) four-flow, non-reheat steam turbine has 38-inch, last stage buckets. The turbine spins at 1,800 rotations per minute.

The generator has a 632,000 kilovolt amphere rating, with an output of 22,000 volts, three-phase, alternating current. Hydrogen gas at 45 pounds per square inch pressure cools the generator.

The substation produces 345,000 volts, 230,000 volts and 115,000 volts, alternating current.

Employees. NSP employs 336 people full time at Monticello. The utility requires all employees to receive annual General Employee Training. The program includes training in security and energy procedures, radiation, respiratory and fire protection practices, and general safety.

Monticello employees utilize the plant's training center and controlroom simulator. NSP spent four years planning and building the simulator, which came into use in 1985.

Radiation protection. Annually, the average Monticello employee receives about 360 millirem of radiation beyond natural background levels. This represents only seven percent of the maximum dose allowed by the NRC.

Monticello's radiation protection department monitors radiation in all areas where people work. The goal of the plant's radiation program is ALARA—As Low As Reasonably Possible.

NSP health physics professionals control worker's exposure to radioactivity and monitor radiation throughout the plant. The utility's radiation protection specialists establish safe procedures, perform radiation surveys, and test and monitor the plant for radiation.

Environmental monitoring. Monticello uses two major systems to protect the environment from increased radioactivity in the air and from the release of warm water into the Mississippi River.

In 1979, NSP added a special system to reduce atmospheric radioactivity from Monticello. The $10-million off-gas treatment system reduces the level of public exposure by about 99 percent and contains a recombiner which eliminates about 90 percent of the gas emissions by converting hydrogen and oxygen into water and recycling it to the reactor. The remaining gas is filtered and stored in hold-up tanks for 50 hours to permit the radioactivity to decay.

NSP began monitoring river temperature, aquatic life and natural background radioactivity about three years before Monticello went into operation. Monitoring data has demonstrated that the plant has not had any adverse effect on the river or the surrounding area.

Waste disposal. Historically, Monticello annually disposes about 6,000 cubic feet of radioactive waste, which falls below the national average of 10,000 cubic feet per year for similar reactors. The plant has received recognition from the Institute of Nuclear Power Operations (INPO) for its agressive waste volume reduction program.

Under the Low-level Radioactive Waste Policy Amendment Act of 1985, states are required to form regional compacts to build and operate disposal sites for their members. Minnesota belongs to the Midwest Compact, and Michigan was chosen as host state for the regional disposal site in 1993. As of 1991, a site had not been found, and the opening date was moved up to 1997 or 1998.

NSP estimated that its own storage space for spent fuel would fill up as early as 1990. In 1984, the utility began shipping spent fuel to a facility in Morris, Illinois.

NSP has a long-term contract with General Electric to store, at no cost until 2002, a 10-year supply of spent fuel from Monticello. The arrangement is predicted to save NSP customers $17 million to $24 million in storage costs. The additional space in the plant's pool would allow the utility to continue operating the plant until the early 21st century.

Emergency planning. NSP makes a booklet available to the public called the *Emergency Planning Guide for Neighbors of Monticello Nuclear Generating Plant*. People within the plant's 10-mile emergency planning zone receive the booklet annually.

Another booklet, *Emergency Radiological Instructions & Information*, is directed to Minnesota farmers, food processors, and distributors with instructions on necessary activity in the event of an accident at Monticello.

Shutdown and renovation. In 1984, NSP shut down Monticello for refueling, pipe repairs, basic maintenance and NRC inspections. After the shutdown, the NRC described the plant as one of the best of the 23 reactors at 15 sites in its eight-state region.

Decommissioning. The cost of decommissioning and building Monticello is reflected in NSP's rates, and spread over the life of the plant as depreciation expenses. Based on a 1990 study, NSP estimated decommissioning costs of both its nuclear plants at roughly $630 million, in 1990 dollars. During 1990, the company recorded $175 million in its accumulated provision for depreciation, and $13 million of the balance was designated for deposit into an external fund in 1991.

Honors and achievements. In 1986, the National Academy for Nuclear Training Accrediting Board honored Monticello for its operator and maintenance training programs, as well as its nonlicensed and licensed operator, shift supervisor, electrical maintenance and mechanical maintenance positions.

In 1987, the board honored Monticello for its engineering and technical training.

Both the INPO and the NRC gave Monticello high ratings during 1990. For the second time, the INPO granted the plant its highest rating, citing its experienced work force, excellent plant material conditions, and impressive long-term performance.

Monticello received the Minnesota Safety Council Award of Honor during 1990. From December 1985 to May 1990, or 2.4 million hours, no worker sustained a lost work day due to injuries. The plant also received the award in 1988.

The Utility Data Institute (UDI) ranked Monticello eighth on its list of lowest-cost producers of electricity in the United States during 1990. According to the UDI, the plant had a production cost of $14.88 per megawatt-hour of electricity while producing 569 gross megawatts at 4,505,927 megawatt-hours.

High capacity factors. Monticello and Prairie Island had the highest combined capacity factor, 82.3 percent, for plants owned by the same utility during the period from January 1, 1987, to December 31, 1989. The two plants also had the highest combined capacity factor from January 1, 1986, to December 31, 1988.

For 1990, *Nucleonics Week* magazine cited Monticello's capacity factor of 92.4 percent as the second highest in the United States, and the fifth highest in the world for boiling water reactors.

In 1992, GE Nuclear Energy recognized Monticello for achieving a capacity factor of 75 percent or higher during 1991.

Sources

"Grass Roots: Washington Watch." *Critical Mass Bulletin*, February 1984.
"Lead Plant Reviews." *U.S. NRC Annual Report 1990*, n.d.
Northern States Power Company 1990 Annual Report: NSP is People. Working for You. Northern States Power Company.
Monticello Plant Receives Safety Council Top Award. (Press release) Northern States Power Company, May 9, 1990, n.d.
NSP Receives Top Ranking For Nuclear Plants' Capacity Factor. (Press release) Northern States Power Company, May 23, 1990.
NSP Nuclear Plants Net Top National Awards. (Press release) Northern States Power Company, November 19, 1990.
"State's Radioactive Waste Becoming Harder to Get Rid of." *Minneapolis-St. Paul CityBusiness*, March 18, 1991.
NSP's Monticello Plant Turns 20. (Press release) Northern States Power Company, June 18, 1991.
"Wisconsin Electric's Point Beach Lowest Cost Nuclear Plant in '90." *INFO*, U.S. Council for Energy Awareness, July 1991.
Nuclear Power at Monticello. Northern States Power Company, July 1991.
"And The Winner Is." *INFO*, U.S. Council for Energy Awareness, May 1992.
NSP's Monticello Plant Turns 20. (Press release) Northern States Power Company, June 18, 1991.

—Charles Ceplecha

---

315 ☆
NEP Nuclear Power Plant, Units 1-2
Charlestown, RI

Basic Facts
Units 1-2 *Status:* Cancelled. *Type of Reactor:* Pressurized light-water reactor. *Megawatts (electric):* 1200. *Megawatts (thermal):* 3425. *Reactor System Supplier:* Westinghouse Electric Corp. (United States). *Owner:* New England Power (NEP). *Operator:* New England Power (NEP).

Key Dates
Units 1-2 *Ordered:* 1974. *Cancelled:* 1979.

---

316 ☆
New Haven Nuclear Power Plant, Units 1-2
New Haven, NJ

Basic Facts
Units 1-2 *Status:* Cancelled. *Type of Reactor:* Pressurized light-water reactor. *Megawatts (electric):* 1240. *Megawatts (thermal):* 3817. *Reactor System Supplier:* Combustion Engineering, Inc. (United States). *Owner:* New York State Electric and Gas. *Operator:* New York State Electric and Gas.

Key Dates
Units 1-2 *Ordered:* 1977. *Cancelled:* 1980.

---

317 ●
Nine Mile Point Nuclear Power Plant, Unit 1

Lake Rd. Phone: (312)343-2110
PO Box 32
Lycoming, NY 13093
Joseph F. Firlit, V.Pres., Nuclear Support

Owning Utility
Niagara Mohawk Power Corporation Phone: (315)474-1511
300 Erie Blvd. W.
Syracuse, NY 13202
B. Ralph Sylvia, Exec.V.Pres., Nuclear

Contact
Robert B. Burtch Jr., Nuc. Commun./
  Pub.Affrs.Dir.
Niagara Mohawk Power Corporation

Basic Facts
Unit 1 *Status:* Operable. *Type of Reactor:* Boiling water reactor. *Megawatts (electric):* 625. *Megawatts (thermal):* 1850. *Reactor System Supplier:* General Electric Co. (United States). *Generator Supplier:* General Electric Co. *Architect Engineer:* Niagara Mohawk Power Corporation.

Key Dates
Unit 1 *Ordered:* 1963. *Construction began:* 1965. *Criticality:* August 1969. *First Power:* November 1969. *Commercial Operation:* December 1969. *Shut Down Expected:* April 2005.

Operating Costs
Units 1-2

Year	Cost	Year	Cost
1970	$1,716,000	1980	$9,404,000
1971	$2,759,000	1981	$26,744,000
1972	$3,575,000	1982	$21,481,000
1973	$4,524,000	1983	$25,517,000
1974	$6,251,000	1984	$26,788,000
1975	$5,810,000	1985	$17,567,000
1976	$5,330,000	1986	$19,693,000
1977	$9,743,000	1987	$10,603,000
1978	$6,382,000	1988	$123,289,000
1979	$11,664,000	1989	$209,434,000

Plant Report

Niagara Mohawk Power Corporation. Nine Mile Point Unit 1 is owned and operated by Niagara Mohawk Power Corporation. The plant is one of two nuclear units operated by the utility. Together they make up 14 percent of the company's generating capacity. Niagara Mohawk's other sources of electricity are

Plant Profiles

hydroelectric and fossil-fueled plants. The power supplier serves 1,500,000 customers in 37 counties covering 24,000 square miles in upstate New York.

Longest run. Nine Mile Point Unit 1 began operation in 1969. It has a generating capacity of 610 megawatts. The plant set a record for the longest run achieved by a U.S. boiling reactor when it completed 415 days of continuous service in 1987. It is one of only 13 nuclear plants of all types in the U.S. to achieve continuous runs in excess of 400 days.

"Troubled Plant". The plant's record with the NRC, however, has not been as exemplary. Niagara Mohawk accumulated fines totaling $340,000 for violations occurring between 1974 and 1987. Defective pipe welds, faulty testing procedures, maintenance irregularities, and radiation-protection problems were listed among the cited infractions. An additional $100,000 fine was assessed because repairs were judged "ineffective."

Beginning of a 2 1/2-year shutdown. On December 19, 1987 a problem with the feedwater system caused the plant to be shut down. Officials at Niagara Mohawk expected a short outage. They moved up the date of a re-fueling, originally scheduled for March 1988, so the procedure could be accomplished during the shutdown.

Their expectations for a short outage were nixed, however, when NRC inspections found 20 other problems at the plaant. These included questions about the operators' training, fire protection measures, and the plants' inservice inspection capability. Allegations of drug use were also under investigation. The NRC placed Nine Mile Point Unit 1 on its "Troubled Plant" list in July 1989.

Cost of supplemental power. During the shutdown, Niagara Mohawk purchased electricity from other sources to make up for power the idled Unit 1 failed to generate. This supplemental power cost an average of $225,000 per day. The utility asked the Public Service Commission for permission to recover the cost of substitute power from the state's electric consumers on a temporary basis.

Radioactive spill cleaned up after 8 years. Also in 1989 while Nine Mile Point Unit 1 was shut down, officials at Niagara Mohawk signed a contract with radioactive waste handlers to clean up a spill at the plant. 150 barrels of highly radioactive waste water had spilled in a sub-basement of the plant in 1981. The utility said the spill was confined to the sub-basement and in stable condition. Radiation measurements outside the room were low, the equivalent of a single chest X-ray, but levels inside the room were too high for workers to enter.

Restarting the reactor. In 1990 the NRC upgraded Nine Mile Point Unit One to "Category 2." This designation indicated that improvements necessary to restart the reactor had been made but that close monitoring would continue. The reactor resumed operation on July 29, 1990 after being out of service for two and a half years. Unit 1 was taken off the NRC "watch list" in June 1991.

Nine | 318

License up for renewal. Unit 1's operating license will be up for renewal in 2,008. Details of the renewal process are uncertain. Some groups claim that licenses should not be renewed because older plants can never be as safe as they were when new. In addition, the NRC has not yet decided whether old plants seeking license renewals should have to meet requirements established after they were first built. In addition, some of the safety equipment at Nine Mile Point Unit 1 has come under scrutiny because operating temperatures have been 20 to 30 °F higher than the equipment was originally designed to withstand.

Visitors welcome. A visitors center, "The Energy Center" is operated jointly by Niagara Mohawk and the New York Power Authority. It serves visitors to Nine Mile Point Unit 1 and Unit 2 as well as the adjacent FitzPatrick Nuclear Power Plant. The center features a model of the Nine Mile Point Unit 1 plant and hands-on exhibits about electricity and nuclear power. Guided tours of the site are available in July and August.

For more information contact: Director, The Energy Center, P.O. Box 81, Lycoming, New York, 13093, (315) 342-4117.

Sources

The Power of Nine Mile Point. Niagara Mohawk, n.d.
Nuclear Focus on Results. Niagara Mohawk, n.d.
"U.S. Grants Upstate A-Plant A License to Run at Full Power." *The New York Times,* July 2, 1987.
"Niagara Mohawk is Fined $100,000 by Nuclear Panel." *The Wall Street Journal,* March 16, 1988.
"8-Year-Old Nuclear Plant Spill is Investigated." *The New York Times,* August 27, 1989.
"Nuked but Alive." *Barron's,* September 18, 1989.
"A-Plant Nominee Admits Mistakes." *The New York Times,* October 13, 1989.
"'Watch List' Pared by NRC." *Nuclear Industry,* Summer 1990.
NRC Annual Report, U.S. Nuclear Regulatory Commission, 1990.
"Due Up for License Renewal: The Future of Nuclear Power." *The New York Times,* June 24, 1991.
"St. Lucie 2 Completes Record Run, Leads World in Efficiency." *Info,* U.S. Council for Energy Awareness, April 1992.
"'Watch List' Pared by NRC." *Nuclear Industry,* Summer 1990.

—Karen Bellenir

318 •

Nine Mile Point Nuclear Power Plant, Unit 2

Lake Rd.
PO Box 32
Lycoming, NY 13093
Joseph F. Firlit, V Pres., Nuclear Support

Phone: (312)343-2110

Owning Utility

Niagara Mohawk Power Corporation
300 Erie Blvd. W.
Syracuse, NY 13202
B. Ralph Sylvia, Exec.V.Pres., Nuclear

Phone: (315)474-1511

Contact

Robert B. Burtch Jr., Nuc. Commun./
 Pub.Affrs.Dir.
Niagara Mohawk Power Corporation

Nine

UNITED STATES

Basic Facts

Unit 2 *Status:* Operable. *Type of Reactor:* Boiling water reactor. *Megawatts (electric):* 1166. *Megawatts (thermal):* 3323. *Reactor System Supplier:* General Electric Co. (United States). *Generator Supplier:* General Electric Co. *Architect Engineer:* Stone & Webster Engineering Corp. (United States).

Key Dates

Unit 2 Ordered: 1972. Construction began: 1975. Criticality: May 1987. First Power: July 1987. Commercial Operation: April 1988. Shut Down Expected: October 2026.

Operating Costs

Units 1-2

Year	Cost	Year	Cost
1970	$1,716,000	1980	$9,404,000
1971	$2,759,000	1981	$26,744,000
1972	$3,575,000	1982	$21,481,000
1973	$4,524,000	1983	$25,517,000
1974	$6,251,000	1984	$26,788,000
1975	$5,810,000	1985	$17,567,000
1976	$5,330,000	1986	$19,693,000
1977	$9,743,000	1987	$10,603,000
1978	$6,382,000	1988	$123,289,000
1979	$11,664,000	1989	$209,434,000

Plant Report

A third nuclear plant begins operation. Nine Mile Point Unit 2, located on the shore of Lake Ontario in Scriba, New York, began operation in 1988. It is adjacent to Nine Mile Point Unit 1, owned by Niagara Mohawk, and the FitzPatrick nuclear plant, owned by the New York Power Authority.

Unit 2 is owned by five New York utilities. Niagara Mohawk Power Corporation holds the largest interest (41 percent) and operates the plant. The other owner utilities are: New York State Electric & Gas (18 percent), Long Island Lighting Company (18 percent), Rochester Gas & Electric Corporation (14 percent), and Central Hudson Gas & Electric Corporation (9 percent).

Fairness Doctrine suit. Controversy arose during the plant's construction. Critics claimed that New York's electric consumers would save money if the plant were abandoned before being completed. The utilities building Unit 2 countered with claims that the plant was vital for employment in the state. They ran supportive advertisements on television stations in five cities near the plant.

The Syracuse Peace Council, an antinuclear group, sued under the "Fairness Doctrine" for free air time to present an opposing viewpoint. In 1984 the FCC sided with the Syracuse Peace Council against WTVH-TV. The ruling was the first time in five years the FCC had censured a broadcaster for violating the Fairness Doctrine.

Expensive plant. When the NRC granted the five utilities a construction permit in 1974 to build the second unit at Nine Mile Point, cost estimates stood at $600 million. The plant was supposed to be finished in five years. Early estimates, however, fell far short of actual experience. Unit Two took 14 years to build and came at a cost of $6.3 billion.

In 1986, because of concerns about rising costs, the New York State Public Service Commission and the plant's builders agreed to put a cap of $4.16 billion on the amount the utilities would be able to pass on to rate payers when the plant was finished. Continued cost overruns forced all the utilities involved to cut dividends to their stock holders. In an effort to recoup some of the excessive construction costs, they filed suit against Stone & Webster Engineering Corporation, the construction firm hired to build the facility. The suit claimed that the cost overruns were incurred as a result of design deficiencies and mismanagement during construction. The suit was settled out of court for an undisclosed amount.

Faulty valves delay opening and result in law suit. Nine Mile Point Unit 2 achieved its first nuclear chain reaction in May of 1987. By August it had been brought up to 50 percent of capacity and was scheduled to begin commercial operation in September. Problems arose, however, when some valves leaked more water during temperature changes than allowed under federal regulations. Valve replacement forced the plant's opening to be delayed until the following April.

The cost-cap agreement again prevented the utilities from passing on additional costs to their customers. A $500 million lawsuit was filed against the companies who supplied the valves. The suppliers denied responsibility saying that the valves had been sold to the plant 10 years earlier and had been modified.

On "Troubled Plant" list. Nine Mile Point Unit 2 finally began commercial operation April 1988. It shut down six months later for maintenance and inspections. In December 1988, the NRC placed Unit 2 on its "Troubled Plant" list. Problems cited included 17 automatic shutdowns, failure to identify and solve problems, poor worker attitudes, and inadequate personnel. NRC officials blamed the instability on management claiming that Unit 2 was one of the 10 worst managed nuclear plants in the country.

Better days short-lived. In 1990 Unit Two achieved its longest run: 135 days. In June 1991 it was taken off the NRC "Troubled Plant" list. The plant operated 123 consecutive days before the next mishap occurred.

"Site Area Emergency". On August 13, 1991, at 5:48 a.m. electrical power in the control room went out. At 6:00 a.m. officials declared a "Site Area Emergency." Under federal guidelines a "Site Area Emergency" is the third highest in a series of four ascending alert levels. First is an "Unusual Event" which means that there is a possibility that the plant's safety will be reduced. The second, an "Alert" means that a potential exists for substantial reduction in plant safety or that plant safety has actually been reduced. The third, a "Site Area Emergency" means there has been or is likely to be a major failure in the plant functions that are necessary for public protection. The highest alert level, a "General Emergency," is defined as one that threatens "actual or imminent serious damage to the reactor."

The alert levels were instituted after the Three Mile Island accident in 1979. There has never been a "General Emergency," declared at a U.S. nuclear station, although Three Mile Island would have qualified if the regulations had been in place. The "Site Area Emergency" declared at Nine Mile Point Unit Two was only the third such emergency in the United States.

When Unit Two experienced a total loss of power to crucial indicators in the control room, operators began working immediately to bring the plant to "cold shutdown" status. A Cold Shutdown means that the reactor temperature is reduced to 200 ° F, below the temperature necessary to boil water at normal pressure. Cold shutdown was reached at 6:46 p.m. and the emergency alert was cancelled at 7:43 p.m.

Investigators attributed the power outage to a transformer failure that knocked out five of the plant's 10 electrical systems. The blacked-out systems provided electricity to many critical pieces of equipment including the plant's main computer, safety monitors, gauges used to control the reactor and the position of control rods, the plant's radio and paging systems, essential lighting, and monitors needed to measure any release of radiation.

Power was restored to the control room from off-site sources after 22 minutes. Although other nuclear plants had experienced power failures before, no plant had ever lost five supposedly uninterruptible power supplies at the same time. Operators had not developed procedures for such an event because it was not considered a possible scenario. Fortunately, the power to the plant's most vital safety-related equipment remained intact.

Much of the investigation following the incident concerned the question of why the back up power had not come on. The "uninterruptible" electrical systems had two separate back up sources of power. In the event of a power failure, the state power grid was supposed to have supplied electricity to the plant. If that system failed, the plant should have switched to emergency back up batteries.

One theory offered was that the transformer malfunction caused an instability in the circuits required to switch over to back up power. NRC investigators discovered, however, that the outage also blacked out a logic board responsible for re-routing power and that the back up batteries in the logic board were dead. They were designed to last four years and had been in service for five years. The batteries had not been included in plant maintenance schedules.

Nine Mile Point Unit 2 returned to service on September 19, 1991, after the faulty transformer was replaced. Other repairs included re-wiring uninterruptible power supplies and instituting a procedure for battery replacement in the logic board.

Sources

The Power of Nine Mile Point. Niagara Mohawk, n.d.
Nuclear Focus on Results. Niagara Mohawk, n.d.
"Grassroots." *Critical Mass Bulletin*, March 1984.
"The Noise from Syracuse." *Channels of Communication*, January/February 1985.
"Nine Mile Point 2 Settles, Skirts Prudence." *Electrical World*, October 1986.
"Niagara Mohawk Sets Valve Replacements At Nine Mile 2 Plant." *The Wall Street Journal*, March 13, 1987.
"Niagara Mohawk, Others Sue 3 Firms Over Valve Designs." *The Wall Street Journal*, April 20, 1987.
"After 15 Years of Problems, Nuclear Plant Warms Up." *The New York Times*, May 28, 1987.
"U.S. Grants Upstate A-Plant A license to Run at Full Power." *The New York Times*, July 2, 1987.
"Nuclear Plant Restarted." *The New York Times*, August 22, 1987.
"New York State E&G Reduces Payout 24 percent." *The Wall Street Journal*, January 18, 1988.
"3 Reactors In Northeast Among 10 Worst-Managed." *The New York Times*, December 23, 1988.
"Niagara Mohawk Says Suit Over Construction of Power Plant Settled." *The Wall Street Journal*, March 12, 1991.
Press Releases Nos. 1, 10. Niagara Mohawk, August 13, 1991.
"New York A-Plant is Shut Down After Monitoring Instruments Fail." *The New York Times*, August 14, 1991.
"What the Alert Levels Mean." *The New York Times*, August 14, 1991.
"U.S. Team Uncovers New Failures in Nine-Mile Nuclear Disruption." *The New York Times*, August 23, 1991.
"When a Nuclear Accident Doesn't Go by the Book." *The New York Times*, August 25, 1991.
"Wiring Flaw and Battery Failure Cited in Nuclear-Plant Accident." *The New York Times*, September 5, 1991.
Nine Mile Two Restart News Briefing. Niagara Mohawk, September 18, 1991.
News Release. Niagara Mohawk, September 27, 1991.
"If You Hate People Looking Over Your Shoulder, Don't Even Think About Working in a Nuclear Power Plant." *Nuclear Industry*, Fourth Quarter 1991.
"Wiring Flaw and Battery Failure Cited in Nuclear-Plant Accident." *The New York Times*, September 5, 1991.
Nine Mile Two Restart News Briefing. Niagara Mohawk, September 18, 1991.

—Karen Bellenir

North Anna Nuclear Power Station, Units 1-2

PO Box 402   Phone: (703)894-5151
Mineral, VA 23117
G.E. Kane, Stat.Mgr., Nuclear

Owning Utility

Virginia Power   Phone: (804)771-3000
PO Box 26666
Richmond, VA 23261
W.L. Stewart, Sr.V.Pres., Nuclear

Contact

W.N. Curry, Mgr., Corp. Commun.
Virginia Power

Basic Facts

Units 1-2 *Status:* Operable. *Type of Reactor:* Pressurized light-water reactor. *Megawatts (electric):* 947. *Megawatts (thermal):* 2893. *Reactor System Supplier:* Westinghouse Electric Corp. (United States). *Generator Supplier:* Westinghouse Electric Corp. *Architect Engineer:* Stone & Webster Engineering Corp. (United States).

Key Dates

Unit 1 *Ordered:* 1967. *Construction began:* 1971. *Criticality:* April 1978. *First Power:* April 1978. *Commercial Operation:* June 1978. *Shut Down Expected:* April 2018.

Unit 2 *Ordered:* 1967. *Construction began:* 1971. *Criticality:* June 1980. *First Power:* June 1980. *Commercial Operation:* December 1980. *Shut Down Expected:* August 2020.

Operating Costs

Units 1-2

1978	$6,521,000	1984	$66,502,000
1979	$19,519,000	1985	$55,490,000
1980	$25,389,000	1986	$71,732,000
1981	$28,856,000	1987	$103,658,000
1982	$43,494,000	1988	$47,377,000
1983	$45,068,000	1989	$123,653,000

Plant Report

An introduction to the North Anna Nuclear Power Station. The North Anna Nuclear Power Station, located 45 miles northwest of Richmond in Mineral, Virginia, is owned by Virginia Electric & Power Company and produces approximately 20 percent of Virginia's total electric needs. The plant, along with the Surry Nuclear Power Station, also owned by Virginia Power, supplies 40 to 45 percent of the utility's power needs and with the Surry Plant has a combined capacity of 3,400 megawatts. The nuclear reactors were completed for just over $500 a kilowatt and generate power for about a half-cent per kilowatt hours versus approximately two cents for coal and five and one-half cents for oil.

Trouble hits North Anna. On July 16, 1987 a mishap occurred at North Anna. At 6:30 a.m. it was discovered that one of thousands of stainless steel tubes carrying super-heated water to a nuclear power plant steam generator ruptured, releasing a tiny amount of radioactivity into the atmosphere. Plant employees immediately took the reactor off line, but the radioactive gas was vented outside for approximately 90 minutes while they went through the procedure.

It was determined, by using the measurement standard for radioactivity absorbed by living tissue, that up to five one-thousandths of a millirem of radioactivity was released in this incident, which is less than one percent of the amount that the Nuclear Regulatory Commission (NRC) allows Virginia Power to release from its plant. In comparison, a diagnostic x-ray may expose a person to about 10 millirems of radioactivity.

During this incident the reactor was operating at full power. However, the reactor had been restarted the previous weekend, after being out of service for several weeks. It was also found that the tube that ruptured was among thousands inspected in a recent refueling of the plant. According to utility and Federal officials, there were no employees injured and the public was not endangered. However, this incident prompted the Nuclear Regulatory Commission (NRC) to issue three minor safety violations to Virginia Power.

Some critics saw this incident as more than just an operating problem. A nuclear analyst for Public Citizen, a consumer interest group, said these leaks implied a premature and potentially dangerous aging process of nuclear power plants. Other critics of nuclear power charged that this incident was an indication that older nuclear power plants are deteriorating to a dangerous degree.

Trouble strikes for the second time at North Anna. On February 25, 1989, another mishap occurred at North Anna. At 2:26 p.m. a cloud of radioactive gas was released into the atmosphere when a tube carrying water with radioactive particles developed a leak, allowing more than 50 gallons a minute of contaminated water to flow into a secondary water system that is normally not radioactive.

The leak occurred in the "C" steam generator of Unit 1, the same generator that developed a leak in 1987, but it was one-tenth of what was released in the previous incident. The leak was not noticed right away by the automatic monitors due to the lower level of radiation. The technicians manually shut down the flow of gases to the area where the tube was leaking when they noticed the leak.

The leaking tube, one of 10,000 tubes in three steam generators, carries hot, radioactive water which, in turn, heats the secondary water system. This water can reach temperatures up to 630 degrees. The steam created from heating the secondary water system turns large turbines that drive the electrical generator, producing electricity.

There were no injuries due to this incident and officials said the release of radioactive gas in Louisa County, about 90 miles southwest of Washington, was harmless and not a threat to the public during the 44-minute period when the radiation was released.

North Anna seems to be improving with age. From the late 1980s to the early 1990s the North Anna Nuclear Power Station was performing at above average levels according to the watchdogs of nuclear power. Some of the achievements of North Anna are described below.

In 1990 North Anna was one of the 10 nuclear plants that showed "consistently superior performance" during the five years between 1986 and 1990 by achieving a capacity factor greater than 70 percent and by turning in below-average costs each year, according to the 1991 Nuclear Industry Performance Compendium, published by Temple, Barker & Sloane.

North Anna was one of the top four out of the top 20 nuclear power plants that produced the cheapest electricity for the five year period between 1986 and 1990. The analysis of electricity production costs was done by Utility Data Institute (UDI), an independent data-gathering subsidiary of Halliburton Company. It was determined that North Anna had an average production cost of 1.234 cents per kilowatt-hour.

In 1990 North Anna 2 was named on of the U.S. nuclear plants with longest non-stop operating runs having completed a 469-day run. Any run of longer than 365 days is exceptional for a power plant.

In 1990 the North Anna plant was included on the Utility Data Institute's (UDI) top 10 list of low-cost producers of nuclear electricity. According to UDI, North Anna had a production cost of $12.88 per megawatt-hour electricity, generated 1,959 gross megawatts of power and produced 13,210,187 megawatt hours.

In 1991 North Anna 2 was one of the 15 U.S. nuclear plants that had capacity factors above 90 percent, 97.4 percent to be precise.

Plant Profiles   Oconee | 321

North Anna strives to help the environment. North Anna is working hard to keep nature undisturbed. As of the summer of 1990, the plant had 40 to 50 wood duck nesting boxes located on the plant's premises.

Sources

"In Praise of Nuclear Power." *Forbes*, November 5, 1984.
"Tube Rupture Releases Radioactivity in Virginia." *The New York Times*, July 16, 1987.
"VA Nuclear Power Plant Releases Radioactive Gas." *The Washington Post,*, February 26, 1989.
"Virginia Power Fined Heavily for Violations of Nuclear Plant Safety." *The Washington Post*, May 20, 1989.
"Wisconsin Electric's Point Beach Lowest Cost Nuclear Plant in '90." *Info*, U.S. Council for Energy Awareness, July 1991.
"Around The States: Another Rave Review for 1990." *Nuclear Industry*, Fourth Quarter 1991.
"Virginia Power Nuclear Units Generate Big Savings." *INFO*, U.S. Council for Energy Awareness, February 1992.
"U.S. Nuclear Plants Shatter Records for Output, Performance." *INFO*, U.S. Council for Energy Awareness, February 1992.
"St. Lucie 2 Completes Record Run, Leads World in Efficiency." *INFO*, U.S. Council for Energy Awareness, April 1992.
"Low-Cost Producers." *INFO*, U.S. Council for Energy Awareness, May 1992.
"Virginia Power Nuclear Units Generate Big Savings." *INFO*, U.S. Council for Energy Awareness, February 1992.
"U.S. Nuclear Plants Shatter Records for Output, Performance." *INFO*, U.S. Council for Energy Awareness, February 1992.

—Sharon Miscavage

320 ☆
North Anna Nuclear Power Station, Units 3-4

PO Box 402   Phone: (703)894-5151
Mineral, VA 23117
G.E. Kane, Stat.Mgr., Nuclear

Owning Utility

Virginia Power   Phone: (804)771-3000
PO Box 26666
Richmond, VA 23261
W.L. Stewart, Sr.V.Pres., Nuclear

Contact

W.N. Curry, Mgr., Corp. Commun.
Virginia Power

Basic Facts

Units 3-4 *Status:* Cancelled (Unit 3, construction 7% complete; Unit 4, construction 4% complete). *Type of Reactor:* Pressurized light-water reactor. *Megawatts (electric):* 950. *Megawatts (thermal):* 2631. *Reactor System Supplier:* The Babcock & Wilcox Co. (United States). *Owner:* Virginia Power. *Operator:* Virginia Power.

Key Dates

Unit 3 *Ordered:* 1971. *Cancelled:* 1982. *Construction began:* 1974.
Unit 4 *Ordered:* 1971. *Cancelled:* 1980. *Construction began:* 1973.

Plant Report

See Plant Report for North Anna Nuclear Power Station, Units 1-2.

321 •
Oconee Nuclear Station, Units 1-3

PO Box 1439   Phone: (803)885-3000
Seneca, SC 29679
J.W. Hampton, V.Pres., Oconee Site

Owning Utility

Duke Power Company   Phone: (704)373-4011
PO Box 33189
Charlotte, NC 28242
H.B. Tucker, Sr.V.Pres., Nuclear
  Generation

Contact

S.C. Adams, Mgr., Community Relations
Duke Power Company

Basic Facts

Units 1-3 *Status:* Operable. *Type of Reactor:* Pressurized light-water reactor. *Megawatts (electric):* 934. *Megawatts (thermal):* 2568. *Reactor System Supplier:* The Babcock & Wilcox Co. (United States). *Generator Supplier:* General Electric Co. (United States). *Architect Engineer:* Duke Power Company; Bechtel.

Key Dates

Unit 1 *Ordered:* 1966. *Construction began:* 1967. *Criticality:* 1973. *First Power:* May 1973. *Commercial Operation:* July 1973.

Unit 2 *Ordered:* 1966. *Construction began:* 1967. *Criticality:* 1973. *First Power:* December 1973. *Commercial Operation:* September 1974.

Unit 3 *Ordered:* 1967. *Construction began:* 1967. *Criticality:* 1974. *First Power:* September 1974. *Commercial Operation:* December 1974.

Operating Costs

Units 1-3

Year	Cost	Year	Cost
1973	$911,000	1982	$88,016,000
1974	$6,982,000	1983	$82,851,000
1975	$12,449,000	1984	$93,024,000
1976	$16,735,000	1985	$123,707,000
1977	$25,038,000	1986	$135,342,000
1978	$29,600,000	1987	$158,650,000
1979	$40,177,000	1988	$157,658,000
1980	$52,004,000	1989	$153,211,000
1981	$58,788,000		

Plant Report

Duke Power's oldest nuclear plant. The Oconee Nuclear Station is the oldest of three stations operated by Duke Power Company, a utility serving the Piedmont region of the Carolinas. The station sits on the shore of Lake Keowee, approximately 26 miles west of Greenville, South Carolina. It has no cooling tower. Instead, vital cooling functions are performed by circulating water through the lake.

All three units at Oconee are 847-megawatt (net) pressurized water reactors. The plant cost $574 million, which is considered below industry averages for similar projects.

Duke Power attributes its cost containment accomplishments to a policy of using employees rather than outside contractors. The utility claims that in-house construction helped save money on contractor and consultant fees. Many of the workers who built the Oconee station also helped build Duke's subsequent nuclear plants, Catawba and McGuire. By using its own employees, Duke says it was able to develop a crew committed to long-term quality, safety, reliability and efficiency, as well as overcome communication problems that have hindered nuclear construction efforts made by other utilities.

Training. Duke Power began its own nuclear training program in 1968. Members of the first class of 25 students became Oconee's initial operators. In August 1983, the Institute of Nuclear Power Operations (INPO) honored the Oconee instruction program by making it the first one to receive accreditation.

Records set. The Oconee plant has also received industry distinction for setting production records. On July 28, 1991, at 10:39 a.m., Oconee-1 became the first plant in U.S. history to generate a total of 100 billion kilowatt-hours of electricity. That same year, Oconee-2 was listed among the top five U.S. nuclear plants, as rated by capacity factor. (Capacity Factor is a ratio of a plant's actual output compared to its maximum ability.) Unit 2 achieved 97.2 percent.

A steam leak forces "Alert". On November 23, 1991, officials were forced to declare an "Alert" at the station. In a series of four ascending warning categories established by the NRC, an "Alert" is the second highest. During the alert, Oconee's operators successfully brought Unit 3 to cold shutdown status following a leak of approximately 70,000 gallons of radioactive water from the primary coolant system. Examiners blamed the incident on a compression fitting that had separated. The NRC praised Oconee's operators for their performance during the crisis.

Visitors Welcome. Duke Power Company operates the World of Energy Visitor Center at the Oconee site. World of Energy features hands-on exhibits, interactive computer displays, and a control-room simulator. For more information contact: World of Energy, 7812 Rochester Highway, Seneca, SC 29678. Phone: (803)885-4600 or (800)777-1004.

Sources

Making Nuclear Power Work: A Special Report. Duke Power Company, n.d.
"Oconee-3 Steam Leak Prompts Cold Shutdown, 15-Hour Alert." *Inside N.R.C.*, December 2, 1991.
"First U.S. Nuclear Plants Join the 100 Billion Kilowatt-Hour Club." *Nuclear Industry*, Fourth Quarter 1991.
"U.S. Nuclear Plants Shatter Records for Output, Performance." *INFO*, U.S. Council for Energy Awareness, February 1992.
Duke Power Company. Corporate Communications Office. 1992.
"Explore the World of Energy." Duke Power Company, n.d.

—*Karen Bellenir*

Oyster Creek Nuclear Generating Station

Rte. 9 Phone: (609)971-4000
Forked River, NJ 08731
John J. Barton, V.Pres.

Owning Utility

GPU Nuclear Corporation Phone: (201)316-7000
1 Upper Pond Rd. Telex: 136-482
Parsippany, NJ 07054
Phillip R. Clark, Pres. & COO

Contact

K.M. Bromery, Mgr., Pub.Info.
Oyster Creek Nuclear Generating Station

Basic Facts

Status: Operable. *Type of Reactor:* Boiling water reactor. *Megawatts (electric):* 670. *Megawatts (thermal):* 1930. *Reactor System Supplier:* General Electric Co. (United States). *Generator Supplier:* General Electric Co. (United States). *Architect Engineer:* Burns & Roe, Inc. (United States); General Electric. *Owner:* Jersey Central Power and Light Company. *Operator:* GPU Nuclear Corporation.

Key Dates

Ordered: 1963. *Construction began:* 1964. *Criticality:* 1969. *First Power:* September 1969. *Commercial Operation:* December 1969. *Shut Down Expected:* 2009.

Operating Costs

Year	Cost	Year	Cost
1970	$1,953,000	1980	$37,529,000
1971	$3,097,000	1981	$45,253,000
1972	$3,877,000	1982	$60,810,000
1973	$6,311,000	1983	$73,246,000
1974	$10,678,000	1984	$91,175,000
1975	$12,310,000	1985	$101,788,000
1976	$10,399,000	1986	$122,574,000
1977	$14,833,000	1987	$113,530,000
1978	$15,898,000	1988	$95,888,000
1979	$13,055,000	1989	$102,626,000

Plant Report

Old plant undergoes refurbishing. The Oyster Creek Nuclear Generating Station in Forked Creek, New Jersey is owned by the Jersey Central Power and Light Company (JCP&L) and operated by GPU Nuclear Corporation. It provides about 25 percent of the electricity used by JCP&L's customers. The 650 megawatt boiling water reactor began commercial operation in December 1969 and is the second oldest large-scale commercial nuclear plant operating in the United States.

During the 1980's major additions and modifications were completed to modernize the plant. The first phase, begun in 1983, was the most extensive overhaul of a plant in the history of commercial nuclear power. The plant underwent 50 modifications and thousands of maintenance tasks during the 20-month outage. The second phase, completed during 1986, included improvements to fire prevention, detection, and suppression systems. In September 1988, the plant shut down for the third phase. Earlier modifications were completed as workers modernized plant systems, improved plant safety and efficiency, and extended the operating life of some equipment. In total, 115 major projects at a cost of $628 million were completed. Oyster Creek returned to commercial production in May 1989.

Operating errors and violations. An incident on April 25, 1987 led the NRC to levy a $205,000 fine for the improper operation of two valves. The violation of NRC regulations occurred when employees tied the valves open with ropes. The plant had been shut down at the time so workers could replace a steam pressure electromagnetic relief valve monitor. The plant resumed operation May 16, 1987.

During an outage the following September, control room operators erroneously closed two valves critical to the plant cooling system. The error caused an alarm to go off in the control room and was corrected in two minutes. An investigation of the mistake uncovered evidence showing that someone had tampered with the recorded log of the incident. Five people were relieved of their duties during the subsequent inquest; three were reinstated. The plant resumed operation November 24, 1987.

Sets records for capacity factor and longest run. The plant completed its longest run, 229 days, between November 1987 and July 9, 1988. As a result Oyster Creek achieved the third highest quarterly capacity factor of all boiling water reactors in the world, 102.5 percent, during the first quarter of 1988. Because of its many outages, however, the plant's lifetime capacity factor is only about 51 percent.

License questions. Oyster Creek, like many nuclear plants, originally received a 40-year license to operate. Its license will expire in 2009. Details of the renewal process are uncertain. Some groups claim that licenses should not be renewed because older plants can never be as safe as they were when new. In addition, the NRC has not yet decided whether old plants seeking license renewals should have to meet requirements established after they were first built.

Sources

"Nuclear Agency Investigates Actions at New Jersey Plant." *The New York Times*, September 20, 1987.
"Oyster Creek in Profile." *Backgrounder*, GPU Nuclear Communications Division, July 1989.
"Eventful Year for Area's Nuclear Power Plants." *Philadelphia Business Journal*, October 1, 1990.
"Oyster Creek Overview: Preparing for a New Century." *Backgrounder*, GPU Nuclear Communications Division, January 1991.
"Due Up for License Renewal: The Future of Nuclear Power." *The New York Times*, June 24, 1991.

—Karen Bellenir

323 ●

Palisades Nuclear Plant

27780 Blue Star Highway　　　　Phone: (616)764-8913
Covert, MI 49043　　　　　　　　Fax: (616)764-8258
Gerald B. Slade, Plant Gen.Mgr.

Owning Utility

Consumers Power Company　　　Phone: (517)788-0550
212 W. Michigan Ave.　　　　　　Fax: (517)788-0045
Jackson, MI 49201　　　　　　　　Telex: 223454
David P. Hoffman, V.Pres., Nuclear Operations

Contact

Mark A. Savage, Pub.Aff. Dir.
Palisades Nuclear Plant

Basic Facts

Status: Operable. *Type of Reactor:* Pressurized light-water reactor. *Megawatts (electric):* 845. *Megawatts (thermal):* 2530. *Reactor System Supplier:* Combustion Engineers, Inc. (United States). *Generator Supplier:* Westinghouse Electric Corp. (United States). *Architect Engineer:* Bechtel. *Owner:* Consumers Power Company, Bechtel, Westinghouse. *Operator:* Consumers Power Company.

Key Dates

Ordered: 1966. *Construction began:* 1967. *Criticality:* May 1971. *First Power:* December 1971. *Commercial Operation:* December 1971.

Operating Costs

1972	$753,000	1981	$44,139,000
1973	$3,160,000	1982	$38,450,000
1974	$11,778,000	1983	$57,029,000
1975	$9,601,000	1984	$51,568,000
1976	$9,848,000	1985	$58,496,000
1977	$6,569,000	1986	$65,672,000
1978	$15,393,000	1987	$63,607,000
1979	$26,345,000	1988	$76,002,000
1980	$19,251,000	1989	$74,716,000

Plant Report

Power for southern lower Michigan. The Palisades Nuclear Plant occupies a site in South Haven, Michigan, on the eastern shore of Lake Michigan. The operating utility for the plant is Consumers Power, which is owned by the CMS Energy Corporation. Palisades accounts for 15 percent of the electrical power that Consumers Power provides to some 3.5 million residential and business customers in southwestern lower Michigan.

Licensing. Palisades got its 40-year operating license in 1967, which was slated to expire in 2012. In June 1991, the NRC was considering whether to grant a 20-year renewal for the plant.

In December 1990, the Advisory Committee on Reactor Safeguards (ACRS) issued a report on Consumers Power's request for a full-term operating license for Palisades. The plant had been operating under a provisional operating license for 19 years. The ACRS believed there was reasonable assurance that the plant could continue to operate without undue risk to public health and safety.

Shutdown. In May 1986, three safety-related valves failed to open properly at Palisades, and triggered an automatic shutdown of the plant. The failure of the valves, coupled with turbine problems, caused the plant to remain closed for 10 months while safety work was done. The NRC allowed the plant to return to service in March 1987.

Shared ownership. In October 1987, Consumers Power made plans to sell a stake in Palisades to the Bechtel Corporation. The two agreed that the utility would receive $365 million and 44 percent of equity in the venture, and would own the plant. The remaining equity would go to Bechtel and other investors. In return, the utility agreed to drop legal claims against Bechtel, involving its work on Consumers Power's plant in Midland, Michigan. That plant was written off as a $4.2 billion loss.

In August 1988, Bechtel agreed to pay an additional $100 million for the deal, bringing the price to $465 million, and making the corporation a 33 percent owner in the plant.

Inspections and fines. In 1989, the NRC began a five-month inspection of Palisades. The inspection revealed that the plant made several pipe support modifications without adequate design reviews and engineering analyses. Pipe supports help to protect the plant's pipes in the event of an earthquake or other jolting movement. The NRC also found that Consumers Power did not react quickly enough or sufficiently when they discovered the design problems.

The NRC said the individual design flaws were not of major significance, but the number of problems indicated a "breakdown in the utility's control of design activities for the pipe support modifications." In March 1990, the commission proposed a $75,000 fine regarding the flaws.

In November 1989, there were a spurious power-operated relief valve opening and a failure of a motor-operated block valve at Palisades. The NRC investigated these events as well.

Replacement project. In November 1989, Consumers Power made a presentation to the NRC which detailed plans to replace both of Palisades' 420-ton steam generators during the next refueling outage. The utility formed the Palisades Steam Generator Replacement Project (SGRP). The Palisades SGRP was the first such replacement for a plant designed by Combustion Engineering. Palisades was also the first large plant constructed by the company.

The Palisades SGRP differed in two respects from previous similar projects. First, in order to facilitate replacement of the steam generators, a temporary construction access opening through the containment wall was required. Secondly, an improved gas tungsten arc welding process, using a "narrow groove" technique, was to be used on the primary coolant system piping cuts.

To prepare for the Palisades SGRP, the NRC staff visited the Ringhals Nuclear Power Plant in Sweden to discuss the containment opening and other aspects of successful steam generator replacement. The Ringhals plant had previously had a similair process done. The commission also reviewed the narrow groove welding process at contractor's testing laboratory in Tennessee.

Palisades used coordinated phosphate secondary side chemistry until 1975, when it switched to an all volatile treatment. The use of phosphate chemistry control, coupled with carbon steel tube sheets and drilled tube supports, led to severe steam generator tube denting and corrosion, thus necessitating the replacement.

The Palisades SGRP constituted the largest single repair project since the plant was constructed. Palisades was shut down on September 15, 1990, for a five-month period, during which the reactor was slated to be defueled, the steam generators to be replaced, the reactor to be refueled, and testing to be performed on both the primary coolant system and the containment building.

The NRC conducted meetings to assess Consumers Power's progress in planning and evaluating all aspects of the replacement. They also performed inspections, both at Palisades and at the offices of the utility's contractor. The plant was scheduled to restart on February 15, 1991.

Another owner. In July 1990, Consumers Power announced that it was selling a 22 percent stake in Palisades to the Pittsburgh-based Westinghouse Corporation, in return for $20 million, making the corporation the third owner of the plant.

The partnership between Consumers Power, Bechtel, and Westinghouse was to be known as Palisades Generating Co., and was to buy the plant for book value, after a replacement closure in September 1990.

Although Westinghouse would own a 22 percent equity interest in the partnership, its voting share would be only 4.9 percent. By keeping its voting interest under 5 percent, the corporation would avoid the restrictions of the 1935 Public Utility Holding Company Act, which would restrict its ability to enter similar arrangements in other states.

The deal was opposed by the Association of Businesses Advocating Tariff Equity (ABATE), a group of industrial firms. ABATE planned to file objections with the Michigan Public Service Commission, which regulates state utilities. The commission was also fighting the deal because it believed that it would transfer regulatory power over how much money electric customers would pay for Palisades' electricity to the federal government.

The deal was also opposed by Michigan Attorney General Frank Kelley who accused Consumers Power of trying to circumvent state laws by asking the federal government for approval for electricity generated by Palisades. The utility denied the allegations.

A spokesman for Consumers Power said that the deal would allow the utility to spread the risk of running a plant among three owners, without creating financial risks to customers. The utility would run Palisades for the partnership and purchase its entire output, while removing the plant from its rate base. According to the spokesman, customers were eager to find partners to help improve Palisades, which traditionally had run at 47 percent capacity. The national average was 65 percent at the time.

Sources

"Business Briefs: Consumers Power." *The Wall Street Journal*, March 23, 1987.
"Consumers Power Gets Bechtel Pact." *The New York Times*, August 16, 1988.
U.S. NRC 1990 Annual Report. U.S. Regulatory Commission, n.d.
"NRC Seeks $75,000 Fine Against CMS Energy Unit." *The Wall Street Journal*, March 6, 1990.
"Bechtel, Westinghouse buy Into Nuclear Plant." *The Washington Post*, July 20, 1990.
"Nuclear Power Plant Deal Draws Criticism." *The Detroit News*, July 23, 1990.

Plant Profiles

"ACRS Favors Full-term OLs for Palisades, Dresden-2." *Nuclear News*, February 1991.

"Due up for License Renewal: The Future of Nuclear Power." *The New York Times*, June 24, 1991.

—Charles Ceplecha

324 •
Palo Verde Nuclear Generating Station, Units 1-3

PO Box 52034 Phone: (602)393-5000
Phoenix, AZ 85072 Fax: (602)932-1695
James M. Levine, V.Pres., Nuclear Production

Owning Utility

Arizona Public Service Company Phone: (602)250-1000
400 N. Fifth St. Fax: (602)250-2061
Phoenix, AZ 85004
William F. Conway, Exec.V.Pres., Nuclear

Contact

Donald B. Andrews, Mgr., Commun./Pub.Info.
Palo Verde Nuclear Generating Station

Basic Facts

Units 1-3 *Status:* Operable. *Type of Reactor:* Pressurized light-water reactor. *Megawatts (electric):* 1303. *Megawatts (thermal):* 3817. *Reactor System Supplier:* Combustion Engineering, Inc. (United States). *Generator Supplier:* General Electric Co. (United States). *Architect Engineer:* Bechtel. *Owner:* Arizona Nuclear Power Project, involving Arizona Public Service, Public Service of New Mexico, Southern California Edison, Southern California Public Power Authorities, El Paso Electric, Salt River Project. *Operator:* Arizona Public Service.

Key Dates

Unit 1 *Ordered:* 1973. *Construction began:* 1976. *Criticality:* May 1985. *First Power:* June 1985. *Commercial Operation:* January 1986. *Shut Down Expected:* December 2024.

Unit 2 *Ordered:* 1973. *Construction began:* 1976. *Criticality:* May 1986. *First Power:* May 1986. *Commercial Operation:* September 1986. *Shut Down Expected:* December 2025.

Unit 3 *Ordered:* 1973. *Construction began:* 1976. *Criticality:* 1987. *First Power:* March 1987. *Commercial Operation:* January 1988. *Shut Down Expected:* July 2027.

Operating Costs
Units 1-3
1986 $119,252,000
1987 $171,774,000
1988 $238,369,000
1989 $334,925,000

Plant Report

Palo Verde overview. The Palo Verde Nuclear Generating Station near Wintersburg, Arizona, is the largest and most powerful commercial nuclear plant in the United States. The plant, which consists of three 1,270-megawatt, pressurized water reactor units for a total generating capacity of 3,810 megawatts of electricity, serves four million people. Only the Paluel nuclear plant in France and a plant near Leningrad in Russia are capable of generating more power. Palo Verde, located on a 4,080-acre site in the desert about 55 miles west of downtown Phoenix, currently employs about 3,100 people.

The Arizona Nuclear Power Project (ANPP) filed an application with the U.S. Nuclear Regulatory Commission (NRC) on July 11, 1974, to build the three units. Following a 22-month review of safety, environmental, financial and antitrust considerations, the NRC issued a construction permit on May 25, 1976. Unit 1 began commercial operation in January 1986, Unit 2 in September 1986, and Unit 3 in January 1988.

Ownership and utility implications. Palo Verde is owned by ANPP, a consortium of seven Southwestern utilities spread throughout Arizona, New Mexico, Texas, and California. ANPP was organized to engineer, design, construct and operate the plant.

The Arizona Public Service Co. (APS) is the plant's operator and primary owner at 29.1 percent. APS is trailed by the Salt River Project (17.49), Southern California Edison (15.8), El Paso Electric Co. (15.8), Public Service Co. of New Mexico (10.2), the Southern California Public Power Authority (5.91), and the Los Angeles Department of Water & Power (5.7). APS is owned by Pinnacle West Capital Corp. (PWCC), a Phoenix-based holding company that lost $186.9 million in 1991.

In January 1992, the El Paso Electric Co. (EPE) became the second major investor-owned utility to seek Bankruptcy Court protection after the Public Service Co. of New Hampshire's claim in 1988. EPE, which serves 248,000 customers in El Paso, western Texas and southern New Mexico, invested $1.5 million in Palo Verde. But the utility sought Chapter 11 protection after being beset by investment problems in Palo Verde, real estate, thrifts and a steel company. EPE was $2 billion in debt.

In 1989, PacifiCorp, an Oregon-based utility, attempted to acquire APS and a hefty share of Palo Verde for $1.7 billion. In addition, PacifiCorp offered to buy PWCC.

APS was involved in a nearly 2-year-old rate case that ended with utility rates increasing an average of 5.2 percent by early December 1991. The increase, which affected the utility's 605,000 customers, was expected to raise an additional $66.5 million in annual revenues for APS. Originally, APS requested a 21 percent increase.

Regulatory and legal implications. Although Palo Verde has been saddled with various litigation proceedings, NRC officials cite continued signs of improvement at the facility in recent years.

In the mid-1980s, the NRC characterized security measures at Palo Verde as among the region's worst. The agency also questioned the plant's integrity. However, an NRC assessment covering Dec. 1, 1990, to Feb. 29, 1992, indicated that Palo Verde improved its overall performance since its previous evaluation 18 months prior. "You're doing good, certainly a lot better than a couple of years ago, and generally satisfactory," said

John Martin, the NRC's regional administrator, in *The Arizona Republic*.

In the most recent evaluation, Palo Verde earned a Category 1—the NRC's highest rating—for radiological controls and emergency preparedness. However, APS officials were criticized for occurrences such as the backward installation of check valves in a safety system and problems that caused the Unit 3 reactor to be placed on alert status.

ANPP has developed a track record of being fined by the NRC. From December 1983 to February 1992, the agency fined ANPP 10 times. Some fines and descriptions follow.

In 1988, the NRC fined Palo Verde $250,000 for a number of safety violations, including one worker's excessive exposure to radiation, the inadvertent shutdown of part of a reactor unit's air-conditioning system, "and a radiation oversight committee not fulfilling its responsibilities."

In October 1990, the NRC fined Palo Verde $125,000 for failing to maintain an operable emergency lighting system.

In 1991, the NRC fined Palo Verde operators $200,000 for emergency-lighting and medical reporting problems.

In February 1992, the NRC proposed a civil penalty of $162,500 against Palo Verde for two safety violations involving a partial loss of off-site power and a refueling problem.

In other legal matters, two California men repaid Palo Verde $1.3 million in restitution in 1990 for a scheme involving the sale of counterfeit switches and circuit breakers to the facility as well as to the Diablo Canyon nuclear power plant in California. The men were sentenced on two felony counts of using counterfeit labels to falsely identify circuit breakers and switches.

Allan Mitchell, an electrical engineer who helped monitor activities at Palo Verde, filed a lawsuit against APS, claiming that he was fired for complaining about plant problems. Mitchell argued that plant breakdowns were the result of human error, and that ratepayers should not have to bear the cost. However, a local court dismissed the suit in 1991, citing insufficient evidence that APS was responsible for Mitchell's ouster.

APS was also scrutinized several times by the National Labor Relations Board (NLRB). Once, NLRB investigated charges in 1991 against APS of unfair labor practices against plant security guards.

Then, in May 1992, the U.S. Department of Labor ordered APS to upgrade its performance appraisal of Linda Mitchell, another whistleblower/electrical engineer who was punished for complaining about gross safety problems at Palo Verde.

Social implications. Palo Verde has not been void of social controversy. Various organizations—from radical to conservative—have raised questions and protested operation of the plant. For example, Arizonans for Safe Energy gathered 79,000 signatures on stop-Palo Verde petitions in 1976. The Palo Verde Intervention Fund (PVIF) was established in 1981 to study the facility's operating license proceedings. One of PVIF's objectives was to notify the NRC of potential problems in Palo Verde's construction. And in 1989, five members of Earth First!, a radical antinuclear organization, were arrested for conspiring to sabotage Palo Verde.

Public relations may be a weak point for APS and Palo Verde. In 1985, APS President Mark DeMichele acknowledged to a group of people that building Palo Verde was a mistake. DeMichele was unaware that a reporter from *The Arizona Republic* was present and the comment was printed. In addition, APS was criticized by the NRC for delaying problem reports.

Education has benefited from Palo Verde's presence. Property-rich Arizona utilities generate a lot of money for school districts, and the Ruth Fisher Elementary School owes virtually all of its $2 billion in assessed property value to Palo Verde. Palo Verde has contributed to the community with acts such as donating light poles to the Arizona Lutheran School.

A NRC inspection team determined that there is an "adequate level" of emergency preparedness at Palo Verde. Hundreds of millions of dollars were invested in safety systems for the facility to prevent an emergency that would necessitate evacuation.

Approximately 3,100 people—a relatively small number—live within the plant's 10-mile radius or "emergency preparedness zone (EPZ)." The radius is surrounded by 37 sirens to alert residents.

Palo Verde has reached "alert status," a second level emergency, just once. In May 1992, the loss of three monitoring systems caused the "alert," which is one level worse than an "unusual event," but two levels from a "meltdown," a worst-case nuclear scenario.

In 1982, a study by the Sandia National Laboratories for the NRC estimated that a full-scale meltdown at Palo Verde could cause 19,000 deaths and $90 billion in property damage. Subsequently, radio, television and newspaper reports noted that those figures were "worst-case" estimates derived from the study.

Sandia and nuclear proponents reacted by indicating that the media misinterpreted the results, creating a distorting and confusing picture of nuclear power accident probabilities and consequences. A NRC scientist noted that even a few hundred deaths would be unlikely from a severe accident at Palo Verde because of the sparsely populated EPZ.

Power output. Palo Verde, a 3,810-megawatt facility, supplies power to approximately four million customers in Arizona, Texas, New Mexico and California. Up to 60 percent of the power is exported out of Arizona, with about 30 percent going to California utilities.

The facility has lived up to being the largest nuclear plant in the United States, recording gargantuan power outputs. In 1988, for instance, Palo Verde was No. 1 in the nation for power production and Unit 3 generated more electricity than any other nuclear unit in the free world. Unit 3 produced 10,865,800 gross megawatt-hours of electricity, enough for about 1.3 million people.

Then, in back-to-back years, 1990 and 1991, Palo Verde again produced enormous power. In 1991, the plant generated 26.7 million hours, almost 34 percent more than its nearest competitor, the Oconee plant in South Carolina. In February 1992, Unit 2 was the nation's top power producer and Unit 3 ranked second.

Palo Verde has also earned high marks in efficiency. For the first eight months of 1991, the operating United States plants, then 111, enjoyed an average capacity factor (the percentage of maximum output assuming continuous full capacity production) at 70.4 percent, better than France and Japan. Palo Verde Units 1, 2, and 3 contributed to the mark with leading capacity factors of 84.2, 92.8, and 68.7 percent, respectively.

Palo Verde experienced a poor year in 1989, generating power just 23 percent of the time, a performance regulators called abysmal. However, after turning in one of the country's worst performance records for the first six months of 1990, the plant rebounded with a 58 percent rate that same year, slightly under the industry average of 66 percent. Palo Verde is the biggest and most expensive engineering project in Arizona's history. Construction costs totalled $5.9 billion and financing charges created a plant value of $9.3 billion.

The NRC considered overall construction of Palo Verde to be "generally satisfactory." Also, a four-year, $7 million audit by an independent accounting firm determined that construction of Palo Verde was "reasonable and prudent." The study found that, of the $5.9 billion project cost, approximately one percent was unreasonable.

Palo Verde experienced project shortcomings. *The Arizona Republic* reported cost overruns of at least 118 percent, in addition to a 3 1/2-year delay in construction. Also, a major mechanical failure boosted the plant's final cost by $1 million, and problems with pressure steam valves and primary coolant pumps contributed $800,000 to the price tag and caused a six-month scheduling delay in operations for all three units.

On the positive side, APS estimated that it took 15.35 man-hours to build each kilowatt of generating capacity, under the industry average of 18.6. In addition, The Bechtel Power Corp., the project's prime contractor, indicated that maximum standardization and design of equipment saved $300 million because all three units are identical.

Plant engineers originally intended to build five reactors, but after the 1979 Three Mile Island accident, they opted to construct three. Most Palo Verde construction workers and employees praised the plant's quality of design and construction. "This is the safest and best built plant I've ever seen," according to one engineer who had worked in at least six plants including Palo Verde.

Environmental and health considerations. Because Palo Verde, the only nuclear plant in the middle of a desert, is far from a natural body of water, a special condenser cooling system was built.

The plant's $220 million Water Reclamation Facility—an advanced wastewater treatment plant and the world's largest—provides the cooling water. Palo Verde's distance from a natural body of water makes the cooling system unique to the nuclear industry.

Palo Verde consumes approximately 20.8 billion gallons and 64,000 acre-feet of water annually when all three reactors are operating. Sewage effluent, which is used for the cooling water, is purchased from the nearby cities of Phoenix, Glendale, Mesa, Scottsdale, Tempe, Youngtown, and Tolleson. The effluent is pumped to Palo Verde through an underground pipeline.

APS and the United States Atomic Safety & Licensing Board, an arm of the NRC, believe that there will always be an adequate supply of water for Palo Verde. They contend that conservation plans and importation of water from the Colorado River provide that assurance. However, an "escape clause" in the 1973 contract between APS and the Municipal Water Users Association (MWUA) allows five cities that will sell effluent to Palo Verde to withhold it in time of emergency.

"If it ever comes down to water for power or water for people, we will send the water to the people," according to MWUA Executive Director Bill Stephens as quoted in *The Phoenix Gazette*.

The future of Palo Verde. Palo Verde's license expires in the year 2025. Sewage effluent will be supplied through 2034.

The United States Department of Energy is seeking a nuclear waste disposal site to store spent fuel from all of the country's reactors. However, Arizona Gov. Fife Symington is opposed to any such project in his state primarily for environmental reasons.

In February 1992, a coalition from financially ailing Apache County tried to obtain a $100,000 federal grant to study the handling of nuclear waste, the first step in being considered as the site of a temporary facility for nuclear-waste storage 220 miles from Phoenix. According to the United States Department of Energy, the county would be eligible for a $10 million cash bonus from the Federal government if it were selected as the site for the $1 billion facility.

Sources

"Palo Verde Estimate Called Incredible,' $1 Billion Low." *The Arizona Republic*, April 25, 1984.
"Drought Could Turn off the Taps to Palo Verde." *The Phoenix Gazette*, June 27, 1984.
"Palo Verde: Focus of a Troubled Nuclear Industry." *The Phoenix Gazette*, June 27, 1984.
"Plant Construction A-OK, Workers Claim." *The Phoenix Gazette*, June 27, 1984.
"Problems Add to Final Cost of Palo Verde." *The Phoenix Gazette*, June 27, 1984.
"Thirsty Desert Plant Has Unique Water System." *The Phoenix Gazette*, June 27, 1984.
"Anything But The Truth." *New Times*, April 23-29, 1986.
"Large Atom Plant Dedicated." *The New York Times*, December 6, 1987.
"Atom Plant in Arizona is Fined for Safety Violations." *The New York Times*, December 4, 1988.
"Pinnacle West Finds PacifiCorp's Bid Less Than Friendly." *The Business Journal*, December 18, 1989.
"Arizona Nuclear Power Project Reports $1.3 Million Restitution Ordered in Circuit Breaker Fraud." *Business Wire*, May 1, 1990.
"Palo Verde Rated Among Worst Plants in U.S." *The Phoenix Gazette*, July 20, 1990.
"1990 Efforts Garner Praise for Palo Verde." *The Phoenix Gazette*, January 7, 1991.
"Palo Verde Turnaround." *The Phoenix Gazette*, January 15, 1991.
"A-Plant Gets Good' Grades." *The Arizona Republic*, February 5, 1991.

Palo

"U.S. Panel to Examine APS' Labor Practices." *The Arizona Republic*, June 29, 1991.

"Earth First! Informer Tells Jury About Drug History." *The Phoenix Gazette*, July 17, 1991.

"Utilities Mean Big Bucks for a Few Districts School Funding/Have We Failed?" *The Phoenix Gazette*, October 7, 1991.

"Texas Utility Forced to Borrow $208 Million." *The Arizona Republic*, December 28, 1991.

"Utility Files Chapter 11; El Paso Electric's Woes Linked to Palo Verde A-Plant." *The Arizona Republic*, January 9, 1992.

"Utility Chief Blames Filing on Creditors." *The Arizona Republic*, January 12, 1992.

"Governor Denies Bid for Nuclear Waste Site." *The Arizona Republic*, January 26, 1992.

"Palo Verde Faces $162,500 Fine for Safety Violations." *The Phoenix Gazette*, February 5, 1992.

"Palo Verde is Top Producer." *The Arizona Republic*, February 15, 1992.

"Rural Program for Studying Nuclear Waste; Symington Is Asked Not to Bar Bid." *The Arizona Republic*, February 15, 1992.

"Efficiency of Nuke Plants Beats Both France and Japan." *The Arizona Republic*, February 25, 1992.

"Palo Verde Wins High Marks in Report." *The Phoenix Gazette*, April 14, 1992.

"At A Glance." *The Arizona Republic*, April 23, 1992.

"3 Systems Fail at Palo Verde, Plant on Alert." *The Arizona Republic*, May 5, 1992.

"Arizona Companies Saw Red Last Year as Recession Ruled." *The Arizona Republic*, May 10, 1992.

"APS Under Fire for Action on Whistle-Blower." *The Phoenix Gazette*, May 14, 1992.

"Palo Verde Given Better Grades: Recent Events Mar Improved Report by NRC." *The Arizona Republic*, May 27, 1992.

—Michael Richman

325 Palo Verde Nuclear Generating Station, Units 4-5

PO Box 52034
Phoenix, AZ 85072
James M. Levine, V.Pres., Nuclear Production

Phone: (602)393-5000
Fax: (602)932-1695

Owning Utility

Arizona Public Service Company
400 N. Fifth St.
Phoenix, AZ 85004
William F. Conway, Exec.V.Pres., Nuclear

Phone: (602)250-1000
Fax: (602)250-2061

Contact

Donald B. Andrews, Mgr., Commun./Pub.Info.
Palo Verde Nuclear Generating Station

Basic Facts

Units 4-5 *Status:* Cancelled. *Type of Reactor:* Pressurized light-water reactor. *Megawatts (electric):* 1303. *Megawatts (thermal):* 3817. *Reactor System Supplier:* Combustion Engineering, Inc. (United States). *Generator Supplier:* General Electric Co. (United States). *Architect Engineer:* Bechtel. *Owner:* Arizona Nuclear Power Project, involving Arizona Public Service, Public Service of New Mexico, Southern California Edison, Southern California Public Power Authorities, El Paso Electric, Salt River Project. *Operator:* Arizona Public Service.

Plant Report

See Plant Report for Palo Verde Nuclear Generating Station, Units 1-3.

326 Pathfinder Nuclear Power Plant

Souix Falls, SD

Owning Utility

Northern States Power Company
414 Nicollet Mall
Minneapolis, MN 55401
Leon Eliason, V.Pres., Nuclear Generation

Phone: (612)330-5500
Fax: (612)330-2900

Basic Facts

Status: Shut down. *Type of Reactor:* Boiling water reactor. *Megawatts (electric):* 63. *Megawatts (thermal):* 189.

Key Dates

Ordered: 1957. *Construction began:* March 1964. *Criticality:* February 1966. *First Power:* July 1966. *Commercial Operation:* July 1966. *Shut Down:* 1967. *Decommissioned:* 1968-1991.

Plant Report

A former South Dakota power center. The Pathfinder Nuclear Power Plant occupied a site along the Big Sioux River, near Sioux Falls, South Dakota.

The Northern States Power Company (NSP) owned and operated Pathfinder, which housed one boiling water reactor, until it was decommissioned beginning in 1968.

A project of the Power Demonstration Reactor Programme. Realizing that utilities could not justify to their shareholders the investment costs required to build nuclear power plants, the Atomic Energy Commission and Congress devised the Power Demonstration Reactor Programme (PDRP) in 1955 to help launch the commercial nuclear industry in the United States. The PDRP offered cooperating utilities financial incentives such as research and development assistance, financing of the reactor system, and a waiver of normal fuel charges during the first five years of plant operation. In 1957, Pathfinder was among the third round of projects undertaken in response to the PDRP.

Decommissioning begins. When NSP began decommissioning Pathfinder in 1968, it became the first investor-owned utility to undertake the process on its own plant.

During the first phase of decommissioning, work included the removal of virtually all electric service to the nuclear poortion of the plant, as well as changes to the reactor vessel. The three-inch-thick steel vessel measured about 32 feet high and 11 feet wide. Workers removed the fuel core, which was shipped to several sites around the United States for reprocessing, and filled the vessel with six tons of gravel.

NRC approval. After a roughly 20-year hiatus, NSP sent a dismantlement plan for Pathfinder to the NRC in July 1989. In

June 1990, the NRC sent the utility a license amendment, which allowed it to continue the decommissioning of the nuclear portion of the plant.

Minimum workforce. The project manager for the decommissioning of Pathfinder was Al Kuroyama. Kuroyama considered staying within budget to be a top priority of his team, so he kept his staff to the minimum number of workers needed.

Noticeable changes. A tour through Pathfinder in October 1990 revealed the extent of changes during the decommissioning process. The beginning of full-scale decommissioning signaled the need for tighter security, which entailed closing off the gate to the plant, and keeping all vehicles outside it. Access was granted only by checking in at a newly added security desk.

On the plant's exterior, the outer surface insulation on a large area around the containment equipment hatch was removed, and a heavy steel platform sat in front of the closed hatch. Workers had removed the welding that sealed the hatch since the first phase of decommissioning took place. The platform was designed to ease the removal of large components from the containment later in the process.

Inside the plant, the nuclear portion was fenced off from the conventional side to allow decommissioning to proceed without interfering with the operation of the rest of the plant. The decommissioning area contained a fully equipped and staffed access control center, set up in former storage areas of the plant. No one could get in or out without passing through access control. Also, although radiation levels were extremely low, a full radiation control program was in effect.

The most dramatic accomplishment of the decommissioning team was evident in the containment building. Almost all the asbestos insulation, which once covered the entire reactor vessel and most if its piping, was removed. These areas were cleaned and coated with a sealant material to encapsulate any remaining asbestos fibers.

Getting adequate electric service back into the decommissioning area was an early priority. Workers accomplished this by installing large, brand-new breaker panels in the former spent fuel storage area.

On the main floor of the containment building, a new duct work system was installed to keep the decommissioning area under a slight vacuum. During the process, mildly radioactive materials, such as concrete dust, could get into the air and reach the environment outside the plant. The duct system insured that any air leakage occurred from the outside in, with all air drawn in passing through high-efficiency filters to remove contaminated particles.

A worker decontamination shower stood in an area where low-level waste was once handled. The low-tech, low-cost shower was set up in case of any worker exposure.

Preparing to remove the vessel. Preparing and lifting the vessel out of Pathfinder's containment building was a formidable task for the decommissioning team. The vessel was considered low-level radioactive waste because its radioactivity decayed over the roughly 20 years that the plant was closed. At the time of removal, the contained about 562 curies of radioactivity. Because the metal of the vessel and its support components contained almost all of the radioactivity, there was no danger for it to escape. Surface radiation readings ranged from about four to 30 millirem per hour. Workers in close proximity to the vessel wore monitoring devices and were supervised by NSP radiation protection specialists, but the low-level surface radiation readings showed that no increased risk existed.

To prepare the vessel for the lift, workers filled it with a cement/grout mixture to fix the gravel and internal components. This would prevent any shifting during handling and shipping. The mixture also provided additional shielding which lowered external radiation readings on the vessel's surface. The bare vessel originally weighed about 100 tons, but with the internal components, gravel, and grout mixture, its weight rose to about 245 tons.

Workers also cut all piping outlets from the vessel, and welded cover plates over the pipe openings. Workers removed some of the bolts holding the vessel to its support columns, and attached a special lifting rig to the top of the vessel. The rest of the bolts were removed on the day of the lift.

Lifting the vessel. NSP contracted Truck Crane Service/Lakehead to execute the lift on May 14, 1991. NSP provided the news media with maps to Pathfinder, instructions on gaining access and a viewing area on plant property. The public was not allowed on the site during the lift.

To get the vessel out of the containment building, workers cut a hole in the roof above it. A larger guy derrick with a lifting capacity of 275 tons at maximum boom radius lifted the vessel through the hole and located it above wooden cribbing. A second crane, attached to the lower end of the vessel, assisted the larger one in rotating the vessel to a horizontal position and lowering it on to the cribbing.

Shipping preparations. After the lift, the decommissioning staff prepared the vessel for placement in a specialized railcar. They removed support pads that would interfere with the shield plate to be attached to vessel; the saddles and lower portions of the shield plate were installed in the railcar. The contractor then made a second lift to position the vessel in the railcar. Once the vessel was in place, workers installed the almost two inches of remaining shielding, and attached special impact-limiting material around the vessel.

Two satellites track the move. The vessel was to be moved, via the Burlington Northern Railroad, 1,650 miles, through six states, to a burial trench at the U.S. Ecology disposal facility near Richland, Washington. To facilitate the move, NSP enlisted the services of a satellite tracking and communications system called TRANSCOM, which is operated by the U.S. Department of Energy (DOE).

DOE had a good deal of experience using the system for truck shipments, but less for rail shipments. Generally, the system was used to track cargo with high radioactivity levels or weapons shipments. Tracking the Pathfinder shipment presented an opportunity for DOE to further field test and provide user training for TRANSCOM, and for NSP to pinpoint the location of the train at all times.

TRANSCOM uses two satellites to locate the shipment's latitude and longitude, and to communicate with a commercial ground satellite dish in San Diego, California. From there, the signal travels by phone to DOE's TRANSCOM control center in Oak Ridge, Tennessee. The system provides position updates at regular intervals automatically. DOE-approved TRANSCOM users for a particular shipment can use an ordinary PC with a modem to call up maps showing the location of the shipment. The location can be shown on a national, state or county map, and the system will show either highways or railroads or both.

A compact terminal unit, traveling with the shipment, allowed NSP staff on the train to exchange messages with the DOE control center in Oak Ridge, and provided a back-up communication system. NSP personnel on the train also had cellular phones, phones using the IMTS system and Burlington Northern's private phone system. If the remoteness of some areas along the route made cellular or IMTS communication impossible, TRANSCOM took over.

NSP staff had 24-hour access to the shipment's location to respond to calls inside and outside NSP about the vessel's progress. Staff at the Richland disposal sight had the same access to predict the shipment's arrival time more easily.

Two other systems were utilized for the move. The first system operated off the train's electric current with battery-powered back up, and the second system was a new solar-powered system that DOE wanted to field test. Each system had an antenna; the solar-powered system's antenna was mounted on the car with the vessel package, and the second antenna was mounted on the personnel car. DOE was interested in seeing how the new solar-powered system withstood the rigors of rail traveling over a long distance.

A successful decommissioning. The vessel reached its destination on August 21, 1991, and all aspects of the journey went according to plan. Project manager Kuroyama accredited the smooth move to advanced public relations work. A year before the move, the decommissioning team discussed the trip with a variety of regional, state and city officials. Several meetings were held along the route, and team members communicated regularly with local and regional officials, as well as the governor's designees in every state through which the train traveled. Any delay could have doubled or tripled the cost of the move, so the extent of the communication was important.

NSP's self-designed communication network also played an important role in the success of the move. The phone systems, along with automatic paging, voice messaging, and laptop computers helped facilitate the move, but the TRANSCOM system was credited with the vast majority of the communication.

Back at the Pathfinder site, project members had to complete a final cleanup and survey of the area, demolish the containment building and demonstrate to the NRC that the site was free of radioactivity.

Sources

Northern States Power Company 1990 Annual Report. Northern States Power Co., n.d.
U.S. NRC 1990 Annual Report. U.S. Nuclear Regulatory Commission, n.d.
"Plant Anatomy." *NSP News,* Northern States Power Company, October 1990.
Mounfield, Peter. *World Nuclear Power.* Routledge, 1991.
"20/20." *NSP News,* Northern States Power Company, April 1991.
Pathfinder Reactor Vessel Lift Q & A. Northern States Power Company Press Release, May 1991.
Media Advisory. Northern States Power Company, May 6, 1991.
"The Last Path." *NSP News,* Northern States Power Company, September 1991.

—*Charles Ceplecha*

327 ★

Peach Bottom Nuclear Power Plant, Unit 1

R.D. 1
Box 208
Delta, PA 17314
Donald B. Miller Jr., V.Pres.

Phone: (717)456-7014
Fax: (717)456-4891

Owning Utility

Philadelphia Electric Company (PECO)
2301 Market St.
PO Box 8699
Philadelphia, PA 19101
Dickinson M. Smith, Sr.V.Pres., Nuclear

Phone: (215)841-4000
Fax: (215)841-4188

Contact

Neil J. McDermott, Dir., Media & PR
Philadelphia Electric Company

Basic Facts

Unit 1 *Status:* Shut down. *Type of Reactor:* High temperature gas reactor. *Megawatts (electric):* 42. *Megawatts (thermal):* 115. *Reactor System Supplier:* General Electric Co. (United States). *Generator Supplier:* General Electric Co. *Architect Engineer:* Bechtel.

Key Dates

Unit 1 *Ordered:* 1953. *Construction began:* 1962. *Criticality:* March 1966. *First Power:* January 1967. *Commercial Operation:* June 1967. *Shut Down:* November 1974.

Plant Report

See Plant Report for Peach Bottom Nuclear Power Plant, Units 2-3.

328 ●

Peach Bottom Nuclear Power Plant, Units 2-3

R.D. 1
Box 208
Delta, PA 17314
Donald B. Miller Jr., V.Pres.

Phone: (717)456-7014
Fax: (717)456-4891

Owning Utility

Philadelphia Electric Company (PECO)
2301 Market St.
PO Box 8699
Philadelphia, PA 19101
Dickinson M. Smith, Sr.V.Pres., Nuclear

Phone: (215)841-4000
Fax: (215)841-4188

Contact

Neil J. McDermott, Dir., Media & PR
Philadelphia Electric Company

Basic Facts

Units 2-3 *Status:* Operable. *Type of Reactor:* Boiling water reactor. *Megawatts (electric):* 1098. *Megawatts (thermal):* 3293. *Reactor System Supplier:* General Electric Co. (United States). *Generator Supplier:* General Electric Co. (United States). *Architect Engineer:* Bechtel. *Owner:* Atlantic City Electric; Delmarva Power and Light; Public Service Electricity and Gas; Philadelphia Electric Company. *Operator:* Philadelphia Electric Company.

Key Dates

Unit 2 *Ordered:* 1966. *Construction began:* 1968. *Criticality:* 1973. *First Power:* February 1974. *Commercial Operation:* July 1974. *Shut Down Expected:* January 2008.

Unit 3 *Ordered:* 1967. *Construction began:* 1969. *Criticality:* 1974. *First Power:* September 1974. *Commercial Operation:* December 1974. *Shut Down Expected:* January 2008.

Operating Costs

Units 2-3

Year	Cost	Year	Cost
1973	$1,605,000	1982	$81,669,000
1974	$1,791,000	1983	$105,285,000
1975	$12,619,000	1984	$97,208,000
1976	$30,601,000	1985	$145,786,000
1977	$46,674,000	1986	$140,272,000
1978	$39,306,000	1987	$183,806,000
1979	$40,005,000	1988	$311,459,000
1980	$56,875,000	1989	$239,702,000
1981	$72,615,000		

Plant Report

Early commitment to nuclear power. The Peach Bottom site, in Delta PA, was a pioneer in U.S. nuclear involvement. Philadelphia Electric ordered its first reactor in 1958 for Peach Bottom 1, an experimental high-temperature gas-cooled 40 megawatt plant. The plant began producing commercial power in 1967 and operated 88 percent of the time, excluding planned shutdowns, until it was closed in 1974. Peach Bottom 1 was a small prototype for the reactors that were to follow. Although it no longer produces electricity, Unit 1 houses the control room training simulator for Units 2 and 3.

Peach Bottom's newer units. Peach Bottom 2 and Peach Bottom 3 came on line five months apart in 1974. They are operated by Philadelphia Electric (PECO) and owned by four utilities. PECO and a New Jersey utility, Public Service Electric & Gas Company (PSE&G) are the major partners. Another New Jersey utility, Atlantic City Electric Company and a Delaware utility, Delmarva Power and Light Company, each own a 7 percent interest.

In addition to the two plants at Peach Bottom, PECO owns and operates the two Limerick nuclear plants and owns an interest in the two Salem units which are operated by PSE&G. Nuclear power typically generates about 65 percent of the electricity supplied by PECO to its 2,475-square-mile service area. The remaining capacity is generated from coal, oil, and hydroelectric sources or purchased from other utilities. world records for performance and earned recognition as exemplary plants. Their reputation, however, changed as the nuclear industry changed following the accident at Three Mile Island. New regulations and new procedures affected employee attitudes at Peach Bottom sending the facility into decline. In an ironic twist, PECO's considerable experience with nuclear power was blamed for its problems. Critics charged that it lulled plant management into accepting a complacent attitude.

Early warning signs of brewing trouble. In 1984 the Institute of Nuclear Power Operation (INPO) told PECO that it had discovered evidence of declining performance at Peach Bottom. In 1985 an NRC inspector found one of Peach Bottom's operators with his head back and eyes closed. Disciplinary action was not pursued because the inspector could not prove that the operator had been sleeping, but the NRC did lower the plant's safety rating. The following year the NRC fined PECO $200,000 for operating procedure violations and poor management supervision. An INPO report in 1986 found standards of performance unacceptably low.

Between 1982 and 1987, Peach Bottom amassed a total of $635,000 in NRC fines for various infractions. Declining performance and worker attitudes led to an incident that culminated in the largest fine ever imposed by the NRC against a corporate operator of a nuclear power plant.

Sleepy operators. On March 31, 1987 the NRC ordered the plant shut down when it found control room operators sleeping on duty. The incident marked the first time the agency had closed a plant for a non-mechanical reason. The NRC had been tipped off by engineers working for General Electric who had frequent access to control room. The engineers reported the sleeping episodes to the NRC because their complaints to PECO management had failed to produce any results.

An investigation following the incident uncovered a history of complacency. Reactor operators passed the time reading magazines, playing video games, and sleeping. Occasionally everyone on a shift in the control room was napping. Problems at Peach Bottom were not limited to sleeping operators. Inspectors also discovered radioactive contamination on one third of the surface area inside the plants, 3,000 uncompleted maintenance tasks, and improperly identified low-level radioactive waste.

The NRC identified numerous causes for the problems. At the time PECO was building and staffing its Limerick Generating Station. The new plant drew personnel from Peach Bottom, creating a shortage of operators and a need for excessive overtime. There were communication breakdowns in PECO's upper management so that critical information did not get to headquarters. Relationships between Peach Bottom operators and GE engineers were poor.

Allegations of drug use surfaced. Four PECO employees and two contract employees were indicted during the fall of 1987 for drug dealing. (Two were subsequently acquitted.) The following spring PECO discharged 17 employees for using drugs. In May 1988, the FBI charged two more employees with drug-related activities at the plant.

The NRC fined PECO $1.25 million for infractions at Peach Bottom and also proposed fines against 33 individual workers. This marked the first time such penalties were imposed against individuals. The agency also barred three PECO employees from being placed in operational positions.

PECO found that nearly all control room operators had slept on duty, but none were discharged. Twelve of the 36 were assigned to other jobs or retired. The remaining underwent extensive retraining as part of the process to restart Peach Bottom's reactors.

Plans to restart. PECO's original plans for restarting Peach Bottom were rejected. The NRC called for a complete change in the "corporate culture" of the utility. Improvements necessary to restart Peach Bottom included:

- Reassignment of essentially all plant management
- Less comfortable chairs with low backs in the control room
- Supeervisor's chair elevated so he can see better
- New rule instituted: Anyone caught sleeping would be immediately suspended and fired
- Elimination of planned overtime
- 13,000 maintenance tasks performed
- 178 equipment modifications made
- Radiation contamination reduced.

Joseph F. Paquette, Jr. replaced James Lee Everett III as PECO's Chairman in the aftermath of the sleeping incident. He summarized the lesson his organization had learned: "TMI exposed the technical risks of nuclear power. Chernobyl exposed the international risks. What Peach Bottom did was expose the management risks."

The NRC finally approved the restart plans at Peach Bottom. Unit 2 achieved criticality on April 27, 1989 and went to full power on August 4. Unit 3 was restarted on November 20, 1989 and achieved full power January 5, 1990. In February 1990, the NRC removed Peach Bottom from its list of troubled plants, but PECO's legal trouble was not over.

Law suits. During the shut down PECO spent $5 million per month to replace the power Peach Bottom failed to generate. The cost to all four owner utilities totaled $15 million per month. PSE&G, Atlantic Electric, and Delmarva Power all filed suits against PECO, the operator at Peach Bottom, seeking to recover their costs of replacing electricity during shutdown. The New Jersey utilities faced additional expenses as a result of rate reductions incurred when Peach Bottom was removed from their rate base. PECO filed a contingent counterclaim against PSE&G for outages and NRC fines at the Salem plant operated by PSE&G and partially owned by PECO.

Improved capacity. PECO continues to make improvements at Peach Bottom. In 1990 the utility opened a new 70,000-square-foot training center. A Nuclear Maintenance Division was created to service both the Limerick and Peach Bottom facilities in a more efficient manner. The improvements helped Peach Bottom achieve a capacity factor of 78 percent in 1990, higher than industry averages.

PECO management also takes pride in its reliance on nuclear power. The utility boasts that during the summer months of 1991 nuclear generation enabled it to meet record demands for electrical power while saving up to 178.4 million pounds of sulfur dioxide emissions associated with fossil-fueled power plants.

Sources

Today and Tomorrow. Philadelphia Electric Company, n.d.
"Philadelphia Electric's Everett Says Peach Bottom Has Put Job on the Line." *The Wall Street Journal,* April 10, 1987.
"Nuclear-power Industry Gets a Wake-up Call." *U.S. News & World Report,* April 13, 1987.
"PECO's Problem and Pride." *Focus,* February 17, 1988.
"The Peach Bottom Syndrome." *The New York Times,* March 27, 1988.
"More Employees Arrested for Drugs at Nuclear Plant." *The Wall Street Journal,* May 13, 1988.
"$1.25 Million Fine Is Sought for Napping at Pennsylvania Reactor." *The New York Times,* August 12, 1988.
"Peach Bottom Fine Set at $1,250,000." *Focus,* August 24, 1988.
"Paradise Lost . Paradise Regained." *Nuclear Industry,* First Quarter 1990.
"PE Summer Nuclear Output Saves 89,200 Tons of Sulfur Dioxide Emissions." Philadelphia Electric Company Press Release, September 30, 1991.
"Philadelphia Electric Lauds Nuclear's Contribution on 25th Anniversary of Peach Bottom 1 Plant." *INFO,* U.S. Council for Energy Awareness, November/December 1991.
"Philadelphia Electric Customers Will Pay Less." *INFO,* U.S. Council for Energy Awareness, February 1992.

—*Karen Bellenir*

329 ☆

Pebble Springs Nuclear Power Generating Plant, Units 1-2

Arlington, OR

Owning Utility

Portland General Electric Company
121 S.W. Salmon
Portland, OR 97204
James C. Cross, V.Pres./Chief Nuclear Officer

Phone: (503)464-8000
Fax: (503)464-2233
Telex: 62934967

Contact

David W. Heintzman, Pub.Aff. Rep.
Portland General Electric Public Affairs Office

Phone: (503)464-2354

Basic Facts

Units 1-2 *Status:* Cancelled. *Type of Reactor:* Pressurized light-water reactor. *Megawatts (electric):* 1314. *Megawatts (thermal):* 3600. *Reactor System Supplier:* The Babcock & Wilcox Co. (United States). *Owner:* Portland General Electric, Puget Sound Power and Light Company. *Operator:* Portland General Electric.

Plant Profiles

Plant Report

Milestones. Following are important events in the history of the Pebble Springs plant:

- 1972: Portland General Electric (PGE) filed notice with the Oregon Nuclear and Thermal Energy Council of intent to apply for certification for three sites for a thermal plant. Preferred site was located at the Carty reservoir approximately 12 miles southwest of Boardman in north-central Oregon. Another was approximately three miles southeast of Arlington at Pebble Springs, and the third was just under 13 miles east of The Dalles, Oregon.

- February 15, 1973: PGE executed an Agreement effective February 15 under which Babcock and Wilcox would furnish the nuclear steam supply system and nuclear core fabrication services for the initial core load and six reload batches (refueling cycles).

- December 1973: Site application filed with the state Nuclear and Thermal Energy Council for both a nuclear and a coal-fired plant at the Boardman site. Alternate site application was also made for a nuclear plant at Pebble Springs; this site would be used if the Boardman project site approval was delayed. The Boardman site was preferred because of a large cooling lake planned. Additionally, this location could accommodate a long-range total of four nuclear plants as well as two coal fired plants. The alternate Pebble Springs site would accommodate two nuclear plants.

- 1974: Boardman site approval is delayed. PGE decided to relocate the initial plant to Pebble Springs. Power load and resource forecasting indicated an energy deficit which would be partially filled by Pebble Springs Unit 1 in 1980-81; a second Pebble Springs unit was expected to provide energy in 1982-83. Both units would be 1260 net megawatt capacity plants, sponsored by PGE with a 47.1 percent majority share.

- 1974-1978: Series of site application hearings at both the federal and state level continued to delay the approval of a site for a nuclear plant. Completion date altered to 1987 for unit 1 and 1989 for unit 2. The Nuclear Thermal Energy Council recommended the site certificate be issued. However, the Oregon Supreme Court reversed the NTEC order in March 1977. Construction permit federal hearings underway with hoped late 1979 permit issuance.

- December 6, 1978: Evidentiary hearings before the Oregon Energy Facilities Siting Council (EFSC) completed.

- January 19, 1979: EFSC hearing reopened for additional testimony on the issue of risk following the NRC policy statement regarding the Rasmussen report on nuclear safety. All testimony turned over to an Administrative Law Judge with hopes of his recommendation in the Spring of 1979 and issuance of site certificate during the summer of 1979.

- November 1979: Oregon Department of Energy issued reports in response to storage of radioactive waste and Three Mile Island issues.

- March 1980: EFSC suspended hearings on site certification proceedings until July 1, 1981, with automatic 6-month extensions.

- November 1980: Oregon voters passed a statewide referendum prohibiting the siting of a nuclear power plant in the state of Oregon until the Federal government had licensed a permanent disposal facility for spent fuel. The referendum also required statewide voter approval before a site certificate is issued.

- 1981: Minimal licensing efforts continued.

- April 1982: The Oregon Public Utility Commissioner (OPUC) issued a directive to write down investment in the proposed facility.

- October 1982: The Pebble Springs Project was abandoned. Decision was made as a result of licensing difficulties, restrictive legislation, and an OPUC order to write-off investment.

- 1985: Puget Sound Power and Light Company, who had invested $47 million into the Pebble Springs nuclear project, requested to recover their investment through higher rates. Although the Washington Utilities and Transportation Commission had allowed Puget Sound to recover the investment through rates over a 10-year period, the decision was appealed to the Washington State Supreme Court by the consumer affairs division of the attorney general's office.

Sources

History of Portland General Electric's Pebble Springs Nuclear Power Generating Plant. Portland General Electric Public Affairs Office, n.d.
"Puget Power Wiping Nuclear Exposure from Its Books" *Puget Sound Business Journal*, February 4, 1985.
"Utilities Gear for Diversification Push." *Puget Sound Business Journal*, June 17, 1985.

330 ☆

Perkins Nuclear Power Plant, Units 1-3

Mocksville, NC

Owning Utility

Duke Power Company Phone: (702)373-4011
PO Box 33189
Charlotte, NC 28242
H.B. Tucker, Sr.V.Pres., Nuclear
  Generation

Contact

Andy Thompson Thompson, Dir. of
  Issues Mgmt.
Duke Power Company

Basic Facts

Units 1-3 *Status:* Cancelled. *Type of Reactor:* Pressurized light-water reactor. *Megawatts (electric):* 1345. *Megawatts (thermal):* 3800. *Reactor System Supplier:* Combustion Engineering, Inc. (United States). *Owner:* Duke Power. *Operator:* Duke Power.

Key Dates

Units 1-3 *Ordered:* 1973. *Cancelled:* 1982.

331 Perry

Perry Nuclear Power Plant, Unit 1

10 Center Rd.
Perry, OH
Robert A. Stratman, Gen.Mgr., Perry Plant Operations
Phone: (216)259-3737
Fax: (216)259-3554

Owning Utility

Cleveland Electric Illuminating Co.
PO Box 5000
Cleveland, OH 44101
Michael D. Lyster, V.Pres., Nuclear
Phone: (216)622-9800
Fax: (650)479-6540

Contact

Michael J. Lumpe, Mgr., Pub.Aff. Sect.
Cleveland Electric Illuminating Co.

Basic Facts

Unit 1 *Status:* Operable. *Type of Reactor:* Boiling water reactor. *Megawatts (electric):* 1250. *Megawatts (thermal):* 3579. *Reactor System Supplier:* General Electric Co. (United States). *Generator Supplier:* General Electric Co. (United States). *Architect Engineer:* Gilbert. *Owner:* Cleveland Electric Illuminating, Toledo Edison, Ohio Edison Co., Duquesne Light, Pennsylvania Power. *Operator:* Cleveland Electric Illuminating Co.

Key Dates

Unit 1 *Ordered:* 1972. *Construction began:* 1974. *Criticality:* June 1986. *First Power:* December 1986. *Commercial Operation:* November 1987. *Shut Down Expected:* March 2027.

Plant Report

Perry's owners. The Perry Nuclear Power Plant is a double reactor facility located on the shore of Lake Erie, about 35 miles east of Cleveland, Ohio. Only one of the units, however, is operational. Perry-1, a 1,205 megawatt boiling water reactor, went into commercial production on November 18, 1987. Construction on the unfinished Perry-2, was halted in 1985.

Five utilities, Cleveland Electric Illuminating (CEI), Toledo Edison, Ohio Edison Co. of Akron, Duquesne Light (a Pittsburgh utility), and Pennsylvania Power, share ownership of the Perry complex. The three Ohio utilities supply most of the power in northern Ohio.

CEI and Toledo Edison merged to form a holding company, Centerior Power Corporation, in April 1986. Centerior owns 51 percent of Perry-1; the remaining 49 percent is divided among the other three utilities. The partially constructed second unit is owned by the Central Area Power Coordination Group, a company comprising the five utilities, however Duquesne abandoned its interest in Perry-2 in 1986.

Controversy shakes plant. The construction of the Perry complex was surrounded with controversy. As early as 1983 groups began asking that Perry-2 be cancelled. They said the unit was not needed because the utilities already had sufficient generating capacity. Antinuclear activists, politicians, and religious groups spoke out against the plant. Questions also arose concerning the containment-vessel, built to prevent radioactive materials from being released in case of an accident. Necessary changes in the design eventually led to a lawsuit being filed by the five owner utilities against the supplier, General Electric.

Political opposition increased when an earthquake rocked the site on January 31, 1986. It measured 5.0 on the Richter scale and was centered 10 miles from the reactor. Opponents to the plant claimed that the strength of the quake was uncomfortably close to the 5.3 limit Perry-1 had been built to withstand.

Although the plant itself did not sustain any damage, roads and bridges along planned evacuation routes were destroyed. Officials in Cuyahoga County (greater Cleveland) asked that the licensing process be stopped until evacuation plans could be re-evaluated. Governor Richard Celeste withdrew his approval of the plant and appointed a task force to study the proposed evacuation plans.

Costly licensing delays. When antinuclear forces sued the NRC for refusing to hold hearings on the earthquake's impact, the U.S. Circuit Court of Appeals barred the NRC from granting a full-power license to Perry-1. Each day's delay cost Perry's owners $2.3 million in interest expenses. CEI launched a "Pro-Perry" advertising campaign and 3,000 employees formed the "Perry Employees Committee." They demonstrated aggressively in favor of the plant and spoke out to vouch for its safety.

In October 1986, the stay against licensing was withdrawn and the NRC voted to bar state intervention in licensing procedure. Perry-1 started generating power in December 1986, and went to 100 percent capacity in mid-1987. Official commercial operation began on November 19, 1987.

Financial impact. The Perry complex came with a $6 billion price tag, making it one of the most expensive nuclear plants built. Public Utility Commission regulators refused to allow the utilities to recover any of their construction costs through rate hikes until the plant attained commercial status. CEI had spent $1.5 billion, a full 25 percent of its assets. When Perry-1 began producing electricity, they requested a $120 million rate increase (9 percent). Toledo Edison received a similar rate increase.

The Perry Public School District has been a beneficiary of the nuclear plant. The system, with a student population of 1,500, expects to receive $44 million annually from tax revenues associated with the plant. Their plans call for constructing new schools and sharing funds with neighboring districts.

Centerior Fuel Corp. In 1989 as part of an ongoing effort to reduce expenses, Centerior Power formed a fuel leasing company, Centerior Fuel Corporation. Under the plan, Centerior Fuel buys the raw material for nuclear fuel (mainly uranium oxide) and leases it to CEI and Toledo Edison. The two utilities then pay Centerior Fuel as the fuel is consumed at Perry-1 and the other nuclear plants in which Centerior has an interest. By consolidating the fuel financing, the two utilities expect to pass on a $10 million per year savings to their customers.

Effects of the Clean Air Act of 1990. Although the nuclear venture has been an expensive one, Centerior's officials expect their nuclear capacity to be financially justified in light of the Clean Air Act of 1990. They point out that nuclear power does

Plant Profiles

not contribute to acid rain or global warming. Rather than spend the money needed to limit sulfur dioxide emissions, Centerior plans to shut down some of its older coal-fired plants and rely more heavily on nuclear power. They expect their power prices to drop in the 1990's as competing utilities have to raise rates to pay for complying with acid rain regulations.

Questionable future for Perry 2. The potential cost effects of acid rain legislation have caused the utilities involved at the Perry complex to renew their interest in the possibility of increasing nuclear capacity. This may effect the status of Perry-2.

Perry-2 was more than 50 percent complete (including facilities shared with Perry-1) when construction was suspended in 1985 after a down turn in the economy raised questions as to whether more power was necessary. The partners currently spend $1 million per year to maintain the facility. Cost estimates to finish it are $2 billion. The original construction permit was to have expired on November 20, 1991, but the owners filed for a 10-year extension to keep their options open.

Perry 1 receives recognition. In 1991 Perry 1 placed on the list of top U.S. nuclear plants, as determined by capacity factor. Capacity factor is a measurement of a plant's electrical output stated as a percentage of the theoretical maximum. The average capacity factor of U.S. reactors in 1991 was 69.3 percent. Perry 1 achieved 85.9 percent.

Sources

"Grassroots." *Critical Mass Bulletin*, December 1983.
"Troubles in Nuclear Power Industry Hurt Centerior." *Crains Cleveland Business*, May 12, 1986.
"Analysts Turn Bearish on Centerior Due to Heavy Nuclear Commitment." *Crains Cleveland Business*, September 8, 1986.
"Centerior Takes Offensive in Selling Perry to Public." *Crains Cleveland Business*, October 13, 1986.
"Where 'No-Nukes' are Fighting Words." *Business Week*, November 10, 1986.
"Nukes of the Ohio." *The Nation*, November 15, 1986.
"A Regulator's Nuclear-Age Balancing Act." *The Wall Street Journal*, August 31, 1987.
"Agency Rejects $627.8 million of Perry Costs." *The Wall Street Journal*, January 13, 1988.
"Duquesne Could Suffer Earnings Hit." *Pittsburgh Business Times and Journal*, September 26, 1988.
"Acid Rain Curbs Would Cost Centerior $1 billion." *Crains Cleveland Business*, September 4, 1989.
"Centerior Sees $10 Million Saving by Refinancing N-Fuel Debt." *Crains Cleveland Business*, October 9, 1989.
"Centerior Expects to Lean on Nuclear Power in '90's." *Crains Cleveland Business*, April 30, 1990.
"Centerior Reports Progress to Share Owners." *Business Wire*, April 23, 1991.
"Five Operating Utilities are Suing GE Over a Nuclear Plant Near Cleveland." *The Wall Street Journal*, August 22, 1991.
"Utilities Ask NRC to Extend Perry-2 Construction Permit." *Nucleonics Week*, November 28, 1991.
"U.S. Nuclear Plants Shatter Records for Output, Performance." *INFO*, February 1992.

—*Karen Bellenir*

332 △
Perry Nuclear Power Plant, Unit 2

10 Center Rd. Phone: (216)259-3737
Perry, OH Fax: (216)259-3554
Robert A. Stratman, Gen.Mgr., Perry Plant Operations

Owning Utility

Cleveland Electric Illuminating Co. Phone: (216)622-9800
PO Box 5000 Fax: (650)479-6540
Cleveland, OH 44101
Michael D. Lyster, V.Pres., Nuclear

Contact

Michael J. Lumpe, Mgr., Pub.Aff. Sect.
Cleveland Electric Illuminating Co.

Basic Facts

Unit 2 *Status:* Indefinitely deferred (construction 57% complete as of January 1992). *Type of Reactor:* Boiling water reactor. *Megawatts (electric):* 1250. *Megawatts (thermal):* 3579. *Reactor System Supplier:* General Electric Co. (United States). *Generator Supplier:* General Electric Co. (United States). *Architect Engineer:* Gilbert. *Owner:* Cleveland Electric Illuminating, Toledo Edison, Ohio Edison Co., Duquesne Light, Pennsylvania Power. *Operator:* Cleveland Electric Illuminating Co.

Key Dates

Unit 2 *Ordered:* 1974. *Construction began:* 1974. *Commercial Operation Expected:* Indefinite.

Plant Report

See Plant Report for Perry Nuclear Power Plant, Unit 1.

333 ☆
Phipps Bend Nuclear Power Plant, Units 1-2

Surgionsville, TN

Owning Utility

Tennessee Valley Authority Phone: (615)751-0011
6N 38A Lookout Pl. Fax: (615)751-4904
1101 Market St. Telex: 361951
Chattanooga, TN 37402-2801
J. R. Bynum, V.Pres., Nuclear Operations

Basic Facts

Units 1-2 *Status:* Cancelled (Unit 1, construction 27% complete; Unit 2, construction 5% complete). *Type of Reactor:* Boiling water reactor. *Megawatts (electric):* 1269. *Megawatts (thermal):* 3579. *Reactor System Supplier:* General Electric Co. (United States).

Key Dates

Units 1-2 *Ordered:* 1974. *Cancelled:* 1982.

● Operable   ○ Under Construction   △ Indefinitely Deferred   ▲ Decommissioning   □ Planned   ■ On Order   ☆ Cancelled   ★ Shut Down

334 Pilgrim Station, Unit 1

Rocky Hill Rd.
Plymouth, MA 02360
Roy A. Anderson, Sr.V.Pres., Nuclear

Phone: (508)747-8000
Fax: (508)747-8921

Owning Utility

Boston Edison Company
800 Boylston St.
Boston, MA 02199
Ralph G. Bird, Exec.V.Pres., Operations

Phone: (617)424-2000
Fax: (508)747-8357

Contact

David Tarantino, District Mgr., Nuclear Info.
Pilgrim Station

Basic Facts

Unit 1 *Status:* Operable. *Type of Reactor:* Boiling water reactor. *Megawatts (electric):* 691. *Megawatts (thermal):* 1998. *Reactor System Supplier:* General Electric Co. (United States). *Generator Supplier:* General Electric Co. *Architect Engineer:* Bechtel.

Key Dates

Unit 1 *Ordered:* 1967. *Construction began:* 1967. *Criticality:* 1971. *First Power:* July 1972. *Commercial Operation:* December 1972. *Shut Down Expected:* June 2012.

Operating Costs

Unit 1

1972	$144,000	1981	$34,995,000
1973	$4,787,000	1982	$42,437,000
1974	$9,527,000	1983	$47,277,000
1975	$7,340,000	1984	$73,609,000
1976	$16,633,000	1985	$61,245,000
1977	$15,320,000	1986	$72,031,000
1978	$14,187,000	1987	$112,482,000
1979	$18,386,000	1988	$113,518,000
1980	$27,785,000	1989	$104,911,000

Plant Report

A power settlement in Massachusetts. Occupying a 1600-acre site in Rocky Point, Massachusetts—approximately 40 miles south of Boston—Pilgrim station is an extensive nuclear power center. The complex, owned by Boston Edison, employees nearly 950 people who contribute to the boiling water reactor's production of 670 megawatts of electricity. Constructed in the late 1960s and early 1970s at a cost of $231 million, the facility began commercial operation in 1972 and is licensed to serve Boston Edison's 1,500,000 customers until June 8, 2012.

In addition to the Rocky Point site, the overall operation of Pilgrim Station includes an engineering complex in Braintree, as well as two training facilities and an emergency preparedness division in Plymouth.

Taking into account the payment of wages and taxes, along with the supporting businesses generated by the nuclear industry, Pilgrim Station is valued around $25 million to the community.

Problems with the NRC: 1978-1985. Since 1982, Pilgrim Station had been getting average to below average marks on the National Regulatory Commission's (NRC) annual report card, the Systematic Assessment of Licensee Performance (SALP), which rates nine areas of responsibility. Internal reports and assessments by Boston Edison had also indicated sub-par performance at the plant.

However, Pilgrim's problems go back further than the 1982 NRC evaluation. In 1978, difficulties in communication with the NRC began to emerge at Boston Edison. That year the NRC asked Boston Edison to run tests at the Pilgrim Station according to specified guidelines. After the results were submitted, the NRC discovered that Boston Edison had not complied with the guidelines indicated by the commission. A similar problem occurred in 1984 when the NRC made a request for tests to be run on concrete reinforcement. The tests were never carried out.

In 1982, the NRC fined Boston Edison $550,000, the largest fine ever proposed up to that time, for lack of management review and attention and a substandard gas-control system at Pilgrim Station. Prior to 1982, Boston Edison had paid only $41,000 in fines.

In response to the fine, a Continuous Improvement Program was set up to increase awareness at Pilgrim Station, $300 million was spent to upgrade it, and management was restructured.

By 1985, two more fines, totalling $90,000, were assessed for Pilgrim Station: one for incorrect labeling of highly radioactive material and inadequate instructions to workers, and a second for security lapses. During 1985, the plant was responsible for 40 percent of the power produced by Boston Edison.

Two-and-a-half year shut down: 1986-1988. On April 5, and 12, 1986, mechanical failures caused two shutdowns of Pilgrim Station. After the second, the NRC instructed Boston Edison to keep the plant closed until the Commission was satisfied that the plant was in acceptable running condition.

The NRC generally concluded that the plant was understaffed. The list of other complaints includes:

- Less than adequate protection for workers from radiation.
- Supervisors' failure to follow instructions on radiation work permits.
- Improper or missed equipment testing.
- Failure to recalibrate test equipment.
- Lack of up-to-date emergency plans.
- Failure to file required reports.
- Unauthorized overtime.

Boston Edison spent $200 million to bring Pilgrim Station back to NRC standards, including the replacement of all the mechanical components responsible for the shutdown.

Boston Edison expected a 3-month shutdown, but modifications and restructuring lengthened it to over two years. Some factors that contributed to the prolonged length were:

- Extensive equipment, procedure and training improvements.
- Physical modifications and management restructuring.

Plant Profiles   Pilgrim | 334

- Fire protection, regulatory requirements and refueling which had to be completed before permission to restart was granted.
- A general rise in industry and NRC requirements and standards.
- A need to address emergency preparedness issues before restart.
- A $50,000 fine in November 1987 for security violations.

In September 1988, the NRC found Pilgrim Station suitable for a restart, and Boston Edison proposed a 4-month supervisory plan to oversee the restart process. In December 1988, the NRC was satisfied, and the gradual restart began.

Dealing with the economic effects of the shutdown: 1988-1990. Because of the shutdown, customers of Boston Edison had to obtain power from other sources. In September 1988, the company refunded $7.6 million in replacement power costs from the outage, with the refunds reflected in third quarter financial statements.

In August 1989, the Massachusetts State Energy Office estimated that Boston Edison spent between $400 and 500 million of ratepayers money on the outage and the purchase of power from other sources. Instead of a rate increase, a settlement was reached between the company and the Massachusetts Department of Public Utility, calling for a freeze of rates for three years, an incentive that ties earnings to performance and safety record, and a $75 million investment in energy-efficiency programs.

The next year, 1990, after a two-year lawsuit regarding the shutdown, Boston Edison reached a $24 million out of court settlement with customers of the Commonwealth Electric Company who normally received power generated from Pilgrim Station. Later in the year, the company reached another settlement with a group of 13 municipalities who purchase about three percent of the output from Pilgrim Station.

More shutdowns, but good NRC ratings: 1990-1991. By 1990, the Pilgrim Station was off the NRC watchlist. Even so, early in the year, the plant was temporarily shutdown again, this time for malfunctioning water valves in a reactor cooling system coupled with the failure of the backup safety system. Repairs were made and operations resumed without further incident.

In the NRC's 1991 SALP report, Pilgrim Station was rated as operating in a "safety-conscious manner." The plant achieved a "superior" level of performance in radiological controls, emergency preparedness, and security. The plant personnel achieved a mark of "good" in all other areas of the report. In the same year, NRC figures showed Pilgrim Station in the top twenty percent of thirty-seven boiling water reactors in the United States for average radiation dosage to workers during a three-year period ending in 1990.

Despite the positive reports, a planned seventy-day shutdown for refueling and maintenance in April 1991 was started five days early because of a leaky pump seal.

Pilgrim 2 and 3 cancelled. Anticipating a high level of growth in the levels of demand for electricity in the region, several New England utilities planned in the early 1970s to build more nuclear power plants. Many of these plants were opposed by nuclear industry regulators and utility financial advisors, forcing the utilities to eventually cancel the plants. Among the plants that were cancelled were Pilgrim 2 and 3, Seabrook 2, Sears Island, and Montague 1 and 2.

According to Paul F. Levy, 1987 chairman of the Massachusetts Department of Public Utilities, had all these plants been built it would have given the region an extra 8000 or 9000 megawatts of capacity and caused an immense strain on the regional economy.

Information and programs. A third-party panel of scientists, educators, and physicians is available to provide information to the media about Pilgrim Station. Contact with this panel can be obtained through Pilgrim Station's "directory for the media" from the Nuclear Information Office.

Pilgrim Station has a Nature Trail-Shorefront area on Cape Cod Bay. The area is open from April 1 to November 20 for fishing, hiking, and observation. Also located at Pilgrim Station is a Marine Monitoring Program designed to determine whether the plant's operations have any effect on the marine ecology. Information about both of these programs can be obtained from the Nuclear Information Office.

Requests for information should be directed to the Nuclear Information Office at (617) 746-0912.

Sources

Pilgrim Station. Boston Edison Company, n.d.
Marine Monitoring Program at Pilgrim Nuclear Power Station, Boston Edison Company, n.d.
"Crossroads." *Critical Mass Bulletin,* 1984.
"Boston Edison Gets the Word: Fix the Nuke or Fold It." *Business Week,* June 30, 1986.
"Utility Chief Not a Man Hungry for Power: Paul F. Levy." *Boston Business Journal,* January 26, 1987.
"Pilgrim's Progress." *New England Business,* February 2, 1987.
"The Glow is Gone: Will Pilgrim N-Plant Ever Be On-Line Again?" *Boston Business Journal,* May 4, 1987.
"Study in Contrasts: Pilgrim and Millstone, Two Nuclear Power Plants Have Disparate Fates." *The Wall Street Journal,* July 28, 1987.
"Inquiry into Boston Edison Management of Nuclear Plant Is Asked By Customer." *The Wall Street Journal,* December 30, 1987.
"Restart Approved at Nuclear Plant." *The New York Times,* September 13, 1988.
"Boston Edison to Refund Customers $7.6 Million in Replacement Power Costs." *Business Wire,* September 15, 1988.
"NRC Proposes Fine At Pilgrim Replacement Reactor for Security Lapse." *The Wall Street Journal,* November 3, 1988.
"Power Plant Workers Trying to Stop Leak of Radioactive Water." *The Wall Street Journal,* May 22, 1989.
"Boston Edison Nuclear Plant." *The Wall Street Journal,* June 27, 1989.
"Pilgrim Plant Aims to Alter Image." *Christian Science Monitor,* August 18, 1989.
"More Ratepayers Get a Break." *Christian Science Monitor,* November 9, 1989.
Boston Edison Annual Report 1990. Boston Edison Company, 1990.
"NRC Assessment of Emergency Preparedness Plan Not Thorough." *U.S. NRC 1990 Annual Report,* U.S. Nuclear Regulatory Commission, 1990.
"NRC Removes Three From Watchlist'." *Nuclear Industry,* Fourth Quarter 1990.
"Pilgrim Shutdown Nets Utility $24M." *Boston Globe,* January 12, 1990.
"Boston Edison Company Reaches Settlement Agreements for Municipal Wholesale Contract Disputes." *Boston Globe,* April 26, 1990.
"NRC Faulted for Reopening of Massachusetts Power Plant." *The Washington Post,* July 26, 1990.
"Pilgrim Plant May Restart by Saturday." *Boston Globe,* September 13, 1990.
"Leukemia Link to Pilgrim Plant Found for '72-'79." *Boston Globe,* October 10, 1990.

335 Pilgrim

UNITED STATES

"Edison Disputes Leukemia Study." *Boston Globe*, October 12, 1990.
"Edison Sued in Pilgrim Worker's Death." *Boston Globe* October 13, 1990.
"7 Suits Seek $29m over Pilgrim Plant Radiation Problems." *Boston Globe*, October 19, 1990.
"NRC Sets Stricter Radiation Limits for Atom Workers, Plant Neighbors." *Boston Globe*, December 14, 1990.
"Leaky Seal Hastens Plan to Close Pilgrim Plant." *Boston Globe*, April 30, 1991.
"The Good, the Bad—and the Economy." *Boston Globe* , June 11, 1991.
"Pilgrim Station Fact Sheet." *Backgrounder*, Boston Edison Company, 1992.
"NRC Sets Stricter Radiation Limits for Atom Workers, Plant Neighbors." *Boston Globe*, December 14, 1990.

—Charles Ceplecha, Linda M. Ross

335 ☆ Pilgrim Station, Unit 2

Rocky Hill Rd. Phone: (508)747-8000
Plymouth, MA 02360 Fax: (508)747-8921
Roy A. Anderson, Sr.V.Pres., Nuclear

Owning Utility

Boston Edison Company Phone: (617)424-2000
800 Boylston St. Fax: (508)747-8357
Boston, MA 02199
Ralph G. Bird, Exec.V.Pres., Operations

Contact

David Tarantino, District Mgr., Nuclear
   Info.
Pilgrim Station

Basic Facts

Unit 2 *Status:* Cancelled. *Type of Reactor:* Pressurized light-water reactor. *Megawatts (electric):* 1240. *Megawatts (thermal):* 3629. *Reactor System Supplier:* Combustion Engineering, Inc. (United States).

Key Dates

Unit 2 *Ordered:* 1972. *Cancelled:* 1981.

Plant Report

See Plant Report for Pilgrim Station, Unit 1.

336 ★ Piqua Nuclear Power Plant

Piqua, OH

Basic Facts

Status: Shut down. *Type of Reactor:* Organic Moderated Reactor. *Megawatts (electric):* 12. *Megawatts (thermal):* 45.5. *Reactor System Supplier:* Atomics International (United States). *Architect Engineer:* Holmes and Narver. *Owner:* U.S. Department of Energy. *Operator:* City of Piqua.

Key Dates

Ordered: 1959. *Construction began:* 1960. *Criticality:* 1963. *First Power:* July 1963. *Commercial Operation:* November 1963. *Shut Down:* January 1966.

Plant Report

A project of the Power Demonstration Reactor Programme (PDRP). Realizing that utilities could not justify to their shareholders the investment costs required to build nuclear power plants, the Atomic Energy Commission and Congress devised the Power Demonstration Reactor Programme in 1955 to help launch the commercial nuclear industry in the United States. The PDRP offered cooperating utilities with financial incentives such as research and development assistance, financing of the reactor system, and a waiver of normal fuel charges during the first five years of plant operation.

The small, organically-cooled plant in Piqua, Ohio, was initiated under the PDRP in the late 1950s, shut down in 1966, and subsequently entombed.

Sources

Mounfield, Peter. *World Nuclear Power*. Routledge, 1991.

337 ● Point Beach Nuclear Plant, Units 1-2

6610 Nuclear Rd. Phone: (414)755-2321
Two Rivers, WI 54241 Fax: (414)755-2321
Gregory J. Maxfield, Mgr., Nuclear
   Operations

Owning Utility

Wisconsin Electric Power Company Phone: (414)221-2345
231 W. Michigan St. Fax: (414)221-2010
Milwaukee, WI 53201
Robert E. Link, V.Pres., Nuclear Power

Contact

Jeff Rauh, Nuclear Info. Coord.
Wisconsin Electric Power Company

Basic Facts

Units 1-2 *Status:* Operable. *Type of Reactor:* Pressurized light-water reactor. *Megawatts (electric):* 524. *Megawatts (thermal):* 1518. *Reactor System Supplier:* Westinghouse Electric Corp. (United States). *Generator Supplier:* Westinghouse Electric Corporation. *Architect Engineer:* Bechtel.

Key Dates

Unit 1 *Ordered:* 1965. *Construction began:* 1966. *Criticality:* November 1970. *First Power:* November 1970. *Commercial Operation:* December 1970. *Shut Down Expected:* 2010.

Unit 2 *Ordered:* 1966. *Construction began:* 1968. *Criticality:* May 1972. *First Power:* August 1972. *Commercial Operation:* October 1972. *Shut Down Expected:* 2013.

Operating Costs

Units 1-2

1971	$1,309,000	1981	$26,820,000
1972	$2,305,000	1982	$31,950,000
1973	$3,647,000	1983	$36,666,000
1974	$5,229,000	1984	$42,054,000
1975	$6,159,000	1985	$41,412,000
1976	$6,592,000	1986	$43,126,000
1977	$8,014,000	1987	$47,424,000
1978	$7,395,000	1988	$58,205,000
1979	$12,462,000	1989	$53,220,000
1980	$17,906,000		

Plant Report

Electric power for Wisconsin. The Point Beach Nuclear Plant is located on a site in Two Creeks, Wisconsin, about 100 miles north of Milwaukee on Lake Michigan.

Spent fuel storage. Point Beach's spent fuel rods are stored at the plant, which uses about 28 fuel rods a year. It was predicted that the plant would have enough capacity on site to maintain its spent fuel rods until the mid-1990s.

Licensing change. Point Beach went into operation in 1967. In 1987, rule changes by the NRC pushed the expiration dates for the plant's 40-year federal licenses from 2007 to 2010 for Unit 1, and from 2008 to 2013 for Unit 2. Although Point Beach's licenses were extended, Wisconsin Electric had already begun making plans for the decommissioning of the plant. The utility hired Irving Trust and National Investment Services of America to manage its decommissioning fund, which was worth about $107.86 million by June 1987. The utility planned to add $13 million to the fund each year until Point Beach's licenses expired.

It was predicted that the dismantling of Point Beach would produce 400,000 cubic feet of low-level radioactive waste, including contaminated plant materials such as pumps, pipes and broken concrete. In 1986, the cost of decommissioning was estimated at $190 million, which expected total of the fund.

Upon decommissioning, Wisconsin Electric planned on either turning Point Beach into a park, or returning it to its natural environment.

Accomplishments. Between 1986 and 1990, Point Beach was ranked 10th on the list of nuclear power plants that produced the lowest-cost electricity in the United States during that five-year period. The plant had an average production cost of 1.205 cents per kilowatt-hour. The list was compiled by the Utility Data Institute (UDI), an independent data-gathering subsidiary of the Halliburton Company.

During 1990, Point Beach was the lowest-cost producer of electricity in the United States. According to the 1991 Nuclear Industry Performance Compendium, the plant generated power at an operating and maintenace cost of 0.73 cents per kilowatt-hour.

In a separate study, done by the UDI for the same period, Point Beach's production cost was estimated at $12.08 per megawatt-hour, or 1.2 cents per kilowatt-hour, while producing 7,323,591 total megawatts of electricity. This study also ranked the plant as the lowest-cost producer in the United States.

In 1991, Unit 1 had a capacity factor of 84.1 percent, and Unit 2 had one of 82.8 percent. Along with all the other U.S. nuclear plants, Point Beach helped in accounting for nearly 22 percent of the electricity generated by electrical utilities in 1991.

Sources

"A Proper Burial." *The Business Journal-Milwaukee*, July 27, 1987.
"Wisconsin's Point Beach Lowest Cost Nuclear Plant in '90." *INFO*, U.S. Council for Energy Awareness, July 1991.
"Another Rave Review for 1990." *Nuclear Industry*, Fourth Quarter 1991.
"U.S. Nuclear Plants Shatter Records for Output, Performance." *INFO*, U.S. Council for Energy Awareness, February 1992.
"Low-Cost Producers." *INFO*, U.S. Council for Energy Awareness, May 1992.

—Charles Ceplecha

338

Prairie Island Nuclear Plant, Units 1-2

1717 Wakonade Dr. E. (Rt. 2) Phone: (612)388-1121
Welch, MN 55089 Fax: (612)330-5743
E.L. Watzl, Site Gen.Mgr.

Owning Utility

Northern States Power Company Phone: (612)330-5500
414 Nicollet Mall Fax: (612)330-2900
Minneapolis, MN 55401
Leon Eliason, V.Pres., Nuclear Generation

Contact

John Bousquet, Dir., Pub.Info.
Northern States Power Company

Basic Facts

Units 1-2 *Status:* Operable. *Type of Reactor:* Pressurized light-water reactor. *Megawatts (electric):* 534. *Megawatts (thermal):* 1650. *Reactor System Supplier:* Westinghouse Electric Corp. (United States). *Generator Supplier:* Westinghouse Electric Corp. *Architect Engineer:* Fluor Power Services, Inc.

Key Dates

Unit 1 *Ordered:* 1967. *Construction began:* 1968. *Criticality:* 1973. *First Power:* 1973. *Commercial Operation:* December 1973. *Shut Down Expected:* August 2008.
Unit 2 *Ordered:* 1967. *Construction began:* 1968. *Criticality:* 1974. *First Power:* 1974. *Commercial Operation:* December 1974. *Shut Down Expected:* October 2014.

Operating Costs

Units 1-2

1973	$101,000	1982	$28,168,000
1974	$4,216,000	1983	$31,252,000
1975	$7,261,000	1984	$33,298,000
1976	$15,574,000	1985	$49,014,000
1977	$17,090,000	1986	$49,548,000
1978	$14,214,000	1987	$5123922,000
1979	$15,346,000	1988	$63,839,000
1980	$23,174,000	1989	$41,757,000
1981	$26,791,000		

Plant Report

Electricity for Minnesota. The Prairie Island Nuclear Plant occupies a 560-acre site on Prairie Island, a two-by-eight-mile peninsula in the Mississippi River near Red Wing, Minnesota. The site is 28 miles southwest of Minneapolis-St. Paul. Prairie Island's construction cost about $413 million.

Community impact. The Northern States Power Company (NSP) owns and operates Prairie Island. The utility employs about 350 people at the plant, most of which live in the neighboring town of Red Wing, Minnesota. NSP's $12,565,935 annual tax payment for the plant helps keep taxes low for the city's 13,000 residents.

Fuel facts. Uranium-235 pellets fill Prairie Islands's fuel rods, which measure 12 feet long. Groups of 174 rods form one fuel assembly that contains enough uranium to produce more than 100 million kilowatt-hours of electricity. Maximum fuel-pellet centerline temperature measures 3,500 degrees Fahrenheit.

Each reactor holds 121 fuel assemblies, or 120,000 pounds of uranium, and each assembly produces power for about three years. Every 18 months, one reactor refuels 48 assemblies, while the other remains operating.

Features

- A reinforced concrete shield building, measuring 2 1/2 feet thick, surrounds the steel containment vessel.
- Water temperature in each reactor averages 560 degrees Fahrenheit at 2,235 pounds per square inch pressure.
- There are 29 movable control rods in the core to control absorption of free neutrons. The rods are assemblies of cadmium-indium-silver alloy encased in stainless steel, plus liquid boron injected into the primary coolant in controllable amounts.
- The plant uses Mississippi River water. The water condenses to steam at a rate of 10,000 gallons per minute, and drives the steam generator, which spins at 1,800 revolutions per minute.
- The Minnesota Pollution Control Agency requires the plant to maintain its circulating cooling water below 86 degrees Fahrenheit. Until 1983, NSP cooled the water only by using its five-story cooling towers. During the 1983-84 winter season, the utility began using an open-cycle system, allowing a bypass of the towers. NSP uses the system all year, and the cooling towers from April to November. This system allows the plant to use less of its own power to run cooling tower pumps and fans, and to decrease its maintenance expense on the towers.
- The plant's substation output measures 345,000 volts and 161,000 volts.

Water intake and discharge system. Prairie Island's water intake and discharge project uses a screen system for filtering river water, an environmental laboratory for biological testing, a de-icing system for keeping intake water ice-free, and a water discharge for improved mixing.

The eight-unit traveling intake screen system and screen house are unique. The system resembles large conveyors, and consists of 220 screens that continuously filter aquatic life and debris from river water. The system contains some 1,760 interchangeable screens.

Fish pass through an 18-inch diameter pipe to either the plant's biological laboratory or back to the river. The fish are held for 48 hours in specialized tanks, and studied for survivability. The annual cost of the study runs about $75,000.

A series of pumps in the de-icing system force warm discharge water into the intake water to keep the river water that enters the plant ice-free. This reduces the impact of the discharge water on the aquatic environment, and increases plant efficiency when intake water reenters the cooling water system.

In December 1984, NSP completed construction of two eight-foot-high concrete mixing barriers in the plant's water canal, and one 12-foot-high barrier at the outlet of the recyle-water canal. These barriers interrupt the flow so that warm recycled water mixes with cool river water, and attains a uniform temperature before going to the four 150,000-gallon-per-minute circulating pumps and cooling water pumps. The barriers cost $100,000, and the uniform-temperature water enables Unit 2 to produce an additional 10 megawatts.

Spent fuel storage. NSP stores Prairie Island's spent fuel assemblies in a 40-foot-deep water pool at the site. The pool is made of reinforced concrete, lined with stainless steel.

In 1977, NSP modified the arrangement of its fuel assemblies to accommodate more spent fuel. The storage racks were modified again in 1982, which increased the pool capacity from 687 to 1,582 stored assemblies. This additional capacity gave the plant storage room until 1994.

In 1990, NSP initiated a plan to utilize an air-cooled storage system, consisting of 8-foot-wide, 16-foot-long, 120-ton steel casks. The utility planned on investing up to $40 million on the plan, which was designed to transfer older and cooler spent fuel from the pool into the casks to make room to store newer spent fuel. At the time, the pool was expected to fill up in three years, causing the plant to shut down.

In November 1990, the Minnesota Environmental Quality Board released a draft environmental impact statement which generally supported the new sttorage method.

Low-level waste disposal. As of 1991, NSP transported low-level waste from Prairie Island to a disposal site in Hanford, Washington. Because of effective low-level waste management and efforts to reduce contamination, the plant generates about one-third as much low-level waste as the nuclear industry average.

In 1991, questions arose among NSP officials concerning storage space, amidst predictions that the Washington site would close its doors in 1993. The utility planned to store its low-level waste in Michigan from 1993 to 1996, but a site was not yet found and not expected to open until 1997 or 1998.

Radiation and the environment. Annually, the average employee at Prairie Island gets about 200 millirem of radiation

Plant Profiles

beyond natural background levels, only four percent of the maximum dose allowed by the Environmental Protection Agency.

NSP employs 19 radiation protection specialists at the plant, whose duties include monitoring the water systems for radioactive contamination, testing for airborne radioactive materials and testing workers for exposure.

NSP began radiation monitoring two years before the plant opened. At that time, the utility measured the area's background radioactivity at about 125 millirem a year, and compares those findings with weekly ones at four locations.

NSP estimates that the plant's closest neighbors receive less than one millirem of radiation annually from the plant. However, in late 1975 and 1977, the utility detected an increased radiation in Minnesota's air, vegetation, and milk. NSP blamed this on Chinese nuclear weapons testing, citing that monitoring data revealed no environmental radiation attributable to the plant.

Around 1988, NSP installed a Radiation Dose Assessment Computer System (RDACS), a sophisticated weather monitoring and radiation dose projection system, at the plant. The RDACS emergency operations involve monitoring radiation released from the plant in order to track the releases, and to plan emergency response activities.

Emergency planning. NSP publishes a booklet called the *Emergency Planning Guide for Neighbors of the Prairie Island Nuclear Plant*, and mails it periodically to people living within the plant's 10-mile emergency planning zone. Residents also receive an annual report detailing information about plant events and projects during each year.

In 1982, NSP installed 63 emergency warning sirens in a 10 mile radius of the plant. The sirens can be used for plant emergencies, as well as threatening weather or non-nuclear emergencies in the community.

The plant has two NRC inspectors on site, which allows for immediate communication with the commission about any abnormal occurence. The NRC also performs unannounced inspections of the plant and its emergency plan.

Problems. In December 1989, a reactor trip and partial loss of off-site power occurred at Prairie Island's Unit 2. In 1990, the NRC assigned augmented inspection teams to investigate the event.

Decommissioning and plant life extension. Throughout the life of Prairie Island, the cost of both building and decommissioning the plant is reflected in NSP's rates and as depreciation costs. Based on a 1990 study, the utility estimated the decommissioning cost for both its nuclear plants at approximately $630 million, in 1990 dollars. During 1990, NSP recorded $175 million in its accumulated provision for depreciation, with $13 million of this balance designed for deposit into an external fund in 1991.

In 1988, NSP explored the possibility of "plant life extension" to extend the use of Prairie Island from a normal 40 years to a 60- or 70-year span.

Low-cost producer. According to the Utility Data Institute (UDI), Prairie Island was the second lowest-cost producer of electricity in the United States from 1982 to 1986, and the 10th lowest-cost producer from 1986 to 1990.

For 1990, UDI ranked the plant 10th on its list of lowest-cost producers, citing its production cost of $15.03 per net megawatt-hour, while producing 1,186 gross megawatts in 7,633,375 megawatt-hours.

Outstanding capacity factors. From January 1, 1986, to December 31, 1988, Prairie Island and its sister plant, Monticello, had the highest combined capacity factor among U.S. nuclear plants.

Up to 1988, Unit 2's lifetime capacity factor was 84 percent. From 1987 to 1989, Unit 2 ranked second in the world, and Unit 1 ranked seventh. The units ranked first in the United States, and were in the world's top 10 for combined capacity factor during the same period.

In 1989, Unit 1's capacity factor of 94.94 percent ranked it among the top 25 plants in the world.

In 1990, Unit 1's capacity factor of 86.1 percent and Unit 2's of 96.6 percent fell well above the U.S. average of 69.3 percent.

General accomplishments. In 1987, the U.S. Council of Energy Awareness named Prairie Island's Unit 2 as its top U.S. nuclear power performer.

When Unit 2 went out of service for scheduled maintenance and refueling on January 6, 1988, it ended a 406-day run. This set an NSP record, and was the fourth-longest run by a pressurized water reactor in the world.

Also in 1988, Westinghouse awarded Unit 2 for its lifetime cumulative availability rating of 86.8 percent, and its availability rating of 95.28 percent for 1986 and 1987.

In February 1990, the NRC named the plant one of the best-run nuclear plants in the United States. The commission based its rating on several factors, including overall operating performance and safety records. Together, Prairie Island and Monticello, produced 36 percent of NSP's electricity during the year.

In November 1990, the INPO gave the plant its third "Award of Excellence," based on receiving the institute's highest performance rating.

In January 1991, the NRC ranked Units 2 and 3 among the top five nuclear reactors in the United States. This ranking was also based on operational performance and safety records, as well as consistently excellent Systematic Assessment of Licensee Performance reports.

In June 1991, the NRC cited Prairie Island for "achieving outstanding safety performance."

Sources

Nuclear Power at Prairie Island. Northern States Power Company, October 1988.
Northern States Power Company 1990 Annual Report: NSP is people.working for you. Northern States Power Company, n.d.
"World Class." *Nuclear Industry,* First Quarter 1990.
"Operational Safety Assessment." *U.S. NRC 1990 Annual Report,* U.S. Nuclear Regulatory Commission, n.d.

339 PSEG

NRC Names Prairie Island One Of Top Four In U.S.. (Press release) Northern States Power Company, February 20, 1990.

NSP Receives Top Ranking For Nuclear Plants' Capacity Factor. (Press release) Northern States Power Company, May 23, 1990.

NSP Nuclear Plants Net Top National Awards. (Press release) Northern States Power Company, November 19, 1990.

"NSP Plans to Store Waste Clears Initial Environmental Hurdle." *Minneapolis-St. Paul CityBusiness*, December 3, 1990.

NSP's Prairie Island Plant Makes NRC's Top Five. (Press release) Northern States Power Company, January 30, 1991.

"State's Radioactive Waste Becoming Harder to Get Rid of." *Minneapolis-St. Paul CityBusiness*, March 18, 1991.

Prairie Island Unit Ranks Number Two In The World. (Press release) Northern States Power Company, March 26, 1991.

"Wisconsin Electric's Point Beach Lowest Cost Nuclear Plant In '90." *INFO*, U.S. Council for Energy Awareness, July 1991.

"Four Nuclear Plant Win Kudos From NRC." *INFO*, U.S. Council for Energy Awareness, July 1991.

"U.S. Nuclear Plants Shatter Records for Output, Performance." *INFO*, U.S. Council for Energy Awareness, February 1992.

"Low-Cost Producers." *INFO*, U.S. Council for Energy Awareness, May 1992.

"Wisconsin Electric's Point Beach Lowest Cost Nuclear Plant In '90." *INFO*, U.S. Council for Energy Awareness, July 1991.

—Charles Ceplecha

339 ☆
PSEG Offshore Nuclear Power Plant, Units 1-2

Owning Utility

Public Service Electric and Gas Company
80 Park Plaza
Newark, NJ 07101
Steven E. Miltenberger, V.Pres. & Chief Nuclear Officer

Phone: (201)430-7000

Contact

Michaele L. Camp, Mgr., Nuclear Commun.
Public Sevice Electric and Gas Company

Basic Facts

Units 1-2 *Status:* Cancelled. *Type of Reactor:* Pressurized light-water reactor. *Megawatts (electric):* 1212. *Reactor System Supplier:* Offshore Power Systems. *Owner:* Public Service Electric and Gas. *Operator:* Public Service Electric and Gas.

Key Dates

Units 1-2 *Ordered:* 1973. *Cancelled:* 1978.

341 ●
Quad Cities Station, Units 1-2

22710 206th Ave. No.
Cordova, IL 61242
Richard Bax, Stat.Mgr.

Phone: (309)654-2241

Owning Utility

Commonwealth Edison Company
One First National Plaza
PO Box 767
Chicago, IL 60690
Cordell Reed, Sr.V.Pres., Nuclear Operations

Phone: (312)294-4321
Fax: (312)294-2995

Contact

John F. Hogan, Dir., Commun. Svcs.
Commonwealth Edison Company

Basic Facts

Units 1-2 *Status:* Operable. *Type of Reactor:* Boiling water reactor. *Megawatts (electric):* 833. *Megawatts (thermal):* 2511. *Reactor System Supplier:* General Electric Co. (United States). *Generator Supplier:* General Electric Company. *Architect Engineer:* Sargent & Lundy Engineers (United States). *Owner:* Commonwealth Edison Company; Iowa-Illinois Gas and Electric Co. *Operator:* Commonwealth Edison Company.

Key Dates

Unit 1 *Ordered:* 1966. *Construction began:* 1967. *Criticality:* January 1971. *First Power:* April 1972. *Commercial Operation:* February 1973.

Unit 2 *Ordered:* 1966. *Construction began:* 1966. *Criticality:* April 1972. *First Power:* May 1972. *Commercial Operation:* March 1973.

Operating Costs

Units 1-2

Year	Cost	Year	Cost
1972	$2,033,000	1981	$37,271,000
1973	$6,290,000	1982	$43,606,000
1974	$9,210,000	1983	$44,939,000
1975	$14,777,000	1984	$71,231,000
1976	$18,268,000	1985	$64,359,000
1977	$17,756,000	1986	$66,680,000
1978	$22,168,000	1987	$74,742,000
1979	$32,417,000	1988	$93,954,000
1980	$38,685,000	1989	$108,553,000

Plant Report

A power center for Iowa and Illinois. Quad-Cities Station is located on a 600-acre site on the east bank of the Mississippi River, four miles north of Cordova, Illinois. The plant got its name by being situated near the "quad cities" of Davenport, Rock Island, Moline and East Moline.

The operating utility of Quad-Cities is Commonwealth Edison, and it is the utility's only jointly owned generating station. The Iowa-Illinois Gas and Electric Co. of Davenport, Iowa shares one-quarter interest in the plant, but Commonwealth Edison is responsible for its operation.

The initial project cost for the plant was $262 million, but an additional $180 million was spent on modifications, required by the NRC to meet newer safety and environmental regulations. At full power, Quad Cities can generate enough electricity to serve a city with a population of about 800,000 people.

Features. The pressure vessels at Quad-Cities are 68 1/2 feet high, with a diameter of 20 feet 4 inches. The walls have a 6 1/8 inch minimum thickness, and the design pressure is 1250 pounds per square inch gauge.

Plant Profiles   Rancho | 342

The core consists of 724 fuel assemblies, containing 62 fuel rods per assembly. Eaach rod is 145.42 inches in length and 0.483 inches in diameter. The total core represents approximately 128.7 metric tons of uranium fuel at 2.12 percent enrichment. Between one-third and one-fourth of the total core is replaced approximately every 18 months during scheduled outages.

The turbine generator is driven by steam at 1800 revolutions per minute, and produces electricity at 18,000 volts, stepped up by the transformer to 345,000 volts for transmission.

Investigation. In 1989, a probe by the NRC's Office of Investigations confirmed that a licensed senior reactor operator (SRO) at Quad-Cities deliberately failed to report and tried to cover up an erroneous fuel movement during a refueling outage. As a result, his SRO license was suspended, pending completion of an NRC-approved remediation program.

Modification project. In 1991, Commonwealth Edison awarded a $1 million contract to the Nutech Engineers Division of Pacific Nuclear Systems Inc. to provide structural and engineering services for Quad-Cities. The project was scheduled for completion in 1992, and called for the engineers to provide modification design packages for one of the plant's safety systems. The modifications were designed to improve system reliability and facilitate maintenance and testing.

Award winner. In 1992, Quad-Cities' Unit 1 was honored by GE Nuclear Energy as being one of 23 U.S.-supplied boiling water reactors that operated at 75 percent efficiency or above during 1991.

Sources

Quad-Cities Nuclear Station Fact Sheet, Commonwealth Edison, May 1988.
"Enforcement Actions/Civil Penalties." *U.S. NRC 1990 Annual Report*, U.S. Nuclear Regulatory Commission, n.d.
Mounfield, Peter. *World Nuclear Power*. Routledge, 1991.
"Pacific Nuclear Systems Inc. Announces $1 Million Engineering Services Contract." *Business Wire*, April 9, 1991.
"And the Winner is." *INFO*, U.S. Council for Energy Awareness, May 1992.
"World List of Nuclear Power Plants." *Nuclear News*, August 1991.

—Charles Ceplecha

---

342 | ★

Rancho Seco Nuclear Generating Station

14440 Twin Cities Rd.   Phone: (916)732-4737
Herald, CA 95638-9799   Fax: (916)732-6185
Jim Shetler, Dep.Asst.Gen.Mgr., Nuclear

Owning Utility

Sacramento Municipal Utility District   Phone: (916)452-3211
    (SMUD)
6201 S St.
PO Box 15830
Sacramento, CA 95813
Leo Fassler, Asst.Gen.Mgr. & COO

Contact

Dace Udris, Pub.Info.
Sacramento Municipal Utility District

Basic Facts

Status: Shut down. *Type of Reactor:* Pressurized light-water reactor. *Megawatts (electric):* 966. *Megawatts (thermal):* 2772. *Reactor System Supplier:* The Babcock & Wilcox Co. (United States). *Generator Supplier:* Westinghouse Electric Corp. (United States). *Architect Engineer:* Bechtel.

Key Dates

Ordered: 1966. *Construction began:* 1969. *Criticality:* 1974. *First Power:* October 1974. *Commercial Operation:* April 1975. *Shut Down:* 1989.

Operating Costs

1975	$11,607,000	1983	$52,588,000
1976	$7,193,000	1984	$57,961,000
1977	$14,000,000	1985	$93,534,000
1978	$11,834,000	1986	$88,927,000
1979	$13,720,000	1987	$123,997,000
1980	$28,408,000	1988	$128,044,000
1981	$35,541,000	1989	$58,559,000
1982	$36,329,000		

Plant Report

Overview. The Rancho Seco Nuclear Generating Station is owned by the Sacramento Municipal Utility District (SMUD), who serves more than 400,000 customers in Sacramento, California. The plant was built for $375 million.

Midway through construction, building costs exceeded projections of $200 million. Electricity rates jumped 10 percent before the formal dedication in October 1974, when valve problems forced a shutdown for repairs. The plant was closed for 13 of its first 18 months in operation due to malfunctioning equipment.

Power production at Rancho Seco continued to decline from 57 percent in 1980 to 25 percent in 1985. As of 1987, Rancho Seco had been in operation 47 percent of the time since it first went on line. Between 1975 and 1986 average rates in the district increased by 184 percent, and 1987 brought another 20 percent increase. In 1987, the NRC and industry experts named Rancho Seco one of the most problem-plagued plants in the country.

In its 15 year history, Rancho Seco has had some 100 unplanned shutdowns, more than twice the national average. The plant's lifetime average was only 38 percent capacity.

Carbon copy of the Three Mile Island plant?. Rancho Seco is one of nine electric-generating plants in the U.S. that uses nuclear-reactor equipment designed and manufactured by Babcock & Wilcox. Its image was marred by the severe core-melting accident in March 1979 at Pennsylvania's Three Mile Island nuclear plant, whose reactor design is nearly identical to Rancho Seco.

Apparently, Rancho Seco and Three Mile Island have more than just their design in common. In comparing an accident on March 20, 1978 at Rancho Seco and the 1979 accident at Three Mile Island, Daniel Whitney, a Rancho Seco nuclear engineer, pointed out that "for about the first seven minutes, the two events looked almost identical."

A study in November 1987 by the NRC found that "despite improvements in the Babcock & Wilcox reactors after the Three Mile Island accident, the number and complexity of events in plants with reactors designed by Babcock & Wilcox have not decreased as expected."

Accidents. In October 1974, an operator error destroyed one of the emergency feedwater pumps by failing to open a valve that would have supplied water to cool it while the pump was being used to transfer water from a storage tank to the steam generators.

In June 1975, one of the two emergency feedwater pumps that had been started during a test suddenly stopped due to electrical failures resulting from a loose nut.

On February 19, 1977, an operator discovered one of the emergency feedwater pumps tested the day before was disabled because an operator failed to reset the switches properly so it could be turned on again.

On March 20, 1978, at 4:25 a.m., a worker dropped a light bulb behind an instrument panel in the control room that caused a short circuit and blackout, and activated the emergency pumps. After one hour and nine minutes, the operators restored power to their instruments and brought the plant, which was operating at 70 percent capacity, back under control.

On December 26, 1985, from 4:30 a.m. to 8:41 a.m., all power to the plant's computerized control system was lost for 26 minutes. A faulty power supply cut electricity to the control room, triggering a series of equipment failures and operator errors that resulted in an over-cooling of the reactor vessel.

The system that supplies water to be boiled into steam, and keeps the reactor from overheating, malfunctioned by pumping too fast. Instead of using controls in the control room to slow down the water flow, workers attempted to close valves manually, but the valves had not been lubricated in the plant's 11-year history and were rusted open.

Operators cut the water supply to the pumps, which led to a spill of 1,200 gallons of slightly radioactive water inside the plant and the release of a small amount of radioactive gas into the atmosphere. Two workers received small exposures to excess radiation. SMUD later found cracks in the steel, too small to be of concern, it concluded.

The plant was shut down for 27 months. Repairs and improvement costs came to over $400 million, more than the plant's original cost, bringing Rancho Seco's price tag to $1 billion. In addition, it cost $200 million more to buy replacement power until the plant reopened in April 1988 and caused a 70 percent rate hike for residential customers. This marked Rancho Seco's fifth over-cooling accident.

SMUD tries to sell Rancho Seco. In 1985 SMUD suffered an operating loss of more than $9 million and an estimated loss of $2.5 million in 1986. As a result, in 1987 SMUD began talking with several utilities in an effort to sell off equity interests in Rancho Seco to provide them with greater access to alternative energy sources, greater management expertise, and more savings. Also, to help pay for repairs and get the plant back on line, SMUD issued $390 million worth of Series S bonds.

SMUD received the following offers—none of which were accepted—to their proposal to sell ownership and/or operation of Rancho Seco:

On September 3, 1987 San Francisco-based Pacific Gas & Electric Company offered to take over SMUD and shut down Rancho Seco.

In December 1987, Duke Power, the nation's seventh-largest investor-owned utility, was considering buying the plant, an idea that was later dropped. In April 1988, Duke offered to operate Rancho Seco for $25 million a year plus $1 million for each percentage point reactor reliability rose above 50 percent. This bid was rejected by SMUD.

On July 20, 1988, Quadrex Corporation of Campbell, California submitted a proposal to run Rancho Seco for $5 million a year—$20 million less than Duke Power. In August 1988 Quadrex proposed operating Rancho Seco during the first phase of any agreement for a base fee of $5 million, plus incentives of up to $5 million based on operating improvements. A second phase called for the firm to take over part or all of the plant's operating costs.

On March 29, 1989, Babcock & Wilcox and Bechtel Power Corporation offered to become partners in operating Rancho Seco for an unspecified price, that was later estimated at $30 million. Supposedly ratepayers would benefit from this agreement because the firms' take would be based in part on plant performance.

In April 1989, Irvine-based Fluor Daniel Inc., an $8 billion worldwide contractor, offered to help SMUD run Rancho Seco for a base fee, adjusted according to how well the plant performed.

Asbestos lawsuit. On June 11, 1987, approximately 100 laborers involved in the construction and maintenance of Rancho Seco filed a $300 million lawsuit—$3 million in exemplary damages for each worker—against the SMUD and six of its contractors, claiming they were needlessly exposed to asbestos which is known to cause certain types of cancer. It was found that Kaylo, a pipe covering which contains 15 percent asbestos materials, was used to insulate pipes at the plant.

The suit charges SMUD and its co-defendant with creating dangerous work conditions, failing to warn workers of asbestos hazards, and failing to take reasonable precautions to protect workers. As a result of their exposure to asbestos, the suit says many of the workers have been rendered "sore, lame and disabled" and now complain of internal and external injuries and shock to their nervous systems. At this time, only one man had contracted a form of cancer that could be traced to asbestos exposure.

Rancho Seco put to a vote: will it stay or will it go?. Six hundred volunteers from Sacramentans for Safe Energy (SAFE) worked for six months to gather more than 50,000 signatures that put the issue of whether to shut down Rancho Seco on the June 1988 ballot.

The group was able to use California's initiative and referendum process to accomplish this task because the plant was run by a publicly owned utility, which meant it was owned by and for the benefit of its customers, not stockholders.

Two measures appeared on the ballot relating to Rancho Seco. One called for the shutdown of the plant, and the other, Measure C, authorized an 18-month trial run, dictated a follow-up public vote, and required that SMUD divest itself of all or part of the ownership of the plant.

In June 1988 the 530,000 voters decided to allow Rancho Seco to remain open, 51.6 percent to 48.4 percent, a narrow 2,000 vote margin. The 1988 vote followed a 27-month shutdown when the utility spent several hundred million dollars for repairs, retraining, ongoing maintenance, and replacement power.

Prior to the election, voters were told there were no planned rate increases and that the plant would operate at 70 percent capacity. However, within one month after the election, SMUD announced 16.7 percent rate increases and during the trial period the plant's overall performance was 38 percent.

Improvement. A 40-page report based on 15,427 hours of on-site inspection by the NRC concluded that Rancho Seco had improved in every area analyzed from July 1986 through December 1988 even though the report was released while Rancho Seco struggled through its third unplanned shutdown since December 12, 1988, and its 56th day without producing power in 1989.

The report graded the plant in nine areas on a scale of one to three, with three being the worst. In every area but one Rancho Seco received a two, up from the near-solid threes it received when last graded in 1986. In one area, start-up testing, the plant received a one.

The report also criticized plant management for not completing repairs suggested months before, for not communicating enough between managers, ffor weaknesses in handling radioactive water, and for 58 other specific problems.

NRC fines. On January 17, 1989, the NRC fined Rancho Seco $100,000 for violating liquid radioactive waste-disposal rules between 1983 and 1986. The series of six violations, some of which resulted in the discharge of radioactive materials into the environment, were given a Level Three severity ranking on a scale of one to five, with Level One being the most serious. The levy brought to $695,000 the total penalties by SMUD from 1980 to January 1989.

A second vote and major victory for antinuclear activists. In June 1989 Rancho Seco faced another election and another chance that it might be closed down. According to a telephone poll published May 1989 by the *Sacramento Bee*, 46.2 percent of those surveyed were in favor of closing Rancho Seco, 36.5 percent favored keeping it open, and 17.4 percent had not yet formed an opinion.

By the end of May 1989, more than $300,000 had been donated in support of Rancho Seco with most of it coming from the nuclear industry; plant opponents had raised more than $170,000.

Both sides promised lower rates. Plant supporters claimed nuclear power was still the cheapest source of electricity and pointed out that its likely replacement, gas, had jumped 15 percent in 1988. However, opponents said Rancho Seco is cheap only if it works, which is infrequent, and SMUD should instead invest in alternative power sources. Opponents cited the plant's poor performance record from the first election, June 1988, to the second election, June 1989, which was only 44 percent capacity.

In May 1989 SMUD analysts estimated that closing the plant would mean a 7.5 percent average rate increase starting in January 1990 while keeping it running would mean only a 4.3 percent increase.

On June 7, 1989 at 10:45 p.m. the results were in. The Sacramento-area residents voted to close Rancho Seco, marking the first time voters decided to abandon a licensed and operating generating station and the first time in 15 elections nationwide that antinuclear activists won a ballot measure against atomic energy.

The count was 111,867 in favor of Measure K, the shutdown of Rancho Seco, and 97,460 opposed, a 53.4 percent to 46.6 percent margin, with about 40 percent of registered voters casting ballots.

Although the vote was not legally binding, SMUD directors had announced in March 1989 that, if voters chose to shut down the plant, it would be closed.

According to David A. Boggs, SMUD's general manager, it will cost $210 million to close Rancho Seco with tens of millions more each year to secure it until at least 2009 when its radioactive fuel and machinery could be safely disposed of.

Decommissioning. On June 9, 1989, SMUD released a 127-page plan that called for actual dismantling of the plant to begin in 20 years, and for spending as much as $30,000 to make a video of the plant's death and $12,000 for brochures and pamphlets. The plan envisions that nuclear fuel will be stored on the site until 2010 when actual dismantling and demolition will begin. Final details will be worked out in 2005.

At 10:40 a.m. on June 8, 1989 Rancho Seco slipped into "hot shutdown," where the plant does not generate electricity but continues to generate heat from its nuclear pile. This process lasted 72 hours. When the interior temperature dropped below 200 degrees Fahrenheit, the plant went into cold shutdown and the nuclear chain reaction stopped. This mode lasted for several months.

Within the first two or three years, the building housing the nuclear reactor will be thoroughly scrubbed to remove radioactive dust. After that SMUD will still have to monitor and maintain the facility for at least 20 years.

Once safe enough, the nuclear fuel will be removed and stored on site until a federal nuclear repository opens, probably not before 2008. The coolant water will then be drained and treated to eliminate radioactivity. Finally, the plant will be sealed up and abandoned for up to 30 years to allow radioactivity to diminish. It will then sit idle until part of the facility is re-used with natural gas or until radiation levels have dropped low enough to let crews cut up the plant and haul it away to the nuclear repository.

At a minimum, experts say it is likely to take at least 30 years to clean up, mothball, dismantle, and truck away the pieces of a power plant. The NRC allows a total of 60 years for the four-step process.

Most of the power generated by Rancho Seco will be replaced by electricity that SMUD will buy from Pacific Gas and Southern California Edison under agreements that became final on July 31, 1989.

SMUD tries to sell Rancho Seco, again. SMUD directors voted three to two at the end of June to try to sell Rancho Seco to another operator, claiming the referendum did not explicitly bar others from running the plant.

As a result, five bidders stepped forward on June 30, 1989, with offers to buy Rancho Seco and run it independently. Quadrex Corporation's bid of $232 million was the only offer to reopen the facility as a nuclear plant. Other offers came from Methacoal Corporation, Consolidated Power Company, Sacramento Delta Power Company, and CMS-Rancho Seco Company.

On July 20, 1989 the board of directors of SMUD exclusively selected Quadrex to begin negotiating the purchase of Rancho Seco. However, on September 11, 1989 the board voted five to zero to drop Quadrex's proposal.

SMUD attempts to cut their losses. As of July 1989 SMUD had paid $6.4 million for a $10 million replica of the Rancho Seco control room they ordered in 1986 from CAE Electronics Ltd. who still intended to deliver the simulator by June 1990 and collect the rest of the money. SMUD was attempting to negotiate out of the contract because it no longer can use the simulator.

SMUD needs more power. As of June 1991, SMUD was weighing proposals from a dozen power companies to sell SMUD electricity. SMUD must pick a mix of projects that would cost more than $1 billion to develop and would give it from 800 megawatts to 1,000 megawatts of power over the next 10 years.

One of the 28 energy plans being considered by SMUD was an offer by CMS Energy who said it could generate up to 610 megawatts of electricity by converting Rancho Seco to a gas fired plant with a life span of 30 years. The conversion would cost between $500 and $600 million to develop.

Sacramento citizens take a different approach to energy. As a result of their expensive experience with nuclear energy, many residents of the Sacramento area have decided to conserve energy. Their goal is to replace 700-megawatts of energy lost with Rancho Seco through conservation efforts. The Sacrmento community has taken the following steps toward their goal.

Local utilities will now pay residents of the Sacramento area to trade in their old energy wasting refrigerator if they buy a more efficient model.

Urban neighborhoods give away trees so that 10 years from now houses with adequate shade will need less air-conditioning during Sacramento's 100-degree summers.

Consumers are buying more low-flow shower heads, low-watt light bulbs, and high efficiency window panes.

Utility officials have reclaimed two megawatts at the abandoned power plant by setting up an orchard of solar collection panels in the shadow of Rancho Seco's cooling towers.

By June 1992, two years into their conservation efforts, Sacramentans were saving 300-megawatts, halfway to their goal of 700-megawatts.

Sources

Network Earth. Turner Multimedia, June 7, 1992.
"Holding Company Criticized." *Sacramento Bee*, June 20, 1991.
Initial Study & Proposed Negative Declaration: Rancho Seco Proposed Decommissioning Plant. Sacramento Municipal Utility District, June 19, 1991.
"Michigan Firm Wants to Convert Seco to Gas." *Sacramento Bee*, February 17, 1991.
"SMUD Mulls Spate of Factors in Eyeing New Power Sources." *The Business Journal-Sacramento*, December 24, 1990.
"Rancho Seco Decked Again." *Bulletin of the Atomic Scientists*, December 1989.
"SMUD Votes 5-0 to Dump Quadrex." *Sacramento Bee*, September 12, 1989.
"SMUD Study Casts Doubt on Seco Plant." *Sacramento Bee*, August 1, 1989.
"5 Companies Present Bids to Buy Rancho Seco Nuclear Plant." *Los Angeles Times*, July 1, 1989.
"Golden State." *The Nation*, June 26, 1989.
"Utility Seeking Sale of Plant." *The New York Times*, June 22, 1989.
"Seco Boss Quits as Contractors Get Pink Slips." *Sacramento Bee*, June 9, 1989.
"Rancho Seco Nuclear Plant Shutdown Begins." *Los Angeles Times*, June 8, 1989.
"Vote Fallout—Seco Begins Shutdown." *Sacramento Bee*, June 8, 1989.
"Voters Force Shutdown of Nuclear Power Plant." *The Washington Post*, June 8, 1989.
"Voters, in a First, Shut Down Nuclear Reactor." *The New York Times*, June 8, 1989.
"Voters Reject Rancho Seco Nuclear Plant." *Los Angeles Times*, June 7, 1989.
"Rancho Seco A-Plant Facing a Live-or-Die Decision From Voters." *Los Angeles Times*, June 4, 1989.
"South State Behemoth Seeks Role with Seco." *Sacramento Bee*, April 19, 1989.
"Seco Bidders Want Bigger Stake; Foes Cry Foul." *Sacramento Bee*, March 31, 1989.
"NRC Gives Seco Good Grades in New Report." *Sacramento Bee*, March 30, 1989.
"NRC to Fine Rancho Seco Nuclear Plant $100,000." *Los Angeles Times*, January 18, 1989.
"Calif. Firm Lowballs Duke Bid to Run Rancho Seco." *The Business Journal-Charlotte*, August 8, 1988.
"Nuclear-Plant Vote Worries Utilities." *The New York Times*, June 6, 1988.
Sacrificing Safety: A Profile of the Safety Records of Rancho Seco and Duke Power's Nuclear Reactors. Public Citizen, May 1988.
"Duke Power Gets Nuclear Plant Job." *The New York Times*, April 30, 1988.
"Headed for the Last Shutdown." *Sierra*, September-October 1987.
"Duke Power Says it Could Revive Calif. Nuke Plant." *The Business Journal-Charlotte*, September 14, 1987.
"Pacific Gas Chairman Sees Growth Prospect." *The New York Times*, September 9, 1987.
"SMUD Slapped with $300 Million Asbestos Lawsuit." *The Business Journal-Sacramento*, June 22, 1987.
"America's Big Risk." *Newsweek*, April 27, 1987.
"SMUD Hunts for Seco Partners." *The Business Journal-Sacramento*, February 2, 1987.
"Rancho Seco Reactor Suffers Another Mishap." *Science*, January 24, 1986.
"A Reporter at Large: Three Mile Island II-The Paper Trail." *The New Yorker*, April 13, 1981.
"America's Big Risk." *Newsweek*, April 27, 1987.

Plant Profiles River 343

"SMUD Hunts for Seco Partners." *The Business Journal-Sacramento,* February 2, 1987.

—Sharon Miscavage

343 •
River Bend Station, Unit 1

PO Box 220
St. Francisville, LA 70775
J.C. Deddens, Sr.V.Pres., Nuclear

Phone: (504)635-6094
Fax: (504)381-4872

Owning Utility

Gulf States Utilities Company
350 Pine St.
PO Box 2951
Beaumont, TX 77704
J.E. Booker, Mgr., Nuclear Industry Relations

Phone: (409)838-6631
Fax: (409)839-3077
Telex: 779312

Contact

W.L. Benedetto, Pub.Info.
River Bend Station

Basic Facts

Unit 1 *Status:* Operable. *Type of Reactor:* Boiling water reactor. *Megawatts (electric):* 991. *Megawatts (thermal):* 2894. *Reactor System Supplier:* General Electric Co. (United States). *Generator Supplier:* General Electric Co. (United States). *Architect Engineer:* Stone & Webster Engineering Corp. (United States). *Owner:* Gulf States Utilities Company; Cajun Electric Power Cooperative. *Operator:* Gulf States Utilities Company.

Key Dates

Unit 1 *Ordered:* 1972. *Construction began:* 1979. *Criticality:* October 1985. *First Power:* December 1985. *Commercial Operation:* June 1986.

Operating Costs

Unit 1
1986 $62,676,000
1987 $120,891,000
1988 $109,374,000
1989 $127,640,000

Plant Report

The site. River Bend Station occupies a 3,300-acre site on the Mississippi River, south of St. Francisville, Louisiana. The site is about 29 miles northwest of Baton Rouge.

Ownership. The operating utility for River Bend is the Gulf States Utilities Company (GSU). GSU owns 70 percent of the plant, and the Cajun Electric Power Cooperative, which consists of 13 member companies, owns the remaining 30 percent.

Construction delays. In August 1971, GSU announced its plans to build River Bend, its first nuclear plant. The original commercial operation date was to be October 1979. In 1973, the date was changed to March 1980, mainly due to constantly changing regulations.

In September 1975, a Limited Work Authorization was approved for River Bend, and site development began. In May 1975, the commercial operation date was changed to November 1981, due to changes in the licensing of nuclear plants and a better forecast of the personnel work involved in construction.

By January 1976, site development was 11 percent complete. In September, a moratorium was placed on nuclear plant construction permits, which caused GSU to seek an extension of its Limited Work Authorization. When the extension was granted in November 1976, site development was 64 percent complete.

In December 1976 and June 1977, site development work was delayed to update existing drawings to include new federal safety regulations and vendor equipment design information.

On March 25, 1977, the construction permit was finally issued for River Bend. In May, the commercial operation date was changed to October 1983, after more regulatory and construction delays. For the next two years, engineering and construction were held to a minimum because of financial restraints.

In June 1979, an updated, 50-month construction schedule was instated which showed the new commercial operation date to be May 1984. The schedule was based on the Nuclear Power Construction Stabilization Agreement. River Bend was the first plant to be built under the agreement, which was created in 1979, to provide innovative procedures for the building of nuclear plants.

River Bend actually began commercial operation in June 1986. Construction required 53.6 million manhours by over 4,000 workers. The total payroll was $712 million.

About one-third of the total cost of River Bend was for the expense of money borrowed for construction. Another 46 percent of added costs were attributed to regulatory changes.

Financial troubles. Due to increased construction costs, and perhaps some unwise spending habits by GSU, the utility had found itself unable to cover $4 billion in debt and preferred stock.

In 1979, when construction of the GE Mark III reactor began, GSU was under the impression that it's major customers were going to require additional generated power in the near future. Also, in 1978, President Carter signed the Fuel Use Act which stated that by 1990, all utilities would have to stop using natural gas for the production of electricity. Since all GSU's plants were gas powered, it chose to make the needed changes as soon as possible.

GSU blamed it's financial troubles on the oil and gas markets which had bottomed out. The Gulf Coast, which relied heavily on these commodities, was hit by hard economic times. GSU's $4 billion investment in River Bend became the scapegoat of the utility's ailments. Once the nuclear plant had been completed, GSU found itself strapped for money. In order to cover cost overruns, GSU proposed to the Louisiana and Texas regulators, which has jurisdiction over the area served by River Bend, to increase rates paid by consumers. Specifically, the utility was looking for a 17 percent rate hike in Texas, and a 39 percent hike in Louisiana.

● Operable   ○ Under Construction   △ Indefinitely Deferred   ▲ Decommissioning   □ Planned   ■ On Order   ☆ Cancelled   ★ Shut Down   **p. 431**

Regional recovery plan. In what became known as the Regional Recovery Plan, GSU filed an action with the Public Utility Commission of Texas (PUCT) in March 1989 which asked for a $67.5 million net first-year increase in rates. The proposal also included the consideration of the fuel refund and a cost reconciliation study on part of GSU's investment in River Bend.

In the major portion of the plan, GSU proposed to place $1.4 billion of it's River Bend investment, which had been placed in abeyance for further examination by the PUCT the previous Spring, into inventory in order to avoid a big rate increase. The PUCT would then decide when GSU could bring the plant into the rate base—hopefully after depreciation had lowered the book value of the plant so as to avoid an additional rate impact.

Prudent or imprudent. GSU hired it's own consulting firm, Sandlin Associates, in order to justify the cost overruns in River Bend. The company found that 82 percent of the cost increase was the result of additional nuclear safety requirements which were beyond GSU's control. However, in May 1989, the PUCT in a 2-1 vote denied the appeal on grounds that the prudency of the $1.4 billion was decided in the last GSU rate case of 1988. Compounding GSU's worries, a Texas district judge stated that the PUCT would be barred from considering the additional costs of River Bend in the current rate case.

Problems compound for GSU. In July 1990, due to a delay in a decision on the Regional Recovery Plan, which by this time was under consideration by the Texas Supreme Court, the GSU filed a motion with the PUCT to ask for a temporary $56.5 million rate increase. At about the same time, the GSU was also having problems in attempting to add the imprudently spent $1.4 billion into the rate base in Louisiana. In April 1991, in a five to two decision, the Louisiana Supreme Court upheld the 1987 Louisiana Public Service Commission (LPSC) order that the additional cost was imprudently incurred. The high court stated that a lower-cost lignite plant should have been built at River Bend.

The Louisiana Supreme Court suggested that the LPSC could implement a "deregulated asset plan" so that the financial impact of the imprudent spending could be minimized for GSU customers. The high court also chastised a state district court for ordering the LPSC to implement this type of plan. The Supreme Court did help the GSU by rejecting the Louisiana attorney general's assertions that almost all of the investment in River Bend be disallowed. It also sympathized with GSU by agreeing that it did not receive due process from the LPSC. As of the end of 1991, Gulf States Utilities was planning to take its case to the U.S. Supreme Court.

The uncertain future of Gulf States Utilities. Although GSU has been able to obtain numerous interim rate hikes in both Texas and Louisiana, the utility is running out of time. At one point, it considered filing bankruptcy. By hiring a bankruptcy attorney, GSU planned to be prepared for all possibilities. However, the utility has said repeatedly that it doesn't want to declare Chapter 11. In a 1986 interview, Louisiana Attorney General William J. Guste, Jr. said that he felt that bankruptcy would be the best thing for both GSU and its customers, because shock treatment would be better than dealing with GSU's financial difficulties for decades.

Internal GE report accidently released. In June 1987, an internal report, prepared by engineers from General Electric in 1975, was accidently released. The "Reed Report" dealt with the company's Mark III reactors, one of which is used at River Bend, and raised serious safety and design questions. The uncertainties about the Mark III and the Mark II reactors were so profound that the engineers advised the company to stop selling them. Safety concerns cited in the report included metal deterioration, inadequate cooling systems, earthquake hazards, and potential radiation dangers to workers.

Since 1976, the U.S. Congress had been trying to obtain a copy of the controversial report to investigate the allegations. Up to 1987, all attempts to view the report were unsuccessful. According to General Electric and the NRC, the issues surrounding the reactors had been adressed and resolved.

Public information. The River Bend Energy Center is located within the training center at the entrance to the plant. The center offers mechanical demonstrations, audio/visual presentations and graphic displays about nuclear power.

The center is open Monday through Friday, from 8 a.m. to 4 p.m., to individuals, families, groups and organizations. More information can be obtained by calling the center during regular business hours at: (504) 635-5004.

Sources

River Bend. The Gulf States Utilities Company, n.d.
The River Bend Energy Center. The Gulf States Utilities Company, n.d.
"GSU: Bankruptcy Is Not A Solution." *Baton Rouge Business Report*, August, 1986.
"Time May Be Running Out For Gulf States Utilities." *Business Week*, November 17, 1986.
"GE Questioned Safety of Its Own Reactors." *New Orleans City Business*, June 22, 1987.
"Gulf States Utilities Files Rate Case Proposal." *Business Wire*, March 21, 1989.
"Gulf States Utilities Co. Said Utility Commission Decision Helps." *Business Wire*, May 11, 1989.
"Gulf States Utilities Co., Files Motion With Public Utility Commission of Texas." *Business Wire*, July 23, 1990.
"Gulf States Utilities Co. Supreme Court Decision." *Business Wire*, April 8, 1991.
River Bend Station News Media Manual, The Gulf States Utilities Company, May 1991.
"Gulf States Utilities Co.: U.S. Supreme Court Appeal Planned on Louisiana Ruling." *The Wall Street Journal*, June 24, 1991.
"Gulf States Seeks High Court Relief." *Nuclear Industry*, Third Quarter 1991.

—*Yvette M. Costa, Charles Ceplecha*

River Bend Station, Unit 2

PO Box 220
St. Francisville, LA 70775
J.C. Deddens, Sr.V.Pres., Nuclear

Phone: (504)635-6094
Fax: (504)381-4872

Plant Profiles   Robinson 345

Owning Utility

Gulf States Utilities Company   Phone: (409)838-6631
350 Pine St.   Fax: (409)839-3077
PO Box 2951   Telex: 779312
Beaumont, TX 77704
J.E. Booker, Mgr., Nuclear Industry
   Relations

Contact

W.L. Benedetto, Pub.Info.
River Bend Station

Basic Facts

Unit 2 *Status:* Cancelled. *Type of Reactor:* Boiling water reactor. *Megawatts (electric):* 991. *Megawatts (thermal):* 2894. *Reactor System Supplier:* General Electric Co. (United States). *Generator Supplier:* General Electric Co. (United States). *Architect Engineer:* Stone & Webster Engineering Corp. (United States). *Owner:* Gulf States Utilities Company; Cajun Electric Power Cooperative. *Operator:* Gulf States Utilities Company.

Key Dates

Unit 2 *Ordered:* 1972. *Cancelled:* 1984. *Construction began:* 1975.

Plant Report

See Plant Report for River Bend Station, Unit 1.

---

345 ●

H.B. Robinson Electric Power Plant, Unit 2

PO Box 790   Phone: (803)383-4524
Hartsville, SC 29550   Fax: (803)383-1319
C.R. Dietz, Mgr., Robinson Nuclear
   Project

Owning Utility

Carolina Power and Light Company   Phone: (919)546-6111
PO Box 1551   Fax: (919)546-7678
Raleigh, NC 27602
R.A. Watson, Sr.V.Pres., Nuclear
   Generation

Contact

R.J. White, V.Pres., Corp. Commun.
Carolina Power and Light

Basic Facts

Unit 2 *Status:* Operable. *Type of Reactor:* Pressurized light-water reactor. *Megawatts (electric):* 769. *Megawatts (thermal):* 2300. *Reactor System Supplier:* Westinghouse Electric Corp. (United States). *Generator Supplier:* Westinghouse Electric Corp. *Architect Engineer:* Ebasco. *Owner:* Carolina Power and Light; North Carolina Municipal Power Authority. *Operator:* Carolina Power and Light.

Key Dates

Unit 2 *Ordered:* 1966. *Construction began:* 1967. *Criticality:* September 1970. *First Power:* September 1970. *Commercial Operation:* March 1971. *Shut Down Expected:* April 2007.

Operating Costs
Unit 2

1971	$1,918,000	1981	$21,789,000
1972	$1,780,000	1982	$43,164,000
1973	$4,609,000	1983	$38,176,000
1974	$4,780,000	1984	$66,077,000
1975	$6,360,000	1985	$42,533,000
1976	$5,903,000	1986	$57,657,000
1977	$6,859,000	1987	$52,657,000
1978	$14,355,000	1988	$62,224,000
1979	$15,142,000	1989	$52,782,000
1980	$22,086,000		

Plant Report

Unusual combination plant. The Robinson Electric Power Plant, named for H. Burton Robinson, a native of South Carolina and retired executive vice president of Carolina Power and Light (CP&L), is a double unit generating station. Robinson 1, a coal-fired unit, began operation in May 1960. It uses 21 rail cars of coal per day when operating at full capacity. Robinson 2, 665 megawatt pressurized water reactor, began commercial operation in March 1971. Together the two units have a total generating capacity of 854 megawatts. The Robinson station is one of only a few generating plants in the United States with two separate operations using different fuels.

The Robinson plant, located in Darlington County, South Carolina, is owned and operated by CP&L. CP&L, founded in 1908, provides electricity to 986,000 customers in parts of North and South Carolina. Its service area covers 30,000 square miles. To meet customer demands, the utility owns an interest in 16 power plants including fossil, nuclear, and hydroelectric plants. CP&L's total generating capacity is 9,613 megawatts. Robinson 2 was CP&L's first nuclear unit.

Fuel information. Robinson 2 uses uranium dioxide fuel pellets and requires one delivery of nuclear fuel per year. The individual fuel pellets are approximately as thick as a pencil and one-half inch long. The reactor core contains 90 tons of pellets. The pellets are configured into fuel rods and the fuel rods are bundled together into fuel assemblies. At Robinson 2, there are 204 fuel rods per assembly and 157 assemblies in the reactor. One third of the fuel must be replaced annually. Used fuel is stored on site in deep pools of water.

Water supply. Water is vital to the operation of the plant. An impoundment was created to supply water for necessary cooling functions. The impoundment, behind a 5,000-foot dam, covers 2,250 acres. It is 7 1/2 miles long and has 25 miles of shore line. Warm water from the plant is circulated back to the impoundment through a 4 1/2-mile long canal. As the water recycles, heat dissipates into the air.

Steam generators and cost questions. In 1984 the steam generators at Robinson 2 had to be replaced because of tube degradation. The Hartsville Group, intervenors contesting CP&L's request to replace the steam generators, claimed that South Carolina's electric customers would save $50 million over fifteen years if the plant were shut down instead of repaired. The generators were replaced and CP&L filed a lawsuit against Westinghouse, the supplier. Utility officials claimed that Robinson's nuclear unit has saved its customers $780 million in fuel costs during the plant's first twenty years in operation.

Service and decommissioning. Robinson 2 was taken out of service for four months in 1989. Workers replaced suction piping in the auxiliary feedwater system and performed vital maintenance tasks. During the outage, other necessary modifications were made.

In 1990 the NRC issued regulations requiring utilities operating nuclear power units to set aside funds for eventual decommissioning. CP&L established an external decommissioning trust to set aside funds which will be collected through electric rates. The utility plans to decommission the plant using the prompt dismantlement method. The project is expected to cost $143.4 million in 1987 dollars.

Visitor's Center. CP&L operates a visitor's center at the Robinson plant. The center, which underwent complete renovation in 1992, is open 8:30 a.m. to 4:30 p.m. Monday through Friday, except holidays. Contact: H.B. Robinson Information Center at (803) 332-2633.

Sources

"Grassroots." *Critical Mass Bulletin*, March 1984.
Robinson Electric Power Plant, Carolina Power and Light Company, February 1985.
"Visit the Information Center: H. B. Robinson Electric Power Plant." Carolina Power and Light Company, November 1990.
Carolina Power & Light Company 1990 Annual Report. Carolina Power and Light Co., n.d.
"Nuclear Notes." *Info*, July 1991.
Carolina Power & Light Company 1991 Annual Report. Carolina Power and Light Co., n.d.

—Karen Bellenir

St. Lucie Nuclear Power Station, Units 1-2

PO Box 128 Phone: (407)465-3550
Fort Pierce, FL 33454 Fax: (407)468-4119
G.J. Boissy, Plant Mgr.

Owning Utility

Florida Power and Light Company Phone: (407)694-4000
700 Universe Blvd. Fax: (407)694-4311
PO Box 14000
Juno Beach, FL 33408-0420
J.H. Goldberg, Pres., Nuclear Div.

Contact

R.R. Golden, Nuc. Corp. Commun. Coord.
Florida Power and Light Company

Basic Facts

Units 1-2 *Status:* Operable. *Type of Reactor:* Pressurized light-water reactor. *Megawatts (electric):* 872. *Megawatts (thermal):* 2700. *Reactor System Supplier:* Combustion Engineering, Inc. (United States). *Generator Supplier:* Westinghouse Electric Corp. (United States). *Architect Engineer:* Ebasco. *Owner:* Florida Power and Light Company; Florida Municipal Power Agency; Orlando Utilities Commission. *Operator:* Florida Power and Light Company.

Key Dates

Unit 1 *Ordered:* 1968. *Construction began:* 1969. *Criticality:* April 1976. *First Power:* May 1976. *Commercial Operation:* December 1976. *Shut Down Expected:* July 2010.
Unit 2 *Ordered:* 1972. *Construction began:* 1977. *Criticality:* June 1983. *First Power:* June 1983. *Commercial Operation:* August 1983. *Shut Down Expected:* April 2023.

Operating Costs

Units 1-2

1976	$3,249,000	1983	$36,039,000
1977	$7,528,000	1984	$64,700,000
1978	$15,814,000	1985	$70,289,000
1979	$14,392,000	1986	$73,385,000
1980	$16,380,000	1987	$92,578,000
1981	$23,240,000	1988	$84,281,000
1982	$21,854,000	1989	$82,881,000

Plant Report

Background. The St. Lucie Nuclear Power Station, located on Hutchinson island near West Palm Beach, Florida, is a two-unit operation. Florida Power & Light Co. (FP&L), which operates both plants, has an 85 percent ownership stake in St. Lucie. The remaining 15 percent is owned by the Florida Municipal Power Agency and the Orlando Utilities Commission.

Failed evacuation attempt. St. Lucie encountered problems during its first full-scale evacuation test for handling a nuclear disaster. The three-day exercise, which began on March 9, 1984, had developed so many problems that the plan was considered useless. Problems encountered included: 1) the wind pattern didn't fit the plan; 2) sirens couldn't be heard within buildings; 3) rescue helicopters were flying outside the exercise zone; and 4) only 21 of 50 Red Cross volunteers showed up to assist "victims."

Florida Power's shining light. Both reactors have ranked among the top performers in terms of capacity utilization at U.S. nuclear power plants. For 1989, St. Lucie 1 finished in the top 25 by operating at 95.89 percent of capacity. In 1981, it operated at 80.2 percent of capacity. St. Lucie 2 achieved an even higher honor by operating at 101.33 percent of capacity in 1991. It also set a U.S. record for operating 488 consecutive days. It achieved this milestone on April 6, 1991 and the scheduled shutdown wasn't until April 29, 1992.

World's most efficient plant. This record run for St. Lucie 2 earned the plant the distinction of the world's most efficient nuclear plant for 1991. Any run that is longer than 365 consecutive days is considered exceptional by industry standards. The 1991 goal was a culmination of five years of better than aver-

Plant Profiles

age operations at St. Lucie 2. During the five years, the plant operated at greater than 70 percent while also turning in below-average costs. In 1989, St. Lucie 2 operated for 427 consecutive days, a run still ranked among the top 20 U.S. nuclear plants with the longest consecutive runs.

Sources

"Grassroots." *Critical Mass Bulletin*, April 1984.
"The Eureka File." *Nuclear Industry*, First Quarter 1990.
"Nuclear Economics." *Tampa Tribune*, June 10, 1991.
"Another Rave Review for 1990." *Nuclear Industry*, Fourth Quarter 1991.
"FP&L's St. Lucie 2 Operates at 100 Percent." *INFO*, U.S. Council for Energy Awareness, February 1992.
"U.S. Nuclear Plants Shatter Records for Output, Performance." *INFO*, U.S. Council for Energy Awareness, February 1992.
"St. Lucie 2 Completes Record Run, Leads World in Efficiency." *INFO*, U.S. Council for Energy Awareness, February 1992.

—Yvette M. Costa

347 ☆
St. Rosalie Nuclear Power Plant, Units 1-2

Alliance, LA

Basic Facts

Units 1-2 *Status:* Cancelled. *Type of Reactor:* Gas-cooled reactor. *Megawatts (electric):* 1300. *Reactor System Supplier:* General Atomic. *Owner:* Louisiana Power and Light. *Operator:* Lousiiana Power and Light.

Key Dates

Units 1-2 *Ordered:* 1974. *Cancelled:* 1975.

348 ★
Fort St. Vrain Nuclear Power Plant

16805 WCR 19-1/2   Phone: (303)785-6471
Platteville, CO 80651
Don W. Warembourg, Nuclear Opns. Mgr./Station Mgr.

Owning Utility

Public Service Company of Colorado   Phone: (303)571-7511
1224 17th St.   Fax: (303)294-8120
Denver, CO 80202
Delwin D. Hock, Chairman, Pres., & CEO

Contact

Mark E. Severts, Dir., Media Relations
Public Service Company of Colorado

Basic Facts

Status: Shut down. *Type of Reactor:* High temperature gas reactor. *Megawatts (electric):* 342. *Megawatts (thermal):* 842. *Reactor System Supplier:* General Atomic, later GA Technologies. *Generator Supplier:* General Electric Co. (United States).

Key Dates

Ordered: 1965. *Construction began:* 1968. *Criticality:* January 1974. *First Power:* December 1976. *Commercial Operation:* July 1979. *Shut Down:* August 1989.

Plant Report

Overview. Construction began in 1968 on a nuclear power generating station at Fort St. Vrain in Colorado, about 35 miles north of Denver. The reactor became operational in 1976 as the only commercially operated helium-cooled nuclear reactor in the United States. But after only thirteen years of operation, Fort St. Vrain was permanently shut down by its owners, Public Service Company (PSC) of Colorado, due to the plant's egregious operating record. The plant was inoperative 85 percent of the time primarily due to difficulties with the reactor's support system. Although the plant had the best safety record in the industry and its builders continue to assert that high-temperature gas-cooled reactors are the wave of the future, Fort St. Vrain remains a one-of-a-kind nuclear plant in the United States.

One of the most significant factors contributing to the demise of the nuclear power facility at Fort St. Vrain is the fact that it was the demonstration case for high temperature gas-cooled reactors in the U.S. The engineers at St. Vrain were forced to learn as they went along, through trial and error setting precedents for this type of nuclear power plant operation. Announcing the permanent shutdown of nuclear operations at Fort St. Vrain, a spokesperson for PSC remarked: "We pioneered new technology. We feel good that it's a technology that will work in the future. But because so many other unrelated systems had problems it has not been cost-effective for us, and it was a disappointment to Public Service Co."

Paying for Fort St. Vrain. Although Fort St. Vrain cost nearly $1 billion in research and contruction, Public Service Company of Colorado assumed only about $200 million of that cost, including millions spent on repairs after the plant came on line. General Atomic (GA), later known as GA Technologies, a subsidiary of Chevron Oil, devised the technology for the nation's first Gas-Cooled Reactor in the 1960s, and along with the federal government, absorbed the majority of the costs associated with building the reactor at St. Vrain. Construction of the plant was completed in 1976 and St. Vrain went into commercial operation in 1979 despite continuing uncertainty that the reactor could operate at its 330-megawatt capacity. The design firm returned $180 million in cash and services to PSC in 1979 and reimbursed the plant's owners for additional construction on the plant shortly after it came on-line.

As is usual with utility companies, PSC had boosted its rates to customers to offset expenditures on Fort St. Vrain. But ratepayers felt that they had been forced to pay twice for the cost of the reactor through federal taxes and raised rates for their electricity. The Colorado Public Utilities Commission eventually agreed, ordering PSC to offer rebates and reduced rates in the amount of $130 million to its customers. In 1986, PSC agreed to remove

Fort St. Vrain from its rate base, used to calculate electric bills to its customers.

Fort St. Vrain similar to Chernobyl. The nuclear reactor at Fort St. Vrain is a rare helium-cooled unit encased in graphite, somewhat similar to the reactor that caught fire at Chernobyl in the Soviet Union in 1986. The significant difference between the Chernobyl plant and the one at Fort St. Vrain is that the latter utilized helium rather than water as a coolant. Because helium does not absorb radioactivity, the possibility of radiation contamination to workers is significantly lower than in conventional nuclear plants. The graphite encasing which caught fire at Chernobyl would give Fort St. Vrain engineers several hours to solve problems before the core was in danger of meltdown; in water-cooled reactors it only takes minutes for coolant to boil away and core meltdown to begin. While St. Vrain's designers insist that their design prevented the instability that doomed the Chernobyl facility, the American plant suffered a number of other design and management related difficulties that frequently brought electrical production to a halt.

Count down to shut down. In 1988, PSC announced its plans to shut down the troubled Fort St. Vrain plant when the extant fuel supply ran out in 1991. The plant was under repair at the time and start-up was scheduled for the fall of 1988. The reactor was still inoperative that December, however, and the start-up date was pushed back to January 1989 while employees removed water from the reactor vessel (a recurring problem at Fort St. Vrain). The date for permament shutdown was moved up to June 1990 and PSC developed a reserve fund in the amount of $160 million to offset the cost of decommissioning.

During routine testing as a prelude to start-up in January, 1989, a back-up shutdown system was accidentally activated, causing a further five-month delay while the system was restored. In August 1989, Fort St. Vrain was shut down again, this time due to problems with the control rod in the reactor. During repairs, inspectors found 37 hairline cracks in the pipe system that carried steam to the generator. In the face of further costly repairs, PSC decided to give up the possibility of further nuclear operations at Fort St. Vrain, and announced the immediate shutdown of the reactor. Ironically, the August 1989 decision to close the plant came on the heels of a rare three-month stretch of continuous operations, including the best month for electrical generation in the plant's history—178,000 megawatt hours net generation.

Disposing of nuclear waste. Dismantling Fort St. Vrain will yield an estimated 140,000 cubic feet of low-level radioactive waste, which will be shipped to a site in Washington state for shallow burial. Disposal of the reactor's high-level waste, while much smaller in bulk, has proved to be more difficult. When PSC shut down the reactor, it intended to transfer the highly radioactive spent fuel to the Department of Energy, which would place this material in permanent storage at its facility in Idaho. However, the state's governor, Cecil Andrus, has consistently refused to allow the radioactive waste into the state, stalling the decommissioning process throughout 1990. In February 1991, PSC filed suit against Andrus, charging that only the federal government has the authority to dispose of hazardous waste, thus the ban interferes with a contract between the government and PSC. The issue had not yet been settled in the federal courts by June 1992.

Faced with the possibility of an indefinite delay in decommissioning Fort St. Vrain as a result of the ban, PSC applied to the NRC for permission to build a temporary spent fuel storage facility on the plant site, at an estimated cost of $47 million. The Independent Spent Fuel Storage Installation (ISFSI) is PSC's solution to its high-level nuclear waste disposal problem. Unlike most nuclear facilities in the U.S., which store spent fuel in pools of water, PSC has devised a unique low-maintenance method of keeping high-level waste cool by natural air convection. Every twelve hours, six fuel elements are plucked from St. Vrain's reactor core and deposited into a concrete vault in the ISFSI. Air enters the ISFSI at ground level, circulates around fuel encased in carbon steel containers, and exits through the chimney. The NRC has certified such dry-cask systems as being safe for up to 100 years.

Decommissioning Fort St. Vrain. In 1990, PSC began reading proposals from engineering companies who responded to the call for bids on decommissioning and converting Fort St. Vrain from nuclear to gas-fueled power generation. Fort St. Vrain is considered ideal for conversion to gas-fired power generation because its support systems are equipped to handle the high temperatures and pressures associated with conventional power generation; water-cooled nuclear reactors operate at much lower temperatures and therefore require more extensive retro-fitting for conversion to conventional power generation.

Conversion to gas-fired generation is a three-fold operation: defueling the plant, which entails removing the spent radioactive material from the containment core (estimated cost at Fort St. Vrain, $116 million); decommissioning the plant, which entails disassembling irradiated components of the plant (estimated cost $80 million); and converting the fuel system to natural gas (estimated cost $50 million). By 1990, PSC had reserved $233 million against the total cost of the decommissioning process, and announced a sale of bonds amounting to another $200 million to offset further costs associated with disposing of the nuclear reactor at Fort St. Vrain and bringing the plant back on-line for conventional power generation.

The bid went to three companies who joined forces after negotiation with PSC: Westinghouse Electric Corporation, Morrison Knudsen Corporation, and Black & Veatch. The $350 million project called for the formation of a new coporate entity which would own the rejuvenated Fort St. Vrain. The natural-gas fired facility was projected to go on-line in 1995 with 350 megawatts of generating power as an independent power producer under the purview of the Federal Energy Regulatory Commission (rather than the state public utilities commission). It was assumed that PSC, whose total share in the new company was $63 million, equalling the net worth of its assets in Fort St. Vrain, would be one of the new company's largest customers.

Planning and engineering work began at Fort St. Vrain in 1991, with actual decommissioning scheduled to begin by mid-1992, when all concerned hoped that the disposition of high-level waste at Fort St. Vrain would be resolved. By 1992, plans for the new generating plant called for a 350-megawatt natural gas-burning facility with 35 acres of photovoltaic cells for generating an additional 14 megawatts of solar power to come on-line by 1998.

Sources

"PSC Officials Believe St. Vran 'Over the Hump.'" *Rocky Mountain Business Journal*, July 22, 1985.
"St. Vrain Shutdown Cost Hiked." *Denver Post*, September 28, 1988.
"Public Service Co. of Colorado Announces Accelerated Plans to Discontinue Nuclear Operations." *Business Wire*, Decmeber 5, 1988.
"PSC Will End Nuclear Use at St. Vrain Plant." *Denver Post*, December 6, 1988.
"Safest Reactor Is Closing Because It Rarely Runs." *The New York Times*, December 8, 1988.
"Public Service Co. of Colorado Financial Results." *Business Wire*, January 30, 1989.
"Public Service Co. of Colorado Ends Fort St. Vrain Operations Earlier than Anticipated." *Business Wire*, August 29, 1989.
"PSC Gives Up Battle, Closes Out St. Vrain." *Denver Post*, August 30, 1989.
"PSCo. Studies Bids to Convert St. Vrain in $250 Million Job." *The Denver Business Journal*, May 14, 1990.
"PSC Seeks Partners in St. Vrain Plan." *Denver Post*, July 6, 1990.
"Public Service Co. of Colorado." *The Wall Street Journal*, July 11, 1990.
"Fort St. Vrain." *NRC Annual Report*, U.S. Nuclear Regulatory Commission, 1990
"PS Colorado Files Suit." *The Wall Street Journal*, February 11, 1991.
"Westinghouse Electric Corp.: Company, Morrison Knudsen Get Colorado Reactor Job." *The Wall Street Journal*, May 7, 1991.
"Westinghouse Takes Lead in Dismantling Nuclear Plants." *Pittsburgh Business Times & Journal*, May 20, 1991.
"One More Renewable Energy Source: Nuke Plants." *Business Week*, September 16, 1991.
"Dismantling Nuclear Reactors." *Garbage: The Practical Journal for the Environment*, March/April 1992.
"A Mountain and a Mission." *Garbage: The Practical Journal for the Environment*, May/June 1992.
"Westinghouse Takes Lead in Dismantling Nuclear Plants." *Pittsburgh Business Times & Journal*, May 20, 1991.
"One More Renewable Energy Source: Nuke Plants." *Business Week*, September 16, 1991.

—Mary Gillis

Salem Nuclear Generating Station, Units 1-2

PO Box 236
Hancocks Bridge, NJ 08038
Calvin Vondra, Gen.Mgr., Salem Operations

Phone: (609)935-6000
Fax: (609)339-2956

Owning Utility

Public Service Electric and Gas Company
80 Park Plaza
Newark, NJ 07101
Steven E. Miltenberger, V.Pres. & Chief Nuclear Officer

Phone: (201)430-7000

Contact

Michaele L. Camp, Mgr., Nuclear Commun.
Salem Nuclear Generating Station

Basic Facts

Units 1-2 *Status:* Operable. *Type of Reactor:* Pressurized light-water reactor. *Megawatts (electric):* 1132 (Unit 1); 1158 (Unit 2). *Megawatts (thermal):* 3411. *Reactor System Supplier:* Westinghouse Electric Corp. (United States). *Generator Supplier:* Westinghouse Electric Corp. (Unit 1); General Electric Co. (United States) (Unit 2). *Architect Engineer:* Public Service Electric and Gas Company. *Owner:* Public Service Electric and Gas Company; Atlantic City Electric; Delmarva Power and Light; Philadelphia Electric Company. *Operator:* Public Service Electric and Gas Company.

Key Dates

Unit 1 *Ordered:* 1966. *Construction began:* 1968. *Criticality:* December 1976. *First Power:* December 1976. *Commercial Operation:* June 1977. *Shut Down Expected:* September 2008.

Unit 2 *Ordered:* 1966. *Construction began:* 1968. *Criticality:* August 1980. *First Power:* June 1981. *Commercial Operation:* October 1981. *Shut Down Expected:* September 2008.

Operating Costs

Units 1-2

1977	$12,708,000	1983	$175,553,000
1978	$22,311,000	1984	$200,872,000
1979	$47,511,000	1985	$166,838,000
1980	$59,684,000	1986	$174,444,000
1981	$77,500,000	1987	$155,034,000
1982	$156,614,000	1989	$143,557,000

Plant Report

Part of complex power picture. The Salem Nuclear Generating Station is located in southwestern New Jersey on Artificial Island in Lower Alloways Creek. Salem shares the three-mile long island with the Hope Creek Nuclear Generating Station.

Four utilities own Salem's two 1,172 megawatt pressurized water reactors. Public Service Electric & Gas (PSE&G), a New Jersey utility, and Philadelphia Electric are the major partners. Atlantic Electric and Delmarva Power each own 7.41 percent.

Salem's operator, PSE&G, is New Jersey's largest utility. The company supplies electricity to approximately 75 percent of the state's population. Nuclear power and coal-generated power are both used to meet the needs of residential, commercial and industrial customers. The PSE&G expected its diverse generating capability to enable it to provide a stable supply of electricity in spite of disruptions in the fossil fuel market.

Inconsistent plant. Salem, however, has failed to generate a consistent supply of electricity. The plant's output has fluctuated widely. For example, Unit One experienced two separate eight-month outages between 1982 and 1984. Then in 1985, Unit 1 set a record for the yearly gross electricity produced by a single nuclear unit. The record stood for three years.

1985 was also a significant year for Unit 2. It resumed operation in April after being shut down for 19 of the previous 27 months due to generator failures. As a result of many outages, capacity factors at both Salem units have been in the middle 50's, below industry averages. In 1987 the New Jersey Board of Regulatory Commissioners (BRC) set performance standards for nuclear generating stations owned by New Jersey utilities. Under the regulations a utility receives a reward or is assessed a penalty based on its annual capacity factor. In 1991 PSE&G's aggregate capacity factor, 71 percent, was at a level where neither rewards nor penalties applied.

The 1991 capacity factor was achieved in spite of shut downs at both Salem units. An electrical storm in June knocked Unit 1 off line when lightning struck the main power transformer. The Unit was returned to service within a week and performed near peak capacity during a record-setting demand period. Unit 2 shut down automatically in November 1991 when protection devices failed and a fire caused severe damage to the turbine generator. No injuries were sustained during the incident and damage was confined to the non-nuclear portion of the station. PSE&G estimated equipment repairs would cost between $65 and $75 million.

Extraterrestrial interference. For reasons not completely understood, the Salem plants have experienced more than average interference from geomagnetic storms, caused by flare ups on the sun's surface. Geomagnetic radiation resulting from these storms can overload transmission lines. A powerful storm in March 1989 knocked out three transformers at the Salem station. Engineers at the plants compensate by turning down the power when intense storm activity is expected, however early warning systems are only 13 percent accurate. Scientists continue to study the situation. One theory suggests that Salem's problems may stem from the fact that it sits on solid rock and attracts storm particles.

NRC impact. In addition to technical problems and natural phenomena, Salem's units have been affected by NRC regulations and corrections. In 1983 the NRC levied against PSE&G the highest fine it had imposed at the time—$850,000—for problems at Salem 1. The plant's protection system failed on two occasions only days apart. The NRC censured PSE&G for restarting the Salem 1 without full comprehension of the incident.

The NRC also issued new industry-wide regulations following the Three Mile Island accident in 1979. A ruling requiring utilities operating nuclear stations to study "human engineering deficiencies" in control rooms had a significant impact on the Salem station. "Human engineering deficiencies" refer to design faults such as the placement of gauges in hard-to-read locations. In response to the order, PSE&G identified 168 deficiencies. Modifications to the control room for Salem 2 were made during an outage in September 1988; modifications for Unit 1 began during an outage in March 1989.

Environmental concerns. Both the NRC and the New Jersey Department of Environmental Protection enforce regulations to keep harmful effects to the environment at a minimum. PSE&G monitors air and water quality paying close attention to a 20-mile stretch of the Delaware River where the nuclear plants draw and discharge a continuous supply of water.

Salem's water intake system originally caught and killed a large number of fish. In 1981 a traveling screen system consisting of panels that measure two feet by 10 feet with a bucket to catch fish was installed. The new system returns fish to the river and has reduced the fish kill by about 40 percent.

In 1990 New Jersey Department of Environmental Protection and Energy issued an order for the construction of cooling towers at Salem. The agency argued in favor of the towers, citing the need to eliminate any chance of harming aquatic life in the Delaware River. PSE&G opposed the towers because the utility believed less costly alternatives would satisfactorily meet the Department's objectives. The issue is currently unresolved.

PSE&G is proud of its environmental record at Salem. The utility expects public support of nuclear power to increase because of environmental awareness. Officials point out that nuclear generation does not contribute to acid rain or the green house effect.

Sources

Facts About Salem Nuclear Generating Station. Public Service Electric & Gas, n.d.
"NRC Strikes Out." *Critical Mass Bulletin,* September 1984.
"Future Energy Needs Well Within Utilities' Capacity, Experts Note." *Focus,* November 26, 1986.
PSE&G News. PSE&G, April 1, 1989.
"Utilities Worry About Catching Rays." *Boston Business Journal,* December 3, 1990.
Public Service Enterprise Group Incorporated Annual Report for 1991. Public Service Enterprise Group, n.d.

—Karen Bellenir

San Onofre Nuclear Generating Station, Units 1-3

PO Box 128
San Clemente, CA 92674-0128
H.E. Morgan, V.Pres. and Site Mgr.

Phone: (714)368-6291
Telex: 677268

Owning Utility

Southern California Edison Company
2244 Walnut Grove Ave.
Rosemead, CA 91770
Harold B. Ray, Sr.V.Pres., Nuclear Organization

Phone: (818)302-1212
Fax: (818)302-7893
Telex: 677268

Contact

R.W. Kreiger, Stat.Mgr.
San Onofre Nuclear Generating Station

Basic Facts

Unit 1 *Status:* Operable. *Type of Reactor:* Pressurized light-water reactor. *Megawatts (electric):* 456. *Megawatts (thermal):* 1347. *Reactor System Supplier:* Westinghouse Electric Corp. (United States). *Generator Supplier:* Westinghouse Electric Corp. *Architect Engineer:* Bechtel. *Owner:* Southern California Edison; San Diego Gas and Electric Company. *Operator:* Southern California Edison.

Units 2-3 *Status:* Operable. *Type of Reactor:* Pressurized light-water reactor. *Megawatts (electric):* 1127. *Megawatts (thermal):* 3390. *Reactor System Supplier:* Combustion Engineering, Inc. (United States). *Generator Supplier:* General Electric Co. (United Kingdom). *Architect Engi-*

Plant Profiles

neer: Bechtel. *Owner:* Southern California Edison; San Diego Gas and Electric Company. *Operator:* Southern California Edison.

Key Dates

Unit 1 *Ordered:* 1963. *Construction began:* 1964. *Criticality:* 1967. *First Power:* June 1967. *Commercial Operation:* January 1968. *Shut Down Expected:* 2008.

Unit 2 *Ordered:* 1970. *Construction began:* 1974. *Criticality:* 1980. *First Power:* September 1982. *Commercial Operation:* August 1983. *Shut Down Expected:* 2023.

Unit 3 *Ordered:* 1970. *Construction began:* 1974. *Criticality:* 1983. *First Power:* September 1983. *Commercial Operation:* April 1984. *Shut Down Expected:* 2024.

Operating Costs
Units 1-3

1968	$1,481,000	1979	$11,668,000
1969	$1,975,000	1980	$31,089,000
1970	$2,236,000	1981	$24,397,000
1971	$2,412,000	1982	$36,829,000
1972	$3,518,000	1983	$97,044,000
1973	$5,839,000	1984	$290,160,000
1974	$5,559,000	1985	$264,681,000
1975	$8,668,000	1986	$210,382,000
1976	$10,490,000	1987	$199,649,000
1977	$8,123,000	1988	$208,934,000
1978	$14,517,000	1989	$231,308,000

Plant Report

Background. Located 60 miles south of Los Angeles on the coast of California, the San Onofre Nuclear Generating Station is comprised of three pressurized water reactors built by Southern California Edison and San Diego Gas and Electric at a cost of $4.5 billion. The plant occupies federal land inside the Camp Pendleton Marine Corps Base, and is four miles from the earthquake fault known as the Offshore Zone Deformation. In October 1986, the California Public Utilities Commission voted to disallow $344 million of the $4.5 billion cost of building the San Onofre facility. The disallowance reflects costs deemed "excessive" and therefore not eligible to be recuperated by Southern California Edison (which owns 75 percent of the facility) from ratepayers through higher rates. In an unusual move that was viewed as a direct result of new appointments made to the Commission by Republican Governor George Deukmejian, the Commission voted three-to-two in July, 1987 to reduce the disallowance by $79 million, bringing the total to $265 million that San Onofre's owners would have to absorb.

Earthquake protection. Located within four miles of an offshore earthquake fault, the San Onofre Generating Facility has been subject to strict regulations by the NRC regarding the plant's ability to withstand an earthquake. All three reactors at San Onofre are designed to withstand an earthquake emitting from the Offshore Zone Deformation measuring 7.0 on the Richter Scale. Utilities measure ground movement such as that associated with earthquakes in terms of how it compares to the force of gravity; an earthquake occuring at the Offshore Zone Deformation measuring 7.0 on the Richter scale at its epicenter would register .67 G at San Onofre. As an additional safety measure, San Onofre Units 2 and 3 are designed to shut down manually at .33 G and automatically at .48 G. Reactor 1, which is older, had no specific G-level at which it would shut down, either manually or automatically, though in 1987 Southern California Edison proposed that such standards be devised for that reactor as well.

By contrast, the earthquake which occured in October 1987 at Whittier fault in Southern California killiing six people, measured 6.1 on the Richter scale but only .03 G at San Onofre. To date, San Onofre has never shut down or suffered any damage due to earthquake activity, despite its location in quake-prone California. Further, San Onofre has never been subject to the kind of consistent, strong opposition by antinuclear groups that has marked the history of another major nuclear plant, Diablo Canyon, also located in California.

Drugs at San Onofre. In 1988, the drug testing policy in force at the San Onofre Nuclear Generating Station came under attack from two sides. The union at the plant filed a grievance against the company, citing random testing of employees for drug use as a violation of the collective bargaining agreement. On the other side of the issue, the plant's medical director, Harold Leider, claimed that he was fired by Southern California Edison for criticizing the company's lax attitude toward drug use by plant employees.

In 1986, when San Onofre began random drug testing of its employees, union local 246 filed a grievance and obtained a temporary restraining order from a Superior Court judge. The company switched to annual rather than random testing in an attempt to appease the union. In May of 1987, a federal appeal court ruled that unions cannot seek protection from drug testing under the state constitution. By 1988, when Harold Leider filed suit for losing his job as medical director at the plant, the case had not yet been resolved.

At issue in Dr. Leider's case was San Onofre's "fitness for duty" program, which Leider claimed was too permissive in part because the plant's management did not want to upset the union. Leider criticized such practices as giving advance notice of testing, which allowed employees the opportunity to substitute or dilute specimens, and sending home intoxicated employees rather than subjecting them to testing. Leider also contended that he was asked to alter test results, and that in one case, an employee who failed a drug test was still employed in the plant.

Leider claimed that the findings of a 1986 NRC inspection upheld his view that the fitness for duty program at San Onofre was ineffective. Leider's suit quoted the NRC report issued in 1987, which criticized the program because it "permitted too many repeated failures before management concluded that the employee was not trustworthy and should be terminated." A spokesperson for Southern California Edison retorted that the comment had been taken out of context and that the NRC had approved the fitness for duty program in operation at San Onofre.

San Onofre and federal intervention. Documents released under the Freedom of Information Act in the 1980s revealed that an antinuclear demonstration staged at San Onofre in the 1970s illustrates how nuclear power may be an issue that allows the distinction between civilian and military authority to be blurred in the United States. When a peaceful demonstration by antinuclear activists was planned for August 1977 at San Onofre Nuclear Generating Station located on the Camp Pendleton Marine Corps Base, news of it reached the U.S. Marshalls Office and the Justice Department. Utilizing the San Diego Police force in conjunction with the plant's own security force was not deemed sufficient to protect the facility from possible damage caused by the demonstrators. The U.S. Marshalls office agreed to intervene only if they were given authority over other law enforcement officials during the demonstration, but Associate Deputy Attorney General Larry Gibson remarked in a letter to Southern California Edison that "the U.S. Marshall Service simply does not possess the capability to provide either continuing protective service or protection against a major armed assault on the plant." He concluded, "the Marines of Camp Pendleton are the only practical Federal alternative..."

The peaceful demonstration in which more than 1,000 activists participated that August, was attended by U.S. Marshalls, San Diego County deputies, state highway patrol and U.S. Border Patrol officers, and Marine Corps military police. No arrests were made and a spokesperson for the San Diego County sheriff's office told reporters that the demonstrators themselves were "doing all the crowd control for us." Nonetheless, many perceive this type of military intervention in a civilian action a dangerous precedent.

Looking toward the future. In the 1990s, San Onofre Unit 3 was cited as one of the top 10 nuclear producers of electricity in the United States. Southern California Edison's response to President George Bush's plan to streamline the licensing procedures for future nuclear power plants was cautious, however. A spokesperson for the utility remarked that Southern California Edison would wait and see how the Tennessee Valley Authority fared under the revised regulations before planning any new nuclear facilities itself. Growth in the demand for energy in the United States slowed during the 1980s, and a utilities analyst predicted that the surplus capacity would not be depleted until the year 2000.

Sources

"To the Shores of San Onofre." *The Progressive*, January 1984.
"Southern California Edison Co." *The Wall Street Journal*, July 30, 1987.
"Fears of Nuclear Power Are Intensified by Latest California Quake." *The New York Times*, October 11, 1987.
"San Onofre Plant's Drug-Test Policy Put on Trial." *San Diego Business Journal*, September 26, 1988.
"Despite Bush Push, Nuclear Not in L.A.'s Outlook." *The Los Angeles Business Journal*, February 25, 1991.
"Palo Verde Units 3 & 2 Top Producers in United States for January: Unit 3 to Be Refueled Starting March 16." *Business Wire*, March 13, 1991.

—Mary Gillis

351 ★

Santa Susana Sodium-Graphite Reactor Experiment (SRE)

Santa Susana, CA

Owning Utility

Southern California Edison Company
2244 Walnut Grove Ave.
Rosemead, CA 91770
Harold B. Ray, Sr.V.Pres., Nuclear Organization

Phone: (818)302-1212
Fax: (818)302-7893
Telex: 677268

Basic Facts

Status: Shut down. *Type of Reactor:* Sodium Graphite Reactor. *Megawatts (electric):* 8. *Megawatts (thermal):* 30. *Architect Engineer:* Atomics International. *Owner:* U.S. Department of Energy. *Operator:* Southern California Edison Co.

Key Dates

Ordered: 1954. *Criticality:* 1957. *First Power:* July 1957. *Commercial Operation:* August 1957. *Shut Down:* February 1964.

352 ●

Seabrook Station, Unit 1

PO Box 300
Seabrook, NH 03874
William A. DiProfio, Stat.Mgr.

Phone: (603)474-9521
Fax: (603)474-3808

Owning Utility

New Hampshire Yankee
PO Box 300
Seabrook, NH 03874
Bruce L. Drawbridge, Exec. Dir., Nuclear Production

Phone: (603)474-9521
Fax: (603)474-2987

Contact

Bruce Gretter, Mgr., Community Relations
Seabrook Station

Basic Facts

Unit 1 *Status:* Operable. *Type of Reactor:* Pressurized light-water reactor. *Megawatts (electric):* 1194. *Megawatts (thermal):* 3411. *Reactor System Supplier:* Westinghouse Electric Corp. (United States). *Generator Supplier:* General Electric Co. (United States). *Architect Engineer:* United Engineers & Constructors (United States).

Key Dates

Unit 1 *Ordered:* 1972. *Construction began:* 1976. *Criticality:* June 1989. *First Power:* June 1989. *Commercial Operation:* July 1990.

Plant Report

Seabrook overview. The U.S. nuclear industry scored a victory with the opening of Seabrook Station in Seabrook, N.H. The plant and its principal owner, the Public Service of New Hampshire (PSNH), became snarled in a 17-year struggle of countless political tangles, regulatory snags, legal fights and cost overruns before a green light was granted for operation in

1990. A review process resulted in thousands of hours of hearings and millions of pages of documents.

In 1973, PSNH applied for a construction permit with the Atomic Energy Commission for two reactors on marshland in Seabrook at $1 billion. But delays arising from environmental concerns stymied the project and, after years of financial reverses and cost overruns, PSNH scrapped Unit Two following an $800 million investment.

Unit One was completed in 1986 at $4.3 billion, but it was blocked from being licensed by disputes over evacuation procedures led by Massachusetts Gov. Michael Dukakis. Seabrook sits on New Hampshire's southeastern coastline, two miles from the Massachusetts border and 40 miles north of Boston. The Massachusetts government argued that the area around Seabrook, a popular beach resort, could not be evacuated safely during an accident.

However, Seabrook was jumpstarted by the U.S. Federal Government. In December 1988, President Ronald Reagan issued an executive order for the Federal Emergency Management Agency (FEMA) to develop emergency response plans when state or local governments are unable or unwilling to develop their own as in the case with Massachusetts. Thereafter, the Nuclear Regulatory Commission (NRC) authorized startup, low-power, and full-power testing at Seabrook. On March 1, 1990, the NRC licensed the plant, which began full-power commercial operation on August 19 of that year. In 1991, Massachusetts, with newly elected Gov. William Weld, cooperated in emergency planning.

Political impact. The Seabrook controversy evolved into a battle between federal, state, and local governments. Political resistance played a major role in delaying Seabrook's opening.

Former governors Michael Dukakis of Massachusetts and John Sununu of New Hampshire held opposite views toward Seabrook. Dukakis, a Democrat, opposed the plant despite not coming out publicly, making it clear that his fears for public safety and evacuation procedures outweighed the financial consequences of delays. On the contrary, Sununu, a Republican and long-time advocate of nuclear power, was passionately pro-Seabrook.

Edward Kennedy and John Kerry, both Democratic senators from Massachusetts, were longtime foes of Seabrook, as was Massachusetts Attorney General James Shannon. Kerry blocked Senate confirmation of Forrest Remick, a Seabrook supporter, to be a NRC member.

The NRC, as part of the process to grant an operating permit for Seabrook, required New Hampshire and Massachusetts to submit evacuation plans to FEMA in 1986. Despite opposition from several towns, New Hampshire submitted its plan to protect the approximately 140,000 people in a 10-mile radius of the plant and carried out a test. Massachusetts, however, with six towns within the 10-mile zone, refused to submit a plan.

Ironically, Dukakis was prepared to present an evacuation plan long before the Chernobyl disaster in the Soviet Ukraine in 1986. The governor wanted Seabrook closed for the first two summers of operation until fallout shelters on the area's beaches could be built, and he advocated off-site monitoring stations to insure early detection of a radiation release. But after Chernobyl, Dukakis experienced tremendous pressure from Massachusetts and New Hampshire legislators urging him to stop the plant's opening, and he halted all emergency planning.

Induced by the lack of cooperation from Dukakis and New York Gov. Mario Cuomo, who adamantly refused to submit an evacuation plan for the Shoreham reactor on Long Island, New York, President Ronald Reagan in December 1988 issued the executive order for FEMA to develop emergency plans when states are unwilling to do so.

Some believe the road for Seabrook's licensing was cleared when George Bush, a nuclear power advocate, easily defeated Dukakis in the 1988 Presidential election. Bush went on to appoint Sununu as the White House chief of staff, emphasizing the administration's stand on nuclear power.

New Hampshire's Democratic alliance opposed Seabrook until the end; since it was conceived in 1973, the power plant dominated state Democratic politics. Sununu battled the state Democratic Party, which opposed Seabrook from the beginning. Party leaders such as chairman Chris Spirou and House Democratic Leader Mary P. Chambers were prominent in the anti-Seabrook movement. Spirou was quoted by *The Boston Sunday Globe* in August 1991, saying, "Since 1974, we've been fighting Seabrook and its effect on the health and pocketbooks of New Hampshire residents." Edward Shumaker, a lawyer from Concord, New Hampshire, said in *The Globe*, "Seabrook was the Achilles' heel of the Democratic Party."

Despite the Democrats' solidarity against Seabrook in New Hampshire, plant opposition was bipartisan. For instance, former New Hampshire Senator Gordon Humphrey, a conservative Republican, opposed Seabrook. Humphrey ordered a congressional investigation of the NRC for its decision-making with Seabrook.

Regulatory impact. The NRC was the subject of much criticism during the period leading to Seabrook's opening, probably the most grueling licensing battle in U.S. nuclear industry history.

After Seabrook was completed in 1986, the process dragged on until the NRC voted 3-0 to license Seabrook in March 1990. But the commission was not without opponents; anti-Seabrook demonstrators and politicians believed that government regulators such as the NRC and the New Hampshire Public Utilities Commission were biased in favor of the reactor.

Some prominent members of Congress were displeased with NRC licensing decisions. In particular, Representative Peter Kostmayer of Pennsylvania, former chairman of a House subcommittee which examined the agency and possible bias, cited seven major incidents where the NRC changed or ignored its own rules to accommodate the Seabrook license. Senator Edward Kennedy denounced the NRC as a "rogue agency."

"Much as Dr. Frankenstein gave up his own humanity to breathe life into his monster, so the NRC has compromised its duty to the safety of the Seabrook facility and our nation," said Kostmayer as reported by *Environmental Action* in 1990.

During negotiations for the plant's licensing, the NRC was responsible for issuing a Facility Operating License with conditions that allow fuel load but not reactor startup (1986); approving a rule change permitting its review and approval of offsite emergency plans developed by utilities without local participation (1987); and authorizing a low-power testing permit contingent upon the owners providing $72.1 million for decommissioning if a full-power license was denied (1988).

However, *Environmental Action* reported several cases where the NRC may have overstepped its authority. For instance, in 1989, when the agency's Atomic Safety and Licensing Appeals Board ruled Seabrook's evacuation plan inadequate, NRC commissioners overrode the board's jurisdiction.

Before its decision to license Seabrook in March 1990, the NRC was buoyed by Ronald Reagan's 1988 executive order for FEMA to develop evacuation plans. One month later, George Bush, who had defeated Dukakis, became President with John Sununu as his White House chief of staff.

The NRC was not all pro-Seabrook. The agency surprised many critics by suspending the plant's low-power testing license of 1988 after operators ignored plant procedures and NRC suggestions, failing to manually close the reactor after a pressure rise. The NRC also proposed fining PSNH $100,000 for shortcomings in the company's Seabrook weld X-ray program.

Utility impact. Stagnant negotiations, demonstrations, and numerous NRC-enforced regulations that were originally unexpected delayed Seabrook's opening and caused expenses to mushroom, wreaking havoc on PSNH, the reactor's principal owner.

Originally envisioned as a two-unit, 2,300-megawatt complex costing $1 billion, the price tag soared to $6.35 billion for only a single unit. PSNH became the first investor-owned utility since the Depression to seek bankruptcy protection. Owner of a 35.6 percent stake in Seabrook, PSNH filed a Chapter 11 bankruptcy claim in 1988, the fourth largest in U.S. corporate history.

In 1989, the U.S. Supreme Court denied PSNH the right to raise rates 15 percent to recover Seabrook's cost. Economists determined that the ruling set a precedent, telling financially troubled utilities they could no longer rely on regulators to bail them out with rate increases.

Two other utilities holding smaller claims in Seabrook also sank into bankruptcy, EUA Power Corp., a 12.2 percent owner and subsidiary of Eastern Utilities Associates, and New Hampshire Electric Cooperative, Inc., the owner of a 2.2 percent stake.

In 1991, PSNH emerged from bankruptcy when it was acquired by Connecticut-based Northeast Utilities (NU) for $2.3 billion. The acquisition turned NU, already the largest electric company in New England, into a regional powerhouse. NU also owns the three units at Millstone in Connecticut and the lone Connecticut Yankee unit.

Currently, Seabrook Station is jointly owned by 12 entities, the largest being North Atlantic Energy Corp., a NU subsidiary, at 35.6 percent.

To pay for Seabrook's decommissioning after it stops producing electricity, New Hampshire residents are now required to pay an itemized charge on their electric bills. The average PSNH customer experienced an initial 13-cent charge, which was included in the 5.5 percent rate hikes NU negotiated with the state after PSNH went bankrupt.

However, the increased charge came under siege by critics, who argued the plan is underfunded and is not enough to decommission Seabrook if it closes before 40 years. Decommissioning is estimated to cost $242 million in 1987 dollars, but will exceed $1.2 billion in 40 years, according to the Nuclear Decommissioning Finance Committee, a New Hampshire group in charge of financial planning for the plant's old age. Seabrook's owners insist that the plant's steel-and-concrete shell and other expensive features make it the world's safest reactor.

Nonetheless, the facility is dwarfed by various concerns. The most pressing matter appears to be evacuation procedures for the tens of thousands of people who flock each summer to at least a half dozen beaches near Seabrook. As reported in *The Boston Globe* in 1990, surveys of residents and summertime visitors found that many believe it would be impossible to leave quickly on already congested narrow roads if there were an accident. Overall, 23 towns sit within 10 miles of the plant.

The possible discharges of super-heated cooling water that would kill nearby marine life has sparked concern. Seabrook has also been questioned for alcohol and drug use among workers, alleged deficiencies in welds and other construction work and supposed security breaches.

Plant officials stress that Seabrook will not become a nuclear waste dump and that the reactor will last longer than 40 years. Decommissioning will include decontaminating radioactive parts, dismantling the 400-ton reactor vessel, carving up its 400,000 cubic yards of concrete and 40,000 tons of reinforcing rods and structural steel, and trucking it all to radioactive waste dumps and landfills.

Social impact. Public opposition was a major factor in delaying Seabrook's start. The Seabrook station was the target of massive protests during a harrowing 17-year period of construction and licensing. Protestors contended in part that the New Hampshire coastline site was unsuitable for a technology that could dump radiation on crowded beaches two miles from the plant. On a typical summer day, up to 70,000 New Englanders storm the beaches.

Thousands of people were arrested in dozens of demonstrations that made Seabrook a symbol of antinuclear activism. The largest demonstration happened in 1977, when 1,414 protestors were arrested. Just prior to the reactor's first low-power test in June 1989, hundreds of protestors, including children and handicapped people, converged on the station and 627 were arrested. Some organizations opposed to Seabrook were the Clamshell Alliance, Citizens Against a 10-mile Radius, Republicans Against Seabrook Station, and the Seacoast Anti-Pollution League. The Alliance was a major organizer of demonstrations.

Plant Profiles

The plant also earned a historical footnote; what was noted as the nation's first group civil disobedience against a nuclear reactor occurred at Seabrook in 1976.

Congressional and regulatory hearings were staged to critique Seabrook. At a House hearing in 1978, then Massachusetts Representative Paul Tsongas said, "To many citizens around the state, the plant is viewed as an intrusion upon their way of life."

Seabrook opposition appeared to fade after the NRC issued an operating license in March 1990. In a *Boston Globe*/WBZ-TV poll in June 1990, just six percent of 402 New Hampshire residents surveyed identified Seabrook as the state's most pressing problem, and 47 percent said the plant should be allowed to open. Comparatively, in 1987, 33 percent of those responding to a similar poll described Seabrook as most important and 58 percent opposed the opening.

Seabrook's current status. Seabrook Station is a single unit, 1,150-megawatt reactor that, with eight other New England nuclear plants, provides more than one-third of the region's electricity. Operated by North Atlantic Energy Service Corp., Seabrook alone supplies power to one million New England homes, creating the region's biggest source of power since Millstone III opened in 1987.

The reactor has significantly reduced New England's need for costly foreign oil. Since operation started on Aug. 19, 1990, Seabrook has displaced the need to burn more than 27 million barrels of foreign oil, the fuel source used by many of New England's conventional power plants. Also, as of August 1, 1992, Seabrook generated 16.8 billion kilowatt-hours of electricity, operating at a capacity factor of more than 80 percent.

Sources

Nuclear Siting and Licensing Process (Seabrook Station, N.H.). U.S. House of Representatives, Subcommittee on Energy and the Environment, 95th Congress, August 7, 1978.
"Failure at Seabrook Could Set Off a Chain Reaction." *Business Week,* October 1, 1984.
"Seabrook Agonistes." *New England Business,* October 7, 1985.
"Seabrook Operations Postponed to 1987 as Bay State Stalls on Emergency Plan." *New England Business,* June 16, 1986.
"We Are in a Heap of Trouble." *Time,* October 26, 1987.
"Buried Under a Nuclear Pile." *Time,* February 8, 1988.
"After Seabrook, It's a Cold World for Utilities." *Business Week,* February 15, 1988.
"Order By Reagan Seeks The Opening of A-Power Plants." *The New York Times,* November 19, 1988.
"Order on A-Plants Concerns Experts." *The New York Times,* November 21, 1988.
"That Rumbling Could Be Seabrook Warming Up." *Business Week,* December 5, 1988.
"Nuclear Plant's Plans for Emergency Backed." *The New York Times,* December 21, 1988.
"Panel Approves Evacuation Plan for Area Around Seabrook Plant," *The New York Times,* January 3, 1989.
"High Court Lets Ruling on Seabrook Cost Stand." *The New York Times,* January 24, 1989.
"Another Setback for Seabrook: New Hampshire Will Seek to Block Low-Power Testing." *The New York Times,* March 9, 1989.
"Despite Cost at Polls, Governor Fought for Seabrook Reactor." *The Washington Post,* March 12, 1989.
"Seabrook." *Science for the People,* May/June 1989.
"Output at Bustling Seabrook: Frustration, Not Electricity." *The New York Times,* August 15, 1989.

"N.R.C. Panel Supports a License for Seabrook." *The New York Times,* November 14, 1989.
"Seabrook Nuclear Plant Is Cleared for Full-Power License; Foes Vow Fight." *The Washington Post,* November 14, 1989.
"The Seabrook Follies." *The Washington Post,* November 19, 1989.
"Allegations Spark Dispute." *The Boston Sunday Globe,* February 18, 1990.
"Northeast Eyes Plant Takeover." *The Boston Sunday Globe,* February 18, 1990.
"Presidential Election Is Called Decisive Factor." *The Boston Globe,* March 2, 1990.
"Seabrook Gets Gift From Uncle Sam." *The Boston Sunday Globe,* March 4, 1990.
"Opponents Go to Court in Fight Over Seabrook," *The New York Times,* March 8, 1990.
"Another Hurdle to Seabrook License Is Lifted." *The New York Times,* March 15, 1990.
"Seabrook's Demise a Topic of Debate." *The Boston Sunday Globe,* March 25, 1990.
"Seabrook A-borning." *National Review,* April 30, 1990.
"Frankenstein's Monster Lives!" *Environmental Action,* May/June 1990.
"Seabrook Issue Shown Fading." *The Boston Sunday Globe,* July 15, 1990.
"Full Power From Seabrook." *Nuclear Industry,* First Quarter 1990.
"Another Utility Files for Chapter 11 Status Over Seabrook Debt." *The Wall Street Journal,* May 7, 1991.
"Democrats Move to Bury Ghost of Seabrook." *The Boston Sunday Globe,* August 11, 1991.
"Seabrook Weld Issue Leads to $100,000 Proposed Fine." *Nucleonics Week,* November 28, 1991.
"Full Power From Seabrook." *Nuclear Industry,* First Quarter 1990.
"Another Utility Files for Chapter 11 Status Over Seabrook Debt." *The Wall Street Journal,* May 7, 1991.

—Michael Richman

Seabrook Station, Unit 2

PO Box 300
Seabrook, NH 03874
William A. DiProfio, Stat.Mgr.

Phone: (603)474-9521
Fax: (603)474-3808

Owning Utility

New Hampshire Yankee
PO Box 300
Seabrook, NH 03874
Bruce L. Drawbridge, Exec. Dir., Nuclear Production

Phone: (603)474-9521
Fax: (603)474-2987

Contact

Bruce Gretter, Mgr., Community Relations
Seabrook Station

Basic Facts

Unit 2 *Status:* Cancelled (construction 25% complete). *Type of Reactor:* Pressurized light-water reactor. *Megawatts (electric):* 1194. *Megawatts (thermal):* 3411. *Reactor System Supplier:* Westinghouse Electric Corp. (United States). *Generator Supplier:* General Electric Co. (United States). *Architect Engineer:* United Engineers & Constructors (United States).

Key Dates

Unit 2 *Ordered:* 1972. *Cancelled:* 1986. *Construction began:* 1976.

Plant Report

See Plant Report for Seabrook Station, Unit 1.

UNITED STATES

354 ☆

Sears Island Nuclear Power Plant

Searsport, ME

Basic Facts

Status: Cancelled. *Type of Reactor:* Pressurized light-water reactor. *Megawatts (electric):* 1194. *Megawatts (thermal):* 3411. *Reactor System Supplier:* Westinghouse Electric Corp. (United States). *Generator Supplier:* Westinghouse Electric Corp. (United States). *Architect Engineer:* United Engineers & Constructors (United States).

Plant Report

New England utilities cancel planned plants. Anticipating a high level of growth in the levels of demand for electricity in the region, several New England utilities planned in the early 1970s to build more nuclear power plants. Many of these plants were opposed by nuclear industry regulators and utility financial advisors, forcing the utilities to eventually cancel the plants. Among the plants that were cancelled were Pilgrim II and III, Seabrook II, Sears Island, and Montague I and II.

According to Paul F. Levy, 1987 chairman of the Massachusetts Department of Public Utilities, had all these plants been built it would have given the region an extra 8000 or 9000 megawatts of capacity and caused an immense strain on the regional economy.

Sources

"Utility Chief Not a Man Hungry for Power: Paul F. Levy." *Boston Business Journal,* January 26, 1987.

355 ●

Sequoyah Nuclear Plant, Units 1-2

PO Box 2000 Phone: (615)843-6000
Soddy-Daisy, TN 37379 Fax: (615)843-8750
J.L. Wilson, Site V.Pres.

Owning Utility

Tennessee Valley Authority Phone: (615)751-0011
6N 38A Lookout Pl. Fax: (615)751-4904
1101 Market St.
Chattanooga, TN 37402-2801
J.R. Bynum, V.Pres., Nuclear Operations

Contact

Kay Wittenburgh, Pub.Info.
Sequoyah Nuclear Plant

Basic Facts

Units 1-2 *Status:* Operable. *Type of Reactor:* Pressurized light-water reactor. *Megawatts (electric):* 1183. *Megawatts (thermal):* 3411. *Reactor System Supplier:* Westinghouse Electric Corp. (United States). *Generator Supplier:* Westinghouse Electric Corporation. *Architect Engineer:* Tennessee Valley Authority.

Key Dates

Unit 1 *Ordered:* 1968. *Construction began:* 1969. *Criticality:* July 1980. *First Power:* October 1980. *Commercial Operation:* July 1981.

Unit 2 *Ordered:* 1968. *Construction began:* 1969. *Criticality:* November 1981. *First Power:* December 1981. *Commercial Operation:* June 1982.

Operating Costs

Units 1-2

1981	$19,217,000	1986	$122,015,000
1982	$47,756,000	1987	$198,798,000
1983	$68,588,000	1988	$197,637,000
1984	$76,755,000	1989	$145,745,000
1985	$88,978,000		

Plant Report

Background. The Sequoyah Nuclear Plant is located on Chickamauga Lake in Soddy Daisy, Tennessee, about 18 miles northeast of Chattanooga. The plant, which is the second nuclear station built by Tennessee Valley Authority (TVA), houses two pressurized water reactors capable of generating enough electricity to supply about 475,000 homes a day or two cities the size of Chattanooga.

Facts and Features. Each reactor vessel contains 193 nuclear fuel assemblies with a steel pressure vessel, about 44 feet high and 14 feet in diameter, supported by concrete walls that are eight feet thick.

The dome-shaped containment that encloses each reactor is about 151 feet high and 134 feet in diameter. The outer concrete structure has walls about 3 feet thick. Racks within the containment hold two and one half million pounds of ice.

Each turbogenerator receives steam at more than 800 pounds per square inch pressure and at more than 500 degrees Fahrenheit. Three low-pressure and one high-pressure steam turbines spin a generator at 1800 revolutions per minute. The total weight of these rotating parts is more than one million pounds.

The condensers are cooled by river water in a separate system. This water can be cooled further in Sequoyah's two 450-foot cooling towers, before being returned to Chickamauga Lake.

$1.66 billion over budget. At first, Sequoyah was scheduled for completion in 1973-74 at an approximate cost of $336 million. After its completion in 1981, the cost totalled almost $2 billion.

A great part of the cost increase was a result of a study of Sequoyah done by the General Accounting Office (GAO) in 1975. The GAO found 23 instances where structures or components had to be torn out and rebuilt, or added because of required changes. Eleven of these changes increased the construction costs more than $2 million.

Another earlier contributor to the escalated cost was the use of a new containment building that had not been tested. Design weaknesses were discovered and repaired in 1972, but more weaknesses were found, prompting TVA to redesign the containment in 1974.

Plant Profiles

Bad press. In 1980, Critical Mass, a nuclear watchdog group, released a study on Sequoyah which found that the plant had more licensee event reports than any other plant in the country, even though it had only been operating 11 percent of the time.

In 1982, about a year after Sequoyah's official start, a private research group went door-to door, handing out potassium iodide tablets to residents within five miles of the plant. The tablets would saturate the human thyroid gland with iodine, thus preventing the absorption of cancer-causing radioactive iodine in the event of a leak from the plant. The action was covered by large newspapers across the country.

Shutdown. In the early 1980s, the NRC and the FBI investigated Sequoyah, finding weaknesses in design, management, and operations, as well as safety violations and improperly installed equipment. These investigations, coupled with higher operating costs and public uneasiness with the Three Mile Island accident in 1979, caused TVA to shutdown Sequoyah in 1985.

During the shutdown, Sequoyah underwent an expensive repair program, most of which was completed on Sequoyah 2.

Restart. TVA had originally intended to restart Sequoyah in 1987, to improve capacity margins, and to reduce purchased power and coal-fired generations. However, the NRC did not issue a new permit to operate Sequoyah until 1988. Before the restart was allowed, two issues had to be resolved: faulty readings from crucial temperature-sensing devices, and last-minute concerns about fire protection standards. These issues were resolved, and the permit was granted. In 1989, Sequoyah was removed from the NRC's list of troubled plants.

Contamination and contracts. In August 1990, the NRC did an investigation at Sequoyah Fuels in Oklahoma, where spent fuel from the Sequoyah plant is stored. The investigation concerned an instance of uraniun contaminated water seepage into an excavation near a solvent extraction building.

In 1991, TVA signed a five-year, $116 million contract with Westinghouse to provide refueling, outage, engineer, and licensing support services for Sequoyah. The contract was designed to shorten the seven fueling outages that it covered and to enhance the availabilty of Sequoyah, which was the only TVA plant in operation at the time.

Accomplishments

- In January 1989, Sequoyah 2 set a new record for large generating units by operating continuously for 209 days before shutting down for a scheduled refueling outage.
- In December 1989, Sequoyah 1 broke the record by operating 299 days before automatically shutting down.
- In June 1990, Sequoyah had the third highest capacity factor and was rated second in the United States and eighth in the world for total generation, producing over 12 billion kilowatt hours of electricity in fiscal year 1989.
- During July 1990, the plant ranked in the top 10 nuclear producers of electricity. Sequoyah 2 was ranked ninth, producing 838,650 megawatt-hours, and Sequoyah 1 was ranked 10th, producing 838,200 megawatt-hours.

- In December 1990, Sequoyah completed two refueling outages, making 212 modifications.
- During 1991, Sequoyah 2 was ranked fourth in the United States and 14th in the world for total generation, producing about 9.7 billion kilowatts of electricity.
- In January 1991, Sequoyah 1 produced 871,480 megawatt-hours of electricity, the eighth highest amount in the United States.
- In May 1991, Sequoyah 1 produced 865,450 megawatt-hours of electricity, the 10th highest amount in the United States.
- In August 1991, Sequoyah 2 operated for 210 consecutive days, breaking its old record.
- In November 1991, Sequoyah 2 operated for 306 consecutive days, breaking TVA's longest continuous run record.
- During 1991, Sequoyah 2 had a capacity factor of 93.3 percent, the eighth highest factor among U.S. plants.

Sources

Sequoyah Nuclear Plant, The Tennessee Valley Authority, n.d.
"Why Do You Need Something That Will Kill You for Heat You Can't Afford?" *The Progressive*, February 1982.
"The Cult of the Atom." *The New Yorker*, November 1, 1982.
"Lost in the Dark: TVA Stumbling Blindly Over its Nuclear Program." *Public Citizen*, January 1987.
"TVA Reactor to be Restarted." *The New York Times*, March 23, 1988.
"Generating Woe: if TVA were Private, it'd be Bankrupt." *Barron's*, May 2, 1988.
"Operational Safety Assessment." *U.S. NRC Annual Report 1990*, U.S. Nuclear Regulatory Commission, n.d.
"Nine U.S. Nuclear Plants Made the World's top 25 Last Year, and That's not all the Good News." *Nuclear Industry*, First Quarter 1990.
"TVA Looking Toward Nuclear Future Again." *Nuclear Industry*, Summer 1990.
"Palo Verde Unit 3 is No. 1 in U.S. Electric Generation." *Business Wire*, September 19, 1990.
"Palo Verde Units 3 & 2 Top Producers in United States for January; Unit 3 to be Refueled Starting March 16." *Business Wire*, March 13, 1991.
"TVA Nuclear Plant Contract Is Estimated at $116 Million." *The Wall Street Journal*, April 11, 1991.
"U.S. Gives Backing To Start Reactor Shut For 6 Years." *The New York Times*, May 3, 1991.
"Palo Verde Units 1 & 2 Top American Producers in May; Unit 2 Continues Record Run." *Business Wire*, July 12, 1991.
"U.S. Nuclear Plants Shatter Records for Output, Performance." *INFO*, U.S. Council for Energy Awarenss, February 1992.
"TVA Nuclear Plant Contract Is Estimated at $116 Million." *The Wall Street Journal*, April 11, 1991.

—*Charles Ceplecha*

Shippingport Atomic Power Station

Shippingport, PA

Owning Utility

Duquesne Light Company
301 Grant St.
One Oxford Centre
Pittsburgh, PA 15279
John D. Sieber, V.Pres., Nuclear

Phone: (412)393-6000
Fax: (412)393-6448

Shippingport

Contact
J.M. Sasala, Dir., Nuclear Commun.
Duquesne Light Company

Basic Facts
Status: Shut down. *Type of Reactor:* Pressurized light-water reactor. *Megawatts (electric):* 68. *Megawatts (thermal):* 238. *Reactor System Supplier:* Westinghouse Electric Corp. (United States). *Generator Supplier:* Westinghouse Electric Corp. *Architect Engineer:* Stone & Webster Engineering Corp. (United States). *Owner:* Department of Energy. *Operator:* Duquesne Light.

Key Dates
Ordered: 1953. *Construction began:* 1955. *Criticality:* December 1957. *First Power:* December 1957. *Commercial Operation:* December 1957. *Shut Down:* October 1982. *Decommissioned:* 1985-1990.

Plant Report

The first commercial nuclear power plant. On October 22, 1953, the Atomic Energy Commission (AEC) announced that a 60,000 kilowatt nuclear power plant was to be built at Shippingport, Pennsylvania, through a cooperative effort of Duquesne Light Company of Pittsburgh, Pennsylvania, and the Atomic Energy Commission. It was to become the world's first nuclear power plant dedicated solely to the purpose of producing electricity.

After approximately two and one-half years of construction in which building costs had increased by at least 50 percent, the Shippingport Atomic Power Station went critical on December 18, 1957, and within a few days, was operating at its 60,000 kilowatt capacity. At a budgeted cost of $47,700,000, the final cost for constructing Shippingport came to $84,000,000 with another $36,000,000 allocated for research and development.

The pressurized water reactor (PWR) at Shippingport operated efficiently until 1977 when it was converted to an experimental light-water system which bred a different fissile isotope of uranium. In 1982, Shippingport was retired after generating 6.5 billion kilowatts of electricity, and although the reactor was efficient, it was never economical. The pressurized light-water reactor design is not, by today's standards, the best model for the U.S. reactor industry.

The first reactor to be decommissioned. Perhaps the most significant contribution of Shippingport, other than being known as the first commercial electricity reactor, is that it was also the first nuclear reactor to be fully decommissioned. When reactors were first built during the 1950's and '60's, power companies knew that 30 or 40 years down the road, the reactors would have to be shut down. However, no thought was given to how that would be accomplished.

Now, with the decommissioning of Shippingport, much thought has been given to the process of disassembling a nuclear reactor. Congressional Representatives Harold Volkmer and Marilyn Lloyd, both chairmen of two subcommittees of the House's Committee on Science and Technology, had agreed to work with the Department of Energy (DOE) on using Shippingport as a Decommissioning Model. They wanted to show that a nuclear power plant could be dismantled without causing any harm to the workers of the surrounding neighborhood. The goal of federal officials was to effectively remove the reactor so that the land could be returned to unrestricted use.

The major task facing officials for dismantling Shippingport was finding a solution for the disposal of radioactive waste. While most of the waste had been removed from the reactor by the time it was ready to be taken apart, there was still some low-level contamination in the concrete and steel. However, Shippingport did have a major advantage over other commercially-owned reactors in that it was partially owned by a government agency. This fact allowed the contaminated materials to be buried at a government disposal site.

Burying a reactor. The radioactive waste from Shippingport was transported and buried at a nuclear waste dump located at Hanford Military Reservation in Washington state. In the Spring of 1989, the 1,000-ton reactor vessel was loaded on a barge, the Paul Bunyan, to make the voyage to its burial site. The 7,800 mile trip was made by water because the vessel was too heavy to travel long-distance on land. The Paul Bunyan mapped a course down the Ohio River, to the Mississippi River, into the Gulf of Mexico, through the Panama Canal and eventually up the Pacific coast to Hanford Military Reservation.

The availability of a disposal site for Shippingport has raised many questions, but has not provided many answers, about locating burial land for future decommissioning as well as entombing the extremely radioactive spent fuel rods from active reactors. Currently, there are only three operational disposal sites, but these locations are approaching capacity as well as nearing the end of their lifespans. However, these sites are not currently available to the private sector. Therefore, since most of the nuclear power plants are owned by the private sector and there aren't any disposal sites available, spent fuel rods are stored in cooling pools at the power plants, but space at the plants for storage of these rods is also approaching capacity.

Size and cost may affect future decommissioning. The decommissioning, which began in September 1985, was expected to be complete in 1990 at a cost of $98.3 million. Compared to the dollar amount of electricity generated over the reactor's lifetime, it is felt by Mark Hills, president of Science Concepts, that "squeezing" a couple of hundred million at the end of its life is not a big deal. However, this view may not be widely shared. Current estimates of future decommissioning range from $100 million to $3 billion.

Using Shippingport as a model has also received much criticism due to the size of the reactor. Critics feel that Shippingport's relatively small size cannot provide an ideal pattern for future decommissioning. Shippingport's 72-megawatt pressurized light-water reactor reactor is very small compared to the 1000-megawatt reactors operating today. Some experts feel that while Shippingports reactor was removed in one piece, the giant reactors of today would require carving the vessel into pieces which could expose workers to more radiation.

Sources

"Adding to Electric Costs: the Retirement of Brittle, Old Reactors." *Discover*, November 1986.

"A Demonstration at Shippingport: Coming of Line." *American Heritage*, June/July 1981.

"Death of a Nuclear Plant." *U.S. News & World Report*, November 2, 1987.

"Reactor Unit Is Removed From Plant." *The New York Times*, December 15, 1988.

"Retiring a Reactor." *Technology Review*, July 1989.

—Yvette M. Costa

Shoreham Nuclear Power Plant

PO Box 618
Wading River, NY 11792

Owning Utility

Long Island Power Authority
200 Garden City Plaza
Garden City, NY 11530
Stanley B. Klimberg, Pres. of Shoreham/
 Gen. Counsel

Phone: (516)742-2200
Fax: (516)742-2084

Basic Facts

Status: Shut down. *Type of Reactor:* Boiling water reactor. *Megawatts (electric):* 880. *Megawatts (thermal):* 2436. *Reactor System Supplier:* General Electric Co. (United States). *Generator Supplier:* General Electric Co. (United States). *Architect Engineer:* Stone & Webster Engineering Corp. (United States). *Owner:* 1966-1989, Long Island Lighting Company (LILCO); 1989-present, Long Island Power Authority (LIPA).

Key Dates

Ordered: 1967. *Construction began:* 1973. *Criticality:* February 1985. *First Power:* August 1986. *Shut Down:* February 1989. *Decommissioned:* 1992.

Plant Report

Shoreham overview. The Shoreham nuclear power plant on Long Island, New York, is considered one of the greatest financial disasters in U.S. industry history.

Conceived in 1966, the reactor was seen by nuclear power proponents as a cheap, clean and safe alternative to Long Island's messy, aging fossil fuel plants. Originally priced at $70 million, the plant endured numerous legal entanglements and cost overruns before its completion in 1984 at approximately $5.5 billion. Still, the Long Island Lighting Company (LILCO), Shoreham's original owner, was not granted a full-power license for the plant until 1989. After more than 20 years of campaigning, the power facility was given the green light to exercise its 809 megawatt operating potential, enough power for one-third of Long Island's 934,000 homes.

However, because of political interference from New York and Suffolk County, where the plant sits, Shoreham never began commercial operation. New York Gov. Mario Cuomo led an intense campaign against the reactor's opening. Opponents relied on two arguments: Long Island, with 2.7 million residents, could not be evacuated safely in the event of a nuclear disaster at Shoreham, which is located 60 miles east of Manhattan; and the plant would cause crippling electricity rates because of its enormous price tag.

In 1989, LILCO turned the plant over to the state for $1 in return for five percent rate increases in each of the next 10 years. Currently, the idle reactor is owned by the state-run Long Island Power Authority (LIPA), which is responsible for decommissioning.

Issues leading to the decision to not open Shoreham. Although Shoreham had proponents other than LILCO, public opposition was too much for LILCO to overcome. In all, tremendous skepticism toward LILCO's evacuation plans and concerns about suspected increased electricity rates led to the reactor's downfall.

In 1973, LILCO received a construction permit from the Atomic Energy Commission to build Shoreham. Then, for about a decade, the plant's construction was supported by the state and Suffolk County, according to LILCO President Anthony F. Earley, Jr. The New York State Public Service Commission (NYSPSC) expressly approved Shoreham's construction on several occasions.

Nonetheless, in 1983, Suffolk County decided to oppose the plant's operation and the state later adopted the same position. State and local governments collaborated to prevent Shoreham from receiving an operating license from the Nuclear Regulatory Commission (NRC), citing the absence of a government-sponsored emergency plan. That same year, a state-authorized study criticized LILCO's emergency preparedness and environmental impact assessments for Shoreham in the event of a nuclear disaster. LILCO representatives fired back that the plant was built to federal safety specifications with the full approval of state officials. "total disregard for the safety of the people of Long Island" by LILCO and Atomic Energy Commission (AEC) officials. The report also indicated that LILCO's $265 million estimated cost for Shoreham, after the March 1969 announcement to upgrade the plant, led state officials to believe that "the facility would be able to produce very inexpensive electricity, which made the project attractive to the ratepayers."

LILCO mounted an effort massive in scope and intensity to license Shoreham. But all of its evacuation plans were rejected and the NRC licensing process remained in holding, forcing the company to repeatedly seek timely decisions by the Federal government to break the Shoreham stalemate.

By 1988, LILCO and the state were in a crisis over Shoreham. The crisis came to a head when the NYSPSC ordered LILCO to present a plan for operating its system without Shoreham. The utility then entered into its second negotiated settlement with the state on February 28, 1989 selling the plant to LIPA for $1, and the NRC belatedly issued a full-power license. The agreement was approved by the company's shareholders, NYSPSC, LIPA, and the New York Power Authority (NYPA).

Abbreviated history of Shoreham. Shoreham made history as the first fully licensed nuclear plant in the United States to be dismantled without ever operating commercially. Until LILCO signed over ownership of Shoreham to LIPA, the projected nuclear station had been a 26-year wrestling match over politics, energy policy, and money. Through Shoreham's lingering controversy, cost estimates and designated dates for completion each changed four times. A brief chronology is as follows:

- April 1966: LILCO announced plans to build a 500-megawatt plant on the north shore of Long Island.
- March 1969: LILCO approved a plan to upgrade the plant to 820 megawatts; it was predicted to cost $265 million with completion in 1975.
- April 1973: LILCO began construction while utility and county officials drafted emergency plans. The plant was estimated to cost $506 million with completion in 1977.
- February 1983: The county withdrew its own evacuation plans and Mario Cuomo did not approve LILCO's evacuation plans.
- March 1983: Major construction ended and LILCO admitted that Shoreham cost $4 billion. A state panel said Shoreham's location made emergency planning impossible.
- July 1986: Cuomo signed legislation establishing LIPA, which has since taken over Shoreham.
- February 1989: LILCO sold Shoreham to the state for $1 in return for rate increases.
- June 1991: The NRC granted a possession-only license for Shoreham, allowing LILCO to transfer ownership to the state or LIPA.
- July 1991: After a request by Shoreham supporters, the Bush Administration asked that the possession-only license be stayed until an environmental review was completed. However, a U.S. Court of Appeals refused the stay.
- June 1992: Shoreham's final chapter opened as work crews began demolishing the plant.

Financial implications for LILCO. LILCO incurred enormous financial costs in pursuing Shoreham's operating license from the NRC. Those costs, plus the financial burden of the plant's construction the cost of which increased rapidly, brought LILCO to the brink of bankruptcy several times in the mid-1980s.

According to LILCO President Anthony F. Earley, Jr., the company suspended dividend payments on both preferred and common stocks; it faced the loss of investor confidence and access to the financial markets; it suffered a series of downgradings by the financial ratings agencies, resulting in securities being rated below investment grade; and relations with customers were poisoned by the bitterness of the Shoreham dispute.

LILCO is now regaining financial health, primarily because of the five percent rate hikes it received in the agreement to abandon Shoreham. The company is still an independent, investor-owned utility and has resumed payments to common-stock shareholders who failed to receive dividends for several years during the crisis. LILCO's common stock rose from $3.25 in 1984 to more than $20 in 1991. As of 1988, LILCO still owed $2.5 billion for construction costs, part of which is now being paid by customers.

Economic implications. If Shoreham opened, customers suspected alarming electricity increases in a region that already had the fifth-highest utility rates in the country as of 1989. Long Island is the only geographic enclave in the United States totally dependent on foreign oil for electricity.

New York Representative George J. Hochbrueckner, who oversees the district in which Shoreham is located, was opposed to the plant's opening for economic reasons. With the nuclear facility, he claimed that Long Island residents would experience a 100 percent increase in utility rates up to the year 2000, creating the country's highest rates. Otherwise, the settlement between the state and LILCO results in 63 percent increases, he noted.

Because Shoreham is being decommissioned, savings for LILCO customers are estimated at approximately $48,000 daily or $17.5 million yearly, according to a LILCO spokesman.

Some companies made surprising moves to deal with Long Island's electricity rates. In 1989, Grumman Corp., then LILCO's largest customer, announced plans to bypass the utility's electrical system and use power from a $50 million generator being built nearby.

Social and political implications. During the 1970s and 1980s, Shoreham was Long Island's most controversial issue and one of New York's most dominant political issues.

According to a 1983 state report on Shoreham, the public was deeply outraged at LILCO and the NRC, and found great offense at both organizations "for inflicting this Shoreham debacle on them." The report noted the following: "The vast majority feel that Shoreham poses a grave threat to their physical well-being and financial stability. They feel the power plant was ill-conceived and arrogantly constructed."

Anti-Shoreham demonstrations grew and diversified as activists became more creative. Protestors constantly appeared at Shoreham's gates; 785 people were arrested from 1979-1980 for disrupting construction and blocking plant entrances.

Opponents were likely alarmed by a 1982 Federal study, which estimated that a core meltdown at Shoreham would cause 40,000 early fatalities, 75,000 early injuries, 35,000 cancer deaths, and $157 billion in property damage. An expert on nuclear plant accidents noted that those projections were conservative, estimating that deaths would total nearly a million if easterly winds sent a plume of radioactive poisons from Shoreham directly into New York City.

Nevertheless, Shoreham enjoyed formidable backers. The White House, the Department of Energy (DOE), and the NRC supported LILCO in its attempt to open the plant. In addition, the nuclear industry, scientists from the Brookhaven National Laboratory on Long Island, Citibank (which functioned as a financial overseer of LILCO) and major media (*The New York Times* and *Newsday*) wanted the plant to start. Scientists and Engineers for Secure Energy, Inc., and the Shoreham-Wading River Central School District, which depended on plant-generated taxes, petitioned for Shoreham's opening.

The Ritter amendment, the most favorable piece of Congressional legislation toward Shoreham, was introduced by Pennsylvania Representative Don Ritter in the late 1980s. Ritter's intention was to prevent New York from dismantling Shoreham. The legislation was criticized though, for concerning Congress with an issue that some say was purely local.

Ronald Reagan approved Shoreham's opening; George Bush and the DOE fought in favor of the plant until the end. The Bush administration went to federal court to block the plant's closing, but a U.S. Court of Appeals cleared the way for Shoreham's destruction. Overall, the final settlement between the state and LILCO resulted in a stunning victory for grass-roots, antinuclear activists.

Judicial implications. Litigation became frequent during the Shoreham crisis. LILCO took legal action and was targeted by the same.

LILCO earned the distinction as the first American utility to seek both an injunction and money damages, contending that antinuclear demonstrators who disrupted construction at Shoreham infringed on the company's private property rights. The company argued that those rights superseded First Amendment rights to demonstrate.

LILCO also took the offensive against the state and General Electric Co. (GE). In 1987, one day before LIPA came into existence, the utility asserted that Mario Cuomo's attempts to block Shoreham's opening violated the constitutional and civil rights of the company and its stockholders. LILCO filed a $400 million lawsuit against GE over the construction of the reactor, claiming the contractor covered up potentially catastrophic safety defects in Shoreham's containment system.

In 1989, Suffolk County was awarded $10 million in damages in a class-action, racketeering suit against LILCO. LILCO was found guilty of misleading state regulators about Shoreham's cost and schedule in order to get approval for rate hikes to continue plant construction.

Court rulings on Shoreham's decommissioning were made in New York's State Court of Appeals, the state's highest court, a U.S. Court of Appeals, and the U.S. Supreme Court.

Shoreham's impact on the nuclear industry. The inability of LILCO, the federal government, and nuclear proponents to open Shoreham was a vital wakeup call to the industry, which poured funds into a drive to save the plant.

Most importantly, the Shoreham dilemma demonstrated a questionable licensing system for nuclear reactors. In Shoreham's case, officials staged hearings after the plant was completed to discuss final NRC approval. Shoreham was ready for operation in 1984, but failed to receive licensing until 1989, during which time New York State and Suffolk County refused to participate in emergency planning.

However, part of the National Energy Security Act, which Congress is currently deliberating, calls for NRC pre-construction certification of reactors and sites. State and local governments may still express displeasure with plant proposals, as in the Shoreham case, but all judgements will be made beforehand. "We want all the input up front so we can make decisions on plants without fighting it," said a spokesman for the U.S. Council on Energy Awareness, a trade association representing the nuclear industry.

Future of Shoreham. The decommissioning of Shoreham, now commonly referred to as Long Island's "great white elephant" and "albatross," is underway and rests with LIPA. NRC officials were reluctant at first to approve a transfer license to LIPA because of its inexperience in decommissioning a nuclear plant, but that move was made on February 29, 1992.

Various cost estimates exist regarding decommissioning. The process is expected to cost $186 million over 27 months.

Because of low-level testing in 1985 and 1987, Shoreham's 66-foot, 600-ton reactor vessel is contaminated with 600 curies of radiation, a small amount for a nuclear plant. The contaminated pieces are to be disposed in a hazardous-waste site. The fuel, contaminated with 150,000 curies, will be shipped to a reprocessing facility, given on-site storage, given off-site storage at a licensed plant, or sold to another utility for further use.

LIPA is also studying the prospects of converting Shoreham into a gas-fired plant, having commissioned a $300,000 feasibility study for that purpose. However, a LIPA spokesman indicated that there are no details on the conversion, and the LIPA board hasn't even determined if it is feasible or likely. Two other nuclear facilities, Rancho Seco (California) and Fort St. Vrain (Colorado) are also being considered for gas conversion.

For more information. Long Island Power Authority, current owner of Shoreham, can be contacted with the information given above. To contact the Long Island Lighting Co. (LILCO) write J.W. McDonnell, Vice President of Public Affairs, at 175 E. Old Country Rd, Hicksville, NY, 11801; or phone 516-933-4590.

Sources

"Legal Showdown at Shoreham." *The Nation*, March 21, 1981.
Report of the New York State Fact Finding Panel on the Shoreham Nuclear Power Facility. Shoreham Commission, Stony Brook, N.Y., December 1983.
"Nuclear Fallout." *Time*, March 19, 1984.
"Something Rotten in Suffolk?" *Forbes*, February 11, 1985.
"Governor Sued By L.I. Utility on Atom Plant." *The New York Times*, January 15, 1987.
"The $5 Billion Nuclear Waste." *Time*, June 6, 1988.
"As Shoreham Goes." *The Nation*, June 11, 1988.
"The Details of the Agreement." *The New York Times*, June 17, 1988.
"Shoreham Chickens, Home to Roost." *The New York Times*, July 25, 1988.
"LILCO Wins Key Rulings in Two Shoreham Suits." *Newsday*, September 7, 1988.
"3 Legislators Seek LILCO Role." *Newsday*, January 19, 1989.
"Watkins Vows to Prevent Dismantling of A-Plant." *The Washington Post*, April 28, 1989.
The Agreement Between the State of New York and the Long Island Lighting Company to Close the Shoreham Nuclear Powerplant. Subcommittee on Energy and the Environment, U.S. House of Representatives, November 9, 1989.
"LILCO, GE Talks Break Off." *Newsday*, December 18, 1990.
"Rate Hikes Not Enough: LILCO." *LI Business News*, January 14, 1991.
"LI Lighting Comes to Life." *LI Business News*, May 20, 1991.
"As Shoreham's Power Fades, State Moves to be a Supplier." *The New York Times*, June 23, 1991.
"U.S. Is Suing to Salvage Shoreham." *The New York Times*, July 12, 1991.
"A Chronology of Shoreham's 25-Year Political Meltdown." *The New York Times*, July 20, 1991.
"U.S. Appeals Court Clears Way to Dismantle Shoreham A-Plant." *The New York Times*, July 20, 1991.

"Way Cleared for Dismantling of Shoreham Nuclear Plant." *The Washington Post*, July 20, 1991.
"One More Renewable Energy Source: Nuke Plants." *Business Week*, September 16, 1991.
"Court Upholds Closing Shoreham Nuclear Plant." *The New York Times*, October 23, 1991.
"At Shoreham, A Somber Beginning of the End." *The New York Times*, October 25, 1991.
"The Fallout of the Demise of Shoreham." *The New York Times*, November 12, 1991.
"Shoreham A-Plant is Sold to Agency." *The New York Times*, March 1, 1992.
"Shoreham's Options Concentrate on Gas." *The New York Times*, March 1, 1992.
"Nearly New A-Plant Going to Scrap Heap." *The Washington Post*, June 18, 1992.

—Michael Richman

358 ☆

Skagit/Hanford Nuclear Project, Units 1-2

Richland, WA

Basic Facts

Units 1-2 *Status:* Cancelled. *Type of Reactor:* Boiling water reactor. *Megawatts (electric):* 1335. *Megawatts (thermal):* 3800. *Reactor System Supplier:* General Electric Co. (United States); Bechtel. *Generator Supplier:* Westinghouse Electric Corp. (United States). *Architect Engineer:* Bechtel. *Owner:* Puget Sound Power and Light Company. *Operator:* Puget Sound Power and Light Company.

Key Dates

Units 1-2 Ordered: January 1973. Cancelled: 1983.

Plant Report

Proposed plant. The Skagit/Hanford Nuclear Project began as a proposal for a twin 1,275 megawatt reactors in Skagit County, Washington. The first public announcement was made in January 1973. Four utilities participated in the plans. The project sponsor, Puget Sound Power & Light Company, would have owned 40 percent of the plant. The other partners were Portland General Electric Company (30 percent), Pacific Power & Light Company (20 percent) and Washington Water Power Company (10 percent).

Legal challenges. Governor Dan Evans signed the Site Certification Agreement on January 5, 1977, and immediately SCANP (Skagitonians Against Nuclear Power) filed suit challenging the legality of the agreement. In November 1978, the owner utilities prevailed in the courtroom.

Four months later, the nation's worst nuclear mishap occurred at Three Mile Island in Pennsylvania. The accident raised concerns among local residents about the safety of nuclear power. An advisory measure (Proposition 4) was placed on the November 6, 1979 ballot. The public voted 71 1/2 percent against construction of the proposed plant. As a result, the project lost its zoning permit.

A new location. In July 1980, Puget Power began site studies for the purpose of moving the project to the Hanford Reservation in eastern Washington. The project name was officially changed to Skagit/Hanford Nuclear Project in October 1981.

Costly abandonment. In 1983, the Northwest Power Planning Council refused to consider the Skagit/Hanford Nuclear Project as part of the regional 20-year power plan. Without assurances of regional support, Puget Power President John Ellis announced his company's intention to abandon the project. In a news conference on August 30, 1983, listed uncertainties about construction costs, questions about future energy demands, the political situation, and the post-TMI regulatory environment as reasons.

Although Puget Power never entered the construction phase of the project, the utility had invested $128 million. The Washington Utilities and Transportation Commission allowed the company to recover only $82 million through electric rates. The WUTC disallowed costs incurred after June 1980. The commission ruled that Puget Power had not considered the cost effectiveness of proceeding after being forced to quit the original site.

Sources

"Puget Power Wiping Nuclear Exposure from Its Books." *Puget Sound Business Journal*, February 4, 1985.
"Puget Power Seeks Termination of Skagit/Hanford Nuclear Power Project." (News release) Puget Sound Power & Light Company, August 30, 1983.
"Skagit/Hanford Information Sheet: Chronology." Puget Sound Power & Light Company, n.d.
Puget Power 1991 Annual Report. Puget Power, n.d.

—Karen Bellenir

359 ☆

Somerset Nuclear Power Plant, Units 1-3

Somerset, NY

Basic Facts

Units 1-3 *Status:* Cancelled. *Type of Reactor:* Boiling water reactor. *Megawatts (electric):* 1247. *Megawatts (thermal):* 3579. *Reactor System Supplier:* General Electric Co. (United States); United Engineers & Constuctors (United States). *Generator Supplier:* General Electric Co. *Architect Engineer:* United Engineers & Constructors. *Owner:* New York State Electric and Gas Corporation. *Operator:* New York State Electric and Gas Corporation.

Plant Report

Planned nuclear power plant turns to coal. Somerset is a vivid example of the revival of coal energy when the U.S. shied away from new sources of commercial nuclear power in the late 1970s and the decade of the 80s. Operated by the New York State Electric & Gas Corp., the Somerset station was originally designed to be powered by three nuclear reactors.

A state-imposed ban on nuclear power plants combined with the site's proximity to the Niagara Escarpment, a geological fault, led the utility to modify the plant's original 1973 plans from nuclear to coal.

Plant Profiles South

The plant, which overlooks the shore of Lake Ontario in Niagara County, generates 650 megaawatts of electricity. The $1 billion station uses approximately 1.5 million tons of coal a year.

Sources

"Ol' Reliable Coal Still Blazing Its Own Path." *Business First Buffalo*, February 3, 1986.

360 ☆
South Dade Nuclear Power Plant, Units 1-2

South Dade, FL

Owning Utility

Florida Power and Light Company
700 Universe Blvd.
PO Box 14000
Juno Beach, FL 33408-0420
J.H. Goldberg, Pres., Nuclear Div.

Phone: (407)694-4000
Fax: (407)694-4311

Contact

R.R. Golden, Nuc. Corp. Commun. Coord.
Florida Power and Light Company

Basic Facts

Units 1-2 *Status:* Cancelled. *Type of Reactor:* Pressurized light-water reactor. *Megawatts (electric):* 1210. *Megawatts (thermal):* 3411. *Reactor System Supplier:* Westinghouse Electric Corp. (United States). *Owner:* Florida Power and Light Co. *Operator:* Florida Power and Light Co.

Key Dates

Units 1-2 *Ordered:* 1975. *Cancelled:* 1977.

361 ●
South Texas Project Electric Generating Station, Units 1-2 (STP)

PO Box 289
Wadsworth, TX 77482
M.R. Wisenburg, Plant Mgr.

Phone: (512)972-3611

Owning Utility

Houston Lighting and Power Company
PO Box 1700
Houston, TX 77251
D.P. Hall, Group V.Pres., Nuclear

Phone: (713)228-9211
Fax: (713)220-5016

Contact

C. Glen Walker, Mgr., Pub.Info.
South Texas Project Electric Generating Station

Basic Facts

Units 1-2 *Status:* Operable. *Type of Reactor:* Pressurized light-water reactor. *Megawatts (electric):* 1312. *Megawatts (thermal):* 3817. *Reactor System Supplier:* Westinghouse Electric Corp. (United States). *Generator Supplier:* Westinghouse Electric Corporation. *Architect Engineer:* Bechtel. *Owner:* Houston Lighting and Power Company; City Public Services of San Antonio; Central Power and Light Company (CPL); and the City of Austin. *Operator:* Houston Lighting and Power Company.

Key Dates

Unit 1 *Ordered:* 1973. *Construction began:* 1976. *Criticality:* March 1988. *First Power:* March 1988. *Commercial Operation:* August 1988.
Unit 2 *Ordered:* 1973. *Construction began:* 1976. *Criticality:* March 1989. *First Power:* April 1989. *Commercial Operation:* June 1989.

Operating Costs

Units 1-2
1988 $28,182,000
1989 $133,026,000

Plant Report

A first for Texas. The South Texas Project Electric Generating Station (STP) occupies a 12,200-acre site on the Colorado River in Matagorda County, Texas. The site is located between Bay City and Palacios, Texas, about 90 miles southwest of Houston.

STP was Texas' first nuclear power plant. Its two pressurized water reactors are capable of producing approximately 15 billion kilowatt hours of electricity per year, or enough to serve 500,000 homes.

Ownership. STP is jointly owned by the Houston Lighting & Power Company (HL&P), City Public Services of San Antonio, the Central Power and Light Company (CPL), and the City of Austin. These utilities share the electricity STP produces in proportion to their degree of ownership. HL&P has 30.8 percent ownership; San Antonio has 28 percent; CPL has 25.2 percent; and the city of Austin has 16 percent.

Customers and economics. Electricity from STP flows to Houston, Austin, San Antonio, Corpus Christi, and many other Texas communities.

Because nuclear fuel costs only about one-fourth as much as conventional power plant fuels, STP saves over $1.01 million per day in fuel costs. The plant is expected to save its customers $40 billion over its 40-year life, and to replace the equivalent of 25 million barrels of imported oil, or 8.4 million tons of coal in a year's time.

Facts and features. The containment buildings at the plant are dome-like structures, made of steel-reinforced concrete, with walls measuring four feet thick. The buildings rest on concrete and steel foundations, measuring 18 feet thick.

Each reactor vessel is surrounded by two concrete shield walls. One wall is seven and one-half feet thick, and the other is three and one-half feet thick. The vessels have steel walls which are six inches thick.

Each reactor houses 193 fuel assemblies. A fuel assembly consists of 264 fuel rods, approximately 14 feet long, and a single rod contains more than 300 uranium dioxide pellets. While one assembly costs about $400,000, it can supply the energy equivalent of more than $400 million worth of oil.

● Operable   ○ Under Construction   △ Indefinitely Deferred   ▲ Decommissioning   □ Planned   ■ On Order   ☆ Cancelled   ★ Shut Down

The plant has a 7,000-acre reservoir for plant cooling, which is the largest above-ground reservoir in the world.

Spent fuel. About one-third of STP's fuel rods are replaced every 18 months. Spent fuel is stored in underwater storage pools at the plant's fuel handling buildings. The fuel will ultimately be sent to U.S. government depositories, but STP has space to store it for the length of its licensed operating life.

Plant safety. STP is the only nuclear power plant in the United States that features three independent and redundant safety systems. All other U.S. plants have only two. These multiple control systems provide plant operators with complete control of the speed of the nuclear reaction, as well as an ability to halt the reaction at any time.

Emergency preparedness. STP's emergency preparedness plan includes the use of periodic practice drills to keep Matagorda County, plant employees, and state officials in a state of readiness at all times.

Residents living within 10 miles of STP are notified of any emergency by 12 warning sirens, tone alert radios, and a computerized telephone calling system, provided and maintained by the plant. The notification system is activated by county officials to broadcast emergency information on nuclear matters, as well as any other kind of emergency.

Employees. STP employs over 1,600 people on a full-time basis, making it the largest employer and taxpayer in Matagorda County. All total, there are over 2,200 employees on the site, including contract labor.

Plant operators participate in an operator training program at STP, which includes hands-on instruction in a state-of-the-art control room simulator. Operators learn how to handle both normal and emergency situations, and participate in a weeklong "refresher" class about every six weeks.

Problems. When construction began on STP in 1975, HL&P had originally hired Brown & Root, a construction company known for its oil refineries, to design and build the plant. STP was the company's first original nuclear power plant design and was scheduled for completion by 1982 at a cost of $900 million.

Delays and cost overruns began to pile up at STP, and within five years, the final cost estimate had risen to over $2.7 billion. According to the NRC, plant workers began to harass Brown & Root employees who were monitoring construction and making sure it was up to NRC standards. NRC investigators had found "unacceptable defects" in welds, and improper consolidation practices in pouring concrete for the walls of the reactors' containment vessels. The commission also found that HL&P had failed to conduct audits of Brown & Root's construction practices, and ordered the utility to conduct a thorough check of all safety-related construction. The utility responded by hiring the Bechtel Corporation as a consultant on quality control.

By 1980, the NRC inspectors at STP had filed nearly 3,000 reports of deficiencies in builders' work. After a three-month investigation and a report which cited 22 violations of federally mandated construction procedures, the NRC recommended a $100,000 fine for HL&P.

Legal questions. In July 1989, a Dallas, Texas jury ruled in favor of HL&P in a suit brought against the utility by the city of Austin. The suit alleged deceit and mismanagement of STP by HL&P, and sought $419 million in damages for Austin.

In January 1990, legal and ethical questions were raised when HL&P planned to fly some of the jurors from the trial to STP for a free plant tour. A spokesman from the utility said that the jurors had expressed an interest to see the plant, and that the action was not improper because they would not be seated in another case involving HL&P. Legal ethics experts felt that the utility's actions, while not actually being jury tampering, would give jurors the impression that they would be awarded a vacation for voting a certain way.

HL&P settlements. In June 1990, the Texas Public Utility Commission (PUC) affirmed a settlement which allowed HL&P to raise its rates by $227 million and include a part of STP's construction and operating costs in its rate base.

In February 1991, the increase was stepped up to $313 million, allowing HL&P to recover its remaining investment in STP. At the time, the settlement still needed regulatory approval, but was scheduled to take effect on May 16, 1991.

In a related 1990 decision, the Texas PUC disallowed $375.5 million of HL&P's $2.8 billion investment in STP. The utility was permitted to reduce the disallowance by the value of proceeds from a legal settlement with the plant's original architect/engineer and constructor, which resulted in an after-tax write-off of $15 million. The ruling followed an investigation into the prudence of plant expenditures.

As of December 31, 1990, HL&P's investments (net of accumulated depreciation and amortization) in STP and in nuclear fuel, including allowance for funds used during construction, were $2.3 billion and $139 million, respectively.

Decommissioning fund. In July 1990, HL&P and the other owners of STP filed a certificate of financial assurance with the NRC for the decommissioning of the plant. This certificate assures that HL&P will meet the minimum decommissioning funding requirements mandated by the commission. During 1990, the utility was funding STP's Unit 1 at an annual amount of about $1.9 million, the amount which the commission allowed to be collected through electric rates in a previous rate order.

As required by the NRC, decommissioning funds are being accumulated in an external trust. There is no assurance that all decommissioning costs will be recoverable through rates, or that the amounts recoverable will be adequate, but a 1990 rate proceeding was to provide $6 million of annual funds for STP. Over the useful life of the plant, this funding level was to provide approximately $136 million, in 1989 dollars, an amount which exceeds the NRC minimum but is less than the HL&P estimate of $168 million which was based on a site-specific study in 1989. The difference in the two estimates is due primarily to a reduction in the contingency factor used in the study.

Plant Profiles

CPL settlements. On October 9, 1990, the Texas PUC affirmed a two-step settlement that allowed CPL to collect $144 million of interim rates to pay for the cost of STP's Unit 1. CPL had been collecting the rates since March, and the decision made the collection permanent. The commission permitted CPL to include in its rate base all of its investment in Unit 1 and to recognize income due to the deferral of carrying costs to the extent interest charges were incurred for Unit's 1 and 2. CPL is owned by the Central South West Corporation, a Dallas-based public utility holding company.

In December 1990, the Texas PUC approved the second step of the settlement which allowed CPL to increase its base rates by $120 million on January 1, 1991, to begin recovering its cost for STP's Unit 2. As a part of the settlement, CPL had agreed not to seek additional relief until after December 31, 1994, except in the case of specified unusual circumstances.

Accomplishments. For 1990, STP achieved an average capacity factor of 56.8 percent. The plant also received high marks in annual evaluations by the NRC and the Institute of Nuclear Power Operations, and became a fully-accredited member of the National Academy for Nuclear Training.

During January 1991, STP's Unit 2 produced 888,140 megawatts of electricity, which gave the plant the sixth highest total among U.S. electricity producers.

During May 1991, STP's Unit 2 produced 866,060 megawatts of electricity, which gave the plant the ninth highest total among U.S. electricity producers.

International technology trade. The NRC chose STP to represent the United States in an exchange of nuclear power safety information with the Soviet Union. The commission provided the Soviets with "the design and licensing requirements of a state-of-the-art American nuclear power station," while the Soviet Union reciprocated with information on a pressurized water reactor in the southern Ukraine. In addition to the Soviet Union, nuclear specialists from around the world have toured STP, and examined its design and operation.

STP wildlife. The American alligator has become the unofficial mascot of STP. Approximately 50 of the reptiles live and feed at the site, which has been declared a wildlife sanctuary. The site is also inhabited by deer and several species of birds.

Public access. The STP Visitor Center, which is located near the plant, offers educational displays, films, and presentations about nuclear power. The center is open from 9 a.m. to 5 p.m., Monday through Saturday. Information can be obtained and drive-through plant tours can be scheduled by calling: (512) 972-5023.

Speakers are available, free of charge, to address clubs or organizations on STP or nuclear energy in general. Arrangements can be made by calling the Central Power and Light Company at (512) 881-5504.

Sources

The South Texas Project Electric Generating Station. Houston Lighting & Power Company, n.d.
"Another Nuclear Scandal." *Newsweek*, May 26, 1980.
Houston Industries Incorporated 1990 Annual Report. Houston Industries, Inc., n.d.
"Legal Beat: Texas Utility." *The Wall Street Journal*, January 12, 1990.
Fact Sheet: South Texas Project Electric Generating Station. Houston Lighting & Power Company, September 1990.
"Central and South West Corp. Unit Gets $120 Million Rate Increase for South Texas Project Unit 2." *Business Wire*, December 11, 1990.
"Palo Verde Units 3 & 2 Top Producers in United States for January; Unit 3 to Be Refueled Starting March 16." *Business Wire*, March 13, 1991.
"Palo Verde Units 1 & 2 Top American Producers in May; Unit 2 Continues Record Run." *Business Wire*, July 12, 1991.

—Charles Ceplecha

---

362 ☆

Sterling Nuclear Power Plant

Sterling, NJ

Owning Utility

Rochester Gas and Electric Corporation
89 E Ave.
Rochester, NY 14649
Roger W. Kober, Pres.

Phone: (716)546-2700
Fax: (716)546-1511

Contact

John W. Edmunds, Dept. Mgr., Pub.Aff.
Rochester Gas and Electric Corporation

Basic Facts

Status: Cancelled (construction less than 1% complete). *Type of Reactor:* Pressurized light-water reactor. *Megawatts (electric):* 1192. *Megawatts (thermal):* 3411. *Reactor System Supplier:* Westinghouse Electric Corp. (United States). *Owner:* Rochester Gas and Electric. *Operator:* Rochester Gas and Electric.

Key Dates

Ordered: 1973. *Cancelled:* 1980. *Construction began:* 1978.

---

363 ●

Virgil C. Summer Nuclear Station

PO Box 88
Jenkinsville, SC 29065
J.L. Skolds, V.Pres., Nuclear

Phone: (803)345-5209
Fax: (803)345-4020

Owning Utility

South Carolina Electric and Gas Company
1426 Main St.
Columbia, SC 23065
B.D. Kenyon, Pres. & COO

Phone: (803)748-3000
Fax: (803)748-3667

Basic Facts

Status: Operable. *Type of Reactor:* Pressurized light-water reactor. *Megawatts (electric):* 933. *Megawatts (thermal):* 2775. *Reactor System Supplier:* Westinghouse Electric Corp. (United States). *Generator Supplier:* General Electric Co. (United States). *Architect Engineer:* Gilbert. *Owner:* South Carolina Electric and Gas Company; South Carolina

Public Service Authority. *Operator:* South Carolina Electric and Gas Company.

Key Dates

Ordered: 1971. *Construction began:* 1973. *Criticality:* October 1982. *First Power:* November 1982. *Commercial Operation:* January 1984. *Shut Down Expected:* January 2024.

Operating Costs

1984	$65,041,000
1985	$70,657,000
1986	$61,496,000
1987	$63,562,000
1988	$68,856,000
1989	$75,829,000

Plant Report

A power center in South Carolina. The Virgil C. Summer Nuclear Station occupies a site in Fairfield County near Jenkensville, South Carolina. The site is about 26 miles southwest of Columbia.

Summer Station is jointly-owned by the South Carolina Electric & Gas Company (SCE&G) and the South Carolina Public Service Authority (Santee Cooper). SCE&G operates the plant and owns two-thirds, and Santee Cooper owns one-third.

Summer Station cost $1.3 billion to build, and was named after Virgil C. Summer, a former chief executive officer and chairman of the board of SCE&G.

Features of Summer Station. The reactor vessel is 40 feet high, and weighs 360 tons. It has an inside diameter of 13 feet, with an average wall thickness of 8 1/2 inches.

The reactor is fueled with 157 fuel assemblies. Each assembly is made up of 264 fuel rods, containing about 272 uranium dioxide pellets each, for a total of approximately 11,275,000 pellets. The assemblies are 144 inches high and 8 1/2 inches in diameter, with a weight of about 1/2 tons each. The uranium has been enriched to increase the amount of the fissionable isotope, U235, from the .7 percent that is found in natural uranium, to about 3.0 percent in the fuel. Each pellet contains as much energy as 1780 pounds of coal.

The three steam generators weigh 340 tons (dry). They are 68 feet high, with an upper diameter of 15 feet and a lower diameter of 11 feet, and contain 4674 tubes each. The pressurizer weighs 77 tons, and is 42 feet high.

Cooling is provided by Lake Monticello, a 6800-acre manmade lake that eliminates the need for cooling towers. There are three reactor cooling pumps, with a design capacity of 100,700 gallons per minute. The pump motors have a rating of 7000 horsepower each.

There are 246,000 cubic yards of concrete and 3,900,000 linear feet of cable at the plant. The exclusion zone is an area within a one mile radius around the plant, or about 2200 acres. Approximately 22 acres of the plant are fenced in.

Emergency preparedness. SCE&G makes an emergency preparedness calendar available to residents of Fairfield, Lexington, Newberry, and Richland counties, which surround Summer Station. The calendar provides information on evacuation routes, reception centers, and instructions to follow in case of an accident at the plant.

Accomplishments. In November 1985, SCE&G was honored by *Electric Light and Power Magazine* as its 1985 Utility of the Year. The basis for the award was the utility's completion of Summer Station, and its inclusion into SCE&G's rate base. The magazine recognized Summer Station's 1982 completion cost of $1,426 per kilowatt, which was 24 percent below the national average. Also noted was the South Carolina Public Service Commission's March 1984 decision to grant the SCE&G a rate increase of $133 million, which included the plant in its rate base. The utility had originally asked for $193 million.

In 1992, the NRC recognized Summer Station for its high level of safety performance. Only three other plants in the United States received the commendation. The NRC said its senior managers had reviewed performance in areas including: operational safety, self-assessment, problem resolution, and plant management and oversight.

Sources

General Facts (Summer Station). South Carolina Electric & Gas Company, n.d.
Nuclear Systems Data (Summer Station). South Carolina Electric & Gas Company, n.d.
Milestones (Summer Station). South Carolina Electric & Gas Company, n.d.
Electricity: Energy From SCE&G. South Carolina Electric & Gas Company, n.d.
"Magazine Recognizes SCE&G." *South Carolina Business Journal,* January 1986.
SCE&G Calendar. South Carolina Electric & Gas Company, January 1992.
SCE&G News Release. South Carolina Electric & Gas Company, February 7, 1992.

—*Charles Ceplecha*

Summit Nuclear Power Plant, Units 1-2

Middletown, DE

Basic Facts

Units 1-2 *Status:* Cancelled. *Type of Reactor:* Gas-cooled reactor. *Megawatts (electric):* 781. *Megawatts (thermal):* 2000. *Reactor System Supplier:* General Atomic. *Owner:* Delmarva Power and Light. *Operator:* Delmarva Power and Light.

Key Dates

Units 1-2 *Ordered:* 1971. *Cancelled:* 1975. *Construction began:* 1974.

Sundesert Nuclear Power Plant, Units 1-2

Blythe, CA

Plant Profiles

Surry 366

Basic Facts

Units 1-2 *Status:* Cancelled. *Type of Reactor:* Pressurized light-water reactor. *Megawatts (electric):* 978. *Megawatts (thermal):* 2785. *Reactor System Supplier:* Westinghouse Electric Corp. (United States). *Owner:* San Diego Gas and Electric. *Operator:* San Diego Gas and Electric.

Key Dates

Units 1-2 *Ordered:* 1975. *Cancelled:* 1978.

366 •

Surry Nuclear Power Station, Units 1-2

PO Box 315 Phone: (804)357-3184
Surry, VA 23883
M.R. Kansler, Stat.Mgr., Nuclear

Owning Utility

Virginia Power Phone: (804)771-3000
PO Box 26666
Richmond, VA 23261
W.L. Stewart, Sr.V.Pres., Nuclear

Contact

W.N. Curry, Mgr., Corp. Commun.
Virginia Power

Basic Facts

Units 1-2 *Status:* Operable. *Type of Reactor:* Pressurized light-water reactor. *Megawatts (electric):* 848. *Megawatts (thermal):* 2441. *Reactor System Supplier:* Westinghouse Electric Corp. (United States). *Generator Supplier:* Westinghouse Electric Corp. *Architect Engineer:* Stone & Webster Engineering Corp. (United States).

Key Dates

Unit 1 *Ordered:* 1966. *Construction began:* 1968. *Criticality:* July 1972. *First Power:* July 1972. *Commercial Operation:* December 1972. *Shut Down Expected:* May 2012.

Unit 2 *Ordered:* 1966. *Construction began:* 1968. *Criticality:* March 1973. *First Power:* March 1973. *Commercial Operation:* May 1973. *Shut Down Expected:* January 2013.

Operating Costs
Units 1-2

1972	$607,000	1981	$31,184,000
1973	$5,102,000	1982	$33,087,000
1974	$9,878,000	1983	$57,159,000
1975	$15,270,000	1984	$59,146,000
1976	$14,796,000	1985	$56,877,000
1977	$15,977,000	1986	$85,515,000
1978	$19,323,000	1987	$76,638,000
1979	$23,313,000	1988	$117,571,000
1980	$29,459,000	1989	$89,369,000

Plant Report

Background information. Surry Nuclear Power Station is located 200 miles southeast of Washington D.C. on the James River, halfway between Norfolk and Richmond in Surry, Virginia. Owned by Virginia Electric & Power Company (VEPCO), also known as Virginia Power, Surry's two units supply more than one-fifth of Northern Virginia's electric power and generate 1,562-megawatts of power.

The Surry Nuclear Power Station, along with the North Anna Nuclear Power Station, also owned by Virginia Power, supply 40 percent to 45 percent of the utility's power needs and have a combined capacity of 3,400 megawatts. The nuclear reactors were completed for just over $500 a kilowatt and generate power for about a half-cent per kilowatt hours versus approximately two cents for coal and five and one-half cents for oil.

Surry in the 70s. The 1970s were a time of soaring costs and regulatory barriers for the Surry Nuclear Power Station. From the plant's onset in 1972 and throughout the 70s, the two 800-megawatt units performed below the national average for nuclear reactors, somewhere in the range of a 50 percent capacity factor.

By 1978, the facility's six steam generators had deteriorated and needed replacing. According to company officials, in the early days of nuclear power plant operation there was little known about the chemistry of the water that circulates through the steam generators. This water is carried via tubes from the reactor vessel where it is heated by the nuclear fuel. It was found that impurities in the water caused a multitude of problems that damaged several of the more than 3,000 tubes in each of the three steam generators in each unit. Virginia Power had to plug many of the tubes, which reduced the ability of the generators to produce steam and turn the turbines. The replacement was complete in 1980.

On top of the technical problems Surry faced in the 70s, in 1977 the slowdown in electricity demand growth that followed the Arab oil embargo had forced Virginia Power to cancel two more planned units at Surry.

After the cancellation, Virginia Power decided to implement a reorganization which included separating their nuclear operations from their other fossil generating stations. Vice President William Stewart was put in charge of this function. He immediately built up the staff, improved working conditions, and increased training. The result: The plant's capacity factor improved and Surry ranked among the lowest-cost producers in the country throughout the 80s.

Surry in the 80s. The 1980s were not without adversity for the Surry plant. In December 1986, four contract employees died when a steam pipe carrying hot water ruptured, bathing eight workers in scalding water. The 350-degree water flashed to steam burning four of the men to death and injuring two others. This accident occurred on the non-nuclear side of Unit 2, thus no radioactivity was released. Virginia Power immediately shut down both units and began inspecting all pipes at the Surry station as well as those at its sister plant, North Anna.

The Nuclear Regulatory Commission's (NRC) investigation confirmed the company's finding: The rupture was the result of a previously unknown form of erosion-corrosion that thinned the pipe walls and not by any problem related to the age of the plant. The NRC absolved Virginia Power of any blame stating there was no way the utility could have anticipated the event.

Virginia Power then took matters into their own hands. They immediately dispatched a description of the problem to the Institute of Nuclear Power Operations (INPO), the industry association responsible for setting standards of excellence for the nation's nuclear operators, notified every company with units of similar design and vintage of the problem by telephone, and held a series of detailed seminars to warn nuclear operators worldwide of the problem.

Despite this slight setback, both Surry units continued to run smoothly until the latter half of 1988 when minor problems transpired at the plant. In May 1988, 30,000 gallons of water leaked from a reactor cavity when an inflatable seal suddenly failed. No radiation was released and there were no injuries, but the NRC was concerned about the potential hazard involved and the failure of Virginia Power to report the incident promptly. The NRC was not informed of this accident until August, three months after it had happened.

Another violation occurred in September 1988 when the plant operated with its emergency service water system pumps unable to function at the required capacity. Each of the three pumps in the backup system, which are responsible for cooling the reactor if the regular pumping system goes out, had a capacity of 12,000 gallons a minute, 20 percent less than the NRC required amount of 15,000 gallons a minute.

These incidents prompted intensive inspections from the NRC that spanned from September 1988 until March 1989. Many basic questions were raised about design inadequacies at the plant, defects that would not be allowed in facilities built today. Critics argued that Surry was a typical illustration of an aging atomic plant whose deterioration makes it expensive to operate, inefficient, and unsafe. As a result, Virginia Power acted on its own accord and shut down both units in September 1988, vowing not to start them up again until all necessary corrections were made, the plant had undergone a complete inspection and refurbishment.

However, the violations were not the initial reasons for the shutdown. Unit 1 was originally taken out of service for emergency repairs to its diesel generator and Unit 2 was turned off for a normal refueling outage.

The 1988 investigation. At the start of the inspection Virginia Power created a committee to look at every aspect of Surry's engineering. They tested, analyzed, and replaced or restored thousands of parts in key operational and safety systems. They also conducted a comprehensive review to ensure the individual systems, many of which had been upgraded as a result of post-Three Mile Island safety requirements, operated smoothly together.

However, during the investigation more prevalent problems were uncovered. They included the following:

- Errors and missing information were found throughout the plant. Some of these stemmed from lower documentation standards that were in effect when Surry began operation.
- In December, reactor employees found that power supplies for two critical motor-operated valves were reversed which meant that in an emergency both could fail. This error also dated from the plant's construction.
- When tests began on the motor-operated valves throughout the facility, employees could not find any manufacturer's standards to tell them at what force the motor should be set to open the valves.
- The NRC expressed concern that the control room air conditioning had been installed at lower power levels than design called for which in a worst-case situation could force a reactor shutdown.

Results of the inspection. Following the investigation some of the major repair work included:

- Upgrading the emergency pumps that supply water if the plant loses its normal power supply. At that time they had a capacity 20 percent less than the NRC required.
- Installing more than 100 new sections of pipe to replace prematurely aging equipment in the steam generator in the non-nuclear section of the plant, the same location where four men died after a pipe exploded in 1986.
- Upgrading its emergency diesel generators. The NRC questioned whether the generators would supply enough electricity to run Surry if the station lost its normal power.

Virginia Power spent $12 million on repairs in 1988 alone. Besides the cost of repairs, from February 1988 to July 1989 the utility paid $800,000 in NRC penalties, more than the plant had been assessed in the previous decade. The largest fine, $500,000 for 12 safety violations found at the plant over a six-month period, stemmed from the leak of 30,000 gallons of water in May 1988. This fine marked the sixth time the NRC had imposed fines of $500,000 or more at one time. Virginia Power was also fined for other violations including:

- Potential worker overexposure to radiation.
- The presence of debris in emergency cooling wells which could have hampered the proper operation of two safety-related systems. The utility had no program for knowing the foreign materials were there, for cleaning it up, or for keeping it out of the reactor sumps. This cost the utility $50,000.
- Careless maintenance.

In August 1988 Virginia Electric was fined $100,000 for the following violations:

- Failing to provide radiological monitoring devices to personnel entering high radiation areas.
- Not posting areas where respirators were needed.
- Not performing adequate surveys to evaluate the extent of airborne radioactive materials.

As a result of the shutdown, Surry became one of the nation's 13 least efficient nuclear plants in 1988. One of its reactors produced less than 40 percent of its capacity and the other barely half. The industry average is about 60 to 66 percent.

Finally, in July 1989 Unit 1 came back on line, followed by Unit 2 in September. But Surry was not free from doubt. In June 1989, the Surry plant was put on the NRC's watch list of troubled plants for stricter oversight, which included the hiring of a third on-site inspector.

Plant gets back on track. In the summer of 1990 the Surry units were taken off the NRC's watch list. In 1990 the two units combined for a 77.5 percent capacity factor, well above the industry average of 67.5 percent. In addition, the Surry plant was included on the Utility Data Institute's (UDI) top 10 list of low-cost producers of nuclear electricity in 1990. The UDI, a subsidiary of Halliburton Company, is an independent research and data base publishing company in Washington, DC. According to UDI, Surry had a production cost of $15.71 per megawatt-hour of electricity and produced 10,609,965 megawatt-hours.

But the improvement didn't stop here. In 1991 Unit, 1 had a 96.1 percent capacity factor, making it one of the 15 U.S. nuclear plants that had capacity factors above 90 percent. Unit 2, which was undergoing refueling, had a 75.7 percent factor. Also, in June 1991, the NRC found Surry had improved in four of seven areas following a 12-month evaluation.

Surry undergoes individual plant examination process. An Individual Plant Examination (IPE) analysis at the Surry plant in 1991 showed a core damage frequency (CDF) from internal flooding of a one chance in 1,000 per reactor year. As a result of the IPE, the plant made some physical changes that included:

- Adding flow shields to all expansion joints in the two units.
- Installing check valves on the charging pump cubicle drain line.
- Adding four diesel-driven sump pumps to the nine already in the turbine building.
- Relocating the power source for turbine building sump pumps out of a potential flood path.

Besides these physical changes, Virginia Power took additional measures above what was required of them, to ensure safety including:

- A new procedural emphasis on preventive maintenance of flood protection devices.
- Employing additional compensatory measures to avoid a maintenance-induced flood.
- Modifying instructions to operators for responding to a flooding event in terms of the priority of mitigative actions.

The outcome of the actions taken by Virginia Power resulted in a reduction of the CDF from internal flooding from a one chance in 1,000 per reactor year to nearly a one chance in 10,000, determined in the re-analysis.

Virginia Power gives back to the community. Virginia Power contributes to the welfare of the community and the environment in many different ways.

Virginia Power paid nearly $5.6 million in taxes in 1990, approximately 80 percent of Surry's county budget. Since the company began paying taxes, it has added nearly $49 million into the county's funds, enabling Surry to rank in Virginia's top 10 in per capita local school aid, and by the time its 40-year license is up for renewal in 2010, it will have paid more than $100 million.

Virginia Power maintains a bald eagle nesting site at the Surry nuclear plant.

In 1985, Virginia Power began the Foundations Program which gives scholarships to a company nuclear training program. Graduates of the program receive college credits from one of two colleges that participate in the program and so far every graduate has been offered a job.

Virginia Power supports a formal volunteer program. In 1990, they had more than 150,000 hours of worker volunteer time and the company earned a medal from President Bush.

The Surry nuclear plant maintains an information center and offers tours of the plant.

The Surry plant has a policy of hiring from the local populace whenever possible and they have also donated used vehicles to the county.

Sources

"In Praise of Nuclear Power." *Forbes*, November 5, 1984.
"Tube Rupture Releases Radioactivity in Virginia." *The New York Times*, July 16, 1987.
"NRC Staff Proposes Fine for Dominion's Virginia Electric Unit." *The Wall Street Journal*, November 15, 1988.
"Virginia Power Fined Heavily for Violations of Nuclear Plant Safety." *The Washington Post*, May 20, 1989.
"Surry Nuclear Plant is Fired Up to Run." *The Washington Post*, July 9, 1989.
"'Watch List' Pared by NRC." *Nuclear Industry*, Summer 1990.
"Wisconsin Electric's Point Beach Lowest Cost Nuclear Plant in '90." *Info*, U.S. Council for Energy Awareness, July 1991.
"New Awareness, Not Plant Changes, Major Consequence of Surry IPE." *Inside N.R.C.*, December 2, 1991.
"Surry Nuclear Plant: Improving With Age." *Nuclear Industry*, Third Quarter 1991.
"High-Tech Revolution in Historic Surry County." *Nuclear Industry*, Third Quarter 1991.
"Virginia Power Nuclear Units Generate Big Savings." *INFO*, U.S. Council for Energy Awareness, February 1992.
"U.S. Nuclear Plants Shatter Records for Output, Performance." *INFO*, U.S. Council for Energy Awareness, February 1992.

—*Sharon Miscavage*

Surry Nuclear Power Station, Units 3-4

PO Box 315 Phone: (804)357-3184
Surry, VA 23883
M.R. Kansler, Stat.Mgr., Nuclear

Owning Utility

Virginia Power Phone: (804)771-3000
PO Box 26666
Richmond, VA 23261
W.L. Stewart, Sr.V.Pres., Nuclear

Contact

W.N. Curry, Mgr., Corp. Commun.
Virginia Power

Susquehanna

Basic Facts
Units 3-4 *Status:* Cancelled. *Type of Reactor:* Pressurized light-water reactor. *Megawatts (electric):* 928. *Megawatts (thermal):* 2631. *Reactor System Supplier:* The Babcock & Wilcox Co. (United States).

Key Dates
Units 3-4 *Ordered:* 1972. *Cancelled:* 1977.

Plant Report
See Plant Report for Surry Nuclear Power Station, Units 1-2.

Susquehanna Steam Electric Station, Units 1-2

PO Box 467
Berwick, PA 18693
H. Gene Stanley, Plant Supt.

Phone: (717)542-2181
Fax: (717)542-3177

Owning Utility
Pennsylvania Power and Light Company
Two N. Ninth St.
Allentown, PA 18101
Harold W. Keiser, Sr.V.Pres., Nuclear

Phone: (215)774-5151
Fax: (215)774-5019

Contact
James Marsh, Dir., Corp. Commun.
Pennsylvania Power and Light

Basic Facts
Units 1-2 *Status:* Operable. *Type of Reactor:* Boiling water reactor. *Megawatts (electric):* 1100. *Megawatts (thermal):* 3293. *Reactor System Supplier:* General Electric Co. (United States). *Generator Supplier:* General Electric Company. *Architect Engineer:* Bechtel. *Owner:* Pennsylvania Power and Light Company; Allegheny Electric Cooperative Inc.

Key Dates
Unit 1 *Ordered:* 1970. *Construction began:* 1973. *Criticality:* 1982. *First Power:* November 1982. *Commercial Operation:* June 1983.

Unit 2 *Ordered:* 1970. *Construction began:* 1973. *Criticality:* 1984. *First Power:* June 1984. *Commercial Operation:* February 1985.

Operating Costs
Units 1-2

1983	$34,305,000	1987	$133,580,000
1984	$70,502,000	1988	$157,250,000
1985	$123,031,000	1989	$160,974,000
1986	$131,241,000		

Plant Report

A Pennsylvania power center. The Susquehanna Steam Electric Station occupies a 1,700-acre site in Salem Township, Lucerne County, about five miles northeast of Berwick, Pennsylvania. The site is near the east and west banks of the Susquehanna River. Susquehanna's two reactors are capable of producing nearly 50 million kilowatt-hours of electricity a day.

Susquehanna Station is jointly-owned by the Pennsylvania Power & Light Company (PP&L) and Allegheny Electric Cooperative Inc. (AE). In 1978, PP&L sold a 10 percent interest in the plant to AE, which includes 13 rural electric cooperatives in Pennsylvania and one in New Jersey. AE purchased 210,000 kilowatts of the plant's total capacity, and distributes electricity to more than 155,000 member-consumers in 47 counties. The plant is predicted to produce about one-third of PP&L's required power in the 1990s.

The building of the plant cost $4.1 billion and involved more than 5,000 workers.

Features of Susquehanna Station. The containment structures are made of reinforced concrete and are 161 feet high. The walls are six feet thick with a .25 inch steel lining. The volume of the structures is 519,450 cubic feet, and the design pressure is 53 pounds per square inch. The structures were designed to withstand earthquakes, and studies made prior to construction showed no movement in the site's bedrock in at least 200 million years.

The pressure vessels are made of carbon clad steel with a stainless steel lining. They are 73.5 feet high and weigh 750 tons. The vessels have an inside diameter of 20.9 feet, and walls measuring four to nine inches thick. The design pressure is 1,250 pounds per square inch at 575 degrees Fahrenheit.

The reactors have a core coolant flow rate of 264,000 gallons per minute. The feedwater inlet temperature is 383 degrees, and the steam outlet temperature is 543 degrees Fahrenheit. The reactors have a coolant pressure (inlet) of 1,053 pounds per square inch, and a steam capacity of 13,464,000 pounds per hour. The heat output is 11,915,000,000 British thermal units per hour.

The fuel cores contain 32,000,000, or 132.2 metric tons of uranium oxide pellets per reactor. The pellets have an enrichment of .71 to 4.91 percent, and are .275 inches long and .424 inches in diameter.

There are 60,356 fuel rods, made of Zircalloy 2, per reactor. The rods are 14.66 feet long, with a cladding thickness of .030 inches and an outside diameter of .424 inches.

The generators run at 1,800 revolutions per minute and are cooled by hydrogen-activated motors. They produce 24,000 volts of electricity, stepped up by transformer to 230,000 volts in Unit 1, and 500,000 volts in Unit 2. The transformers are oil-cooled and have a capacity of 1,214,000 kilovolt-amperes.

The turbine-generators are 208 feet long. The tandem compound, six flow turbines have a steam pressure of 965 pounds per square inch at 540 degrees Fahrenheit. The steam flow is 13,464,500 pounds per hour.

The condensers are made of 81,500 stainless steel tubes. The tubing length is 616 miles, and the condensing surface is 880,000 square feet. The cooling water flow is 448,000 gallons per minute, and the steam flow is 8,650,339 gallons per hour.

The natural draft, evaporative cooling towers are 540 feet high with a base diameter of 415 feet and a top diameter of 301 feet.

Water flows in from the Susquehanna River at 448,000 gallons per minute, and the cooling range is 35 degrees Fahrenheit.

Storage and disposal of waste and fuel. Susquehanna Station has a radwaste building where the plant's low-level radioactive waste is processed and packaged. The plant was one of the first in the nation to construct a low-level waste holding facility. This facility would be used to store the waste if disposal sites become unavailable.

The spent fuel pools at the plant have enough capacity for more than 5,000 fuel assemblies, or enough to last through the late 1990s. Spent fuel will be stored at the plant until a federal repository is built under guidelines established by the national Nuclear Waste Policy Act of 1983. Under that act, the Department of Energy has the responsibility to develop, schedule, site, construct, and operate a deep-mine geologic waste repository.

Plant staff and training. Susquehanna Station has a workforce of about 1,200. The plant's training program was the first in the nation to be fully accredited by an independent panel, appointed by the Institute for Nuclear Power Operations. Workers are trained at the plant's Susquehanna Center, which became the first member of the National Academy for Nuclear Training in 1986. The academy will eventually include "campuses" at all of the nation's nuclear power plants.

News and accomplishments of 1990. During 1990, Susquehanna Station produced 14.727 million kilowatt-hours of electricity, a six percent increase over the PP&L's Corporate Employee Incentive Award Program Goal. This level of generation was the second highest in plant history.

Unit 2 set a plant record for completing a refueling and inspection outage in the shortest time since the plant began operating. The 60-day outage, which included the removal and replacement of about one-quarter of the unit's fuel assemblies, beat the previous record of 66 days set by Unit 1.

The plant's Nuclear Department personnel suffered only 17 injuries during the year, a reduction of 19 percent from the 1989 total, making 1990 the safest year since the plant began operating. Several significant safety milestones were reached during the year, including:

- 3,000,000 safe manhours in Nuclear Plant Engineering
- 2,000,000 safe manhours in the Nuclear Department
- 1,500,000 safe manhours in Nuclear Services Nuclear Quality Assurance
- 1,000,000 safe manhours in the Nuclear Department's Operations Section.

In radiological safety, total radiation exposure of department personnel was 440 manrem, which slightly exceeded the department limit. This was the lowest total manrem since the start of commercial operation. There were 315 personnel contaminations during the year.

Operational review. During 1990, PP&L performed an Operational Effectiveness Review of the plant, which took 175 manhours to complete. Out of 25,000 comments from 1,300 people, 55 major recommendations were chosen and implemented.

News and accomplishments of 1991. Susquehanna Station was honored by GE Nuclear Energy for having a 95.6 percent capacity factor during 1991. The plant placed first among boiling water reactors world-wide.

Construction began on a new 15,000 square-foot addition to the Susquehanna Training Center, which was to house an upgraded control room simulator. The new simulator was expected to be operational by late 1992, and predicted to provide more effective training of plant operators and those studying for the federal NRC operator license examination.

In June, work began at Susquehanna Station to remove a number of used reactor components, stored underwater in the plant's spent fuel pools. After the removal, the components were to be shipped to a federally approved, low-level radioactive waste storage facility in Barnwell, South Carolina. The removal concerned a total of 98 control rod blades and 72 local power range monitors which were taken out of service over the past eight years during refueling outages for both units. The event marked the first time that the PP&L disposed of used reactor components. Work was to be completed in October 1991.

Problems. The NRC inspected Susquehanna Station in the summer of 1990, after PP&L informed them of problems with polyurethane damper actuator seals. During the inspection, the NRC found that while PP&L had also identified some concerns with certain Limitorque valve actuators in 1988, the utility had not established, by 1990, that the actuators were environmentally qualified. Because of this, the NRC fined the utility $25,000 in February 1991. The fine was mitigated by 50 percent from the base because of PP&L's later corrections and overall good past performance, and was paid.

In January 1992, a small blast occurred at Susquehanna Station, contaminating a worker with radioactive dust. The plant continued to operate, and no radiation was released into the atmosphere. The worker underwent decontamination procedures, and was to undergo tests to determine if his system absorbed any radioactivity.

Emergency preparedness planning. The average background radiation in the area around Susquehanna Station is about 150 millirems each year, and the plant adds less than one millirem annually. There are a system of in-plant monitors to detect and measure any radiation that might be released from the site. In addition, PP&L has established more than 170 environmental monitoring locations in areas outside the plant boundary. Federal and state officials also maintain radiation monitors around the plant. If an emergency occurs, plant and field monitors would be supplemented by mobile teams, made up of trained personnel, to measure and track radiation releases.

PP&L makes an emergency planning pamphlet available to the public. The pamphlet provides instructions, information and evacuation routes for people living in Columbia and Lucerne counties, which are in close proximity to Susquehanna Station.

Environmental studies. PP&L's radiological environmental monitoring program has been analyzing food, water and air samples for radioactivity in the area surrounding Susquehanna Station since 1971. These data provide a catalog of normal environmental conditions so that any irregularity can be pinpointed. The monitoring will continue throughout the life of the plant.

PP&L began nonradiological studies on aquatic life in the Susquehanna River more than 10 years before the projected startup date. There are a laboratory and a collecting station on the west bank of the river, and study areas include: water chemistry, algae, macroinvertebrates, fish life, and terrestrial activities.

Meteorological monitoring is also done around the site. Data is collected on wind velocity and direction, geology, hydrology, seismology, geography, and demography.

Recreation and public tours. The 400-acre Susquehanna Riverlands recreation area is open to the public seven days a week. The area has day-use facilities for fishing, picnicking, canoeing, bicycling, hiking, ball games, and other recreation. The Riverlands staff is available for special group programs, and inquiries or arrangements can be made by calling (717) 542-2306, or writing to Susquehanna Riverlands, R.R. 1, Box 1797, Berwick, PA 18603.

The Susquehanna Energy Information Center provides perimeter bus tours for groups of 21 people, and inside-the-plant tours for groups of 20 people. The center is located along Route 11, about 11 miles north of Berwick, Pennsylvania, and is open seven days a week. Special programs and tours can be arranged by calling: (717) 542-2131, or writing to: Susquehanna Energy Information Center, R.R. 1, Box 1797, Berwick, PA 18603.

Sources

Susquehanna Nuclear Power Plant. Pennslyvania Power and Light Company, n.d.
Susquehanna Steam Electric Station. Pennslyvania Power and Light Company.
Susquehanna Energy Information Center. Pennsylvania Power and Light Company, December 1990.
Understanding Emergency Planning: Susquehanna Steam Electric Station. Pennsylvania Power and Light, 1990-1991.
"Four Fines Proposed, One Imposed, One Fought." *Nuclear News,* February 1991.
1990 Nuclear Department Performance Report. Pennsylvania Power and Light, June 1991.
Inside Susquehanna. Pennsylvania Power and Light, July 1991.
"Susquehanna Nuclear Plant Earns High Marks in Evaluation." *Pennsylvania Power and Light News Release,* November 1, 1991.
"Small Blast at Nuclear Plant." *The Toronto Sun,* January 19, 1992.
"And the Winner is." *INFO,* U.S. Council for Energy Awareness, May 1992.
Inside Susquehanna. Pennsylvania Power and Light, July 1991.
"Susquehanna Nuclear Plant Earns High Marks in Evaluation." *Pennsylvania Power and Light News Release,* November 1, 1991.

—Charles Ceplecha

Three Mile Island Nuclear Power Plant, Unit 1

Rte. 441 S
Londonderry Township
PO Box 480
Middletown, PA 17057
T. Gary Broughton, V.Pres.
Phone: (717)944-7621

Owning Utility

GPU Nuclear Corporation
1 Upper Pond Rd.
Parsippany, NJ 07054
Carol Clawson, V.Pres., Commun.
Phone: (201)316-7000
Telex: 136-482

Contact

Douglas H. Bedell, Mgr., Pub.Info.
Three Mile Island 1 Nuclear Power Plant

Basic Facts

Unit 1 *Status:* Operable. *Type of Reactor:* Pressurized light-water reactor. *Megawatts (electric):* 824. *Megawatts (thermal):* 2535. *Reactor System Supplier:* The Babcock & Wilcox Co. (United States). *Generator Supplier:* General Electric Co. (United States). *Architect Engineer:* Stone & Webster Engineering Corp. (United States). *Owner:* Metropolitan Edison; Jersey Central Power and Light Company; Pennsylvania Electric Company. *Operator:* GPU Nuclear Corporation.

Key Dates

Unit 1 *Ordered:* 1966. *Construction began:* 1967. *Criticality:* 1974. *First Power:* June 1974. *Commercial Operation:* September 1974. *Shut Down:* 2014.

Operating Costs

Unit 1

1974	$3,351,000
1975	$14,226,000
1976	$17,840,000
1977	$13,287,000
1978	$17,954,000

Plant Report

Unit 1 at Three Mile Island impacted by Unit 2 accident. The Three Mile Island site in Londonderry Township, about 10 miles south of Harrisburg, Pennsylvania, is best known as the location of the worst mishap in the history of U. S. commercial nuclear power. The accident occurred in Unit Two. At the time, Unit One was shut down for refueling. Although not directly involved, Unit One suffered many repercussions.

A six-year shut down begins. The NRC ordered Unit One to remain shut down until the agency completed a comprehensive investigation of both units. The shutdown lasted a total of six and a half years. More than half a dozen groups, including the Union of Concerned Scientists, and the Three Mile Island Alert, filed lawsuits to prevent the plant from restarting. They raised objections about the plant's design, safety, and emergency planning. Management integrity was also questioned.

Incriminations against Unit One's management stemmed from a cheating incident in 1981. The NRC had ordered TMI-1's operators to be re-examined as part of the process to restart the plant. During the examination, a senior reactor operator copied answers from another. Although both operators were fired, rumors about more extensive cheating led the NRC to retest all Unit 1 operators.

The possibility of restarting Unit One was further delayed when a federal court ordered psychological surveys of the community to be conducted. One study done for the NRC reported high levels of anxiety symptoms in the local population. Other studies reported increases in alcohol and tranquilizer use.

Steam generator tubes cause problems. Equipment trouble forced more delays. Tubes in both of Unit One's steam generators showed signs of excessive corrosion and many were found to have serious leaks. Thousands of steam generator tubes, like the one that ruptured at the Ginna plant in New York, had to be modified to prevent further cracking. The improvements took a year to complete.

Ready to re-start. On June 8, 1985, plant operators brought the plant into "hot standby" readiness after the NRC gave its approval for TMI-1 to restart. Fuel oil and electricity costs to maintain hot-standby status totaled $40,000 per day. Costs mounted as legal maneuvering forced further delays.

Four months later on October 3, 1985, TMI-1 reinitiated criticality at 1:30 p.m. On October 9, operating at 15 percent power, the plant went back on line. Full power was achieved on January 6, 1986.

GPU Nuclear Corporation. During the long outage, Metropolitan Edison, Uniit One's original operator, was succeeded by GPU Nuclear Corporation. The NRC transferred TMI's operating license to GPUN in 1982 stating that the newly formed corporation represented a significant improvement in organizational structure.

GPUN is a subsidiary of General Public Utilities Corporation (GPU), a holding company for Metropolitan Edison, Pennsylvania Electric Company, and Jersey Central Power and Light Company. Together the three utilities serve more than 4 million people in Pennsylvania and New Jersey. In addition to TMI-1, GPUN operates the Oyster Creek Nuclear Generating Station in New Jersey and the permanently shut down TMI-2.

Improvements made after restart. In 1986 TMI-1 switched to a more enriched form of uranium fuel and changed its fuel bundle arrangements. These modifications put the plant on an 18-month fuel cycle instead of a 12-month cycle which boosted productivity. Further efficiency was obtained through the implementation of an infrared thermal imaging process used to diagnose mechanical problems without cutting pipes or taking equipment apart. These improvements helped TMI-1 in 1989 to achieve a world record capacity factor of 100.03 percent.

In 1990 Three Mile Island received top ratings when the NRC granted it Category 1 status in seven areas on its Systematic Assessment of Licensee Performance (SALP) report. *Nuclear News* Magazine credited the plant for having the most improved efficiency rating over a three year period. Three Mile Island set another world record in 1991 by completing a continuous operating run of 479 days. At the time, it was the longest run achieved by any light water reactor.

The future at TMI-1. In spite of the achievements at TMI-1, GPUN is uncertain about its future. The 1990 GNU annual report cautiously points out that operations at TMI-1 and its Oyster Creek plant "for the full term of their now assumed lives cannot be assured."

The plant, which began commercial operation in September 1974, is scheduled to be retired in 2014. In accordance with NRC requirements, GPU established a sinking fund to provide cash for eventual decommissioning, estimated to cost $130 million (in 1990 dollars). TMI-2, currently in long term storage, will be decommissioned at the same time.

Sources

"Mopping Up at Three Mile Island." *Maclean's*, March 29, 1982.
"Steam Tubes Further Stay TMI-1 Restart." *Science News*, April 24, 1982.
"A Fresh Start at Three Mile Island?" *Science News*, June 8, 1985.
"False Start at TMI." *Science News*, September 7, 1985.
"The Eureka File." *Nuclear Industry*, First Quarter 1990.
General Public Utilities Corporation 1990 Annual Report. General Public Utilities Corp., n.d.
"GPU Nuclear Corporation — In Profile." *Backgrounder*. GPU Nuclear Communications, September 1991.
"Three Mile Island Unit 1: Unit 1 in Profile." *Backgrounder*. GPU Nuclear Communications, September 1991.
"You Can't Keep a Good Plant Down." *Nuclear Industry*, First Quarter 1992.
General Public Utilities Corporation 1990 Annual Report. General Public Utilities Corp., n.d.
"GPU Nuclear Corporation — In Profile." *Backgrounder*. GPU Nuclear Communications, September 1991.
"Three Mile Island Unit 1: Unit 1 in Profile." *Backgrounder*. GPU Nuclear Communications, September 1991.

—*Karen Bellenir*

370 ★
Three Mile Island Nuclear Power Plant, Unit 2

Rte. 441 S
Londonderry Township
PO Box 480
Middletown, PA 17057

Owning Utility

GPU Nuclear Corporation
1 Upper Pond Rd.
Parsippany, NJ 07054
Carol Clawson, V.Pres., Commun.

Phone: (201)316-7000
Telex: 136-482

Basic Facts

Unit 2 *Status:* Shut down. *Type of Reactor:* Pressurized light-water reactor. *Megawatts (electric):* 906. *Megawatts (thermal):* 2772. *Reactor System Supplier:* The Babcock & Wilcox Co. (United States). *Generator Supplier:* Westinghouse Electric Corp. (United States). *Architect Engineer:* Burns & Roe, Inc. (United States). *Owner:* Metropolitan Edison; Jersey Central Power and Light Company; Pennsylvania Electric Company. *Operator:* GPU Nuclear Corporation.

Key Dates

Unit 2 *Ordered:* 1969. *Construction began:* 1970. *Criticality:* March 1978. *First Power:* April 1978. *Commercial Operation:* December 1978. *Shut Down:* March 1979.

Plant Report

What happened. The worst accident in the history of American commercial nuclear power generation occurred near Middletown, Pennsylvania, on March 28, 1979.

At 4 a.m. that morning, Three Mile Island Unit 2, operating at full power, began experiencing a partial meltdown and suffered a Loss of Cooling Accident (LOCA) so dangerous that it is a "worst-case scenario" in the industry's safety manuals. The plant automatically shut down when a pump bringing cooling water to the reactor core stopped functioning.

The accident lasted for about five hours. It triggered the release of 2.5 million curies of radioactive noble gases and 15 curies of radioiodines, according to the *Rogovin Report*, the Nuclear Regulatory Commission's (NRC) inquiry into the accident. Those figures are less than one percent of the annual dose from both natural background radiation and medical practice, the report indicated.

All radiation monitors in TMI-2's vent stacks, where 80 percent of the radiation escaped, went off scale during the accident and thus were inoperable for three days.

The nuclear fuel core was uncovered for approximately two-and-a-half hours as water was leaving the reactor through a stuck-open valve. Approximately 32,000 gallons of contaminated coolant water that had covered the core, more than one-third of the reactor coolant system's capacity, escaped from the system and spread across the reactor's basement floors and auxiliary buildings. Workers shut the valve, but not in time to prevent much of the core from melting and radiation from escaping. The intensely hot and highly radioactive fuel rods were exposed during the accident.

The sequence of events included a loss of feedwater flow, a primary relief valve sticking open, the premature turning off of the emergency core cooling system, and later the turning off of all reactor coolant pumps.

The following day, Metropolitan Edison Corporation (Met Ed), a subsidiary of TMI-owner General Public Utilities (GPU) and the plant operator when the accident happened, said that less than one percent of the reactor's fuel was damaged. However, technicians later estimated a 50 percent fuel melt and internal temperatures that exceeded 5,000 degrees, 3,000 more than originally cited. Hydrogen "bubbles" of gas formed in the reactor's coolant system apparently causing the threat of an explosion and traces of radioactive iodine were detected nearby.

Pennsylvania Gov. Richard Thornburgh then advised evacuation of an estimated 3,500 pregnant women and children living within five miles of the plant. Instead, about 200,000 people fled their homes. On April 9, NRC officials declared that the crisis had ended, and Thornburgh lifted the order to close nearby schools, saying it was safe for pregnant women and children to return.

Social impact. The TMI-2 accident ignited a chain reaction of private, political and scientific criticism of the handling of the crisis and of the U.S. commercial nuclear industry as a whole.

Three Mile Island Alert (TMIA), formed two years before the accident, became the lead citizen intervenor and is still at the forefront of Three Mile Island public opposition.

In addition to TMIA, as many as six public interest groups formed after the accident. For instance, Concerned Mothers and Women began fighting for health and safety rights of families near TMI. The group routinely met with NRC and public officials to express concern over GPU and NRC practices at TMI. The Susquehanna Valley Alliance (SVA) successfully blocked GPU from dumping the accident-generated water into the Susquehanna River.

Protests, which were organized immediately after the accident and ensued thereafter, included the following: Between 5,000 and 10,000 people participated in a "Women and Children" march to Met Ed headquarters; an estimated 100,000 people marched in Washington, D.C., in response to the accident and to demand an end to the United States' reliance on nuclear power; on March 28, 1980, the accident's first anniversary, peaceful rallies were staged at nuclear plants and utilities in several states; and the Rev. Jesse Jackson and consumer advocate Ralph Nader attended a vigil and a rally, respectively, on the fifth anniversary.

Many local residents went into a frenzy after the accident. On March 30, then Pennsylvania Gov. Richard Thornburgh ordered the evacuation of pregnant women and children within a five-mile radius from the plant. As many as 200,000 people fled the area.

Three Mile Island sits in the middle of the Susquehanna Valley. About two million people lived within a 50-mile radius of the plant when the accident occurred.

The Rogovin report concluded that nobody was completely informed about the incident itself or the course of events that followed, a major cause of chaos. As a result, according to the report, "The residents around Three Mile Island were unduly confused and alarmed, and the level of anxiety nationwide about the safety of nuclear plants was unnecessarily raised."

The report noted that the accident-related information Met Ed and the NRC provided to the news media was "often inaccurate, incomplete, overly optimistic, or ultraconservative. Errors in judgement by Met Ed and NRC officials were major contributors to the inadequate public information effort at TMI." The study also blamed the NRC for the shoddy information that reached Harrisburg, Pennsylvania, the state capitol located 10 miles from TMI.

Congress was also swift in reacting to TMI. Congressional hearings and investigations into the accident were plentiful, and federal laws impacting on the nuclear industry have either been passed or amended. Most notably, the Price-Anderson Act, passed as a temporary measure in 1957, was reauthorized and modified in 1988 to grant the industry no-fault insurance and a cap on liability for personal injury and property damage resulting from a nuclear accident. Now, in the event of an accident, the industry needs to pay annually up to a cap of about $7 billion. The cap was originally $695 million. Three Mile Island was not serious enough to trigger Price-Anderson's provisions.

Today, Three Mile Island is a fascinating tourist attraction in a region that has seen tremendous economic and population growth since the accident. GPU welcomes travelers at both a Three Mile Island visitors center and on the island itself.

Environmental impact. Governmental studies such as the Rogovin report and the Kemeny Commission revealed no negative environmental affects from the accident. However, it is unclear if they made a concerted effort to search for environmental changes.

Public activists and residents of the area's farm towns have since complained of crop failures, of livestock unable to bear young, and of giant mutated vegetation.

But unlike the 1986 Chernobyl disaster in the Soviet Union and the 1957 Windscale disaster in Britain, there was no major release of radioactive iodine 131. The Three Mile Island accident triggered the release of 2.5 million curies of radioactive noble gases and 15 curies of radioiodines, according to the Rogovin Report. Those figures are less than one percent of the annual dose from both natural background radiation and medical practice.

In 1980, the NRC ignited a storm of controversy, allowing GPU to release 43,000 curies of radioactive Krypton gas from the plant into the atmosphere for 11 days. TMI-2 was designed to release approximately 770 curies of Krypton a year.

Virtually all of the iodine 131 that remained in the reactor at the time of the accident, an estimated 66 million curies, has decayed away.

Cleanup efforts. Shortly after the accident, authorities began cleanup operations that would eventually total more than $1 billion for TMI-2, which was decommissioned and is now in long-term monitored storage. Some nuclear experts originally estimated that the cleanup would cost around $40 million and the facility would be repaired and put back into service.

The TMI-2 Safety Advisory Board, under the auspices of General Public Utilities Nuclear (GPUN), a GPU subsidiary formed after the accident to oversee TMI-1 and TMI-2 and its prime contractor, Bechtel Corporation, coordinated the cleanup. The Board was composed of more than 1000 international specialists in nuclear science, engineering, chemistry, physics, mathematics, medicine, and government.

The cleanup's most integral part was the removal and shipment of the damaged reactor fuel. More than 150 tons of damaged fuel and other debris were removed from the reactor vessel.

The initial step occurred in August 1979, when the first low-level, accident-generated waste was shipped to Richland, Washington. Afterward, accomplishments included the removal of the plant's damaged core in 1985, defueling, and shipment of about 308,000 pounds of fuel debris and core structural components to the Idaho National Engineering Laboratory in April 1990.

Through cleanup efforts, new materials were developed that are now used by the Germans, the French, and the Japanese in their nuclear power endeavors. Incidentally, Japan contributed approximately $18 million to GPUN in the cleanup just for the privilege of studying it. Some other major contributors included insurers ($300 million), customers through electric rates ($257 million), and the U.S. nuclear industry ($152 million).

In February 1991, the National Society of Professional Engineers named the TMI-2 cleanup, a 12-year project, one of the top engineering achievements in the United States in 1990. Nevertheless, several fires prompted the NRC to cite GPU for violations. In addition, TMIA found the cleanup tainted, highlighting information about two liquid radioactive waste spills, and a cleanup worker who was contaminated after falling into a pool of radioactive water.

Overall, the accident altered the development of nuclear facilities and led to electricity from nuclear power eventually exceeding coal in cost. Regulators began requiring utilities to make billions of dollars in changes and many plants were abandoned, some in the late stages of construction.

Judicial impact. The TMI-2 accident ignited various types of lawsuits, ranging from the local citizens versus the utility, the utility versus the manufacturer, and the government versus the utility.

More than 2,200 area residents have filed lawsuits because of health problems that they attribute to radiation released during the accident. So far, about 280 claims have been settled for a total of about $14 million.

A look at some other litigation provides further insight into the legal entanglement promulgated by the accident:

- In October 1979, the NRC fined Met Ed $155,000.
- After GPU vented 43,000 curies of radioactive Krypton gas into the atmosphere in 1980, the U.S. Court of Appeals for the District of Colombia ruled that the venting was illegal, but another court later found that NRC officials were immune from suit for approving the venting.
- A 1981 landmark class-action suit resulted in a $25 million settlement for TMI-area residents against the plant's owners. Five million dollars went toward creating the Three Mile Island Public Health Fund, which studies the health effects of radiation on people living near TMI.
- GPU sued Babcock & Wilcox, which manufactured TMI-2, for $4 million in 1979. In 1983, the parties settled for $37 million out of court on the condition that Babcock & Wilcox pay in the form of rebates on services to GPU.
- In 1983, the NRC fined GPU $140,000 for submitting a false statement regarding the license certification of Jim Floyd, the TMI-2 supervisor of operations who had cheated on his license requalification exam four months after the accident.
- The U.S. Department of Justice indicted Met Ed in 1983 for falsifying leak rate data and destroying documents before the accident. In a U.S. District Court in 1984, Met Ed pleaded

guilty to one count and no contest to six counts of an 11-count indictment.

- In 1983, three Three Mile Island employees submitted affidavits, charging GPUN and Bechtel with cutting corners in cleanup efforts and circumventing necessary safety. In turn, the NRC claimed the companies violated regulatory procedures.
- In 1984, GPU agreed to an out-of-court settlement with cleanup worker William Pennsyl, who was fired in 1982 for insisting he be allowed to wear a respirator during the cleanup.
- In 1985, a local Dauphin County court approved GPU's $4 million settlement offer to 70 families with children who had been injured by accident-related radioactive contamination, including $1,095,000 for a 5-year-old boy born nine months after the accident with Down's syndrome.

Health impact. Extensive studies conducted to document health-related problems causedd by radioactive gas leaks during the TMI-2 accident indicate that, from a public health standpoint, the accident was not a disaster.

However, Three Mile Island was a serious accident in terms of the severe mental stress caused among tthe population near the plant, which was greatly exacerbated by the press coverage. A 1991 study by Columbia University and National Audubon Society researchers reported that cancer rates increased among people living within about three-and-a-half miles of the plant in 1982 because of stress, not radiation.

The Pennsylvania State Health Department found no evidence of increased cancer through radiation, according to a report issued in 1985. In 1990, the Three Mile Island Public Health Fund announced that, while rates of certain cancers rose among the 160,000 people within a 10-mile radius of the plant, the increases were not statistically significant.

Two more studies by the state health department in 1991, one of 1,000 people living within a five-mile radius of the plant and another of 5,292 women of childbearing age living within a 10-mile radius, not only found no association between radiation and cancer, but no association between psychological stress and cancer.

Public interest groups such as TMIA are quick to discredit any government-founded results. "Because of certain government biases, such as the Nuclear Regulatory Commission's position that little radiation was released during the accident, and because no one knows precisely how much actually got out, the government has not conducted responsible health studies," according to TMIA in 1989.

A state Health Department study in the mid-80s found that the nuclear accident did not affect the incidence of fetal and infant deaths, congenital defects, and premature births within a 10-mile radius of TMI. But Three Mile Island opponents were alarmed when denied access to the data, contending that it was unscientific and did not support the evidence.

Government officials have also labeled private studies unscientific. In 1984, a local couple, Norman and Marjorie Aamodt, did a survey and concluded that there was much more cancer around the reactor than would be expected. However, state officials argued that the Aamodts counted cancer victims who moved into the area after the accident, and that it often takes more than 10 years for many radiation-induced cancers to show.

Gordon MacLeod, Pennsylvania's Secretary of Health at the time of the accident, admitted that the state health department "lacked the tools to deal with the extraordinarily serious health problems facing Pennsylvania during and after the accident."

Regulatory impact. The NRC was shaken by the TMI-2 accident and the events which followed. NRC Chairman Joseph Hendrie was moved aside and the agency's executive director for operations resigned. Numerous investigations, reports, and resulting and related actions wrought change in the NRC and the way it operates.

Special inquiries were conducted in the agency, for the agency, and of the agency to discover the accident's cause and meaning.

Some investigations are extremely notable, particularly the Rogovin report, the Kemeny Commission report, and congressional hearings staged by various Senate and House committees. Rogovin and Kemeny passed strongly negative judgements on the NRC.

Shortly after the accident, the NRC hired the Washington, D.C., law firm of Rogovin, Stern & Huge to conduct its own investigation. Mitchell Rogovin headed the study, which noted, "We have found in the Nuclear Regulatory Commission an organization that is not so much badly managed as it is not managed at all. In our opinion, the Commission is incapable, in its present configuration, of managing a comprehensive national safety program for existing nuclear powerplants adequate to ensure public health and safety."

The Kemeny Commission report was authorized by President Jimmy Carter and chaired by John Kemeny, the Dartmouth College president. In its overall conclusion, the report stated, "To prevent nuclear accidents as serious as Three Mile Island, fundamental changes will be necessary in the organization, procedures, and practices—and above all—in the attitudes of the Nuclear Regulatory Commission and, to the extent that the institutions we investigated are typical, of the nuclear industry."

Other studies and congressional hearings shed light on the incident, its impact on the NRC and the industry, and the future of TMI. For instance, the House Subcommittee on Energy, Research and Production compared the accident's impact on the industry to how the 1967 Apollo fire, in which three astronauts died, sent notice to the America's space program.

In the accident's aftermath, the NRC and the nuclear industry took decisive steps to improve the safety of nuclear power plants. A NRC report released 10 years after the accident noted an improved safety commitment within the agency and the industry.

Nevertheless, public and private skepticism of the NRC remained steady. To this day, TMIA and the Nuclear Information and Resource Service in Washington, D.C., continue to criticize

Plant Profiles

the NRC for what they perceive as negligence during and after the accident.

For example, as a result of the accident, congressional mandate required the NRC to have utilities and state governments organize evacuation plans for a 10-mile zone around every nuclear facility. But once Massachusetts Governor Michael Dukakis and New York Governor Mario Cuomo refused to license plants in their jurisdictions, the NRC rescinded the rule, saying utilities could implement their own evacuation plans without state intervention.

The future of TMI-2. TMI-2, which is not included in GPU's present energy supply plans, is in monitored storage after defueling was completed in fiscal year 1990. No funds are presently being expended to preserve the plant or equipment for future use.

TMI-2 is expected to remain in monitored storage until sometime in the next century, when it will be decommissioned along with TMI-1.

Sources

"Kemeny Commission Releases TMI Report." *Groundswell*, Nuclear Information and Resource Service, Washington, D.C., November 1979.
Report of The President's Commission on The Accident at Three Mile Island. Washington, D.C., 1979.
Three Mile Island: A Report to the Commissioners and to the Public. Nuclear Regulatory Commission Special Inquiry Group, Washington, DC, 1980.
Nuclear Powerplant Safety After Three Mile Island. Subcommittee on Energy, Research and Production, House of Representatives, March 1980.
Some Public Health Lessons From Three Mile Island: A Case Study in Chaos. Human Environment Research and Management, Volume 10, Number 1, 1981.
"A Reporter At Large: Three Mile Island." (Part 1), *The New Yorker*, April 6, 1981.
"A Reporter At Large: Three Mile Island." (Part 2), *The New Yorker*, April 13, 1981.
Financing the Cleanup of Three Mile Island Unit 2 Nuclear Powerplant. Subcommittee on Energy and the Environment, House of Representatives, April 23 and 27, 1982.
"$380 Million and Counting." *Newsweek*, October 10, 1983.
"TMI Opponents Denied Access to State Health Data." *The Philadelphia Inquirer*, January 7, 1986.
"Three Mile Island Cleanup Uncovers New Details." *The New York Times*, January 26, 1988.
A Decade of Delay, Deceit and Danger: Three Mile Island 1979-1989. Three Mile Island Alert, Inc., Harrisburg, Pa., 1989.
The Three Mile Island Accident, 10th Anniversary: A Decade of Delay, Deceit and Danger. Three Mile Island Alert, Inc., Harrisburg, PA, 1989.
"Cleansing the Atom." *Life*, March 1989.
"TMI: Ten Years Later." *Newsday*, March 26, 1989.
"A Decade After Accident, Legacy at TMI Is Mistrust." *The Washington Post*, March 28, 1989.
"Three Mile Island - From Disaster to a Fascinating Tourist Attraction." *The Boston Globe*, February 11, 1990.
Three Mile Island - Unit 2 Safety Advisory Board. General Public Utilities Nuclear, March 1990.
"Litigation Is in the Air Over Hanford." *Seattle Times*, August 7, 1990.
"Study of Three Mile Island Accident Finds Negligible Increase in Cancers." *The New York Times*, September 1, 1990.
"Stress May Have Boosted Cancer Rates Near Damaged Nuclear Plant, Study Says." *The Washington Post*, May 27, 1991.
Myth Busters 7. Safe Energy Communication Council, Washington, D.C., Winter 1992.
"Litigation Is in the Air Over Hanford." *Seattle Times*, August 7, 1990.

"Study of Three Mile Island Accident Finds Negligible Increase in Cancers." *The New York Times*, September 1, 1990.

—Michael Richman

Trojan Nuclear Plant

71760 Columbia River Hwy. Phone: (503)556-3713
Rainier, OR 97048 Fax: (503)556-7888
William R. Robinson, Gen.Mgr., Trojan Plant

Owning Utility

Portland General Electric Company Phone: (503)464-8000
121 S.W. Salmon Fax: (503)464-2233
Portland, OR 97204 Telex: 62934967
James C. Cross, V.Pres. & Chief Nuclear Officer

Contact

Steve P. Sautter, Nuclear Energy Info.
Portland General Electric Company

Basic Facts

Status: Operable. *Type of Reactor:* Pressurized light-water reactor. *Megawatts (electric):* 1178. *Megawatts (thermal):* 3411. *Reactor System Supplier:* Westinghouse Electric Corp (United States). *Generator Supplier:* General Electric Co. (United States). *Architect Engineer:* Bechtel. *Owner:* Portland General Electric Company; Pacific Power and Light; Eugene Water and Electric Board. *Operator:* Portland General Electric.

Key Dates

Ordered: 1968. *Construction began:* 1971. *Criticality:* December 1975. *First Power:* December 1975. *Commercial Operation:* May 1976. *Shut Down Expected:* 1996.

Operating Costs

1976	$5,921,000	1983	$30,344,000
1977	$13,628,000	1984	$46,089,000
1978	$15,204,000	1985	$46,172,000
1979	$16,958,000	1986	$52,017,000
1980	$25,790,000	1987	$79,131,000
1981	$32,205,000	1988	$79,131,000
1982	$30,629,000	1989	$102,294,000

Plant Report

Background. Located on the Columbia River 45 miles northwest of Portland in Rainier, Oregon, the Trojan Nuclear Plant began producing electricity commercially on May 20, 1986 after almost five years of construction. In 1977, Trojan's first complete year of operation, the plant produced 6.5 kilowatt-hours of electricity, more than any other plant in the United States. Trojan cost PGE $4.460 million which was higher than the original estimate. Throughout the construction process, PGE obtained approvals from 60 county, state, regional, and federal agencies for different areas of operation.

Portland General Electric Co., a subsidiary of Portland General Corp., owns 67.5 percent of Trojan. The plant provides approximately 25 to 30 percent of PGE's total electricity load and approximately 15 to 20 percent of the power load statewide.

History of the Trojan site. The site chosen for construction of Trojan offered archaeologists some interesting historical information. It was discovered that the area was used by Indians for camping and fishing. Later, the area was settled by pioneers who farmed and logged the land. During World War I, the site was purchased by Trojan Powder Co., where it was used as a depot, but after World War II, the company phased out all of its operations. Up until 1968, when Portland General Electric purchased the site, it had been vacant. After a six-year intensive study, PGE purchased the site from Trojan Powder Co. for $750,000.

Fuel and waste disposal. Trojan Nuclear Plant utilizes uranium-235 fuel for the necessary chain reaction that produces electricity. The 1130-megawatt plant houses 193 fuel assemblies, each composed of 264 fuel rods. Each rod contains 271 uranium dioxide pellets. In all, there are 13.8 million fuel pellets contained within 50,952 fuel rods. After Trojan produces approximately 7.5 billion kilowatt-hours of electricity, it must be shut down for refueling and maintenance. Generally, this procedure is done on an annual basis.

The by-products of the chain reaction slowly build up in the fuel and will eventually block the process. During refueling, spent fuel rods are removed and replaced, and remaining rods are moved to a new position. Approximately one-third of the 193 fuel assemblies are replaced each year.

Currently, Trojan stores these spent fuel rods on-site in a specially designed storage facility. The underground pool can store up to 1,408 used fuel assemblies. Eventually, as prescribed in the Nuclear Waste Policy Act of 1982, the U.S. Department of Energy will open a permanent underground repository for the nation's high-level nuclear waste.

Trojan also produces some low-level nuclear waste which is generally found in paper products and filters, protective clothing, tools, and radioactive filters and filtering agents. These contaminated materials are sealed in steel containers and transported and stored at a disposal site in Hanford Nuclear Reservation located in Washington State. However, plant employees do their best to minimize the amount of low-level waste that is produced.

Violations and fines. Since the Trojan Plant opened in 1976, it has incurred numerous Nuclear Regulatory Commission (NRC) fines and violations, although not every violation at Trojan has led to a fine by the NRC.

The NRC uses a grading system to judge how severe a violation is and how much the corresponding penalty will be. Violations are divided into five categories. A Level I violation is the most severe because these transgressions are viewed as having a direct impact on the safety of the public. Level V violations are the least severe and they only cause minor concern. Generally, a Level IV or V violation will not incur a fine.

The following represents some of Trojan's violations and associated fines:

- July 1983: Control rods were inadequately tested three times in a three-year period.
- December 1983: PGE discovered that inexperienced and untrained supervisors were signing welding and cutting permits. This same problem was found in January 1987. In September 1987, Portland General was cited.
- In 1983 and 1984, Trojan was one of 20 plants out of 105 total U.S. nuclear plants that had the most significant safety violations.
- In 1983 and 1986, a total of 71 mishaps had resulted in federal citations. Three of these violations resulted in a fine of $180,000.
- September 1987: Officials at Trojan failed to have a fire protection engineer review all plant design changes after remodeling.
- September 1987: Inadequate training was provided to both the on-site fire brigade and the volunteer fire department.
- July 1989: First identified by plant personnel, debris was found in and around a recirculation pump in the Containment Building. It was also discovered that some screens meant to keep debris out were either damaged or missing. This violation resulted in a $280,000 fine (an NRC Level II infringement).
- March 1991: Medical records at Trojan were found to be in violation of established requirements for workers with operator licenses. The NRC fined PGE $50,000 for this violation (an NRC Level III infraction).

Extensive closures. Another controversial issue associated with Trojan is the extensive periods of time that the plant has remained shut down. A history of Trojan's major shutdowns follows:

- 1978: Trojan was closed for most of the year for refueling, maintenance, and construction modifications.
- 1986: The plant was closed just under three months for unplanned repairs.
- 1987: The facility was shut down for more than five months for scheduled and unscheduled repairs. Also during 1987, the plant was closed four separate times due to unplanned repairs.
- 1989: Trojan was closed for approximately five months due to federal safety and maintenance violations.
- 1991: Trojan was closed most of the year due to minuscule cracks and corrosion that were discovered in tubes that converted heat to steam in the turbine generator.

High costs of Trojan to Portland General. The financial impact of Trojans closures on PGE and Portland General Corp. has been expensive. During the first half of 1989, Portland General Corp. spent $44 million on Trojan for operation, repair and maintenance, which is the highest amount paid for a half-year operation since the plant opened in 1976. This amount represents approximately $15 million more in costs on Trojan than in the same period for the previous year.

Plant Profiles

For the year 1989, a $26.9 million loss on $797 million in revenues was posted by Portland General. In comparison to a $97.3 million profit on $756.3 million in revenues for 1988, the loss in 1989 was significant for Portland General Corp. By the end of 1989, Trojan cost the company an additional $19 million due to cost overruns found during a routine shutdown. As of third quarter 1991, due to an extensive closure of Trojan in 1991, Portland General incurred a loss of $438 million compared to the same period in 1990 when it reported net income of $31.9 million.

Voters keep Trojan open. Trojan's many problems have caused great concern for local activists. They feel that the violations and emergency shutdowns may indicate deeper problems. They also feel that PGE has been too slow to correct problems. Activists of the Don't Waste Oregon Committee, represented by Portland attorney Greg Kafoury, have tried twice to close the plant. Most recently, in 1990, the group obtained the necessary 62,568 petition signatures needed to place the measure on the November ballot.

In 1986, PGE defeated a measure to close Trojan by spending $2.5 million in advertising and public relations. However, in 1990, activists had more ammunition. First, questions regarding Trojan's ability to withstand a major earthquake were causing great concern among voters. Second, California voters had successfully closed Rancho Seco Nuclear Power Plant. And third, the unavailability of a federal nuclear waste depository site in the United States until at least the year 2011 was contributing to the increasing negative view of nuclear power plants.

The measure stated that PGE would be forced to close Trojan until a federal repository was built, the plant was proven cost effective, and Trojan was shown to be able to withstand a major earthquake. Opponents of the measure argued that if Trojan wasn't meeting safety standards, then the NRC and/or the Department of Energy would have shut the plant long ago. They also cited that the expense PGE would have to incur in order to decommission Trojan would be astronomical.

The Oregon Public Utility Commission (PUC) estimated that a decommissioning would probably cost PGE $270 to $900 million. Northwest Power Planning Council estimated this cost to be higher at $1.5 to $3 billion. Proponents of the measure argued that safety should be the deciding factor, not cost.

In case of a closing, PGE had made call-back provisions with two California electric companies to supply power. A spokesperson for PGE stated that with expected population growth of 3-4 percent a year, consumers couldn't afford to close Trojan because of the unpredictability of water supplies and the hydroelectric dam system. Proponents argued that since output from Trojan only produces an average of 425-megawatts, PGE could make up the difference through conservation. Only after a hotly contested battle on both sides were voters able to defeat the measure.

PGE makes efforts to change. PGE made general staff changes by hiring 200 contractors into full-time positions to increase employee stability and loyalty. Experienced consultants were also hired so that PGE could learn from the success and mistakes of other plants. In 1989, Trojan opened a training center to provide PGE employees, as well as contract personnel, necessary training for safe operation of the plant. This program followed the December 1987 accredited status Trojan achieved with the recently formed National Academy for Nuclear Training.

Although Trojan encountered many operational and management problems over the years, PGE has seemed to work to rectify these issues. The fines that Trojan incurred have been paid without protest, and management has proceeded with reorganization to avoid future incidents. Through management promotions and reassignments, Portland General is hoping to prosper in a more competitive environment while at the same time improve communication among managers for problem solving.

Trojan to shut down in 1996. PGE recently decided to close the Trojan Nuclear Plant five years sooner than originally planned. The company concluded that continued operation of the plant through 2011 would be economically infeasible, with cost estimates as high as $125 million, due to a local abundance of inexpensive natural gas and hydropower. PGE plans to replace some of the energy produced by the plant with 600 megawatts of combustion turbines, to begin operation when Trojan closes.

Plant tours. Plant tours are available on specific days. Please call (503) 464-7200 for tour information.

Sources

"Feds Demand an Answer to Trojan Safety Flaws." *The Business Journal-Portland*, December 28, 1987.

"Costly Trojan Repairs Torpedo Utility's Profits." *The Business Journal-Portland*, August 14, 1989.

"Utility Revamps Management and Strategy." *The Business Journal-Portland*, November 6, 1989.

"Portland General Electric Notified of Proposed Civil Penalty by NRC." *Business Wire*, October 9, 1989.

"Management Changes Faces Problems at Trojan." *The Business Journal-Portland*, December 11, 1989.

"PGE Addresses Trojan Ills, Critics Remain Unconvinced." *The Business Journal-Portland*, March 19, 1990.

"PGE's Trojan Nuclear Unit Shows Improved NRC SALP Rating." *Business Wire*, June 7, 1990.

"Measure 4 Puts Trojan Power Questions to the People." *The Business Journal-Portland*, October 29, 1990.

"PGE Pays $50,000 Fine, Tries to Correct Trojan Violations." *The Business Journal-Portland*, March 4, 1991.

"Kay Stepp: Electric Utility Exec Shuns Limelight, Seeks Results as Head of PGE." *The Business Journal-Portland*, March 11, 1991.

"Nuclear Regulatory Commission Issues Annual Report Card for Trojan Plant." *Business Wire*, June 20, 1991.

The Trojan Nuclear Plant, Portland General Electric, n.d.

—*Yvette M. Costa*

Turkey Point Nuclear Station, Units 3-4

PO Box 4332
Princeton, FL 33032
L.W. Pearce, Plant Mgr.

Phone: (305)246-1300
Fax: (305)246-6225

373 Tyrone

UNITED STATES

Owning Utility

Florida Power and Light Company
700 Universe Blvd.
PO Box 14000
Juno Beach, FL 33408-0420
J.H. Goldberg, Pres., Nuclear Div.

Phone: (407)694-4000
Fax: (407)694-4311

Contact

R.R. Golden, Nuc. Corp. Commun. Coord.
Florida Power and Light Company

Basic Facts

Units 3-4 *Status:* Operable. *Type of Reactor:* Pressurized light-water reactor. *Megawatts (electric):* 699. *Megawatts (thermal):* 2200. *Reactor System Supplier:* Westinghouse Electric Corp. (United States). *Generator Supplier:* Westinghouse Electric Corp. (United States). *Architect Engineer:* Bechtel. *Owner:* Florida Power and Light Company; Florida Municipal Power Agency; Orlando Utilities Commission. *Operator:* Florida Power and Light Company.

Key Dates

Unit 3 *Ordered:* 1965. *Construction began:* 1967. *Criticality:* 1972. *First Power:* November 1972. *Commercial Operation:* December 1972. *Shut Down Expected:* April 2007.

Unit 4 *Ordered:* 1965. *Construction began:* 1967. *Criticality:* June 1973. *First Power:* June 1973. *Commercial Operation:* September 1973. *Shut Down Expected:* April 2007.

Operating Costs

Units 3-4

Year	Cost	Year	Cost
1972	$247,000	1981	$30,275,000
1973	$4,059,000	1982	$32,067,000
1974	$9,660,000	1983	$47,777,000
1975	$15,492,000	1984	$60,054,000
1976	$18,602,000	1985	$73,203,000
1977	$15,109,000	1986	$118,338,000
1978	$18,602,000	1987	$123,474,000
1979	$22,510,000	1988	$155,106,000
1980	$30,831,000	1989	$159,867,000

Plant Report

Background. Operated by Florida Power & Light Co. (FP&L), the twin reactors at Turkey Point Nuclear Station are located in south Dade County, Florida. Since the 1300-megawatt station went into operation, it has displaced more the 190 million barrels of oil, and has saved FP&L customers more than $3 billion.

NRC infractions to an extensive shutdown. Turkey Point has had to overcome some obstacles during its twenty year existence. It has fallen short of federal standards for fracture toughness, and has been placed on the NRC "watch list" because of poor performance and management.

During 1991, both plants were shut down for an extensive period of time so workers could increase backup power capacity and modernize the security system. However, work at both plants was finished just before the scheduled completion date of November 4, 1991.

Alternative storage facilities. The Turkey Point plants have also been the subject of a U.S. Department of Energy study. In an effort to solve the problem of developing a permanent repository for spent fuel rods, the DOE is experimenting with the possibility of storing nuclear waste in dry storage vaults or wells. Advocates of this method point to the corrosion and cracking of fuel cladding that could occur if the storage pool method isn't changed. Opponents of dry storage cite the environmental and work hazards as well as the possible threat of terrorism.

Sources

"Comment." *Critical Mass Bulletin*, January 1984.
"NRC Removes Three Plants from 'Watch List.'" *Nuclear Industry*, First Quarter 1990.
"Nuclear Economics." *Tampa Tribune*, June 10, 1991.
"FPL's Turkey Point Resumes Service." *INFO*, U.S. Council for Energy Awareness, November/December 1991.

—Yvette M. Costa

373 ☆

Tyrone Nuclear Power Plant, Units 1-2

Durand, WI

Owning Utility

Northern States Power Company
414 Nicollet Mall
Minneapolis, MN 55401
Leon Eliason, V.Pres., Nuclear Generation

Phone: (612)330-5500
Fax: (612)330-2900

Contact

John Bousquet, Dir., Pub.Info.
Northern States Power Company

Basic Facts

Units 1-2 *Status:* Cancelled. *Type of Reactor:* Pressurized light-water reactor. *Megawatts (electric):* 1192. *Megawatts (thermal):* 3411. *Reactor System Supplier:* Westinghouse Electric Corp. (United States). *Owner:* Northern States Power Co. *Operator:* Northern States Power Co.

Key Dates

Unit 1 *Ordered:* 1973. *Cancelled:* 1979.
Unit 2 *Ordered:* 1973. *Cancelled:* 1974.

374 ★

Vallecitos Nuclear Power Plant

VBWR
Pleasonton, CA

Owning Utility

Pacific Gas and Electric Company
77 Beale St.
San Francisco, CA 94106
Gregory M. Rueger, Sr.V.Pres., Nuclear Power Generation

Phone: (415)973-7000

Plant Profiles Vermont 376

Basic Facts

Status: Shut down. *Type of Reactor:* Boiling water reactor. *Megawatts (electric):* 5. *Megawatts (thermal):* 50. *Reactor System Supplier:* Bechtel. *Generator Supplier:* General Electric Co. (United States). *Owner:* General Electric; Pacific Gas and Electric. *Operator:* Pacific Gas and Electric.

Key Dates

Ordered: 1952. *Criticality:* 1957. *First Power:* October 1957. *Commercial Operation:* October 1957. *Shut Down:* December 1963.

Plant Report

A project of the Power Demonstration Reactor Programme (PDRP). Realizing that utilities could not justify to their shareholders the investment costs required to build nuclear power plants, the Atomic Energy Commission and Congress devised the Power Demonstration Reactor Programme in 1955 to help launch the commercial nuclear industry in the United States. The PDRP offered cooperating utilities with financial incentives such as research and development assistance, financing of the reactor system, and a waiver of normal fuel charges during the first five years of plant operation. Vallecitos was among the initial projects undertaken in response to the PDRP.

Vallecitos was shut down and Mothballed in 1963—a method of decommissioning where buildings with radioactive materials or components inside are fenced off to allow much of the radioactivity to fade away in place. After radioactivity declines for several decades, the plant can be dismantled.

Sources

Mounfield, Peter. *World Nuclear Power.* Routledge, 1991.

---

375 ☆

Vandalia Nuclear Power Plant

Prairie City, IA

Owning Utility

Iowa Electric Light & Power Company Phone: (319)398-4411
PO Box 351 Fax: (319)398-8192
Cedar Rapids, IA 52406
John F. Franz Jr., V.Pres., Nuclear

Contact

Colleen R. Dykes, Mgr., Corp. Commun.
Iowa Electric Light & Power Company

Basic Facts

Status: Cancelled. *Type of Reactor:* Pressurized light-water reactor. *Megawatts (electric):* 1300. *Reactor System Supplier:* The Babcock & Wilcox Co. (United States). *Owner:* Iowa Electric Light and Power Co. *Operator:* Iowa Electric Light and Power Co.

Key Dates

Ordered: 1976. *Cancelled:* 1982.

---

376 ●

Vermont Yankee Nuclear Power Plant

Governor Hunt Rd. Phone: (802)257-7711
PO Box 157 Fax: (802)254-7711
Vernon, VT 05354
Donald A. Reid, Plant Mgr.

Owning Utility

Vermont Yankee Nuclear Power Corp. Phone: (802)257-5271
Ferry Rd. Fax: (802)254-2127
Brattleboro, VT 05301
J. Gary Weigand, Pres.

Contact

Mary Schneider, Mgr. of Commun.
Vermont Yankee Nuclear Power Corp.

Basic Facts

Status: Operable. *Type of Reactor:* Boiling water reactor. *Megawatts (electric):* 540. *Megawatts (thermal):* 1593. *Reactor System Supplier:* General Electric Co. (United States). *Generator Supplier:* General Electric Co. (United States). *Architect Engineer:* Ebasco.

Key Dates

Ordered: 1966. *Construction began:* 1966. *Criticality:* 1972. *First Power:* September 1972. *Commercial Operation:* November 1972. *Shut Down Expected:* 2007.

Operating Costs

1972	$414,000	1981	$26,795,000
1973	$4,957,000	1982	$33,764,000
1974	$5,692,000	1983	$46,310,000
1975	$7,682,000	1984	$43,203,000
1976	$7,912,000	1985	$46,416,000
1977	$9,775,000	1986	$52,025,000
1978	$11,191,000	1987	$49,760,000
1979	$14,208,000	1988	$42,188,000
1980	$22,588,000	1989	$53,813,000

Plant Report

Background. Located in Vernon, Vermont on the Connecticut River, the Vermont Yankee nuclear power plant has withstood some shaky grounds. Owned by a coalition of 13 utilities including Green Mountain Power Corporation, Central Vermont Public Service Corporation, New England Power Company, Connecticut Light & Power Company, Central Maine Power Company, and the Public Service Commission of New Hampshire the 540 megawatt boiling water reactor produces approximately 75 percent of Vermont's electricity.

Vermont Yankee is operated by the Vermont Yankee Nuclear Power Corporation which exists solely to provide daily management services at the plant for the owning utilities.

The twenty-year old plant has turned in some impressive numbers, including a 1991 operating capacity of 91.1 percent. That same year, Vermont Yankee had a higher availability factor than any other boiling water reactor in the world.

Infractions at Vermont Yankee. The Vermont Yankee plant has had difficulties in its history as well. Continuing problems with faulty fuel rods and cracks in the emergency cooling system

---

● Operable ○ Under Construction △ Indefinitely Deferred ▲ Decommissioning □ Planned ■ On Order ☆ Cancelled ★ Shut Down

have led to several unplanned and extended shutdowns. During the first eighteen months of operation, thirty-four such incidents occurred. In addition, the plant was shutdown from September 1985 to May 1986 because of corroded cooling pipes.

The most serious infraction at Vermont Yankee occurred in 1973 during a routine shutdown. In order to save time in maintenance checks, plant operators violated safety procedures. As a result, too many control rods were removed from the exposed core which caused a potentially serious situation. Although the error was discovered, and the rods were replaced immediately, the plant incurred a $15,000 federal fine.

Later, on July 18, 1976, Vermont Yankee was ordered to pay the state of Vermont an out-of-court settlement fine of $30,000 for dumping 83,000 gallons of radioactive water into the Connecticut River. Furthermore, the plant's safety record has received several below average ratings from the Nuclear Regulatory Commission (NRC).

In April 1986, Vermont residents filed a complaint with the Public Service Board against the plant for purchasing uranium from South Africa. The complainants requested assurance that no electricity would be generated using South African uranium.

Despite many of its problems, Vermont Yankee achieved a 145-day record run for continuous operation on September 20, 1977, and it has amassed a good record of electrical production and low radiation releases in comparison with other nuclear plants.

Prompt removal and dismantling. Since Vermont Yankee's operating license is due to expire in the year 2007, the company has begun plans for its decommissioning. To contain costs once the plant closes and enable a speedy dissolution of the company, officials have chosen the controversial prompt removal dismantling program. This method is viewed as favorable because the operating company does not need to remain in business to oversee a long term dismantling program. The difficulty is that the method is considered unsafe because the core will be dismantled while it is still "hot" with radioactivity. As a result, extreme and expensive safety precautions such as special shields and remote control devices are required.

Although this method does have some drawbacks, it also offers some advantages. It may be the most expensive method, but in the long run, it makes good economic and perhaps environmental sense. For example, prompt removal of the core will mean that the site will be safe and reusable within 10 years.

Also, since the plant is the only income generating asset of Vermont Yankee Nuclear Power, the company wants to avoid costs associated with mothballing the plant to give the radioactive core a chance to cool—including maintaining and paying a security staff, and supplying managers to oversee the eventual dismantling.

Dispute over dismantling costs. The biggest obstacle facing Vermont Yankee with regards to dismantling the plant is obtaining the needed funds. According to a 1983 study completed by officials at Yankee with the help of industry consultants, the cost to decommission the plant would be $459 million. This figure was to be adjusted yearly for inflation, and every four years as new studies were conducted. As of June 1987, the $459 million figure had jumped an additional $36 million.

The June 1987 figure, however, is the subject of debate. Antinuclear forces such as the Vermont Public Interest Research Group of Montpelier placed the figure as high as $2 billion, citing the fact that there is no basis in the industry for the $459 million cost estimate.

Saving for the future. Vermont Yankee began planning for the plant's decommissioning back in 1981 when a study by Nuclear Energy Services was conducted. Currently, customers are paying an additional 10 percent for using electricity produced by Yankee which is placed in an escrow account. Some criticism about this prepayment plan has been voiced since the actual cost of decommissioning is impossible to project. Vermont Yankee defends its program pointing out that even if its estimates are significantly lower than actual cost, any amount saved will be beneficial. Also, the costs may actually be underestimated, claims Vermont Yankee, because of the advancements in technology that may help cut costs even beyond projections.

A risky plan to increase the amount of money available in the fund at the time of decommissioning was proposed by Vermont Yankee to the Federal Energy Regulatory Commission (FERC). The plan would have used the funds to invest in the stock market rather than in pension-grade return markets. The benefit was clear if the stocks had done well, but if poor investments occurred, Vermont Yankee suggested that the customers absorb the loss through increased rates. This plan was rejected by the FERC.

NRC response to decommissionings. The area of plant decommissionings is not one in which the NRC has had great involvement. According to federal officials, there are no legal guidelines for establishing or funding a decommissioning program (although plans for such guidelines are being pursued). NRC spokesperson Karl Abraham asserted that by accepting an NRC permit, the owners take full responsibility for their plants from cradle to grave. However, the NRC has approved Vermont Yankee's decommissioning plan and funding mechanism.

Extending Yankee's life. In 1989, Vermont Yankee sought approval from the NRC to extend its operating license another four years to March 2012. According to NRC policy regarding granting extensions, the facility must show that the additional operational period will not have a significant affect on the health or safety of the public and the environment.

Those who oppose the granting of a license extension include former Vermont Governor Madeline Kunin. In 1989, Kunin ordered the NRC to review each request before granting the extension. This marked the first time the NRC had been challenged regarding extension. As a result of this mandated investigation, it was discovered that Vermont Yankee's maintenance and quality assurance programs were inadequate to justify an extension.

Vermont residents also filed an opposition with the Atomic Safety and Licensing Board regarding the extension on nine different points. The only objection accepted by the board concerned maintenance and the age of the plant's components.

Plant Profiles   Vogtle 378

When Vermont Yankee finally does close its doors, the question of where to store the spent fuel rods and low-level radioactive waste may be answered. Currently, the rods are stored on-site, but no site has been designated for their disposal after the plant dismantling occurs. Clearly this will present a problem for the prompt dismantling plan as the radioactive waste cannot remain on the site.

Consideration of a new venting process. In 1987, Vermont Yankee decided to explore the advantages and disadvantages of installing a controlled ventilation system to update the facility. In the event of an accident, this feature would allow the timely release of radiation versus the explosion which may occur from keeping the containment vessel under pressure.

The new process was under consideration because of reports from the NRC in 1986 that the Mark I design of Vermont Yankee had a ninety percent chance of bursting when the fuel rods overheat and melt. Later in 1987, another NRC senior official put the probability of a containment failure at 10 to fifty percent. Vermont Yankee believed that this figure could be reduced to seven percent or less if the reactor were vented.

Should an accident occur, the venting system allows operators to take advantage of favorable wind conditions to minimize the risk to the public. Also, containing a potential disaster within the vessel could provide time in which corrective action could be taken or evacuation begun. If, however, an accident occurred in which the radiation was prematurely released or the vents would not close, the disaster would not be averted.

Sources

"Big Power in a Small Town." *Blair & Ketchum's Country Journal*, April 1980.
"Special Vermont Yankee Workers Help Buoy Local Economy." *Vermont Business*, January 1986.
"A Vermont Yankee in South Africa's Court." *The Progressive*, April 1986.
"Decommissuining: To Pay Now or Pay Later." *Vermont Business*, June 1987.
"Shift in Safety Strategy Gains at Nuclear Plants." *The New York Times*, August 29, 1987.
"Balancing Cost and Risk at VT Yankee." *Vermont Business*, October 1990.
"First Generation Nuclear Retirement." *The New York Times*, June 2, 1991.
"U.S. Nuclear Plants Shatter Records for Output, Performance." *INFO*, U. S. Council for Energy Answers, February 1992.
"Vermont Nuclear Plant Goes to Head of Class." *Info*, U.S. Council for Energy Answers, February 1992.
"Balancing Cost and Risk at VT Yankee." *Vermont Business*, October 1990.
"First Generation Nuclear Retirement." *The New York Times*, June 2, 1991.
"U.S. Nuclear Plants Shatter Records for Output, Performance." *INFO*, U. S. Council for Energy Answers, February 1992.

Yvotto M. Costa

377 ☆
Vidal Nuclear Power Plant, Units 1-2
Vidal, CA

Owning Utility
Southern California Edison Company   Phone: (818)302-1212
2244 Walnut Grove Ave.   Fax: (818)302-7893
Rosemead, CA 91770   Telex: 677268
Harold B. Ray, Sr.V.Pres., Nuclear
  Organization

Basic Facts
Units 1-2 *Status:* Cancelled. *Type of Reactor:* High temperature gas reactor. *Megawatts (electric):* 779. *Megawatts (thermal):* 2000. *Reactor System Supplier:* General Atomic. *Owner:* Southern California Edison Co. *Operator:* Southern California Edison Co.

Key Dates
Unit 1 *Ordered:* 1971. *Construction began:* 1974. *Criticality:* March 1987. *First Power:* March 1987. *Commercial Operation:* June 1987.
Unit 2 *Ordered:* 1974. *Construction began:* 1976. *Criticality:* March 1989. *First Power:* April 1989. *Commercial Operation:* May 1989.

378 ●
Vogtle Nuclear Power Station, Units 1-2
PO Box 1600   Phone: (404)554-9961
Waynesboro, GA 30830   Fax: (404)826-3762
William B. Shipman, Gen.Mgr.

Owning Utility
Georgia Power Company   Phone: (404)526-6526
PO Box 4545
Atlanta, GA 30302
C. Ken McCoy, V.Pres., Vogtle Project,
  Southern Nuclea

Contact
Ed Crosby, Mgr., Pub.Aff., Southern
  Power
Georgia Power Company

Basic Facts
Units 1-2 *Status:* Operable. *Type of Reactor:* Pressurized light-water reactor. *Megawatts (electric):* 1160. *Megawatts (thermal):* 3411. *Reactor System Supplier:* Westinghouse Electric Corp. (United States). *Generator Supplier:* General Electric Co. (United States). *Architect Engineer:* Georgia Power Company; Bechtel. *Owner:* Municipal Electric Authority of Georgia; Oglethorpe Power Corp; City of Dalton. *Operator:* Georgia Power Company.

Key Dates
Unit 1 *Ordered:* 1971. *Construction began:* 1974. *Criticality:* March 1987. *First Power:* March 1987. *Commercial Operation:* June 1987.
Unit 2 *Ordered:* 1974. *Construction began:* 1976. *Criticality:* March 1989. *First Power:* April 1989. *Commercial Operation:* May 1989.

Operating Costs
Units 1-2
1987 $67,667,000
1988 $113,531,000
1989 $97,085,000

● Operable   ○ Under Construction   △ Indefinitely Deferred   ▲ Decommissioning   □ Planned   ■ On Order   ☆ Cancelled   ★ Shut Down

Vogtle

Plant Report

Background. The Vogtle power plants in Burke County near Augusta, Georgia have been plagued by controversy after controversy. From soaring costs to serious allegations made by former executives relating to safety at the plant, Vogtle management has had its hands full with defending its actions.

Jointly owned by Georgia Power, the Municipal Electric Authority of Georgia (MEAG), Oglethorpe Power Corp., and the City of Dalton, the Vogtle plants are operated by Georgia Power, a subsidiary of Southern Company. Recently, Georgia Power and Southern Company obtained approval from the Securities and Exchange Commission (SEC) to transfer daily operations and technical support to a newly formed subsidiary, Southern Nuclear Operating Company (SONOPCO).

Southern Company formed the new subsidiary in order to bring the nuclear operations of the company into a single unit where it could focus solely on the requirements of the nuclear power industry. SONOPCO assumed plant operation of Vogtle at the end of 1992. (Contact information for SONOPCO is P.O. Box 1295, Birmingham, AL 35201; phone 205-868-5000.)

Skyrocketing costs. Georgia Power's difficulties with Plant Vogtle started when construction costs began to skyrocket. In 1984 alone, costs had increased by 8.4 percent, and by mid-1985, Vogtle had cost $7.2 billion and wasn't slated for completion until 1987.

Georgia Power had estimated the final cost of Vogtle to be $8.4 billion whereas consultants figured it to be much higher at $9.2 billion. Although Georgia Power claimed that Vogtle 1 would be operational by June 1987 and Vogtle 2 by the fall of 1988, consultants predicted that Vogtle 1 wouldn't come on line until December 1987 and Vogtle 2 until July 1989.

Vogtle's final price tag came in at $8.87 billion—a far cry from Georgia Power's originial estimate of $660 million.

No relief in sight. In order to manage the increasing costs at Plant Vogtle, Georgia Power sought relief from the Georgia Public Service Commission (PSC). In 1987, while Vogtle was still under construction, Georgia Power asked that Georgia legislature for the authority to begin billing customers for costs associated with Plant Vogtle. The company was also looking for $650 million in long-term financing and $1 billion in term loans. The issue of allowing Georgia Power to include Vogtle costs in the rate case became a case of prudency.

In July 1987, the Georgia Court of Appeals voted 5-4 to uphold a Georgia PSC decision that disallowed $951 million of Georgia Power's investment to Vogtle on grounds that expenses were imprudently incurred. Southern Company asked the Court of Appeals to reconsider it's decision so that it could avoid a $190 million charge against it's earnings. In the end, the Georgia PSC did allow the issuance of bonds and preferred stock, but stated that shareholders would have to bear costs that were deemed imprudently incurred.

Cost overruns result in charge against earnings. In order to keep the construction process on track, Georgia Power's parent, Southern Company, stated that it would absorb any and all costs above the projected $8.4 billion. By the time Plant Vogtle was complete, it had cost $8.87 billion. Because actual costs had exceeded expected costs, Southern Company was required to make-up the difference.

In 1987, Southern Company was forced to take a $229 million charge against it's earnings because of cost overruns. As a result, the board at Southern Company voted not to increase its quarterly dividend but to maintain it at the previous rate. Also, another large write-off of $226 million was slated for January 1988.

Although Southern Company did bear some of the responsibility for cost overruns at Plant Vogtle, the company did attempt to recover some of its losses through consumer rate hikes. The write-offs were the result of the 1987 Georgia PSC rate case decision. In early 1990, Southern Company again filed a $413 million rate case increase with the Georgia PSC to recover its investment in Vogtle 2 as well as to cover costs of providing other services.

Alleged violations at Vogtle. In a rush to complete the Vogtle plants, Georgia Power incurred two NRC civil penalties for alleged violations totaling $250,000. Of the total, $200,000 was for infractions of NRC security requirements, which represented the largest fine of this nature imposed by the NRC. The remaining $50,000 was in response to Georgia Power's failure to comply with technical specifications on two separate occasions.

A spokesperson for the NRC stated that the violations represented a "programmatic breakdown" in the physical security program at Vogtle. Georgia Power said that in the rush to complete the plants, Vogtle management didn't pay sufficient attention to the completion of the security system or the implementation of security programs. Georgia Power didn't intend to fight the proposed fine.

Site Area Emergency. On March 20, 1990 at 9:20 in the morning, the Vogtle 1 nuclear power plant narrowly averted a potential disaster. A truck delivering fuel oil and lubricants backed into a utility pole on the grounds of the plant and knocked out an electrical transformer. After the loss of power was detected by the system, a back-up generator kicked on to replace the power loss. However, problems really began when, after only 70 seconds of emergency power, the generator shut off.

By the time workers restored power, the temperature of the coolant water in the core had risen 46 degrees Fahrenheit to 136 degrees Fahrenheit. If the water had continued to heat, it could have boiled and melted the uranium-filled rods thereby releasing radiation into the containment building. From there, if the containment building hadn't been sealed off, radiation could have escaped into the countryside. During the emergency, workers took 80 minutes to seal off the containment building instead of the required 57 minutes. The delay was due to the fact that a hatch in the building was barricaded with maintenance items.

Workers became confused in the commotion because they were unclear as to how to get the generator running again. Thirty-six minutes passed before workers restored power to the generator.

One-time failure or negligence. The level-2 Site Area Emergency raised many questions regarding the safety of Plant Vogtle. The NRC investigation revealed that the generator shut down was attributed to a malfunctioning Calcon sensor device. The sensor is designed to measure the water temperature so as to turn the generator off if it overheats. The Site Emergency was declared only after the emergency generator repeatedly failed.

The NRC investigation revealed that the Calcon sensors had failed 69 times since 1985 and not one incident had been reported to the agency. The NRC indicted Georgia Power for failure to correct repeated equipment breakdowns at Vogtle, but the company defended itself by claiming that the problem had been overstated because the sensors hadn't failed since the last series of testing which had taken place one month prior. Georgia Power also stated that the failures it did encounter were not serious enough to be reported.

However, the tendency for the sensors to fail was noted by the industry seven months prior to the accident at Vogtle. This fact contributed to the chastising Georgia Power received from the NRC. The agency reprimanded Vogtle management for not heeding the industry-wide warning which requested that plants using the sensor should implement a surveillance plan to detect any problems. During that time, Vogtle experienced 29 failures but took little action to determine the cause, review past occurrences, or log succeeding problems.

Even though the plant was shut down for maintenance during the emergency, the core was still highly radioactive. If emergency measures at Vogtle had not been successful, the highly radioactive core could have been exposed within three and a half hours.

Former exec accuses Vogtle of unsafe operations. A few months after the accident, insiders at Vogtle accused the plant of lying to federal regulators. Allen L. Mosbaugh, who once held the third highest management position at Vogtle, claimed that the company concealed safety problems, and lied about the reliability of the generator. Mosbaugh cited other technical problems and equipment failures that were never reported to the NRC.

Mosbaugh also made serious allegations that senior management at Vogtle promoted risk-taking, avoided reprimanding managers for questionable compliance practices of investigating possible infractions, and sought reprisal against managers who voiced their concerns to Georgia Power. Mosbaugh's claims prompted a surprise inspection by the NRC. Although the regulators did not find any serious problems, Mosbaugh contends that the agency may have found extraordinary items that could have resulted in fines to Vogtle.

The investigation continues. After Mosbaugh pointed out deficiencies in safety-related issues, he believed his job was in jeopardy. Therefore, he began secretly recording conversations with plant personnel. When plant officials learned about the recordings, Mosbaugh was placed on paid leave and banned from the plant.

Mosbaugh was not the only person firing allegations against Vogtle. Marvin Hobby, Jr., who formerly held a general manager position at the plant, said he was fired because of his claims that Plant Vogtle may have been illegally transferred to SONOPCO from Georgia Power.

In October 1990, several members of Vogtle's brass were called to NRC headquarters for a highly confidential meeting. As told by some of the participants of the meeting to Mosbaugh, the NRC accused Vogtle of having a relaxed view toward safety-related matters. However, the NRC would not comment on the proceedings of the meeting. As of October 1990, both complaints were still pending with the Labor Department. However, in both cases, the preliminary investigation has been in favor of Georgia Power.

Exceptional track record at Vogtle. Despite the Plant Vogtle's problems over the years, both units have performed quite well. For 1989, Vogtle 1 ranked among the Top 25 produces of nuclear power. In January 1991, the plant continued its outstanding performance by producing 865,800 megawatt-hours of electricity. Vogtle 2 has also turned in a shining rating by operating at 92 percent of capacity in 1991.

Sources

"BellSouth Leads 'Year of Profits.'" *Business Atlanta*, July 1985.
"Southern Shareholders Bear Some Vogtle Burdens" *Atlanta Business Chronicle*, June 16, 1986.
"Southern Company." *The New York Times*, March 10, 1987.
"Southern Company to Post Charge of $226 Million." *The Wall Street Journal*, March 10, 1987.
"Georgia Power Plant is Slated to be Granted a Full-Power License." *The Wall Street Journal*, March 13, 1987.
"Nuclear Plant to be at Full Power" *The New York Times*, March 14, 1987.
"Eclipse Across the Peach State." *Business Atlanta*, July 1987.
"NRC Staff Sets Fines for Alleged Violations at Georgia Power Plant." *The Wall Street Journal*, September 8, 1987.
"Georgia Power to Appeal Ruling" *The New York Times*, November 5, 1987.
"Southern Company Plans to Form New Unit for Nuclear Plants." *The Wall Street Journal*, May 19, 1988.
"Paying Dividends." *Georgia Trend*, July 1988.
"The Eureka File." *Nuclear Industry*, First Quarter 1990.
"Corrections & Amplifications." *The Wall Street Journal*, May 4, 1990.
"Southern Company Presidents Addresses the Company's 44th Annual Meeting of Stockholders." *Business Wire*, May 23, 1990.
"Southern Co. Plans to Appeal Ruling to Plant Vogtle Costs." *Atlanta Constitution*, July 24, 1990.
"Sensing Trouble at Vogtle." *Atlanta Business Chronicle*, July 30, 1990.
"Ga. Power Accused of Lying to NRC." *Atlanta Constitution*, September 18, 1990.
"NRC is Investigating Vogtle Safety Again." *Atlanta Business Chronicle*, October 1, 1990.
"Southern Announces Approval From Securities and Exchange Commission to Form a New Subsidiary." *Business Wire*, December 18, 1990.
"Palo Verde Units 3 & 2 Top Producers in United States for January; Unit 3 to be Refueled Starting March 16." *Business Wire*, March 13, 1991.
"Power Briefs." *Nuclear News*, February 1991.
"U.S. Nuclear Plants Shatter Records for Output, Performance." *INFO*, U.S. Council for Energy Awareness, February 1992.
"Southern Announces Approval From Securities and Exchange Commission to Form a New Subsidiary." *Business Wire*, December 18, 1990.
"Palo Verde Units 3 & 2 Top Producers in United States for January; Unit 3 to be Refueled Starting March 16." *Business Wire*, March 13, 1991.

—Yvette M. Costa

Washington Nuclear Project, Units 1 & 3 (WNP)

PO Box 968
Richland, WA 99352-0968
J.W. Baker, Plant Mgr., WNP-2
Phone: (509)377-8000
Fax: (509)377-2476
Telex: 185220

Owning Utility

Washington Public Power Supply System (WPPSS)
PO Box 968
Richland, WA 99352-0968
G.A. Tupper, Commun./Ext.Affrs.Mgr.
Phone: (509)372-5000
Fax: (509)372-5328
Telex: 185220

Contact

J.C. Britton, Public Informaiton
WNP-2

Basic Facts

Unit 1 *Status:* Indefinitely deferred (construction 65% complete as of January 1992). *Type of Reactor:* Pressurized light-water reactor. *Megawatts (electric):* 1339. *Megawatts (thermal):* 3760. *Reactor System Supplier:* The Babcock & Wilcox Co. (United States). *Generator Supplier:* Westinghouse Electric Corp. (United States). *Architect Engineer:* United Engineers & Constructors (United States).

Unit 3 *Status:* Indefinitely deferred (construction 75% complete as of January 1992). *Type of Reactor:* Pressurized light-water reactor. *Megawatts (electric):* 1324. *Megawatts (thermal):* 3800. *Reactor System Supplier:* Combustion Engineering, Inc. (United States). *Generator Supplier:* Westinghouse Electric Corp. (United States). *Architect Engineer:* Ebasco.

Key Dates

Unit 1 *Ordered:* 1973. *Construction began:* 1975. *Commercial Operation Expected:* Indefinite.

Unit 3 *Ordered:* 1973. *Construction began:* 1977. *Commercial Operation Expected:* Indefinite.

Plant Report

WPPSS overview. The Washington Public Power Supply System (WPPSS) is a municipal corporation and joint operating agency of 13 publicly owned utilities. It operates two electrical generating facilities in Washington state. WPPSS's commercial nuclear reactor, Washington Nuclear Project (WNP-2), is located at the U.S. Department of Energy's (DOE) Hanford Nuclear Reservation in Richland, Washington.

WNP-2 produces approximately six billion kilowatt-hours of power annually, enough to meet the needs of 375,000 all-electric homes. The electricity is delivered to the Bonneville Power Administration (BPA)—a U.S. Government agency that sells hydroelectric and thermal power to towns, utilities, and industries in the Pacific Northwest—and is pooled with power from other region-wide plants.

Initially, WPPSS, comprised of 88 member utilities, intended to finance five nuclear plants in Washington in one of the most ambitious nuclear projects ever. The supply system hoped to provide an abundance of cheap energy in a region where electricity rates were already ranked with the nation's least expensive. Construction plans for WNP-1, WNP-2, and WNP-3 began in 1972. In 1976, BPA warned that power demand was likely to outstrip supply in the 1980s and 1990s, prompting construction of WNP-4 and WNP-5. WPPSS wanted to triple the Northwest's power supply in 20 years.

However, WPPSS, which used an inexperienced nuclear board and three different designs to coordinate the five projects, suffered repeated delays and huge cost overruns. By 1982, the plan's total projected price had skyrocketed from $4.1 billion to $23.8 billion, more than $5,000 for every person in the state. On top of that, forecasts of power shortages in the region dwindled because of surging oil prices in the late 1970s and a series of economic recessions that stunned the Northwest. By some estimates, the Pacific Northwest experienced a 17-percent surplus above peak power demand by early 1984.

In all, two major factors led to the suspension or termination of four projects. WPPSS and BPA got swept away by overoptimistic projections of regional electricity demands, while WPPSS suffered serious financial woes incurred through the skyrocketing price tag on the power plants. Other problems included BPA's perennial desire to expand, labor strife, regulatory delays, sparing interest rates, and what many critics called inept management.

In January 1982, WNP-4 was terminated at 23 percent completion, and work ended on WNP-5 at 17 percent. Thereafter, Projects 1 and 3 were suspended at 63-percent and 75-percent completion, respectively. WNP-1 is at Hanford and WNP-3 is in western Washington at Satsop.

Located in eastern Washington, WNP-2 began operation in December 1984, after 12 years of construction and $3.2 billion in costs originally estimated at $450 million. Power from the facility goes into a large pool of electricity produced by six large thermal plants and 24 hydroelectric plants in Washington, Oregon, Idaho, Montana, and Wyoming. WPPSS holds the infamous distinction of suffering the largest default in municipal bond market history.

In 1973, WPPSS started borrowing $8.3 billion in bonds to finance construction of five nuclear plants. In August 1983, though, the utility defaulted on $2.25 billion in high-yielding, tax-exempt securities sold for WNP-4 and WNP-5. The default followed a court ruling that 28 municipal utilities didn't have to pay WPPSS for power they promised to buy but no longer needed. WPPSS did not default on the remaining $6 billion in bonds for Projects 1, 2, and 3 because, unlike 4 and 5, they were backed by BPA.

In all, though, WPPSS created one of the greatest financial tragedies in nuclear industry history by failing to make a $15.6 million monthly debt-service payment to Chemical Bank—a trustee for thousands of bondholders—on the $2.25 billion in bonds. The default, which happened on the 50th anniversary of the Federal government beginning wholesale electric-power generation with the Tennessee Valley Authority, earned WPPSS the nickname, "Whoops."

Concern about WPPSS's problems depressed the municipal bond market. Still, the market was not severely shaken because the default had been expected for weeks, and some observers were hoping to avoid it. Most bond prices held fairly steadily.

In the aftermath, questions rose regarding the construction of power plants by public utilities, which were forced to promise exceptionally high tax-exempt interest rates to sell new bonds. Some municipal bond dealers felt frustration over the WPPSS blunder and its potentially negative effects on the industry. "I feel threatened and so does the whole industry," said a popular municipal bond dealer as quoted in *Time* magazine of August 8, 1983. "I feel the shame of Whoops."

The default cheated WPPSS bondholders, many of whom were elderly people depending on bonds for retirement income. Securities that were bought for $5,000 just two years prior to the default were sold for as little as $700 afterward. One Idaho couple suffered after an $80,000 investment in WPPSS bonds.

Insurance companies, which owned about 15 percent of Projects 4 and 5, were probably the biggest losers. Heading into 1983, Aetna Life & Casualty, American Express unit Fireman's Fund, and Kemper, respectively, had $50 million, $48.9 million, and $24 million invested in WPPSS securities. None of the companies experienced financial trouble because WPPSS investments were minimal compared to other corporate ventures.

Eighty-eight public and private utilities from six Pacific Northwest states joined in the project, originally agreeing to pay for the plants. On January 25, 1983, they began paying off the $4.8 billion in interest due over 30 years on the $2.25 billion in bonds sold to finance WNP-4 and WNP-5.

Putting the size of the WPPSS debacle in perspective, each of the three largest municipal bond defaults prior to the utility's bailout was approximately one-twentieth the size. The West Virginia Turnpike Commission defaulted on $133 million in 1958; the Chesapeake Bay Bridge and Tunnel Commission defaulted on $100 million in 1978; and Chicago's Calumet Skyway collapsed under bond indebtedness of $101 million in 1963. Had New York City defaulted in 1975, it would have been slightly ahead of WPPSS's record at $2.4 billion. However, the city secured moratoriums on short-term interest payments until it could begin paying creditors.

Some utilities with investments in the nuclear project were jolted. For instance, Pacific Power & Light, Co., put forth a $292 million pre-tax write-off of WPPSS. Oreas Power and Light, Co., declared bankruptcy; its share of the debt was $14.5 million plus interest estimated at $45 million over 35 years.

Down but not out, WPPSS re-entered the municipal bond market to refinance $2 billion in tax-exempt securities on WNP-1 and WNP-3. The bonds carried rates as high as 15 percent and—as reported in September 1989—WPPSS never missed an interest payment on them. Again backed by BPA, the bonds garnered high marks from investment ratings services.

U.S. government and other reactions. The Securities and Exchange Commission (SEC) investigated the default. The SEC found that WPPSS failed to provide investors with complete and consistent disclosures on construction cost estimates, load forecasts, its financing program, and the project's legal status. Specific accusations arose that WPPSS and Wall Street brokers knew the project had problems, but failed to inform bondholders, included the following.

Chemical Bank contended that WPPSS constantly created an overly optimistic picture for investors. A WPPSS study in the late 1970s found the project had less than a 20 percent chance of meeting schedule and cost goals, something investors were not told.

WPPSS didn't acknowledge a 1978 U.S. government report that the region's future electrical needs may be lower than originally predicted.

WPPSS kept secret a 1980 revelation by brokerage firm Blyth Eastman that WPPSS "might be unable to raise funds it needs to maintain construcction cash flow at acceptable interest rate levels, or at any interest rate at all."

WPPSS indicated that a law firm examined and approved 72 of the 88 agreements signed by utilities sponsoring Projects 4 and 5. However, WPPSS failed to note that the firm found legal problems with the remaining 16 utilities, mostly from Idaho and Oregon.

The SEC didn't accuse WPPSS of criminal fraud, but recommended that more stringent disclosure standards be maintained by municipal financing companies.

Legal ramifications. WPPSS's inability to repay bonds used to finance WNP-4 and WNP-5 provoked numerous immediate lawsuits.

Originally, bondholders eager to reduce their losses sued virtually everyone with a connection to the default. Suits were targeted at WPPSS, the 88 Northwest utilities for welshing, and prominent brokerage houses such as Merrill Lynch, Prudential-Bache Securities, Smith Barney, Blyth Eastman, Pain Webber, Goldman Sachs and Salomon Bros., all of which had underwritten WPPSS securities. The 88-utility consortium also faced multiple lawsuits from unpaid construction workers.

Bondholders apparently were able to recoup losses. Decisions such as an out-of-court settlement paid investors $20 million from three Washington state utilities involved in Projects 4 and 5. In February 1992, a Federal appeals court approved an $800 million settlement of the default, clearing the way for payments to attorneys and investors who accused the utility of fraud and securities law violations.

Although the city of Seattle was not in the group of 88, it was sued and paid $50 million to bondholders of WNP-4 and WNP-5. A WPPSS member, Seattle designated individuals who served on the supply system's board of directors and executive board when bonds were issued.

BPA was sued by utilities such as the Portland General Electric Co. (PGE), the Washington Water Power Co. (WWP), PacifiCorp, and Puget Sound Power & Light Co. In a settlement, the utilities agreed to trade their combined 30 percent stake in WNP-3 for about 161,000 kilowatts from BPA that could be resold to customers. BPA was also released from any obligation to build WNP-3. PGE, WWP and PacifiCorp took write-downs of $75.5 million, $36 million and $318 million, respectively, to end their investments in the project.

● Operable   ○ Under Construction   △ Indefinitely Deferred   ▲ Decommissioning   □ Planned   ■ On Order   ☆ Cancelled   ★ Shut Down

WPPSS and the General Electric Co. (GE) were embroiled in seven years of litigation. Originally, WPPSS sued GE, which was involved in the design and construction of WNP-2, for fraud and breach of contract, but the case ended in a mistrial. Next, WPPSS filed a $1.2 billion claim against GE for negligent misrepresentation. The two parties settled in March 1992, when GE was ordered to provide free goods and services for Plant 2 and provide substantial discounts on others.

Utility rate implications. Electric rates in the Pacific Northwest skyrocketed, particularly during construction of the project in the late 1970s and after the default. Customers' rates increased to pay for Projects 1, 2 and 3, and consumers also faced hikes in tax levies.

In 1985, it was reported that the rates charged by BPA to utilities and industry had increased 600 percent in six years. In 1979, the average Northwest retail utility paid 1.9 cents per kilowatt hour compared to 3.6 cents in 1985.

BPA purchased the future output of WNP-1, WNP-2, and WNP-3 from Northwestern utilities. In turn, the power authority paid interest on the debt for all three units, although only Plant 2 was operable, causing a dramatic rise in electricity costs.

Economic implications. The WPPSS default was a major factor that the Pacific Northwest region fell into a recession—and what some called a depression.

In the mid-1970s, the region was seen to be poised for boundless demographic and economic expansion. By 1985, however, the area was saddled with an immense debt and owed millions of dollars because of bondholders' lawsuits.

The default had a more negative impact on Eastern Washington, where the economy is less diverse than the other side of the state. In particular, the Tri-Cities of Richland, Kennewick, and Pasco in Southeastern Washington took a precipitous economic drop.

When Projects 1, 2, and 4 were being built, the Tri-Cities became one of America's booming metropolitan areas. Nuclear construction fueled the area's economy, making it the country's seventh fastest-growing region and apparently guaranteeing years of steady employment and residence. In the 1970s, the population nearly doubled to around 100,000. Nevertheless, the default left the economy reeling, wiped out more than 16,000 jobs, and forced a mass exodus of people seeking work in other areas. Units 1 and 4 employed 6,300 people prior to cancellation, a figure which plummeted to 1,700 afterward.

The status of WPPSS reactors and other endeavors. Both Projects 1 and 3 are in a preserved state with maintenance costs of about $25 million annually. Work on those plants was suspended indefinitely until BPA, which intends to purchase 100 percent of WNP-1's power and 70 percent from WNP-3, makes a decision.

Projects 1 and 3 are probably the least expensive increments of new generating capacity available in the Pacific Northwest. However, public opinion and an estimated price tag of $1.5 million each may not permit their completion. Some studies indicate that electricity from both units has only a 16 percent chance of being needed by the late 1990s.

After citing "significant deficiencies" at WNP-2 in 1983, the U.S. Nuclear Regulatory Commission noted improved operations in a 1990 report.

DOE considered acquiring Project 1 in a move where interest now appears to be waning. The government wanted to convert it to a nuclear-weapon facility to replace DOE's N-Reactor, a tritium producer which closed in 1986. The U.S. General Accounting Office conducted two studies on the transition, which Rep. Ron Wyden (D-Ore.) introduced legislation to prevent.

In an effort to recoup some of the approximately $2.25 billion it defaulted on, WPPSS negotiated with China to sell major assets from Project 5 in 1986. WPPSS considered selling nuclear reactors, turbine generators, additional hardware, and nearly completed engineering designs. After talks lasted several years, they were stymied when the United States became more restrictive on sales to China in 1989. The Supply System has sold components from WNP-4 and WNP-5 to other nuclear plants such as the Palo Verde Nuclear Generating Station in Arizona.

Other contact information. WNP-1 contact information same as above. WNP-1 Site Manager, G.K. Dyekman; WNP-1 Public Information, L.G. Woehle. WNP 3-5 mailing address: Satsop/Elma, P.O. Box 1223, Elma, WA 98541. WNP 3-5 phone, 206-482-4428; WNP 3-5 facsimile, 206-482-5970. WNP 3-5 Site Manager, Charles M. Butros; WNP 3-5 Public Information, Michael Louisell.

Sources

"Workers vs. Nukes." *The Progressive*, October 1981.
"A Win Against Whoops." *The Progressive*, January 1982.
"Tough Times in the Tri-Cities." *The Washington Post*, August 1, 1982.
"Hell or High Water." *Barron's*, August 30, 1982.
"A Fiasco That May Rock Municipal Bonds." *Business Week*, February 7, 1983.
"If Northwest Power Network Goes Belly Up." *U.S. News & World Report*, April 4, 1983
"Public Power Fiasco." *Barron's*, May 23, 1983.
"Whoops Woes." *Time*, May 30, 1983.
"Whoops! A $2 Billion Blunder." *Time*, August 8, 1983.
"Significant Deficiencies Found in Nuclear Plant 2." *Seattle Times*, August 31, 1983.
"Pacific Power & Light: Back to Basics After Swallowing its Losses From Whoops'." *Business Week*, March 5, 1984.
"After Default, the Questions of Blame and Duty Linger." *The Washington Post*, December 5, 1984.
"After the Fall." *Sierra,*, May/June 1985.
"WPPSS Asks China to Buy Reactors." *The Business Journal-Portland*, March 10, 1986.
"DOE Studying Conversion of WPPSS Reactor for Defense Production," *Nucleonics Week*, August 21, 1986.
"Two Utilities Taking Charges on Nuclear Unit." *The Wall Street Journal*, January 9, 1987.
"Pannel Urges Closer Look at Safety at Nuclear Plant in Washington." *The New York Times*, August 29, 1987.
"Power Agency to Seek Ruling." *The New York Times*, January 9, 1988.
"Seattle Agrees to Settlement in Suit on WPPSS Default." *The Wall Street Journal*, April 13, 1988.
"Trial in Utility Case Begins" *The New York Times*, September 15, 1988.
"The SEC is Coming!" *American City & County*, January 1989.
"Oregon Lawmaker Introduces Bill That Would Bar Nuclear-Plant Conversion." *Associated Press*, January 4, 1989.
"Whoops is Back." *Barron's*, March 6, 1989.
"The Primary Aluminum Industry and Electric Power in the Pacific Northwest." *Pacific Northwest Executive*, April 1989.
"Miller, Canfield Tries $1B Lawsuit." *Crains Detroit Business*, April 30, 1990.

Plant Profiles | Waterford | 382

"Fool Me Twice." *Barron's*, August 21, 1989.
"Whoops: Investors May Let Bygones Be Bygones." *Business Week*, September 4, 1989.
"A Word to the Muni-Mad." *Newsweek*, September 25, 1989.
"WPPSS Makes a Comeback." *Institutional Investor*, January 1990.
"Born-Again Bonds." *Forbes 400*, October 22, 1990.
"Washington Utility Accord." *The New York Times*, February 6, 1 992.
"A Word to the Muni-Mad." *Newsweek*, September 25, 1989.

—Michael Richman

380 •
Washington Nuclear Project, Unit 2 (WNP)

PO Box 968 Phone: (509)377-8000
Richland, WA 99352-0968 Fax: (509)377-2476
J.W. Baker, Plant Mgr., WNP-2 Telex: 185220

Owning Utility
Washington Public Power Supply System (WPPSS) Phone: (509)372-5000
PO Box 968 Fax: (509)372-5328
Richland, WA 99352-0968 Telex: 185220
G.A. Tupper, Commun./Ext.Affrs.Mgr.

Contact
J.C. Britton, Public Informaiton
WNP-2

Basic Facts
Unit 2 *Status:* Operable. *Type of Reactor:* Boiling water reactor. *Megawatts (electric):* 1154. *Megawatts (thermal):* 3323. *Reactor System Supplier:* General Electric Co. (United States). *Generator Supplier:* Westinghouse Electric Corp. (United States). *Architect Engineer:* Burns & Roe, Inc. (United States).

Key Dates
Unit 2 *Ordered:* 1971. *Construction began:* 1972. *Criticality:* 1984. *First Power:* May 1984. *Commercial Operation:* December 1984.

Operating Costs
Unit 2
1985 $40,299,000
1986 $67,358,000
1987 $67,106,000
1988 $72,213,000
1989 $84,114,000

Plant Report
See Plant Report for Washington Nuclear Project, units 1-3.

381 ☆
Washington Nuclear Project, Units 4-5 (WNP)

PO Box 968 Phone: (509)377-8000
Richland, WA 99352-0968 Fax: (509)377-2476
J.W. Baker, Plant Mgr., WNP-2 Telex: 185220

Owning Utility
Washington Public Power Supply System (WPPSS) Phone: (509)372-5000
PO Box 968 Fax: (509)372-5328
Richland, WA 99352-0968 Telex: 185220
G.A. Tupper, Commun./Ext.Affrs.Mgr.

Contact
J.C. Britton, Public Informaiton
WNP-2

Basic Facts
Unit 4 *Status:* Cancelled (construction 23% complete). *Type of Reactor:* Pressurized light-water reactor. *Megawatts (electric):* 1339. *Megawatts (thermal):* 3760. *Reactor System Supplier:* The Babcock & Wilcox Co. (United States). *Generator Supplier:* Westinghouse Electric Corp. (United States). *Architect Engineer:* United Engineers & Constructors (United States).

Unit 5 *Status:* Cancelled (construction 17% complete). *Type of Reactor:* Pressurized light-water reactor. *Megawatts (electric):* 1324. *Megawatts (thermal):* 3800. *Reactor System Supplier:* Combustion Engineering, Inc. (United States). *Generator Supplier:* Westinghouse Electric Corp. (United States). *Architect Engineer:* Ebasco.

Key Dates
Unit 4 *Ordered:* 1974. *Cancelled:* January 1982. *Construction began:* 1975.

Unit 5 *Ordered:* 1974. *Cancelled:* January 1982. *Construction began:* 1977

Plant Report
See Plant Report for Washington Nuclear Project, units 1-3.

382 •
Waterford Power Station, Unit 3

PO Box B Phone: (504)467-8211
Highway 18 Fax: (504)464-3262
Kilona, LA 70066
Dan Packer, Gen.Mgr., Plant Operations

Owning Utility
Entergy Operations, Inc. Phone: (601)984-9000
PO Box 31995 Fax: (601)984-9817
Jackson, MS 39286-1995
F.B. Rives, Dir., Nuclear Fuels

Contact
Phil R. Miracle, Dir., Corp. Commun.
Entergy Operations, Inc.

Basic Facts
Unit 3 *Status:* Operable. *Type of Reactor:* Pressurized light-water reactor. *Megawatts (electric):* 1153. *Megawatts (thermal):* 3410. *Reactor System Supplier:* Combustion Engineering, Incorporated (United States). *Generator Supplier:* Westinghouse Electric Corporation (United States). *Architect Engineer:* Ebasco. *Owner:* Louisiana Power and Light. *Operator:* Entergy Operations, Inc.

● Operable   ○ Under Construction   △ Indefinitely Deferred   ▲ Decommissioning   □ Planned   ■ On Order   ☆ Cancelled   ★ Shut Down

UNITED STATES

Key Dates

Unit 3 *Ordered:* 1970. *Construction began:* 1974. *Criticality:* March 1985. *First Power:* March 1985. *Commercial Operation:* September 1985.

Operating Costs

Unit 3
1985 $13,788,000
1986 $88,118,000
1987 $86,265,000
1988 $90,551,000
1989 $97,924,000

Plant Report

Louisiana's first nuclear reactor. Waterford 3, Louisiana's first nuclear plant, is owned by Louisiana Power & Light (LP&L) and operated by Entergy Operations. Both LP&L and Entergy Operations are wholly owned subsidiaries of Entergy Corporation, a public utility holding company. LP&L supplies power to 582,000 customers in 46 of Louisiana's 64 parishes. Entergy Operations was formed in 1990 for the purpose of operating the three nuclear stations (Waterford 3, ANO in Arkansas, and Grand Gulf in Mississippi) owned by utilities in the Entergy System.

The reactor is located on the west bank of the Mississippi River 25 miles upriver from New Orleans. It occupies a 3,600 acre site adjacent to Waterford 1 & 2, two of LP&L's fossil-fueled generating plants. The Mississippi River supplies 1 million gallons of water per minute to cool the plant's three condensers.

A decade of delays. In 1970 all of LP&L's electrical production relied on natural gas. Faced with the expiration of its long-term natural gas contracts, the utility announced plans for a nuclear generating station in September 1970. The original cost estimate was $233 million. Because of a lengthy antitrust review, LP&L was not granted a construction permit until November 14, 1974. LP&L faced additional delays when changing requirements necessitated a revamping of the Environmental Impact Statement. Community fears about the health effects of nuclear power and unanswered concerns about nuclear waste management deepened the controversy surrounding plant.

In 1978 the Federal Fuel Use Act complicated the power generation picture in Louisiana. The Act mandated that by 1990 utilities could not rely on natural gas for electrical generation. LP&L maintained that complying with the Act would be impossible without Waterford 3. When the utility applied for an operating license in 1978 three intervenors, The Oyster Shell Alliance, Save Our Wetlands, Inc., and Louisiana Consumer League, blocked the process. The Louisiana state legislature responded by appointing a Joint Legislative Study Committee to investigate the issues surrounding the plant. By the end of the 1970's, the plant was only 74 percent complete and cost estimates exceeded $1 billion. In the aftermath of the Three Mile Island accident, changing regulations added $500 million more in additional costs.

More problems surface. Further problems surfaced in May 1983 when allegations of safety violations began appearing in the local press. The NRC responded by appointing a special task force to review the plant's safety. A two-month investigation discovered cracks in the foundation underneath the reactor and falsified safety records. Although the licensing process was delayed until 1985, critics charged that the NRC expedited the plant's licensing for economic reasons without adequate resolution of the problems it had identified.

Commercial operation finally achieved. A full power license was granted on March 16, 1985 and commercial operation began commercial the following September. In spite of numerous cost overruns and construction delays utility officials claimed that the plant cost less per kilowatt of capacity than the national average for similar plants constructed during the same period. Waterford 3's price came to $2,500 per kilowatt of capacity; the United States average was $3,000.

Since it began commercial operation, Waterford 3 has received recognition for its performance. In 1989 the plant achieved the second highest annual capacity factor in the U.S., 96.82 percent. Capacity factor is the ratio of a plant's actual output compared with its maximum capability. Waterford 3 was one of only nine United States plants to rank on the list of the top 25 performers worldwide.

Visitors welcome. Entergy operates a visitors center, the Energy Education Center, at the Waterford 3 site. The center features exhibits, hands-on displays, and information about conservation. The Energy Education Center is open Tuesday through Friday from 8:00 a.m. until 4:30 p.m. Guided tours, which include the control room simulator, are also available. For more information call the Energy Education Center at (504) 739-6075.

Sources

"Waterford 3 On the Eve of the Atom." *New Orleans*, June 1980.
"Middle South CEO Effectively Defends Grant Gulf." *Jackson Journal of Business*, May 1985.
"Are Regional Reactors Safe?" *New Orleans CityBusiness*, January 19, 1987.
"Waterford 3 Nuclear Unit Fact Sheet." Middle South Utilities System, n.d.
"The Eureka File." *Nuclear Industry*, First Quarter 1990.
"NRC Okays Nuclear Merger at Entergy Corp." *Electrical World*, February 16, 1990.
"Generate Some Family Fun." Entergy Operations, n.d.
Face-to-Face '92: Building for the Future. Entergy Corporation, n.d.

—Karen Bellenir

383

Watts Bar Nuclear Plant, Units 1-2

PO Box 2000
Spring City, TN 37381
Bill Museler, Site V.Pres.

Phone: (615)365-8100
Fax: (615)365-1924

Owning Utility

Tennessee Valley Authority
6N 38A Lookout Pl.
1101 Market St.
Chattanooga, TN 37402-2801
J. R. Bynum, V.Pres., Nuclear Operations

Phone: (615)751-0011
Fax: (615)751-4904
Telex: 361951

Contact

Russ Greene, Pub.Info.
Watts Bar Nuclear Station

Basic Facts

Units 1-2 *Status:* Under construction (Unit 1, 89% complete as of January 1992; Unit 2, 58% complete as of January 1992). *Type of Reactor:* Pressurized light-water reactor. *Megawatts (electric):* 1218. *Megawatts (thermal):* 3411. *Reactor System Supplier:* Westinghouse Electric Corp. (United States). *Generator Supplier:* Westinghouse Electric Corp. *Architect Engineer:* Tennessee Valley Authority.

Key Dates

Unit 1 *Ordered:* 1970. *Construction began:* 1972. *Commercial Operation Expected:* 1994.

Unit 2 *Ordered:* 1970. *Construction began:* 1972. *Commercial Operation Expected:* Indefinite.

Plant Report

Power for Tennessee. The Watts Bar Nuclear Plant is located on the Tennessee River, about seven miles southwest of Spring City, Tennessee. The plant, which is not yet in operation, will be able to provide enough electricity to supply approximately 488,000 houses a day or two cities the size of Knoxville. Its generating capacity will represent about seven percent of the total capacity of the operating utility, the Tennessee Valley Authority (TVA).

Inspection scandal. Construction began at Watts Bar in 1972. By March 1984, major construction was nearing completion when an inspector from the Hartford Steam Boiler Inspection and Insurance Co., a Connecticut firm that checks nuclear plants, discovered unsafe welds at the plant. By the time the inspector found the welds, TVA had installed 56 pipes that had them. The inspector refused to sign a report edorsing the work and was allegedly pressured by TVA and his own superiors to sign against his will.

In 1985, TVA certified that Watts Bar was ready for its operating license, but the inspection scandal arose. TVA staff received phone calls, telling them to raise the pay of the inspectors from Hartford Steam Boiler, or face exposure of the scandal. Indopendent investigators could not tie the calls to Hartford Steam Boiler and the issue remained unresolved.

Workers reveal construction flaws. After the scandal, employees of Watts Bar made charges that management had overlooked construction flaws at the plant. The employees took their charges to congressional aides, and, in a confidential employee survey, hundreds of workers identified 1,700 potential safety and operational problems at Watts Bar. TVA countered by spending $3.6 million to have an outside firm interview employees, but the NRC eventually denied the utility's request for an operating permit, and work was halted.

More troubling charges. Work on Watts Bar continued in 1986, and TVA hired a new nuclear power manager, Steven A. White. In a letter to the NRC, White said that safety procedures at the plant were in compliance with NRC standards, and the NRC disagreed. They accused White of lying, noting that Watts Bar was riddled with defects. White claimed his words were misinterpreted, and in December 1987, the U.S. Justice Department investigated the NRC's charges.

Programs work towards a startup. By September 1987, TVA had spent $4.632 billion on Watts Bar, including the replacement of some 300 miles of cables. In November, the utility formed an independent Watts Bar Program Team (WBPT) made up of TVA personnel assisted by recognized nuclear power experts. The WBPT was responsible for defining the scope of necessary corrective actions and special programs at the plant, as well as developing a Watts Bar Nuclear Performance Plan (WBNPP). The WBNPP was designed to describe the corrective action planned to qualify for the licensing of Unit 1.

In 1988, the NRC approved TVA's licensing approach, and in May, the WBNPP was submitted. In June 1990, the NRC issued a Safety Evaluation Report (SEV) approving all but two of the 29 specific programs suggested in the plan, and reinitiated the licensing review for Unit 1.

TVA increased staffing at Watts Bar by about 700 short-term positions due to renewed construction. But even with all the staffing and programs, TVA stopped work in December because of quality assurance problems with the new workforce.

Still working towards a startup. In June 1991, TVA dismissed 5,800 hourly construction workers on its payroll at Watts Bar and sought outside bids on replacements. In August, the utility signed a $112-million contract with Ebasco Constructors for the work on Unit 1.

From an August/September inspection at Watts Bar, the NRC issued TVA a Notice of Violation (NOV) in October. The NOV dealt with three instances of "proceedural non-compliances," involving revisions of drawings of Unit 1.

In November, the NRC approved a conditional construction restart at Watts Bar. Conditions of the restart included slowing down the speed of the work, coordinating all work done with the NRC, and discussing any significant changes to procedures or work with the NRC before making them.

The restart only applied to Unit 1. The NRC planned to make another assessment of Unit 1 at the end of January 1992.

Sources

"Can a Chastened TVA Reform its Nuclear Ways?" *Business Week*, July 15, 1985.
"Nuclear Scandal Shakes the TVA." *Fortune*, October 27, 1986.
"Charges Against a TVA Official Become an Issue on Capitol Hill." *The New York Times*, December 7, 1987.
"Generating Woe: If TVA Were Private, it'd be Bankrupt." *Barron's*, May 2, 1988.

"TVA Announces Nuclear Organization and Staffing Plan." *Business Wire*, April 26, 1989.
"TVA Projects." *U.S. NRC Annual Report 1990*. U.S. Nuclear Regulatory Commission, n.d.
"NRC Grants Conditional Construction Restart Permission for Watts Bar." *Inside NRC*, December 2, 1991.
Watts Bar Nuclear Plant, The Tennessee Valley Authority, 1992.

—*Charles Ceplecha*

Wolf Creek Nuclear Power Plant

PO Box 411
Burlington, KS 66839
Otto Maynard, Plant Mgr.
Phone: (316)364-8831
Fax: (316)364-4146

Owning Utility

Wolf Creek Nuclear Operating Corporation
PO Box 411
Burlington, KS 66839
John Bailey, Mgr. of Nuclear Fuels
Phone: (316)364-8831
Fax: (316)364-4146

Contact

Ronn Smith, Info. Admin.
Wolf Creek Nuclear Operating Corporation

Basic Facts

Status: Operable. *Type of Reactor:* Pressurized light-water reactor. *Megawatts (electric):* 1192. *Megawatts (thermal):* 3411. *Reactor System Supplier:* Westinghouse Electric Corp. (United States). *Generator Supplier:* General Electric Co. (United States). *Architect Engineer:* Bechtel; Sargent & Lundy Engineers (United States). *Owner:* Kansas Gas and Electric Company; Kansas City Power & Light Company; Kansas Electric Power Cooperative. *Operator:* Wolf Creek Nuclear Operating Corporation.

Key Dates

Ordered: 1973. *Construction began:* 1977. *Criticality:* 1985. *First Power:* June 1985. *Commercial Operation:* September 1985.

Operating Costs

1985 $13,643,000
1986 $88,118,000
1987 $86,265,000
1988 $90,551,000
1989 $97,924,000

Plant Report

Serves three utilities. The Wolf Creek Nuclear Power Plant is located near Burlington, Kansas, about 90 miles southwest of Kansas City. Wolf Creek Operating Corporation operates the plant which is owned by three utilities. Kansas City Power & Light Company and Kansas Gas and Electric Company each own 47 percent. A Topeka cooperative owns the remaining 6 percent.

The plant is one of two plants in the United States built using a standardized design (the other is located in Callaway County, Missouri.) The final cost of the plant was $3.05 billion.

Rate battles. In 1984 anti-nuclear groups citing the high cost of nuclear power called for the cancellation of the Wolf Creek plant. According to them, electric consumers would have saved 4.1 cents per kilowatt-hour if the plant and cited statistics claiming that both Kansas Gas and Electric and Kansas City Power and Light had generating capacities above the recommended 18 percent reserve margin. The utilitiies countered that a 22 percent margin was necessary.

As predicted, when the plant came on line in 1985, rates went up. Kansas Gas & Electric was granted a 29 percent increase in 1986 to guarantee adequate cash flow, although small decreases followed in subsequent years.

Some problems. In its early years the plant had trouble during refueling, and some concerns were expressed about safety systems and training. One "significant event" (an NRC defined category indicating a serious problem) occurred in December 1990 when problems developed with pumps in the emergency cooling system. Improvement in the ability to recognize and act on problems has resulted in the plant receiving high ratings from the NRC.

Wolf Creek, like other nuclear plants in the United States, has no permanent solution to the problem of nuclear waste. Approximately 357.4 curies of low-level waste are removed from the plant annually. In 1989 there were 81.2 million metrics tons of high-level nuclear waste stored at the site.

Honors of efficiency. Wolf Creek's efficient operation has resulted in the plant's 81 percent availability, much higher than industry averages. In 1989 the plnat produced more than 10 billion kilowatt hours of electricity, a record for the amount of power produced at any U.S. power plant, nuclear or fossil-fueled.

Wolf Creek also holds the record for the number of days in continuous operation. On September 19, 1991 Wolf Creek shut down for a scheduled re-fueling after 487 days of continuous operation. This was the longest run achieved by any light water reactor in the world.

Sources

"Grassroots." *Critical Mass Bulletin*, March 1984.
"The Legacy of Seabrook I." *New England Business*, April 7, 1986.
"Glasnost at KCP&L." *The Corporate Report-Kansas City*, April 1988.
"The Eureka File." *Nuclear Industry*, First Quarter 1990.
"Union Electric's Callaway Nuclear Plant." *Union Electric*, May 1991.
"TMI 1, Wolf Creek Plants Pass 400 Days on Line." *INFO*, U.S. Council for Energy Awareness, July 1991.
"Consumer Group's Report Flatters Region's Utilities." *Kansas City Star*, July 19, 1991.
"Nuclear Plants Perform Well." *Kansas City Star*, July 22, 1991.
"Around the States." *Nuclear Industry*, Fourth Quarter 1991.
"TMI 1, Wolf Creek Plants Pass 400 Days on Line." *INFO*, U.S. Council for Energy Awareness, July 1991.
"Consumer Group's Report Flatters Region's Utilities." *Kansas City Star*, July 19, 1991.
"Nuclear Plants Perform Well." *Kansas City Star*, July 22, 1991.

—*Karen Bellenir*

385 ★
Yankee Rowe Nuclear Power Plant

Star Rte. Phone: (413)625-6147
Rowe, MA 01367
Norman. L. St. Laurent, Plant Supt.

Owning Utility

Yankee Atomic Electric Company Phone: (508)779-6711
580 Main St.
Bolton, MA 01740
Rudy Grube, Nuclear Fuels Mgr.

Contact

William J. McGee Pub.Af, Yankee
  Atomic Electric Company

Basic Facts

Status: Shut down. *Type of Reactor:* Pressurized light-water reactor. *Megawatts (electric):* 185. *Megawatts (thermal):* 600. *Reactor System Supplier:* Westinghouse Electric Corp. (United States). *Generator Supplier:* Westinghouse Electric Corp. *Architect Engineer:* Stone & Webster Engineering Corp. (United States). *Owner:* Northeast Utilities; New England Electric System; Boston Edison. *Operator:* Yankee Atomic Electric Co.

Key Dates

Ordered: 1956. *Construction began:* 1957. *Criticality:* 1960. *First Power:* November 1960. *Commercial Operation:* July 1961. *Shut Down:* February 1992.

Plant Report

The nation's oldest commercial nuclear power plant. The Yankee Rowe nuclear power plant is located in the northwest corner of Massachusetts, one mile south of the Vermont boarder and 35 miles east of Albany in Rowe, Massachusetts. The 186-megawatt plant went into operation in 1960, cost $40 million to build, and had one of the best operating records in the U.S. nuclear industry. The plant had only been shut down 38 times in its 31 years of operation.

Before its closing in 1992, Yankee Rowe provided jobs for over half of the 357 residents of Rowe, paid for the roads and schools of the town, and, along with a major hydroelectric project, paid 85 percent of Rowe's taxes.

The facility is owned by a small group of New England Utilities including Northeast Utilities in Hartford, Connecticut with 37 percent, New England Electric System in Westborough, Massachusetts with 30 percent, and Boston Edison Company with 10 percent.

1988 vote keeps Yankee Rowe open. On November 8, 1988 Massachusetts voters had a chance to shut down the state's two nuclear plants, Yankee Rowe and the Pilgrim plant in Plymouth. If passed, the referendum would have banned generating electricity by commercial nuclear power plants "by means which result in the production of nuclear waste" after July 4, 1989.

The referendum was opposed by the state's utilities. They contributed more than $7 million for a lobbying effort aimed at defeating the voter-initiated referendum while supporters only raised $330,000 from 14,500 contributors.

The dominant group backing the referendum was Massachusetts Citizens for Safe Energy. Their primary concern was safety. They alleged that both plants had outdated designs and that every day the plants operated they produced 150 pounds of high-level radioactive waste. They also claimed the utilities had not focused on other power sources such a hydro-electric and conservation efforts.

On the other side, opponents of the referendum formed the "No On 4 Committee: Massachusetts Citizens Against the Shutdown Initiative." The group argued the shutdown would worsen the region's electricity supply problems and increase air pollution by forcing utilities to increase their dependence on plants that burn oil and coal.

They also asserted that the state might have to compensate the plants' owners for the shutdown at a cost to taxpayers of as much as $2 bbillion and that the state's economy could be damaged by the closing which would result in the loss of more than 1,000 jobs.

The 1988 referendum was defeated and Yankee Rowe remained operational.

Recent safety assessments and incidents. In an August 1990 safety assessment, the NRC noted a number of deficiencies but concluded that Yankee Rowe could operate safely until the end of its current full cycle, estimated to be April 1992.

On June 15, 1991, nearly one year after the safety assessment, Yankee Rowe demonstrated that its emergency system was indeed functional when a powerful lightning bolt struck an electric line serving Yankee Rowe forcing a shutdown. The charge went through the cables to a transformer next to the reactor, shattering two four-foot-high lightning arresters. An automatic shutdown was triggered and the plant remained closed for one week during an extensive inspection. Fortunately, the investigators discovered that the lightning had not damaged the reactor.

A re-licensing controversy emerges. As the use of nuclear power in the United States enters its ffifth decade, many plants will have reached the end of their operating licenses. In fact, between 2000 and 2016 the licenses of 66 reactors in the U.S. will expire. Because no new nuclear plant has been ordered in the U.S. since 1974, and because building a new plant could cost $5 billion or more as of 1991, extending the life of these nuclear facilities has become critical to the industry and was a priority of the Bush administration's energy policy.

As a result, the nuclear industry and the NRC have been working to draft regulations that would allow the licenses to be extended by up to 20 years. Yankee Rowe's operating license was scheduled to expire in 2000, making it the first plant to go through the NRC's re-licensing procedure that started in 1992.

Although it is a relatively small plant, producing less than one percent of New England's electricity, Yankee Rowe was key to both sides of the controversy in setting a precedent for future re-licensing requests. Because of this, in 1990 former President

Bush called Yankee Rowe "the model for the future of nuclear power."

The pros and cons of license renewal. In a June 1991 Gallup poll of 846 adults, sponsored by the U.S. Council for Energy Awareness (USCEA) and Yankee Atomic Electric Co., it was found that 73 percent of Americans thought it was a "good idea" to issue license renewals to nuclear energy plants that meet federal safety standards, 20 percent said it was "not a good idea," and seven percent "didn't know."

Yankee Rowe provides a pilot case to work out the issues involved with license renewal. Like other plants that will come up for evaluation in the near future, Yankee Rowe was beginning to show its age. The specific issue for Yankee Rowe was whether the steel of the reactor has become so brittle that it would crack like glass from the shock of a sudden change in temperature or pressure. Critics of nuclear power say the aging vessel would do just that and help turn an accident into a meltdown, sending radioactive steam into the air and rendering parts of Massachusetts and Vermont unlivable.

The nuclear power industry contends, however, that aging isn't a problem because atomic plants are closely regulated and carefully maintained. The U.S. Council on Energy Awareness, the public communications arm of the nuclear industry, says plants are being continuously "renewed from the inside out" with new components, thus eliminating any dangers associated with age.

According to Andrew J. Kadak, president of Yankee Atomic Electric Co., electronic technology has advanced substantially since the plant began commercial operation in 1961, and so much has been replaced or added since then, that "it's not the same plant."

Still, critics maintain that the reactors and their parts are worn and were never designed to last forever.

The real issues for Yankee Rowe. In addition to concern about the brittle vessel, Yankee Rowe has some unique and complicating issues. Unlike most pressurized water reactors, Yankee Rowe's containment vessel is not made of reinforced concrete and is not embedded in the earth. Instead it is a small, spherical steel shell mounted on stilts of metal piping. It is the only U.S. nuclear plant where the reactor and its radiation containment structure are above ground. Although replacement of the reactor vessel sounds like a simple solution to the controversy, that would cost at least $50 million, according to Kadak.

A petition filed for an immediate shutdown of Yankee Rowe. In June 1991 a petition requesting the immediate shutdown of the Yankee Rowe nuclear power plant was filed by the Union of Concerned Scientists and the New England Coalition on Nuclear Pollution. They claimed the Yankee Rowe reactor vessel wasn't safe because it had been weakened by years of radiation exposure and that the containment vessel didn't meet the NRC's flexibility and safety standards to justify continued operation.

Yankee Atomic acknowledged the fact that the vessel had become brittle, but said the problem wasn't as bad as portrayed by the Union of Concerned Scientists. The company was studying the problem and had already done several things to diminish it, including heating the emergency cooling water to 130 degrees to lessen the shock when it floods the 500-degree reactor core.

Nevertheless, internal documents published by the *Boston Globe* in June 1991 indicated that the vessel might fracture when subjected to a force measures at 35 foot pounds when NRC regulations specify that it be able to withstand 50 foot pounds.

The petition alleged that this condition renders the vessel "subject to cracking and rupture and if such cracking and rupture occur, they would almost certainly lead to a meltdown and uncontrolled release of radioactivity into the environment."

The NRC rejected the petition even though the vessel did not meet their current standards for fracture strength stating that the "vessel condition continues to provide adequate protection of the public health and safety."

Political debate begins. Once the NRC statement had been made, political opposition began to surface. Rep John D. Dingell (D-Mich.), Chairman of the House Energy and Commerce Committee, questioned the decision to allow the plant to keep operating. In a letter to the NRC, he said he was "concerned" that there is "no clear consensus within the scientific community and indeed within the commission itself" that the decision was proper.

Eventually, at the prodding of key members of Congress, the NRC scheduled a special meeting July 11, 1991 to consider charges that the Yankee Rowe was unsafe and should be shut down immediately.

Once again, the NRC voted to reject the Union's petition. In its decision it cited what it called "low" probability, which means about a one in 1,000 chance each year of the type of pipe break or other coolant-loss accident that could suddenly stress the vessel.

Yankee Atomic takes action to address problem. Even though the petition was denied, official at Yankee Rowe decided to address the problem because of the serious ramifications of an accident. Studies were conducted by metallurgists at the University of Michigan and by metal fatigue experts from Europe to determine whether the steel could withstand the pressure of an accident that would overheat the reactor.

As a result of the findings, the Yankee Atomic Electric Company's board voted in July 1991 to delay, for at least a year, its effort to win re-licensing for Yankee Rowe. The decision came just days before the NRC was scheduled to vote on whether to shut down the plant until the safety of its reactor vessel could be assessed.

Yankee Rowe voluntarily shuts down. On October 1, 1991, expressing concern regarding the integrity of the containment vessel, the NRC recommended that Yankee Rowe be shut down immediately. Within a few hours, the closing procedures had begun. Although the closing was not considered permanent, the event marked the first time in nearly five years that the NRC had recommended closing a plant for safety reasons.

Yankee Atomic Electric Company disagreed with the NRC staff recommendation and hoped to restart the plant in the future. The company had planned to present its findings to the regulatory commission October 2, 1991, but a plant spokesman, William McGee, said that under the circumstances, "it seemed like the prudent thing to do for us to voluntarily initiate the shutdown" rather than wait for a vote. The plant's owners said the plant would remain closed until the commission gave permission to reopen.

The NRC's decision to close Yankee Rowe was seen as significant because it marked a change in the traditionally pro-nuclear stance of the agency. Some observers have suggested that the NRC's handling of Yankee Rowe could signal the formation of a tough policy regarding license renewal.

Nevertheless, the U.S. Council for Energy Awareness, a pro-nuclear group, insisted that the closing of Yankee Rowe was not a negative reflection on the nuclear power industry. "We look upon this as a step in the right direction, that in fact we, as a nation, can say we have a responsible regulator in government and responsible nuclear operators," said Phillip Bayne, president of the Council.

Plans for reopening permanently abandoned. Yankee Atomic formulated a plan to recondition the metal of the vessel by the wet anneal method in which the metal is heated to 650 degrees Fahrenheit and slow cooled thus increasing its strength and reducing its brittleness. The objective of the process was to prepare for a 1992 reopening of the plant. Neither of these events occurred and, in February 1992, officials announced that the closing was permanent.

Cost of decommissioning to be high. Citing estimated decommissioning costs at $247 million, the Yankee Atomic Electric Co. has requested approval for a $28 million annual rate increase until the year 2000. The high price-tag for decommissioning is largely due to increased costs for low-level waste disposal and spent fuel storage.

Sources

"Yankee Rowe Decommissioning Estimate Alarms Industry." *Power Engineering*, August 1992.
"A Year of Shakeups and Quips at the NRC." *The New York Times*, June 28, 1992.
"Yankee Atomic Electric Co. Plans to Wet Anneal." *Inside NRC*, December 2, 1991.
"Turning Off the Juice." *The New York Times*, October 6, 1991.
"A-Plant's Closing Brings Shrugs as Well as Doubt." *The New York Times*, October 5, 1991.
"A-Plant to Close Over Safety Issue." *The New York Times*, October 2, 1991.
"Massachusetts Nuclear Plant is Shut Down." *The Wall Street Journal*, October 2, 1991.
"Yankee Rowe: Aging Nuclear Plants Become a Hot Issue as Re-Licensing Near." *The Wall Street Journal*, September 12, 1991.
"Board Votes to Delay Effort to Seek Plant Re-licensing." *The Wall Street Journal*, July 26, 1991.
"Near Old Nuclear Plant Town Keeps the Faith." *The New York Times*, July 20, 1991.
"NRC Schedules Hearing on Safety of Oldest Nuclear Power Plant." *The Washington Post*, July 3, 1991.
"NRC Rule to Allow Renewal of Nuclear Plant Licenses." *INFO*, July 1991.
"NRC Rejects Move to Close Yankee Rowe Nuclear Plant." *The Wall Street Journal*, June 26, 1991.
"Lightning Forces Closing of Oldest Reactor in U.S." *The Wall Street Journal*, June 17, 1991.
"Massachusetts to Vote on Nuclear Plants." *The New York Times*, October 23, 1988.
"Not Worth the Risk." *Public Citizen*, November 1988.

—Sharon Miscavage

386 ☆

Yellow Creek Nuclear Power Plant, Units 1-2

Corinth, MS

Owning Utility

Tennessee Valley Authority
6N 38A Lookout Pl.
1101 Market St.
Chattanooga, TN 37402-2801
J. R. Bynum, V.Pres., Nuclear Operations

Phone: (615)751-0011
Fax: (615)751-4904
Telex: 361951

Basic Facts

Units 1-2 *Status:* Cancelled (Unit 1, construction 35% complete; Unit 2, construction 3% complete). *Type of Reactor:* Pressurized light-water reactor. *Megawatts (electric):* 1339. *Megawatts (thermal):* 3800. *Reactor System Supplier:* Combustion Engineering, Inc. (United States). *Generator Supplier:* General Electric Co. (United States). *Architect Engineer:* Stone & Webster Engineering Corp. (United States).

Key Dates

Units 1-2 *Ordered:* 1974. *Cancelled:* 1984. *Construction began:* 1978.

Plant Report

High cost projections lead to plant cancellation. The Yellow Creek Nuclear Plant occupied a site near Iuka, Mississippi until it was cancelled in 1984 by the operating utility, the Tennessee Valley Authority (TVA). High cost projections for operations and maintenace were partial factors for the cancellation.

Yellow Creek was to house reactors. At the time construction was suspended, Yellow Creek 1 was 35 percent completed, and Yellow Creek 2 was three percent completed.

It is estimated that Yellow Creek Plant would have cost $10 billion to complete. Along with two other units, Hartsville A-1 and A-2, which were also cancelled in 1984, TVA had invested $2.71 billion. To compensate, TVW charged $800 million of the investments against their 1984 fiscal earnings. The rest was to be written off over the next 11 years.

Late in 1984, TVA, Combustion Engineering, and Ebasco tried to sell both of the Yellow Creek reactors to interested parties from China.

Sources

"China Syndrome 2." *Critical Mass Bulletin*, June 1984.
"The End of Four TVA Nuclear Plants." *Business Week*, July 30, 1984.

"Lost in the Dark: TVA Stumbling Over its Nuclear Program." *Public Citizen,* January 1987.

—*Charles Ceplecha*

387 ☆
William H. Zimmer Nuclear Power Plant, Units 1-2

Moscow, OH

Basic Facts

Units 1-2 *Status:* Cancelled (Unit 1, construction 97% complete). *Type of Reactor:* Boiling water reactor. *Megawatts (electric):* Unit 1, 840; Unit 2, 1200. *Megawatts (thermal):* Unit 1, 2436. *Reactor System Supplier:* General Electric Co. (United States). *Owner:* Grouping of Cincinatti Gas and Electric, Columbus and Southern Ohio Electric, Dayton Power and Light . *Operator:* Cincinnatti Gas and Electric.

Key Dates

Unit 1 *Ordered:* 1969. *Cancelled:* August 1984. *Construction began:* 1972.
Unit 2 *Ordered:* 1974. *Cancelled:* 1978.

Plant Report

Nuclear plant brings prosperity and fear. Cincinnati Gas & Electric made a costly error in June 1976. By mishandling a community information meeting, the utility and the Ohio Disaster Services created a middle-class antinuclear movement that eventually stalled the nearly completed William H. Zimmer power plant.

Before that meeting, the 326 residents of Moscow, Ohio had expressed faith in government and company officials. The plant project had returned prosperity to their decaying rural center. Located 25 miles south east of Cincinnati on the Ohio River, Moscow had become a commuter community for that city.

The power plant meant permanent jobs and a bonanza of tax dollars both during construction and later when the plant began operation. By 1976, the town embraced nuclear energy.

However, as Zimmer's 479-foot cooling tower grew, so did the apprehension of some of the residents. To allay their fears, they requested the June 1976 information meeting, expecting to hear detailed reassurances of safety and well-conceived evacuation plans. Instead, they discovered that no evacuation plan existed and the Ohio Disaster Services felt none was needed. An informal group of concerned Moscow residents evolved from that meeting to form the Zimmer Area Citizens of Ohio (ZAC).

Fuel loading completed despite opposition. ZAC unsuccessfully opposed the loading of fuel rods into the plant in 1979. The NRC issued a low-power test license that summer, in preparation for the plant to go to full power in 1981. Although public demonstrations attracted thousands of protesters, fuel loading went as scheduled.

In October 1981, ZAC intervened in Zimmer's federal licensing hearing. Ever since Three Mile Island, the Nuclear Regulatory Commission (NRC) required proper evacuation planning. ZAC was able to demonstrate serious flaws in the plan devised by the Federal Emergency Management Agency. The NRC denied the plant's license application.

Over the next year, numerous allegations of intimidation of quality-control workers surfaced. Former employees told of some 15,000 altered official records. The NRC levied a $200,000 fine on CG&E for safety violations.

Reactor designer may have sold unsafe design. In addition, design flaws in the General Electric Mark III boiling water reactor necessitated costly changes during construction. Later CG&E would sue GE for $1 billion for knowingly selling the utility a flawed design.

Original plans called for Zimmer to go on-line in 1980 at a cost of $230 million. By 1982, construction costs had climbed to $1.7 billion with the plant 97 percent finished. However, questions of weld safety meant completion of the reactor would cost another $1.4 to $1.7 billion. (Plans for a second nuclear unit at Zimmer were announced in 1974 and later cancelled in 1978.)

In 1982, the NRC issued a stop-work order because of safety concerns.

Delays cost utilities millions. The Ohio Public Utilities Commission allowed CG&E and its two partners, Dayton Power and Light Company, and Columbus and Southern Ohio Electric (C&SOE) to recapture $861 million of the construction cost through rate increases. In 1984, the Ohio Supreme Court ordered C&SOE to refund its customers $12.9 million collected during construction of Zimmer nuclear power plant.

The three utilities now faced a financial loss on a nearly completed plant at a time when acceptable sites for any kind of power plant were difficult to find. In addition, increasing demand for electricity and aging of Ohio's existing plants foreshadowed serious shortfalls in generating capacity over the next decade.

Desparate utilities opt for risking fuel change operation. In a move that surprised industry experts, the consortium decided not to complete the nuclear reactor, or to abandon the project entirely. Instead, on August 1, 1984, the utilities announced they would spend $1.9 billion to change the plant to a coal-fired station, making Zimmer the world's first nuclear-to-coal conversion.

As a bonus, the coal-fired plant would produce 1,300 megawatts electric instead of the 834 megawatts electric expected from the nuclear system. The cost remained equivalent—roughly 2 cents per kilowatt hour.

Zimmer will consume 3.5 million tons of Ohio's bituminous coal at a cost of $30 per ton. That translates as 10 percent of the state's coal production.

State and federal regulations required state-of-the-art pollution control devices. Obtaining the necessary permits took three years, pushing construction three months behind schedule before it began.

The 305-acre site, bisected by a meandering stream, worked well for a nuclear facility, but a coal plant with sophisticated

Plant Profiles   Zion 388

scrubbers and pulverizing equipment, needed over 1,000 acres. To gain the needed space, the contractor diverted the stream into a semicircle around the perimeter of the site. Four million cubic yards of sand and gravel dredged from the river raised and levelled the terrain.

A two-lane highway provided the only truck access to the site. No rail link existed. The river provided the only viable means of transporting construction material.

A final incentive for conversion lay in the make-up of the consortium. C&SOE's parent company, American Electric Power Company Inc. (AEP), had already built six 1,300 megawatts electric coal-fired power plants. With that experience in hand, the corporation devised an innovative plan to turn the Zimmer disaster into a success.

Modular construction makes project work. AEP's decision to use modular components built off-site made the project uniquely innovative. Zimmer became the first power plant constructed using this method.

Specialized manufacturing facilities built key components, then shipped them by barge to the site, where a 1,200 ton crane lifted the pieces onto prepared foundations. One project, a 30-piece precipitator built in Mobile, Alabama and transported by ocean-going barge, represented the largest volume shipment ever on the Mississippi and Ohio Rivers. The largest piece weighed over 500 tons. Two turbines weighing 130 and 150 tons and a 450-ton generator stator came from Europe.

The method reduced the amount of land needed for assembly and improved the quality of the components. About one million man-hours of labor shifted from on-site construction to various manufacturing facilities where better working conditions and experience enhanced productivity.

Informing the community becomes priority. Most importantly, AEP initiated a program of public speeches and meetings. It distributed an information newsletter quarterly to 30,000 residents detailing construction activity.

As its final construction act, AEP repainted the highly visible sky-blue nuclear portion of the plant. Now the entire facility wears a more modest coat of clay-brown, allowing it to blend more easily into the background.

AEP completed the conversion on March 30, 1990, two months ahead of schedule and more than 20 percent below budget.

The station generated electricity for the first time on December 31, 1990. Utility officials expect approval for a 20 percent rate increase to pay for the $3.6 billion total construction bill.

Sources

"Fuel-Change Operation." *Progressive*, March 1984.
"Grassroots." *Critical Mass Bulletin*, June 1984.
"Zimmer Conversion Proceeding, Bids Under Way." *Cincinnati Business Courier*, March 10, 1986.
"Moscow Radicals' Stop a Nuclear Plant." *Sierra*, January 1987.
"Zimmer Conversion Work Could Start Next Month." *Cincinnati Business Courier*, February 23, 1987.
"A Nuclear Cloud Hangs Over GE's Reputation." *Business Week*, June 15, 1987.

"A Salvage Operation." *Forbes*, September 1988.
"Plugging Zimmer In." *Ohio Business Journal*, December 13, 1989.
"The Zimmer Plant Generated its First Electricity." *Nuclear News*, February 1991.
"World's First Nuclear-to-Coal Conversion Goes Commercial." *Electrical World*, April 1991.
"One More Renewable Energy Source: Nuke Plants." *Business Week*, September 16, 1991.
"Plugging Zimmer In." *Ohio Business Journal*, December 13, 1989.

—Al Cook

388 ● Zion Station, Units 1-2

101 Shiloh Blvd.   Phone: (708)746-2084
Zion, IL 60099
Thomas Joyce, Stat.Mgr.

Owning Utility

Commonwealth Edison Company   Phone: (312)294-4321
One First National Plaza   Fax: (312)294-2995
PO Box 767
Chicago, IL 60690
Cordel Reed, Sr.V.Pres., Nuclear Operations

Contact

John F. Hogan, Dir., Commun. Svcs.
Commonwealth Edison Company

Basic Facts

Units 1-2 *Status:* Operable. *Type of Reactor:* Pressurized light-water reactor. *Megawatts (electric):* 1908. *Megawatts (thermal):* 3250. *Reactor System Supplier:* Westinghouse Electric Corp. (United States). *Generator Supplier:* Westinghouse Electric Corp. *Architect Engineer:* Sargent & Lundy Engineers (United States).

Key Dates

Unit 1 *Ordered:* 1967. *Construction began:* 1968. *Criticality:* June 1973. *First Power:* June 1973. *Commercial Operation:* December 1973.

Unit 2 *Ordered:* 1967. *Construction began:* 1968. *Criticality:* December 1973. *First Power:* December 1973. *Commercial Operation:* September 1974.

Operating Costs

Units 1-2

Year	Cost	Year	Cost
1973	$444,000	1982	$52,616,000
1974	$9,234,000	1983	$48,670,000
1975	$12,735,000	1984	$56,860,000
1976	$18,268,000	1985	$69,753,000
1977	$18,104,000	1986	$70,774,000
1978	$20,383,000	1987	$83,743,000
1979	$26,953,000	1988	$129,377,000
1980	$37,656,000	1989	$113,951,000
1981	$44,864,000,		

Plant Report

A power station for Illinois. Zion Station is located on a 250-acre site near Zion, Illinois, approximately half way between

Chicago and Milwaukee, Wisconsin. The Station's initial project cost was $583 million.

Bad evaluations. Zion station has had a history of poor evaluations from the Nuclear Regulatory Commission (NRC). In February 1990, senior managers of the NRC appraised several indications of overall declining performance at Zion Station. In the area of operator training, five of 12 operators examined by the NRC in September 1989 failed, resulting in an unsatisfactory rating for a Zion requalification program. In the area of maintenance, a maintenance team inspection in June and July 1989 cited lack of management involvement, insufficient engineering support, poor preparation of work packages, inadequate procedures, and poor adherence to procedure. Performance had been particularily weak in the area of engineering and technical support.

During June 1990, a 16-member NRC team was sent to Zion Station and the corporate offices of Commonwealth Edison. The team's findings are summarized below.

Operator knowledge and skills were sufficient to insure safe plant operation, however, interviews and observations revealed that management control over the operator activities was weak, teamwork within the operating department was poor, operator overtime was not well managed or controlled, and both operators and managers accepted equipment degradation and cumbersome administrative requirements. The team also determined that the training department was not well prepared to support the NRC requalificaton examinations scheduled for September 1990.

Safety-related systems were poorly maintained—corrective maintenance was not timely, and, in several instances, preventive maintenace was limited, root causes of equipment failures were not indentified, and motor operated valves were poorly maintained. Testing of safety-related components was not adequate to detect degradation; valves, service heat exchangers, and water pumps were also not tested adequately.

Poor material condition had been caused by years of operation without an effective maintenance program. Poor maintenance and testing were caused by deficient programs without clear assignments of responsibility. Lack of effective direction and oversight from the corporate office was also a contributing factor.

In the areas of engineering design and technical support, several main weaknesses were found, including weak in-house design engineering expertise, a failure to develop a safety analysis or operating procedure for post-accident cooling when one unit was operating and the other shut down, and the discovery of accident vulnerabilities.

In the area of management and organization, the team found insufficient corporate support for timely resolutions of problems, unclear assignments of responsibility at the plant, weak management control and support for operator activities, failure to accept recommended improvements from outside, and site and corporate engineering staffs not exercising sufficient foresight in problem resolution.

Although Commonwealth Edison was aware of some of the weaknesses found by the team, the extent and impact of them had not been understood. The utility took corrective actions following 1990 NRC visit.

In 1991, however, an NRC review of Zion Station found more operating problems. The NRC ordered Commonwealth Edison to correct them, but allowed the plant to remain open during the process.

Pump failure. In December 1990, the failure of a cooling pump bearing in Zion Station's Unit 1 contributed to a five-month unplanned outage at the plant. The failure prompted a utility scramble for a replacement pump, which Commonwealth Edison purchased for about $2.5 million through the Pooled Inventory Management System.

An analysis of the bearing by the Argonne National Laboratory failed to establish the cause of the failure. It was thought that cause may have been found if the journal was sent to the laboratory. As of November 1991, the utility was contemplating this action.

Because the other three pumps at Unit 1 were the same age and manufacture as the failed one, Commonwealth Edison re-instrumentated all the pumps at Unit 1 with sophisticated analysis equipment.

Sources

Zion Nuclear Station Fact Sheet, Commowealth Edison Co., May 1988.
U.S. NRC 1990 Annual Report. U.S. Nuclear Regulatory Commission, n.d.
"NRC Review Finds Problems at Zion Plant." *The Chicago Tribune*, January 5, 1991.
"Reactor Coolant Pump Failure Remains a Mystery a Year Later." *Nucleonics Week*, November 28, 1991.
"Commonwealth to Make Equipment Change at Zion." *Inside NRC*, December 2, 1991.

Appendix I: NRC Report Cards

The following tables provide assessments prepared by the U.S. Nuclear Regulatory Commission (NRC) on U.S. nuclear power plants profiled in this directory. Assessments are presented in chronological order beginning with the most recent data. The dates in the left-hand column indicate the month and year the assessment was published by the NRC. Refer to the legend at the bottom of each page for an explanation of the NRC coding system. Each table also includes a "See" reference to the plant profile in the main section of this directory that corresponds to the assessment and, if necessary, contains explanatory text regarding the data. For additional information about the assessments, contact the NRC's Reactor Inspection and Licensee Performance Division at (301) 504-2903.

Table 1: Arkansas Nuclear One Steam Electric Station, Units 1-2 (See Entry 238)

	Plant operations	Radiological controls	Maintenance	Surveillance	Maintenance/surveillance	Fire protection	Emergency preparedness	Security	Outages	Quality programs	Licensing activities	Training & qualification effectiveness	Engineering/technical support	Safety assessment/quality verification
04/91	2	2	*	*	2	*	1	1	*	*	*	*	2	2
02/90	2	2	*	*	3	*	1	2	*	*	*	*	2	3
03/89	1	1	*	*	2	*	1	2	*	*	*	*	2	2
05/87	1	1	2	2	*	1	2	3	1	2	1	1	*	*
09/85	2	1	2	2	*	2	2	2	1	2	1	2	*	*
09/84	2	2	3	3	*	2	2	1	1	2	1	N	*	*
11/83	2	3	3	2	*	3	2	1	2	2	2	N	*	*
03/83	3	2	3	3	*	2	3	2	2	3	2	N	*	*
09/82	2	2	2	3	*	2	1	2	2	1	N	N	*	*
09/80	2	2	2	2	*	2	2	3	2	2	N	N	*	*

1 = superior performance/NRC attention reduced; 2 = good performance/NRC attention maintained at normal levels;
3 = acceptable performance/NRC inspection levels increased; N = no rating given; * = inspection category not used in current year.

Table 2: Beaver Valley Power Station, Units 1-2 (See Entry 242)
Ratings for 1981-88 are for Unit 1 only.

	Plant operations	Radiological controls	Maintenance	Surveillance	Maintenance/surveillance	Fire protection	Emergency preparedness	Security	Outages	Quality programs	Licensing activities	Training & qualification effectiveness	Engineering/technical support	Safety assessment/quality verification
05/91	1	2	*	*	1	*	1	1	*	*	*	*	2	1
02/90	2	2	*	*	2	*	1	1	*	*	*	*	2	2
11/88	1	2	*	*	2	*	1	1	*	*	*	*	2	2
12/87	2	2	2	2	*	2	1	1	2	2	2	2	*	*
05/86	2	2	1	2	*	1	1	1	3	N	1	N	*	*
08/84	2	1	1	2	*	2	1	1	2	N	1	N	*	*
06/83	2	1	2	2	*	2	2	1	2	N	1	N	*	*
05/82	2	1	3	2	*	2	2	3	2	N	2	N	*	*
02/81	3	2	3	3	*	3	3	3	2	3	N	N	*	*

Table 3: Big Rock Point Nuclear Power Plant (See Entry 244)

	Plant operations	Radiological controls	Maintenance	Surveillance	Maintenance/surveillance	Fire protection	Emergency preparedness	Security	Outages	Quality programs	Licensing activities	Training & qualification effectiveness	Engineering/technical support	Safety assessment/quality verification
08/90	1	2	*	*	1	*	1	2	*	*	*	*	2	2
05/89	1	2	*	*	2	*	1	2	*	*	*	*	2	2
03/88	1	2	2	1	*	2	2	2	2	2	1	2	*	*
02/87	2	2	2	2	*	2	1	1	3	2	1	1	*	*
07/85	1	2	2	1	*	2	1	1	1	2	2	N	*	*
10/83	1	2	1	1	*	2	1	1	N	2	2	N	*	*
12/82	1	3	1	2	*	2	2	2	1	N	2	N	*	*
07/82	1	2	2	2	*	2	2	1	1	2	2	N	*	*
02/81	2	3	1	1	*	2	2	2	2	2	N	2	*	*

1 = superior performance/NRC attention reduced; 2 = good performance/NRC attention maintained at normal levels; 3 = acceptable performance/NRC inspection levels increased; N = no rating given; * = inspection category not used in current year.

Appendix I: NRC Report Cards

Table 4: Braidwood Station, Units 1-2 (See Entry 248)

	Plant operations	Radiological controls	Maintenance	Surveillance	Maintenance/surveillance	Fire protection	Emergency preparedness	Security	Outages	Quality programs	Licensing activities	Training & qualification effectiveness	Engineering/technical support	Safety assessment/quality verification
05/90	2	2	*	*	1	*	1	2	*	*	*	*	2	2
06/89	2	2	*	*	2	*	1	2	*	*	*	*	2	2
05/88	2	2	2	2	*	2	1	2	N	2	2	1	*	*
05/87	1	2	N	N	*	2	2	2	N	1	2	2	*	*

Table 5: Browns Ferry Nuclear Plant, Units 1-3 (See Entry 249)

	Plant operations	Radiological controls	Maintenance	Surveillance	Maintenance/surveillance	Fire protection	Emergency preparedness	Security	Outages	Quality programs	Licensing activities	Training & qualification effectiveness	Engineering/technical support	Safety assessment/quality verification
08/90	?	1	*	*	3	*	2	2	*	*	*	*	2	3
09/85	3	2	3	3	*	3	?	3	N	3	3	2	*	*
06/84	3	3	3	2	*	N	2	3	3	3	2	N	*	*
06/83	3	3	3	2	*	2	2	3	1	3	2	N	*	*
11/82	3	3	2	2	*	3	N	2	2	3	N	N	*	*
01/81	2	3	2	2	*	2	2	2	2	3	N	N	*	*

1 = superior performance/NRC attention reduced; 2 = good performance/NRC attention maintained at normal levels;
3 = acceptable performance/NRC inspection levels increased; N = no rating given; * = inspection category not used in current year.

Table 6: Brunswick Nuclear Plant, Units 1-2 (See Entry 250)

	Plant operations	Radiological controls	Maintenance	Surveillance	Maintenance/surveillance	Fire protection	Emergency preparedness	Security	Outages	Quality programs	Licensing activities	Training & qualification effectiveness	Engineering/technical support	Safety assessment/quality verification
01/91	2	2	*	*	2	*	2	1	*	*	*	*	3	2
12/89	2	2	*	*	2	*	2	1	*	*	*	*	2	2
12/88	2	2	*	*	2	*	2	1	*	*	*	*	3	3
01/88	2	2	2	1	*	2	2	1	2	2	2	2	*	*
01/86	2	2	2	2	*	2	2	1	2	2	2	N	*	*
08/84	2	1	2	2	*	2	1	1	1	2	2	N	*	*
06/83	3	2	3	3	*	3	1	1	3	3	3	N	*	*
09/82	3	3	3	2	*	3	2	2	N	3	N	N	*	*
01/81	2	3	2	2	*	2	2	2	2	2	N	N	*	*

Table 7: Byron Station, Units 1-2 (See Entry 251)

Ratings for 1984 are for Unit 1 only.

	Plant operations	Radiological controls	Maintenance	Surveillance	Maintenance/surveillance	Fire protection	Emergency preparedness	Security	Outages	Quality programs	Licensing activities	Training & qualification effectiveness	Engineering/technical support	Safety assessment/quality verification
07/90	1	1	*	*	1	*	1	1	*	*	*	*	2	2
03/89	2	2	*	*	1	*	1	1	*	*	*	*	2	2
03/88	2	2	2	2	*	2	1	1	2	2	1	1	*	*
05/87	2	2	2	2	*	2	1	2	1	2	2	2	*	*
04/86	3	3	2	3	*	2	1	3	N	2	2	N	*	*
09/84	N	2	N	N	*	3	2	2	N	2	2	N	*	*

1 = superior performance/NRC attention reduced; 2 = good performance/NRC attention maintained at normal levels; 3 = acceptable performance/NRC inspection levels increased; N = no rating given; * = inspection category not used in current year.

Appendix I: NRC Report Cards

Table 8: Callaway Nuclear Plant, Unit 1 (See Entry 252)

	Plant operations	Radiological controls	Maintenance	Surveillance	Maintenance/surveillance	Fire protection	Emergency preparedness	Security	Outages	Quality programs	Licensing activities	Training & qualification effectiveness	Engineering/technical support	Safety assessment/quality verification
05/90	1	1	*	*	1	*	1	1	*	*	*	*	1	1
12/88	2	2	*	*	1	*	1	1	*	*	*	*	2	1
10/87	2	2	1	2	*	N	1	1	1	1	1	1	*	*
12/86	2	2	2	1	*	1	2	1	1	2	2	1	*	*
10/85	2	2	2	2	*	1	2	2	N	2	1	N	*	*
04/84	N	1	N	N	*	2	N	2	N	2	1	N	*	*
06/83	N	2	2	1	*	N	2	2	N	1	1	N	*	*

Table 9: Calvert Cliffs Nuclear Power Plant, Units 1-2 (See Entry 254)

	Plant operations	Radiological controls	Maintenance	Surveillance	Maintenance/surveillance	Fire protection	Emergency preparedness	Security	Outages	Quality programs	Licensing activities	Training & qualification effectiveness	Engineering/technical support	Safety assessment/quality verification
06/91	2	2	*	*	2	*	2	1	*	*	*	*	2	2
06/90	3	2	*	*	3	*	2	1	*	*	*	*	2	3
05/89	2	1	*	*	2	*	2	1	*	*	*	*	2	3
03/88	2	1	2	2	*	N	2	1	1	2	2	2	*	*
09/86	2	1	2	1	*	N	1	1	2	2	1	2	*	*
01/85	1	1	2	2	*	1	1	1	1	N	1	N	*	*
02/84	2	2	3	3	*	1	2	1	2	N	2	N	*	*
02/83	2	1	2	1	*	1	1	2	1	N	2	N	*	*
12/80	3	2	3	2	*	2	2	3	2	3	N	N	*	*

1 = superior performance/NRC attention reduced; 2 = good performance/NRC attention maintained at normal levels;
3 = acceptable performance/NRC inspection levels increased; N = no rating given; * = inspection category not used in current year.

Table 10: Catawba Nuclear Station, Units 1-2 (See Entry 257)

	Plant operations	Radiological controls	Maintenance	Surveillance	Maintenance/surveillance	Fire protection	Emergency preparedness	Security	Outages	Quality programs	Licensing activities	Training & qualification effectiveness	Engineering/technical support	Safety assessment/quality verification
05/91	2	2	*	*	2	*	1	1	*	*	*	*	2	2
03/90	2	2	*	*	2	*	1	2	*	*	*	*	2	2
12/88	2	2	*	*	2	*	2	2	*	*	*	*	2	2
01/88	2	2	2	2	*	1	2	2	N	2	2	2	*	*
12/85	2	2	2	2	*	1	2	1	N	2	2	2	*	*

Table 11: Clinton Nuclear Power Station, Unit 1 (See Entry 260)

	Plant operations	Radiological controls	Maintenance	Surveillance	Maintenance/surveillance	Fire protection	Emergency preparedness	Security	Outages	Quality programs	Licensing activities	Training & qualification effectiveness	Engineering/technical support	Safety assessment/quality verification
06/91	2	2	*	*	2	*	1	2	*	*	*	*	2	2
03/90	2	2	*	*	2	*	1	2	*	*	*	*	3	2
01/89	2	2	*	*	2	*	2	1	*	*	*	*	2	2
01/88	2	2	2	2	*	2	2	2	N	2	1	2	*	*
04/87	2	2	3	N	*	2	2	2	N	3	2	N	*	*

Table 12: Comanche Peak Nuclear Power Plant, Unit 1 (See Entry 262)

	Plant operations	Radiological controls	Maintenance	Surveillance	Maintenance/surveillance	Fire protection	Emergency preparedness	Security	Outages	Quality programs	Licensing activities	Training & qualification effectiveness	Engineering/technical support	Safety assessment/quality verification
05/91	2	2	*	*	2	*	1	1	*	*	*	*	2	2

1 = superior performance/NRC attention reduced; 2 = good performance/NRC attention maintained at normal levels; 3 = acceptable performance/NRC inspection levels increased; N = no rating given; * = inspection category not used in current year.

Appendix I: NRC Report Cards

Table 13: Comanche Peak Nuclear Power Plant, Unit 2 (See Entry 263)

	Plant operations	Radiological controls	Maintenance	Surveillance	Maintenance/surveillance	Fire protection	Emergency preparedness	Security	Outages	Quality programs	Licensing activities	Training & qualification effectiveness	Engineering/technical support	Safety assessment/quality verification
05/91	2	2	*	*	2	*	1	1	*	*	*	*	2	2

Table 14: Donald C. Cook Nuclear Plant, Units 1-2 (See Entry 264)

	Plant operations	Radiological controls	Maintenance	Surveillance	Maintenance/surveillance	Fire protection	Emergency preparedness	Security	Outages	Quality programs	Licensing activities	Training & qualification effectiveness	Engineering/technical support	Safety assessment/quality verification
12/90	1	2	*	*	3	*	1	1	*	*	*	*	3	2
10/89	1	2	*	*	2	*	1	2	*	*	*	*	2	2
07/88	2	2	2	2	*	2	2	2	1	2	1	2	*	*
04/87	2	2	2	2	*	2	?	2	1	2	2	2	*	*
05/86	2	2	2	3	*	2	2	3	?	3	1	N	*	*
08/84	2	2	2	2	*	2	2	2	1	3	2	N	*	*
08/83	2	2	3	2	*	3	2	2	2	3	2	N	*	*
10/82	2	2	2	3	*	3	3	2	2	3	3	N	*	*
02/81	2	2	2	2	*	3	3	2	2	2	N	N	*	*

1 = superior performance/NRC attention reduced; 2 = good performance/NRC attention maintained at normal levels; 3 = acceptable performance/NRC inspection levels increased; N = no rating given; * = inspection category not used in current year.

Table 15: Cooper Nuclear Power Plant (See Entry 265)

	Plant operations	Radiological controls	Maintenance	Surveillance	Maintenance/surveillance	Fire protection	Emergency preparedness	Security	Outages	Quality programs	Licensing activities	Training & qualification effectiveness	Engineering/technical support	Safety assessment/quality verification
12/90	1	1	*	*	2	*	2	2	*	*	*	*	2	2
08/89	1	1	*	*	2	*	2	2	*	*	*	*	2	3
06/88	1	1	2	2	*	1	2	3	2	2	1	2	*	*
10/86	1	2	2	2	*	2	2	3	1	2	1	2	*	*
04/85	2	2	2	2	*	1	2	2	1	2	1	N	*	*
12/83	2	2	1	1	*	1	2	2	1	2	1	N	*	*
11/82	1	1	1	1	*	1	3	1	1	2	1	N	*	*
06/82	1	2	2	1	*	2	3	2	1	1	N	N	*	*
08/80	2	2	2	1	*	2	2	2	1	2	N	N	*	*

Table 16: Crystal River Energy Complex, Unit 3 (See Entry 266)

	Plant operations	Radiological controls	Maintenance	Surveillance	Maintenance/surveillance	Fire protection	Emergency preparedness	Security	Outages	Quality programs	Licensing activities	Training & qualification effectiveness	Engineering/technical support	Safety assessment/quality verification
06/90	2	1	*	*	2	*	1	1	*	*	*	*	2	2
05/89	2	2	*	*	2	*	2	1	*	*	*	*	3	2
03/88	2	2	1	3	*	1	2	2	2	2	2	2	*	*
06/86	2	2	1	2	*	2	2	3	2	3	2	3	*	*
02/85	2	2	1	3	*	N	2	3	N	2	2	N	*	*
12/83	2	1	2	2	*	2	2	2	1	2	2	N	*	*
01/83	2	2	3	1	*	2	2	2	2	N	3	N	*	*
01/81	2	2	2	2	*	2	3	2	2	2	N	N	*	*

1 = superior performance/NRC attention reduced; 2 = good performance/NRC attention maintained at normal levels; 3 = acceptable performance/NRC inspection levels increased; N = no rating given; * = inspection category not used in current year.

Appendix I: NRC Report Cards

Table 17: Davis-Besse Nuclear Power Plant, Unit 1 (See Entry 267)

	Plant operations	Radiological controls	Maintenance	Surveillance	Maintenance/surveillance	Fire protection	Emergency preparedness	Security	Outages	Quality programs	Licensing activities	Training & qualification effectiveness	Engineering/technical support	Safety assessment/quality verification
11/90	2	2	*	*	2	*	1	1	*	*	*	*	2	2
07/89	2	2	*	*	2	*	1	1	*	*	*	*	2	2
06/88	2	1	2	1	*	3	2	1	1	2	2	2	*	*
09/87	N	N	N	N	*	N	N	N	N	N	N	N	*	*
04/85	2	1	3	2	*	3	3	2	1	3	2	3	*	*
10/83	2	1	3	2	*	2	2	2	1	2	2	N	*	*
09/82	2	1	3	2	*	2	1	2	1	3	2	N	*	*
12/80	2	2	2	2	*	2	3	3	2	2	N	N	*	*

Table 18: Diablo Canyon Nuclear Power Plant, Units 1-2 (See Entry 269)

	Plant operations	Radiological controls	Maintenance	Surveillance	Maintenance/surveillance	Fire protection	Emergency preparedness	Security	Outages	Quality programs	Licensing activities	Training & qualification effectiveness	Engineering/technical support	Safety assessment/quality verification
05/90	1	1	*	*	2	*	1	2	*	*	*	*	2	2
01/89	2	1	*	*	2	*	1	2	*	*	*	*	2	2
10/87	2	1	2	2	*	2	1	2	1	2	2	1	*	*
10/86	2	2	1	2	*	1	1	1	2	2	1	1	*	*
10/85	2	2	1	2	*	1	1	1	N	1	1	N	*	*
11/84	2	2	2	2	*	2	2	1	N	2	2	N	*	*
03/83	2	2	2	2	*	2	1	1	N	2	2	N	*	*

1 = superior performance/NRC attention reduced; 2 = good performance/NRC attention maintained at normal levels;
3 = acceptable performance/NRC inspection levels increased; N = no rating given; * = inspection category not used in current year.

Dresden

Table 19: Dresden Nuclear Station, Unit 1 (See Entry 271)

	Plant operations	Radiological controls	Maintenance	Surveillance	Maintenance/surveillance	Fire protection	Emergency preparedness	Security	Outages	Quality programs	Licensing activities	Training & qualification effectiveness	Engineering/technical support	Safety assessment/quality verification
02/86	2	2	2	2	*	2	1	2	1	2	1	N	*	*
02/85	2	3	3	1	*	2	1	2	1	2	2	N	*	*
07/83	2	2	3	2	*	2	2	1	1	2	1	N	*	*
12/82	3	3	2	2	*	3	1	1	1	3	1	N	*	*
11/80	2	3	2	2	*	2	3	2	2	2	N	2	*	*

Table 20: Dresden Nuclear Station, Units 2-3 (See Entry 272)

	Plant operations	Radiological controls	Maintenance	Surveillance	Maintenance/surveillance	Fire protection	Emergency preparedness	Security	Outages	Quality programs	Licensing activities	Training & qualification effectiveness	Engineering/technical support	Safety assessment/quality verification
08/90	1	2	*	*	2	*	1	1	*	*	*	*	2	2
05/89	1	2	*	*	2	*	1	2	*	*	*	*	2	2
06/88	2	2	3	2	*	N	2	2	N	2	2	2	*	*
06/87	2	2	2	2	*	3	2	2	2	3	1	2	*	*
02/86	2	2	2	2	*	2	1	2	1	2	1	N	*	*
02/85	2	3	3	1	*	2	1	2	1	2	2	N	*	*
07/83	2	2	3	2	*	2	2	1	1	2	1	N	*	*
12/82	3	3	2	2	*	3	1	1	1	3	1	N	*	*
11/80	2	3	2	2	*	2	3	2	2	2	N	2	*	*

1 = superior performance/NRC attention reduced; 2 = good performance/NRC attention maintained at normal levels; 3 = acceptable performance/NRC inspection levels increased; N = no rating given; * = inspection category not used in current year.

Appendix I. NRC Report Cards

Table 21: Duane Arnold Energy Center (See Entry 273)

	Plant operations	Radiological controls	Maintenance	Surveillance	Maintenance/surveillance	Fire protection	Emergency preparedness	Security	Outages	Quality programs	Licensing activities	Training & qualification effectiveness	Engineering/technical support	Safety assessment/quality verification
06/91	2	2	*	*	2	*	1	2	*	*	*	*	3	3
04/90	1	2	*	*	2	*	2	2	*	*	*	*	2	2
03/89	2	2	*	*	2	*	3	2	*	*	*	*	2	2
01/88	1	2	2	2	*	1	2	2	1	3	1	2	*	*
06/86	1	2	2	2	*	1	1	2	2	2	1	2	*	*
03/85	1	2	1	3	*	1	2	3	N	N	1	N	*	*
07/83	2	2	1	2	*	2	2	2	2	2	2	N	*	*
10/82	3	2	2	2	*	3	2	2	2	3	2	N	*	*
12/80	2	3	2	2	*	2	3	2	3	2	N	2	*	*

Table 22: Joseph M. Farley Nuclear Power Station, Units 1-2 (See Entry 275)

	Plant operations	Radiological controls	Maintenance	Surveillance	Maintenance/surveillance	Fire protection	Emergency preparedness	Security	Outages	Quality programs	Licensing activities	Training & qualification effectiveness	Engineering/technical support	Safety assessment/quality verification
05/91	1	1	*	*	1	*	2	2	*	*	*	*	2	1
12/89	1	1	*	*	1	*	2	1	*	*	*	*	2	1
09/88	1	1	2	2	*	1	2	2	1	2	1	2	*	*
02/87	1	1	1	1	*	1	2	2	1	2	1	2	*	*
03/85	1	1	1	2	*	1	1	2	1	2	1	N	*	*
04/84	1	1	1	1	*	N	2	1	1	1	2	N	*	*
03/83	1	1	1	1	*	2	1	1	1	N	2	N	*	*
06/82	1	1	1	2	*	2	2	2	2	1	N	N	*	*

1 = superior performance/NRC attention reduced; 2 = good performance/NRC attention maintained at normal levels;
3 = acceptable performance/NRC inspection levels increased; N = no rating given; * = inspection category not used in current year.

Table 23: Enrico Fermi 2 Nuclear Power Plant (See Entry 277)

	Plant operations	Radiological controls	Maintenance	Surveillance	Maintenance/surveillance	Fire protection	Emergency preparedness	Security	Outages	Quality programs	Licensing activities	Training & qualification effectiveness	Engineering/technical support	Safety assessment/quality verification
05/91	2	1	*	*	2	*	2	1	*	*	*	*	2	2
04/90	2	1	*	*	3	*	1	1	*	*	*	*	2	2
05/89	2	2	*	*	3	*	1	1	*	*	*	*	3	2
08/88	3	2	3	3	*	2	1	1	2	3	2	2	*	*
01/88	3	2	2	3	*	N	1	2	2	2	2	3	*	*
05/87	N	N	N	N	*	N	N	N	N	N	N	N	*	*
02/86	2	2	2	2	*	3	1	2	1	2	2	N	*	*

Table 24: James A. Fitzpatrick Nuclear Power Plant (See Entry 279)

	Plant operations	Radiological controls	Maintenance	Surveillance	Maintenance/surveillance	Fire protection	Emergency preparedness	Security	Outages	Quality programs	Licensing activities	Training & qualification effectiveness	Engineering/technical support	Safety assessment/quality verification
05/91	2	3	*	*	2	*	1	1	*	*	*	*	2	3
05/90	1	2	*	*	2	*	1	1	*	*	*	*	2	2
09/88	1	2	2	2	*	N	1	1	N	2	2	N	*	*
07/87	2	2	2	2	*	N	1	1	2	2	2	2	*	*
06/86	2	2	2	2	*	1	1	1	2	2	2	2	*	*
12/84	2	2	2	2	*	1	1	1	1	N	2	N	*	*
07/83	2	2	2	2	*	2	1	1	N	N	2	N	*	*
09/82	3	3	2	2	*	2	2	1	2	N	2	N	*	*
02/81	2	3	2	2	*	3	3	3	2	3	N	N	*	*

1 = superior performance/NRC attention reduced; 2 = good performance/NRC attention maintained at normal levels; 3 = acceptable performance/NRC inspection levels increased; N = no rating given; * = inspection category not used in current year.

Appendix I: NRC Report Cards — Ginna

Table 25: Fort Calhoun Generating Station (See Entry 281)

	Plant operations	Radiological controls	Maintenance	Surveillance	Maintenance/surveillance	Fire protection	Emergency preparedness	Security	Outages	Quality programs	Licensing activities	Training & qualification effectiveness	Engineering/technical support	Safety assessment/quality verification
09/90	2	1	*	*	2	*	2	2	*	*	*	*	2	1
08/89	2	2	*	*	2	*	2	2	*	*	*	*	2	2
11/88	2	3	2	2	*	2	2	2	2	3	2	3	*	*
12/86	1	2	1	1	*	1	2	3	2	2	1	2	*	*
05/85	1	1	1	1	*	1	2	3	1	2	1	N	*	*
02/84	1	2	2	1	*	2	2	2	1	3	2	N	*	*
12/82	1	2	2	1	*	2	2	2	2	3	1	N	*	*
08/82	2	2	2	2	*	2	2	2	N	2	N	N	*	*
08/80	2	2	2	2	*	2	2	2	2	2	N	N	*	*

Table 26: Robert Emmet Ginna Nuclear Power Plant (See Entry 283)

	Plant operations	Radiological controls	Maintenance	Surveillance	Maintenance/surveillance	Fire protection	Emergency preparedness	Security	Outages	Quality programs	Licensing activities	Training & qualification effectiveness	Engineering/technical support	Safety assessment/quality verification
12/90	2	2	*	*	2	*	1	2	*	*	*	*	2	2
11/89	2	2	*	*	2	*	1	2	*	*	*	*	2	2
05/88	2	2	2	1	*	N	1	1	N	2	1	1	*	*
12/86	2	1	2	1	*	N	2	1	1	2	1	2	*	*
05/85	2	2	1	1	*	1	2	1	1	3	1	N	*	*
09/83	2	2	2	2	*	1	2	1	1	N	1	N	*	*
03/83	1	2	1	2	*	2	1	1	1	N	1	N	*	*
03/82	1	2	1	2	*	2	2	2	1	2	N	N	*	*
07/80	2	2	2	2	*	2	2	2	2	2	N	N	*	*

1 = superior performance/NRC attention reduced; 2 = good performance/NRC attention maintained at normal levels;
3 = acceptable performance/NRC inspection levels increased; N = no rating given; * = inspection category not used in current year.

Table 27: Grand Gulf Nuclear Station (See Entry 284)

	Plant operations	Radiological controls	Maintenance	Surveillance	Maintenance/surveillance	Fire protection	Emergency preparedness	Security	Outages	Quality programs	Licensing activities	Training & qualification effectiveness	Engineering/technical support	Safety assessment/quality verification
05/91	1	1	*	*	1	*	1	1	*	*	*	*	1	2
01/90	1	1	*	*	1	*	1	1	*	*	*	*	2	2
09/88	1	1	2	2	*	1	2	1	1	2	2	2	*	*
04/87	2	1	2	2	*	2	2	1	1	2	2	2	*	*
07/85	2	2	2	2	*	2	3	1	N	3	3	N	*	*
01/84	3	2	3	3	*	2	1	2	N	3	3	N	*	*
04/83	3	2	N	3	*	N	2	2	N	2	2	N	*	*
11/82	2	N	2	3	*	N	N	N	N	1	N	N	*	*

Table 28: Haddam Neck Nuclear Power Plant (See Entry 288)

	Plant operations	Radiological controls	Maintenance	Surveillance	Maintenance/surveillance	Fire protection	Emergency preparedness	Security	Outages	Quality programs	Licensing activities	Training & qualification effectiveness	Engineering/technical support	Safety assessment/quality verification
07/90	1	1	*	*	1	*	1	1	*	*	*	*	1	1
12/88	1	1	*	*	2	*	1	1	*	*	*	*	1	1
09/87	2	2	2	2	*	3	2	1	2	2	2	2	*	*
08/86	1	2	2	2	*	N	2	1	2	2	2	2	*	*
08/85	1	2	1	2	*	2	2	1	1	2	1	N	*	*
01/83	1	1	1	1	*	1	1	1	1	N	1	N	*	*
07/82	1	1	1	2	*	1	2	1	1	1	N	N	*	*
10/80	2	2	2	2	*	2	2	2	2	2	N	N	*	*

1 = superior performance/NRC attention reduced; 2 = good performance/NRC attention maintained at normal levels; 3 = acceptable performance/NRC inspection levels increased; N = no rating given; * = inspection category not used in current year.

Appendix I: NRC Report Cards

Table 29: Shearon Harris Nuclear Power Plant, Unit 1 (See Entry 290)

	Plant operations	Radiological controls	Maintenance	Surveillance	Maintenance/surveillance	Fire protection	Emergency preparedness	Security	Outages	Quality programs	Licensing activities	Training & qualification effectiveness	Engineering/technical support	Safety assessment/quality verification
04/90	1	1	*	*	2	*	1	2	*	*	*	*	1	1
11/88	2	2	1	1	*	2	2	1	1	2	2	1	*	*
01/88	2	2	1	1	*	1	2	2	N	2	1	2	*	*

Table 30: Edwin I. Hatch Nuclear Power Plant, Units 1-2 (See Entry 293)

	Plant operations	Radiological controls	Maintenance	Surveillance	Maintenance/surveillance	Fire protection	Emergency preparedness	Security	Outages	Quality programs	Licensing activities	Training & qualification effectiveness	Engineering/technical support	Safety assessment/quality verification
05/91	1	2	*	*	2	*	1	2	*	*	*	*	2	1
01/90	1	2	*	*	1	*	1	2	*	*	*	*	2	1
11/88	2	2	2	2	*	2	2	2	2	2	1	2	*	*
03/87	2	2	2	2	*	1	1	3	2	2	2	1	*	*
10/85	2	1	3	2	*	2	2	2	2	2	2	N	*	*
03/84	3	2	3	2	*	N	1	1	N	3	2	N	*	*
03/83	2	2	2	2	*	2	2	2	N	2	2	N	*	*
10/82	2	2	2	2	*	2	N	1	2	2	N	N	*	*
01/81	2	2	2	2	*	2	2	2	2	2	N	N	*	*

Table 31: Hope Creek Nuclear Generating Station, Unit 1 (See Entry 295)

	Plant operations	Radiological controls	Maintenance	Surveillance	Maintenance/surveillance	Fire protection	Emergency preparedness	Security	Outages	Quality programs	Licensing activities	Training & qualification effectiveness	Engineering/technical support	Safety assessment/quality verification
11/90	1	1	*	*	2	*	1	1	*	*	*	*	1	1
09/89	1	1	*	*	2	*	2	1	*	*	*	*	2	2
06/88	2	2	1	2	*	N	1	1	N	2	2	1	*	*
04/87	2	2	1	2	*	N	1	1	N	2	1	2	*	*

1 = superior performance/NRC attention reduced; 2 = good performance/NRC attention maintained at normal levels;
3 = acceptable performance/NRC inspection levels increased; N = no rating given; * = inspection category not used in current year.

Table 32: Indian Point Nuclear Generating Station, Unit 2 (See Entry 299)

	Plant operations	Radiological controls	Maintenance	Surveillance	Maintenance/surveillance	Fire protection	Emergency preparedness	Security	Outages	Quality programs	Licensing activities	Training & qualification effectiveness	Engineering/technical support	Safety assessment/quality verification
08/90	1	2	*	*	2	*	2	1	*	*	*	*	2	2
06/89	2	2	*	*	2	*	2	1	*	*	*	*	3	2
05/88	2	2	2	2	*	N	2	1	1	2	2	2	*	*
01/87	2	2	2	1	*	1	2	2	1	2	2	2	*	*
12/85	2	3	2	1	*	2	1	1	2	N	2	N	*	*
02/85	2	3	1	1	*	3	1	2	2	2	2	N	*	*
08/83	2	2	2	2	*	2	N	3	1	3	1	N	*	*
09/82	2	2	2	3	*	2	2	1	2	N	2	N	*	*
04/81	3	2	3	2	*	2	2	3	2	3	N	N	*	*

Table 33: Indian Point Nuclear Power Plant, Unit 3 (See Entry 300)

	Plant operations	Radiological controls	Maintenance	Surveillance	Maintenance/surveillance	Fire protection	Emergency preparedness	Security	Outages	Quality programs	Licensing activities	Training & qualification effectiveness	Engineering/technical support	Safety assessment/quality verification
07/90	2	1	*	*	2	*	1	3	*	*	*	*	2	2
01/89	2	1	*	*	2	*	1	1	*	*	*	*	3	2
02/88	2	1	1	1	*	N	1	1	N	2	1	2	*	*
06/86	2	1	1	1	*	N	1	1	1	2	2	2	*	*
12/84	1	1	1	1	*	1	1	1	1	2	2	N	*	*
08/83	1	1	1	1	*	1	1	1	1	N	3	N	*	*
08/82	2	1	1	1	*	1	2	2	N	N	2	N	*	*
04/81	2	3	3	2	*	2	2	3	2	2	N	N	*	*

1 = superior performance/NRC attention reduced; 2 = good performance/NRC attention maintained at normal levels; 3 = acceptable performance/NRC inspection levels increased; N = no rating given; * = inspection category not used in current year.

Appendix I: NRC Report Cards — Lasalle

Table 34: Kewaunee Nuclear Plant (See Entry 303)

	Plant operations	Radiological controls	Maintenance	Surveillance	Maintenance/surveillance	Fire protection	Emergency preparedness	Security	Outages	Quality programs	Licensing activities	Training & qualification effectiveness	Engineering/technical support	Safety assessment/quality verification
04/91	1	1	*	*	1	*	1	2	*	*	*	*	2	2
03/89	1	1	*	*	1	*	1	1	*	*	*	*	1	1
09/87	1	1	1	1	*	1	1	2	1	1	1	1	*	*
08/86	1	1	1	2	*	1	1	2	1	2	1	N	*	*
11/84	1	1	1	1	*	1	2	2	1	N	2	N	*	*
11/83	2	2	1	2	*	1	2	2	1	N	2	N	*	*
07/82	1	1	1	2	*	2	2	2	1	N	2	N	*	*
12/80	2	2	2	2	*	2	3	2	2	2	N	2	*	*

Table 35: Lasalle County Nuclear Station, Units 1-2 (See Entry 304)

	Plant operations	Radiological controls	Maintenance	Surveillance	Maintenance/surveillance	Fire protection	Emergency preparedness	Security	Outages	Quality programs	Licensing activities	Training & qualification effectiveness	Engineering/technical support	Safety assessment/quality verification
02/91	1	2	*	*	2	*	2	2	*	*	*	*	2	2
10/89	1	2	*	*	2	*	1	2	*	*	*	*	2	2
08/88	?	?	2	2	*	2	2	2	2	2	1	2	*	*
04/87	2	2	2	?	*	2	2	2	2	2	2	2	*	*
04/86	3	2	3	3	*	1	2	2	N	3	2	N	*	*
02/85	3	2	2	2	*	2	2	2	1	2	2	N	*	*
07/83	2	2	1	3	*	2	2	3	1	N	2	N	*	*
11/82	N	2	2	3	*	2	2	3	N	2	2	N	*	*

1 = superior performance/NRC attention reduced; 2 = good performance/NRC attention maintained at normal levels;
3 = acceptable performance/NRC inspection levels increased; N = no rating given; * = inspection category not used in current year.

Table 36: Limerick Generating Station, Units 1-2 (See Entry 305)
Ratings for 1986-88 are for Unit 1 only.

	Plant operations	Radiological controls	Maintenance	Surveillance	Maintenance/surveillance	Fire protection	Emergency preparedness	Security	Outages	Quality programs	Licensing activities	Training & qualification effectiveness	Engineering/technical support	Safety assessment/quality verification
02/91	1	1	*	*	1	*	2	1	*	*	*	*	2	1
02/90	1	1	*	*	1	*	3	1	*	*	*	*	2	1
08/88	1	1	*	*	1	*	2	1	*	*	*	*	2	1
09/87	1	1	1	1	*	N	1	2	N	1	2	1	*	*
12/86	1	2	2	2	*	N	1	3	N	1	1	2	*	*

Table 37: Maine Yankee Nuclear Power Plant (See Entry 306)

	Plant operations	Radiological controls	Maintenance	Surveillance	Maintenance/surveillance	Fire protection	Emergency preparedness	Security	Outages	Quality programs	Licensing activities	Training & qualification effectiveness	Engineering/technical support	Safety assessment/quality verification
06/91	1	2	*	*	1	*	2	2	*	*	*	*	2	2
05/90	1	2	*	*	1	*	2	2	*	*	*	*	2	2
01/89	1	3	*	*	1	*	1	2	*	*	*	*	2	2
08/87	1	2	2	1	*	N	1	1	N	2	2	1	*	*
04/86	2	2	2	1	*	1	1	1	2	2	N	2	*	*
01/85	3	2	2	2	*	2	1	1	1	N	2	N	*	*
10/83	3	2	2	2	*	1	2	2	1	2	2	N	*	*
01/83	3	2	2	2	*	1	2	2	N	3	2	N	*	*
04/82	2	2	2	1	*	1	1	3	1	2	N	2	*	*
11/80	2	2	2	2	*	2	2	2	3	3	N	N	*	*

1 = superior performance/NRC attention reduced; 2 = good performance/NRC attention maintained at normal levels; 3 = acceptable performance/NRC inspection levels increased; N = no rating given; * = inspection category not used in current year.

Appendix I: NRC Report Cards — Millstone

Table 38: McGuire Nuclear Station, Units 1-2 (See Entry 310)
Ratings for 1982 are for Unit 1 only.

	Plant operations	Radiological controls	Maintenance	Surveillance	Maintenance/surveillance	Fire protection	Emergency preparedness	Security	Outages	Quality programs	Licensing activities	Training & qualification effectiveness	Engineering/technical support	Safety assessment/ quality verification
05/91	2	1	*	*	2	*	1	1	*	*	*	*	1	1
03/90	2	1	*	*	2	*	1	2	*	*	*	*	2	2
12/88	2	2	*	*	2	*	1	2	*	*	*	*	2	2
01/88	2	2	2	2	*	2	1	2	2	2	2	2	*	*
06/86	3	2	2	2	*	2	2	1	1	2	2	2	*	*
12/84	3	2	2	2	*	N	1	2	1	2	1	N	*	*
10/83	1	1	1	2	*	2	1	1	1	2	1	N	*	*
12/82	2	1	1	1	*	2	1	2	1	N	2	N	*	*

Table 39: Millstone Nuclear Power Plant, Unit 1 (See Entry 312)

	Plant operations	Radiological controls	Maintenance	Surveillance	Maintenance/surveillance	Fire protection	Emergency preparedness	Security	Outages	Quality programs	Licensing activities	Training & qualification effectiveness	Engineering/technical support	Safety assessment/ quality verification
05/91	1	2	*	*	1	*	1	1	*	*	*	*	2	2
04/90	1	2	*	*	1	*	1	1	*	*	*	*	2	1
07/88	1	2	1	2	*	N	1	2	1	2	2	1	*	*
12/86	1	3	1	1	*	N	1	1	1	2	1	2	*	*
08/85	1	2	1	1	*	1	1	1	2	N	1	N	*	*
03/84	2	1	1	2	*	1	2	1	1	N	1	N	*	*
01/83	1	1	1	1	*	1	1	1	1	N	1	N	*	*
07/82	1	2	1	1	*	1	1	2	1	1	N	N	*	*
10/80	2	2	2	2	*	2	2	3	2	2	N	N	*	*

1 = superior performance/NRC attention reduced; 2 = good performance/NRC attention maintained at normal levels;
3 = acceptable performance/NRC inspection levels increased; N = no rating given; * = inspection category not used in current year.

Table 40: Monticello Nuclear Generating Plant (See Entry 314)

	Plant operations	Radiological controls	Maintenance	Surveillance	Maintenance/surveillance	Fire protection	Emergency preparedness	Security	Outages	Quality programs	Licensing activities	Training & qualification effectiveness	Engineering/technical support	Safety assessment/quality verification
09/90	1	2	*	*	1	*	1	2	*	*	*	*	2	1
06/89	1	1	*	*	1	*	1	3	*	*	*	*	2	2
04/88	2	1	1	1	*	1	1	2	1	2	1	2	*	*
11/86	1	1	1	2	*	1	1	1	1	2	1	2	*	*
04/85	1	2	1	2	*	2	1	1	1	2	1	N	*	*
10/83	2	2	2	1	*	2	1	1	1	N	2	N	*	*
12/82	1	2	2	1	*	2	1	1	1	N	1	N	*	*
05/82	1	2	1	1	*	2	2	2	1	N	N	N	*	*
01/81	2	3	2	2	*	2	3	2	2	2	N	2	*	*

Table 41: Nine Mile Point Nuclear Power Plant, Unit 1 (See Entry 317)

	Plant operations	Radiological controls	Maintenance	Surveillance	Maintenance/surveillance	Fire protection	Emergency preparedness	Security	Outages	Quality programs	Licensing activities	Training & qualification effectiveness	Engineering/technical support	Safety assessment/quality verification
06/91	2	2	*	*	2	*	1	1	*	*	*	*	2	2
08/90	3	2	*	*	3	*	1	1	*	*	*	*	2	3
07/89	3	2	*	*	3	*	1	1	*	*	*	*	3	3
07/88	2	2	2	2	*	N	1	1	N	3	2	2	*	*
07/87	2	2	3	2	*	N	1	1	2	3	1	2	*	*
09/85	1	1	2	1	*	1	1	1	N	N	1	N	*	*
09/84	3	2	2	2	*	1	1	1	1	N	1	N	*	*
08/83	2	2	1	N	*	1	N	1	1	N	1	N	*	*
09/82	2	3	2	2	*	2	2	1	1	N	1	N	*	*
05/81	2	3	2	2	*	2	3	2	2	2	N	N	*	*

1 = superior performance/NRC attention reduced; 2 = good performance/NRC attention maintained at normal levels; 3 = acceptable performance/NRC inspection levels increased; N = no rating given; * = inspection category not used in current year.

Appendix I: NRC Report Cards

Table 42: Nine Mile Point Nuclear Power Plant, Unit 2 (See Entry 318)

	Plant operations	Radiological controls	Maintenance	Surveillance	Maintenance/surveillance	Fire protection	Emergency preparedness	Security	Outages	Quality programs	Licensing activities	Training & qualification effectiveness	Engineering/technical support	Safety assessment/quality verification
06/91	2	2	*	*	2	*	1	1	*	*	*	*	2	2
08/90	3	2	*	*	3	*	1	1	*	*	*	*	2	3
07/89	3	2	*	*	3	*	1	1	*	*	*	*	3	3
07/88	2	2	2	2	*	N	1	1	N	3	2	2	*	*
08/87	2	2	N	2	*	N	1	1	N	2	3	2	*	*

Table 43: North Anna Nuclear Power Station, Units 1-2 (See Entry 319)

	Plant operations	Radiological controls	Maintenance	Surveillance	Maintenance/surveillance	Fire protection	Emergency preparedness	Security	Outages	Quality programs	Licensing activities	Training & qualification effectiveness	Engineering/technical support	Safety assessment/quality verification
12/90	1	2	*	*	2	*	1	2	*	*	*	*	1	1
10/89	1	2	*	*	2	*	2	2	*	*	*	*	2	2
09/88	2	2	2	2	*	1	2	2	3	2	1	1	*	*
12/86	2	2	2	2	*	1	2	2	2	2	1	2	*	*
05/85	2	1	2	2	*	2	2	1	1	2	?	N	*	*
02/84	1	1	2	2	*	N	1	1	1	2	1	N	*	*
04/83	2	1	2	2	*	2	2	1	1	2	2	N	*	*
11/82	2	1	2	2	*	2	2	2	2	3	N	N	*	*
12/80	2	2	2	2	*	2	2	2	2	2	N	N	*	*

1 = superior performance/NRC attention reduced; 2 = good performance/NRC attention maintained at normal levels;
3 = acceptable performance/NRC inspection levels increased; N = no rating given; * = inspection category not used in current year.

Table 44: Oconee Nuclear Station, Units 1-3 (See Entry 321)

	Plant operations	Radiological controls	Maintenance	Surveillance	Maintenance/surveillance	Fire protection	Emergency preparedness	Security	Outages	Quality programs	Licensing activities	Training & qualification effectiveness	Engineering/technical support	Safety assessment/quality verification
11/90	1	1	*	*	2	*	1	2	*	*	*	*	1	2
06/89	1	1	*	*	2	*	1	2	*	*	*	*	2	2
01/88	1	2	1	1	*	2	1	1	2	1	2	2	*	*
06/86	1	2	2	1	*	1	1	2	1	2	2	2	*	*
12/84	1	2	1	1	*	N	1	1	1	2	2	N	*	*
10/83	2	2	1	3	*	N	2	1	1	2	2	N	*	*
10/82	2	2	1	2	*	2	1	1	1	N	2	N	*	*
10/80	2	2	2	2	*	2	2	2	2	2	N	N	*	*

Table 45: Oyster Creek Nuclear Generating Station (See Entry 322)

	Plant operations	Radiological controls	Maintenance	Surveillance	Maintenance/surveillance	Fire protection	Emergency preparedness	Security	Outages	Quality programs	Licensing activities	Training & qualification effectiveness	Engineering/technical support	Safety assessment/quality verification
09/90	2	3	*	*	2	*	2	2	*	*	*	*	2	2
09/89	3	3	*	*	2	*	2	2	*	*	*	*	2	2
05/88	3	2	2	2	*	N	2	1	N	2	2	2	*	*
06/87	2	2	2	1	*	N	1	1	2	2	2	1	*	*
03/86	2	1	3	2	*	N	1	2	2	N	2	N	*	*
10/84	1	1	2	1	*	2	2	2	2	N	2	N	*	*
06/83	2	2	2	2	*	2	2	1	2	N	2	N	*	*
06/82	2	2	3	3	*	2	2	2	2	N	2	N	*	*
03/81	2	2	2	2	*	2	2	3	2	2	N	N	*	*
10/80	2	3	2	3	*	2	2	2	2	3	N	N	*	*

1 = superior performance/NRC attention reduced; 2 = good performance/NRC attention maintained at normal levels; 3 = acceptable performance/NRC inspection levels increased; N = no rating given; * = inspection category not used in current year.

Appendix I: NRC Report Cards

Table 46: Palisades Nuclear Plant (See Entry 323)

	Plant operations	Radiological controls	Maintenance	Surveillance	Maintenance/surveillance	Fire protection	Emergency preparedness	Security	Outages	Quality programs	Licensing activities	Training & qualification effectiveness	Engineering/technical support	Safety assessment/quality verification
04/91	1	2	*	*	1	*	2	1	*	*	*	*	2	2
01/90	2	2	*	*	2	*	1	1	*	*	*	*	2	2
09/88	2	2	2	1	*	3	1	1	1	2	2	2	*	*
12/87	2	2	3	2	*	2	2	2	2	3	2	2	*	*
05/86	2	2	3	3	*	2	2	2	N	3	2	N	*	*
08/85	2	2	2	2	*	2	2	2	1	2	2	N	*	*
10/83	1	2	2	1	*	2	1	2	N	1	2	N	*	*
01/83	2	3	2	1	*	2	2	2	2	2	2	N	*	*
07/82	2	2	2	3	*	1	2	2	2	3	3	N	*	*
02/81	3	3	2	3	*	2	2	2	2	2	2	N	2	*	*

Table 47: Palo Verde Nuclear Generating Station, Units 1-3 (See Entry 324)
Ratings for 1985 and 1987 are for Units 1-2 only.

	Plant operations	Radiological controls	Maintenance	Surveillance	Maintenance/surveillance	Fire protection	Emergency preparedness	Security	Outages	Quality programs	Licensing activities	Training & qualification effectiveness	Engineering/technical support	Safety assessment/quality verification
03/91	2	2	*	*	2	*	1	2	*	*	*	*	2	2
01/90	2	2	*	*	3	*	2	2	*	*	*	*	3	3
03/89	3	3	*	*	2	*	2	2	*	*	*	*	2	3
02/88	2	2	2	2	*	1	1	2	1	2	2	2	*	*
01/87	2	2	2	2	*	1	2	3	1	2	2	1	*	*
12/85	2	2	2	2	*	1	2	2	1	2	2	N	*	*

1 = superior performance/NRC attention reduced; 2 = good performance/NRC attention maintained at normal levels;
3 = acceptable performance/NRC inspection levels increased; N = no rating given; * = inspection category not used in current year.

Table 48: Peach Bottom Nuclear Power Plant, Units 2-3 (See Entry 328)

	Plant operations	Radiological controls	Maintenance	Surveillance	Maintenance/surveillance	Fire protection	Emergency preparedness	Security	Outages	Quality programs	Licensing activities	Training & qualification effectiveness	Engineering/technical support	Safety assessment/quality verification
10/90	2	2	*	*	2	*	1	1	*	*	*	*	2	2
02/90	2	2	*	*	2	*	1	2	*	*	*	*	2	2
12/88	2	2	*	*	2	*	2	3	*	*	*	*	1	2
12/87	N	2	2	2	*	3	2	2	N	N	2	N	*	*
12/86	2	2	2	2	*	2	2	3	1	3	2	2	*	*
08/85	2	3	1	2	*	2	2	3	1	N	1	N	*	*
05/84	2	2	2	2	*	2	2	1	2	N	1	N	*	*
09/83	2	3	2	3	*	3	1	1	2	N	2	N	*	*
10/82	2	3	2	2	*	3	2	2	2	N	1	N	*	*
09/81	2	2	2	1	*	3	2	2	1	2	N	N	*	*
07/80	2	N	3	2	*	2	2	3	2	3	N	N	*	*

Table 49: Perry Nuclear Power Plant, Unit 1 (See Entry 331)

	Plant operations	Radiological controls	Maintenance	Surveillance	Maintenance/surveillance	Fire protection	Emergency preparedness	Security	Outages	Quality programs	Licensing activities	Training & qualification effectiveness	Engineering/technical support	Safety assessment/quality verification
11/90	2	2	*	*	2	*	1	1	*	*	*	*	2	2
10/89	2	2	*	*	2	*	1	1	*	*	*	*	2	2
09/88	2	2	2	2	*	N	1	1	N	2	2	2	*	*
01/88	2	2	2	2	*	2	1	1	N	2	2	2	*	*
08/86	2	2	2	2	*	2	1	1	N	2	1	2	*	*

1 = superior performance/NRC attention reduced; 2 = good performance/NRC attention maintained at normal levels; 3 = acceptable performance/NRC inspection levels increased; N = no rating given; * = inspection category not used in current year.

Appendix I: NRC Report Cards

Table 50: Pilgrim Station, Unit 1 (See Entry 334)

	Plant operations	Radiological controls	Maintenance	Surveillance	Maintenance/surveillance	Fire protection	Emergency preparedness	Security	Outages	Quality programs	Licensing activities	Training & qualification effectiveness	Engineering/technical support	Safety assessment/quality verification
11/90	2	1	*	*	2	*	2	1	*	*	*	*	1	2
12/89	2	2	*	*	2	*	2	1	*	*	*	*	1	2
12/88	2	3	2	2	*	2	2	2	N	2	2	2	*	*
06/87	2	3	2	3	*	3	2	3	1	3	2	2	*	*
05/86	3	3	2	2	*	N	3	2	1	N	1	N	*	*
06/85	2	3	1	1	*	2	3	2	1	N	1	N	*	*
01/84	2	2	2	1	*	1	1	2	N	N	1	N	*	*
11/82	3	2	2	2	*	3	1	2	2	N	2	N	*	*
10/81	3	2	3	2	*	2	1	2	2	3	N	N	*	*
04/81	2	3	2	2	*	2	3	2	3	3	N	N	*	*

Table 51: Point Beach Nuclear Plant, Units 1-2 (See Entry 337)

	Plant operations	Radiological controls	Maintenance	Surveillance	Maintenance/surveillance	Fire protection	Emergency preparedness	Security	Outages	Quality programs	Licensing activities	Training & qualification effectiveness	Engineering/technical support	Safety assessment/quality verification
01/91	1	2	*	*	1	*	2	2	*	*	*	*	2	2
07/89	2	1	*	*	1	*	2	2	*	*	*	*	2	2
03/88	2	2	1	1	*	1	2	1	1	2	2	1	*	*
11/86	1	2	1	1	*	2	2	1	1	2	2	1	*	*
04/85	1	2	2	1	*	2	2	1	2	2	1	N	*	*
08/83	1	2	1	1	*	2	2	2	2	2	N	N	*	*
09/82	2	1	2	2	*	3	2	2	2	N	2	N	*	*

1 = superior performance/NRC attention reduced; 2 = good performance/NRC attention maintained at normal levels;
3 = acceptable performance/NRC inspection levels increased; N = no rating given; * = inspection category not used in current year.

Table 52: Prairie Island Nuclear Plant, Units 1-2 (See Entry 338)

	Plant operations	Radiological controls	Maintenance	Surveillance	Maintenance/surveillance	Fire protection	Emergency preparedness	Security	Outages	Quality programs	Licensing activities	Training & qualification effectiveness	Engineering/technical support	Safety assessment/quality verification
08/89	1	1	*	*	1	*	1	2	*	*	*	*	2	1
04/88	2	1	1	1	*	2	1	1	1	2	1	2	*	*
12/86	2	1	1	2	*	2	1	1	1	2	1	2	*	*
04/85	2	1	1	2	*	1	1	1	1	2	1	N	*	*
10/83	2	1	1	1	*	1	1	1	1	N	1	N	*	*
12/82	1	1	1	1	*	1	2	1	1	2	1	N	*	*
05/82	1	1	1	1	*	1	2	2	1	2	N	N	*	*
01/81	2	3	2	2	*	2	3	2	2	2	N	2	*	*

Table 53: Quad Cities Station, Units 1-2 (See Entry 341)

	Plant operations	Radiological controls	Maintenance	Surveillance	Maintenance/surveillance	Fire protection	Emergency preparedness	Security	Outages	Quality programs	Licensing activities	Training & qualification effectiveness	Engineering/technical support	Safety assessment/quality verification
06/91	3	2	*	*	2	*	1	1	*	*	*	*	2	2
03/90	2	1	*	*	2	*	1	1	*	*	*	*	2	2
01/89	2	1	*	*	2	*	2	1	*	*	*	*	2	2
09/87	2	1	2	1	*	N	2	1	2	2	1	2	*	*
03/86	2	2	2	2	*	2	1	1	1	2	1	N	*	*
02/85	3	2	2	1	*	2	1	1	1	2	1	N	*	*
07/83	2	2	2	1	*	1	2	1	1	N	1	N	*	*
11/82	1	2	2	2	*	2	2	2	1	N	1	N	*	*
11/80	2	3	2	2	*	3	3	2	2	2	N	2	*	*

1 = superior performance/NRC attention reduced; 2 = good performance/NRC attention maintained at normal levels; 3 = acceptable performance/NRC inspection levels increased; N = no rating given; * = inspection category not used in current year.

Appendix I: NRC Report Cards

Table 54: Rancho Seco Nuclear Generating Station (See Entry 342)

	Plant operations	Radiological controls	Maintenance	Surveillance	Maintenance/surveillance	Fire protection	Emergency preparedness	Security	Outages	Quality programs	Licensing activities	Training & qualification effectiveness	Engineering/technical support	Safety assessment/quality verification
07/89	2	2	*	*	2	*	2	2	*	*	*	*	2	2
11/86	3	3	3	2	*	1	3	3	2	3	3	2	*	*
11/85	2	3	2	2	*	2	2	1	1	2	2	2	*	*
06/84	3	2	2	3	*	2	2	1	1	N	3	N	*	*
03/83	2	2	2	2	*	2	2	1	1	N	1	N	*	*
04/82	2	2	2	2	*	1	2	2	1	2	N	N	*	*
08/80	2	2	2	1	*	2	2	2	2	3	N	N	*	*

Table 55: River Bend Station, Unit 1 (See Entry 343)

	Plant operations	Radiological controls	Maintenance	Surveillance	Maintenance/surveillance	Fire protection	Emergency preparedness	Security	Outages	Quality programs	Licensing activities	Training & qualification effectiveness	Engineering/technical support	Safety assessment/quality verification
06/91	1	2	*	*	2	*	1	2	*	*	*	*	2	2
05/90	1	1	*	*	2	*	1	2	*	*	*	*	2	2
03/89	2	2	*	*	2	*	1	3	*	*	*	*	1	1
06/87	1	2	2	2	*	2	2	2	N	1	2	1	*	*
04/86	2	2	2	2	*	1	2	2	N	2	2	2	*	*
02/85	2	2	N	N	*	N	2	2	N	2	2	N	*	*

1 = superior performance/NRC attention reduced; 2 = good performance/NRC attention maintained at normal levels;
3 = acceptable performance/NRC inspection levels increased; N = no rating given; * = inspection category not used in current year.

Table 56: H.B. Robinson Electric Power Plant, Unit 2 (See Entry 345)

	Plant operations	Radiological controls	Maintenance	Surveillance	Maintenance/surveillance	Fire protection	Emergency preparedness	Security	Outages	Quality programs	Licensing activities	Training & qualification effectiveness	Engineering/technical support	Safety assessment/quality verification
05/91	2	1	*	*	2	*	2	1	*	*	*	*	2	2
05/90	2	1	*	*	2	*	2	1	*	*	*	*	2	2
02/89	2	2	*	*	2	*	2	1	*	*	*	*	2	2
01/88	2	2	2	1	*	2	2	1	1	2	2	2	*	*
01/86	2	2	2	2	*	2	1	1	1	2	2	N	*	*
08/84	2	2	2	1	*	1	1	2	1	2	2	N	*	*
09/82	2	3	2	2	*	2	2	2	2	2	N	N	*	*
01/81	2	2	2	2	*	2	2	2	2	2	N	N	*	*

Table 57: St. Lucie Nuclear Power Station, Units 1-2 (See Entry 346)
Ratings for 1981 and January 1983 are for Unit 1 only.

	Plant operations	Radiological controls	Maintenance	Surveillance	Maintenance/surveillance	Fire protection	Emergency preparedness	Security	Outages	Quality programs	Licensing activities	Training & qualification effectiveness	Engineering/technical support	Safety assessment/quality verification
01/91	1	1	*	*	2	*	1	1	*	*	*	*	1	1
09/89	1	2	*	*	1	*	1	2	*	*	*	*	1	1
03/88	1	2	1	1	*	N	1	2	1	1	1	1	*	*
08/86	1	2	1	1	*	2	2	2	2	2	1	1	*	*
02/85	1	1	1	1	*	2	1	1	1	2	1	N	*	*
12/83	1	1	1	1	*	2	1	2	1	2	2	N	*	*
01/83	1	1	2	1	*	2	2	2	1	N	2	N	*	*
01/81	2	2	2	2	*	2	2	2	2	2	N	N	*	*

1 = superior performance/NRC attention reduced; 2 = good performance/NRC attention maintained at normal levels; 3 = acceptable performance/NRC inspection levels increased; N = no rating given; * = inspection category not used in current year.

Appendix I: NRC Report Cards **Salem**

Table 58: Fort St. Vrain Nuclear Power Plant (See Entry 348)

	Plant operations	Radiological controls	Maintenance	Surveillance	Maintenance/surveillance	Fire protection	Emergency preparedness	Security	Outages	Quality programs	Licensing activities	Training & qualification effectiveness	Engineering/technical support	Safety assessment/quality verification
04/89	2	2	*	*	3	*	2	1	*	*	*	*	2	2
08/87	2	1	2	2	*	1	3	2	2	2	2	2	*	*
07/86	2	1	3	2	*	1	3	3	3	3	3	2	*	*
05/85	3	1	3	2	*	2	2	2	1	3	3	2	*	*
06/84	3	2	2	2	*	N	1	2	2	2	3	2	*	*
01/83	3	2	1	2	*	1	2	2	N	2	1	N	*	*
06/82	3	2	2	2	*	2	2	2	2	2	N	N	*	*
10/80	2	2	2	2	*	2	2	3	2	2	N	N	*	*

Table 59: Salem Nuclear Generating Station, Units 1-2 (See Entry 349)

	Plant operations	Radiological controls	Maintenance	Surveillance	Maintenance/surveillance	Fire protection	Emergency preparedness	Security	Outages	Quality programs	Licensing activities	Training & qualification effectiveness	Engineering/technical support	Safety assessment/quality verification
11/90	2	2	*	*	2	*	1	1	*	*	*	*	2	2
09/89	3	2	*	*	2	*	2	1	*	*	*	*	2	2
06/88	2	2	1	2	*	N	1	1	1	1	2	2	*	*
02/87	2	1	1	2	*	N	1	1	2	2	2	2	*	*
02/86	2	1	2	2	*	2	2	1	2	N	2	N	*	*
02/85	3	2	2	2	*	3	2	1	2	N	2	N	*	*
01/84	3	2	2	2	*	2	1	2	1	N	2	N	*	*
01/83	2	1	1	1	*	2	2	3	1	N	2	N	*	*
10/81	1	2	1	1	*	2	2	3	1	1	N	N	*	*

1 = superior performance/NRC attention reduced; 2 = good performance/NRC attention maintained at normal levels;
3 = acceptable performance/NRC inspection levels increased; N = no rating given; * = inspection category not used in current year.

Table 60: San Onofre Nuclear Generating Station, Units 1-3 (See Entry 350)

Ratings for 1981 are for Unit 1 only; ratings for 1982 are for Units 1-2 only.

	Plant operations	Radiological controls	Maintenance	Surveillance	Maintenance/surveillance	Fire protection	Emergency preparedness	Security	Outages	Quality programs	Licensing activities	Training & qualification effectiveness	Engineering/technical support	Safety assessment/quality verification
05/90	2	1	*	*	1	*	1	1	*	*	*	*	2	2
01/89	1	1	*	*	2	*	1	1	*	*	*	*	3	3
01/88	1	2	2	2	*	2	1	2	1	2	2	1	*	*
08/86	2	1	2	2	*	1	2	2	1	2	2	2	*	*
11/84	3	3	2	2	*	2	1	2	2	N	2	N	*	*
12/82	2	2	2	2	*	2	1	2	1	2	2	N	*	*
09/81	2	3	3	1	*	2	1	3	1	2	N	N	*	*

Table 61: Seabrook Station, Unit 1 (See Entry 352)

	Plant operations	Radiological controls	Maintenance	Surveillance	Maintenance/surveillance	Fire protection	Emergency preparedness	Security	Outages	Quality programs	Licensing activities	Training & qualification effectiveness	Engineering/technical support	Safety assessment/quality verification
01/91	2	2	*	*	2	*	1	1	*	*	*	*	2	2

Table 62: Sequoyah Nuclear Plant, Units 1-2 (See Entry 355)

	Plant operations	Radiological controls	Maintenance	Surveillance	Maintenance/surveillance	Fire protection	Emergency preparedness	Security	Outages	Quality programs	Licensing activities	Training & qualification effectiveness	Engineering/technical support	Safety assessment/quality verification
07/90	2	2	*	*	2	*	2	2	*	*	*	*	2	2
06/89	2	2	*	*	2	*	2	2	*	*	*	*	3	2
09/85	2	2	3	2	*	2	2	2	2	3	2	2	*	*
06/84	2	1	1	1	*	1	3	2	1	3	2	N	*	*
06/83	2	2	2	1	*	N	2	3	2	3	2	N	*	*
11/82	3	2	2	2	*	2	2	2	N	3	N	N	*	*

1 = superior performance/NRC attention reduced; 2 = good performance/NRC attention maintained at normal levels; 3 = acceptable performance/NRC inspection levels increased; N = no rating given; * = inspection category not used in current year.

Appendix I: NRC Report Cards

Table 63: Shoreham Nuclear Power Plant (See Entry 357)

	Plant operations	Radiological controls	Maintenance	Surveillance	Maintenance/surveillance	Fire protection	Emergency preparedness	Security	Outages	Quality programs	Licensing activities	Training & qualification effectiveness	Engineering/technical support	Safety assessment/quality verification
02/88	1	1	2	2	*	N	1	1	N	1	1	2	*	*
11/86	2	3	2	N	*	N	1	1	2	2	3	3	*	*
07/85	1	1	1	1	*	2	N	1	N	N	2	N	*	*
09/83	2	2	2	N	*	2	2	2	N	N	2	N	*	*
07/82	N	2	2	N	*	2	N	2	2	N	2	N	*	*
05/81	2	N	2	2	*	N	N	N	N	2	N	N	*	*

Table 64: South Texas Project Electric Generating Station, Units 1-2 (See Entry 361)

Ratings for 1988 are for Unit 1 only.

	Plant operations	Radiological controls	Maintenance	Surveillance	Maintenance/surveillance	Fire protection	Emergency preparedness	Security	Outages	Quality programs	Licensing activities	Training & qualification effectiveness	Engineering/technical support	Safety assessment/quality verification
06/90	1	2	*	*	1	*	2	1	*	*	*	*	2	1
05/89	2	2	*	*	2	*	2	2	*	*	*	*	2	1
08/88	2	2	2	2	*	2	2	3	N	2	2	2	*	*

1 = superior performance/NRC attention reduced; 2 = good performance/NRC attention maintained at normal levels;
3 = acceptable performance/NRC inspection levels increased; N = no rating given; * = inspection category not used in current year.

Table 65: Virgil C. Summer Nuclear Station (See Entry 363)

	Plant operations	Radiological controls	Maintenance	Surveillance	Maintenance/surveillance	Fire protection	Emergency preparedness	Security	Outages	Quality programs	Licensing activities	Training & qualification effectiveness	Engineering/technical support	Safety assessment/quality verification
08/90	1	1	*	*	1	*	2	1	*	*	*	*	2	2
05/89	2	1	*	*	1	*	2	2	*	*	*	*	2	2
12/87	2	1	1	1	*	2	2	1	1	2	2	1	*	*
05/86	3	1	1	2	*	3	1	1	2	2	2	N	*	*
10/84	2	1	1	2	*	2	1	1	N	2	2	N	*	*
05/83	2	1	1	2	*	3	1	2	N	N	2	N	*	*
11/82	2	N	2	2	*	N	1	2	2	2	N	N	*	*
01/81	2	2	2	2	*	2	2	2	2	2	N	N	*	*

Table 66: Surry Nuclear Power Station, Units 1-2 (See Entry 366)

	Plant operations	Radiological controls	Maintenance	Surveillance	Maintenance/surveillance	Fire protection	Emergency preparedness	Security	Outages	Quality programs	Licensing activities	Training & qualification effectiveness	Engineering/technical support	Safety assessment/quality verification
05/91	2	1	*	*	2	*	1	1	*	*	*	*	2	2
07/90	2	2	*	*	3	*	1	1	*	*	*	*	2	2
11/89	3	3	*	*	3	*	3	1	*	*	*	*	2	3
09/88	2	2	2	2	*	2	2	2	2	2	1	1	*	*
12/86	1	2	2	2	*	1	2	2	2	1	2		*	*
05/85	2	1	2	3	*	N	2	1	1	2	1	N	*	*
02/84	2	3	2	1	*	2	1	1	1	2	1	N	*	*
04/83	1	2	2	1	*	2	2	1	2	2	2	N	*	*
11/82	3	2	2	2	*	2	N	2	2	3	N	N	*	*
12/80	3	2	2	2	*	2	2	2	2	2	N	N	*	*

1 = superior performance/NRC attention reduced; 2 = good performance/NRC attention maintained at normal levels; 3 = acceptable performance/NRC inspection levels increased; N = no rating given; * = inspection category not used in current year.

Appendix I: NRC Report Cards

Table 67: Susquehanna Steam Electric Station, Units 1-2 (See Entry 368)
Ratings for 1983-84 are for Unit 1 only.

	Plant operations	Radiological controls	Maintenance	Surveillance	Maintenance/surveillance	Fire protection	Emergency preparedness	Security	Outages	Quality programs	Licensing activities	Training & qualification effectiveness	Engineering/technical support	Safety assessment/quality verification
03/91	1	2	*	*	1	*	1	1	*	*	*	*	2	1
01/90	1	2	*	*	1	*	2	1	*	*	*	*	1	1
08/88	1	2	1	1	*	N	1	1	1	1	1	1	*	*
03/87	1	1	1	1	*	N	1	1	1	2	1	1	*	*
09/85	2	1	1	2	*	2	1	1	1	1	1	N	*	*
08/84	2	2	1	2	*	1	1	1	N	N	1	N	*	*
08/83	2	2	2	1	*	1	1	1	N	N	1	N	*	*

Table 68: Three Mile Island Nuclear Power Plant, Unit 1 (See Entry 369)

	Plant operations	Radiological controls	Maintenance	Surveillance	Maintenance/surveillance	Fire protection	Emergency preparedness	Security	Outages	Quality programs	Licensing activities	Training & qualification effectiveness	Engineering/technical support	Safety assessment/quality verification
10/90	1	1	*	*	1	*	1	1	*	*	*	*	1	1
06/89	1	1	*	*	2	*	1	1	*	*	*	*	2	2
05/88	2	1	1	1	*	2	1	1	2	2	2	1	*	*
05/87	2	1	1	1	*	N	1	2	N	2	1	1	*	*
10/86	2	1	2	1	*	N	1	2	N	2	1	1	*	*
05/86	2	1	2	1	*	N	N	N	N	1	N	1	*	*
05/85	1	1	1	1	*	1	1	1	N	N	2	N	*	*
07/84	2	1	1	1	*	1	1	1	N	1	2	N	*	*
01/83	1	2	2	1	*	1	1	1	N	1	1	N	*	*
06/81	2	2	2	2	*	2	N	2	N	2	N	N	*	*

1 = superior performance/NRC attention reduced; 2 = good performance/NRC attention maintained at normal levels;
3 = acceptable performance/NRC inspection levels increased; N = no rating given; * = inspection category not used in current year.

Trojan

Table 69: Trojan Nuclear Plant (See Entry 371)

	Plant operations	Radiological controls	Maintenance	Surveillance	Maintenance/surveillance	Fire protection	Emergency preparedness	Security	Outages	Quality programs	Licensing activities	Training & qualification effectiveness	Engineering/technical support	Safety assessment/quality verification
08/90	1	2	*	*	3	*	2	2	*	*	*	*	2	2
06/89	2	2	*	*	2	*	1	3	*	*	*	*	2	3
05/88	2	2	2	2	*	3	2	2	2	2	2	2	*	*
12/86	2	2	2	2	*	2	2	2	1	2	2	2	*	*
01/86	2	1	2	2	*	2	1	2	1	2	2	N	*	*
01/85	3	1	2	1	*	2	1	1	1	3	2	N	*	*
04/84	2	2	2	1	*	2	1	1	1	3	1	N	*	*
12/82	2	1	1	2	*	2	1	1	1	2	1	N	*	*
04/82	1	2	2	2	*	1	1	2	1	2	N	N	*	*
10/80	2	2	2	2	*	2	2	2	2	2	N	N	*	*

Table 70: Turkey Point Nuclear Station, Units 3-4 (See Entry 372)

	Plant operations	Radiological controls	Maintenance	Surveillance	Maintenance/surveillance	Fire protection	Emergency preparedness	Security	Outages	Quality programs	Licensing activities	Training & qualification effectiveness	Engineering/technical support	Safety assessment/quality verification
10/90	2	2	*	*	2	*	2	2	*	*	*	*	2	1
12/89	2	2	*	*	2	*	2	3	*	*	*	*	2	2
11/88	3	2	3	2	*	2	2	3	2	2	2	2	*	*
11/87	2	2	2	2	*	N	1	3	2	2	2	3	*	*
08/86	3	2	3	2	*	2	1	2	2	3	2	3	*	*
02/85	3	2	3	2	*	2	2	2	2	3	1	N	*	*
12/83	3	1	2	2	*	N	2	2	2	1	2	N	*	*
01/83	2	1	2	2	*	2	2	2	1	N	2	N	*	*
01/81	2	2	2	2	*	2	2	2	2	2	N	N	*	*

1 = superior performance/NRC attention reduced; 2 = good performance/NRC attention maintained at normal levels; 3 = acceptable performance/NRC inspection levels increased; N = no rating given; * = inspection category not used in current year.

Appendix I: NRC Report Cards

Table 71: Vermont Yankee Nuclear Power Plant (See Entry 376)

	Plant operations	Radiological controls	Maintenance	Surveillance	Maintenance/surveillance	Fire protection	Emergency preparedness	Security	Outages	Quality programs	Licensing activities	Training & qualification effectiveness	Engineering/technical support	Safety assessment/quality verification
05/91	1	2	*	*	1	*	1	2	*	*	*	*	1	1
03/90	1	2	*	*	1	*	1	2	*	*	*	*	1	1
01/89	1	2	2	1	*	N	2	1	N	2	1	1	*	*
06/87	1	2	2	1	*	N	2	1	1	2	1	1	*	*
03/86	1	2	1	1	*	N	2	2	1	2	1	N	*	*
05/85	1	2	2	1	*	2	1	1	1	2	1	N	*	*
11/83	2	1	1	1	*	1	N	1	2	N	1	N	*	*
03/83	1	1	1	1	*	1	2	1	1	N	1	N	*	*

Table 72: Vogtle Nuclear Power Station, Units 1-2 (See Entry 378)

Ratings for 1988-89 are for Unit 1 only.

	Plant operations	Radiological controls	Maintenance	Surveillance	Maintenance/surveillance	Fire protection	Emergency preparedness	Security	Outages	Quality programs	Licensing activities	Training & qualification effectiveness	Engineering/technical support	Safety assessment/quality verification
01/91	2	1	*	*	2	*	3	3	*	*	*	*	2	2
01/90	2	2	*	*	1	*	2	2	*	*	*	*	2	2
01/89	2	2	*	*	2	*	1	2	*	*	*	*	2	2
03/88	3	2	1	2	*	2	1	3	N	2	2	1	*	*

1 = superior performance/NRC attention reduced; 2 = good performance/NRC attention maintained at normal levels;
3 = acceptable performance/NRC inspection levels increased; N = no rating given; * = inspection category not used in current year.

Washington

Table 73: Washington Nuclear Project, Unit 2 (See Entry 380)

	Plant operations	Radiological controls	Maintenance	Surveillance	Maintenance/surveillance	Fire protection	Emergency preparedness	Security	Outages	Quality programs	Licensing activities	Training & qualification effectiveness	Engineering/technical support	Safety assessment/quality verification
12/90	1	2	*	*	2	*	2	2	*	*	*	*	2	1
10/89	2	2	*	*	2	*	1	2	*	*	*	*	2	2
10/88	3	2	*	*	2	*	2	1	*	*	*	*	3	2
07/87	1	2	2	2	*	2	1	2	2	1	2	1	*	*
07/86	2	2	2	2	*	3	2	2	1	2	2	2	*	*
05/85	2	2	2	1	*	2	2	2	N	2	2	N	*	*
12/83	2	2	N	N	*	N	2	2	N	N	2	N	*	*

Table 74: Waterford Power Station, Unit 3 (See Entry 382)

	Plant operations	Radiological controls	Maintenance	Surveillance	Maintenance/surveillance	Fire protection	Emergency preparedness	Security	Outages	Quality programs	Licensing activities	Training & qualification effectiveness	Engineering/technical support	Safety assessment/quality verification
03/90	1	1	*	*	2	*	1	1	*	*	*	*	2	2
02/89	2	2	*	*	2	*	1	1	*	*	*	*	2	2
04/87	2	2	2	1	*	1	2	2	1	2	1	1	*	*
03/86	3	2	3	2	*	2	2	1	N	2	2	2	*	*
07/82	3	N	N	2	*	N	N	2	N	3	3	N	*	*

1 = superior performance/NRC attention reduced; 2 = good performance/NRC attention maintained at normal levels; 3 = acceptable performance/NRC inspection levels increased; N = no rating given; * = inspection category not used in current year.

Appendix I: NRC Report Cards   Yankee

Table 75: Wolf Creek Nuclear Power Plant (See Entry 384)

	Plant operations	Radiological controls	Maintenance	Surveillance	Maintenance/surveillance	Fire protection	Emergency preparedness	Security	Outages	Quality programs	Licensing activities	Training & qualification effectiveness	Engineering/technical support	Safety assessment/ quality verification
10/90	2	2	*	*	2	*	1	1	*	*	*	*	2	2
07/89	2	2	*	*	2	*	1	1	*	*	*	*	2	2
09/88	2	2	2	2	*	1	2	1	3	3	2	2	*	*
05/87	2	2	1	2	*	2	2	2	2	2	1	1	*	*
04/86	2	2	1	1	*	2	2	2	N	1	1	2	*	*
02/85	N	2	2	N	*	1	3	2	N	2	1	N	*	*

Table 76: Yankee Rowe Nuclear Power Plant (See Entry 385)

	Plant operations	Radiological controls	Maintenance	Surveillance	Maintenance/surveillance	Fire protection	Emergency preparedness	Security	Outages	Quality programs	Licensing activities	Training & qualification effectiveness	Engineering/technical support	Safety assessment/ quality verification
05/91	2	1	*	*	2	*	1	1	*	*	*	*	2	1
12/89	1	2	*	*	1	*	1	1	*	*	*	*	1	1
09/88	1	2	2	1	*	N	2	2	1	1	1	2	*	*
03/87	1	1	1	1	*	1	2	2	1	1	1	2	*	*
05/85	1	2	1	1	*	1	1	2	1	2	1	N	*	*
02/84	1	2	1	1	*	1	1	2	1	2	1	N	*	*
01/83	1	1	1	1	*	1	1	1	1	N	1	N	*	*
03/82	1	2	1	1	*	1	1	2	1	2	N	N	*	*
07/80	2	2	2	2	*	2	2	2	2	2	N	N	*	*

1 = superior performance/NRC attention reduced; 2 = good performance/NRC attention maintained at normal levels;
3 = acceptable performance/NRC inspection levels increased; N = no rating given; * = inspection category not used in current year.

Table 77: Zion Station, Units 1-2 (See Entry 388)

	Plant operations	Radiological controls	Maintenance	Surveillance	Maintenance/surveillance	Fire protection	Emergency preparedness	Security	Outages	Quality programs	Licensing activities	Training & qualification effectiveness	Engineering/technical support	Safety assessment/ quality verification
03/91	3	2	*	*	3	*	2	2	*	*	*	*	3	2
02/90	2	2	*	*	2	*	2	2	*	*	*	*	3	2
10/88	2	2	2	2	*	2	1	1	1	2	1	2	*	*
08/87	2	2	2	2	*	2	2	1	1	2	1	1	*	*
04/86	2	2	2	2	*	2	2	1	1	2	1	N	*	*
02/85	2	2	1	2	*	2	2	3	1	2	1	N	*	*
08/83	2	2	2	2	*	2	2	2	1	N	2	N	*	*
11/82	2	3	2	2	*	N	2	2	1	N	2	N	*	*
11/80	2	3	2	2	*	2	3	2	2	2	N	2	*	*

1 = superior performance/NRC attention reduced; 2 = good performance/NRC attention maintained at normal levels; 3 = acceptable performance/NRC inspection levels increased; N = no rating given; * = inspection category not used in current year.

Appendix II: Glossary

Arranged alphabetically, the Glossary provides definitions of the terms, concepts, and acronyms commonly used within the nuclear power industry, with particular emphasis on the terminology employed in the country reports and plant profiles. The Glossary also includes See and See Also references as appropriate.

accident A not reasonably expected event that causes or threatens a rupture of a radioactive material barrier.

AGR (Advanced Gas-cooled Reactor) An improvement on the Magnox type of reactor, it operates at a higher temperature to give a greater burnup. The fuel consists of slightly enriched uranium oxide pellets canned in stainless steel.

alert Second least serious of the four classifications of a nuclear emergency, this is an event that could reduce the plant's level of safety. An alert is considered to pose no danger to the public and the public is not required to take any action. Federal, state, and local officials are notified.

alpha particle A positively charged particle ejected spontaneously from the nuclei of some radioactive elements. Short range and low penetrating power, generally failing to penetrate the skin. Hazardous when introduced to the body internally. Alpha particles can be stopped by a sheet of paper.

Anticipated Transient Without Scram (ATWOS) See **transient**

atom The smallest particle of an element that cannot be divided by chemical means. Consists of a central core called the nucleus, which contains protons and neutrons. Electrons revolve in orbits in the region surrounding the nucleus.

atomic energy Energy released when a neutron initiates the breaking up or fissioning of an atom's nucleus into smaller pieces. Synonymous with nuclear energy.

auxiliary feedwater Backup feedwater supply used during nuclear plant start-up and shutdown.

availability factor The ratio of gross available generation to gross maximum generation, expressed as a percentage. Available generation is the energy that can be produced if the nuclear unit is operated at the maximum power level permitted by equipment and regulatory limitations. Maximum generation is the energy that can be produced by a nuclear unit in a given period if operated continuously at maximum capacity. Availability factor is used rather than capacity factor because it does not penalize nuclear units for fuel loading or reserve shutdown and, thus, allows more realistic comparisons among units.

background radiation The radiation in the natural environment, including cosmic rays and radiation from the naturally radioactive elements, both outside and inside the bodies of humans, animals, and plants.

baseload electricity The supply of electricity that is available 24 hours a day; it contrasts with peak-load electricity, a supply called upon to meet needs during times of greatest usage.

Becquerel (Bq) Has replaced the Curie (Ci) as the unit used for measuring quantity in terms of the number of radioactive disintegrations taking place in a material. Defined simply as the transformation of one atom per second in any quantity of any radioactive substance.

beta particle A negatively charged particle emitted from a nucleus during radioactive decay. A beta particle is an electron that has a mass equal to 1/1837 that of a proton. Beta particles can be stopped by a thin sheet of metal or plastic. Large amounts of beta radiation may cause skin burns and beta emitters are harmful if they enter the body.

BeV American for GeV (1,000 million electron volts).

bioassay Analysis of human hair, tissue, nasal smears, urine, or fecal samples to determine the amount of radioactive material in the body.

biological shield A barrier placed around a reactor or radioactive source.

Boiling Water Reactor (BWR) A nuclear reactor in which water is boiled in the reactor vessel and the resulting steam drives a turbine to generate electricity.

Bq See **Becquerel**

breeder reactor (See Also **breeding**) A nuclear reactor that produces fissionable fuel while consuming it, generating more fuel than it consumes.

breeding (See Also **breeder reactor**) The process of generating nuclear fuel, e.g. plutonium from uranium, by absorption of neutrons.

BTU (British Thermal Unit) The amount of heat required to change the temperature of one pound of water one degree Fahrenheit at sea level.

burnup The fraction or percentage of atoms in a reactor fuel that has undergone fission; also the total amount of heat released per unit mass of fuel (usually expressed in megawatt days per ton).

BWR See **Boiling Water Reactor**

capacity factor The amount of electricity produced by a given power plant over a specified length of time (usually a year), expressed as a percentage of what the plant could have produced if it had operated at full power, 24 hours a day, over that length of time. For example, a plant operating at full power for a full year has a capacity factor of 100 percent. At half power for a full year, or full power for half the year, its capacity factor is 50 percent.

cerenkov radiation Visible light emitted when charged particles pass through a transparent material at a velocity greater than that of light in the material. It can be seen, for example, as a blue glow in the water around fuel elements in the spent fuel pool.

chain reaction (nuclear) A process in which one nuclear transformation sets up conditions that permit a similar nuclear transformation to take place in another atom. Thus, when fission occurs in uranium atoms, neutrons are released that in turn produce fission in neighboring uranium atoms.

China Syndrome Vernacular term describing a core accident where molten fuel melts through the reactor containment building and, theoretically, through the earth to China.

Ci See **Curie**

cladding The thin-walled metal tube that forms the outer jacket of a nuclear fuel rod. The cladding serves as a barrier by preventing the release of radioactivity into the coolant.

coastdown A gradual decrease in reactor power.

cold shutdown A reactor condition in which the coolant temperature has been reduced below 212 degrees Fahrenheit.

collective dose equivalent The sum of the products of the individual dose equivalents and the number of individuals receiving the dose in each group in an exposed population. Also known as effective dose equivalent.

collective effective dose equivalent This is a composite figure in which all the doses from different types of radiation to different body organs are combined together according to their capacity to cause fatal cancers. This quantity is often simply called 'dose'.

condenser A heat exchanger that cools non-nuclear steam from the turbine to water.

containment The provision of a pressurized shell or other enclosure around a reactor to confine fission products that might be released to the environment as the result of an accident.

contamination See **radioactive contamination**

control rods Rods, plates, or tubes of steel or aluminum-containing boron, cadmium, or some other strong absorber of neutrons. They are used to hold a reactor at a given power level; by absorbing neutrons, they prevent the neutrons from causing further fission.

coolant A fluid, usually water, that is circulated through or about the core of a reactor to maintain a low temperature and prevent the fuel from overheating.

core The central portion of a nuclear reactor containing the fuel assemblies.

critical mass The smallest amount of fuel necessary to sustain a chain reaction.

criticality A reactor is "critical" when it achieves a self-sustaining nuclear chain reaction.

Curie (Ci) A unit of radioactivity. It is the quantity of a radioactive isotope that disintegrates at a rate of 37,000 million disintegrations per second.

Appendix II: Glossary

daughter product The product formed in the radioactive decay of a nucleus (called the parent). Synonym for decay product.

decay When a radioactive atom disintegrates it is said to decay. What remains is a different element. For example, an atom of polonium decays to form lead, ejecting an alpha particle in the process.

decay heat The heat produced by radioactive atoms in the reactor after the reactor has been shut down.

decay product see **daughter product**

decontamination The reduction or removal of contaminating radioactive material using soap, water, and/or cleaning solvents.

depleted uranium Uranium having less than the natural content (i.e., 0.7 percent), of the easily fissionable uranium U-235. An example of depleted uranium is the residue from a diffusion plant or a reactor.

deuterium (See Also **heavy water**) The isotope of hydrogen of mass 2, often called heavy hydrogen.

diffusion (1) A method of isotope separation (see **gaseous diffusion**). (2) In nuclear physics, the passage of particles through matter in such a way that the probability of scattering is large compared with that of capture.

dose (See Also **collective dose**) The amount of radioactive exposure received or potentially received. Measured in rems or millirems. For any ionizing radiation it is the energy, measured in grays or rads, that is imparted to matter by the ionizing particles per unit mass of irradiated material.

dose limits The limits recommended by the International Commission on Radiological Protection (ICRP) for controlling the exposure of members of the public to ionizing radiations.

dosimeter A portable instrument for measuring and registering exposure to radiation.

effective dose equivalent See **collective dose**

electrical generator An electromagnetic device that converts mechanical (rotational) energy into electrical energy.

electron A particle that "orbits" the nucleus of an atom.

emergency core cooling system A backup cooling system that removes heat from the plant's reactor core, thus preventing a core meltdown if the normal core cooling system fails.

enriched fuel Nuclear fuel that has been enriched in the fissile component, e.g. uranium containing more than 0.7 percent U-235.

excursion A sudden, very rapid rise in the power level of a reactor caused by supercriticality. Excursions are usually quickly suppressed by the reactor's negative temperature coefficient and/or by control rods.

exposure The absorption of radiation or ingestion of a radionuclide.

fast reactor A nuclear reactor in which most of the fissions are caused by neutrons moving at high speeds. Such reactors contain little or no moderator.

feedwater Water in the steam generators that removes heat from the reactor and becomes steam that drives the turbine generator.

fission The splitting of a nucleus into at least two other nuclei, accompanied by the release of a relatively large amount of heat energy.

fission gases Fission products (primarily noble gases like xenon, krypton and radon) that exist in a gaseous state. Most fission products are radioactive.

flux (heat) Rate of heat flow across a boundary (i.e., cladding).

flux (neutron) The intensity of neutron radiation. It is expressed as the number of neutrons passing through one square centimeter in one second.

fuel assembly A cluster of fuel rods that make up a reactor core.

fuel cycle The period of time a reactor can operate without needing new fuel.

fuel pellet A small cylinder of uranium fuel approximately 3/8 inches in diameter and 5/8 inches long.

fuel rod A long slender tube that holds fuel or fissionable material.

fusion The process of building up more complex nuclei by the combination, or fusion, of simpler ones. The formation is usually accompanied by the release of energy.

gamma ray Electromagnetic radiation emitted by the nuclei of radioactive substances during decay, similar in nature to X-rays.

gas centrifuge process A method of isotopic separation in which heavy gaseous atoms or molecules are separated from lighter ones by centrifugal force.

gaseous diffusion A method of separating isotopes, for example those of uranium, by causing a gaseous compound to diffuse through a porous membrane; the lighter molecules diffuse faster than the heavy ones and consequently their concentration increases on passing through the membrane, for the initial portions of gas. In practice, one allows a certain amount to diffuse through a cell membrane, and the rest, somewhat depleted in the light component, to flow past.

Geiger-Mueller Counter An instrument for detecting and measuring beta and gamma radiation. It was named for Hans Geiger and W. Mueller who invented it in the 1920s. It is sometimes called simply a Geiger counter or a G-M counter.

general emergency This is the most serious of the four classifications of a nuclear emergency. A large amount of radioactive material is being or could be released from the plant into the environment. Federal, state, local, and power plant officials take actions to protect the public. EBS stations broadcast information and instructions.

generator See electrical generator

graphite A form of carbon used as a moderator in nuclear reactors. It is made from purified petroleum coke compressed into bricks and heated to high temperatures.

half-life The time in which half the atoms of a particular radioactive substance disintegrate to another nuclear form. Measured half-lives vary from millionths of a second to billions of year.

heat exchanger A device that transfers heat from one material, such as water or gas, to another substance with no direct contact between the two materials. Two examples are steam generators and feedwater heaters.

heat sink Anything that absorbs heat.

heavy water (See Also **deuterium**) Water consisting of molecules in which the hydrogen is replaced by deuterium, or heavy hydrogen. It is present in ordinary water as about 1 part in 5,000.

high radiation area Any area of the power plant where a human could receive a radiation dose of 100 millirem (0.1 rem) in one hour. Access into these areas is maintained under strict control.

IAEA (International Atomic Energy Agency) An independent United Nations organization that promotes the peaceful uses of atomic energy and establishes international standards of safety.

incident Any event not considered as part of planned operations.

ingestion exposure pathway An area in which there could be ingestion of contaminated water or foods such as milk or fresh vegetables. The duration of the principal exposures could range in length from hours to months.

internal radiation Radiation resulting from radioactive substances in the body.

International Atomic Energy Agency See **IAEA**

ion (1) An atom that has too many or too few electrons, causing it to be chemically active; (2) an electron that is not associated (in orbit) with a nucleus.

ionizing radiation Any radiation capable of displacing electrons from atoms or molecules, thereby producing ions. Examples: alpha, beta, gamma, X-rays, neutrons and ultra-violet light. Doses of ionizing radiation may produce skin or tissue damage.

irradiation Exposure to radiation.

isotope Two atoms are said to be isotopes if they are the same chemical element but have different masses. This means that isotopic nuclei contain the same number of protons but different numbers of neutrons. Synonymous with nuclide.

kilowatt A unit of power equal to one thousand watts. Most electric plants express their generating capacity in kilowatts or megawatts (one million watts).

light-water reactor Reactors that use ordinary water as coolant, including boiling water reactors and pressurized water reactors.

load factor The load factor of a generating plant reflects its availability and usage. It is the ratio of the units (kWH) actually generated to those that could have been generated if the plant had worked continuously at full power.

LOCA (Loss Of Coolant Accident) A LOCA can result from an opening in the primary cooling system, such as due to a pipe break or a stuck open relief valve (as occurred at Three Mile Island). At the first sign of a LOCA, the reactor would shut down automatically. Although the reactor is shut down, the fuel assemblies would continue to generate heat and so cooling water must continue to circulate through the reactor. If there is an interruption in the main flow of cooling water because a LOCA has occurred, then backup cooling water is needed. This is pumped automatically from the independent water sources referred to as the Emergency Core Cooling Systems (ECCS).

magnox The magnesium alloy used for sheathing the uranium in Calder Hall type fuel elements. Nuclear power stations using such fuel elements are often referred to as Magnox stations.

mega (M) Prefix meaning one million.

megawatt (MW) One million watts or a thousand kilowatts, where a watt is the unit of power. In MWe the "e" signifies "electrical;" in MWt the "t" means "thermal power" or heat output.

meltdown A buildup of heat in the core caused by insufficient cooling, which causes the fuel to melt.

MeV One million electron volts.

millirem A unit of radiation dosage equal to 0.001 of a rem. An individual member of the public can receive up to 500 millirems per year according to U.S. federal standards. This limit does not include radiation received for medical treatment, nor does the limit include the 100 to 250 millirems people receive annually from background radiation.

moderator The material in a reactor used to reduce the energy (by scattering), and hence the speed, of fast neutrons, as far as possible without capturing them.

mothball A method of decommissioning where buildings with radioactive materials or components inside are fenced off to allow much of the radioactivity to fade away in place. The plant can be dismantled several decades later after radioactivity declines.

NEA (Nuclear Energy Agency) An agency of the Organization for Economic Co-operation and Development (OECD); comprises countries whose objective is the orderly development of the peaceful uses of atomic energy.

neutron An uncharged particle found in the nucleus of every atom heavier than hydrogen. Neutrons sustain the fission chain reaction in a nuclear reactor.

noble gas An inert gas or gaseous chemical element that does not readily enter into chemical combination with other elements. The noble gases are helium, neon, argon, krypton, xenon, and radon.

NRC See **Nuclear Regulatory Commission**

nuclear chain reaction See **chain reaction**

nuclear energy See **atomic energy**

Nuclear Energy Agency See **NEA**

Nuclear Non-Proliferation Treaty A treaty with the aim of preventing the diversion of nuclear materials from peaceful uses to nuclear weapons or other nuclear explosive devices. It came into force on March 5, 1970, and was initially ratified by 47 countries.

nuclear reactor A structure in which a fission chain reaction can be maintained and controlled. It usually contains a fuel, coolant, moderator, and control absorbers and is most often surrounded by a concrete biological shield to absorb neutron and gamma ray emission.

Nuclear Regulatory Commission (NRC) The independent civilian agency of the U.S. federal government with the authority to regulate, inspect, and oversee the nuclear industry to assure the safe operation of U.S. nuclear power plants.

nucleus The small, central, positively charged region of an atom that carries essentially all its mass.

nuclide Any species of atom that exists for a measurable length of time. A nuclide can be distinguished by its atomic weight, atomic number, and energy state. Synonymous with isotope.

outage A situation in which the plant is temporarily out of service because of extensive inspection, repairs, and/or refueling. An outage is distinguished from a "shutdown" which refers to the reactor and/or turbine generator being out of operation.

personnel monitoring Determination by either physical or biological measurement of the amount of ionizing radiation to which an individual has been exposed, such as by using a film badge or a TLD (thermo-luminescent dosimeter).

plume A "cloud" of airborne radioactive particles moving away from a damaged nuclear plant in a direction and at a speed determined by prevailing wind.

plume exposure pathway An area in which the whole body external exposure to gamma radiation could be from a plume and/or from deposited materials, or from inhalation of a passing radioactive plume. The duration of the principal potential exposures could range in length from hours to days.

plutonium The element, atomic number 94, produced by neutron irradiation of U-238. The isotope Pu-239 is an important fissionable material and is usually made in reactors. It is used as a nuclear fuel, usually as plutonium oxide.

poison Any material that readily absorbs neutrons and hence removes them from the fission chain reaction in a reactor.

Pressurized Water Reactor (PWR) A reactor in which heat is transferred from the core to steam generators by high temperature water kept under high pressure. Steam is generated in a secondary system.

pressurizer A tank that controls the pressure in the primary or nuclear system of a pressurized water reactor.

primary loop A closed system of piping that provides cooling water to the reactor and transfers heat energy to the secondary loop.

rad The basic unit of an absorbed dose of radiation.

radiation Energy from alpha, beta, neutron particles, or gamma photons emitted from the nucleus of an atom.

radiation area Any area in a power plant where the dose of radiation to an individual during one hour would exceed five millirems.

radiation burns If radiation is sufficiently intense or prolonged it causes surface burns on the skin that redden and blister, not unlike heat burns or severe sunburn. Patients subject to intense irradiation, e.g. in cancer therapy by X-rays or gamma rays, sometimes suffer surface or skin radiation burns.

radiation detection instrument A device that detects and records the characteristics of ionizing radiation, such as a Geiger counter.

radiation risk The risk to health from exposure to ionizing radiation.

radiation sickness The symptoms, usually including nausea, induced by acute whole-body overexposure to radiation, i.e. sudden large doses received over the whole body.

radioactive contamination Radioactive particles in a place where they are not wanted, constituting a potential hazard.

radioactive waste Solid, liquid, and gaseous materials from nuclear operations that are radioactive. Wastes are generally classified as high-level or low-level, depending on the amount of radiation emitted by the material. Spent fuel elements are examples of high-level radioactive waste. Contaminated gloves and rags are examples of low-level radioactive waste.

radioactivity Emission of energy by alpha particles, beta particles, or gamma rays from the nucleus of an isotope.

radon A naturally occurring radioactive element.

reactivity A measure of the amount of the possible departure of a reactor from the critical condition, where the reaction is just self-supporting. At any steady rate of operation the reactivity is zero. Addition of positive reactivity causes an increase in power; addition of negative reactivity causes the reaction to die down.

reactor coolant system The primary cooling system used to remove energy from the reactor core.

reactor core The central portion of a nuclear reactor containing the nuclear fuel, such as uranium or plutonium, and the moderator, if any.

reactor vessel A cylindrical, steel vessel that contains the core, control rods, coolant, and structures that support the core. In a pressurized water reactor, the reactor vessel is constructed of steel approximately eight inches thick.

reactor vessel head The removable top section of a reactor.

redundant system Each of the safety systems in a nuclear plant has at least one backup system that automatically takes over if the first system should fail for any reason. Since there are at least two systems to do the same function, they are called redundant.

relief valve A valve that automatically opens to release steam and prevent excessive pressure build-up.

REM (Roentgen Equivalent Man) A measure of radiation exposure that indicates the potential biological impact on human cells.

reprocessing The procedure of removing fission products from fuel before reusing it. One main aim is to remove poisons that would absorb and waste neutrons; another is to remove mechanical stresses due to irradiation, especially in the case of metallic fuels.

Roentgen Equivalent Man See **REM**

safety rod A neutron-absorbing rod that, in an emergency, can be put into and shut down a nuclear reactor in a fraction of a second. It is normally operated by gravity so as to be independent of power supply.

scram A shutdown of the reactor by rapid insertion of control rods, either automatically or manually. Also known as trip.

secondary loop In pressurized water reactors, a system of piping that carries nonradioactive water. Water in the secondary loop absorbs heat from water in the primary loop through the steam generator tubes, is boiled and, as steam, is used to spin the turbines.

shielding Any material or obstruction that absorbs or blocks radiation.

shutdown A decrease in the rate of fission and heat production in the reactor.

site area emergency The second most serious of the four classifications of a nuclear emergency. Radiation could be released from the plant but at low levels that do not require the public to take any action. Federal, state, and local officials are notified and recommend that farmers within two miles of the plant place their milk-producing animals on stored feed.

source Concentrated radioactive matter used as a source of radiation.

spent fuel Nuclear fuel that can no longer effectively sustain a chain reaction.

spent fuel pool An underwater storage and cooling facility for spent fuel.

startup An increase in the rate of fission and heat production in a reactor, usually by the removal of control rods from the core.

steam generator The heat exchanger used to transfer heat from the primary system to the secondary system.

surface contamination The deposition and attachment of radioactive materials to a surface.

Appendix II: Glossary

thermal reactor A nuclear reactor that includes a moderator and therefore uses slow or thermal neutrons for fission of its fuel.

transient A deviation from normal operating conditions that can usually be controlled by minor adjustments without shutting down the reactor. A significant transient can result in activation of emergency systems and/or reactor scram.

trip See **scram**

turbine An engine made with a series of curved vanes on a rotating shaft that is usually turned by water or steam.

unusual event The least serious of the four classifications of a nuclear emergency. It indicates there is a minor problem at the plant. The Nuclear Regulatory Commission (NRC), state, and local agencies are notified.

uranium A radioactive element with the atomic number 92 and, as found in natural ores, an average atomic weight of approximately 238. The two principal natural isotopes are uranium-235 (0.7 percent of natural uranium), which is fissionable, and uranium-238 (99.3 percent of natural uranium). Uranium is the basic fuel of a nuclear reactor.

waste, radioactive See **radioactive waste**

whole-body counter A device used to identify and measure radiation in the body.

whole-body exposure An exposure of the body to radiation in which the entire body is irradiated.

Technical Problems Index

This index lists plants that have had significant accidents, breakdowns, or technical failures. The index is organized by the nine main systems of a nuclear plant, plus one category for radiation exposure (categories are shown below). Each citation includes the name of the plant and the country where it is located, the book entry number corresponding to a plant profile in the main section of the directory, a symbol denoting the plant's operational status, and a brief description of the problem.

Condenser System: Condenser Coolant
Condenser System: Cooling Tower
Condenser System: Tubing

Control Systems: Emergency Water Supply
Control Systems: Fire Safety System
Control Systems: Operations Computer
Control Systems: Safety Control System

Coolant Pipes and Valves

Electrical Engineering Components: Generator
Electrical Engineering Components: Transformer
Electrical Engineering Components: Turbine

Fuel Handling: Cooling Ponds
Fuel Handling: Core Loading Machines and Refueling
Fuel Handling: Waste Handling

Primary Cooling Loop: Coolant Lines
Primary Cooling Loop: Coolant Pump
Primary Cooling Loop: Heat Exchanger
Primary Cooling Loop: Seals
Primary Cooling Loop: Valves

Radiation Exposure

Reactor: Containment Vessel
Reactor: Control/Dampening Rods
Reactor: Core
Reactor: Fuel Rod Assemblies
Reactor: Moderator System
Reactor: Standpipes and Rod Guide Tubes

Reactor Building: Primary Containment
Reactor Building: Secondary Containment

Secondary Cooling Loop: Coolant Lines and Valves
Secondary Cooling Loop: Coolant Pumps
Secondary Cooling Loop: Seals
Secondary Cooling Loop: Steam Generator

Condenser System: Condenser Coolant

Blayais Nuclear Power Plant (France) **37** ●
Water temperature in Gironde River rose too high to allow cooling of reactors under licensing agreement.

Clinton Nuclear Power Station (United States) **260** ●
Three workers suffered burns when they removed a glass cover from an evaporator, releasing a spray of hot water.

Crystal River Energy Complex (United States) **266** ●
In August 1991, a mass of seaweed clogged the plant's water intake pipe on the Gulf of Mexico, forcing it to shut down for lack of coolant.

Fukushima Daiichi Nuclear Power Station (Japan) **107** ●
Critics of the plant claimed leakage from the recirculating condenser water had contaminated fishing grounds.

Condenser System: Cooling Tower

Salem Nuclear Generating Station (United States) **349** ●
In 1990, New Jersey's Department of Environmental Protection and Energy issued an order for the construction of cooling towers at Salem in response to the high number of fish caught and killed in the plant's water intake system.

Condenser System: Tubing

Angra Nuclear Power Station (Brazil) **11** ●
Poorly designed condenser tubes were replaced during the first few years of operation.

Doel Nuclear Power Station (Belgium) **7** ●
Corrosion of condenser steam tubing in 1989 will force replacement of entire condenser components in 1993.

Paluel Nuclear Power Plant (France) **56** ●
Corrosion cracking in Inconel-600 condenser tubes prompted Electricite de France to reduce the operating temperature.

Control Systems: Emergency Water Supply

Dresden Nuclear Station (United States) **272** ●
In October 1989, the high pressure coolant injection system failed.

Kozloduy Nuclear Power Plant (Bulgaria) **14** ●
Leaks in the emergency cooling water system contaminated the work environment, resulting in worker's exposure to radioactivity.

Mihama Power Station (Japan) **118** ●
In February 1991, Unit 2's condenser failed, causing a leak of 20 tons of radioactive water. Operator error almost caused a meltdown, but the automatic core flooding system activated.

Pilgrim Station (United States) **334** ●
In 1990, the emergency backup water injection system proved faulty when leaky valves in the cooling system caused the water level in the cooling system to drop.

Rancho Seco Nuclear Generating Station (United States) **342** ★
Operator error repeatedly knocked out emergency feed water pumps. A short circuit and control-room blackout caused by a dropped light initiated a feedwater accident very similar to that of Three Mile Island.

Wolf Creek Nuclear Power Plant (United States) **384** ●
Two emergency water pumps failed in December 1990.

Control Systems: Fire Safety System

Balakovo Nuclear Power Plant (Russia) **157** ●
A fire caused automatic fire control systems to shut the plant down in March 1992. Plant workers controlled the blaze in 40 minutes.

Bohunice Nuclear Power Plant (Czech Republic) **29** ●
Austrian inspectors have claimed that the plant has an inadequate fire safety system.

Browns Ferry Nuclear Plant (United States) **249** ●
In 1975, a worker checking for air flows with a candle ignited insulation below the control rooms of Units 1 and 2. The fire raged for 7.5 hours, knocking out safety control circuits for Unit 1. Workers could not shut the reactor down. It took 18 months to get the reactors back on line and cost $213 million.

Dukovany Nuclear Power Plant (Czech Republic) **30** ●
Fire broke out in the plant on January 21, 1991, alarming neighboring Austrians, but no injuries or contamination were reported.

Greifswald Nuclear Power Plant (Germany) **71** ★
In 1975, a ductwork fire destroyed both normal and emergency control cables for one of the reactors, stopping the water feed pumps and threatening a loss-of-coolant incident and a probable meltdown. A single water pump on an outside power source supplied coolant for the reactor for six hours, allowing repair crews to install temporary replacement lines.

Ignalina Nuclear Power Plant (Lithuania) **132** ●
In 1988, two fires in control cabling at Unit 2 activated the automatic fire safety system. The plant closed temporarily for minor repairs.

Ignalina Nuclear Power Plant (Lithuania) **132** ●
In 1988, Unit 1 closed for major repairs after more than 40 yearly unplanned power outages.

Kozloduy Nuclear Power Plant (Bulgaria) **14** ●
Numerous oil, water, and steam leaks compounded the fire hazards of exposed electrical wiring and lack of redundant safety circuits.

Vandellos I Nuclear Center (Spain) **189** ★
A fire in October 1989 forced the premature closing of the plant. The Spanish federal government decided to make the closure permanent due to a $315 million estimated cost to re-open the plant.

Windscale Advanced Gas Cooled Reactor (United Kingdom) **234** ★
Operator error allowed the graphite core to overheat, igniting fuel rods and eventually the core itself in Britain's worst nuclear accident.

Control Systems: Operations Computer

Diablo Canyon Nuclear Power Plant (United States) **269** ●
Some of the plant's electronic components were used parts or

defective parts that had been refurbished. After the plant went into full power operation, the utility discovered they had been duped by parts suppliers in a nationwide scam.

Enrico Fermi 2 Nuclear Power Plant (United States) **277** ●
In February 1989, the plant experienced control problems with its heating, air conditioning, and ventilation system.

Ignalina Nuclear Power Plant (Lithuania) **132** ●
In 1992, a senior programmer of Russian descent was accused of trying to plant a virus in the computer responsible for overseeing the plant's cooling system.

Sizewell B Power Station (United Kingdom) **231** ○
The French Controlbloc P-20 computer system proved too complex, prompting Nuclear Electric to opt for a Westinghouse system at the last moment. The change added cost and may delay the planned 1994 opening.

Control Systems: Safety Control System

Berkeley Power Station (United Kingdom) **214** ▲
Outdated safety control circuits would have to be modernized if plant stayed open.

Bradwell Power Station (United Kingdom) **215** ●
Outdated safety control circuits would have to be modernized if plant stayed open.

Calder Hall Nuclear Power Station (United Kingdom) **216** ●
Outdated safety control circuits would have to be modernized if plant stayed open.

Chapelcross Nuclear Power Station (United Kingdom) **217** ●
Outdated safety control circuits would have to be modernized if plant stayed open.

Douglas Point Nuclear Generating Station (Canada) **20** ★
The first case of calandria tube failure led to the development of the "mouse" in 1976. The robotic device can replace defective calandria and pressure tubes.

Dungeness A Power Station (United Kingdom) **219** ●
Outdated safety control circuits would have to be modernized if plant stayed open.

Robert Emmet Ginna Nuclear Power Plant (United States) **283** ●
During a core flooding incident, a pressure relief valve at first failed to open, then stuck open.

Hinkley Point A Power Station (United Kingdom) **224** ●
Outdated safety control circuits would have to be modernized if plant stayed open.

Hunterston A Nuclear Power Station (United Kingdom) **227** ▲
Outdated safety control circuits would have to be modernized if plant stayed open.

Krsko Power Station (Slovenia) **180** ●
The Krsko station has been plagued by a high number of unplanned automatic scrams, unplanned actuations of the plant's safety system, and forced outages.

Nine Mile Point Nuclear Power Plant (United States) **318** ●
Transformer failure robbed plant of power for 10 vital functions including computer control and monitoring. Operators declared a "Site Emergency." Power was restored in 22 minutes from off-site. The safety control system that should have switched immediately and automatically to off-site power failed to initiate. Inspections later revealed that the logic board batteries were dead.

Palisades Nuclear Plant (United States) **323** ●
Failure of turbine and three safety valves closed the plant for 10 months.

Palo Verde Nuclear Generating Station (United States) ● **324**
Failure of three monitoring devices precipitated an "Alert" in May, 1992.

Rancho Seco Nuclear Generating Station (United States) **342** ★
In December 1985, a power failure to the plant control room for 22 minutes caused an overcooling of the reactor vessel. Operators tried to regulate the coolant water flow by manually closing valves, but the valves had rusted open. Cutting the water supply to the pumps resulted in a 1,200-gallon spill of slightly radioactive water in the containment structure and a small release of radiation to the atmosphere.

Salem Nuclear Generating Station (United States) **349** ●
Two failures of the automatic safety system within a few days prompted Public Service Electric and Gas Co. to extensively modify its computerized safety control system. The Nuclear Regulatory Commission blamed much of the problem on "human engineering deficiencies."

Coolant Pipes and Valves See also: (Primary Cooling Loop and Secondary Cooling Loop)

Hunterston B Nuclear Power Station (United Kingdom) **228** ●
In October 1977, 8000 liters of seawater entered the reactor vessel through a temporary wasteduct installed during construction. The pipe was meant to exhaust contaminated demineralized water until repairs to defective circulator seals could be made. When the reactor shut down, pressure in the system allowed the water flow to reverse, channelling seawater into the reactor.

Loviisa Power Plant (Finland) **34** ●
A failure in a water feed pipe closed the plant for two weeks in 1990.

Point Lepreau Nuclear Generating Station (Canada) **25** ●
From July to September 1982, four heavy water spills resulted from equipment failures and, in one case, operator error when a valve was left opened.

Surry Nuclear Power Station (United States) **366** ●
In 1986, a high temperature water line ruptured, killing four workers and injuring four others. Investigators blamed the failure on an unexpected form of corrosion erosion that had thinned the pipe wall.

Electrical Generating Components: Generator

Enrico Fermi 2 Nuclear Power Plant (United States) **277** ●
On March 22, 1990, a small amount of lubricating oil ignited on an emergency diesel generator.

Philippsburg Nuclear Power Station (Germany) **84** ●
　In 1990, excessive condenser coolant temperature damaged the generator.

Salem Nuclear Generating Station (United States) **349** ●
　Generator failures, including a fire, shut the unit down for 19 out of 27 months.

Surry Nuclear Power Station (United States) **366** ●
　In 1988, an emergency diesel generator failed, taking the unit off-line for inspections and upgrades.

Ulchin Nuclear Power Plant (Republic of Korea) **147** ●
　In early 1990, both units shut down for 109 days because of generator failure.

Vogtle Nuclear Power Station (United States) **378** ●
　An emergency generator failed, leaving the plant without control power and beginning a "Site Emergency." By the time power was restored, the coolant water temperature had risen to 136 degrees Fahrenheit. If power had not been restored, the coolant could have boiled, releasing radioactive steam into the containment structure. The structure was not secure and took 80 minutes to seal because a hatch was blocked with maintenance equipment.

Electrical Generating Components: Transformer

Dresden Nuclear Station (United States) **272** ●
　In 1990, a failure in an auxiliary transformer left the reactor site without off-site power.

Maine Yankee Nuclear Power Plant (United States) **306** ●
　Fire and an explosion in the main transformer split generator hydrogen pipes causing a dramatic explosion but no threat to the nuclear portion of the plant.

Nine Mile Point Nuclear Power Plant (United States) **318** ●
　Transformer failure robbed the plant of power for 10 vital functions including computer control and monitoring. Operators declared a "Site Emergency." Power was restored in 22 minutes from off-site.

Pickering Nuclear Generating Station (Canada) **24** ●
　A leaking transformer shut down the reactor in April 1984, costing the utility $2 million for replacement generation.

Salem Nuclear Generating Station (United States) **349** ●
　Lightning struck the transformer, closing the plant for one week. The Salem plant was unusually susceptible to geomagnetic interference with its transformers, causing numerous transformer trips.

Yankee Rowe Nuclear Power Plant (United States) **385** ★
　On June 15, 1991, lightning struck the plant's transmission lines. The surge destroyed a transformer and initiated an automatic safe shutdown of the reactor.

Electrical Generating Components: Turbine

Chernobyl Nuclear Power Station (Ukraine) **205** ☆
　In 1991, a fire in the turbine closed Unit 5 and prompted an order to close the entire complex by 1993.

Enrico Fermi 2 Nuclear Power Plant (United States) **277** ●
　On September 4, 1989, the plant shut down when hydrogen contaminated the water of the stator cooling system of the main turbine generator.

Obrigheim Nuclear Power Plant (Germany) **83** ●
　Administrative Court ordered the plant closed until it could prove the integrity of high pressure steam turbines.

Palisades Nuclear Plant (United States) **323** ●
　Failure of turbine and three safety valves closed the plant for 10 months.

Philippsburg Nuclear Power Station (Germany) **84** ●
　In 1990, the plant shut down for 105 days for extensive backfitting, including repairs and modifications to the steam turbine.

Ulchin Nuclear Power Plant (Republic of Korea) **147** ●
　In late 1989, both units suffered turbine blade cracking.

Fuel Handling: Cooling Ponds

Douglas Point Nuclear Generating Station (Canada) **20** ★
　In January 1986, contaminated water leaked from a fuel storage pond at a rate of 60 liters per hour.

Superphenix Fast Breeder Power Station (France) **64**
　In April 1987, a major sodium leak drained 800 liters daily of coolant through a six-inch crack in a sodium storage tank.

Fuel Handling: Core Loading Machines and Refueling

Douglas Point Nuclear Generating Station (Canada) **20** ★
　Early development work involved thousands of repairs and improvements that included refueling machines.

Haddam Neck Nuclear Power Plant (United States) **288** ●
　The fuel assembly dropped back into the core during refueling. Inspection revealed no damage.

Hinkley Point B Power Station (United Kingdom) **225** ●
　When inserting new rods with the reactor running, the pressure changes in the core caused control problems in the boilers.

Hinkley Point B Power Station (United Kingdom) **225** ●
　On November 19, 1978, a fuel assembly being withdrawn from reactor R4 snagged. After disengaging the assembly, workers discovered damage to the graphite sleeve and damage to the fuel pins caused by lack of coolant.

Indian Point Nuclear Power Plant (United States) **300** ●
　Two fuel bundles were accidentally lifted out of the core when workers used a crane to lift off the top of the reactor. No radiation was released.

Fuel Handling: Waste Handling

Cofrentes Nuclear Power Plant (Spain) **183** ●
　On October 19, 1982, 11 meters of rain fell in 33 hours, flooding the plant and causing the release of 580 liters of radioactive waste.

Hinkley Point A Power Station (United Kingdom) **224** ●
　A 1987 report named Hinkley the worst polluting nuclear power station in the United Kingdom. A 1988 health authority report showed a higher-than-average incidence of childhood

leukemia in an eight-mile radius of the facility.

Sosnovy Bor Nuclear Power Plant (Russia) **179** ●
Critics claimed that storage buildings were inadequate for the volume of waste and had cracks and holes and that radiation surrounding them measured up to 380 times above normal.

Tsuruga Nuclear Power Station (Japan) **130** ●
In March 1981, leaking radioactive material was dumped into Tsuruga Bay after radiation alarms were turned off. The resulting contamination is called Japan's worst nuclear incident. In all, the plant has had 21 unreported leaks, and operators admitted they routinely dumped radioactive material into the bay.

Vermont Yankee Nuclear Power Plant (United States) **376** ●
On July 18, 1976, Vermont Yankee was ordered to pay the state of Vermont an out-of-court settlement fine of $30,000 for dumping 83,000 gallons of radioactive water into the Connecticut River.

Primary Cooling Loop: Coolant Lines

Atucha Nuclear Power Plant (Argentina) **1** ●
Broken pipes caused a heavy water spill.

Diablo Canyon Nuclear Power Plant (United States) **269** ●
An allegedlly sabotaged cooling pipe delayed low power testing.

Dresden Nuclear Station (United States) **272** ★
In 1978, officials shut down the plant after discovering radioactive sediment in the cooling piping.

Hinkley Point A Power Station (United Kingdom) **224** ●
Bolts and fasteners in high pressure ducting corroded from reaction with carbon dioxide gas surfaces in 1968.

Hinkley Point B Power Station (United Kingdom) **225** ●
Ineffective insulation forced a rerouting of some ducting around the core. The resulting bends created noise and vibration greater than a jet's tailpipe.

Laguna Verde Nuclear Power Plant (Mexico) **135** ●
In two separate incidents in October 1988, two months after it was loaded for pre-operational testing, the plant experienced piping ruptures that would have been serious if Laguna Verde had been on-line.

Oconee Nuclear Station (United States) **321** ●
A leaking compression fitting spilled 70,000 gallons of radioactive water onto the containment floor, activating an "Alert."

Sosnovy Bor Nuclear Power Plant (Russia) **179** ●
In March 24, 1992, a steam line ruptured, flooding a machine room with radioactive steam, some of which vented to the atmosphere. Western experts claim more radioactivity was released in this incident than at Three Mile Island.

THTR-300 Demonstration Reactor (Germany) **89** ▲
High operational temperatures corroded bolts and insulation in the high pressure ducting system.

Vermont Yankee Nuclear Power Plant (United States) **376** ●
Cracked fuel rod cladding and cracks in cooling pipes closed the reactor 34 times in its first 18 months of operation. The plant closed again in 1985 because of corroded cooling pipes.

Primary Cooling Loop: Coolant Pumps

Douglas Point Nuclear Generating Station (Canada) **20** ★
Early development work involved thousands of repairs and improvements that included the primary circuit pump.

Enrico Fermi 2 Nuclear Power Plant (United States) **277** ●
In 1985, a coolant circulator pump failed during testing.

Fukushima Daiichi Nuclear Power Station (Japan) **107** ●
Excessive vibration destroyed a coolant pump, spreading debris into the reactor moderator pool. The units closed for two years for repairs.

Lasalle County Nuclear Station (United States) **304** ●
Operator error turned off coolant circulating pumps for the core. The result was an oscillation of the nuclear reaction, something the designers had said could not happen. The Nuclear Regulatory Commission ordered General Electric to review its 60 similar plants and address the problem.

Three Mile Island Nuclear Power Plant (United States) **370** ★
In 1979, a coolant pump failure caused a loss-of-coolant incident that rated as "worst-case scenario." The incident is considered America's worst commercial nuclear accident.

Zion Station (United States) **388** ●
In 1990, the reactor closed for five months when a cooling pump failed. An emergency replacement from the Pooled Inventory Management System brought the plant on-line quickly, at a cost of $2.5 million. Later, the Unit's other three coolant pumps were upgraded.

Primary Cooling Loop: Heat Exchanger

Phenix Prototype Fast Reactor Power Station (France) **58** ●
Twice in 1976, welds on rigid sealing plates failed from thermal expansion, causing liquid sodium to leak onto the reactor floor. Emergency systems contained the resulting fire without release of radioactivity.

Primary Cooling Loop: Seals

Crystal River Energy Complex (United States) **266** ●
In 1987, a weakened seal forced the temporary shutdown of the plant.

Dungeness B Power Station (United Kingdom) **220** ●
In the early 1970s, the circulator seals proved unable to withstand the high pressures and temperatures. Construction was halted for three years while new seals were developed.

Hunterston B Nuclear Power Station (United Kingdom) **228** ●
Circulator seals proved unable to withstand the high pressures and temperatures.

Surry Nuclear Power Station (United States) **366** ●
In 1988, an inflatable seal failed, spilling 30,000 gallons of nonradioactive coolant water into the containment structure.

Primary Cooling Loop: Valves

Biblis Nuclear Power Plant (Germany) **66** ●
Operator error resulted in an open valve between the primary

● Operable ○ Under Construction △ Indefinitely Deferred ▲ Decommissioning ☐ Planned ■ On Order ☆ Cancelled ★ Shutdown

and secondary cooling loops. High pressure radioactive steam flooded low pressure piping. Pipe failure could have resulted in a loss-of-coolant meltdown.

Braidwood Station (United States) **248** ●
A residual heat removal pump suction relief valve became stuck open in December 1990.

Enrico Fermi 2 Nuclear Power Plant (United States) **277** ●
In August 1986, smoking wiring on a motor-operated valve shut down the plant until the design problem was rectified. In several incidents in 1988, circulator pump valves failed to close.

Fort Calhoun Generating Station (United States) **281** ●
On July 3, 1992, an electrical component malfunctioned, activating the emergency shutdown system. During the shutdown, a pressure relief valve failed to close, causing the containment structure to flood with 20,000 gallons of water.

Three Mile Island Nuclear Power Plant (United States) **370** ★
In 1979, a stuck valve compounded the failure of a coolant pump to produce America's worst commercial nuclear accident. A loss-of-coolant incident which rated as a "worst-case scenario", the core remained uncovered for 5.5 hours. During the incident, 32,000 gallons of contaminated water flooded the containment building and adjoining auxiliary buildings.

Radiation Exposure

Browns Ferry Nuclear Plant (United States) **249** ●
In 1984, the Nuclear Regulatory Commission fined Tennessee Valley Authority $100,000 because 13 workers had been exposed to excessive radiation from leaking coolant.

Calvert Cliffs Nuclear Power Plant (United States) **254** ●
In 1990, operator error resulted in an hour-long release of gases from a waste decay tank, exposing operators to a small amount of radioactivity.

Dounreay Prototype Fast Breeder Reactor (PFR) (United Kingdom) **218** ●
As a result of studies showing a high incidence of leukemia in residents living near the facility, especially school children, British Nuclear Fuels Ltd. agreed in 1987 to pay pensions to families of workers who had died of cancer.

Haddam Neck Nuclear Power Plant (United States) **288** ●
Operator error released a small amount of radioactive gas when workers removed a gauge that opened a half-inch-diameter hole in the plant system cooler.

Hinkley Point A Power Station (United Kingdom) **224** ●
A 1987 report named Hinkley the worst polluting nuclear power station in the United Kingdom. A 1988 health authority report showed a higher-than-average incidence of childhood leukemia in an eight-mile radius of the facility.

Susquehanna Steam Electric Station (United States) **368** ●
In January 1992, a small blast exposed one worker to radioactive dust. The worker underwent decontamination and is under observation.

Three Mile Island Nuclear Power Plant (United States) **370** ★
Health studies indicated no statistical evidence of increased cancer resulting from the release of radioactive material in 1979; however, more than 2,200 people filed lawsuits for medical compensation. So far 280 have been settled at a cost of $14 million.

Wolsong Nuclear Power Plant (Republic of Korea) **149** ●
Between 1983 and 1990, worker exposure consistently exceeded the 25 REM limit proscribed by the Korean Atomic Energy Law.

Yonggwang Nuclear Power Plant (Republic of Korea) **152** ●
Between 1986 and 1990, worker exposure consistently exceeded the 25 REM limit proscribed by the Korean Atomic Energy Law.

Reactor: Containment Vessel

Berkeley Power Station (United Kingdom) **214** ▲
Cracks and brittle welds in steel alloy vessel from long exposure to radiation forced the early shutdown of the plant.

Bohunice Nuclear Power Plant (Czech Republic) **29** ●
Brittle welds threatened the integrity of the vessel.

Bradwell Power Station (United Kingdom): **215** ●
Cracks and brittle welds in steel alloy vessel from long exposure to radiation threatened thre early shutdown of the plant.

Calder Hall Nuclear Power Station (United Kingdom) **216** ●
Cracks and brittle welds in steel alloy vessel from long exposure to radiation threatened the early shutdown of the plant.

Calvert Cliffs Nuclear Power Plant (United States) **254** ●
In 1989, reacting to charges that the reactor vessel was one of 12 in the United States that had become dangerously brittle, the Nuclear Regulatory Commission increased its brittleness threshold by 30 percent, allowing the reactor to continue operating.

Chapelcross Nuclear Power Station (United Kingdom) **217** ●
Cracks and brittle welds in steel alloy vessel from long exposure to radiation threatened the early shutdown of the plant.

Dungeness A Power Station (United Kingdom) **219** ●
Cracks and brittle welds in steel alloy vessel from long exposure to radiation threatened the early shutdown of the plant.

Dungeness B Power Station (United Kingdom) **220** ●
Carbon dioxide coolant gas corroded the mild steel pressure vessel liner. Researchers injected methane gas to control the reaction, but that caused carboning on fuel rods. Eventually designers prescribed another liner, this one of stainless steel. To install the liner, a portion of the graphite block core had to be disassembled.

Dungeness B Power Station (United Kingdom) **220** ●
Construction halted on the power station when workers discovered the original vessel liner was too small and had caused deformation at the weld points.

Greifswald Nuclear Power Plant (Germany) **71** ★
Soviet scientists used a high-temperature annealing process to repair cracks in vessel walls. The procedure called for heating the vessel to 480 degrees Celsius over a 120-hour period.

Hinkley Point A Power Station (United Kingdom) **224** ●
Cracks and brittle welds in steel alloy vessel from long exposure to radiation threatened the early shutdown of the plant.

Technical Problems Index

Hinkley Point B Power Station (United Kingdom) 225 ●
In 1966, researchers observed corrosion of the mild steel pressure vessel liner from carbon dioxide coolant gas. They injected methane gas to control the reaction, but that caused carboning on fuel rods. In 1972, designers prescribed another liner, this one of stainless steel. To install the liner, a portion of the graphite block core had to be disassembled.

Hunterston A Nuclear Power Station (United Kingdom) 227 ▲
Cracks and brittle welds in the steel alloy vessel from long exposure to radiation threatened the early shutdown of the plant.

Laguna Verde Nuclear Power Plant (Mexico) 135 ●
The steel reactor vessel was dropped from a crane and fell to the ground. Later, during operation tests, seawater accidentally entered the reactor containment building, which may have led to corrosion of the pressure vessel.

Oldbury Power Station, (United Kingdom) 229 ●
Cracks and brittle welds in steel alloy vessel from long exposure to radiation threatened the early shutdown of the plant.

Sizewell A Power Station (United Kingdom) 230 ●
Cracks and brittle welds in steel alloy vessel from long exposure to radiation threatened the early shutdown of the plant.

Trawsfynydd Power Station (United Kingdom) 233 ●
Cracks and brittle welds in steel alloy vessel from long exposure to radiation threatened the early shutdown of the plant.

Yankee Rowe Nuclear Power Plant (United States) 385 ★
Critics of the plant's relicensing application claimed its aging pressure vessel had become so brittle with age and exposure to radiation that it could crack catastrophically under rapid temperature changes. Replacement could cost $5 million. In October, 1991, the operators began a voluntary shutdown of the reactor at the request of the Nuclear Regulatory Commision. They hoped to prove the integrity of the vessel and reopen later. In 1992, a plan to repair the brittle metal with annealing was abandoned.

Reactor: Control/Dampening Rods

Dukovany Nuclear Power Plant (Czech Republic) 30 ●
Between 1985 and 1990, safety systems dropped the reactor control rods on 28 occasions, halting the nuclear reaction each time.

Vermont Yankee Nuclear Power Plant (United States) 376 ●
In 1973, operators, attempting to save time, withdrew too many control rods from the core while the core was exposed during a maintenance shutdown. The potential meltdown was averted when the rods were replaced.

Reactor: Core

Chernobyl Nuclear Power Station (Ukraine) 204 ★
A series of operator errors led to the world's worst nuclear accident. Bypassing safety controls and cooling systems, researchers ignited the graphite core, setting off a series of explosions that blew the 100-ton roof off the reactor building and spread deadly radioactive gases across much of the northern hemisphere.

Dungeness B Power Station (United Kingdom) 220 ●
Tests indicate the graphite core blocks may deform during thermal expansion, possibly causing emergency control rods to jam during an emergency shutdown.

Enrico Fermi 1 Fast Breeder Reactor Project (United States) 276 ▲
Core temperatures reaching 1000 degrees Fahrenheit caused failures in graphite and metal components, necessitating exotic materials and innovative repair designs.

Hartlepool Power Station (United Kingdom) 221 ●
Tests indicate the graphite core blocks may deform during thermal expansion, possibly causing emergency control rods to jam during an emergency shutdown.

SNR 300 Kalkar Nuclear Power Plant (Germany) 86 ☆
Extremely high core temperatures forced development of exotic metals and fuel configurations.

Three Mile Island Nuclear Power Plant (United States) 370 ★
In 1979, America's worst commercial nuclear accident, a loss-of-coolant incident that rated as a "worst-case scenario," uncovered the core for 5.5 hours, causing 50 percent of the core to melt. Temperatures reached 5,000 degrees Fahrenheit and hydrogen bubbles formed in the pressure vessel, threatening an explosion that would have breached the containment structure.

Reactor: Fuel Rod Assemblies

Dungeness B Power Station (United Kingdom) 220 ●
Researchers injected methane gas to control a corrosive reaction between the carbon dioxide coolant gas and the mild steel of the reactor vessel, resulting in carboning on fuel rods.

Enrico Fermi 1 Fast Breeder Reactor Project (United States) 276 ▲
On October 5, 1966, during low-power testing, two fuel assemblies melted when Zirconium segments broke free and blocked the flow of coolant. Redesign and cleanup took four years.

Haddam Neck Nuclear Power Plant (United States) 288 ●
A thermal shield damaged by debris in the core was discarded.

Vermont Yankee Nuclear Power Plant (United States) 376 ●
Cracked fuel rod cladding and cracks in cooling pipes closed the reactor 34 times in its first 18 months of operation.

Reactor: Moderator System

Crystal River Energy Complex (United States) 266 ●
In 1977, a malfunctioning monitor indicated a low-water condition in the core moderator pool. The automatic protection system began to flood the core with more water; however, the pool was not low. More than 43,000 gallons of contaminated water spilled onto the containment floor, forcing employees to evacuate.

Robert Emmet Ginna Nuclear Power Plant (United States) 283 ●
A steam bubble threatened to uncover the core. Operators tried to control the bubble with pressure, condensing it, but the pressure grew too high and resulted in the escape of radioactive steam into the atmosphere.

Reactor: Standpipes and Rod Guide Tubes

Atucha Nuclear Power Plant (Argentina) 1 ●
A probe became unsoldered and broke fuel rod guides inside

● Operable   ○ Under Construction   △ Indefinitely Deferred   ▲ Decommissioning   □ Planned   ■ On Order   ☆ Cancelled   ★ Shutdown

the moderator pool.

Bruce Nuclear Power Development (Canada) **16** ●
Heavy water leakage into calandria tubes in February 1982 and March 1986 prompted Energy Probe to call for a metallurgical review of the station.

Bugey Nuclear Power Plant (France) **38** ●
A maintenance teardown revealed corrosion cracking in Inconel standpipes. The discovery prompted inspections of other French pressurized water reactors.

Diablo Canyon Nuclear Power Plant (United States) **269** ●
When loading fuel rods into the plant for the first time, workers discovered construction errors had resulted in an interchange of standpipes between Units 1 and 2.

Paluel Nuclear Power Plant (France) **56** ●
Corrosion cracking in Inconel-600 standpipes showed up in hydro-testing of vessel. Electricite de France reduced the operating temperature to eliminate the problem.

Pickering Nuclear Generating Station (Canada): **24** ●
In August 1983, a defective spring allowed a fuel bundle to shift, rupturing a pressure tube. The resulting heavy water leak within the containment structure is considered one of the worst nuclear accidents in Canada's history. As a result, both Units 1 and 2 were completely re-tubed, raising questions about CANDU reliability.

Reactor Building: Primary Containment

Armenia Nuclear Station (Armenia) **4** ★
Earthquakes in the region raised doubts about the integrity of the containment structures. Both Units were closed.

Bohunice Nuclear Power Plant (Czech Republic) **29** ●
No provision was made for containment in case of an accident.

Diablo Canyon Nuclear Power Plant (United States) **269** ●
After the discovery of the Hosgri Fault, the containment structure was redesigned at an additional cost of $100 million to withstand earthquakes registering up to 7.5 on the Richter scale.

Guangdong Nuclear Power Plant (People's Republic of China) **141** ○
Workers misread construction blueprints and left out 316 of 576 reinforcing rods in the foundations.

Obrigheim Nuclear Power Plant (Germany) **83** ○
Administrative Court ordered the plant closed until it could prove the integrity of its old-style containment structure.

Pickering Nuclear Generating Station (Canada) **24** ●
During the 1970s, a hole of more than 100 square centimeters in the containment structure went unnoticed for a year-and-a-half.

Sequoyah Nuclear Plant (United States) **355** ●
The discovery of weaknesses in the containment structure in 1972 and 1974 led to a redesign and construction cost overruns.

Reactor Building: Secondary Containment

Bruce Nuclear Power Development (Canada) **16** ●
In 1989, there were 11 complaints of odor outside the reactor building. In 1992, a senior official at the Atomic Energy Control Board admitted minute radioactive particles had been found both on the floor of the reactor and outside the containment structures.

Douglas Point Nuclear Generating Station (Canada) **20** ★
In 1980-1981 the reactor ran for six months without a proper filter to contain radioactive gases.

Enrico Fermi 2 Nuclear Power Plant (United States) **277** ●
In 1985, a valve and pipe assembly used to test for radioactive leakage was left open for two months. The incident could have allowed radioactive materials to escape from the containment structure.

Three Mile Island Nuclear Power Plant (United States) **370** ★
In 1979, America's worst commercial nuclear accident, a loss-of-coolant incident that rated as a "worst-case scenario," triggered the release of 2.5 million curies of radioactive noble gases and 15 curies of radioiodine.

Secondary Cooling Loop: Coolant Lines and Valves

Blayais Nuclear Power Plant (France) **37** ●
Steam pipe maintenance shut down all units down between 1989 and 1990.

Browns Ferry Nuclear Plant (United States) **249** ●
In 1984, the Nuclear Regulatory Commission fined Tennessee Valley Authority $100,000 when 13 workers were exposed to excessive radiation from leaking coolant.

Calvert Cliffs Nuclear Power Plant (United States) **254** ●
In March 1989, a pinhole-sized leak in a steam pipe forced the closure of Unit 2. Repairs and verification took two years.

Calvert Cliffs Nuclear Power Plant (United States) **254** ●
In March 1991, 1,900 gallons of reactor cooling water were sprayed into the containment building because a valve at Unit 2 had been incorrectly connected.

Diablo Canyon Nuclear Power Plant (United States) **269** ●
A valve stuck during testing, diverting cooling water to a cooling tank and temporarily delaying the test.

Fort St. Vrain Nuclear Power Plant (United States) **348** ★
The discovery of 37 hairline cracks in the steam piping led to the plant's early decommissioning.

Nine Mile Point Nuclear Power Plant (United States) **318** ●
Faulty valves delayed the opening of the new plant for one year.

Philippsburg Nuclear Power Station (Germany) **84** ●
In 1990, the plant shut down for 105 days for extensive backfitting, including repairs and modifications to steam piping.

Vandellos II Nuclear Center (Spain) **190** ●
A leaking valve flooded the containment structure with radioactive steam in 1990. The plant re-opened four days later.

Secondary Cooling Loop: Coolant Pumps

Calvert Cliffs Nuclear Power Plant (United States) **254** ●
Repairs to the coolant circulating pumps, along with steam pipe leaks, extended a routine maintenance shutdown to two years.

Technical Problems Index

Davis-Besse Nuclear Power Plant (United States) **267** ●
In 1985, an operator error turned off a water supply pump for the steam generator, threatening to boil the moderator pool and expose the core. A backup pump had not yet been installed even though mandated by the Nuclear Regulatory Commission in 1979.

Hamaoka Nuclear Power Plant (Japan) **111** ●
A circulating pump failure in April 1991 shut down the plant temporarily.

Hope Creek Nuclear Generating Station (United States) **295** ●
Water feed pump and valve failures closed the plant repeatedly in 1987.

Nine Mile Point Nuclear Power Plant (United States) **317** ●
Feed water pump failure caused a two-and-one-half year shutdown to repair numerous deficiencies, including faulty welds and radioactive spills.

Secondary Cooling Loop: Seals

Fort St. Vrain Nuclear Power Plant (United States) **348** ★
Leaking seals habitually allowed water to leak into the helium gas and accumulate in the reactor vessel.

Pilgrim Station (United States) **334** ●
In 1991, a leaking pump seal forced the plant to close five days before a scheduled shutdown.

Secondary Cooling Loop: Steam Generator

Almaraz Nuclear Center (Spain) **181** ●
All six steam generators will be replaced by the mid-1990s because of corroding and clogging of their tubes.

Asco Nuclear Center (Spain) **182** ●
All six steam generators will be replaced by the mid-1990s because of corroding and clogging of their tubes.

Beaver Valley Power Station (United States) **242** ●
In 1991, Beaver Valley plant operators sued Westinghouse over repeated corrosion failures of steam generator tubes.

Beloyarsk Nuclear Power Plant (Russia) **160** ●
Early problems with steam generator tube breaks resulted in a dangerously volatile mixing of liquid sodium and water. The plant's modular design will allow steam generators to be replaced individually without taking the plant off-line.

Dungeness B Power Station (United Kingdom) **220** ●
The plant's original high pressure ductwork design proved too rigid. High noise and vibration threatened early failure, forcing redesign of the boiler system and ducting layout.

Robert Emmet Ginna Nuclear Power Plant (United States) **283** ●
Ruptured tubes spilled 8,000 gallons of radioactive water onto the containment floor. Tubes ruptured a total of four times in seven years.

Indian Point Nuclear Generating Station (United States) **299** ●
Cracks in the steam generator walls threatened proposed power increase.

Loviisa Power Plant (Finland) **34** ●
Inspections of poor welds in steam generator occasioned continual power outages, reducing operating efficiency of the plant.

Mihama Power Station (Japan) **118** ●
Unit 1 was closed from 1974 to 1983 for repairs and improvements. In February 1991, Unit 2's condenser failed, causing a leak of 20 tons of radioactive water and nearly resulting in a core meltdown. The condenser will be replaced in 1994. An inspection of Unit 3 revealed similar pipe damage.

North Anna Nuclear Power Station (United States) **319** ●
On two occasions steam generator tubes failed, releasing a small amount of radioactive steam into the atmosphere.

Obrigheim Nuclear Power Plant (Germany) **83** ●
Both steam generators were replaced in 1983 to modernize the facility.

Ohi Nuclear Power Station (Japan) **120** ●
Condenser failure at Mihama nuclear plant prompted inspection and repairs at Ohi.

Palisades Nuclear Plant (United States) **323** ●
The use of phosphate in the steam generator secondary side promoted corrosion of tubing. Work is underway to replace both of the 420-ton steam generators.

Phenix Prototype Fast Reactor Power Station (France) **58** ●
In April 1982, a special hydrogen detector activated an emergency shutdown when steam generator tubes failed, mixing liquid sodium and water. The controlled shutdown averted a devastating explosion, successfully proving the plant's safety system.

H.B. Robinson Electric Power Plant (United States) **345** ●
Steam generators were replaced due to corrosion of tubes. Carolina Power & Light sued Westinghouse for premature failures.

Rovno Nuclear Power Plant (Ukraine) **208** ●
Recurring problems with breaks in its steam generator tubes prompted researchers to use Rovno to study hypothetical loss-of-coolant scenarios and situations where operators control a sudden change in temperature without shutting down the reactor.

Surry Nuclear Power Station (United States) **366** ●
Between 1978 and 1980, all six steam generators were replaced after water impurities corroded and clogged many of the system's 18,000 tubes.

Takahama Nuclear Power Station (Japan) **127** ●
An inspection of the steam generator revealed an improperly installed anti-vibration bar.

Three Mile Island Nuclear Power Plant (United States) **369** ●
Severe corrosion of tubes was corrected with individual tube repairs and blocking over a one-year period.

Trojan Nuclear Plant (United States) **371** ●
In 1991, the plant closed to repair corrosion cracks in steam generator tubes.

Zaporozhye Nuclear Power Station (Ukraine) **212** ●
Corrosion problems may force the replacement of all five steam generators in the 1990s.

● Operable   ○ Under Construction   △ Indefinitely Deferred   ▲ Decommissioning   □ Planned   ■ On Order   ☆ Cancelled   ★ Shutdown

Alphabetical Index

This index provides an alphabetical listing of all plant names, company names, and other significant details mentioned within the country reports and plant profiles of this directory. The index also includes inversions on significant keywords appearing in the names of the power plants. Bolded numbers following a citation refer to book entry numbers for the plant profiles; numbers preceded by 'p.' are page numbers and refer to names and terms in the country reports. Each plant name citation is followed by a symbol that indicates its current operating status; refer to the legend at the bottom of each page for more information.

AB Kärnkraftutbildning (AKU) p. 241
AB Vattenbyggnadsbryan (Sweden) **191, 193, 194**
Abalone Alliance **269**
ABB Reaktor **146**
ACEOWEN **6, 7, 10**
Advanced Boiling Water Reactors (ABWR) **115**
Advanced Gas Cooled Reactor (AGR) **219, 220, 221, 225, 234, 235**, p. 277
Advanced Thermal Reactor (ATR) **106, 119**, p. 124
AEA Technology **218, 234, 235**, p. 279
AEG Telefunken (Germany) **91**
AETEA **188**
Aetna Life & Casualty **379**
Alabama Power Co. **241, 275**
Alan R. Barton Nuclear Power Plant, Units 1-4 (United States) **241** ☆
ALARA **305**
Alberto Nuclear Power Station **12**
Alert **283, 321**
All-Union Institute of Nuclear Power Operation p. 212
Allegheny Electric Cooperative Inc. **368**
Allen's Creek Nuclear Power Plant, Units 1-2 (United States) **237** ☆
Allgemeine Elektricitaets-Gessellschaft **91**
Allis-Chalmers (United States) **274, 284**
Allis-Chambers (United States) **262, 263**

Almaraz Nuclear Center, Units 1-2 (Spain) **181** ●
Almirante Alvaro Alberto Nuclear Power Station **12**
Alsthom **10, 44, 45, 50, 51, 61, 62, 141, 147, 155, 189**, p. 60
Alsthom-Atlantique **7**
American Electric Power Company Inc. **387**
American Electrical Power Service Corporation **264**
American Express Unit Fireman's Fund **379**
Amperwerke Elektrizitaets-AG **76**
Angelsportverein Landshut/Bayern **76**
Angra Nuclear Power Station, Unit 1 (Brazil) **11** ●
Angra Nuclear Power Station, Units 2-3 (Brazil) **12** ○
ANO-1 **238**
ANO-2 **238**
Ansaldo Meccanico Nucleare SpA (Italy) **101, 102, 103, 156**
AP600 p. 117
Apsara reactor p. 111
Apulia Nuclear Power Station p. 120
Area of Outstanding Natural Beauty (AONB) **230, 231**
Argentine National Atomic Energy Commission **1, 2**, p. 1
Arizona Nuclear Power Project **324, 325**
Arizona Public Service Co. **324, 325**

Arizonans for Safe Energy **324**
Arkansas Nuclear One Steam Electric Station, Units 1-2 (United States) **238** ●
Arkansas Power & Light Co. **238, 284**
Armenia Nuclear Station, Units 1-2 (Armenia) **4** ★
Armerad-Betong **193**
Asco Nuclear Center, Units 1-2 (Spain) **182** ●
ASEA-Atom (Sweden) **18, 19**
ASEA-Atom (Sweden) **35**
ASEA-Atom (Sweden) **80, 191, 192, 193, 194**
Asgen **101**
Asociacion Nuclear Asco **182**
Asociacion Nuclear Vandellos **190**
Asse p. 81
Associated Electric Industries Ltd. (United Kingdom) **214, 219, 225, 229, 235**
Association of Businesses Advocating Tariff Equity (ABATE) **323**
Association Momentanee de Genie Civil (Belgium) **7**
Astrobel General Contractors **10**
Ateliers de Constructions Electriques de Charleroi S.A. (Belgium) **6, 7, 10, 42**
Atlantic City Electric **239, 295, 296, 328, 349**
Atlantic Nuclear Power Plant, Units 1-2 (United States) **239** ☆
Atombolaget Company p. 239

● Operable   ○ Under Construction   △ Indefinitely Deferred   ▲ Decommissioning   □ Planned   ■ On Order   ☆ Cancelled   ★ Shut Down   **p. 543**

Atomenergoexport (former USSR) **13, 14, 15, 27, 29, 32, 34, 71, 72, 85, 92**
Atomenergyoprojekt (former USSR) **4**
Atomgrad p. 212
Atomic Energy Act of 1948 **276**
Atomic Energy of Canada Ltd. **3, 16, 17, 18, 19, 21, 23, 24, 25, 97, 149, 150, 156,** p. 27, p. 105, p. 190
Atomic Energy Control Board p. 27
Atomic Industrial Forum **306**
Atomic Power Construction Ltd. (United Kingdom) **220, 233**
Atomic Power Development Associates (APDA) **276**
Atomics International (United States) **289, 336, 351**
Atommash **176,** p. 270
Atoms for Peace **217,** p. 1, p. 123, p. 157, p. 165, p. 177
Atucha Nuclear Power Station, Unit 1 (Argentina) **1** ●
Atucha Nuclear Power Station, Unit 2 (Argentina) **2** ○
Auxeltra-Delens-Francois (Belgium) **10**
Babcock-Brown Boveri Reaktor, GmbH (Germany) **80**
Babcock Power Ltd. (United Kingdom) **224, 230, 236**
Babcock & Wilcox Co. (United States) **80, 149, 224, 230, 236, 238, 243, 266, 267, 268, 287, 298, 311, 320, 321, 329, 342, 367, 369, 370, 379, 381**
Badenwerk AG **84**
Bailly Nuclear Power Plant (United States) **240** ☆
Balakovo Nuclear Power Plant, Units 1-3 (Russia) **157** ●
Balakovo Nuclear Power Plant, Unit 4 (Russia) **158** □
Baldwin **260, 261**
Baldwin Lima Hamilton (United States) **289**
Ballot **47, 56**
Baltimore Gas & Electric Co. **254**
Barseback Nuclear Power Plant, Units 1-2 (Sweden) **191** ●
Barton Nuclear Power Plant **241**
Barton Nuclear Power Plant, Units 1-4; Alan R. (United States) **241** ☆
Bataafsche Aanneming Maatschappij (Netherlands) **138**
Batan p. 117
Bayernwerk AG **70, 75, 76,** p. 80
Beaver Valley Power Station, Units 1-2 (United States) **242** ●
Bechtel **5, 99, 146, 152, 182, 185, 190, 200, 201, 238, 244, 252, 253, 254, 267, 268, 269, 271, 273, 275, 284, 285, 287, 289, 293, 295, 297, 305, 311, 314, 321, 323, 324, 325, 327, 328, 334, 337, 342, 350, 358, 361, 368, 371, 372, 374, 379, 380, 381, 384, 385**
Belene Nuclear Power Plant, Units 1-2 (Bulgaria) **13** ☆
Bellefonte Nuclear Power Plant, Units 1-2 (United States) **243** △
Belleville Nuclear Power Plant, Units 1-2 (France) **36** ●
Beloyarsk Nuclear Power Plant, Units 1-2 (Russia) **159** ★
Beloyarsk Nuclear Power Plant, Unit 3 (Russia) **160** ●
Beloyarsk Nuclear Power Plant, Unit 4 (Russia) **161** ○
Berkeley Power Station, Units 1-2 (United Kingdom) **214** ▲
Bernische Kraftwerke AG **198**
Beznau Nuclear Power Plant, Units 1-2 (Switzerland) **195** ●
Bharat Heavy Electrical Ltd. (India) **94, 95, 96, 98**
Biblis Nuclear Power Plant, Units A-B (Germany) **66** ●
Big Rock Point Nuclear Power Plant (United States) **244** ●
Bilibino Nuclear Power Plant, Units 1-4 (Russia) **162** ●
Black Fox Nuclear Power Plant, Units 1-2 (United States) **245** ☆
Black & Veatch **348**
Blaton **10**
Blayais Nuclear Power Plant, Units 1-4 (France) **37** ●
Blue Hills Nuclear Power Plant, Units 1-2 (United States) **246** ☆
Blyth Eastman **379**
BN-350 Nuclear Plant (Kazakhstan) **131** ●
BN-600 **160**
BN-800 **160**
BNR-350 **58**
Bodega Bay Nuclear Power Plant (United States) **247** ☆
Bohunice A1 Nuclear Power Plant (Czech Republic) **28** ★
Bohunice Nuclear Power Plant **5**
Bohunice Nuclear Power Plant, Units 1-4 (Czech Republic) **29** ●
Boliden-WP-Contech **193**
Bonneville Power Administration (BPA) **379**
Borssele Nuclear Power Plant (Netherlands) **137** ●
Boston Edison **25, 334, 335, 385**
Bouygues **38, 43, 60**

Bradwell Power Station, Units 1-2 (United Kingdom) **215** ●
Braidwood Station, Units 1-2 (United States) **248** ●
Brazos Electric Cooperative **262**
Bredero **137**
Breeder Reactor p. 126
Breton-Liberation Front p. 61
British Nuclear Fuels **66, 68, 214, 215, 216, 217, 233,** p. 279
Brokdorf Nuclear Power Plant (Germany) **67** ●
Brown Boveri et Cie. (Switzerland) **10, 18, 19, 80, 89, 193, 195, 197, 198, 243, 264**
Brown & Root, Inc. (United States) **250, 262, 263**
Browns Ferry Nuclear Plant, Units 1-3 (United States) **249** ●
Bruce Energy Center **16**
Bruce Heavy Water Plant **16**
Bruce Nuclear Power Development (BNPD) **20**
Bruce Nuclear Power Development, Units 1-4 (Canada) **16** ●
Bruce Nuclear Power Development, Units 5-8 (Canada) **17** ●
Brunsbuettel Nuclear Power Plant (Germany) **68** ●
Brunswick Nuclear Plant, Units 1-2 (United States) **250** ●
BSN S.A. **188, 189**
Bugey Nuclear Power Plant, Units 1-5 (France) **38** ●
Bulgarian Committee on the Uses of Atomic Energy for Peaceful Purposes p. 23
Bulgarian Energy Committee p. 23
Burgerforum gegen Atomkraftwerken **76**
Burns & Roe, Inc. (United States) **144, 259, 322, 370, 380**
Business and Professional People in the Public Interest **248**
Byron Station, Units 1-2 (United States) **251** ●
C.A. Parsons & Co., Ltd. (United Kingdom) **215, 216, 217, 219, 220, 228, 229**
Cabrera Nuclear Power Plant; Jose (Spain) **184** ●
Cadarache Center **58**
Cajun Electric Power Cooperative **343, 344**
Calder Hall Nuclear Power Station, Units 1-4 (United Kingdom) **216** ●
California Public Utilities Commission **269, 350**
Callaway Nuclear Plant, Unit 1 (United States) **252** ●

Callaway Nuclear Plant, Unit 2 (United States) **253** ☆
Calumet Skyway **379**
Calvert Cliffs Nuclear Power Plant, Units 1-2 (United States) **254** ●
Camp Pendleton Marine Corps Base **350**
Campenon-Bernard (France) **10, 46, 48, 141**
Canadian General Electric **20,** p. 165
Canatom Ltd. (Canada) **22, 25, 149**
CANDU p. 27
Caorso Nuclear Power Plant (Italy) **101** ★
Capacity Factor **321**
Carbon Dioxide Analysis Center p. 13
Carolina Environmental Study Group (CESG) **256**
Carolina Power & Light Co. **250, 290, 291, 345**
Carolinas Virginia Nuclear Power Associates **255**
Carolinas-Virginia Tube Reactor (United States) **255** ▲
Carroll County Nuclear Power Plant, Units 1-2 (United States) **256** ☆
Cattenom Nuclear Power Plant, Units 1-4 (France) **39** ●
Centerior Fuel Corporation **267, 331**
Centerior Power Corporation **242, 267, 331**
Central Area Power Coordination **242, 331**
Central Electric Generating Board (CEGB) **214, 215, 216, 217,** p. 279
Central Hudson Gas & Electric Corporation **318**
Central Iowa Power Cooperative **273**
Central Maine Power Company **376**
Central Nuclear de Almaraz **181**
Central Nuclear de Valdecalballeros **188**
Central Power & Light Co. (CPL) **361**
Central South West Corporation **361**
Central de Trillo **187**
Central Vermont Public Service Corporation **376**
Centrale Nucléaire Européenne à Noutrons Rapides S.A. (NERSA) **64, 86**
Centrale Organisatie Voor Radioactief Afval (COVRA) p. 162
Centrales Nucleares del Norte, S.A. (NUCLENOR) **186**
Centre d'Etude de l'Energie Nucleaire (Belgium) **9**
Cernavoda Nuclear Power Station, Units 1-5 (Romania) **156** ○
Chag **56**
Chalk River Nuclear Laboratories **20, 23,** p. 28
Chantiers Modernes (France) **47, 56, 57**

Chapelcross Nuclear Power Station, Units 1-4 (United Kingdom) **217** ●
Chase Econometrics Inc., **248**
Chashma Nuclear Power Plant (Pakistan) **139** □
Chashma Nuclear Power Project (CHASNUPP) p. 165
CHASNUPP p. 165
Chemical Bank **379**
Chernobyl nuclear accident p. 271
Chernobyl Nuclear Power Station, Units 1-3 (Ukraine) **203** ●
Chernobyl Nuclear Power Station, Unit 4 (Ukraine) **204** ★
Chernobyl Nuclear Power Station, Unit 4 (Ukraine) **192**
Chernobyl Nuclear Power Station, Units 5-6 (Ukraine) **205** ☆
Cherokee Nuclear Power Plant, Units 1-3 (United States) **258** ☆
Chesapeake Bay Bridge and Tunnel Commission **379**
Chevron Oil **348**
Chicago Bridge and Iron **11**
Chicago Department of Planning **248**
China Construction Engineering Corp. (China) **141**
China National Nuclear Corp. **142, 143**
China Syndrome **221**
Chinon A Nuclear Power Plant, Units 1-3 (France) **40** ★
Chinon B Nuclear Power Plant, Units 1-4 (France) **41** ●
Chinshan Nuclear Power Plant, Units 1-2 (Taiwan) **199** ●
Chooz A Nuclear Power Plant (France) **42** ★
Chooz B Nuclear Power Plant, Units 1-2 (France) **43** ○
Chooz Nuclear Power Plant p. 14, p. 60
Chubu Electric Power Co. **111, 112**
Chugoku Electric Power Co, Inc. **126**
Cie d'Enterprises CFE S.A. (Belgium) **10**
Cincinnati Gas & Electric **387**
Cirene Nuclear Power Plant p. 120
Citizens Against a 10-mile Radius **352**
Citizens' Association for Sound Energy (CASE) **262**
Citizens for Fair Utility Regulation **262**
City of Austin **361**
City Public Services of San Antonio **361**
Civaux Nuclear Power Plant, Unit 1 (France) **44** ○
Civaux Nuclear Power Plant, Unit 2 (France) **45** □
CLAB **193**
Clamshell Alliance **352**
Clean Air Act **308, 331**

Cleveland County Education Authority **221**
Cleveland Electric Illuminating Co. **242, 267, 331, 332**
Clinch River Breeder Reactor Project (United States) **259** ☆
Clinton Nuclear Power Station, Unit 1 (United States) **260** ●
Clinton Nuclear Power Station, Unit 2 (United States) **261** ☆
Club VVER 440-V 213 **30**
CMS Energy Corporation **323**
Coalition for Alternatives to Shearon Harris (CASH) **290**
Cobalt-60 **24**
Cockeril Ougree-Providence (Belgium) **6, 7, 9, 10, 42**
Cofrentes Nuclear Power Plant (Spain) **183** ●
COGEMA (France) **66**
Cold Shutdown **318**
Colorado Public Utilities Commission **348**
Columbia University **370**
Columbus & Southern Ohio Electric **387**
Comanche Peak Nuclear Power Plant, Unit 1 (United States) **262** ●
Comanche Peak Nuclear Power Plant, Unit 2 (United States) **263** ○
Combined Heat and Power (CHP) **160,** p. 211, p. 271
Combustion Engineering, Inc. (United States) **148, 153, 238, 254, 258, 276, 280, 281, 306, 312, 323, 324, 325, 335, 346, 350, 379, 381, 386**
Comision Federal de Electricidad (CFEM) **135, 136**
Comision Nacional de Energia Atomica (CNEA) **1, 2, 3**
Comision Nacional de Seguridad Nuclear y Salvaguardas (CNSNS) **135**
Comissao Nacional de Energia Nuclear (CNEN) p. 19
Commissariat a l'Energie Atomique **40, 53, 54, 58,** p. 46
Commission of the European Communities (CEC) p. 273
Commonwealth Edison Co. **248, 251, 256, 271, 272, 304, 341, 388**
Commonwealth Electric Company **334**
Compagnie Electro-Mechanique (France) **54, 58**
Compagnie Industrielle de Travauz (France) **42**
Compania Sevillana de Electricidad S.A. **188**
CON-CLAVE **135**
Concerned Mothers and Women **370**
Connecticut Light & Power Company **288, 312, 376**

Connecticut

Connecticut Yankee Atomic Power Co. **288**
Connecticut Yankee Nuclear Power Plant **288**
Consejo de Seguridad Nuclear (CSN) **181, 182**
Consejo de Sequridad Nuclear (CSN) **186**
Consolidated Edison Co. of New York, Inc. **298, 299, 300**
Construtora Norberto Odebrecht **11**
Consumers Power Co. **244, 311, 323**
Control Data p. 170
Convoy system **69**
Cook Nuclear Plant, Units 1-2; Donald C. (United States) **264** ●
Coolwater Farms Ltd. **24**
Cooper Nuclear Power Plant (United States) **265** ●
Coordinadora Nacional Contra Laguna Verda (CON-CLAVE) **135**
Corn Belt Power Cooperative **273**
COVRA (Centrale Organisatie Voor Radioactief Afval) p. 162
Cowans Ford Hydroelectric Station **309**
Creusot-Loire (France) **10, 42**
Creys-Malville Nuclear Power Plant **64**, p. 61
Cruas Nuclear Power Plant, Units 1-4 (France) **46** ●
Crystal River Energy Complex, Unit 3 (United States) **266** ●
Cuban Ministry of Basic Industries **27**
Czech Disposal Centre of Radioactive Waste **30**
Czech Power Board **30, 32, 33**
Czech Power Board Syndicate Training Centre **30**
Czechoslovak Federal Environment Committee **29**
Czechoslovakia Atomic Energy Commission **30**
Dampierre Nuclear Power Plant, Units 1-4 (France) **47** ●
Daniel Construction (United States) **252, 253, 255, 275, 277, 283, 290, 291, 363, 384**
Darlington Nuclear Generation Station, Units 1-3 (Canada) **18** ●
Darlington Nuclear Generation Station, Unit 4 (Canada) **19** ○
David Gegen Goliath (David versus Goliath) **70**
Davis-Besse Nuclear Power Plant, Unit 1 (United States) **267** ●
Davis-Besse Nuclear Power Plant, Units 2-3 (United States) **268** ☆
Daya Bay Nuclear Power Plant **141**
Dayton Power & Light **387**

Decommissioning Model **356**
Del-AWARE **305**
Delft University p. 161
Delmarva Power & Light **328, 349, 364**
Detroit Edison **276, 277, 278, 287**
Detroit Edison's Nuclear Operations Improvement Plan (NOIP) **277**
Diablo Canyon Nuclear Power Plant, Units 1-2 (United States) **269** ●
Division of Ratepayer Advocates **269**
Dodewaard Nuclear Power Plant (Netherlands) **138** ●
Doel Nuclear Power Station, Units 1-2 (Belgium) **6** ●
Doel Nuclear Power Station, Units 3-4 (Belgium) **7** ●
Doel Nuclear Power Station, Unit 5 (Belgium) **8** ☆
Dolmel **145**
Donald C. Cook Nuclear Plant, Units 1-2 (United States) **264** ●
Dong Ah **147**
Don't Waste Oregon Committee **371**
Douglas Point Nuclear Generating Station (Canada) **16, 20** ★
Douglas Point Nuclear Power Plant, Units 1-2 (United States) **270** ☆
Dounreay Prototype Fast Breeder Reactor (PFR) (United Kingdom) **58, 218** ●
Dow Chemical Company **276**
Dow-Edison Industry Advisory Group **276**
Dragon **235**
Dresden Nuclear Station, Unit 1 (United States) **271** ★
Dresden Nuclear Station, Units 2-3 (United States) **272** ●
Duane Arnold Energy Center (United States) **132, 273** ●
Duke Power Co. **258, 321, 330, 342**
Dukovany Nuclear Power Plant, Units 1-4 (Czech Republic) **30** ●
Dumez **37, 38, 39**
Dungeness A Power Station, Units 1-2 (United Kingdom) **219** ●
Dungeness B Power Station, Units 1-2 (United Kingdom) **220** ●
Duquesne Light Co. **242, 331, 332, 356**
Dyckenhoff & Widmann AG **90**
Dyckerhoff & Widmann AG **73**
Earth First! **324**
Eastern Maine Electric Co-op **25**
Eastern Utilities Associates **352**
Ebasco **102, 107, 128, 130, 135, 136, 144, 186, 199, 278, 290, 291, 312, 345, 346, 361, 376, 379, 381**
Ecco-Asia **144**
Ecology Institute **70**

Edwin I. Hatch Nuclear Power Plant, Units 1-2 (United States) **293** ●
El Paso Electric **324, 325**
Eldorado Nuclear p. 27
Electra de Viesgo **186**
Electrabel **6, 7, 8, 10**, p. 13
Electric Power Development Co., Ltd. (Japan) **106**
Electricite de France **10, 36, 37, 38, 39, 40, 41, 43, 44, 45, 46, 47, 48, 49, 50, 51, 52, 55, 56, 57, 58, 60, 61, 62, 63, 65, 141, 155**
Electrobras **11**, p. 19
Electrowatt Ltd. (Switzerland) **197**
Elekroprivreda Zagreb (Croatia) **180**
Elektrosila **132, 133, 157, 158, 159, 165, 166, 167, 168, 169, 171, 172, 173, 174, 176, 203, 205, 206, 207, 208, 209, 210, 211, 212, 213**
Elektrotyazhmash **160, 161, 174**
Elk River Nuclear Power Plant (United States) **274** ▲
Elliot **274**
Embalse Nuclear Power Plant (Argentina) **3** ●
Emch & Berger (Switzerland) **198**
Emergency Core Cooling System **118**
Empresa Brasileira de Engenharia **11**
Empresa Nacional Bazan (Spain) **182, 187, 190**
Empresa Nacional de Ingenieria y Technologia SA (Spain) **182, 185, 190**
Empresa Nuclear Argentina de Centrales Electricas (ENACE S.A.) **2**
Empresarios Agrupados (Spain) **181, 183, 187, 188**
Emsland Nuclear Power Plant (Germany) **69** ●
ENACE S.A. **2**
ENEL p. 120
Energetika Association **13, 14, 15**
Energie-Versorgung Schwaben AG (EVS) **84**
Energie Werke Norde GmbH (EWN) **71, 72, 85**
Energoprojeckt Skoda Lotep (Czech Republic) **28, 29, 30, 31**
Energoproject **14**
Energy Probe **16**
Energy Workers' Union **14**
Engineering Construction Corp. (India) **95**
Engineering Constructions Brno (Czech Republic) **30**
English Electric Co., Ltd. (Canada) **97**
English Electric Co., Ltd. (United Kingdom) **194, 218, 224, 230, 234, 236**
Enrico Fermi 1 Fast Breeder Reactor Project (United States) **276** ▲

Enrico Fermi 2 Nuclear Power Plant (United States) **277** ●
Enrico Fermi 3 Nuclear Power Plant (United States) **278** ☆
Enserch Corporation **144**
Ente Nazionale per l'Energia Elettrica **101, 102, 103, 104, 105,** p. 120
Entergy Corporation **238, 284, 382**
Entergy Operations, Inc. **238, 284, 285, 382**
Entrecanales y Tavora (Spain) **181, 183, 184, 187**
Equipos Nucleares SA (Spain) **187**
ESKOM **155**
ETA **185**
EUA Power Corp. **352**
Eugene Water & Electric Board **371**
Euratom **79, 83, 89, 138**
European Atomic Energy Community (Euratom) **218**
European Atomic Forum p. 106
European Fast Breeder Reactor **58, 218**
European Joint Venture **79, 89**
EXPO '70 **118**
Export-Import Bank **11,** p. 19
Fairness Doctrine **318**
Farley Nuclear Power Station, Units 1-2; Joseph M. (United States) **275** ●
Fast Breeder Reactor **58, 64, 86, 106, 119, 160, 218, 259,** p. 80, p. 112
Fast Breeder Reactor Project; Enrico Fermi 1 (United States) **276** ▲
Fast Flux Test Facility (FFTF) **276**
FBR **86**
FBR Engineering Co., Ltd. (Japan) **119**
Federal Energy Regulatory Commission (FERC) **280, 284**
Federal University of Rio de Janeiro **11**
Fermi 1 Fast Breeder Reactor Project; Enrico (United States) **276** ▲
Fermi 2 Nuclear Power Plant; Enrico (United States) **277** ●
Fermi 3 Nuclear Power Plant; Enrico (United States) **278** ☆
Fessenheim Nuclear Power Plant, Units 1-2 (France) **48** ●
Filial Leningradense de Atomenergoprojekt (former USSR/Cuba) **27**
FILTRA **191**
Finnish Centre for Radiation and Nuclear Safety (STUK) p. 53
First International Conference on the Peaceful Uses of Atomic Energy **276**
First Nuclear Power Station (Taiwan) **199**
Fitzpatrick Nuclear Power Plant; James A. (United States) **279** ●

Flamanville Nuclear Power Plant, Units 1-2 (France) **49** ●
Florida Municipal Power Agency **346, 372**
Florida Power Corp. **266**
Florida Power & Light Co. **346, 360, 372**
Florida Progress Corporation **266**
Fluor **303**
Fluor Daniel Inc. **342**
Fluor Power Services, Inc. **338**
Forked River Nuclear Power Plant (United States) **280** ☆
Formatom p. 106
Forsmark Kraftgrupp **192,** p. 241
Forsmark Nuclear Power Plant, Units 1-3 (Sweden) **192** ●
Fort Calhoun Generating Station (United States) **281** ●
Fort St. Vrain Nuclear Power Plant (United States) **348** ★
Fougerolle **44, 45, 50, 51**
Four Lakes Regional Industrial Development Authority **292**
Fourth Nuclear Power Station (Taiwan) **202**
FPC vs. Hope Natural Gas Company **280**
FR2 **77**
FRAMACECO **10**
Framateg **155**
Framatome (France) **7, 8, 10, 44, 45, 50, 51, 141, 147, 155, 165,** p. 60
Franco Tosi **105**
Franco Toxi SpA (Belgium) **6**
Francois et Fils **10**
Franki-Engema **6**
Friends of the Earth **226**
Fuel Use Act **343**
Fugen ATR Prototype Nuclear Power Plant (Japan) **106** ●
Fuji **106, 119**
Fukui Environmental Radiation Monitoring Council **130**
Fukui Prefectural Fishery Laboratory **130**
Fukushima Daiichi Nuclear Power Station, Units 1-6 (Japan) **107** ●
Fukushima Daini Nuclear Power Station, Units 1-4 (Japan) **108** ●
Fulton Nuclear Power Plant, Units 1-2 (United States) **282** ☆
Furnas-Centrais Eletricas SA **11, 12**
Fusion reactor **71**
GA Technologies **89, 348**
Ganz Electrical Works (Hungary) **92**
Garigliano Nuclear Power Plant (Italy) **102** ★
Gas-Cooled Reactor **348**
GE Canada **22, 97**

Gemeinschaftskernkraftwerk Grohnde GmbH **73**
Gemeinschaftskernkraftwerk Neckar GmbH **81**
Gemeinschaftskraftwerk Tullnerfeld GmbH **5**
General Atomic **282, 347, 348, 364, 377**
General Electric Canada **17, 23, 140**
General Electric Co. (United Kingdom) **128, 141, 146, 221, 222, 225, 227, 231, 232, 277, 350**
General Electric Co. (United States) **74, 99, 102, 107, 128, 130, 135, 136, 138, 150, 153, 156, 183, 186, 188, 199, 200, 201, 238, 240, 244, 249, 250, 252, 253, 254, 260, 261, 264, 265, 267, 268, 271, 272, 273, 277, 278, 279, 281, 284, 285, 287, 292, 293, 295, 296, 297, 299, 304, 305, 312, 313, 314, 317, 321, 322, 324, 325, 327, 328, 331, 332, 333, 334, 341, 343, 344, 348, 349, 352, 353, 357, 358, 359, 363, 368, 369, 371, 374, 376, 380, 384, 386, 387,** p. 111, p. 123, p. 178
General Electric Technical Services Co. (United States) **101, 197, 198**
General Emergency **283**
General Physics International Engineering & Simulation Inc. (GPI) **32, 179**
General Public Utilities Nuclear Corp. **280, 369, 370**
Generale d'Electricité p. 60
Genkai Nuclear Power Plant, Units 1-2 (Japan) **109** ●
Genkai Nuclear Power Plant, Units 3-4 (Japan) **110** ○
Gentilly Nuclear Generating Station, Unit 1 (Canada) **21** ▲
Gentilly Nuclear Generating Station, Unit 2 (Canada) **22** ●
Georgia Power Co. **293**
Georgia Public Service Commission **377**
German Atomic Energy Law **66**
German Convoy System **69**
German Federal Railways **81**
German Green Movement **66**
German Green Party **86**
German Law of Water Rights **76**
Gesellschaft für Reaktorsicherheit (GRS) **70**
Gibbs & Hill Espanola SA (Spain) **181**
Gibbs & Hill, Inc. (United States) **9, 11, 42, 101, 105, 183, 184, 194, 195, 262, 263, 281**
Gilbert **118, 120, 127, 146, 180, 266, 283, 331, 332, 363,** p. 177
Ginna Nuclear Power Plant; Robert Emmet (United States) **283** ●

GKN-1 **81**
GKN-2 **81**
Goesgen Nuclear Power Plant (Switzerland) **196** ●
Golfech p. 61
Golfech Nuclear Power Plant, Unit 1 (France) **50** ●
Golfech Nuclear Power Plant, Unit 2 (France) **51** ○
Gorky Nuclear Facility, Units 1-2 (Russia) **163** ●
Gorky Nuclear Facility, Unit 3 (Russia) **164** □
Gorleben Interim Storage Facility **66**
Gorleben Nuclear Waste Disposal Site **73**
Gospromatomnadzor (GPAN) p. 212, p. 269
Government Accountability Project (GAP) **269**
GPU Nuclear Corp. **280, 322, 369, 370**
Grafenrheinfeld Nuclear Power Plant (Germany) **70** ●
Grand Gulf Nuclear Station, Unit 1 (United States) **284** ●
Grand Gulf Nuclear Station, Unit 2 (United States) **285** ☆
Grands Travaux de Marseille (France) **36, 41, 61, 62, 63**
Gravelines Nuclear Power Plant, Units 1-6 (France) **52** ●
Green Mountain Power Corporation **376**
Greene County Nuclear Power Plant (United States) **286** ☆
Greenpeace **224, 225**
Greenwood Nuclear Power Plant, Units 2-3 (United States) **287** ☆
Greifswald p. 79
Greifswald Nuclear Power Plant, Units 1-4 (Germany) **71** ★
Greifswald Nuclear Power Plant, Units 5-8 (Germany) **72** △
Grohnde Nuclear Power Plant (Germany) **73** ●
Group of 100 **135**
Group of Seven **204**
Grumman Corp. **357**
Guangdong Nuclear Power Joint Venture Co., Ltd. **141**
Guangdong Nuclear Power Plant, Units 1-2 (People's Republic of China) **141** ○
Gulf General Atomic **89**
Gulf States Utilities Co. **246, 343, 344**
Gundremmingen Nuclear Power Plant, Unit A (Germany) **74** ▲
Gundremmingen Nuclear Power Plant, Units B-C (Germany) **75** ●
Gundremmingen Pilot Plant **76**

Haddam Neck Nuclear Power Plant (United States) **288** ●
Hallam Nuclear Power Plant (United States) **289** ★
Hamburgische Electricitaets-Werke AG (HEW) **67, 68, 78, 87**
Hamaoka Nuclear Power Plant, Units 1-3 (Japan) **111** ●
Hamaoka Nuclear Power Plant, Unit 4 (Japan) **112** ○
Hanford Military Reservation **356**
Harris Energy and Environmental Center **290**
Harris Nuclear Power Plant, Unit 1; Shearon (United States) **290** ●
Harris Nuclear Power Plant, Units 2-4; Shearon (United States) **291** ☆
Hartford Electric Light Company **288, 312**
Hartford Steam Boiler inspection and Insurance Co. **383**
Hartlepool Power Station, Units 1-2 (United Kingdom) **221** ●
Hartsville Group **345**
Hartsville Nuclear Power Plant, Units A1, (United States) **292** ☆
Hatch Nuclear Power Plant, Units 1-2; Edwin I. (United States) **293** ●
Haven Nuclear Power Plant, Units 1-2 (United States) **294** ☆
Hazama Gumi Co. (Japan) **118, 127**
H.B. Robinson Electric Power Plant, Unit 2 (United States) **345** ●
Heavy water **16**
Hedgekamp (Germany) **73, 90**
Helmstedt-Wolmirstedt transmission line **71**
Herdi's Management and Investment Corp. **144**
Heritage Coast **230, 231**
Heysham A Power Station, Units 1-2 (United Kingdom) **222** ●
Heysham B Power Station, Units 1-2 (United Kingdom) **223** ●
Hidroelectrica del Cantabrico **187**
Hidroelectrical Espanola, S.A. **183, 185**
High Temperature Gas Cooled Reactor (HTGR) **220**
High Temperature Gas Reactor (HTR) **89**
High Temperature Incinerator (HTI) **109**
Hindustan Construction Co. (Canada) **97**
Hindustan Construction Co. (India) **94, 96**
Hinkley Inquiry **226**
Hinkley Point A Power Station, Units 1-2 (United Kingdom) **224** ●
Hinkley Point B Power Station, Units 1-2 (United Kingdom) **225** ●
Hinkley Point C Power Station (United Kingdom) **226** □

Hispano-Francesa de Energia Nuclear, S.A. (HIFRENSA) **189**
Hitachi **106, 107, 108, 111, 112, 115, 116, 119, 125, 126, 140,** p. 123
Hochtemperatur-Kernkraftwerk GmbH **89**
Hochtemperatur Reaktorbau (HRB) **89**
Hochtief AG (Germany) **66, 75, 80, 91**
Hokkaido Electric Power Co., Inc. **129**
Hokuriku Electric Power Co. **125**
Holmes & Narver **336**
Hoosiers for Economic Development Committee **308**
Hope Creek Nuclear Generating Station, Unit 1 (United States) **295** ●
Hope Creek Nuclear Generating Station, Unit 2 (United States) **296** ☆
House Subcommittee on Energy, Research and Production **370**
Houston Lighting & Power Co. **237, 361**
HTR-500 **89**
Huaxing (China) **141**
Humboldt Bay Nuclear Power Plant, Unit 3 (United States) **297** ★
Hungarian Atomic Energy Commission **92**
Hungarian Electricity Trust p. 105
Hungarian Nuclear Power Act of 1980 p. 106
Hungarian Power Companies Ltd. **92**
Hunterston A Nuclear Power Station, Units 1-2 (United Kingdom) **227** ▲
Hunterston B Nuclear Power Station, Units 1-2 (United Kingdom) **228** ●
Hydro-Quebec **21, 22**
Hyperphenix **64**
Hyundai **146, 152, 153**
Iberdrola I, S.A. **185, 187**
Iberdrola II, S.A. **183, 188**
Iberduero, S.A. **183, 185, 186**
Idaho National Engineering Laboratory **370**
Ignalina Nuclear Power Plant, Units 1-2 (Lithuania) **132** ●
Ignalina Nuclear Power Plant, Units 3-4 (Lithuania) **133** ☆
Ikata Nuclear Power Station, Units 1-2 (Japan) **113** ●
Ikata Nuclear Power Station, Unit 3 (Japan) **114** ○
Illinois Commerce Commission (ICC) **248**
Illinois Power Co. **260, 261**
Imatran Voima Oy **34, 92**
Incident Investigation Team **267**
Indatom **54**
Independent Spent Fuel Storage Installation **348**
India Atomic Energy Commission p. 111

India Department of Atomic Energy, Nuclear Power Board **93, 94, 95, 96, 97, 98, 99, 100**
Indian Point Nuclear Generating Station, Unit 1 (United States) **298** ★
Indian Point Nuclear Generating Station, Unit 2 (United States) **299** ●
Indian Point Nuclear Power Plant, Unit 3 (United States) **300** ●
Indiana Michigan Power Co. **264**
Indira Gandhi Center for Atomic Research p. 112
Indonesian National Atomic Energy Agency p. 117
Industrial Power Company **35**
Informas y Projectas SA (Spain) **182**
Institute of Nuclear Power Operations **265, 269**, p. 274
Interatom **86**
International Chamber of Commerce **144**, p. 178
International Commission on Radiological Protection (ICRP) **155**
Internationale Natrium Brutreaktro Baiges **86**
Interuniversitair Reaktor Instituut at Delft **138**
Iowa Department of Natural Resources **273**
Iowa Electric Light & Power Co. **273, 375**
Iowa-Illinois Gas & Electric Co. **341**
Iowa Power **265**
Isar-Ampwerke AG **76**
Isar Nuclear Power Plant, Units 1-2 (Germany) **76** ●
Isarwerke AG **76**
Isolte Nuclear Power Plant (United States) **301** ☆
Italimpianti **3**
Italy Power Station p. 120
I.V. Kurchatov Institute of Atomic Energy p. 212, p. 269
Izhorski Works p. 270
Jacksonville Electric Authority **309**
James A. Fitzpatrick Nuclear Power Plant (United States) **279** ●
Japan Atomic Energy Basic Law of 1955 p. 123
Japan Atomic Energy Commission p. 123
Japan Atomic Energy Research Institute (JAERI) p. 124
Japan Atomic Power Co. (JAPCO) **119, 128, 130**
Japan Electric Power Coordinating Council **117**
Japanese Ministry of International Trade & Industry (MITI) p. 124
Jersey Central Power & Light Co. **239, 280, 322, 369, 370**

Jeumont-Schneider (France) **10, 38, 40, 189**
John Laing Construction Ltd. (United Kingdom) **231**
Jones **266**
Jose Cabrera Nuclear Power Plant (Spain) **184** ●
Joseph M. Farley Nuclear Power Station, Units 1-2 (United States) **275** ●
Joyo **119, 276**, p. 126
Jukola **35**
Julich AVR **89**
Juragua Nuclear Complex, Units 1-2 (Cuba) **27** △
Kahl Experimental Station **76**
Kaiga Atomic Power Station, Units 1-2 (India) **93** ○
Kajima **106, 107, 108, 111, 112, 113, 114, 115, 116, 119, 122, 123, 125, 126, 128**
Kakrapar Atomic Power Plant, Units 1-2 (India) **94** ○
Kalinin Nuclear Station, Units 1-2 (Russia) **165** ●
Kalinin Nuclear Station, Unit 3 (Russia) **166** ○
Kalinin Nuclear Station, Unit 4 (Russia) **167** □
Kalkar Nuclear Power Plant **86**, p. 13
Kalkar Nuclear Power Plant; SNR 300 (Germany) **86** ☆
Kansai Electric Power Co., Inc. **118, 120, 120, 121, 127**
Kansas City Power & Light Co. **384**
Kansas Electric Power Cooperative **384**
Kansas Gas & Electric Co. **384**
KANUPP p. 165
Karachi Nuclear Power Plant (KANUPP) p. 165
Karachi Nuclear Power Plant (Pakistan) **140** ●
Karlsruhe Nuclear Research Facility (Germany) **77, 86**
Kashiwazaki Kariwa Nuclear Station, Units 1, 2 (Japan) **115** ●
Kashiwazaki Kariwa Nuclear Station, Units 3, 4 (Japan) **116** ○
Kemeny Commission **370**
Kemper **379**
Kernforschungzentrum Karlsruhe **82**
Kernkraftwerk Brokdorf GmbH (KBR) **67**
Kernkraftwerk Goesgen-Daeniken AG **196**
Kernkraftwerk Isar Gmbh (KKI) **76**
Kernkraftwerk Leibstadt AG **197**
Kernkraftwerk Niederaichbach **82**
Kernkraftwerk Obrigheim GmbH (KWO) **83**

Kernkraftwerk Philippsburg GmbH (KKP) **84**
Kernkraftwerk Stade GmbH (KKS) **87**
Kernkraftwerk Unterweser GmbH (KKU) **90**
Kernkraftwerke Gundremmingen Betriebsgesellschaft mbH (KGB) **74, 75**
Kernkraftwerke Lippe-Ems GmbH **69**
Kewaunee Nuclear Plant (United States) **303** ●
Khmelnitsky Nuclear Station, Unit 1 (Ukraine) **206** ●
Khmelnitsky Nuclear Station, Units 2-4 (Ukraine) **207** ○
KKP-1 **84**
KKP-2 **84**
KKS Stade **87**
KNK **86**
KNK and KNK-II **77**
Koblenz Administrative Court **80**
Koeberg Nuclear Power Station, Units 1-2 (Republic of South Africa) **155** ●
Kola Nuclear Plant, Units 1-4 (Russia) **168** ●
Kola Nuclear Plant, Units 5-6 (Russia) **169** □
Kommunales Elektrizitätswerk Mark AG (Elektromark) **69**
Konrad Iron Mine p. 81
Korea Atomic Energy Research Institute p. 190
Korea Atomic Industrial Forum p. 189
Korea Electric Power Corp. **146, 147, 148, 149, 150, 151, 152, 153, 154**, p. 189
Korea Heavy Industries & Construction Co. **147, 148, 150, 153**, p. 190
Korea Institute of Nuclear Safety p. 190
Korea Nuclear Fuel Co. p. 190
Korea Power Engineering Co., Ltd. **153**, p. 190
Korean Atomic Energy Commission (AEC) p. 190
Korean Nuclear Act p. 191
Kori Nuclear Power Plant, Units 1-4 (Republic of Korea) **146** ●
Kozloduy Nuclear Power Plant **173**
Kozloduy Nuclear Power Plant, Units 1-2 (Bulgaria) **14** ★
Kozloduy Nuclear Power Plant, Units 3-6 (Bulgaria) **15** ●
Kozloduy Workers Union **14**
Kraftwerk Union AG (Germany) **1, 2, 5, 12, 66, 67, 68, 70, 73, 75, 76, 86, 90, 91, 137, 187, 196**, p. 61
Krasnodar Nuclear Power Plant (Russia) **170** ☆
Krsko Nuclear Power Plant (Slovenia) **144, 180** ●

Kruemmel KKK Nuclear Power Plant (Germany) 78 ●
Kumagaya Gumi Co. (Japan) 118, 120, 127, 130
Kuosheng Nuclear Power Station, Units 1-2 (Taiwan) 200 ●
Kurchatov Institute of Atomic Energy p. 212, p. 269
Kursk Nuclear Power Station, Units 1-4 (Russia) 171 ●
Kursk Nuclear Power Station, Unit 5 (Russia) 172 □
KWO Obrigheim Plant 83
Kyushu Electric Power Co., Inc. 109, 110, 124
La Hague p. 61
Laguna Verde Nuclear Power Plant, Unit 1 (Mexico) 135 ●
Laguna Verde Nuclear Power Plant, Unit 2 (Mexico) 136 ○
LaMeuse 10
Larson & Toubro (India) 94, 95, 97
Lasalle County Nuclear Station, Units 1-2 (United States) 304 ●
Latina Nuclear Power Plant (Italy) 103 ★
Leibstadt Nuclear Power Plant (Switzerland) 197 ●
Lemoniz Nuclear Power Plant, Units 1-2 (Spain) 185 ○
Leningrad Nuclear Power Plant 179
Limerick Ecology Action 305
Limerick Generating Station, Units 1-2 (United States) 305 ●
Limerick Nuclear Power Plant 92
Lincoln Electric System 265
Lingen Boiling Water Reactor p. 79
Lingen Nuclear Power Plant (Germany) 79 ▲
Lippe-Ems (KLE) project 69
Liquid-Metal Fast Breeder Reactor (LMFBR) 160, 276
Lithuania Ministry of Energy 132, 133
Lithuanian Institute for Physical and Engineering Problems of Energy Research p. 149
Lombardy Nuclear Power Station p. 120
Long Island Lighting Co. 318, 357
Long Island Power Authority 357
Long Term Safety Review 214
Los Angeles Department of Water & Power 307, 324
Loss of Cooling Accident 370
Louisiana Consumer League 382
Louisiana Power & Light 284, 347, 382
Loviisa Power Plant, Units 1-2 (Finland) 34 ●
Low-level Radioactive Waste Policy Amendment Act of 1985 314

Lubmin Nuclear Power Plant 71
Maanshan Nuclear Power Station, Units 1-2 (Taiwan) 201 ●
Madison Gas & Electric 303
Madras Atomic Power Plant (MAPP) p. 112
Madras Atomic Power Station, Units 1-2 (India) 95 ●
Maeda 106, 118, 127, 141
Magnox reactor 214, p. 277
Maine Nuclear Referendum Committee 306, 306
Maine Yankee Atomic Power Co. 306
Maine Yankee Nuclear Power Plant (United States) 306 ●
Maki Nuclear Power Station (Japan) 117 ■
Malibu Citizens for Conservation Inc. 307
Malibu Citizens Group 307
Malibu Hearings 307
Malibu Nuclear Power Plant (United States) 307 ☆
Marble Hill Nuclear Power Plant 240
Marble Hill Nuclear Power Plant, Units 1-2 (United States) 308 ☆
Marblehead Land Company 307
Marcoule G Nuclear Power Plant, Units 1-3 (France) 53 ★
Marcoule G1 p. 59
Maryland Safe Energy Coalition 254
Massachusetts Citizens Against the Shutdown Initiative 385
Massachusetts Citizens for Safe Energy 385
Massachusetts Municipal Wholesale Electric Company 25
Maurice Delens 10
Mayport Nuclear Power Plant, Units 1-2 (United States) 309 ☆
McAlpine 215, 219, 229
Meralco p. 177
Metropolitan Edison 369, 370
Mexican House of Representatives 135
Michigan Citizens Lobby (MCL) 311
Michigan Public Service Commission 278
Mid-Continent Area Power Pool 265
Middle South Energy (MSE) 284
Middle South Utilities (MSU) 284
Midland Cogeneration Venture (MCV) 311
Midland Nuclear Power Plant, Units 1-2 (United States) 311 ☆
Midwest Compact 277, 314
Mihama Power Station, Units 1-3 (Japan) 118 ●
Millstone Nuclear Power Plant, Units 1-3 (United States) 312 ●
Minatom RF p. 212

Minnesota Pollution Control Agency 338
Mississippi Electric Power Association 284
Mississippi Power and Light (MP&L) 284
Missouri Department of Conservation 252
Mitsubishi 135, 136, p. 117
Mitsubishi Atomic Power Industries, Inc. (Japan) 109, 110, 113, 114, 118, 120, 121, 124, 127, 129, 130
Mitsubishi Electric Corp. (Japan) 109, 110, 113, 114, 118, 120, 121, 124, 127, 129, 130
Mitsubishi Heavy Industries, Ltd. (Japan) 106, 109, 110, 113, 114, 118, 119, 120, 121, 124, 127, 129, 130, p. 124
Mixed Oxide Fuel (MOX Fuel) 68, 86, 118, 119
Mochovce Nuclear Power Plant, Units 1-4 (Czech Republic) 31 ○
Mol BR3 (Belgium) 9 p. 13
Monju 276
Monju Nuclear Power Plant (Japan) 119 ○
Monju Prototype Fast Breeder Reactor p. 126
Montague Nuclear Power Plant, Units 1-2 (United States) 313 ☆
Montalto di Castro Nuclear Power Plant p. 119
Montalto di Castro Nuclear Power Plant, Units 1-2 (Italy) 104 ☆
Monticello Nuclear Generating Plant (United States) 314 ●
Montreal Engineering Co. (Canada) 97
Monts d'Arree Nuclear Power Plant (France) 54 ★
Morrison Knudsen Corporation 348
Moscow Research Institute of Instrument-Making 208
Mothballed 374
Mothers for Peace 269
Mouvement Ecologique p. 153
Muehleberg Nuclear Power Plant (Switzerland) 198 ●
Muelheim-Kaerlich Nuclear Power Plant (Germany) 80 ●
Municipal Electric Authority 293
MZFR 77
N-Reactor 379
Narora Atomic Power Plant, Units 1-2 (India) 96 ●
National Academy of Nuclear Training 250, 273
National Academy of Sciences' Commission on Policy Options for Global Warming p. 169
National Agency for Radioactive Waste Management (ANDRA) p. 61
National Audubon Society 370

National Cooperative for the Storage of Radioactive Waste (NAGRA) p. 247
National Nuclear Corp. (United Kingdom) **221, 222, 223, 231, 232**
National Power (UK) p. 278
National Research Council p. 28
National Society of Professional Engineers **370**
Nature Conservancy Council **219, 220, 221, 230, 231**
Nebraska Public Power District **265, 289**
Neckar Nuclear Power Plant, Units 1-2 (Germany) **81** ●
NEI-Parsons Ltd. (Canada) **16, 24, 25, 149**
NEP Nuclear Power Plant, Units 1-2 (United States) **315** ☆
Nest of Earthquakes **111**
Netherlands Energy Research Foundation ECN p. 161
New Brunswick Electric Power Commission **25, 26**
New England Coalition on Nuclear Pollution **385**
New England Electric System **385**
New England Power Co. **315, 376**
New England Power Pool (NEPOOL) **312**
New Hampshire Electric Cooperative, Inc. **352**
New Hampshire Yankee **352, 353**
New Haven Nuclear Power Plant, Units 1-2 (United States) **316** ☆
New Jersey Board of Regulatory Commissioners **349**
New Jersey Department of Environmental Protection **295, 349**
New Orleans Public Service Inc. **284**
New York Power Authority **279, 286, 300, 357**
New York Public Interest Research Group **299**
New York State Electric & Gas **316, 318, 359**
New York State Public Service Commission **318, 357**
Niagara Mohawk Power Corp. **279, 317, 318**
Niederaichbach Nuclear Power Plant (Germany) **82** ▲
Niederaichbach Prototype **76**
Nine First-Level Utilities p. 80
Nine Mile Point Nuclear Power Plant, Unit 1 (United States) **317** ●
Nine Mile Point Nuclear Power Plant, Unit 2 (United States) **318** ●
No On 4 Committee **385**
Nogent s/Seine Nuclear Power Plant, Units 1-2 (France) **55** ●
Non-Fossil Fuel Obligation (N-FFO) **226**

Nord Nuclear Power Plant **71**
Nordel **191, 194**
Nordostschweizerische Kraftwerke AG **195**
Nordring Grafenrheinfeld-Oberhaid-Redwitz **70**
North Anna Nuclear Power Station, Units 1-2 (United States) **319** ●
North Anna Nuclear Power Station, Units 3-4 (United States) **320** ☆
North Atlantic Energy Corp. **352**
North Atlantic Treaty Organization (NATO) **70, 75**
North Carolina Eastern Municipal Power Agency **250, 290, 291**
North Carolina Municipal Power Authority **345**
North Carolina Wildlife Resource Commission's Game Lands Program **290**
Northeast Utilities **288, 312, 313, 352, 385**
Northern Engineering Industries (United Kingdom) **223**
Northern Indiana Public Service Co. **240**
Northern Michigan Electric Cooperative, Inc. **277**
Northern States Power Co. **314, 326, 338, 373**
Northwest Power Planning Council **371**
Novatome **58**
Novotroitsk Military Nuclear Station p. 269
Novovoronezh Nuclear Power Station **85**
Novovoronezh Nuclear Power Station, Units 1-2 (Russia) **173** ★
Novovoronezh Nuclear Power Station, Units 3-5 (Russia) **174** ●
Nuclear Capital of the World **16**
Nuclear Civil Constructors (United Kingdom) **233**
Nuclear Decommissioning Finance Committee **352**
Nuclear Electric plc **214, 215, 219, 220, 221, 222, 223, 224, 225, 226, 229, 230, 231, 233, 236,** p. 278
Nuclear Industry Performance Compendium **283, 312**
Nuclear Information and Resource Service **254**
Nuclear Installations Inspectorate **214, 215, 216, 217**
Nuclear Non-Proliferation Act p. 170
Nuclear Non-Proliferation Treaty p. 1, p. 111, p. 166, p. 203
Nuclear Power Construction Stabilization Agreement **343**
Nuclear Power Corp. of India Ltd. **93, 98, 100**

Nuclear Power Demonstration Reactor (Canada) **23** ★
Nuclear Power Group (United Kingdom) **103, 214, 215, 218, 219, 225, 228, 229**
Nuclear Power Plants Research Institute **30**
Nuclear Utility Maintenance Experience Exchange (NUMEX) **155**
Nuclear Waste Policy Act **281, 306**
Nuclebras Engenaria SA (Brazil) **12**
NUCLENOR **186**
Nucon International **149**
N.V. Elektriciteits-Produktiemaatschappij Zuid-Nederland (EPZ) **137**
N.V. Gemeenschappelijke Kernenergiecentrale Nederland (GKN) **138,** p. 161
N.V. tot Keuring van Elektrotechnische Materialen (KEMA) **138,** p. 161
N.V. Samenwerkende Elektriciteits Produktiebedrijven (SEP) p. 161
Oak Ridge National Laboratory p. 13
Obayashi Gumi Co. (Japan) **109, 110, 118, 119, 120, 127, 129, 130**
Obninsk APS Nuclear Power Plant **175**
Obninsk BR5 Nuclear Power Plant **175**
Obninsk Nuclear Power Plant, Units 1-2 (Russia) **175** ●
O'Brien-Kreitzburg & Associates **248**
Obrigheim Nuclear Power Plant (Germany) **83** ●
Obrigheim Pressurized Water Reactor p. 79
Occupational Safety and Health Administration **262**
Oconee Nuclear Station, Units 1-3 (United States) **321** ●
Off-Gas Hold-up System **130**
Offshore Power Systems **239, 309, 339**
Offshore Zone Deformation **350**
Oglethorpe Power Corp. **293**
Ohi Nuclear Power Station, Units 1-3 (Japan) **120** ●
Ohi Nuclear Power Station, Unit 4 (Japan) **121** ○
Ohio Edison Co. **242, 331, 332**
Ohma Demonstration ATR **106**
OKG p. 241
OKG Aktiebolag **193**
Oktembryan Nuclear Station **4**
Okumura **126**
Oldbury Power Station, Units 1-2 (United Kingdom) **229** ●
Olkiluoto Nuclear Power Plant, Units 1-2 (Finland) **35** ●
Omaha Public Power District **281**
Onagawa Nuclear Power Station, Unit 1 (Japan) **122** ●

Onagawa Nuclear Power Station, Unit 2 (Japan) 123 ○
Ontario Cancer Treatment and Research Foundation p. 28
Ontario Hydro 16, 17, 18, 19, 20, 23, 24
Operational Safety Review Team (OSART) 60, 70
Oreas Power and Light Co. 379
Oregon Department of Energy 329
Oregon Energy Facilities Siting Council (EFSC) 329
Oregon Nuclear and Thermal Energy Council 329
Oregon Public Utility Commission 371
Organization for Economic Cooperation and Development (OECD) 66, 89, 235
Orgrez 32
Orlando Utilities Commission 346, 372
Oskarshamn Nuclear Power Plant, Units 1-3 (Sweden) 193 ●
Otto Hahn 77
Oura Nuclear Power Plant 120
Oyster Creek Nuclear Generating Station (United States) 322 ●
Oyster Shell Alliance 382
Pacific Gas & Electric Co. 247, 269, 297, 342, 374
Pacific Nuclear Systems Incorporated 146
Pacific Power & Light Co. 358
Pacific Power & Light Co. 371
Pacific Power & Light Co. 379
PacifiCorp 324, 379
Pakistan Atomic Energy Commission 139, 140
Paks Nuclear Power Plant, Units 1-4 (Hungary) 92 ●
Palisades Generating Co. 323
Palisades Nuclear Plant (United States) 323 ●
Palisades Steam Generator Replacement Project 323
Palo Verde Intervention Fund (PVIF) 324
Palo Verde Nuclear Generating Station, Units 1-3 (United States) 324 ●
Palo Verde Nuclear Generating Station, Units 4-5 (United States) 325 ☆
Palo Verde Nuclear Plant 147
Paluel Nuclear Power Plant, Units 1-4 (France) 56 ●
Parsons 103
Pathfinder Nuclear Power Plant (United States) 326 ★
Peach Bottom Nuclear Power Plant, Unit 1 (United States) 327 ★
Peach Bottom Nuclear Power Plant, Units 2-3 (United States) 328 ●
Pebble-Bed Reactor 89

Pebble Springs Nuclear Power Generating Plant, Units 1-2 (United States) 329 ☆
Penly Nuclear Power Plant, Units 1-2 (France) 57 ●
Pennsylvania Electric Co. 369, 370
Pennsylvania Power 242, 331, 332, 368
Pennsylvania Public Utility Commission (PUC) 305
Pennsylvania State Health Department 370
People for Maine Yankee's Electricity 306
Perkins Nuclear Power Plant, Units 1-3 (United States) 330 ☆
Perry Employees Committee 331
Perry Nuclear Power Plant, Unit 1 (United States) 331 ●
Perry Nuclear Power Plant, Unit 2 (United States) 332 △
Pheasant Rearing Program 277
Phenix Prototype Fast Reactor Power Station (France) 58 259, 276
Philadelphia Electric Co. (PECO) 282, 305, 327, 328, 349
Philippine Atomic Energy Commission (PAEC) 144, p. 177
Philippine National Power Corp. 144
Philippine Nuclear Power Plant (Philippines) 144 p. 177
Philippine Science Act of 1958 p. 177
Philippsburg Nuclear Power Station, Units 1-2 (Germany) 84 ●
Phipps Bend Nuclear Power Plant, Units 1-2 (United States) 333 ☆
Pickering Nuclear Generating Station, Units 1-8 (Canada) 24 ●
Pilgrim Station, Unit 1 (United States) 334 ●
Pilgrim Station, Unit 2 (United States) 335 ☆
Pinnacle West Capital Corp. (PWCC) 324
Pioneer Service & Engineering Co. 303
Piqua Nuclear Power Plant (United States) 336 ★
Pixies' Mound 224, 225
Pleasant Creek Park 273
Pleasant Creek Reservoir 273
Plogoff Nuclear Power Plant (France) 59 ☆
PN Services Group (PNS) 146
PNPP-1 p. 177
Point Beach Nuclear Plant, Units 1-2 (United States) 337 ●
Point Lepreau Nuclear Generating Station, Unit 1 (Canada) 25 ●
Point Lepreau Nuclear Generating Station, Unit 2 (Canada) 26 △
Point Pleasant Pumping Station 305

Portland General Corp. 371
Portland General Electric Co. 329, 358, 371, 379
Potomac Electric Power 270
Power Contractors Inc. 144, 262, 263
Power Demonstration Reactor Programme 244, 255, 271, 274, 289, 307, 326, 336, 374
Power Reactor Development Co. 276
Power Reactor & Nuclear Fuel Development Corp. 106, 119
Power Reactor and Nuclear Fuel Development Corporation Law 119
Power Station & Network Engineering Co. (Hungary) 92
PowerGen (UK) p. 278
Praha 145
Prairie Island Nuclear Plant, Units 1-2 (United States) 338 ●
Prestressed Concrete Containment Vessel 109, 120
PreussenElektra 67, 68, 78, 87, 90, 91, p. 80
Price-Anderson Act 370
Professional Security Officers Association 277
Promon Engenharia 11
PSEG Offshore Nuclear Power Plant, Units 1-2 (United States) 339 ☆
PSI Energy, Inc. 308
PSI Resources, Inc. 308
Public Citizen 254, 277, 319
Public Citizen's Critical Mass Energy Project 271
Public Service Commission of New Hampshire 352, 376
Public Service Co. of Colorado 348
Public Service Co. of Indiana 240, 308
Public Service Co. of Oklahoma 245
Public Service Electric & Gas Co. 239, 295, 296, 328, 339, 349
Public Service of New Mexico 324, 325
Public Utility Commission of Texas (PUCT) 343
Puerto Rico Water Resources Authority 301
Puget Sound Power & Light Co. 329, 358, 379
Puno Commission 144
PWR Power Projects (United Kingdom) 231
Qinshan Nuclear Power Co. 142, 143
Qinshan Nuclear Power Plant, Unit 1 (People's Republic of China) 142 ○
Qinshan Nuclear Power Plant, Units 2-3 (People's Republic of China) 143 □
Quad Cities Station, Units 1-2 (United States) 341 ●
Quadrex Corporation 342

● Operable   ○ Under Construction   △ Indefinitely Deferred   ▲ Decommissioning   □ Planned   ■ On Order   ☆ Cancelled   ★ Shut Down

Alphabetical Index

Rådet För Kärnkraftsäkerhet (RKS) p. 241
Radium Institute p. 269
Rajasthan Atomic Power Station, Units 1-2 (India) **97** ●
Rajasthan Atomic Power Station, Units 3-4 (India) **98** ○
Rancho Seco Nuclear Generating Station (United States) **342** ★
Rapsodie **58, 276**
Rateau **10, 38, 42, 53**
RBMK reactor p. 149, p. 211, p. 270
Reed Report **135, 343**
Reeves E. Ritchie Nuclear Training Center **238**
Reform Wildlife Management Area **252**
Rem **305**
Remerschem Nuclear Power Plant (Luxembourg) **134** p. 153
Republicans Against Seabrook Station **352**
Rheinisch-Westfälisches Elektrizitätswerk AG (RWE Energie) **80**
Rheinsberg Nuclear Power Plant (Germany) **85** p. 79
Richardson & Cruddas (India) **96**
Richardsons Westgarth Ltd. (United Kingdom) **233**
Ringhals Nuclear Power Plant **323**
Ringhals Nuclear Power Plant, Units 1-4 (Sweden) **194** ●
River Bend Station, Unit 1 (United States) **343** ●
River Bend Station, Unit 2 (United States) **344** ☆
Robert Emmet Ginna Nuclear Power Plant (United States) **283** ●
Robinson Electric Power Plant, Unit 2; H.B. (United States) **345** ●
Rochester Gas & Electric Corp. **283, 318, 362**
Roentgen-Equivalent-Man **305**
Rogovin Report **370**
Rogovin, Stern & Huge **370**
Rokkasho Facility p. 127
Romanian Electricity Authority (RENEL) **156**
Romanian Institute for Power Studies & Design **156**
Romanian Ministry of the Machine Building Industry **156**
Rostov Nuclear Power Plant, Units 1-4 (Russia) **176** ○
Rotterdamse Drookdok Madtdschappij (Netherlands) **137, 138**
Rovno Nuclear Power Plant, Units 1-3 (Ukraine) **208** ●
Rovno Nuclear Power Plant, Unit 4 (Ukraine) **209** ○
Ruhrstahl **74**

Rural Coooperative Power Association **274**
Rural Electrification Administration (REA) **308**
Russian Academy of Sciences and the Institute of Nuclear Safety p. 212
Russian Federation Ministry for Atomic Energy (Minatom RF) p. 212
Russian Kurchatov Institute p. 212
Russian Ministry of Atomic Energy (MINATOM) **157, 158, 159, 160, 161, 162, 163, 164, 165, 166, 167, 168, 169, 170, 171, 172, 173, 174, 175, 176, 177, 178, 179**
RWE Energie **66, 75, 80, 86**, p. 80
Sacramentans for Safe Energy (SAFE) **342**
Sacramento Municipal Utility District (SMUD) **342**
Safe Integral Reactor (SIR) **235**
Sainrapt et Brice (France) **47**
Saint-Alban Nuclear Power Plant, Units 1-2 (France) **60** ●
Saint-Alban/Saint-Maurice **60**
Saint-Laurent A Nuclear Power Plant, Unit A-1 (France) **61** ★
Saint-Laurent A Nuclear Power Plant, Unit A-2 (France) **62** ●
Saint-Laurent B Nuclear Power Plant, Units B1-B (France) **63** ●
St. Laurent des Eaux A2 p. 59
St. Lucie Nuclear Power Station, Units 1-2 (United States) **346** ●
St. Rosalie Nuclear Power Plant, Units 1-2 (United States) **347** ☆
St. Vrain Nuclear Power Plant; Fort (United States) **348** ★
Salem Nuclear Generating Station, Units 1-2 (United States) **349** ●
Salt River Project **324, 325**
San Diego Gas & Electric **350, 365**
San Onofre Nuclear Generating Station, Units 1-3 (United States) **350** ●
Sandia National Laboratories **324**
Sandlin Associates **343**
Santa Maria de Garona Nuclear Center (Spain) **186** ●
Santa Susana Sodium-Graphite Reactor Experiment (SRE) (United States) **351** ★
Sargent & Lundy Engineers (United States) **148, 153, 240, 248, 251, 260, 261, 272, 274, 304, 308, 341, 384, 388**
Save Maine Yankee Committee **306**
Save Our Wetlands, Inc. **382**
Savske Elektrarne Ljubljana (Slovenia) **180**
Schleswig-Holstein Administrative Court **67**

Schnell-Brueter-Kernkraftwerks-Gesellschaft **86**
Science Concepts **356**
Scientists and Engineers for Secure Energy, Inc. **357**
Scottish Nuclear Ltd. **227, 228, 232**, p. 278
Seabrook Station **5**
Seabrook Station, Unit 1 (United States) **352** ●
Seabrook Station, Unit 2 (United States) **353** ☆
Seacoast Anti-Pollution League **352**
Sears Island Nuclear Power Plant (United States) **354** ☆
Second Nuclear Power Station (Taiwan) **200**
Sellafield **227, 234**, p. 278
Senate Subcommittee on Nuclear Regulation **262**
Sendai Nuclear Power Station, Units 1-2 (Japan) **124** ●
Sener **183, 185**
Seoul Electric Co. p. 189
Sequoyah Fuels **355**
Sequoyah Nuclear Plant, Units 1-2 (United States) **355** ●
Shanghai Boiler Factory **142**
Shanghai Turbine Factory **142**
Shearon Harris plant **165**
Shearon Harris Nuclear Power Plant, Unit 1 (United States) **290** ●
Shearon Harris Nuclear Power Plant, Units 2-4 (United States) **291** ☆
Sheldon Station **265**
Shevchenko Nuclear Power Plant **58, 131**
Shika Nuclear Power Plant (Japan) **125** ○
Shikoku Electric Power Co. **113, 114**
Shimane Nuclear Power Station, Units 1-2 (Japan) **126** ●
Shimizu **108, 110, 115, 116, 126, 128, 129, 130**
Shippingport Atomic Power Station (United States) **356** p. 13
Shoreham Nuclear Power Plant (United States) **5, 357** ★
Shoreham-Wading River Central School District **357**
Siemens AG **1, 7, 70, 75**, p. 45
Siemens Schuckertwerke **83**
Sierra Club **247, 269**
Sievert **155**
Sigma p. 45
Site Area Emergency **283, 318, 377**
Site of Special Scientific Interest (SSSI) **219, 220**
Site of Special Scientific Interest (SSSI) **221**

● Operable   ○ Under Construction   △ Indefinitely Deferred   ▲ Decommissioning   □ Planned   ■ On Order   ☆ Cancelled   ★ Shut Down

Site of Special Scientific Interest (SSSI) **224, 225**
Site of Special Scientific Interest (SSSI) **230, 231**
Sizewell A Power Station, Units 1-2 (United Kingdom) **230** ●
Sizewell B Power Station (United Kingdom) **231** ○
Skagit/Hanford Nuclear Project, Units 1-2 (United States) **358** ☆
Skagitonians Against Nuclear Power **358**
Skanska Cementgjuteriet (Sweden) **193**
Skoda (Czech Republic) **28, 29, 30, 31, 32, 92**, p. 45
Skodaexport **145**
Slovak Power Board **28, 29, 31**
Smolensk Nuclear Power Plant, Units 1-3 (Russia) **177** ●
Smolensk Nuclear Power Plant, Unit 4 (Russia) **178** ○
Snowdon Nuke **233**
Snowdonia National Park **233**
SNR 300 **218, 276**
SNR 300 Kalkar Nuclear Power Plant (Germany) **86** ☆
Societe Alsacienne de Constructions Mecaniques (France) **53**
Societe d'Energie Nucleaire Franco-Belge des Ardennes (SENA) **42**
Societe des Forges et Ateliers du Creusot (Usines Schneider) (France) **189**
Societe General d'Enterprises (France) **39, 49, 58**
Societe pour l'Industrie Atomique (France) **189**
Societe Luxembourgoise d'Energie Nucleaire **134**
Society for Reactor Safety **70**
Somerset Education Authority **224, 225**
Somerset Nuclear Power Plant, Units 1-3 (United States) **359** ☆
Sosnovy Bor Nuclear Power Plant, Units 1-4 (Russia) **179** ●
South African Council for Nuclear Safety (CNS) p. 204
South Carolina Electric & Gas Co. **363**
South Carolina Public Service Authority **363**
South Dade Nuclear Power Plant, Units 1-2 (United States) **360** ☆
South Korea Electric p. 189
South of Scotland Electricity Board (SSEB) **216, 217, 227, 228, 232**, p. 279
South Texas Project Electric Generating Station, Units 1-2 (United States) **361** ●
South Ukraine Nuclear Power Plant, Units 1-3 (Ukraine) **210** ●
South Ukraine Nuclear Power Plant, Unit 4 (Ukraine) **211** ○

Southern California Edison **324, 325, 350, 351**
Southern California Edison **377**
Southern California Public Power Authorities **324, 325**
Southern Company **275, 293, 377**
Southern Nuclear Operating Co. **275, 293, 377**
Soviet Ministry of Atomic Energy (former USSR) **4**
Soviet Ministry of Atomic Power and Industry (MAPI) **132**, p. 149, p. 212
Soviet National Academy of Sciences p. 212
Soviet Subcommittee on Atomic Energy and Nuclear Ecology p. 212
Soyland Power Cooperative **260, 261**
Spie Batignolles SA (France) **10, 37, 42, 155**
Stade Nuclear Power Plant (Germany) **87** ●
Stadwerke Muenchen and Energieversorgung Ostbayern (OBAG) **76**
Stal-Laval Turbin AB (Sweden) **35, 191, 192, 193, 194**
State Committee for the Supervision of Nuclear and Radiation Safety p. 212
Statens Kärnbränslenämnd (SKN) p. 241
Statens Kärnkraftinspektion (SKI) p. 241
Statens Strålskyddsinstitut (SSI) p. 241
Statens Vattenfallsverk (SSPB) **192, 194**
Steam Generating Heavy Water Reactor (SGHWR) **235**
Stendal p. 79
Stendal Nuclear Power Plant, Units 1-2 (Germany) **88** △
Sterling Nuclear Power Plant (United States) **362** ☆
Stone & Webster Engineering Corp. (United States) **242, 255, 279, 288, 306, 312, 313, 317, 318, 319, 343, 344, 356, 357, 366, 369, 385, 386**
Stork **137, 138**
Studiecentrum Kernenfersie (Belgium) **9**
Sumitomo Heavy Industries, Ltd. (Japan) **106**
Summa Insurance Company **144**
Summer Nuclear Station; Virgil C. (United States) **363** ●
Summit Nuclear Power Plant, Units 1-2 (United States) **364** ☆
Sundesert Nuclear Power Plant, Units 1-2 (United States) **365** ☆
Superphenix **58, 86, 160, 218, 259**
Superphenix Fast Breeder Power Station (France) **64**
Surry Nuclear Power Station, Units 1-2 (United States) **366** ●

Surry Nuclear Power Station, Units 3-4 (United States) **367** ☆
Susquehanna Nuclear Power Plant **165**
Susquehanna Riverlands **368**
Susquehanna Steam Electric Station, Units 1-2 (United States) **368** ●
Susquehanna Valley Alliance **370**
Suzu Nuclear Power Plant **120**
Svendk Kärnbränslehantering AB (SKB) p. 241
Swedish Nuclear Power Inspectorate p. 150, p. 241
Sydkraft **191**, p. 240
Syracuse Peace Council **318**
System Energy Resources, Inc. **284**
Taisei **113, 114, 119, 124, 127, 129**
Taiwan Power Co. **199, 200, 201, 202**
Takahama Nuclear Power Station, Units 1-4 (Japan) **127** ●
Takenaka **108, 110, 111, 112, 113, 114, 115, 116, 118, 127, 130**
Tarapur Atomic Power Plant, Units 1-2 (India) **99** ●
Tarapur Atomic Power Plant, Units 3-4 (India) **100** ■
Tarapur Atomic Power Station (TAPS) p. 111
Tata Institute of Fundamental Research p. 111
Taylor Woodrow plc **216, 217**
Taylor Woodrow Construction Ltd. (United Kingdom) **216, 217, 224, 230, 236**
Technical Control Association (TÜV) **66, 84**
Tecnatom, S.A. (Spain) **92**, p. 106
Teesmouth Field Centre **221**
Telefunken **5, 74, 79**
Temelin Nuclear Power Plant, Units 1-2 (Czech Republic) **32** ○
Temelin Nuclear Power Plant, Units 3-4 (Czech Republic) **33** △
Tennessee Valley Authority **243, 249, 259, 292, 333, 355, 383, 386**, p. 170
Teollisuuden Voima Oy (TVO) **35**
Termoli Nuclear Power Station p. 119
Tex-La Electric **262**
Texas Public Utility Commission **262, 361**
Texas Utilities (TU) **262**
Thermal Oxide Reprocessing Plant (THORP) **220, 221**
Third Nuclear Power Station (Taiwan) **201**
THORP **222**
Three Mile Island Alert **369, 370**
Three Mile Island Nuclear Power Plant **92**
Three Mile Island Nuclear Power Plant, Unit 1 (United States) **369** ●

Alphabetical Index — Westinghouse

Three Mile Island Nuclear Power Plant, Unit 2 (United States) **370** ★
Three Mile Island Public Health Fund **370**
THTR-300 Demonstration Reactor (Germany) **89** ▲
Tihange Nuclear Power Plant, Units 1-3 (Belgium) **10** ●
Tijdelijke Vereniging Burgerlijke Bouwkunde (Belgium) **7**
Tlatelolco Treaty p. 1
TMI-2 Safety Advisory Board **370**
Tohoku Electric Power Co., Inc. **117, 122, 123**
Tokai Nuclear Power Station, Units 1-2 (Japan) **128** ●
Tokyo Electric Power Co. **107, 108, 115, 116,** p. 124
Toledo Edison **242, 267, 268, 331, 332**
Tomari Nuclear Power Station, Units 1-2 (Japan) **129** ●
Torness Nuclear Power Station, Units 1-2 (United Kingdom) **232** ●
Torness Power Station **177**
Toshiba **106, 107, 108, 111, 112, 115, 116, 119, 122, 123,** p. 124
Traction-Electricite (Belgium) **6, 7, 194**
Transnuklear p. 81
Travaux **10**
Trawsfynydd Power Station, Units 1-2 (United Kingdom) **233** ●
Tricastin Nuclear Power Plant, Units 1-4 (France) **65** ●
Trillo Nuclear Power Plant (Spain) **187** ●
Trino, Piedmont Nuclear Power Station p. 120
Trino Vercellese Nuclear Power Plant (Italy) **105** ★
Trojan Nuclear Plant (United States) **371** ●
Trojan Powder Co. **371**
Tsuruga Nuclear Power Station, Units 1-2 (Japan) **130** ●
TU Electric **262, 263**
Tullnerfeld Nuclear Power Plant (Austria) **5** ▲
Turkey Point Nuclear Station, Units 3-4 (United States) **372** ●
TVO Nuclear Power Plant **35**
Tyrone Nuclear Power Plant, Units 1-2 (United States) **373** ☆
Uentrop Reactor **89**
UK Atomic Energy Authority p. 111
Ukrainian Academies of Science p. 257
Ukratomenergoprom (Ukrainian Nuclear Energy Concern) **203, 204, 205, 206, 207, 208, 209, 210, 211, 212, 213**
Ulchin Nuclear Power Plant, Units 1-2 (Republic of Korea) **147** ●
Ulchin Nuclear Power Plant, Units 3-4 (Republic of Korea) **148** ○
Union of Concerned Scientists **135, 254, 284, 369, 385**
Union Electric Co. **252, 253**
Union Electrica FENOSA, S.A. **184, 187**
United Engineers & Constructors (United States) **250, 272, 285, 299, 300, 341, 349, 352, 353, 354, 359, 370, 379, 381**
United Kingdom Atomic Energy Authority **214, 216, 216, 217, 218, 223, 234,** p. 278
United States Atomic Energy Commission (AEC) **276**
U.S. Bureau of Reclamation **265**
U.S. Council on Energy Awareness **357**
U.S. Department of Energy **259, 274, 289, 336**
U.S./Soviet Joint Coordinating Committee on Civilian Nuclear Reactor Safety (JCCNRS) **165, 173, 208,** p. 213, p. 258, p. 273
University of British Columbia p. 28
Unterweser Nuclear Power Plant (Germany) **90** ●
Unusual Event **283**
Urban Wildlife Sanctuary **290**
Utility Data Institute (UDI) **248**
Utility of the Year **363**
Vaalputs Radioactive Waste Disposal Facility **155,** p. 204
Valdecaballeros Nuclear Center, Units 1-2 (Spain) **188** ○
Vallecitos Nuclear Power Plant (United States) **374** ★
Vandalia Nuclear Power Plant (United States) **375** ☆
Vandellos I Nuclear Center (Spain) **189** ★
Vandellos II Nuclear Center (Spain) **190** ●
VANEA **190**
Vattenfall p. 241
Veracruz Ecological Group **135**
Veracruz Mothers' Committee **135**
Vereinigte Elektrizitätswerke Westfalen AG (VEW) **69, 79**
Vereinigte Energiewerke AG (VEAG) **71,** p. 80
Vermont Public Interest Research Group **376**
Vermont Yankee Nuclear Power Corp. **376**
Vermont Yankee Nuclear Power Plant (United States) **376** ●
Very High Temperature Reactor (VHTR) p. 124
Vidal Nuclear Power Plant, Units 1-2 (United States) **377** ☆
Virgil C. Summer Nuclear Station (United States) **363** ●
Virginia Electric & Power Company **319, 366**
Virginia Power **319, 320, 366, 367**
Vitkovice p. 45
VLJ Repository for Operating Wastes **35**
Vogtle Nuclear Power Station, Units 1-2 (United States) **378** ●
VVER reactor p. 211, p. 270
VVER-91 p. 271
VVER-92 p. 271
VVER-210 prototype **173,** p. 270
VVER-365 prototype **173,** p. 270
VVER-440 reactor **173,** p. 270
VVER-440 Model V213 reactor **92**
VVER-440 Model V230 reactor **71,** p. 211
VVER-1000 reactor **92, 157, 173, 173**
Wabash Valley Power Association (WVPA) **308**
Wackenhut Corp. **251**
Wackersdorf **69,** p. 81
Wakasa Bay National Park **130**
Walchandnagar Industries Ltd. (India) **94**
Walchandnager Industries Ltd. (India) **96**
Washington Nuclear Project, Units 1 & (United States) **379** △
Washington Nuclear Project, Unit 2 (United States) **380** ●
Washington Nuclear Project, Units 4-5 (United States) **381** ☆
Washington Public Power Supply System (WPPSS) **379, 380, 381,** p. 170
Washington Water Power Co. **358, 379**
Waterford Power Station, Unit 3 (United States) **382** ●
Watts Bar Nuclear Performance Plan **383**
Watts Bar Nuclear Plant, Units 1-2 (United States) **383** ○
Watts Bar Program Team **383**
Wayss & Freitag AG **73, 90**
Wedco, a subsidiary of Westinghouse **299, 300**
West Virginia Turnpike Commission **379**
Western Massachusetts Electric Company **288, 312**
Westfalen Coal Works **89**
Westinghouse **42**
Westinghouse (Belgium) **7, 10**
Westinghouse Electric Corp. (United States) **9, 11, 32, 38, 105, 118, 120, 127, 144, 146, 152, 180, 181, 182, 184, 185, 190, 194, 195, 199, 200, 201, 238, 239, 242, 248, 251, 252, 253, 254, 255, 259, 262, 263, 264,**

● Operable   ○ Under Construction   △ Indefinitely Deferred   ▲ Decommissioning   □ Planned   ■ On Order   ☆ Cancelled   ★ Shut Down   **p. 555**

Westinghouse

265, 266, 269, 275, 283, 288, 289, 290, 291, 298, 299, 300, 303, 306, 308, 312, 319, 323, 337, 338, 342, 345, 346, 349, 350, 352, 353, 354, 355, 356, 358, 361, 363, 366, 370, 371, 372, 379, 380, 381, 383, 384, 385, 386, 388, p. 45, p. 169, p. 178, p. 203

Westinghouse (France) 10
Westinghouse International Projects Company 144
Whoops 379
Wigner release 234
William H. Zimmer Nuclear Power Plant, Units 1-2 (United States) 387 ☆
Windscale plutonium pile accident p. 278
Windscale Advanced Gas Cooled Reactor (United Kingdom) 234 ★
Winfrith Atomic Energy Establishment 235
Winfrith Dragon 235
Winfrith Steam Generating Heavy Water Reactor (United Kingdom) 235 ★
Wisconsin Electric Power Co. 294, 337
Wisconsin Power & Light 294, 303
Wisconsin Public Service Corp. 294, 303
WNP-1 379
WNP-2 379
WNP-3 379
WNP-4 379
WNP-5 379
Wolf Creek Nuclear Operating Corp. 384
Wolf Creek Nuclear Power Plant (United States) 384 ●
Wolsong Nuclear Power Plant, Unit 1 (Republic of Korea) 149 ●
Wolsong Nuclear Power Plant, Unit 2 (Republic of Korea) 150 ○
Wolsong Nuclear Power Plant, Units 3-4 (Republic of Korea) 151 ■
Wolverine Electric Cooperative, Inc. 277
World Association of Nuclear Operators (WANO) 14, 30, 155, p. 274
World Atomic Nuclear Operators (WANO) p. 213
World Health Organization 5, 29, p. 9, p. 46
Worldwatch Institute 204
WPPSS 379
Wuergassen Nuclear Power Plant (Germany) 91 ●
Wylfa Power Station, Units 1-2 (United Kingdom) 236 ●
Yankee Atomic Electric Co. 385
Yankee Rowe Nuclear Power Plant (United States) 385 ★
Yellow Creek Nuclear Power Plant, Units 1-2 (United States) 386 ☆

Yenliao Nuclear Power Station, Units 1-2 (Taiwan) 202 ■
Yonggwang Nuclear Power Plant, Units 1-2 (Republic of Korea) 152 ●
Yonggwang Nuclear Power Plant, Units 3-4 (Republic of Korea) 153 ○
Yonggwang Nuclear Power Plant, Units 5-6 (Republic of Korea) 154 ■
Zamech-Elblag 145
Zaporozhye Nuclear Power Station 118
Zaporozhye Nuclear Power Station, Units 1-5 (Ukraine) 212 ●
Zaporozhye Nuclear Power Station, Unit 6 (Ukraine) 213 ○
Zarnowiec Nuclear Power Plant, Units 1-4 (Poland) 145 ★
Zimmer Area Citizens of Ohio (ZAC) 387
Zimmer Nuclear Power Plant, Units 1-2; William H. (United States) 387 ☆
Zion Station, Units 1-2 (United States) 388 ●
Zschokke 195
ZWIBEZ p. 248

● Operable ○ Under Construction △ Indefinitely Deferred ▲ Decommissioning □ Planned ■ On Order ☆ Cancelled ★ Shut Down